AF608623

Richard H. Cushman
Larry M. Bates

Global Aspects of Classical Integrable Systems

Springer Basel AG

Richard H. Cushman
Mathematics Institute
University of Utrecht
Budapestlaan 6
3508 TA Utrecht
The Netherlands

Larry M. Bates
Department of Mathematics and Statistics
University of Calgary
2500 University Dr. N.W.
Calgary, Alberta T2N 1N4
Canada

1991 Mathematics Subject Classification 70-01, 70E99, 58F05, 58-01

A CIP catalogue record for this book is available from the Library of Congress, Washington D.C., USA

Deutsche Bibliothek Cataloging-in-Publication Data

Cushman, Richard H.:
Global aspects of classical integrable systems / Richard H. Cushman ; Larry M. Bates. - Basel ; Boston ; Berlin : Birkhäuser, 1997
ISBN 978-3-0348-9817-1 ISBN 978-3-0348-8891-2 (eBook)
DOI 10.1007/978-3-0348-8891-2
NE: Bates, Larry M.:

Originally published by Birkhäuser Verlag in 1997
Softcover reprint of the hardcover 1st edition 1997

Printed on acid-free paper produced of chlorine-free pulp. TCF ∞

ISBN 978-3-0348-9817-1

9 8 7 6 5 4 3 2 1

Table of Contents

III. The Euler top

IV. The spherical pendulum

V. The Lagrange top

Appendix A. **Fundamental concepts**

Appendix B. **Systems with symmetry**

Appendix C. **Ehresmann connections**

Appendix D. **Action angle coordinates**

Appendix E. **Basic Morse theory**

Foreword

This book gives a complete global geometric description of the motion of the two dimensional harmonic oscillator, the Kepler problem, the Euler top, the spherical pendulum and the Lagrange top. These classical integrable Hamiltonian systems one sees treated in almost every physics book on classical mechanics. So why is this book necessary? The answer is that the standard treatments are not complete. For instance in physics books one cannot see the monodromy in the spherical pendulum from its explicit solution in terms of elliptic functions nor can one read off from the explicit solution the fact that a tennis racket makes a near half twist when it is tossed so as to spin nearly about its intermediate axis. Modern mathematics books on mechanics do not use the symplectic geometric tools they develop to treat the qualitative features of these problems either. One reason for this is that their basic tool for removing symmetries of Hamiltonian systems, called regular reduction, is not general enough to handle removal of the symmetries which occur in the spherical pendulum or in the Lagrange top. For these symmetries one needs singular reduction. Another reason is that the obstructions to making local action angle coordinates global such as monodromy were not known when these works were written.

The point of view adopted in this book is to start with a somewhat unfamiliar abstract mathematical model of the physical system such as the study of the geodesic flow of a left invariant metric on the three dimensional rotation group. Using the symplectic geometric formulation of Hamiltonian mechanics we then show that the equations of motion agree with those found by more traditional methods for a well known physical system, namely, the force free rigid body or Euler top. This justifies our mathematical model. We do not try to build our model from fundamental physical principles. We have not written a book on mechanics or Hamiltonian particle dynamics. We only discuss five special integrable systems, which is a very small sample of the rich variety of general Hamiltonian systems. Moreover the behavior of the solutions of these integrable systems is much more regular than the nearly unpredictable motion of a general Hamiltonian system such as the three body problem.

Our main goal is to understand the global geometric features of our model integrable systems. The main tool we use is reduction to remove the symmetries and to obtain a system with one degree of freedom. This allows us to determine the range and the topology of every fiber of the energy-momentum mapping of the system. The topology of a fiber corresponding to a singular value of the energy-momentum mapping is of great interest. Physically, these motions are simpler than the general motion and therefore are easier to study experimentally. Mathematically, these fibers contain a relative equilibrium of the system, that is, a motion which is also an orbit of the symmetry group. For instance, in the spherical pendulum the relative equilibria are circular orbits on the 2-sphere which lie in a plane parallel to and below the equator. Other examples are the regular precession and sleeping motions of the Lagrange top. Finally, to complete the qualitative picture, we describe how the fibers of the energy momentum map fit together. Sometimes this involves showing that the monodromy of certain torus bundles are nontrivial. That this phenomenon happens in the spherical pendulum and the Lagrange top was not known until the 1980s.

This book is written from a bottom up approach with examples being given prominence over theory. The examples are treated in a uniform way. First the mathematical model is described and then the equations of motion are derived. Next the symmetries and corresponding integrals are obtained and it is shown that the given problem is Liouville integrable. Finally, the geometry of the level sets of the energy momentum map, which gives a complete geometric description of the motion, are obtained by first using reduction to remove the symmetries and then reconstructing the geometry from the geometry of the reduced system. This program may seem to be excessively lengthy. There are two reasons why we have followed it. First, our procedure gives complete answers, whereas short cut ones do not. Second, in carrying out our program the reader sees enough detail in the text to be able to understand the arguments without having to look at the theory. The theory given in the appendices is what the authors feel is the minimum necessary to justify all the unproved assertions in the examples.

This book was not written to be read in a sequential fashion. We strongly encourage the reader to browse.

Introduction

1 The mathematical pendulum

We begin by looking at the mathematical pendulum.

Let $T^*\mathbf{R}$ be the cotangent bundle of $\mathbf{R}$ which we identify with $\mathbf{R}^2$ and give coordinates (x, y). The canonical symplectic form $\omega = dx \wedge dy$ on $T^*\mathbf{R}$ is the element of oriented area on $\mathbf{R}^2$. Consider the Hamiltonian system $(H, T^*\mathbf{R}, \omega)$ with Hamiltonian

$$H : T^*\mathbf{R} \to \mathbf{R} : (x, y) \to \tfrac{1}{2} y^2 - \cos x.$$

▷ The following argument shows that the Hamiltonian vector field X_H on $T^*\mathbf{R}$ corresponding to the Hamiltonian H is

$$X_H(x, y) = \dot{x}\frac{\partial}{\partial x} + \dot{y}\frac{\partial}{\partial y} = y\frac{\partial}{\partial x} - \sin x \frac{\partial}{\partial y}. \tag{1}$$

(1.1) **Proof**: By definition of Hamiltonian vector field (see appendix A section 3)

$$dH(p)z_p = \omega(p)(X_H(p), z_p) \tag{2}$$

for every $z_p = (v_p, w_p)$ in the tangent space $T_p(T^*\mathbf{R})$ to $T^*\mathbf{R}$ at p. Let $X_H(p) = \big(X(p), Y(p)\big)$. Now

$$dH(p)z_p = \frac{\partial H}{\partial x}v_p + \frac{\partial H}{\partial y}w_p.$$

Moreover, $\omega(p)(X_H(p), z_p)$ is the oriented area spanned by the parallelogram with sides $X_H(p)$ and z_p, that is,

$$\omega(p)(X_H(p), z_p) = \det\begin{pmatrix} X(p) & v_p \\ Y(p) & w_p \end{pmatrix} = X(p)w_p - Y(p)v_p.$$

Therefore (2) is equivalent to

$$\frac{\partial H}{\partial x}v_p + \frac{\partial H}{\partial y}w_p = -Y(p)v_p + X(p)w_p \tag{3}$$

for every $(v_p, w_p) \in \mathbf{R}^2$. In (3) choose $(v_p, w_p) = (1, 0)$. Then $X(p) = \frac{\partial H}{\partial y} = y$. Next choose $(v_p, w_p) = (0, 1)$. Then $Y(p) = -\frac{\partial H}{\partial x} = -\sin x$. □

Note that (1) may be written as the second order differential equation

$$\ddot{x} = -\frac{d}{dx}(-\cos x) = -\sin x. \tag{4}$$

By Newton's second law of motion, an integral curve of (1) describes the motion of a particle of unit mass under a force coming from the potential $V : \mathbf{R} \to \mathbf{R} : x \to -\cos x$.

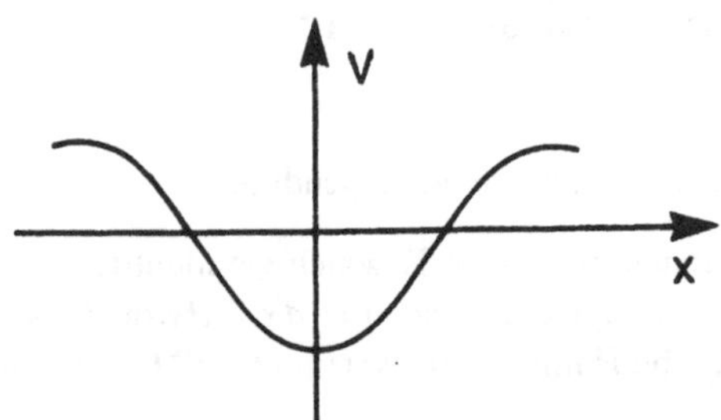

Figure .1. The graph of the potential $V(x) = -\cos x$.

Thus H is the total energy of the particle, namely the sum of the kinetic and potential energy. We now show that H is a Morse function on $T^*\mathbf{R}$.

(1.2) **Proof**: The point $p = (x, y)$ is a critical point of H if and only if $X_H(p) = 0$, that is, if and only if

$$0 = \frac{\partial H}{\partial y} = \sin x$$

$$0 = \frac{\partial H}{\partial x} = y.$$

Thus $\{p = (n\pi, 0) \in \mathbf{R}^2 | n \in \mathbf{Z}\}$ is the set of critical points of H. The corresponding critical value of H is -1 if n is even or 1 if n is odd. Since the Hessian of H at p is

$$D^2H(p) = \begin{pmatrix} \frac{\partial^2 H}{\partial x^2} & \frac{\partial^2 H}{\partial x \partial y} \\ \frac{\partial^2 H}{\partial y \partial x} & \frac{\partial^2 H}{\partial y^2} \end{pmatrix}\Bigg|_p = \begin{pmatrix} \cos n\pi & 0 \\ 0 & 1 \end{pmatrix} = \begin{pmatrix} (-1)^n & 0 \\ 0 & 1 \end{pmatrix},$$

H is a Morse function, because $D^2H(p)$ is nondegenerate. □

When $n = 2k$ the Morse index of $D^2H(n\pi, 0)$ is zero, and so the critical points $(2k\pi, 0)$ are relative minima of H; whereas when $n = 2k + 1$ the Morse index of $D^2H(n\pi, 0)$ is one, and so the critical points $(2k\pi, 0)$ are saddle points of H, see figure .2. By the Morse lemma (see appendix E section 1) there is a neighborhood U_k of $(2k\pi, 0)$ in $((2k-1)\pi, (2k+1)\pi) \times \mathbf{R}$ such that for h slightly greater than -1 the level set $H^{-1}(h) \cap U_k$ is diffeomorphic to a circle. Since H has no critical values in the interval $(-1, 1)$, by the Morse isotopy lemma (see appendix E section 3) we deduce that for every $h \in (-1, 1)$ the level set $H^{-1}(h) \cap U_k$ is diffeomorphic to a circle. Thus for $h \in (-1, 1)$ the whole level set $H^{-1}(h)$ is diffeomorphic to a countable disjoint union of circles. If $h \geq 1$, then $H^{-1}(h)$ is the union of the graphs of two smooth functions

$$y_{\pm} = \pm\sqrt{2(h + \cos x)}.$$

The graphs of $y_{\pm}$ are disjoint if $h > 1$. On the other hand, if $h = 1$, then the graphs of $y_{\pm} = \pm 2 \cos \frac{1}{2} x$ intersect only at the points $((2k + 1)\pi, 0)$. There they intersect transversely as can be seen by applying the Morse lemma at the points $((2k + 1)\pi, 0)$. Thus we have obtained a picture of the level curves of H as given in figure .2.

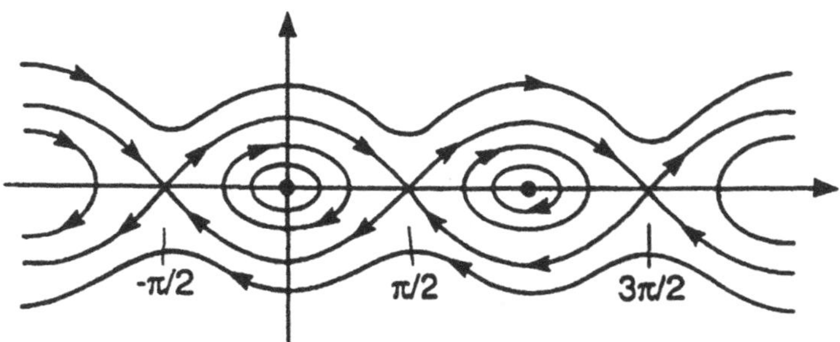

Figure .2. The level sets of $H(x, y) = \frac{1}{2} y^2 - \cos x$.

To simplify the topology of the level sets of H, we make use of the fact that H is invariant under the discrete symmetry

$$\mathbf{Z} \times T^*\mathbf{R} \to T^*\mathbf{R} : \big(n, (x, y)\big) \to (x + 2n\pi, y). \tag{5}$$

Thus H induces a function $\widetilde{H}$ on the space of orbits $T^*\mathbf{R}/2\pi\mathbf{Z}$. Concretely, this orbit space is identified with the cotangent bundle T^*S^1 of the circle S^1. Here S^1 is

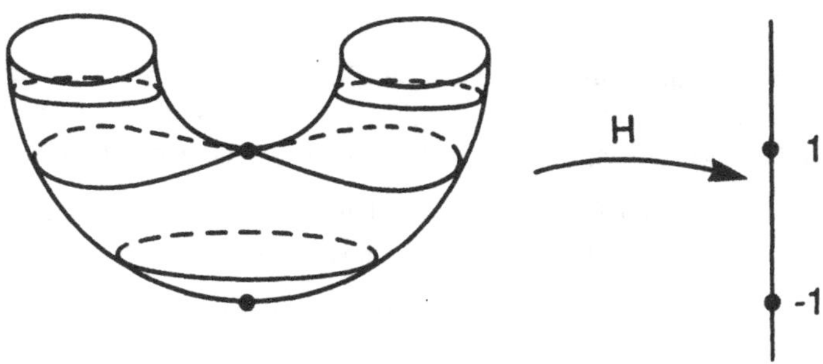

Figure .3. The level sets of $\widetilde{H}(x, y) = \frac{1}{2} y^2 - \cos x$ with $x \bmod 2\pi$ on T^*S^1.

thought of as the orbit space $\mathbf{R}/2\pi\mathbf{Z}$ of the real numbers modulo 2π. Geometrically, T^*S^1 is the cylinder $S^1 \times \mathbf{R}$ which is obtained from $\mathbf{R}^2$ by cutting along the vertical lines $x = 0$

and $x = 2\pi$ and then pasting the edges together. Applying this process to figure .2 gives figure .3 which depicts the level sets of $\widetilde{H}$ and hence the orbits of the induced Hamiltonian vector field $X_{\widetilde{H}}$. A short argument using Newton's second law shows that the second order differential equation

$$\ddot{x} = -\sin x \qquad x \bmod 2\pi$$

describes the motion of a particle of mass one on the unit circle under the influence of a constant vertical downward unit force, see figure .4.

Figure .4. Physical realization of the mathematical pendulum.

From figure .3 we see that the topological circle, defined by the the component of the level set $\widetilde{H}^{-1}(h)$ $(h > 1)$ lying in the upper half cylinder, is very different from the topological circle defined by the level set $\widetilde{H}^{-1}(h)$ $(-1 < h < 1)$. The first circle is not contractible in TS^1 to a point whereas the second circle is. Hence it is impossible to continuously deform the first circle into the second one. This difference in the topological disposition of the two circles corresponds to the physical fact that for small energy the particle oscillates about the bottom of the circle, while for large energy the particle loops over the top of the circle.

2 Exercises

1. Let (x, y) be canonical coordinates on $T^*\mathbf{R} = \mathbf{R}^2$ with symplectic form $\omega = dx \wedge dy$. Suppose that the Hamiltonian $H : T^*\mathbf{R} \rightarrow \mathbf{R}$ is a sum of kinetic and potential energy, that is, $H(x, y) = \frac{1}{2}y^2 + V(x)$, where $V : \mathbf{R} \rightarrow \mathbf{R}$.

   (a). Find a potential function V such that the zero level set of H is connected, compact and has one singular point which is a cusp.

   (b). Construct a polynomial Hamiltonian on $T^*\mathbf{R}$ whose zero level set is an n-leaf clover.

   (c). Show that there is *no* smooth Hamiltonian which is the sum of kinetic and potential energy which has a 3-leaf clover as a level set.

   (d). For smooth V with countable many isolated critical points give a topological characterization of the critical level sets of H.

2. Construct a Hamiltonian function on S^2 which is a Morse function with two critical points. Draw its level sets. Construct a vector field on S^2 with only one equilibrium point and sketch its orbits. Show that this vector field is not Hamiltonian.

3. (a). On $\mathbf{R}^2$ consider the action $\cdot$ of $\mathbf{Z}^2$ defined by $(n, m) \cdot (x, y) = (x + n, y + m)$. The orbit space $\mathbf{R}^2/\mathbf{Z}^2$ is a two dimensional torus T^2, which may be modeled by a square with the opposite sides identified. The symplectic form $\Omega = dx \wedge dy$ on $\mathbf{R}^2$ induces a symplectic form $\widetilde{\Omega}$ on T^2. Show that the vector field $\widetilde{X}$ on T^2 induced by the Hamiltonian vector field $X = \frac{\partial}{\partial x} + \frac{\partial}{\partial y}$ on $\mathbf{R}^2$ is *not* Hamiltonian on $(T^2, \widetilde{\Omega})$.

   (b). Sketch the orbits of a Hamiltonian vector field on $(T^2, \widetilde{\Omega})$ where the Hamiltonian is a Morse function with the fewest number of critical points.

   (c). Construct a vector field X on T^2 with *two* equilibrium points.

   (d)*. Show that a smooth function on T^2 must have at least three critical points. Find a smooth function on T^2 with exactly three critical points. Sketch its level sets.

   (e). Deduce that the vector field X constructed in c) is not Hamiltonian.

4. Let M be a compact connected orientable smooth two dimensional manifold with volume form Ω. In what follows we show that every integral curve of a Hamiltonian vector field X_H of a one degree of freedom Hamiltonian system (H, M, Ω) is either an equilibrium point, a periodic orbit, or is asymptotic to an equilibrium point as $t \rightarrow \pm\infty$. For $m \in M$ let $\gamma : \mathbf{R} \rightarrow M : t \rightarrow \varphi_t^H(m)$ be the integral curve of X_H through m. The *ω-limit set* of γ is

$$\omega(\gamma) = \overline{\bigcap_{T>0} \left\{ \varphi_t^H(m) \middle| t \geq T \right\}}.$$

   (a). Show that $\omega(\gamma)$ is nonempty.

   (b). If γ is an equilibrium point or a periodic orbit of X_H, show that $\omega(\gamma) = \gamma$. Is the converse true?

   (c)*. If γ is not an equilibrium point or periodic orbit of X_H, show that $\omega(\gamma)$ is a critical point of H, that is, an equilibrium point of γ.

5. (Period energy relation for the mathematical pendulum.) When $|h| < 1$ show that the period of an integral curve of the mathematical pendulum which starts at $(x_+, 0)$ where $0 < x_+ = x_+(h) < \pi$ and $h + \cos x_+ = 0$, is given by

$$\tau(h) = 2 \int_{-x_+}^{x_+} \frac{1}{\sqrt{2(h + \cos x)}} \, dx.$$

   Show that $\tau = 4K(\sqrt{(h+1)/2})$, where K is the complete elliptic integral of the first kind (see exercise 1 (b) of chapter 1). Deduce that

   (a). $\tau(-1) = 2\pi$, $\tau(1) = \infty$ and $\tau'(-1) = \pi/4$.

   (b). τ is a real analytic function on $(-1, 1)$.

   (c)*. $\tau' > 0$ on $(-1, 1)$. Hint: show that τ satisfies a differential equation.

6. (a). Suppose that a particle moves on the graph of $y = f(x)$ under the influence of gravity and that the origin is a stable equilibrium point. Determine the shape of the graph of f so that the period of oscillation of the particle about the origin is a constant independent of the energy.

   (b)*. Show that (a) is equivalent to the fact the the derivative of the area enclosed by a level set of the Hamiltonian of the particle with respect to the Hamiltonian itself is a constant. (Hint: see appendix D section 1.)

Chapter I

The harmonic oscillator

1 Hamilton's equations and S^1 symmetry

Physically, the harmonic oscillator in the plane is described by a particle of unit mass acted upon by two linear springs of unit spring constant: one spring acting in the x_1-direction and the other in the x_2-direction.

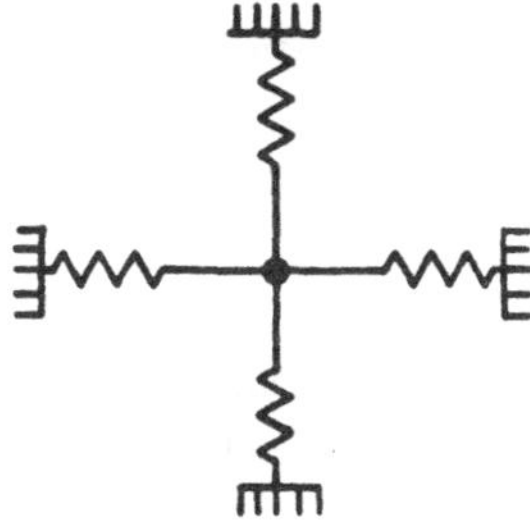

Figure 1.1 A two dimensional harmonic oscillator.

Mathematically, the *configuration space* of the harmonic oscillator is Euclidean 2-space. In other words, the space of positions of the particle is $\mathbf{R}^2$ with coordinates $x = (x_1, x_2)$ and *Euclidean inner product* $(\ ,\ )$ where $(x, x') = x_1 x_1' + x_2 x_2'$. The space of all positions and momenta of the particle is the cotangent bundle $T^*\mathbf{R}^2$ of $\mathbf{R}^2$. This phase space has coordinates (x, y) and a canonical symplectic form $\omega = dx_1 \wedge dy_1 + dx_2 \wedge dy_2$. The Hamiltonian $H : T^*\mathbf{R}^2 \to \mathbf{R}$ of the harmonic oscillator is the sum of *kinetic energy* $\frac{1}{2}(y, y)$ and *potential energy* $\frac{1}{2}(x, x)$. In other words, letting $z = (x, y) \in \mathbf{R}^4$,

$$H(z) = \tfrac{1}{2}(y, y) + \tfrac{1}{2}(x, x) = \tfrac{1}{2}\langle z, z\rangle,$$

where $\langle\,,\,\rangle$ is the Euclidean inner product on $\mathbf{R}^4$, which we have identified with $T^*\mathbf{R}^2$. The motion of the harmonic oscillator is described by Hamilton's equations

$$\frac{d}{dt}\begin{pmatrix} x \\ y \end{pmatrix} = \begin{pmatrix} 0 & I_2 \\ -I_2 & 0 \end{pmatrix}\begin{pmatrix} D_1 H(x,y) \\ D_2 H(x,y) \end{pmatrix} = \begin{pmatrix} y \\ -x \end{pmatrix}.$$

Here I_2 is the 2×2 identity matrix. Since the Hamiltonian vector field $X_H(x,y) = (y,-x)$ is linear, its flow is

$$\varphi^H : \mathbf{R}\times\mathbf{R}^4 \to \mathbf{R}^4 : (t,z) \to (\exp tX_H)\,z = \begin{pmatrix} (\cos t)I_2 & (\sin t)I_2 \\ -(\sin t)I_2 & (\cos t)I_2 \end{pmatrix} z. \qquad (1)$$

Given any initial condition $z \in T^*\mathbf{R}^2$, the *integral curve* of X_H through the point z is $t \to \varphi_t^H(z)$. Thus from a quantitative point of view, we know everything about the vector field X_H, because we have an explicit formula (1) for every integral curve. On the other hand, from a qualitative point of view, the explicit formula is *very* unsatisfactory. For instance, we do not know if the integral curves lie on a lower dimensional invariant manifold or how they fit together. In the rest of this chapter we will describe the *global qualitative* features of the invariant manifolds of X_H.

Claim: The h-level set $H^{-1}(h)$ of H is an invariant manifold of the vector field of the harmonic oscillator.

(1.1) **Proof**: Since

$$\dot{H} = L_{X_H}H = \langle \dot{x}, x\rangle + \langle \dot{y}, y\rangle = \langle y, x\rangle - \langle x, y\rangle = 0,$$

H is constant on the integral curves of X_H, that is, H is an integral (= conserved quantity) of X_H, (see appendix B section 3). In particular the h-level set

$$H^{-1}(h) = \left\{ z \in \mathbf{R}^4 \,\middle|\, \tfrac{1}{2}\langle z, z\rangle = h \right\},$$

which is diffeomorphic to a 3-sphere S^3 when $h > 0$, a point when $h = 0$, and is empty when $h < 0$, is a smooth *invariant manifold* of X_H. In other words, every integral curve of X_H with initial condition in $H^{-1}(h)$ lies in $H^{-1}(h)$ for all time. □

The rotational symmetry of the potential energy $\frac{1}{2}\langle x, x\rangle$ gives rise to another conserved quantity, namely, the *angular momentum*

$$L : T^*\mathbf{R}^2 \to \mathbf{R} : (x,y) \to x_1 y_2 - x_2 y_1. \qquad (2)$$

To see this, consider the $S^1 = \mathbf{R}/2\pi\mathbf{Z}$ action on $\mathbf{R}^2$

$$\psi : S^1 \times \mathbf{R}^2 \to \mathbf{R}^2 : (s,x) \to R_s x = \begin{pmatrix} \cos s & -\sin s \\ \sin s & \cos s \end{pmatrix} x.$$

ψ_s is a counterclockwise rotation through an angle s about the origin. ψ lifts to an action Ψ on $T^*\mathbf{R}^2$ defined by

$$\Psi : S^1 \times T^*\mathbf{R}^2 \to T^*\mathbf{R}^2 : \big(s, (x,y)\big) \to (R_s x, R_s y).$$

Ψ preserves the canonical 1-form $\theta = y_1\,dx_1 + y_2\,dx_2 = (y, dx)$ on $T^*\mathbf{R}^2$, since

$$\Psi_s^*\theta = (R_s y, dR_s x) = (R_s y, R_s dx) = (y, dx) = \theta.$$

Therefore Ψ_s is a symplectic mapping, that is, $\Psi_s^*\omega = \omega$, since

$$\omega = -d\theta = -d\Psi_s^*\theta = -\Psi_s^* d\theta = \Psi_s^*\omega.$$

The infinitesimal generator of the action Ψ is the vector field

$$Y(x,y) = \frac{d}{ds}\bigg|_{s=0} \Psi_s(x,y) = (-x_2, x_1, -y_2, y_1) = \big(D_2L(x,y), -D_1L(x,y)\big).$$

Thus Y is a Hamiltonian vector field X_L corresponding to the angular momentum L (2). (That the lifted S^1 action Ψ preserves the canonical 1-form and is the flow of the Hamiltonian vector field corresponding to the Hamiltonian $L = Y \lrcorner \theta$ is a particular case of a more general set of results, (see appendix B section 3)). Since

$$(\Psi_s^* H)(x,y) = \tfrac{1}{2}(R_s y, R_s y) + \tfrac{1}{2}(R_s x, R_s x) = H(x,y),$$

H is constant on the integral curves of X_L. Therefore L is constant on the integral curves of X_H, that is, L is an integral of X_H, (see appendix A section 4). This implies that the ℓ-level set of L,

$$L^{-1}(\ell) = \left\{ (x,y) \in T^*\mathbf{R}^2 \,\middle|\, x_1y_2 - x_2y_1 = \ell \right\},$$

is an invariant manifold of X_H. When $\ell \neq 0$ the level set $L^{-1}(\ell)$ is diffeomorphic to $S^1 \times \mathbf{R}^2$, while when $\ell = 0$ the level set $L^{-1}(0)$ is homeomorphic but *not* diffeomorphic to $S^1 \times \mathbf{R}^2$ as it is a cone on $S^1 \times S^1$ with vertex at the origin, (see exercise 1).

2 S^1 energy momentum mapping

To organize the qualitative information about the harmonic oscillator which can be obtained from the integrals of energy and angular momentum, we define the S^1 *energy momentum mapping*

$$\begin{aligned} &\mathcal{EM} : T^*\mathbf{R}^2 \to \mathbf{R}^2 : \\ &\quad (x,y) \to \big(H(x,y), L(x,y)\big) = \big(\tfrac{1}{2}(y_1^2 + y_2^2 + x_1^2 + x_2^2),\, x_1y_2 - x_2y_1\big). \end{aligned}$$

Because H and L are polynomial integrals of X_H, the *fiber* $\mathcal{EM}^{-1}(h,\ell)$ is an invariant manifold of X_H which is a real algebraic variety. Other geometric properties of $\mathcal{EM}$ correspond to qualitative properties of X_H. To determine the geometric properties of $\mathcal{EM}$, we shall

1. Find the critical points, critical values and range of $\mathcal{EM}$.

2. Find the topological type of *every* fiber of the energy momentum mapping. This determines the *bifurcation set* of $\mathcal{EM}$, that is, the set of values (h,ℓ) where the *topological type of the fiber* changes.

3. Analyze how the fibers of constant angular momentum *foliate* a given energy level set.

▷ We begin by finding the critical points and corresponding critical values of the energy momentum mapping. A point $z = (x, y) \in T^*\mathbf{R}^2$ is a critical point of $\mathcal{EM}$ if and only if the derivative of $\mathcal{EM}$ at z is not surjective, that is,

$$D\mathcal{EM}(z) = \begin{pmatrix} DH(z) \\ DL(z) \end{pmatrix} = \begin{pmatrix} x_1 & x_2 & y_1 & y_2 \\ y_2 & -y_1 & -x_2 & x_1 \end{pmatrix} \tag{3}$$

has rank less than two. We have two cases to consider.

<u>case</u> I: rank $D\mathcal{EM}(z) = 0$. This can only happen if

$$DH(z) = DL(z) = 0. \tag{4}$$

Then $z = 0$ is the critical point and $\big(H(0), L(0)\big) = (0, 0)$ is the corresponding critical value of $\mathcal{EM}$.

<u>case</u> II. rank $D\mathcal{EM}(z) = 1$. This occurs if and only if $DH(z)$ and $DL(z)$ are linearly dependent and are not both zero. From (3) it follows that $DH(z) = 0$ if and only if $DL(z) = 0$. Therefore we may suppose that $DH(z) \neq 0$, that is, $z \neq 0$. Thus for some $\mu \in \mathbf{R}$

$$0 = DL(z) - \mu\, DH(z) = (y_2, -y_1, -x_2, x_1) - \mu\,(x_1, x_2, y_1, y_2). \tag{5}$$

If $\mu = 0$, then $z = 0$, which is a contradiction. Therefore $\mu \neq 0$. Composing the linear mapping $(x_1, x_2, y_1, y_2) \to \mu^{-1}(y_2, -y_1, -x_2, x_1)$ with itself gives $(x_1, x_2, y_1, y_2) = \mu^{-2}(x_1, x_2, y_1, y_2)$. Thus $\mu^2 = 1$, since $z \neq 0$. Consequently, the solutions of (5) define two punctured 2-planes:

$$\Pi^*_+ = \Big\{ (x_1, x_2, -x_2, x_1) \in T^*\mathbf{R}^2 \big| (x_1, x_2) \in \big(\mathbf{R}^2 - \{(0,0)\}\big) \Big\}$$

and

$$\Pi^*_- = \Big\{ (x_1, x_2, x_2, -x_1) \in T^*\mathbf{R}^2 \big| (x_1, x_2) \in \big(\mathbf{R}^2 - \{(0,0)\}\big) \Big\}.$$

On Π^*_+ the corresponding critical values of $\mathcal{EM}$ are the diagonal ray $\{(h, h) \in \mathbf{R}^2 \,|\, h > 0\}$ since $0 < h = H|\Pi^*_+ = x_1^2 + x_2^2 = L|\Pi^*_+ = \ell$; while on Π^*_- the corresponding critical values of $\mathcal{EM}$ is the antidiagonal ray $\{(h, -h) \in \mathbf{R}^2 \,|\, h > 0\}$. Therefore the critical fiber $\mathcal{EM}^{-1}(h, h)$, $h > 0$ is the circle

$$S^1_{+,h} = H^{-1}(h) \cap \Pi^*_+ = \Big\{ (x_1, x_2, -x_2, x_1) \in T^*\mathbf{R}^2 \big| x_1^2 + x_2^2 = h,\ h > 0 \Big\}; \tag{6}$$

while the critical fiber $\mathcal{EM}^{-1}(h, -h)$, $h > 0$ is the circle

$$S^1_{-,h} = H^{-1}(h) \cap \Pi^*_- = \Big\{ (x_1, x_2, x_2, -x_1) \in T^*\mathbf{R}^2 \big| x_1^2 + x_2^2 = h,\ h > 0 \Big\}. \tag{7}$$

Note that the image of $S^1_{\pm,h}$ under the bundle projection

$$\pi : T^*\mathbf{R}^2 \to \mathbf{R}^2 : (x, y) \to x$$

is the circle $x_1^2 + x_2^2 = h$, which is positively oriented when the sign is + and negatively otherwise. Another way to interpret the critical circles $S^1_{\pm,h}$ is to note that solving (5) subject to the condition that $z \in H^{-1}(h)$, $h > 0$ is equivalent to finding the critical points of L on the smooth manifold $H^{-1}(h)$. Thus L has two critical manifolds $S^1_{\pm,h}$

▷ on $H^{-1}(h)$. The $S^1_{\pm,h}$ are nondegenerate critical manifolds of $L|H^{-1}(h)$ of Morse index 2 for $S^1_{+,h}$ and 0 for $S^1_{-,h}$.

(2.1) **Proof**: To show that $S^1_{+,h}$ is a nondegenerate critical manifold of $L|H^{-1}(h)$ we must verify that at every $p = (x_1, x_2, -x_2, x_1) \in S^1_{+,h}$, the Hessian of $L|H^{-1}(h)$ when restricted to a 2-plane $N_pS^1_{+,h}$ normal to $S^1_{+,h}$ in $T_pH^{-1}(h)$ has Morse index 2. From the fact that p is a critical point of $L|H^{-1}(h)$ with Lagrange multiplier $\mu = 1$, it follows that the Hessian of $L|H^{-1}(h)$ at p is

$$\begin{aligned} Q_+ &= D^2\big(L|H^{-1}(h)\big)(p) = \big(D^2L(p) - D^2H(p)\big)|\, T_pH^{-1}(h) \\ &= \left.\begin{pmatrix} -1 & 0 & 0 & 1 \\ 0 & -1 & -1 & 0 \\ 0 & -1 & -1 & 0 \\ 1 & 0 & 0 & -1 \end{pmatrix}\right|_{\ker DH(p)}, \end{aligned} \tag{8}$$

(see appendix E section 2). Since $S^1_{+,h}$ is an *orbit* of X_H, $T_pS^1_{+,h}$ is spanned by the vector $X_H(p) = (-x_2, x_1, -x_1, -x_2)$. As ker $DH(p)$ is spanned by the linearly independent vectors

$$X_H = (-x_2, x_1, -x_1, -x_2),\ \ f_1 = (x_2, -x_1, -x_1, -x_2),\ \ f_2 = (x_1, x_2, x_2, -x_1),$$

a normal 2-plane $N_pS^1_{+,h}$ is spanned by the vectors $\{f_1, f_2\}$. A calculation shows that the matrix of Q_+ with respect to the basis $\{f_1, f_2\}$ is $-2I_2$. Therefore $Q_+|N_pS^1_{+,h}$ is nondegenerate with Morse index 2. Another calculation shows that at $p \in S^1_{-,h}$ the Hessian $Q_- = D^2\big(L|H^{-1}(h)\big)(p)$ restricted to $N_pS^1_{-,h}$ is equal to $2I_2$. Thus $Q_-|N_pS^1_{-,h}$ is nondegenerate with Morse index 0. □

Consequently, on $S^1_{+,h}$ the function L assumes its maximum value h, while on $S^1_{-,h}$ the function L assumes its minimum value $-h$. Therefore, the closed wedge

$$\Big\{(h,\ell) \in \mathbf{R}^2 \big| 0 \le |\ell| \le h\Big\}$$

is the image of the energy momentum mapping $\mathcal{EM}$.

▷ To find the topology of a fiber of $\mathcal{EM}$ corresponding to a regular value, consider the linear change of coordinates on $\mathbf{R}^4$

$$\begin{pmatrix} x \\ y \end{pmatrix} = \begin{pmatrix} A & -B \\ B & A \end{pmatrix}\begin{pmatrix} \xi \\ \eta \end{pmatrix} = P\begin{pmatrix} \xi \\ \eta \end{pmatrix}, \tag{9}$$

where $A = \frac{1}{\sqrt{2}}\begin{pmatrix} 0 & 0 \\ 1 & -1 \end{pmatrix}$ and $B = \frac{1}{\sqrt{2}}\begin{pmatrix} 1 & 1 \\ 0 & 0 \end{pmatrix}$. Since

$$A^tA + B^tB = I_2 \quad \text{and} \quad A^tB = B^tA,$$

the matrix P is *symplectic* and orthogonal, that is,

$$P^*\omega = \omega \quad \text{and} \quad \langle Pz, Pw\rangle = \langle z, w\rangle$$

for every $z, w \in \mathbf{R}^4$. With respect to the (ξ, η) coordinates, the Hamiltonian H becomes

$$\widetilde{H}(\xi,\eta) = (P^*H)(\xi,\eta) = (H\circ P)(\xi,\eta) = \tfrac{1}{2}(\eta_1^2+\eta_2^2+\xi_1^2+\xi_2^2).$$

Because P is symplectic, the Hamiltonian vector field $X_{\widetilde{H}}$ corresponding to $\widetilde{H}$ is $P^{-1}X_H P$, that is,

$$\frac{d}{dt}\begin{pmatrix}\xi\\ \eta\end{pmatrix} = \begin{pmatrix}\eta\\ -\xi\end{pmatrix}.$$

Moreover the angular momentum L becomes

$$\widetilde{L} = (P^*L)(\xi,\eta) = \tfrac{1}{2}(\eta_2^2-\eta_1^2+\xi_2^2-\xi_1^2).$$

Since P is symplectic, $\widetilde{L}$ is an integral of $X_{\widetilde{H}}$. Therefore the fiber of $\widetilde{\mathcal{EM}} = \mathcal{EM}\circ P$ at (h,ℓ) is the set of (ξ,η) which satisfy

$$\begin{aligned}\tfrac{1}{2}(\eta_1^2+\eta_2^2+\xi_1^2+\xi_2^2) &= h = \widetilde{H}(\xi,\eta)\\ \tfrac{1}{2}(\eta_2^2-\eta_1^2+\xi_2^2-\xi_1^2) &= \ell = \widetilde{L}(\xi,\eta).\end{aligned} \tag{10}$$

From (10) we obtain

$$\eta_1^2+\xi_1^2 = h-\ell \tag{11}$$

$$\eta_2^2+\xi_2^2 = h+\ell. \tag{12}$$

Therefore when (h,ℓ) is a regular value of $\widetilde{\mathcal{EM}}$, that is, when $0 \le |\ell| < h$, (11) and (12) each define a circle. Hence $\widetilde{\mathcal{EM}}^{-1}(h,\ell)$ is a 2-torus $\widetilde{T}^2_{h,\ell}$. Since P is a diffeomorphism, it follows that $\mathcal{EM}^{-1}(h,\ell)$ is a 2-torus when (h,ℓ) is a regular value. □

We now describe geometrically how the orbits of $X_{\widetilde{H}}|\widetilde{T}^2_{h,\ell}$ foliate the 2-torus $\widetilde{T}^2_{h,\ell}$.

▷ Observe that the flow $\varphi_t^{\widetilde{H}}$ of $X_{\widetilde{H}}$ defines a free proper action of $S^1 = \mathbf{R}/2\pi\mathbf{Z}$ on the 2-torus $\widetilde{T}^2_{h,\ell}$.

(2.2) **Proof**: Since S^1 is compact, we need only show that at every point $p \in \widetilde{T}^2_{h,\ell}$ the isotropy group $\{t \in S^1 \mid \varphi_t^{\widetilde{H}}(p) = p\}$ is the identity element of S^1. For $(\xi,\eta) \neq 0$ and $t \in \mathbf{R}$

$$\begin{pmatrix}\xi\\ \eta\end{pmatrix} = \varphi_t^{\widetilde{H}}\begin{pmatrix}\xi\\ \eta\end{pmatrix} = \begin{pmatrix}(\cos t)I_2 & (\sin t)I_2\\ -(\sin t)I_2 & (\cos t)I_2\end{pmatrix}\begin{pmatrix}\xi\\ \eta\end{pmatrix}$$

implies that $t = 2n\pi$ for every $n \in \mathbf{Z}$. These values of t correspond to the identity element in S^1 under the orbit mapping $\mathbf{R} \to \mathbf{R}/2\pi\mathbf{Z}$. □

Thus the orbit space $\widetilde{T}^2_{h,\ell}/S^1$ is a smooth manifold, (see appendix B ((2.9))).

Claim: The orbit space $\widetilde{T}^2_{h,\ell}/S^1$ is diffeomorphic to S^1.

(2.3) **Proof**: To see this we construct a global *cross section* for the flow of $X_{\widetilde{H}}$ on $\widetilde{T}^2_{h,\ell}$. For $p = (\sqrt{h-\ell}, 0, 0, \sqrt{h+\ell}) \in \widetilde{T}^2_{h,\ell}$, the integral curve

$$[0,\pi] \to \widetilde{T}^2_{h,\ell} : s \to \varphi_s^{\widetilde{H}-\widetilde{L}}(p) = \left(\sqrt{h-\ell}\cos 2s, 0, -\sqrt{h-\ell}\sin 2s, \sqrt{h+\ell}\right)$$

of $X_{\widetilde{H}-\widetilde{L}}$ on $\widetilde{T}^2_{h,\ell}$ is closed and hence defines a circle

$$\mathscr{C} = \left\{(\xi,\eta) \in \widetilde{T}^2_{h,\ell} \middle| \xi_1^2 + \eta_1^2 = h-\ell,\ \xi_2 = 0 \ \&\ \eta_2 = \sqrt{h+\ell}\right\}.$$

Next we show that $\mathscr{C}$ is the desired cross section. To do this we have to verify that for every $(\xi,\eta) \in \widetilde{T}^2_{h,\ell} - \mathscr{C}$ the integral curve

$$t \to \varphi_t^{\widetilde{H}}(\xi,\eta) = (\xi_1\cos t + \eta_1 \sin t, \xi_2 \cos t + \eta_2 \sin t, -\xi_1 \sin t + \eta_1 \cos t, -\xi_2 \sin t + \eta_2 \cos t)$$

of $X_{\widetilde{H}}$ intersects $\mathscr{C}$ exactly once at some positive time less than 2π. Since $\xi_1^2 + \eta_1^2 = h-\ell$ and $\xi_2^2 + \eta_2^2 = h+\ell$, there are unique $\chi_1, \chi_2 \in (0, 2\pi)$ such that

$$\begin{aligned} \xi_1 &= \sqrt{h-\ell}\cos\chi_1, \quad \eta_1 = \sqrt{h-\ell}\sin\chi_1 \\ \xi_2 &= \sqrt{h+\ell}\sin\chi_2, \quad \eta_2 = \sqrt{h+\ell}\cos\chi_2. \end{aligned}$$

Thus the integral curve

$$\begin{aligned} &t \to \varphi_t^{\widetilde{H}}(\xi,\eta) = \\ &= \left(\sqrt{h-\ell}\cos(\chi_1 - t), \sqrt{h+\ell}\sin(\chi_2+t), \sqrt{h-\ell}\sin(\chi_1 - t), \sqrt{h+\ell}\cos(\chi_2+t)\right) \end{aligned}$$

intersects $\mathscr{C}$ if and only if

$$\cos 2s = \cos(\chi_1 - t) \tag{13}$$
$$0 = \sin(\chi_2 + t) \tag{14}$$
$$\sin 2s = -\sin(\chi_1 - t) \tag{15}$$
$$1 = \cos(\chi_2 + t). \tag{16}$$

From (14) and (16) and (13) and (15) we see that

$$\chi_2 + t = 2n\pi \quad \text{for } n \in \mathbf{Z} \tag{17}$$
$$2m\pi + t = 2s + \chi_1 \quad \text{for } m \in \mathbf{Z}. \tag{18}$$

Because $\chi_2 \in (0,2\pi)$ and $t \in (0,2\pi)$, from (17) we find that $n=0$ or $n=1$. Because $s \in [0,\pi]$, $t \in (0,2\pi)$ and $\chi_1 \in (0,2\pi)$, from (18) we find that $m=0$ or $m=1$. Suppose that $n=0$. Then $0 = t + \chi_2$, which implies that $t=0$. But this is excluded by hypothesis. Therefore $n=1$ and

$$t = 2\pi - \chi_2. \tag{19}$$

Substituting (19) into (18) gives

$$s = (m+1)\pi - \tfrac{1}{2}(\chi_1 + \chi_2).$$

Since $s \geq 0$ and $0 \leq \frac{1}{2}(\chi_1 + \chi_2) < 2\pi$, we find that $m = 1$. Consequently, when $t \in (0, 2\pi]$ the curve $t \to \varphi_t^{\widetilde{H}}(\xi, \eta)$ intersects $\mathscr{C}$ exactly once at

$$q = \varphi^{\widetilde{H}}_{2\pi-\chi_2}(\xi, \eta) = q(\xi, \eta) = \left(\sqrt{h-\ell}\cos(\chi_1+\chi_2), 0, \sqrt{h-\ell}\sin(\chi_1+\chi_2), \sqrt{h+\ell}\right),$$

where

$$\cos(\chi_1+\chi_2) = \frac{\xi_1\eta_2 - \xi_2\eta_1}{\sqrt{h^2-\ell^2}} \quad \text{and} \quad \sin(\chi_1+\chi_2) = \frac{\xi_1\xi_2 + \eta_1\eta_2}{\sqrt{h^2-\ell^2}}.$$

Since every integral curve of $X_{\widetilde{H}}$ on $\widetilde{T}^2_{h,\ell}$ is periodic of minimal period 2π, even if $(\xi, \eta) \in \mathscr{C}$, the curve $t \to \varphi_t^{\widetilde{H}}(\xi, \eta)$ intersects $\mathscr{C}$ exactly once when $t \in (0, 2\pi]$. Therefore the circle $\mathscr{C}$ is a model for the *orbit space* $\widetilde{T}^2_{h,\ell}/S^1$.

Alternatively, we can give a more abstract proof as follows. Since the S^1 action defined by the flow of $X_{\widetilde{H}}$ is free and proper, we know from results proved in appendix B ((2.9)) that $\widetilde{T}^2_{h,\ell}$ is the total space of a principal bundle with base a one dimensional manifold. Compactness and connectedness of $\widetilde{T}^2_{h,\ell}$ implies that the base must be a circle. □

Suppose that $(\xi, \eta) \in \mathscr{C}$. After time 2π the integral curve of $X_{\widetilde{H}}$ through (ξ, η) intersects $\mathscr{C}$ for the first positive time at $\varphi^{\widetilde{H}}_{2\pi}(\xi, \eta)$, which is in fact (ξ, η). Since the integral curve through (ξ, η) of $X_{\widetilde{H}-\widetilde{L}}$ parameterizes $\mathscr{C}$ and crosses (ξ, η) for the first positive time at 2π, we find that

$$\varphi^{\widetilde{H}-\widetilde{L}}_{2\pi}(\xi, \eta) = \varphi^{\widetilde{H}}_{2\pi}(\xi, \eta).$$

In other words, the integral curve of $X_{\widetilde{H}}$ through a point on $\mathscr{C}$ winds once around $\mathscr{C}$ as its projection winds once around the circle

$$\left\{(\xi, \eta) \in \widetilde{T}^2_{h,\ell} \middle| \xi_1 = 0, \eta_1 = \sqrt{h-\ell},\ \xi_2^2 + \eta_2^2 = h + \ell\right\}.$$

Therefore the rotation number of the flow of $X_{\widetilde{H}}$ on $\widetilde{T}^2_{h,\ell}$ is 1. Applying the linear symplectic coordinate change P^{-1} with P given by (9), we find that $P^{-1}\mathscr{C}$ is a global cross section for the flow of X_H on $T^2_{h,\ell}$ and that every integral curve of $X_H|T^2_{h,\ell}$ has rotation number 1.

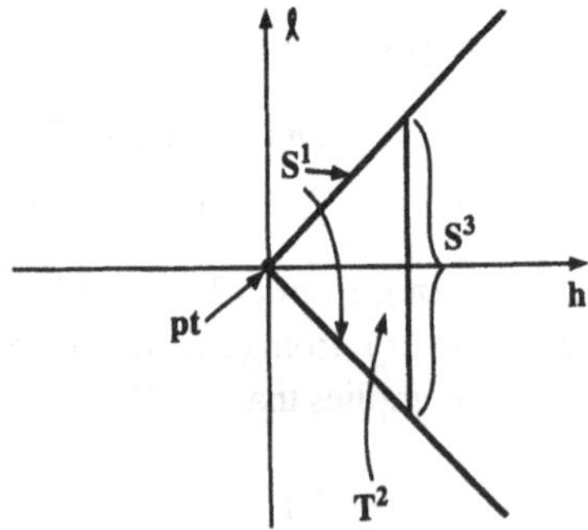

Figure 2.1 The image of the energy momentum mapping $\mathscr{EM}$ is the closed wedge $0 \leq |\ell| \leq h$. The topological type of the fibers of $\mathscr{EM}$ is as indicated.

The information we have obtained so far about the level sets of the energy momentum mapping of the harmonic oscillator is summarized in the *bifurcation diagram* figure 2.1. The set of regular values of $\mathcal{EM}$ is the open wedge $\mathcal{T}$ in the (h, ℓ) plane defined by $0 \le |\ell| < h$, since the critical values of $\mathcal{EM}$ are the two rays $\{(h, \pm h) \in \mathbf{R}_{>} \times \mathbf{R} \,|\, h > 0\}$. Because $\mathcal{T}$ is *simply connected*, the energy momentum mapping $\mathcal{EM}$ defines a *trivial* smooth fibration over $\mathcal{T}$ with fiber T^2, that is, $\mathcal{EM}^{-1}(\mathcal{T})$ is diffeomorphic to $\mathcal{T} \times T^2$, (see appendix C section 2).

To complete the qualitative analysis of the energy momentum mapping we need only understand how an energy surface $H^{-1}(h)$, $h > 0$ (which is diffeomorphic to the 3-sphere S^3) is built up from the fibers of the energy momentum mapping $\mathcal{EM}^{-1}(h, \ell)$ as ℓ varies over $[-h, h]$. Recall that $\mathcal{EM}^{-1}(h, \ell)$ is a circle when $|\ell| = h$ and is a 2-torus T^2 when $|\ell| < h$. This problem will be solved in the next two sections by showing the the S^1 momentum mapping $\mathcal{EM}$ has an extension to a U(2) momentum mapping whose restriction to $H^{-1}(h)$ is the Hopf fibration.

3 U(2) momentum mapping

As with our construction of the S^1 energy momentum mapping, we begin our construction of the U(2) momentum mapping by looking for additional integrals of the harmonic oscillator.

We start by looking for quadratic ones. Suppose that F is a homogeneous quadratic function on $\mathbf{R}^4$. Then F is an integral of X_H if and only if

$$0 = L_{X_H} F = \{F, H\},$$

where $\{\,,\,\}$ is the *standard Poisson bracket* in $C^\infty(\mathbf{R}^4)$, (see appendix A section 4). From $[X_F, X_H] = -X_{\{F,H\}}$, it follows that $[X_F, X_H] = 0$. Conversely, if $[X_F, X_H] = 0$, then $X_{\{F,H\}} = 0$. Thus the function

$$\{F, H\} = (\frac{\partial F}{\partial x}, \frac{\partial H}{\partial y}) - (\frac{\partial F}{\partial y}, \frac{\partial H}{\partial x})$$

is constant. But F is a homogeneous quadratic function on $\mathbf{R}^4$. Hence $\frac{\partial F}{\partial x} = \frac{\partial F}{\partial y} = 0$ at the origin. Therefore $\{F, H\} = 0$, that is, F is an integral of X_H. Thus we have proved

Claim: The homogeneous quadratic function $F : \mathbf{R}^4 \to \mathbf{R}$ is an integral of the harmonic oscillator vector field X_H if and only if $[X_F, X_H] = 0$.

Since F is a homogeneous quadratic function, the Hamiltonian vector field X_F is linear. Thus F is a homogeneous quadratic integral of X_H if and only if the matrices X_F and X_H commute. Now every homogeneous quadratic function F on $\mathbf{R}^4$ is given by a 4×4 symmetric matrix, which can be written as $\begin{pmatrix} -B & A^t \\ A & C \end{pmatrix}$, where B and C are 2×2 symmetric matrices. In other words,

$$2\,F(x, y) = -(Bx, x) + 2\,(Ax, y) + (Cy, y).$$

Therefore the Hamiltonian vector field X_F corresponding to F is has integral curves which satisfy

$$\frac{dx}{dt} = \frac{\partial F}{\partial y} = Ax + Cy$$
$$\frac{dy}{dt} = -\frac{\partial F}{\partial x} = Bx - A^t y.$$

Thus X_F is the 4×4 infinitesimally symplectic matrix $\begin{pmatrix} A & C \\ B & -A^t \end{pmatrix}$, where $B = B^t$ and $C = C^t$, that is, $X_F \in \mathrm{sp}(4, \mathbf{R})$, (see appendix B section 3 example 2). The infinitesimally symplectic matrix corresponding to the harmonic oscillator vector field X_H is $\begin{pmatrix} 0 & I_2 \\ -I_2 & 0 \end{pmatrix}$. A calculation shows that X_F and X_H commute if and only if $X_F = \begin{pmatrix} A & -B \\ B & A \end{pmatrix}$, where $A = -A^t$ and $B = B^t$.

Claim: The set of all 4×4 infinitesimally symplectic matrices which commute with the infinitesimally symplectic matrix X_H corresponding to the harmonic oscillator vector field is isomorphic to the Lie algebra $\mathrm{u}(2)$ of the Lie group $\mathrm{U}(2)$ of 2×2 unitary matrices.

(3.1) **Proof**: Define a mapping

$$\vartheta : \mathrm{u}(2) \to \mathrm{sp}\,(4, \mathbf{R}) : A + iB \to \begin{pmatrix} A & -B \\ B & A \end{pmatrix} = \mathscr{A}.$$

Since $A + iB \in \mathrm{u}(2)$,

$$A^t - iB^t = (\overline{A + iB})^t = -(A + iB).$$

Hence $A = -A^t$ and $B = B^t$, which implies that $\mathscr{A} \in \mathrm{sp}(4, \mathbf{R})$. Thus the image of the mapping ϑ is contained in $\mathrm{sp}(4, \mathbf{R})$. Clearly ϑ is bijective on its image and linear. □

From now on we will consider $\mathrm{u}(2)$ to be a subspace of $\mathrm{sp}(4, \mathbf{R})$. Let

$$E_1 = \begin{pmatrix} 0 & -1 & 0 & 0 \\ 1 & 0 & 0 & 0 \\ 0 & 0 & 0 & -1 \\ 0 & 0 & 1 & 0 \end{pmatrix}, \quad E_2 = \begin{pmatrix} 0 & 0 & 0 & 1 \\ 0 & 0 & 1 & 0 \\ 0 & -1 & 0 & 0 \\ -1 & 0 & 0 & 0 \end{pmatrix},$$
$$E_3 = \begin{pmatrix} 0 & 0 & 1 & 0 \\ 0 & 0 & 0 & -1 \\ -1 & 0 & 0 & 0 \\ 0 & 1 & 0 & 0 \end{pmatrix}, \quad E_4 = \begin{pmatrix} 0 & 0 & 1 & 0 \\ 0 & 0 & 0 & 1 \\ -1 & 0 & 0 & 0 \\ 0 & -1 & 0 & 0 \end{pmatrix}.$$

Then $\{E_1, E_2, E_3, E_4\}$ form a basis for $\mathrm{u}(2)$. The quadratic Hamiltonian function corresponding to the linear Hamiltonian vector field E_i is

$$W_i(z) = \tfrac{1}{2}\,\omega(E_i z, z). \tag{20}$$

Here $z = (x, y)$ and the matrix of the canonical symplectic form ω on $\mathbf{R}^4$ is $-E_4$. Thus we have shown

Claim: The functions

$$\begin{aligned} W_1(x,y) &= x_1y_2 - x_2y_1 \\ W_2(x,y) &= x_1x_2 + y_1y_2 \\ W_3(x,y) &= \tfrac{1}{2}(y_1^2 + x_1^2 - y_2^2 - x_2^2) \\ W_4(x,y) &= \tfrac{1}{2}(x_1^2 + x_2^2 + y_1^2 + y_2^2) \ = \ H(x,y) \end{aligned} \tag{21}$$

are a basis for the vector space of all the quadratic integrals of the harmonic oscillator vector field.

We now give a geometric interpretation of these integrals. Let

$$\gamma : \mathbf{R} \to \mathbf{R}^2 : t \to \big(x_1(t), x_2(t)\big)$$

be the projection onto configuration space of an integral curve of X_H of energy $h > 0$ starting at (x^0, y^0). From (1) we see that $\gamma(t)$ is given by

$$\begin{aligned} x_1(t) &= x_1^0 \cos t + y_1^0 \sin t \\ x_2(t) &= x_2^0 \cos t + y_2^0 \sin t. \end{aligned} \tag{22}$$

Since all the integral curves of X_H of positive energy are closed, it follows that γ is a planar closed curve, called a *Lissajous figure*. To find out what curve γ is we eliminate the variable t from (22) and obtain

$$\xi_4 x_1^2 - 2\xi_2 x_1x_2 + \xi_3 x_2^2 = \xi_3\xi_4 - \xi_2^2, \tag{23}$$

where

$$\begin{aligned} \xi_1 &= W_1(x^0,y^0) = x_1^0y_2^0 - x_2^0y_1^0 \\ \xi_2 &= W_2(x^0,y^0) = x_1^0x_2^0 + y_1^0y_2^0 \\ \xi_3 &= (W_4 + W_3)(x^0,y^0) = (x_1^0)^2 + (y_1^0)^2 \\ \xi_4 &= (W_4 - W_3)(x^0,y^0) = (x_2^0)^2 + (y_2^0)^2. \end{aligned}$$

Note that

$$\xi_3\xi_4 - \xi_2^2 = (x_1^0y_2^0 - x_2^0y_1^0)^2 = (W_1(x^0,y^0))^2 = \xi_1^2.$$

Suppose that $\xi_1 \neq 0$. Then

$$\det A = \det\begin{pmatrix} \xi_4 & \xi_2 \\ \xi_2 & \xi_3 \end{pmatrix} = \xi_1^2 > 0.$$

Since $\xi_3 + \xi_4 = 2h > 0$ and A is symmetric, the matrix A has two positive eigenvalues $\lambda_1 \geq \lambda_2$. Hence (23) defines an ellipse $\mathscr{E}$, see figure 3.1. All the geometric elements describing $\mathscr{E}$ are functions of the ξ_i for $i = 1, \ldots, 4$. In particular $\mathscr{E}$ has its *center* at the origin and its *major semiaxis* has length $\frac{1}{\sqrt{\lambda_2}}$. The eccentricity of $\mathscr{E}$ is $\sqrt{1 - \frac{\lambda_2}{\lambda_1}}$ and the angle the major axis makes with the x_1-axis is $\frac{1}{2}\tan^{-1}\frac{\xi_2}{\xi_3 - \xi_4}$.

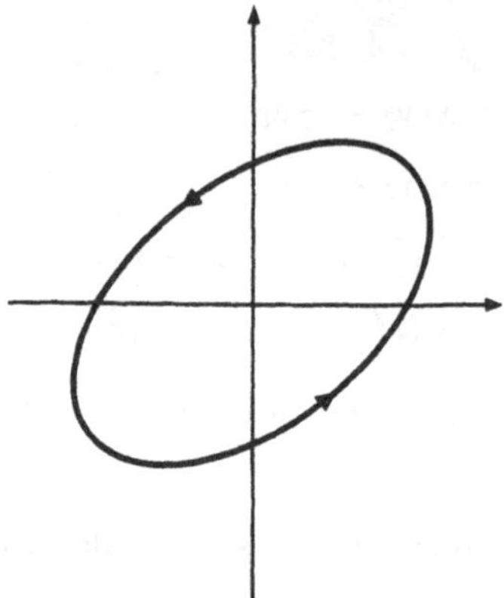

Figure 3.1 A Lissajous figure for the harmonic oscillator.

Now suppose that $\xi_1 = 0$. Then the matrix A has rank 1. Hence (23) defines a line L given by

$$\xi_4 x_1 - \xi_2 x_2 = 0 \quad \text{or} \quad -\xi_2 x_1 + \xi_3 x_2 = 0. \tag{24}$$

Since the energy is fixed, we have

$$x_1^2 + x_2^2 \le x_1^2 + y_1^2 + x_2^2 + y_2^2 = 2h. \tag{25}$$

Therefore γ is a line segment defined by that part of L which lies inside the circle (25).

We now return to the problem of finding an extension of the S^1 momentum map of section 1. Following the construction of the S^1 momentum mapping, we look for a group acting on $T^*\mathbf{R}^2$ which properly contains S^1 and has Hamiltonian vector fields for its infinitesimal generators. In contrast to the S^1 case, the action on $T^*\mathbf{R}^2$ we find will *not* be a lift of an action on the configuration space $\mathbf{R}^2$, but instead will act on the whole phase space $T^*\mathbf{R}^2$. This is reflected in the observation that some of the new integrals are not linear in the momenta, (see appendix B section 3). Consider the *linear action* of the *unitary group*

$$\mathrm{U}(2) = \left\{ \mathrm{U} = \begin{pmatrix} a & -b \\ b & a \end{pmatrix} \in \mathrm{Gl}_4(\mathbf{R}) \,\middle|\, \begin{array}{c} a^t a + b^t b = I_2 \ \&\ a^t b = b^t a \\ a, b \in \mathrm{gl}_2(\mathbf{R}) \end{array} \right\}$$

on $\mathbf{R}^4 = T^*\mathbf{R}^2$ defined by

$$\Psi : \mathrm{U}(2) \times \mathbf{R}^4 \to \mathbf{R}^4 : \big(U, (x, y)\big) \to U \begin{pmatrix} x \\ y \end{pmatrix}.$$

Since

$$\begin{pmatrix} a & -b \\ b & a \end{pmatrix}^t \begin{pmatrix} 0 & I_2 \\ -I_2 & 0 \end{pmatrix} \begin{pmatrix} a & -b \\ b & a \end{pmatrix} = \begin{pmatrix} 0 & I_2 \\ -I_2 & 0 \end{pmatrix},$$

Ψ_U is a linear symplectic map, that is, $\Psi_U^*\omega = \omega$. For $u = \begin{pmatrix} A & -B \\ B & A \end{pmatrix} \in \mathrm{u}(2)$, the infinitesimal generator Y^u corresponding to u is the vector field

$$Y^u(x, y) = \left.\frac{d}{ds}\right|_{s=0} \Psi_{\exp su} \begin{pmatrix} x \\ y \end{pmatrix} = \begin{pmatrix} A & -B \\ B & A \end{pmatrix} \begin{pmatrix} x \\ y \end{pmatrix}$$

with flow

$$\psi_t^u : \mathbf{R}^4 \to \mathbf{R}^4 : \begin{pmatrix} x \\ y \end{pmatrix} \to \exp tu \begin{pmatrix} x \\ y \end{pmatrix}.$$

Because the matrix Y^u is infinitesimally symplectic, the vector field Y^u is linear Hamiltonian with Hamiltonian function

$$J^u : \mathbf{R}^4 \to \mathbf{R} : (x, y) \to \tfrac{1}{2}\omega\Big(u \begin{pmatrix} x \\ y \end{pmatrix}, \begin{pmatrix} x \\ y \end{pmatrix}\Big). \tag{26}$$

For fixed $(x, y) \in \mathbf{R}^4$, the function $u \to J^u(x, y)$ is linear. Therefore by duality it makes sense to define the mapping $J : T^*\mathbf{R}^2 \to \mathrm{u}(2)^*$ by setting $J(x, y)u = J^u(x, y)$.
▷ An important property of J is that it intertwines the linear action of U(2) on $\mathbf{R}^4$ with the coadjoint action of U(2) on $\mathrm{u}(2)^*$, the dual of the Lie algebra u(2).

(3.2) **Proof**: This is a consequence of the calculation

$$\begin{aligned} J(U(x, y))u &= J^u\Big(U\begin{pmatrix} x \\ y \end{pmatrix}\Big) = \tfrac{1}{2}\omega\Big(u\,U\begin{pmatrix} x \\ y \end{pmatrix}, U\begin{pmatrix} x \\ y \end{pmatrix}\Big) \\ &= \tfrac{1}{2}\omega\Big(U^{-1}\,u\,U\begin{pmatrix} x \\ y \end{pmatrix}, \begin{pmatrix} x \\ y \end{pmatrix}\Big), \quad \text{since } U \text{ is symplectic} \\ &= J(x, y)\big(U^{-1}\,u\,U\big) = J(x, y)(\mathrm{Ad}_{U^{-1}}\,u) \\ &= \big(\mathrm{Ad}^t_{U^{-1}} J(x, y)\big)\,u. \qquad \square \end{aligned}$$

Therefore J is the U(2) momentum mapping corresponding to the linear action Ψ, (see appendix B section 3). If we identify $\mathrm{u}(2)^*$ with u(2) using the *Killing metric* k defined by $k(u, v) = \frac{1}{2}\,\mathrm{tr}\, u\,\overline{v}^t$, then the momentum mapping intertwines the linear action of U(2) on $\mathbf{R}^4$ with the adjoint action of U(2) on its Lie algebra u(2). Identifying u(2) with $\mathbf{R}^4$ by choosing the basis $\{E_i\}$, the U(2) momentum mapping J becomes the mapping

$$\mathbf{R}^4 \to \mathbf{R}^4 : (x, y) \to \big(W_1(x, y), W_2(x, y), W_3(x, y), W_4(x, y)\big).$$

▷ Thus the components of the U(2) momentum mapping J are quadratic integrals of the harmonic oscillator.

(3.3) **Proof**: This is just the content of (20). An alternative argument starts by observing that every element $U \in \mathrm{U}(2)$ when written as $\begin{pmatrix} a & -b \\ b & a \end{pmatrix}$ is also an element of SO (4), the group of all orientation preserving linear *isometries* of $(\mathbf{R}^4, \langle\, , \rangle)$. Consequently, Ψ_U preserves H and so

$$0 = L_{Y^u} H = -L_{X_H} J^u. \qquad \square$$

Now consider the 2-torus

$$T^2 = \left\{ \begin{pmatrix} ac & -ad & -bc & bd \\ ad & ac & -bd & -bc \\ bc & -bd & ac & -ad \\ bd & bc & ad & ac \end{pmatrix} \in \mathrm{U}(2) \,\middle|\, \begin{array}{l} a^2 + b^2 = 1 \\ c^2 + d^2 = 1 \end{array} \right\},$$

which is an abelian subgroup of U(2). Restricting the U(2) action on $\mathbf{R}^4$ to a T^2 action gives rise to a momentum mapping $j : T^*\mathbf{R}^2 \to (t^2)^* = \mathbf{R}^2$ where $(t^2)^*$ is the dual of the Lie algebra t^2 of T^2 and $j(x, y) = \big(W_4(x, y), W_1(x, y)\big)$. In other words, j is the S^1 energy momentum map $\mathcal{EM}$ studied in section 2. Therefore the U(2) momentum mapping J is a proper extension of $\mathcal{EM}$.

4 The Hopf fibration

In this section we study the qualitative properties of the *Hopf mapping*

$$\mathcal{H} : \mathbf{R}^4 \to \mathbf{R}^4 : (x, y) \to \big(w_1(x, y), w_2(x, y), w_3(x, y), w_4(x, y)\big), \tag{27}$$

where

$$\begin{aligned} w_1 = 2W_1 &= 2(x_1y_2 - x_2y_1) \\ w_2 = 2W_2 &= 2(x_1x_2 + y_1y_2) \\ w_3 = 2W_3 &= y_1^2 + x_1^2 - y_2^2 - x_2^2 \\ w_4 = 2W_4 &= y_1^2 + x_1^2 + y_2^2 + x_2^2. \end{aligned} \tag{28}$$

The *Hopf variables* w_i, $i = 1, \ldots, 4$ satisfy the relation

$$C(w_1, w_2, w_3, w_4) = w_1^2 + w_2^2 + w_3^2 - w_4^2 = 0, \quad w_4 \geq 0, \tag{29}$$

since

$$w_1^2 + w_2^2 = 4(x_1^2 + y_1^2)(x_2^2 + y_2^2) = w_4^2 - w_3^2.$$

Therefore the image of the Hopf map is contained in the *semialgebraic variety* $\mathcal{C}$ defined ▷ by (29). Topologically $\mathcal{C}$ is a cone on S^2 with vertex at 0. To show that $\mathcal{C}$ is the image of the Hopf map, it suffices to verify that

$$D\mathcal{H}(x, y) : T_{(x,y)}\mathbf{R}^4 \to T_{\mathcal{H}(x,y)}\mathcal{C}$$

is surjective for all $(x, y) \neq (0, 0)$, because $\mathcal{H}(0) = 0$ and $\mathcal{C} - \{0\}$ is a smooth three dimensional manifold.

(4.1) **Proof**: When $(x, y) \neq (0, 0)$ the derivative

$$D\mathcal{H}(x, y) = 2\begin{pmatrix} y_2 & -y_1 & -x_2 & x_1 \\ x_2 & x_1 & y_2 & y_1 \\ x_1 & -x_2 & y_1 & -y_2 \\ x_1 & x_2 & y_1 & y_2 \end{pmatrix}$$

has rank ≥ 3 because its first three rows are nonzero and pairwise orthogonal. It has rank ≤ 3, since

$$\operatorname{im} D\mathcal{H}(x, y) \subseteq T_{\mathcal{H}(x,y)}C = \ker DC(\mathcal{H}(x, y)) = \ker(w_1, w_2, w_3, w_4)$$

and (w_1, w_2, w_3, w_4) is nonzero. □

Restricting $\mathcal{H}$ to the 3-sphere

$$S^3_{\sqrt{2h}} = \left\{(x, y) \in \mathbf{R}^4 \middle| y_1^2 + y_2^2 + x_1^2 + x_2^2 = 2h, \;\; h > 0\right\}$$

and using (29) gives

$$\mathcal{F} : S^3_{\sqrt{2h}} \subseteq \mathbf{R}^4 \to S^2_{2h} \subseteq \mathbf{R}^3 : (x, y) \to \big(w_1(x, y), w_2(x, y), w_3(x, y)\big). \tag{30}$$

Here S^2_{2h} is the 2-sphere $\{w \in \mathbf{R}^3 \mid w_1^2 + w_2^2 + w_3^2 = (2h)^2\}$. The map $\mathcal{F}$ is called the *Hopf fibration*. From a topological point of view the Hopf fibration is quite complicated and will require quite a bit of work to be understood.

Claim: The Hopf fibration $\mathcal{F}$ has the following properties:

I). $\mathcal{F}$ is a proper *submersion*.

II). For every $w \in S^2_{2h}$ the fiber $\mathcal{F}^{-1}(w)$ is a *great circle* on $S^3_{\sqrt{2h}}$ contained in the 2-plane Π^w, (see (31)).

III). For every $w, w' \in S^2_{2h}$ with $w \neq w'$, the circles $\mathcal{F}^{-1}(w)$ and $\mathcal{F}^{-1}(w')$ are linked in $S^3_{\sqrt{2h}}$ with linking number 1, (see (33)).

(4.2) **Proof**:

I). Consider the mapping

$$F : \mathbf{R}^4 \to \mathbf{R}^3 :$$
$$(x, y) \to (w_1, w_2, w_3) = \big(2(x_1y_2 - x_2y_1), 2(x_1x_2 + y_1y_2), y_1^2 + x_1^2 - y_2^2 - x_2^2\big).$$

Clearly F is smooth and has derivative

$$DF(x, y) = 2\begin{pmatrix} y_2 & -y_1 & -x_2 & x_1 \\ x_2 & x_1 & y_2 & y_1 \\ x_1 & -x_2 & y_1 & -y_2 \end{pmatrix}.$$

Since $\mathcal{F} = F|S^3_{\sqrt{2h}}$, the mapping $\mathcal{F}$ is smooth. When $(x, y) \neq (0, 0)$ the rows of $DF(x, y)$ are nonzero and pairwise orthogonal. Hence $D\mathcal{F}(x, y) = DF(x, y)|T_{(x,y)}S^3_{\sqrt{2h}}$ is a surjective linear map from $T_{(x,y)}S^3_{\sqrt{2h}}$ to $T_{\mathcal{F}(x,y)}S^2_{2h}$ for every $(x, y) \in S^3_{\sqrt{2h}}$. In other words, $\mathcal{F}$ is a submersion. In addition, $\mathcal{F}$ is a proper map, because its domain is compact.

II). The following argument shows that every fiber of $\mathcal{F}$ is a great circle on $S^3_{\sqrt{2h}}$. Let Π^w be the 2-plane in $\mathbf{R}^4$ defined by

$$\begin{aligned} \pi_1(x, y) &= (2h + w_3)x_2 - w_2x_1 + w_1y_1 = 0 \\ \pi_2(x, y) &= (2h + w_3)y_2 - w_1x_1 - w_2y_1 = 0, \end{aligned} \tag{31}$$

when $w \in (S^2_{2h} - \{(0, 0, -2h)\})$, and the 2-plane

$$\Big\{(0, x_2, 0, y_2) \in \mathbf{R}^4 \Big| (x_2, y_2) \in \mathbf{R}^2\Big\}$$

when $w = (0, 0, -2h)$. To see that (31) defines a 2-plane we argue as follows. We know that the covectors $\pi_1(x, y)$ and $\pi_2(x, y)$ are linearly dependent if and only if $0 = \pi_1(x, y) \wedge \pi_2(x, y)$, that is, when all the 2×2 minors of

$$\begin{pmatrix} -w_2 & 2h + w_3 & w_1 & 0 \\ -w_1 & w_2 & 0 & 2h + w_3 \end{pmatrix}$$

vanish. In other words, $w_1 = w_2 = 0$ and $w_3 = -2h$.

We now show that $\mathcal{F}^{-1}(w) \subseteq \Pi^w \cap S^3_{\sqrt{2h}}$. Suppose that $w \in (S^2_{2h} - \{(0,0,-2h)\})$. Then for every $(x,y) \in \mathcal{F}^{-1}(w) \subseteq S^3_{\sqrt{2h}}$, we have $x_1^2 + y_1^2 > 0$. To see this note that $(x,y) \in \mathcal{F}^{-1}(w)$ if and only if (28) holds. Adding the last two equations in (28) gives $2(x_1^2+y_1^2) = w_3+w_4$. But $(x,y) \in S^3_{\sqrt{2h}}$, so $w_4 = 2h$. Therefore $2(x_1^2+y_1^2) = w_3+2h$. Since $w_3 \in (-2h, 2h]$, we obtain $x_1^2 + y_1^2 > 0$. Now write the first two equations in (28) as

$$2\begin{pmatrix} -y_1 & x_1 \\ x_1 & y_1 \end{pmatrix}\begin{pmatrix} x_2 \\ y_2 \end{pmatrix} = \begin{pmatrix} w_1 \\ w_2 \end{pmatrix}. \tag{32}$$

Since $x_1^2 + y_1^2 > 0$ we may solve (32) to obtain

$$\begin{pmatrix} -y_1 & x_1 \\ x_1 & y_1 \end{pmatrix}\begin{pmatrix} w_1 \\ w_2 \end{pmatrix} = 2(x_1^2+y_1^2)\begin{pmatrix} x_2 \\ y_2 \end{pmatrix} = (2h+w_3)\begin{pmatrix} x_2 \\ y_2 \end{pmatrix},$$

which is (31). Thus $\mathcal{F}^{-1}(w) \subseteq \Pi^w$. Hence $\mathcal{F}^{-1}(w) \subseteq \Pi^w \cap S^3_{\sqrt{2h}}$. Now suppose that $w = (0,0,-2h)$. Then the last two equations in (28) become

$$\begin{aligned} y_1^2 + x_1^2 - x_2^2 - y_2^2 &= -2h \\ y_1^2 + x_1^2 + x_2^2 + y_2^2 &= 2h, \end{aligned}$$

since $(x,y) \in S^3_{\sqrt{2h}}$. Adding these equations together and dividing by 2 gives $y_1^2+x_1^2 = 0$, that is, $x_1 = y_1 = 0$. Hence $\mathcal{F}^{-1}(0,0,-2h) \subseteq \Pi^{(0,0,-2h)} \cap S^3_{\sqrt{2h}}$. Therefore every fiber $\mathcal{F}^{-1}(w)$ is contained in $\Pi^w \cap S^3_{\sqrt{2h}}$. Because $\mathcal{F}$ is a submersion, $\mathcal{F}^{-1}(w)$ is a smooth compact one dimensional submanifold of $S^3_{\sqrt{2h}}$ without boundary. Therefore $\mathcal{F}^{-1}(w)$ is the great circle $\Pi^w \cap S^3_{\sqrt{2h}}$.

Before proving property III we must define the notion of linking number of two smooth oriented disjoint circles γ_1 and γ_2 in an oriented 3-sphere S^3. Intuitively, the circles γ_i are linked if they cannot be pulled apart without being cut, see figure 4.1. A more precise definition goes as follows. For simplicity we will assume that γ_1 bounds a smooth oriented closed 2-disc in S^3. The general case is handled exercise 3. Since S^3 is simply connected, the circle γ_1 is *null homotopic* and hence is smoothly *contractible* to a point $p \in S^3$. In other words, there is a diffeomorphism

$$F : \overline{D}^2_1 \subseteq \mathbf{R}^2 \to S^3 \subseteq \mathbf{R}^4,$$

called a contraction, such that $F(0) = p$ and such that for every $r \in (0,1]$ the map F restricted to the boundary $\partial\overline{D}^2_r$ of the closed 2-disc $\overline{D}^2_r = \{x \in \mathbf{R}^2 \,|\, (x,x) \le r^2\}$ is a diffeomorphism onto γ_1. Orient the 2-disc $\overline{D}^2_1$ so that $F|\partial\overline{D}^2_1$ is orientation preserving. Furthermore assume that F is *transverse* to γ_2 in S^3, that is, either $\gamma_2 \cap F(D^2_1) = \varnothing$ or for every x in D^2_1 such that $F(x) \in \gamma_2 \cap F(D^2_1)$ we have

$$T_x F(T_x D^2_1) + T_x\gamma_2 = T_x S^3.$$

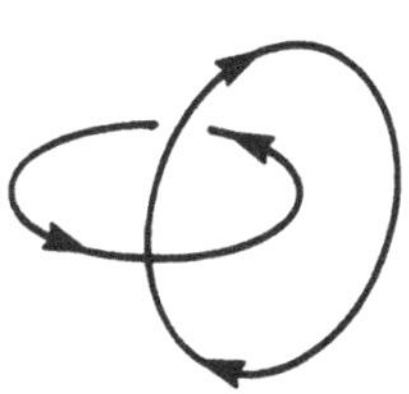

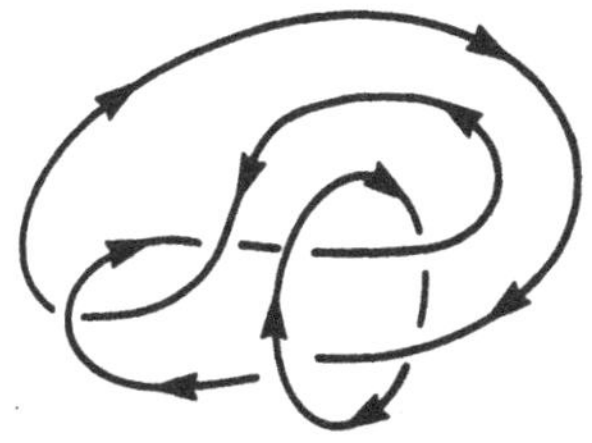

Figure 4.1. In the left hand picture the circles γ_1 and γ_2 have linking number 1. In the right hand picture they have linking number 0. In both cases they can not be pulled apart into two disjoint circles without cutting one or the other.

The *linking number* of the circles γ_1 and γ_2 is the *intersection number* of $F(D_1^2)$ with γ_2, that is,

$$\text{Link}\,(\gamma_1, \gamma_2) = \sum_{x \in \mathcal{T}} \#\big(T_x F(T_x D_1^2), T_x \gamma_2\big) \tag{33}$$

where $\mathcal{T} = \{x \in D_1^2 \mid F(x) \in \gamma_2 \cap F(D_1^2)\}$ and

$$\#\big(T_x F(T_x D_1^2), T_x \gamma_2\big) = \begin{cases} 1, & \text{if the orientation of } T_x F(T_x D_1^2) \oplus T_x \gamma_2 \\ & \text{is the same as } T_x S^3 \\ -1, & \text{otherwise.} \end{cases}$$

▷ Note that the sum in (33) is finite, since F is transverse to γ_2. The following argument shows that the definition of linking number does not depend on the choice of oriented 2-disc with boundary γ_1.

(4.3) **Proof**: Let $G : \widetilde{D}_1^2 \subseteq \mathbf{R}^2 \to S^3$ be another contraction of γ_1. Give the 2-disc $\widetilde{D}_1^2$ the *orientation* opposite to that of the 2-disc D_1^2. Consider the oriented 2-sphere $S^2 \subseteq S^3$ formed from the 2-discs $F(\overline{D}_1^2)$ and $G(\widetilde{D}_1^2)$ by identifying $\partial F(\overline{D}_1^2) = F(\partial \overline{D}_1^2)$ with $\partial G(\widetilde{D}_1^2) = G(\partial \widetilde{D}_1^2)$. The intersection number of S^2 with γ_2 is equal to its intersection number with $F(\overline{D}_1^2)$ minus its intersection number with $G(\widetilde{D}_1^2)$. If we show that the intersection number of S^2 with γ_2 is zero, then we are done. To do this choose a point p in $S^3 - (S^2 \cup \gamma_2)$ and let $\varphi : S^3 - \{p\} \to \mathbf{R}^3$ be *stereographic projection*. Since φ preserves orientation, the intersection number of S^2 and γ_2 is the same as the intersection number of $\varphi(S^2)$ and $\varphi \circ \gamma_2$. Thus we have to show that the intersection number of an oriented smooth circle $\gamma : [0, 1] \to \mathbf{R}^3$ which meets an oriented $S^2 \subseteq \mathbf{R}^3$ transversely is zero. If γ does not intersect S^2 then we are done. Suppose that at $\gamma(t_0) \in S^2$ the curve γ has intersection number 1. Then for some sufficiently small $\varepsilon > 0$, $\gamma(t_0 - \varepsilon)$ lies in the bounded component of $\mathbf{R}^3 - S^2$; while $\gamma(t_0 + \varepsilon)$ is in the unbounded component. Reparametrize γ so that γ is defined on $[0, 1]$ and begins and ends at $p = \gamma(t_0)$. Since γ is a closed curve, there is a $t_1 \in (0, 1)$ such that $\gamma(t_1) \in S^2$. Choose t_1 as small as possible. (There are only finitely many since γ intersects S^2 transversely). Then for every

$t \in (0, t_1)$, $\gamma(t)$ lies in the unbounded component of $\mathbf{R}^3 - S^2$. Now $q_1 = \gamma(t_1) \neq p$ since γ is a diffeomorphism. Since γ is transverse to S^2 at q_1, $\gamma(t_1 + \varepsilon)$ lies in the bounded component of $\mathbf{R}^3 - S^2$ for sufficiently small $\varepsilon > 0$. Therefore the intersection number of γ at q_1 is -1. A similar argument shows that the next intersection point q_2 of γ with S^2 has intersection number 1. If $q_2 = p$ then we are through; otherwise repeat the argument a finite number of times until $t = 1$ is reached. Since γ has an even number of intersections with S^2, its intersection number is 0. Thus the linking number is well defined. □

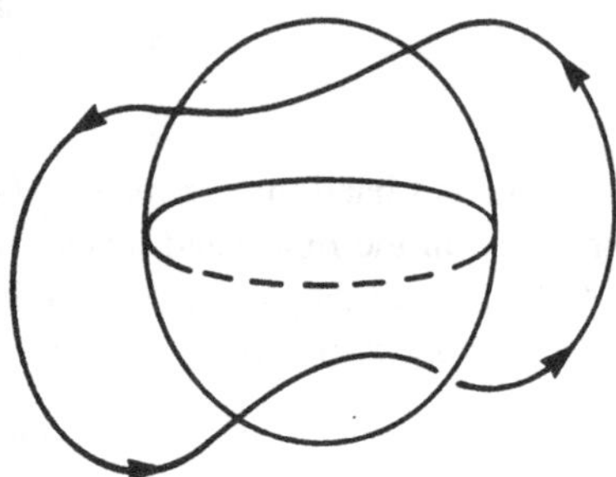

Figure 4.2. The intersection number of an oriented circle with an oriented 2-sphere is zero.

III). With these preliminaries out of the way we are in a position to prove property III of the Hopf fibration $\mathcal{F}$. Suppose that $w, v \in S^2_{2h}$ and $w \neq v$. Then $\mathcal{F}^{-1}(w) \cap \mathcal{F}^{-1}(v) = \varnothing$. Thus the corresponding 2-planes Π^w and Π^v intersect only at 0, (see the proof of property II for the definition of Π^w). Let Π be a 3-plane in $\mathbf{R}^4$ containing Π^w. Then Π^v is not contained in Π. Moreover $\Pi^v \cap \Pi = \ell^v$ is a line in $\mathbf{R}^4$. Let $S^2 = \Pi \cap S^3$ be the great 2-sphere in S^3 cut out by Π and let $S^1_w = \Pi^w \cap S^3$ be the great circle on S^3 cut out by Π^w. Furthermore let H^+ be the closed upper *hemisphere* of S^2 with boundary S^1_w. Then H^+ is diffeomorphic to a closed 2-disc $\overline{D}^2$ with boundary S^1_w (just project points of H^+ onto the equatorial plane containing S^1_w). Since Π^v is not contained in Π, the great circle $S^1_v = \Pi^v \cap S^3$ is not contained in S^2. Because the line ℓ^v intersects the great 2-sphere S^2 in two *antipodal points* p_+ and p_-, the circle S^1_v intersects S^2 at p_+ and p_-. Since $\Pi^w \cap \Pi^v = \{0\}$, p_+ and p_- do not lie on the *equator* S^1_w of S^2. Hence exactly one of the points $p_\pm$, say p_+, lies in the interior of the hemisphere H^+. Therefore the linking number of the two circles $\mathcal{F}^{-1}(w)$ and $\mathcal{F}^{-1}(v)$ in S^3 is ± 1, since we have not been careful about orientations. If we choose orientations properly we can arrange that the linking number is 1. This completes the proof of the properties of the Hopf fibration. □

We now draw some conclusions about the Hopf fibration from the properties we have just proved. From property I we know that $\mathcal{F}$ is a proper submersion. Therefore the Hopf mapping $\mathcal{F} : S^3_{\sqrt{2h}} \rightarrow S^2_{2h} : (x, y) \rightarrow w$ defines a locally trivial bundle with fiber S^1, (see appendix C section 2). To find the local trivializations of $\mathcal{F}$ explicitly we use the identities

$$x_2 = \frac{1}{2h + w_3}(w_2 x_1 - w_1 y_1) \quad \text{and} \quad y_2 = \frac{1}{2h + w_3}(w_1 x_1 + w_2 y_1) \tag{34}$$

which hold for $(x_1, x_2, y_1, y_2) \in \mathcal{F}^{-1}(w)$ when $w \in (S^2_{2h} - \{(0, 0, -2h)\}) = U_1$, (see

(31)), and the identities

$$x_1 = \frac{1}{2h - w_3}(w_2x_2 + w_1y_2) \quad \text{and} \quad y_1 = \frac{1}{2h - w_3}(-w_1x_2 + w_2y_2), \tag{35}$$

which hold when $w \in (S^2_{2h} - \{(0,0,2h)\}) = U_2$. Note that $\{U_1, U_2\}$ form an open covering of S^2_{2h}. Let $S^1 = \{(x,y) \in \mathbf{R}^2 | x^2 + y^2 = 1\}$. Consider the mappings

$$\begin{aligned} &\tau_1 : U_1 \times S^1 \to \mathcal{F}^{-1}(U_1) : (w_1, w_2, w_3, x, y) \to \\ &\tfrac{1}{\sqrt{2}}\Big(x\sqrt{2h + w_3}, \tfrac{1}{\sqrt{2h+w_3}}(w_2x - w_1y), y\sqrt{2h + w_3}, \tfrac{1}{\sqrt{2h+w_3}}(w_1x + w_2y)\Big) \end{aligned} \tag{36}$$

and

$$\begin{aligned} &\tau_2 : U_2 \times S^1 \to \mathcal{F}^{-1}(U_2) : (w_1, w_2, w_3, x, y) \to \\ &\tfrac{1}{\sqrt{2}}\Big(\tfrac{1}{\sqrt{2h-w_3}}(w_2\, x + w_1\, y), x\,\sqrt{2h - w_3}, \tfrac{1}{\sqrt{2h-w_3}}(-w_1\, x + w_2\, y), y\,\sqrt{2h - w_3}\Big) \end{aligned} \tag{37}$$

Using the definition of the Hopf fibration $\mathcal{F}$, it is easy to check that $\mathcal{F} \circ \tau_1 = \pi_1$ and $\mathcal{F} \circ \tau_2 = \pi_2$, where $\pi_1 : U_1 \times S^1 \to U_1$ and $\pi_2 : U_2 \times S^1 \to U_2$ are the projections onto the first factor. A short calculation shows that

$$\begin{aligned} \tau_1^{-1} &: \mathcal{F}^{-1}(U_1) \to U_1 \times S^1 : (x_1, x_2, y_1, y_2) \to (w_1, w_2, w_3, x, y) = \\ &= \Big(2(x_1y_2 - x_2y_1), 2(x_1x_2 + y_1y_2), y_1^2 + x_1^2 - x_2^2 - y_2^2, \tfrac{\sqrt{2}x_1}{\sqrt{2h+w_3}}, \tfrac{\sqrt{2}y_1}{\sqrt{2h+w_3}}\Big), \end{aligned}$$

since $x_1^2 + y_1^2 = \frac{1}{2}(2h + w_3)$ in $\mathcal{F}^{-1}(U_1)$, and also that

$$\begin{aligned} \tau_2^{-1} &: \mathcal{F}^{-1}(U_2) \to U_2 \times S^1 : (x_1, x_2, y_1, y_2) \to (w_1, w_2, w_3, x, y) = \\ &= \Big(2(x_1\, y_2 - x_2\, y_1), 2(x_1\, x_2 + y_1\, y_2), {y_1}^2 + {x_1}^2 - {x_2}^2 - {y_2}^2, \tfrac{\sqrt{2}x_2}{\sqrt{2h-w_3}}, \tfrac{\sqrt{2}y_2}{\sqrt{2h-w_3}}\Big), \end{aligned}$$

since ${x_2}^2 + {y_2}^2 = \frac{1}{2}(2h - w_3)$ in $\mathcal{F}^{-1}(U_2)$. Thus τ_1^{-1} and τ_2^{-1} are continuous. Consequently the mappings τ_1 and τ_2 are local trivializations of the bundle $\mathcal{F}$. The Hopf fibration is *not trivial*, that is, the bundle $\mathcal{F} : S^3_{\sqrt{2h}} \to S^2_{2h}$ is not isomorphic to the trivial bundle $S^2_{2h} \times S^1 \to S^2_{2h}$.

(4.4) **Proof**: This follows from the fact that linking number is a topological invariant and two distinct fibers of the Hopf fibration are linked with linking number one, whereas the fibers of the trivial bundle are unlinked.

Another argument uses the observation that the total space S^3_{2h} of the Hopf fibration is not diffeomorphic to the total space $S^2_{2h} \times S^1$ of the trivial bundle, because the first *homology group* of $S^3_{\sqrt{2h}}$ is 0, whereas the first homology group of $S^2_{2h} \times S^1$ is $\mathbf{Z}$.

For yet another argument see appendix C section 1. □

Claim: The Hopf bundle $\mathcal{F} : S^3_{\sqrt{2h}} \to S^2_{2h}$ is an S^1 principal bundle.

(4.5) **Proof**: We prove this assertion in several steps.

▷ First we show that the S^1 bundle

$$\pi_h : H^{-1}(h) \to M_h = H^{-1}(h)/S^1,$$

where π_h maps each orbit of energy h of the harmonic oscillator vector field X_H to a point, is a principal bundle.

(4.6) **Proof**: $H^{-1}(h)$ is invariant under the flow φ_t^H of X_H. Since $\varphi_{2\pi}^H = id_{S^3_{\sqrt{2h}}}$, the flow φ_t^H defines an action of $S^1 = \mathbf{R}/2\pi\mathbf{Z}$ on $H^{-1}(h) = S^3_{\sqrt{2h}}$ given by

$$S^1 \times S^3_{\sqrt{2h}} \to S^3_{\sqrt{2h}} : \big(t, (x, y)\big) \to \varphi_t^H(x, y) = \begin{pmatrix} (\cos t)\, I_2 & -(\sin t)\, I_2 \\ (\sin t)\, I_2 & (\cos t)\, I_2 \end{pmatrix} \begin{pmatrix} x \\ y \end{pmatrix}.$$

Because every orbit of this action has minimal period 2π, the isotropy group of every point on S^3 is the identity element of $S^1 = \mathbf{R}/2\pi\mathbf{Z}$. Therefore the S^1 action is free. It is also proper, since S^1 is compact. Thus the bundle π_h is a principal S^1 bundle, (see appendix B section 2.2 ((2.12))). □

▷ Next we show that the orbit space M_h is diffeomorphic to a 2-sphere.

(4.7) **Proof**: Because the S^1 action defined by the flow of the harmonic oscillator is free and proper, the orbit space M_h is smooth, (see appendix B section 2.2). We now use Morse theory. Consider the smooth function $L|H^{-1}(h) : H^{-1}(h) \to \mathbf{R}$, which is the restriction of the angular momentum L (2) to the energy level set $H^{-1}(h)$. Because $L|H^{-1}(h)$ is invariant under the flow of X_H, it induces a smooth function $\widetilde{L}_h : M_h \to \mathbf{R}$ on M_h. Since the set of critical points of $L|H^{-1}(h)$ consists of two disjoint circles which are nondegenerate critical submanifolds of $H^{-1}(h)$ of Morse index 0 and 2 (see ((2.1))), the function $\widetilde{L}_h$ is a Morse function on M_h with two critical points, one of Morse index 0 and the other of Morse index 2. Thus, M_h is homeomorphic to a 2-sphere, (see appendix E section 3 ((3.2))). Because M_h is smooth, it is diffeomorphic to a 2-sphere. □

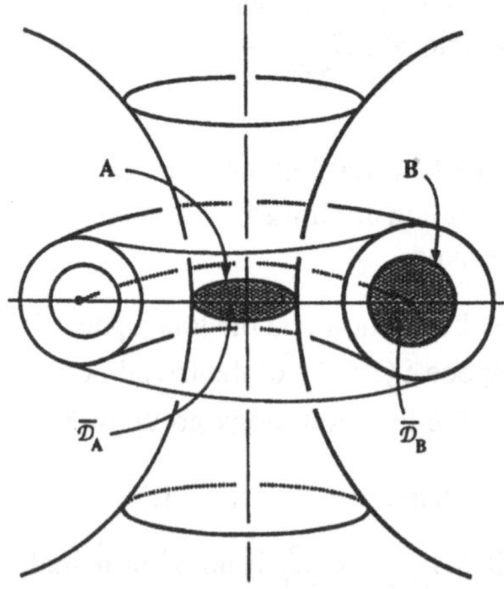

Figure 4.3. The Hopf fibration.

To visualize the orbit space consider figure 4.3. Here the 3-sphere $S^3_{\sqrt{2h}}$ is to be thought of as the *one point compactification* of $\mathbf{R}^3$, the point at infinity having been added. Thus the z-axis is actually a circle. Every S^1 fiber of the Hopf fibration passes transversely through one of the two discs $\overline{D}_A$ or $\overline{D}_B$ which have bounding circles A or B, respectively. Corresponding to each point p on A there is a unique point $\vartheta(p)$ on B where the integral curve of X_H through p meets the circle B. Gluing the disc $\overline{D}_A$ to the disc $\overline{D}_B$ along their boundary by the diffeomorphism ϑ gives a 2-sphere S^2, which is the *orbit space* M_h.

▷ The S^1 bundle $\pi_h : H^{-1}(h) \to M_h$ is *isomorphic* to the *Hopf bundle* $\mathcal{F} : S^3_{\sqrt{2h}} \to S^2_{2h}$.

(4.8) **Proof**: This result follows because the map φ making diagram 4.1 *commutative* is a diffeomorphism. In more detail, the map φ is well defined because each fiber of $\mathcal{F}$ is a single orbit of X_H. Since each fiber of π_h is also a single orbit of X_H, φ is injective. Clearly, φ is surjective. Because $\mathcal{F}$ is a submersion, it is an open mapping. Therefore φ is continuous. Its inverse is continuous since S^2_h is a compact Hausdorff space. Hence φ is a homeomorphism. Because the bundle $\mathcal{F}$ is locally trivial, it has a smooth *local section*. Hence φ is smooth. The inverse of φ is smooth because the bundle π_h is locally trivial and hence has a smooth local section. □

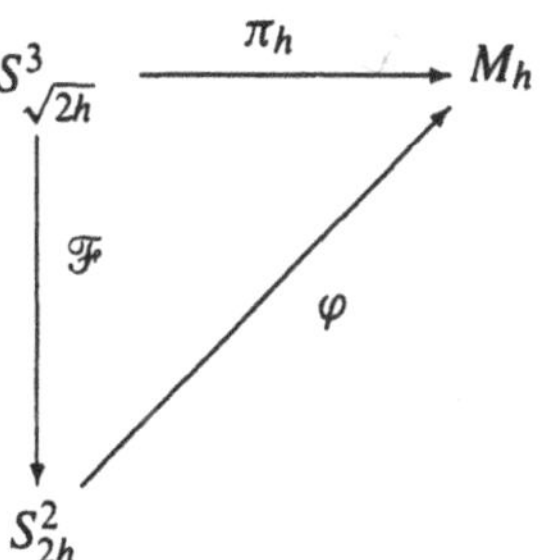

Diagram 4.1

This completes the argument that the Hopf bundle is an S^1-principal bundle. □

To determine which principal bundle $\mathcal{F}$ is, we calculate its classifying map. From the definition of the local trivializations τ_i (36–37) we find that the transition map is given by

$$\tau_2^{-1} \circ \tau_1 : (U_1 \cap U_2) \times S^1 \to (U_1 \cap U_2) \times S^1 : \left((w, \begin{pmatrix} x \\ y \end{pmatrix} \right) \to \left(w, g_{12}(w) \begin{pmatrix} x \\ y \end{pmatrix} \right)$$

where

$$g_{12} : U_1 \cap U_2 \to \mathrm{SO}(2, \mathbf{R}) = S^1 : (w_1, w_2, w_3) \to \frac{1}{\sqrt{4h^2 - w_3^2}} \begin{pmatrix} w_2 & -w_1 \\ w_1 & w_2 \end{pmatrix}.$$

Let

$$S^1_E = \{(w_1, w_2, 0) \in S^2_{2h} \mid w_1^2 + w_2^2 = 4h^2\} \subseteq U_1 \cap U_2$$

be the equator of the 2-sphere S^2_{2h}. By definition, the classifying map of $\mathcal{F}$ is

$$\chi = g_{12}|S^1_E : S^1_E \to \mathrm{SO}(2, \mathbf{R}) : (w_1, w_2) \to \frac{1}{2h} \begin{pmatrix} w_2 & -w_1 \\ w_1 & w_2 \end{pmatrix}. \tag{38}$$

Clearly the mapping χ has degree 1.

We now give a way to visualize geometrically the *fibration* of $H^{-1}(h)$ by level sets of the angular momentum L. When $0 \le \ell < h$ the 2-torus $T^2_{h,\ell} = \mathcal{EM}^{-1}(h,\ell)$ intersects the 2-disc $\overline{D}_B$ in figure 4.3 in a circle concentric with the circle B; while when $-h < \ell \le 0$ the 2-torus $T^2_{h,\ell}$ intersects the 2-disc $\overline{D}_A$ in a circle concentric with the circle A. Note that the torus $T^2_{h,0} = T^2$ contains both of the circles A and B. The X_H orbit through the center of $\overline{D}_B$ is $S^1_{h,+}$ (6); while the X_H orbit through the center of $\overline{D}_A$ is $S^1_{h,-}$ (7). The union of the X_H orbits through $\overline{D}_B$ is the closed *solid torus* $ST_B = \overline{D}_B \times S^1$ with boundary T^2 and the union of the X_H orbits through $\overline{D}_A$ is the closed solid torus $ST_A = \overline{D}_A \times S^1$ with boundary T^2. To understand how the 3-sphere $S^3_{\sqrt{2h}}$ is the union the two solid tori ST_A and ST_B we need to know the map $\psi : T^2 \to T^2$ which *glues* the solid torus ST_B to the solid torus ST_A along their common boundary T^2. Using the local trivializations τ_i we see that $ST_B = \tau_1\big((S^2_{2h} \cap \{w_3 \ge 0\}) \times S^1\big)$, $ST_A = \tau_1\big((S^2_{2h} \cap \{w_3 \le 0\}) \times S^1\big)$ and $T^2 = \tau_1(S^1_E \times S^1) = \tau_2(S^1_E \times S^1)$. Therefore, in the charts provided by the local

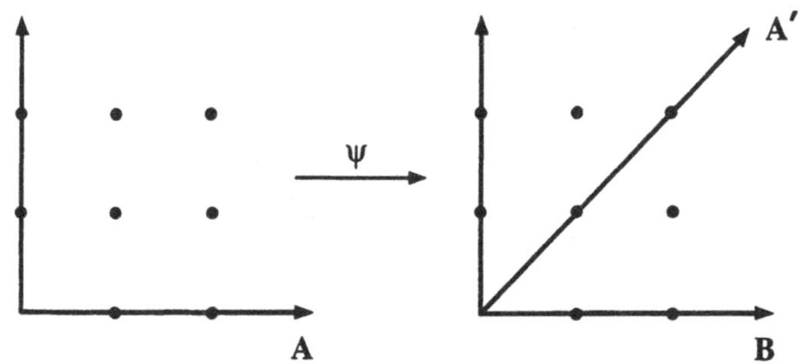

Figure 4.4. The gluing map $\psi : T^2 \to T^2$.

trivializations, the gluing map ψ is the graph of the transition map $\tau_2^{-1} \circ \tau_1$ restricted to $S^1_E \times S^1$, that is,

$$\psi : S^1_E \times S^1 \to S^1_E \times S^1 : \left(w_1, w_2, \begin{pmatrix} x \\ y \end{pmatrix}\right) \to \left(w_1, w_2, \chi(w_1, w_2)\begin{pmatrix} x \\ y \end{pmatrix}\right).$$

To visualize the gluing map ψ, we identify the 2-torus T^2 with the lattice $\mathbf{Z}^2 \subseteq \mathbf{R}^2$. Taking the S^1 orbits of X_H as vertical and the circles A and B as horizontal, the image of A under the mapping ψ is the line $A' : \{(x, -x) \in \mathbf{R}^2 | x \in \mathbf{R}\}$, see figure 4.4. If we identify the 2-torus T with $\mathbf{R}^2/\mathbf{Z}^2$, we see from figure 4.4 that the gluing map ψ is just the map on T^2 induced by the linear map $\widetilde{\psi}$ of $\mathbf{R}^2$ into itself with matrix $\begin{pmatrix} 1 & 0 \\ 1 & 1 \end{pmatrix}$. The map ψ is well defined because $\widetilde{\psi}(\mathbf{Z}^2) = \mathbf{Z}^2$.

The above treatment of the Hopf fibration gives a description of the fibers of the Hopf mapping $\mathcal{H}$ (27) over the $w_4 = 2h > 0$ section of its image cone $\mathcal{C}$ (29). To complete the

Section of $\mathscr{C}$	Topology of section	Topology of fiber
1. $w_4 = k > 0$	S^2	S^3
2. $w_3 = k \neq 0$	$\mathbf{R}^2$	$S^1 \times \mathbf{R}^2$
3. $w_2 = k \neq 0$	$\mathbf{R}^2$	$S^1 \times \mathbf{R}^2$
4. $w_1 = k \neq 0$	$\mathbf{R}^2$	$S^1 \times \mathbf{R}^2$
5. $w_3 = 0$	cone on S^1	cone on T^2
6. $w_2 = 0$	cone on S^1	cone on T^2
7. $w_1 = 0$	cone on S^1	cone on T^2

Table 4.1. The fibers of the Hopf map.

description of the geometry of the Hopf mapping, we look at the fibers of $\mathscr{H}$ over other slices of $\mathscr{C}$. The results are given in table 4.1. We now verify the entries in table 4.1. We will ▷ treat only the $w_3 = 2k < 0$ slice of $\mathscr{C}$. The argument for the other slices is similar and is omitted.

(4.9) **Proof**: We begin by supposing that $k \neq 0$. From (29) we see that the $w_3 = 2k < 0$ slice of $\mathscr{C}$ is a sheet $\mathscr{S}_k^+$ of the *hyperboloid*

$$4k^2 = w_4^2 - w_1^2 - w_2^2, \quad w_4 \geq 0 \ \& \ k \neq 0$$

which is diffeomorphic to $S^1 \times \mathbf{R}^2$. For $\widetilde{w} = (w_1, w_2, w_4) \in \mathscr{S}_k^+$ we determine the topology of the fiber $\mathscr{H}^{-1}(w')$ in a manner analogous to that used to find the topology of the fiber of the Hopf fibration. A straightforward calculation comparable to that used to derive (31) shows that the identities

$$\pi_1'(x, y) = (2k + w_4)y_2 - w_1x_1 - w_2y_1 = 0 \tag{39}$$

and

$$\pi_2'(x, y) = (2k + w_4)x_2 + w_1y_1 - w_2x_1 = 0 \tag{40}$$

hold. Except when $w' = (0, 0, -2k)$ with $k < 0$, $\{\pi_1' = 0\} \cap \{\pi_2' = 0\}$ defines a 2-plane $\Pi^{w'}$:

$$x_2 = \frac{w_2x_1 - w_1y_1}{2k + w_4} \quad \text{and} \quad y_2 = -\frac{w_1x_1 + w_2y_1}{2k + w_4}. \tag{41}$$

Therefore the fiber $\mathscr{H}^{-1}(w')$ is contained in the intersection of the 2-plane $\Pi^{w'}$ with the hyperboloid

$$x_1^2 + y_1^2 = 2k + x_2^2 + y_2^2. \tag{42}$$

Substituting (41) into (42) and simplifying a little gives

$$\big((2k + w_4)^2 - (w_1^2 + w_2^2)\big)(x_1^2 + y_1^2) = 2k(2k + w_4)^2. \tag{43}$$

But $w_1^2 + w_2^2 = w_4^2 - (2k)^2$ and $2k + w_4 \neq 0$. Thus (43) becomes

$$x_1^2 + y_1^2 = \tfrac{1}{2}(2k + w_4) > 0. \tag{44}$$

Therefore, when $w' \neq (0, 0, -2k)$, $\mathscr{H}^{-1}(w')$ is the set of vectors of the form

$$\left(x_1, \frac{1}{2k+w_4}(-w_1y_1+w_2x_1), y_1, \frac{1}{2k+w_4}(w_2y_1+w_1x_1)\right),$$

where $x_1^2+y_1^2 = \frac{1}{2}(2k+w_4) > 0$. Hence $\mathcal{H}^{-1}(w)$ is diffeomorphic to a circle. When $w' = (0,0,-2k)$, $\mathcal{H}^{-1}(w')$ is defined by

$$\begin{cases} x_1x_2+y_1y_2 &= 0 \\ x_1y_2-x_2y_1 &= 0 \\ x_1^2+y_1^2-x_2^2-y_2^2 &= 2k, \quad k<0 \\ x_1^2+y_1^2+x_2^2+y_2^2 &= -2k. \end{cases} \tag{45}$$

Adding and subtracting the last two equations in (45) gives

$$x_1^2+y_1^2=0 \quad \text{and} \quad x_2^2+y_2^2=-2k.$$

Therefore $\mathcal{H}^{-1}(0,0,-2k)$ is the circle in $\mathbf{R}^4$ defined by $x_1=0$, $y_1=0$ and $x_2^2+y_2^2=-2k$ where $k<0$. Hence $\mathcal{H}^{-1}(\mathcal{S}_k^+)$ is a smooth bundle over $\mathbf{R}^2$ with fiber S^1 when $k \neq 0$. Since $\mathbf{R}^2$ is contractible, $\mathcal{H}^{-1}(\mathcal{S}_k^+)$ is diffeomorphic to $S^1 \times \mathbf{R}^2$.
Now suppose that $k=0$. Then $\mathcal{S}_0^+$ is defined by $w_1^2+w_2^2=w_4^2$ where $w_4 \geq 0$. Hence $\mathcal{S}_0^+$ is a cone on a circle with vertex at 0. When $w_4>0$ and $k=0$ (41) and (42) hold. Therefore $\mathcal{H}^{-1}(\widetilde{w})$ is the circle defined by the set of vectors of the form

$$\left(x_1, \frac{1}{w_4}(-w_1y_1+w_2x_1), y_1, \frac{1}{w_4}(w_2y_1+w_1x_1)\right),$$

where $x_1^2+y_1^2=\frac{1}{2}w_4$ and $w_4>0$. If $w_4=0$ and $k=0$, then it follows from (45) that $\mathcal{H}^{-1}(w') = \mathcal{H}^{-1}(0,0,0)=0$. Consequently, $\mathcal{H}^{-1}(\mathcal{S}_0^+)$ is a cone on T^2 with vertex at 0. This completes the verification of the entries in table 4.1. □

5 Invariant theory and reduction

In this section we examine the geometry of the space of orbits of energy h of the harmonic oscillator. We will show that this space is a symplectic manifold. This fact can be exploited in several ways. One way is to gain some insight into the geometry of the foliation of the energy surface $H^{-1}(h)$ by integral curves of X_H and to see how the symplectic structure of this foliation depends on the energy. Another way, which we will not pursue in this book, is to study resonance problems in perturbation theory. The main idea of perturbation theory is to look at Hamiltonian systems which have the harmonic oscillator Hamiltonian as an integral. Using this integral, reduce the original Hamiltonian vector field to a Hamiltonian vector field on the orbit space $H^{-1}(h)/S^1$, which is two dimensions less than the original phase space. We will show how to carry out this reduction process using invariant theory. This procedure has two advantages. For one, it allows us to show that any smooth Hamiltonian, which is invariant under the flow of the harmonic oscillator vector field (and hence has the harmonic oscillator Hamiltonian as an integral), is a smooth function of four quadratic polynomials. Second, using these polynomials, we can explicitly construct the reduced space $H^{-1}(h)/S^1$ together with an embedding in Euclidean space. As a consequence, we obtain the associated Poisson and symplectic structure of the reduced space. These algebraic techniques will be used repeatedly in succeeding chapters of this book because they give a geometrically faithful model of the reduced space, even when it is *not* a smooth manifold.

We may summarize the contents of this section as follows. Let $\mathcal{H}$ be a function which is invariant under the flow of the harmonic oscillator vector field X_H. We reduce the Hamiltonian system $(\mathcal{H}, \mathbf{R}^4, \omega)$ to a Hamiltonian system (K_h, S_h^2, ω_h) on a 2-sphere S_h^2 which is the space formed by associating to each orbit of X_H of energy h a point.

We begin by proving

Claim: The Hamiltonian $\mathcal{H}$ is invariant under the flow of the harmonic oscillator vector field X_H if and only if it is an integral X_H.

(5.1) **Proof**: $\mathcal{H}$ is invariant under the flow of the harmonic oscillator vector field if and only if it is constant on the integral curves of X_H if and only if it is an integral of X_H. More formally, let φ_t^H be the flow of X_H. Since $\mathcal{H}$ is invariant under φ_t^H, it follows that $(\varphi_t^H)^*\mathcal{H} = \mathcal{H}$. Differentiating this condition with respect to t and evaluating the result at $t = 0$ gives $L_{X_H}\mathcal{H} = 0$. In other words, $\mathcal{H}$ is an integral of X_H. Conversely, suppose that $\mathcal{H}$ is an integral of X_H. Then

$$\frac{d}{dt}(\varphi_t^H)^*\mathcal{H} = \frac{d}{ds}\Big|_{s=0}(\varphi_{t+s}^H)^*\mathcal{H} = (\varphi_t^H)^*\Big(\frac{d}{ds}\Big|_{s=0}(\varphi_s^H)^*\mathcal{H}\Big) = (\varphi_t^H)^*(L_{X_H}\mathcal{H}) = 0.$$

Therefore $t \to (\varphi_t^H)^*\mathcal{H}$ is a constant function, that is, $(\varphi_t^H)^*\mathcal{H} = (\varphi_0^H)^*\mathcal{H} = \mathcal{H}$. Thus $\mathcal{H}$ is invariant under the flow of X_H. □

We now describe all the smooth functions which are integrals of X_H. We begin by finding all polynomial integrals of X_H. We show that they are polynomials in the quadratic integrals

$$\begin{aligned} w_1 &= 2(x_1y_2 - x_2y_1) \\ w_2 &= 2(x_1x_2 + y_1y_2) \\ w_3 &= y_1^2 + x_1^2 - y_2^2 - x_2^2 \\ w_4 &= y_1^2 + x_1^2 + y_2^2 + x_2^2. \end{aligned}$$

Claim: The algebra of *polynomials* which are invariant under the S^1 action given by the flow of the harmonic oscillator vector field X_H is *generated* by the quadratic functions w_i, which satisfy the relation

$$w_4^2 = w_1^2 + w_2^2 + w_3^2, \quad \text{where } w_4 \geq 0. \tag{46}$$

(5.2) **Proof**: To prove this assertion we introduce *complex conjugate coordinates*

$$\begin{aligned} \xi_1 &= x_1 + iy_1 \quad & \eta_1 &= x_1 - iy_1 \\ \xi_2 &= x_2 + iy_2 \quad & \eta_2 &= x_1 - iy_2. \end{aligned}$$

Then the algebra $\mathbf{R}[x, y]$ of real polynomials on $\mathbf{R}^4$ becomes the algebra of Hermitian polynomials

$$\mathbf{HP}[\xi, \eta] = \left\{ \sum c_{ij}\xi^i\eta^j \middle| c_{ij} = \overline{c}_{ji} \quad \text{where } c_{ij} \in \mathbf{C} \right\}.$$

Writing the Hamiltonian H in complex conjugate coordinates gives

$$\widetilde{H}(\xi, \eta) = \tfrac{1}{2}(\xi_1\eta_1 + \xi_2\eta_2).$$

Moreover, we obtain the Hamiltonian vector field on $\mathbf{C}^4$

$$\begin{aligned} \dot{\xi} &= -2i\,\frac{\partial \widetilde{H}}{\partial \eta} = -i\xi \\ \dot{\eta} &= 2i\,\frac{\partial \widetilde{H}}{\partial \xi} = i\eta, \end{aligned} \tag{47}$$

where $\xi = (\xi_1, \xi_2)$, $\frac{\partial}{\partial \xi} = \frac{1}{2}(\frac{\partial}{\partial x} - i\frac{\partial}{\partial y})$ and $\eta = (\eta_1, \eta_2)$, $\frac{\partial}{\partial \eta} = \frac{1}{2}(\frac{\partial}{\partial x} + i\frac{\partial}{\partial y})$.

The flow of $X_{\widetilde{H}}$ is the S^1 action $\cdot$ on $\mathbf{C}^4$ given by

$$\big(s, (\xi, \eta)\big) \to (s\,\xi, s^{-1}\eta), \tag{48}$$

where $|s| = 1$ and $s \in \mathbf{C}$. A real polynomial is invariant under the flow of X_H if and only if the corresponding Hermitian polynomial is invariant under the flow of $X_{\widetilde{H}}$. A Hermitian polynomial is invariant if and only if for each of its monomials $M = \xi^i\eta^j = \xi_1^{i_1}\xi_2^{i_2}\eta_1^{j_1}\eta_2^{j_2}$ we have

$$s \cdot M = s^{|i|}\xi^i s^{-|j|}\eta^j = \xi^i\eta^j = M,$$

where $|i| = i_1 + i_2$ and $|j| = j_1 + j_2$. In other words, $|i| = |j|$.

▷ We now show that every S^1 invariant Hermitian monomial M can be written as a product of invariant quadratic monomials

$$\sigma_{\ell k} = \xi_\ell\eta_k, \qquad \text{where } \ell = 1, 2 \text{ and } k = 1, 2.$$

(5.3) **Proof**: The factors in the monomial $M = \xi_1^{i_1}\xi_2^{i_2}\eta_1^{j_1}\eta_2^{j_2}$ can be displayed as two lists

$$\begin{array}{c} \overbrace{\xi_1 \cdots\cdots \xi_1}^{i_1}\;\overbrace{\xi_2 \cdots \xi_2}^{i_2} \\ \underbrace{\eta_1 \cdots \eta_1}_{j_1}\;\underbrace{\eta_2 \cdots\cdots \eta_2}_{j_2}. \end{array}$$

Because $|i| = |j|$, the above two lists have the same length and hence their entries may be paired off. This pairing expresses M as the product of quadratic monomials $\sigma_{\ell k}$ as claimed. □

Since

$$\begin{aligned} \sigma_{11} &= \xi_1\eta_1 = x_1^2 + y_1^2 \\ \sigma_{12} &= \xi_1\eta_2 = (x_1x_2 + y_1y_2) - i(x_1y_2 - x_2y_1) \\ \sigma_{21} &= \xi_2\eta_1 = (x_1x_2 + y_1y_2) + i(x_1y_2 - x_2y_1) \\ \sigma_{22} &= \xi_2\eta_2 = x_2^2 + y_2^2, \end{aligned}$$

every S^1 invariant polynomial is a sum of monomials which are products of w_1, w_2, w_3, w_4 times a real coefficient. From the identity

$$(x_1y_2 - x_2y_1)^2 + (x_1x_2 + y_1y_2)^2 = (x_1^2 + y_1^2)(x_2^2 + y_2^2) \tag{49}$$

it follows that

$$w_1^2 + w_2^2 + w_3^2 = w_4^2.$$

Clearly $w_4 \geq 0$. This proves the claim. □

Claim: The only polynomial relation among the generators w_i of the algebra of polynomials invariant under the flow of X_H is

$$w_1^2 + w_2^2 + w_3^2 - w_4^2 = 0.$$

(5.4) **Proof**: Consider the *complexified* Hopf mapping

$$\begin{aligned}\Phi : \mathbf{C}^4 \to \mathbf{C}^4 : (x_1, x_2, y_1, y_2) \to (w_1, w_2, w_3, w_4) &= \\ = \big(2\,(x_1y_2 - x_2y_1),\ 2\,(x_1x_2 + y_1y_2),\ y_1^2 + x_1^2 - y_2^2 - x_2^2,\ y_1^2 + x_1^2 + y_2^2 + x_2^2\big).\end{aligned}$$

We assert that the image of Φ is equal to the zero set $\mathfrak{Z}_F$ of the polynomial

$$F(w_1, w_2, w_3, w_4) = w_1^2 + w_2^2 + w_3^2 - w_4^2.$$

Since $F \circ \Phi = 0$ (which follows from (49)), the image of Φ is contained in $\mathfrak{Z}_F$. To prove the reverse inclusion, let $(w_1, w_2, w_3, w_4) \in \mathfrak{Z}_F$. Consider the following cases:

1. If $w_4 + w_3 \neq 0$, let

$$\begin{aligned} x_1 &= 0, & x_2 &= -w_1/\sqrt{2(w_4 + w_3)}, \\ y_1 &= \sqrt{(w_4 + w_3)/2}, & y_2 &= w_2/\sqrt{2(w_4 + w_3)}. \end{aligned}$$

2. If $w_4 + w_3 = 0$, let

$$x_1 = y_1 = x_2 = 0 \quad \text{and} \quad y_2 = \sqrt{w_4}.$$

Which branch of the square root one chooses above is immaterial, as long as it is consistent. In all of the above cases, $\Phi(x_1, x_2, y_1, y_2) = (w_1, w_2, w_3, w_4)$. Thus the image of Φ contains $\mathfrak{Z}_F$.

Next we show that F is irreducible. Suppose not. Then F is the product of two factors. Write

$$F = (\alpha w_1 + \beta)(\gamma w_1 + \delta), \tag{50}$$

where $\alpha, \beta, \gamma, \delta$ are polynomials in w_2, w_3, w_4. Clearly we can take $\alpha = \gamma = 1$. Since F has no term which is linear in w_1, we must have $\delta = -\beta$. Therefore (50) becomes

$$-\beta^2 = w_2^2 + w_3^2 - w_4^2. \tag{51}$$

Consequently the degree of β is at most 1. Since $\beta(0, 0, 0) = 0$, we may write

$$\beta = bw_2 + cw_3 + dw_4 \tag{52}$$

for some $b, c, d \in \mathbf{C}$. Squaring (52) and equating coefficients with (51) gives $b = \pm i$, $c = \pm i$ and $bc = 0$, which is a contradiction.

Let $\mathcal{I}$ be the ideal in $\mathbf{C}[w_1, w_2, w_3, w_4]$ generated by F. Since $\mathbf{C}$ is a field, the polynomial ring $\mathbf{C}[w_1, w_2, w_3, w_4]$ is a unique factorization domain and an integral domain. Thus every irreducible element is prime. Since every ideal in an integral domain which is generated by a prime polynomial is a prime ideal, $\mathcal{I}$ is a prime ideal.

Suppose that f is a polynomial in $\mathbf{C}[w_1, w_2, w_3, w_4]$ such that $f \circ \Phi = 0$, that is, f is a polynomial relation among the generators w_i. Since the image of Φ is the zero set of $\mathcal{I}$ and f vanishes on the image of Φ, it follows that the zero set of the ideal generated by f contains the image of Φ. By the Hilbert Nullstellensatz there is a positive integer m such that $f^m \in \mathcal{I}$. Since $\mathcal{I}$ is prime, $f \in \mathcal{I}$. □

The flow of X_H defines an algebraic linear action of $\mathrm{SO}(2, \mathbf{R}) = S^1$ on $\mathbf{R}^4$ given by

$$\varphi^H : \mathrm{SO}(2, \mathbf{R}) \times \mathbf{R}^4 \to \mathbf{R}^4 :$$

$$\left(\begin{pmatrix} a & -b \\ b & a \end{pmatrix}, \begin{pmatrix} x \\ y \end{pmatrix} \right) \to \begin{pmatrix} a\, I_2 & -b\, I_2 \\ b\, I_2 & a\, I_2 \end{pmatrix} \begin{pmatrix} x \\ y \end{pmatrix}$$

where $a^2 + b^2 = 1$. We may use a theorem of Schwarz to conclude that every *smooth* integral of the harmonic oscillator is a smooth function of the quadratic integrals. Thus we have proved

Claim: For every smooth function $\mathcal{H}$ on $\mathbf{R}^4$, which is invariant under the flow of the harmonic oscillator, there is a smooth function K on $\mathbf{R}^4$ such that $\mathcal{H} = J^*K$, where J is the $\mathrm{U}(2)$ momentum mapping of the harmonic oscillator.

We now turn to contructing a Poisson bracket on $\mathbf{R}^3$ with coordinates (w_1, w_2, w_3). Since the Hamiltonian vector fields X_{w_i} for $i = 1, \ldots, 4$ form a Lie algebra which is isomorphic to the Lie algebra $\mathrm{u}(2)$, the quadratic integrals w_i form a Lie algebra under Poisson bracket $\{\,,\,\}$ on $C^\infty(T^*\mathbf{R}^2)$ which is isomorphic to $\mathrm{u}(2)$. The bracket relations for this Lie algebra

$\{A, B\}$	w_1	w_2	w_3	w_4	B
w_1	0	$-4w_3$	$4w_2$	0	
w_2	$4w_3$	0	$-4w_1$	0	
w_3	$-4w_2$	$4w_1$	0	0	
w_4	0	0	0	0	
A					

Table 5.1 The structure matrix of the Poisson algebra $\mathcal{A}$.

are found by calculating

$$\{w_1, w_2\} = \omega(X_{w_1}, X_{w_2}) = -4w_3.$$

The rest of the bracket relations are given in table 5.1. Using the quadratic integrals $w = (w_1, w_2, w_3, w_4)$ as coordinates on $\mathbf{R}^4$, the function space $C^\infty(\mathbf{R}^4)$ can be made into a Poisson algebra $\mathcal{A}$ by defining a Poisson bracket as follows:

$$\{f, g\} = \sum_{i,j=1}^{4} \frac{\partial f}{\partial w_j} \frac{\partial g}{\partial w_i} \{w_j, w_i\}, \tag{53}$$

where $f, g \in C^\infty(\mathbf{R}^4)$, (see appendix A section 4). The bracket $\{\, ,\, \}$ is entirely determined by the bracket relations given in table 5.1.

For $K \in C^\infty(\mathbf{R}^4)$ the corresponding Hamiltonian vector field X_K is

$$\dot{w}_j = \{w_j, K\} = \sum_i \frac{\partial K}{\partial w_i} \{w_j, w_i\} \tag{54}$$

for $j = 1, \ldots, 4$. Note that for any $f \in C^\infty(\mathbf{R}^4)$

$$\{f, w_4\} = 0 \text{ and } \{f, w_4^2 - w_1^2 - w_2^2 - w_3^2\} = 0,$$

because they vanish for $f = w_i$ where $i = 1, \cdots, 4$. Thus the functions

$$w_4 \quad \text{and} \quad w_4^2 - w_1^2 - w_2^2 - w_3^2$$

are Casimir elements of $\mathcal{A}$. Therefore the level set $w_4 = 2h$ for $h > 0$ defines a smooth submanifold $\mathbf{R}^3 \times \{2h\}$ of $\mathbf{R}^4$ diffeomorphic to $\mathbf{R}^3$ which is invariant under the flow of the vector field X_K. Using (54) we see that $X_K|(\mathbf{R}^3 \times \{2h\})$ is given by

$$\begin{aligned}
\dot{w}_1 &= 4\,(\frac{\partial K}{\partial w_3} w_2 - \frac{\partial K}{\partial w_2} w_3) \\
\dot{w}_2 &= 4\,(\frac{\partial K}{\partial w_1} w_3 - \frac{\partial K}{\partial w_3} w_1) \\
\dot{w}_3 &= 4\,(\frac{\partial K}{\partial w_2} w_1 - \frac{\partial K}{\partial w_1} w_2),
\end{aligned} \tag{55}$$

where all the partial derivatives in (55) are evaluated when $\{w_4 = 2h\}$.

Now consider the space $\mathcal{B}$ of smooth functions on $\mathbf{R}^3$ which are restrictions of smooth functions on $\mathbf{R}^4$ to $\{w_4 = 2h\}$. For $K \in \mathcal{A}$ let $K_h \in \mathcal{B}$ be the restriction of K to $\{w_4 = 2h\}$. Because w_4 is a Casimir for $\mathcal{A}$, the space $\mathcal{B}$ is a Poisson subalgebra of $\mathcal{A}$ with bracket relations given in table 5.2.

$\{A, B\}$	w_1	w_2	w_3	B
w_1	0	$-4w_3$	$4w_2$	
w_2	$4w_3$	0	$-4w_1$	
w_3	$-4w_2$	$4w_1$	0	
A				

Table 5.2 The structure matrix of the Poisson algebra $\mathcal{B}$. Here the functions w_i are restricted to $\mathbf{R}^3 \times \{2h\}$.

▷ Using the Poisson algebra $\mathcal{B}$, we show that the smooth 2-sphere S^2_{2h} defined by

$$w_1^2 + w_2^2 + w_3^2 = (2h)^2$$

has a symplectic 2-form

$$\omega_h(w)(u,v) = -\frac{1}{16h^2}(w, u\times v), \tag{56}$$

where $w \in S^2_{2h}$ and $u, v \in T_w S^2_{2h}$. In the preceding formula $(\, ,\,)$ is the Euclidean inner product on $\mathbf{R}^3$.

(5.5) **Proof**: Since the 3×3 matrix $W(w) = (\{w_i, w_j\}(w))$ has rank 2 when $w \neq 0$, its restriction to S^2_{2h} is invertible. (Invertibility follows from the fact that the null space of W is spanned by $(w_1, w_2, w_3)^t$ which is transverse to S^2_{2h}.) Hence $W|S^2_{2h}$ gives rise to a symplectic form ω_h on S^{2h}. □

The Hamiltonian vector field X_{K_h} associated to the function $K_h \in \mathcal{B}$ is the restriction of X_K to $\{w_4 = 2h\}$. In the Poisson algebra $\mathcal{B}$, the function $w_1^2 + w_2^2 + w_3^2$ is a ▷ Casimir. Thus the smooth 2-sphere S^2_{2h} is an invariant manifold for X_{K_h}.

(5.6) **Proof**: Note that (55) can be written in the form

$$\dot{w} = X_{K_h}(w) = 4\, w \times \nabla K_h(w) \tag{57}$$

where $w = (w_1, w_2, w_3)$ and $\times$ is the usual vector product in $\mathbf{R}^3$. Therefore

$$\frac{d(w,w)}{dt} = 2\,(w,\dot{w}) = 2\,(w, X_{K_h}(w)) = 8\,(w, w\times \nabla K_h(w)) = 0. \quad \square$$

▷ In fact X_{K_h} (57) is a Hamiltonian vector field on (S^2_{2h}, ω_h) corresponding to the function $K_h : S^2_{2h} \subseteq \mathbf{R}^3 \rightarrow \mathbf{R}$.

(5.7) **Proof**: From the definition of a Hamiltonian vector field it follows that for every $w \in S^2_{2h}$ and every $u \in T_w S^2_{2h}$

$$\begin{aligned}(\nabla K_h(w), u) &= dK_h(w)u = \omega_h(w)(X_{K_h}(w), u) \\ &= -\frac{1}{16h^2}(w, X_{K_h}(w)\times u), \quad \text{using (56)} \\ &= \frac{1}{16h^2}(X_{K_h}(w)\times w, u).\end{aligned}$$

Therefore for $w \in S^2_{2h}$ we have $\nabla K_h(w) = \frac{1}{16h^2} X_{K_h}(w) \times w$. Consequently

$$\begin{aligned}-16h^2\,\nabla K_h(w)\times w &= w\times (X_{K_h}(w)\times w) \\ &= (w,w)\,X_{K_h}(w) - (X_{K_h}(w), w)\,w \\ &= 4h^2\,X_{K_h}(w),\end{aligned}$$

using $(w,w) = 4h^2$ and $(w, X_{K_h}(w)) = 0$. Dividing both sides of the above equation by $4h^2$ proves the assertion. □

Up to the factor 4 the above equations are *Euler's equations* for the rigid body in momentum coordinates when

$$K = \tfrac{1}{2}\,(I_1^{-1}w_1^2 + I_2^{-1}w_2^2 + I_3^{-1}w_3^2),$$

(see chapter 3 section 3.3). Note that the image under the U (2) momentum mapping J of the integral curves of $X_{\mathcal{H}}|H^{-1}(h)$ are the integral curves of X_{K_h} on the orbit space $H^{-1}(h)/S^1 = S^2_{2h}$ with symplectic form ω_h. This is precisely what the regular reduction theorem says in the case of the harmonic oscillator, (see appendix B section 4). Here (S^2_{2h}, ω_h) is the reduced phase space, K_h is the reduced Hamiltonian and J is the reduction mapping.

6 Exercises

1. Let $Q(x) = x_1^2 + \cdots + x_{p+1}^2 - x_{p+2}^2 - \cdots - x_{p+n+1}^2$ be a quadratic form on $\mathbf{R}^{p+n+1}$. Show that $Q^{-1}(1)$ is diffeomorphic to $S^p \times \mathbf{R}^n$. Show that $Q^{-1}(0)$ is a cone on $S^p \times S^{n-1}$. In other words, $Q^{-1}(0)$ is the space formed from $(S^p \times S^{n-1}) \times \mathbf{R}_{\geq}$ by identifying $(S^p \times S^{n-1}) \times \{0\}$ to a point. Show that $Q^{-1}(0)$ is homeomorphic to $S^p \times \mathbf{R}^n$.

2. (Complex projective 1-space.) *Complex projective 1-space* $\mathbf{CP}^1$ is defined as the set of equivalence classes of vectors in $\mathbf{C}^2 - \{0\}$ under the equivalence relation $\sim$ defined by $(z_1, z_2) \sim (w_1, w_2)$ if there is a $\lambda \in S^1 = \{z \in \mathbf{C} \,|\, |z| = 1\}$ such that $(z_1, z_2) = (\lambda w_1, \lambda w_2)$. Denote the equivalence class of (z_1, z_2) by $[z_1 : z_2]$. In other words $[z_1 : z_2]$ are *homogeneous coordinates* on $\mathbf{CP}^1$. Let

$$U_1 = \{[1 : z_2/z_1] \in \mathbf{CP}^1 \,|\, z_1 \neq 0\} \quad \text{and} \quad U_2 = \{[z_1/z_2 : 1] \in \mathbf{CP}^1 | z_2 \neq 0\}$$

with coordinates $w = z_2/z_1$ and $z = z_1/z_2$, respectively. Show that $\{U_1, U_2\}$ form an atlas for $\mathbf{CP}^1$ with transition function

$$\varphi_{12} : U_1 \cap U_2 \to U_1 \cap U_2 : z \to w = 1/z.$$

Show that $\mathbf{CP}^1$ is diffeomorphic to S^2 by verifying that the function

$$f : \mathbf{CP}^1 \to \mathbf{R} : [z_1 : z_2] \to \tfrac{1}{2}\,(a|z_1|^2 + b|z_2|^2), \qquad a > b > 0$$

is a Morse function with two nondegenerate critical points. Show that

$$\omega = \mathrm{Im}\left(\frac{(z_2 dz_1 - z_1 dz_2) \wedge (\bar{z}_2 d\bar{z}_1 - \bar{z}_1 d\bar{z}_2)}{|z_1|^2 + |z_2|^2}\right)$$

is a symplectic form on $\mathbf{CP}^1$.

3. (Linking number.) Let A and B be two smooth circles in S^3. Suppose that W_A and W_B are disjoint tubular neighborhoods of A and B in S^3. Then W_A and W_B are diffeomorphic to $D_A \times S^1$ and $D_B \times S^1$ for some 2-discs D_A and D_B respectively. There are 2-forms η_A and η_B on S^3 which are nonzero on W_A and W_B and zero elsewhere. To show that η_A is closed we argue as follows. Since the normalized volume 3-form vol_{S^3} generates $H^3(S^3)$, it follows that for some $\lambda \in \mathbf{R}$ $d\eta_A = \lambda \mathrm{vol}_{S^3}$. By *Stokes' theorem* $\int_{S^3} d\eta_A = \int_{\partial S^3} \eta_A = 0$, since $\partial S^3 = \varnothing$. Therefore $\lambda = 0$. A similar argument shows that η_B is closed. Since W_A and W_B are contractible in S^3, the 2-forms η_A and η_B are exact. In other words, there are

1-forms ξ_A and ξ_B on S^3 such that $d\xi_A = \eta_A$ and $d\xi_B = \eta_B$. This also follows because $H^2(S^3) = H^1(S^3) = 0$. Define the *linking number* of A and B as

$$\text{Link}(A, B) = \int_{S^3} \xi_A \wedge \eta_B.$$

Show that Link(A, B) does not depend on the choice of ξ_A. Also show that Link(A, B) does not depend on the representative of $\eta_B \in H^2(S^3)$, that is, if $\eta'_B = \eta_B + d\theta$ for some 1-form θ on S^3, then Link$(A, B) = \int_{S^3} \xi_A \wedge \eta'_B$.

4. (Degree of a map). Let (M, σ) and (N, τ) be two connected compact oriented manifolds of dimension r with volume forms σ and τ, respectively. Suppose that $f : M \to N$ is a smooth map. Let $n = f(m)$ be a regular value of f. For each $p \in f^{-1}(n)$ let

$$\text{sign}_p f = \begin{cases} 1, & \text{if } T_p f : (T_pM, \sigma_p) \to (T_nN, \tau_n) \\ & \text{is orientation preserving} \\ -1, & \text{otherwise} \end{cases}$$

Define the *degree* of f by

$$\deg f = \sum_{p \in f^{-1}(n)} \text{sign}_p f.$$

Show that $\int_M f^*\tau = (\deg f) \int_N \tau$. If $f : M \to N$ and $g : M \to N$ are smoothly homotopic, then show that $\deg f = \deg g$.

5. (Linking number.) Let $f, g : S^1 = \mathbf{R}/2\pi\mathbf{Z} \to \mathbf{R}^3$ be smooth maps whose images are disjoint. Define

$$\alpha_{f,g} : S^1 \times S^1 = T^2 \to S^2 : (s, t) \to \frac{f(s) - g(t)}{\|f(s) - g(t)\|},$$

where $\|x\| = \sqrt{\langle x, x\rangle}$ is the Euclidean length of $x \in \mathbf{R}^3$. Define the linking number of f and g to be Link$(f, g) = \deg \alpha_{f,g}$. Show that

$$\frac{\partial \alpha_{f,g}}{\partial s} = \frac{1}{\|f(s) - g(t)\|^3} \Big[\Big(f(s) - g(t)\Big) \times \Big(f'(s) \times (f(s) - g(t))\Big) \Big].$$

For $v_1, v_2 \in T_xS^2$ let $\text{vol}_{S^2}(x)(v_1, v_2) = \langle x, v_1 \times v_2 \rangle$ be the standard volume form on S^2. Show that

$$\alpha^*_{f,g}\text{vol}_{S^2} = -\frac{1}{\|f(s) - g(t)\|^3} \Big\langle f(s) - g(t), f'(s) \times g'(t) \Big\rangle \, ds \wedge dt.$$

Hence $\int_{T^2} \alpha^*_{f,g}\text{vol}_{S^2} = (\deg \alpha_{f,g}) \int_{S^2} \text{vol}_{S^2}$, that is,

$$\text{Link}(f, g) = -\frac{1}{4\pi} \int_0^{2\pi} \int_0^{2\pi} \frac{\Big\langle f(s) - g(t), f'(s) \times g'(t) \Big\rangle}{\|f(s) - g(t)\|^3} \, ds \, dt.$$

6. (Hopf invariant.) (a). Let $g : S^3 \to S^2$ be a smooth map and let α be a 2-form on S^2 which generates $H^2(S^2)$. Because $H^3(S^2) = 0$, α is closed. Consequently, $g^*\alpha$ is a closed 2-form on S^3. Since $H^2(S^3) = H^1(S^3) = 0$, there is a 1-form β on S^3 such that $g^*\alpha = d\beta$. Define the *Hopf invariant* of g to be

$$\text{Hopf}(g) = \int_{S^3} \beta \wedge d\beta.$$

Show that Hopf(g) does not depend on the choice of β. Moreover, if g and h are homotopic show that Hopf$(g) =$ Hopf(h).

(b). Let p and q be distinct regular values of the map g. Let D_p and D_q be disjoint closed 2-discs on S^2. Let α_p and α_q be 2-forms on S^2 which are nonzero on D_p and D_q, respectively, and are zero elsewhere. Let $\eta_A = g^*\alpha_p$ and $\eta_B = g^*\alpha_q$. From exercise 3 we know that the linking number of the circles $A = g^{-1}(p)$ and $B = g^{-1}(q)$ in S^3 is given by Link$(A, B) = \int_{S^3} \xi_A \wedge \eta_B$, where $\eta_B = d\xi_B$. To show that

$$\text{Link}(A, B) = \text{Hopf}(g) = \int_{S^3} \xi_A \wedge \eta_A$$

we argue as follows. Because dim $H^2(S^2) = 1$, there is a 1-form β on S^2 such that $\alpha_p - \alpha_q = d\beta$. Therefore $\eta_A - \eta_B = g^*(d\beta) = d(g^*\beta)$. Hence

$$\int_{S^3} \xi_A \wedge (\eta_A - \eta_B) = -\int_{S^3} d(\xi_A \wedge g^*\beta) + \int_{S^3} g^*(\alpha_p \wedge \beta) = 0.$$

(c). Consider the map

$$f : S^3 = \{(z_1, z_2) \in \mathbf{C}^2 | \, |z_1|^2 + |z_2|^2 = 1\} \to \mathbf{CP}^1 : (z_1, z_2) \to [z_1 : z_2].$$

Note that $f(1, 0) = [1 : 0]$. Composing f with stereographic projection from the north pole onto the *equatorial* plane show that the result is the Hopf map.

(d). To show that for every $[z_1 : z_2] \in \mathbf{CP}^1 = S^2$ the fiber $f^{-1}([z_1 : z_2])$ is a great circle on S^3 we argue as follows. For $c = (c_1, c_2, c_3) \in S^2 \subseteq \mathbf{R}^3$ we see that $(z_1, z_2) \in f^{-1}(c)$ if and only if

$$|z_1|^2 + |z_2|^2 = 1, \;\; |z_1|^2 - |z_2|^2 = c_3, \;\text{ and } 2z_1\bar{z}_2 = c_1 + i\, c_2.$$

Thus $|z_2|^2 = \frac{1}{2}(1 - c_3) \neq 0$. Choose z_2^0 so that $|z_2^0|^2 = \frac{1}{2}(1 - c_3)$ and z_1^0 satisfies $2z_1^0\bar{z}_2^0 = c_1 + i\, c_2$. Show that

$$f^{-1}(c) = \left\{ \zeta(z_1^0, z_2^0) \in S^3 \big| |\zeta| = 1 \right\}.$$

Thus $f^{-1}(c)$ lies in a complex 1-dimensional subspace of $\mathbf{C}^2$ and hence is a great circle on S^3.

(e). Here we show that the Hopf invariant of the Hopf map is 1. Let p be the north pole of S^2 and D_p be the disc on S^2 containing p and bounded by the equator. Let ω be the 2-form on D_p whose pull back under the chart

$$\{u_1^2 + u_2^2 < 1\} \to S^2 \subseteq \mathbf{R}^3 : (u_1, u_2) \to u_3 = \sqrt{1 - u_1^2 - u_2^2}$$

is $\frac{du_1 \wedge du_2}{2\pi\, u_3}$. Show that $\int_{D_p} \omega = 1$. Using the relation $\sum_{i=1}^{4} x_i dx_i = 0$, which comes from taking the exterior derivative of the defining equation of S^3, show that the pull back of ω by the Hopf map h is $d\beta$ where $\beta = -\frac{1}{\pi}(x_1 dx_2 + x_3 dx_4)$. Finally, using spherical coordinates

$$\begin{cases} x_1 & = \sin\varpi \sin\varphi \cos\theta \\ x_2 & = \sin\varpi \sin\varphi \sin\theta \\ x_3 & = \sin\varpi \cos\varphi \\ x_4 & = \cos\varpi \end{cases}$$

with $0 \le \varpi \le \pi$, $0 \le \varphi \le \pi$ and $0 \le \theta \le 2\pi$, show that

$$\int_{S^3} \beta \wedge d\beta = \frac{2}{\pi^2} \int_{S^3} x_1 dx_2 \wedge dx_3 \wedge dx_4 = 1.$$

7. (n-dimensional harmonic oscillator.) (a). Let $z = (z_1, \ldots, z_n)$ be coordinates for $\mathbf{C}^n$. Put a Kähler structure on $\mathbf{C}^n$ by defining the Kähler form $\Omega = \frac{i}{2} \sum_{j=1}^{n} dz_j \wedge d\overline{z}_j$. Let $h : \mathbf{C}^n \rightarrow \mathbf{R} : z \rightarrow h(z, \overline{z})$ be a smooth real valued function. Define the *Kähler Hamiltonian vector field* X_h associated to the *Kähler Hamiltonian* h by $X_h \,\lrcorner\, \Omega = \overline{d}h$, where $\overline{d}h = \sum_{j=1}^{n} \frac{\partial h}{\partial \overline{z}_j} d\overline{z}_j$. Show that $X_h = -i \sum_{j=1}^{n} \frac{\partial h}{\partial \overline{z}_j} \frac{\partial}{\partial z_j}$. In particular, if $h : \mathbf{C}^n \rightarrow \mathbf{R} : z \rightarrow \frac{1}{2} \sum_{j=1}^{n} z_j \overline{z}_j$ is the harmonic oscillator Hamiltonian, then $X_h = -i \sum_{j=1}^{n} z_j \frac{\partial}{\partial x_j}$, which is a holomorphic vector field on $\mathbf{C}^n$ whose flow is $\varphi : \mathbf{R} \times \mathbf{C}^n \rightarrow \mathbf{C}^n : (t, z) \rightarrow e^{-it} z$.

(b). Let $\langle z, w \rangle = \sum_{j=1}^{n} z_j \overline{w}_j$ be the standard *Hermitian inner product* on $\mathbf{C}^n$. An invertible linear map $U : \mathbf{C}^n \rightarrow \mathbf{C}^n$ is unitary if $\langle Uz, Uw \rangle = \langle z, w \rangle$ for every $z, w \in \mathbf{C}^n$. The set of all unitary matrices forms the Lie group $U(n)$ with Lie algebra $u(n)$ given by the set of all skew Hermitian linear maps $u : \mathbf{C}^n \rightarrow \mathbf{C}^n$, that is, $\langle uz, w \rangle + \langle z, uw \rangle = 0$ for every $z, w \in \mathbf{C}^n$. Show that the $n \times n$ skew Hermitian matrices

$$\begin{cases} i(e_j \otimes \overline{e}_k^t + e_k \otimes \overline{e}_j^t), & 1 \le j < k \le n \\ e_j \otimes \overline{e}_k^t - e_k \otimes \overline{e}_j^t, & 1 \le j < k \le n \\ i(e_j \otimes \overline{e}_j^t), & 1 \le j \le n \end{cases} \tag{58}$$

form a basis for $u(n)$ as a real vector space. Here $\{e_j\}_{j=1}^{n}$ is the standard basis for $\mathbf{C}^n$. Recall that $i(z \otimes \overline{z}^t) \in \mathbf{C} \otimes (\overline{\mathbf{C}^n})^*$ is the skew Hermitian linear map $i(z \otimes \overline{z}^t)(w) = i\langle w, z \rangle z$. Write out the matrices in (58) explicitly. Show that

$$k : u(n) \times u(n) \rightarrow \mathbf{C} : (u, v) \rightarrow \operatorname{tr} u\overline{v}^t$$

is a Hermitian inner product on $u(n)$. Since k is Ad_U-invariant for every $U \in U(n)$, it follows that k is the Killing (Hermitian) metric.

(c). The Lie group $U(n)$ acts on $\mathbf{C}^n$ by

$$\Phi : U(n) \times \mathbf{C}^n \rightarrow \mathbf{C}^n : (U, z) \rightarrow Uz.$$

For every $u = (u_{jk}) \in u(n)$ the vector field infinitesimally generated by u is

$$X^u(z) = \frac{d}{ds}\bigg|_{s=0} \Phi_{\exp su}(z) = \sum_{j,k}^{n} u_{jk}\, z_k \frac{\partial}{\partial z_j}.$$

Show that X^u is the Kähler Hamiltonian vector field associated to the function

$$j^u : \mathbf{C}^n \rightarrow \mathbf{R} : z \rightarrow \frac{i}{2}\langle uz, z\rangle =$$
$$= \frac{i}{2}\Big[\sum_{j<k} \Big(\mathrm{Re}\, u_{jk}\, \mathrm{Re}\, z_j \overline{z}_k - \mathrm{Im}\, u_{jk}\, \mathrm{Im}\, z_j \overline{z}_k\Big) + \sum_{j=1}^{n} u_{jj} z_j \overline{z}_j \Big].$$

Show that j^u is an integral of the harmonic oscillator vector field X_h. Define the map $\mathcal{J} : \mathbf{C}^n \rightarrow u(n)^*$ by $\mathcal{J}(z)u = j^u(z)$ for every $z \in \mathbf{C}^n$ and every $u \in u(n)$. Show that $\mathcal{J}$ intertwines the $U(n)$ action Φ on $\mathbf{C}^n$ with the coadjoint action of $U(n)$ on $u(n)^*$. In other words,

$$\mathcal{J}(Uz)u = \mathrm{Ad}^t_{U^{-1}}(\mathcal{J}(z))u = \mathcal{J}(z)(U^{-1}uU). \tag{59}$$

Thus $\mathcal{J}$ is a *Kähler momentum mapping* for the action Φ. Use the Killing Hermitian metric k to identify $u(n)^*$ with $u(n)$. Using the basis (58) of $u(n)$, show that $\mathcal{J}$ becomes

$$J : \mathbf{C}^n \rightarrow u(n) : z \rightarrow \frac{i}{2}(z \otimes \overline{z}^t) = \frac{i}{2}(z_j \overline{z}_k). \tag{60}$$

Verify directly that J intertwines the $U(n)$ action Φ and the adjoint action of $U(n)$ on $u(n)$.

(d). Show that $U(n)$ acts transitively on the set of all 1-dimensional complex subspaces of $\mathbf{C}^n$ with isotropy group at span$\{e_1\}$ isomorphic to $U(n-1)$. Deduce that the orbit space $U(n)/U(n-1)$ is diffeomorphic to $\mathbf{CP}^{n-1}$, complex projective $(n-1)$-space. For $z \neq 0$ show that $i(z \otimes \overline{z}^t)$ has rank 1 and trace $\langle z, z\rangle$. If $z \neq 0$ deduce that the $U(n)$ adjoint orbit $\mathcal{O}_\zeta$ through $\zeta = J(z)$ is diffeomorphic to $\mathbf{CP}^{n-1}$. In particular, $\mathbf{CP}^{n-1}$ is defined by

$$\begin{cases} (z_j\overline{z}_k)(z_\ell\overline{z}_m) = (z_j\overline{z}_m)(z_\ell\overline{z}_k), \quad 1 \le j,k,\ell,m \le n \\ \sum_{j=1}^{n} z_j\overline{z}_j = 1. \end{cases} \tag{61}$$

Since (61) is the set of all relations among the quadratic (and hence smooth) integrals of the harmonic oscillator restricted to the level set $h^{-1}(\frac{1}{2})$, the space $h^{-1}(\frac{1}{2})/S^1$ of orbits of the harmonic oscillator of energy $\frac{1}{2}$ is diffeomorphic to $\mathbf{CP}^{n-1}$. Show that

$$J|h^{-1}(\tfrac{1}{2}) : h^{-1}(\tfrac{1}{2}) \rightarrow \mathbf{CP}^{n-1} = h^{-1}(\tfrac{1}{2})/S^1$$

is the reduction map for the harmonic oscillator. Use the following argument to find the symplectic form on the reduced space $\mathbf{CP}^{n-1}$. On the orbit $\mathcal{O}_\zeta$ with $\langle z, z\rangle = 1$, there is a symplectic form

$$\omega(\zeta)(v_\zeta, w_\zeta) = k(\zeta, [v_\zeta, w_\zeta]), \tag{62}$$

where $v_\zeta = DJ(z)v$ and $w_\zeta = DJ(z)w$ lie in $T_\zeta \mathcal{O}_\zeta \subseteq u(n)$. Using (62), $\langle z, z \rangle = 1$ and the fact that $v_\zeta = \frac{i}{2}(v_j \overline{z}_k + z_j \overline{v}_k)$, $w_\zeta = \frac{i}{2}(w_j \overline{z}_k + z_j \overline{w}_k)$, show that

$$\omega = \tfrac{1}{4} \operatorname{Im} \big(\langle v, w \rangle \langle z, z \rangle - \langle v, z \rangle \langle w, z \rangle \big). \tag{63}$$

Replacing z by $z/\sqrt{\langle z, z \rangle}$, (63) becomes

$$\omega = \tfrac{1}{4} \operatorname{Im} \left(\frac{\langle v, w \rangle \langle z, z \rangle - \langle v, z \rangle \langle w, z \rangle}{\langle z, z \rangle} \right),$$

which is $\frac{1}{4}$ times the imaginary part of the Fubini-Study Hermitian metric on $\mathbf{CP}^{n-1}$.

8. Let φ be the S^1 action defined by the flow of the harmonic oscillator on $T\mathbf{R}^4$ restricted to the 3-sphere S^3. Show that the orbit space of this S^1 action *cannot* be realized as a cross section for the action φ.

Chapter II

Geodesics on S^3

In this chapter we study the geodesic vector field on the tangent bundle of the 3-sphere. We examine its relation to the Kepler vector field, which governs the motion of two bodies in $\mathbf{R}^3$ under gravitational attraction. We give a method to regularize the flow of the Kepler vector field for all negative energies at once and not energy surface by energy surface.

1 The geodesic and Delaunay vector fields

In this section we find the geodesic vector field on the 3-sphere and give an explicit formula for its flow. Rescaling time gives the Delaunay vector field, which we need when regularizing the Kepler vector field, (see section 3.4).

We begin by discussing the geodesic vector field. Suppose that $\langle\,,\,\rangle$ is the Euclidean inner product on $\mathbf{R}^4$. This induces a Riemannian metric g on $\mathbf{R}^4$ defined by $g(x)^\sharp(y)z = \langle y, z\rangle$ where $x \in \mathbf{R}^4$ and $y, z \in T_x\mathbf{R}^4 = \mathbf{R}^4$. Pulling back the canonical symplectic 2-form Ω on $T^*\mathbf{R}^4$ by the map $g^\sharp$ (see appendix A section 2) we obtain the symplectic form $\omega_4 = -d\langle y, dx\rangle$ on $T\mathbf{R}^4$. On $(T\mathbf{R}^4, \omega_4)$ consider the Hamiltonian function

$$\mathcal{H} : T\mathbf{R}^4 \rightarrow \mathbf{R} : (x, y) \rightarrow \tfrac{1}{2}\langle y, y\rangle. \tag{1}$$

Since an integral curve of the Hamiltonian vector field $X_{\mathcal{H}}$ satisfies

$$\begin{cases} \dot{x} &= y \\ \dot{y} &= 0, \end{cases}$$

it is a straight line on $T\mathbf{R}^4$, except when $y = 0$ where it is a point. Hence $X_{\mathcal{H}}$ describes the *motion* of a particle in $T\mathbf{R}^4$ which is not subject to any *force*. To constrain this free particle so that it moves on the 3-sphere

$$S^3 = \left\{ x \in \mathbf{R}^4 \,\middle|\, \langle x, x\rangle = 1 \right\},$$

we add a force $\lambda(x, \dot{x})x$ which is normal to S^3 at the point x. The particle subjected to this force moves according to

$$\ddot{x} = \lambda(x, \dot{x})\, x. \tag{2}$$

Differentiating the defining equation of S^3 twice gives

$$\langle x, \ddot{x} \rangle + \langle \dot{x}, \dot{x} \rangle = 0. \tag{3}$$

Substituting (2) into (3) and using $\langle x, x \rangle = 1$ gives $\lambda(x, \dot{x}) = -\langle \dot{x}, \dot{x} \rangle$. Hence the *motion* of the free particle constrained to S^3 is governed by the second order equation

$$\ddot{x} = -\langle \dot{x}, \dot{x} \rangle x \tag{4}$$

subject to the constraints $\langle x, x \rangle = 1$ and $\langle \dot{x}, x \rangle = 0$. Written as a first order equation on the tangent bundle

$$TS^3 = \left\{ (x, y) \in T\mathbf{R}^4 \middle| \langle x, x \rangle = 1 \ \& \ \langle x, y \rangle = 0 \right\}$$

of S^3, the constrained system (4) becomes

$$\begin{cases} \dot{x} &= y \\ \dot{y} &= -\langle y, y \rangle x. \end{cases} \tag{5}$$

This defines a vector field Y on TS^3. Note that TS^3 is an invariant manifold of (5), thought of as a vector field on $T\mathbf{R}^4$, since the initial conditions $\langle x, x \rangle = 1$ and $\langle x, y \rangle = 0$ are preserved under its flow.

▷ The above discussion is not at all Hamiltonian. What we want to do is to show that Y is a Hamiltonian vector field on the phase space (TS^3, Ω_4), where Ω_4 is a suitable symplectic form.

(1.1) **Proof**: To do this, we use modified Dirac brackets, (see appendix A section 4). On the open subset $M = T\mathbf{R}^4 - (\{0\} \times \mathbf{R}^4)$ of $T\mathbf{R}^4$ consider the constraint functions

$$c_1 : M \to \mathbf{R} : (x, y) \to \tfrac{1}{2}(\langle x, x \rangle - 1) \ \text{ and } \ c_2 : M \to \mathbf{R} : (x, y) \to \langle x, y \rangle.$$

Let $\{\,,\,\}$ be the standard Poisson bracket on $C^\infty(T\mathbf{R}^4)$, the space of smooth functions on the symplectic manifold $(T\mathbf{R}^4, \omega_4)$, (see appendix A section 4). Since the matrix

$$(\{c_i, c_j\}) = \begin{pmatrix} 0 & \langle x, x \rangle \\ -\langle x, x \rangle & 0 \end{pmatrix}$$

is invertible on M with inverse

$$(C_{ij}) = \frac{1}{\langle x, x \rangle} \begin{pmatrix} 0 & -1 \\ 1 & 0 \end{pmatrix}$$

and since 0 is a regular value of the constraint map

$$\mathscr{C} : M \to \mathbf{R}^2 : m \to \big(c_1(m), c_2(m)\big),$$

the constraint manifold $TS^3 = \mathscr{C}^{-1}(0)$ is a *cosymplectic* submanifold of $(M, \omega_4|M)$. In other words, $\Omega_4 = \omega_4|TS^3$ is a symplectic form on TS^3. For $F \in C^\infty(M)$ let

$$F^* = F - \sum_{i,j=1}^{2} (\{F, c_i\} + F_i) C_{ij} c_j,$$

where the F_i lie in the ideal generated by c_1 and c_2. Define a Poisson bracket $\{\, , \,\}_{TS^3}$ on $C^\infty(TS^3)$ by

$$\{F|T^+S^3, G|T^+S^3\}_{TS^3} = \{F^*, G^*\}|TS^3.$$

Note that the Hamiltonian vector field $X_{F|TS^3}$ of the Hamiltonian F constrained to TS^3 is the Hamiltonian vector field X_{F^*} restricted to TS^3.

Applying these remarks to the unconstrained Hamiltonian $\mathcal{H}$ (1) gives

$$\begin{aligned}
\mathcal{H}^* &= \mathcal{H} - \sum_{i,j}(\{\mathcal{H}, c_i\} + \mathcal{H}_i)C_{ij}c_j \\
&= \tfrac{1}{2}\langle y, y\rangle + \frac{1}{\langle x, x\rangle}\Big\langle\big(\langle x, y\rangle - \mathcal{H}_1, \langle y, y\rangle - \mathcal{H}_2\big), \big(-\langle x, y\rangle, \tfrac{1}{2}(\langle x, x\rangle - 1)\big)\Big\rangle \\
&= \tfrac{1}{2}\big(\langle x, x\rangle\langle y, y\rangle - \langle x, y\rangle^2\big),
\end{aligned}$$

where we have chosen

$$\mathcal{H}_1 = \langle x, y\rangle(1 - \tfrac{1}{2}\langle x, x\rangle) \quad \text{and} \quad \mathcal{H}_2 = -\langle y, y\rangle(\langle x, x\rangle - 1).$$

From Hamilton's equations on $(T\mathbf{R}^4, \omega_4)$ it follows that the integral curves of $X_{\mathcal{H}^*}$ satisfy

$$\frac{d}{dt}\begin{pmatrix} x \\ y \end{pmatrix} = A(x, y)\begin{pmatrix} x \\ y \end{pmatrix} = \begin{pmatrix} -\langle x, y\rangle & \langle x, x\rangle \\ -\langle y, y\rangle & \langle x, y\rangle \end{pmatrix}\begin{pmatrix} x \\ y \end{pmatrix}. \tag{6}$$

Using (6) and the definition of TS^3, it is easy to see that the integral curves of $X_{\mathcal{H}^*}|TS^3$ satisfy (5). Because $X_{\mathcal{H}|TS^3} = X_{\mathcal{H}^*}|TS^3$, the geodesic vector field on TS^3 is the Hamiltonian vector field X_H on (TS^3, Ω_4) corresponding to the Hamiltonian function

$$H = \mathcal{H}^*|TS^3 : TS^3 \to \mathbf{R} : (x, y) \to \tfrac{1}{2}\langle y, y\rangle. \tag{7}$$

Note that H is the free particle Hamiltonian on $T\mathbf{R}^4$ restricted to TS^3. Thus the integral curves of the geodesic vector field X_H on TS^3 satisfy

$$\begin{cases} \dot{x} &= y \\ \dot{y} &= -\langle y, y\rangle x. \end{cases} \qquad \square \tag{8}$$

▷ To find the flow of the geodesic vector field X_H, we first look for integrals (= conserved quantities) of the vector field $X_{\mathcal{H}^*}$. From the construction of the Hamiltonian $\mathcal{H}^*$ on $T\mathbf{R}^4$, we know that TS^3 is an invariant manifold of $X_{\mathcal{H}^*}$. Therefore the functions

$$f_1(x, y) = \tfrac{1}{2}\langle x, x\rangle \quad \text{and} \quad f_2(x, y) = \langle x, y\rangle \tag{9}$$

are integrals of $X_{\mathcal{H}^*}$. A calculation shows that $f_3(x, y) = \frac{1}{2}\langle y, y\rangle$ is also an integral of $X_{\mathcal{H}^*}$. The integrals $\{f_1, f_2, f_3\}$ span a Lie subalgebra of $(C^\infty(T\mathbf{R}^4), \{\, , \,\})$, which is isomorphic to $\mathrm{sl}(2, \mathbf{R})$ since

$$\{f_1, f_2\} = 2f_1, \ \{f_1, f_3\} = f_2, \ \text{and} \ \{f_3, f_2\} = -2f_3.$$

Because the functions f_i are constant along the integral curves of $X_{\mathcal{H}^*}$, so is the matrix $A(x, y)$ (6). Since $A^2(x, y) = -2\,\mathcal{H}^*(x, y)\,I_2$ and $\mathcal{H}^*(x, y) \geq 0$, the flow of $X_{\mathcal{H}^*}$ is

$$\begin{aligned} \varphi_t^{\mathcal{H}^*}(x, y) &= \exp tA(x, y) \begin{pmatrix} x \\ y \end{pmatrix} \\ &= \left(\cos(t\sqrt{2\mathcal{H}^*})I_2 + \left(\sin(t\sqrt{2\mathcal{H}^*})/\sqrt{2\mathcal{H}^*}\right) A(x, y)\right) \begin{pmatrix} x \\ y \end{pmatrix}. \end{aligned}$$

Restricting $\varphi_t^{\mathcal{H}^*}$ to the invariant manifold TS^3 gives

$$\varphi_t^H(x, y) = \begin{pmatrix} \cos(t\sqrt{2H}) & \sin(t\sqrt{2H})/\sqrt{2H} \\ -\sqrt{2H}\sin(t\sqrt{2H}) & \cos(t\sqrt{2H}) \end{pmatrix} \begin{pmatrix} x \\ y \end{pmatrix}, \tag{10}$$

which is the flow of the geodesic vector field X_H on TS^3. □

Clearly, all of the integral curves of X_H on the level set $H^{-1}(h)$ with $h > 0$ are periodic ▷ of *period* $2\pi/\sqrt{2h}$. In fact, when $y \neq 0$, the image of the integral curve $t \to \varphi_t^H(x, y)$ under the bundle projection to S^3 is the geodesic

$$\gamma_{(x,y)} : \mathbf{R} \to S^3 : t \to x\left(\cos(t\sqrt{2H})\right) + y\left(\sin(t\sqrt{2H})/\sqrt{2H}\right). \tag{11}$$

(1.2) **Proof**: To see that $\gamma_{(x,y)}$ is a geodesic on S^3 it suffices to show that

1. $\gamma_{(x,y)}$ is parametrized up to an affine transformation by arc length.
2. The *acceleration* $\ddot{\gamma}_{(x,y)}$ has no tangential component.

From (8) it follows that 2) holds. 1) holds because γ is parametrized. Another argument to prove 1) goes as follows. Differentiating (11) gives

$$\langle \dot{\gamma}_{(x,y)}, \dot{\gamma}_{(x,y)} \rangle = 2H\sin^2(t\sqrt{2H})\,\langle x, x\rangle + \cos^2(t\sqrt{2H})\,\langle y, y\rangle \;=\; 2H,$$

which is a constant of motion. This constant is nonzero, since $y \neq 0$. □

The explicit formula (10) for the flow of the geodesic vector field gives no *qualitative* information about how the integral curves are organized into invariant manifolds. To understand the invariant manifolds, it is useful to explain the role of the obvious symmetry of the problem, namely, the group SO(4) of rigid motions of the 3-sphere. This will be done in section 2.

We now discuss the Delaunay vector field. Let

$$T^+S^3 = \left\{ (x, y) \in TS^3 \middle| y \neq 0 \right\}$$

be the tangent bundle of S^3 less its zero section. Clearly T^+S^3 is an open subset of TS^3. Therefore $\Omega_4 | T^+S^3$ is a symplectic form $\widetilde{\Omega}_4$ on T^+S^3. On $(T^+S^3, \widetilde{\Omega}_4)$ consider the *Delaunay Hamiltonian*

$$\widetilde{\mathcal{H}} : T^+S^3 \subseteq T\mathbf{R}^4 \to \mathbf{R} : (x, y) \to -\tfrac{1}{2}\frac{\mu^2}{\langle y, y\rangle}. \tag{12}$$

The integral curves of the *Delaunay vector field* $X_{\widetilde{\mathcal{H}}}$ satisfy

$$\begin{cases} \frac{dx}{dt} &= \frac{\mu^2}{\langle y, y\rangle^2}\, y \\ \frac{dy}{dt} &= -\frac{\mu^2}{\langle y, y\rangle}\, x. \end{cases} \tag{13}$$

(1.3) **Proof**: As with the geodesic Hamiltonian, we will treat $(\widetilde{\mathcal{H}}, T^+S^3, \widetilde{\Omega}_4)$ as a constrained Hamiltonian system. Give M, which is $T\mathbf{R}^4$ with the coordinate planes $\{0\} \times \mathbf{R}^4$ and $\mathbf{R}^4 \times \{0\}$ removed, the symplectic form $\widetilde{\omega}_4 = \omega_4|M$. Consider the unconstrained system $(\widetilde{H}, M, \widetilde{\omega}_4)$ with unconstrained Hamiltonian

$$\widetilde{H} : M \subseteq T\mathbf{R}^4 \to \mathbf{R} : (x, y) \to -\tfrac{1}{2}\,\frac{\mu^2}{\langle y, y\rangle}.$$

Now constrain this system to T^+S^3, which is a submanifold of M defined by the constraint functions

$$c_1 : M \to \mathbf{R} : (x, y) \to \tfrac{1}{2}\big(\langle x, x\rangle - 1\big), \quad c_2 : M \to \mathbf{R} : (x, y) \to \langle x, y\rangle,$$

and the condition $\langle y, y\rangle > 0$. For every $(x, y) \in T^+S^3$ the 2×2 matrix $(\{c_i, c_j\})$ is invertible with inverse

$$(C_{ij}) = \tfrac{1}{\langle x,x\rangle}\begin{pmatrix} 0 & -1 \\ 1 & 0 \end{pmatrix}.$$

Thus T^+S^3 is a cosymplectic submanifold of $(M, \widetilde{\omega}_4)$ with symplectic form $\Omega_4 = \widetilde{\omega}_4|TS^3$. Since T^+S^3 is an open subset of TS^3, T^+S^3 is a symplectic manifold with symplectic form $\widetilde{\Omega}_4 = \Omega_4|T^+S^3$. Therefore we may use the modified Dirac procedure to compute the Delaunay vector field $X_{\widetilde{\mathcal{H}}}$. Towards this goal let

$$\begin{aligned} \widetilde{\mathcal{H}}^* &= \mathcal{H} - \sum_{i,j}\big(\{\mathcal{H}, c_i\} + \mathcal{H}_i\big)C_{ij}c_j \\ &= -\tfrac{1}{2}\tfrac{\mu^2}{\langle y,y\rangle} - \tfrac{1}{\langle x,x\rangle}\Big\langle\Big(-\tfrac{\mu^2\langle x,y\rangle}{\langle y,y\rangle^2} + \widetilde{\mathcal{H}}_1, \tfrac{\mu^2}{\langle y,y\rangle} + \widetilde{\mathcal{H}}_2\Big), \Big(-\langle x,y\rangle, \tfrac{1}{2}(\langle x,x\rangle - 1)\Big)\Big\rangle \\ &= -\tfrac{1}{2}\frac{\mu^2}{\langle y, y\rangle} - \tfrac{1}{2}\langle x, y\rangle^2 + \tfrac{1}{2}\frac{\mu^2}{\langle y, y\rangle}\big(\langle x, x\rangle - 1\big), \end{aligned}$$

where we have chosen the functions

$$\widetilde{\mathcal{H}}_1(x, y) = \Big(\frac{\mu^2}{\langle y, y\rangle^2} - \tfrac{1}{2}\langle x, x\rangle\Big)\langle x, y\rangle \quad\text{and}\quad \widetilde{\mathcal{H}}_2(x, y) = \frac{\mu^2}{\langle y, y\rangle}\big(\langle x, x\rangle - 1\big)$$

to lie in the ideal generated by c_1 and c_2. On $(M, \widetilde{\omega}_4)$ the Hamiltonian vector field $X_{\widetilde{\mathcal{H}}^*}$ has integral curves which satisfy

$$\begin{cases} \frac{dx}{dt} &= \frac{\partial\widetilde{\mathcal{H}}^*}{\partial y} = \frac{\mu^2}{\langle y, y\rangle^2}\, y - \langle x, y\rangle\, x + \frac{\mu^2}{\langle y, y\rangle^2}\big(\langle x, x\rangle - 1\big)y \\ \frac{dy}{dt} &= -\frac{\partial\widetilde{\mathcal{H}}^*}{\partial x} = -\frac{\mu^2}{\langle y, y\rangle}\, x + \langle x, y\rangle\, y. \end{cases}$$

Since $X_{\widetilde{\mathcal{H}}} = X_{\widetilde{H^*}|T^+S^3} = X_{\widetilde{H^*}}|T^+S^3$, the integral curves of $X_{\widetilde{\mathcal{H}}}$ satisfy (13). □

Before we show that the Delaunay vector field $X_{\widetilde{\mathcal{H}}}$ is a time rescaling of the geodesic vector field X_H, we discuss the general concept of rescaling. Let X be a smooth vector field on a smooth manifold M with flow φ_t and let f be a smooth real valued function on M. Consider the vector field $f \cdot X$, defined by $(f \cdot X)(m) = f(m)X(m)$ for $m \in M$. Define a new time scale s by

$$\frac{dt}{ds} = f(\varphi_t(m)), \tag{14}$$

that is,

$$s(t) = \int_0^t \frac{dt}{f(\varphi_t(m))}.$$

We assume that the right hand side of the above equation exists and that the function s is invertible. In the new time $t = t(s)$ the integral curve $t \to \gamma(t)$ of X with $\gamma(0) = m$ satisfies

$$\frac{d\gamma(t(s))}{ds} = \frac{d\varphi_{t(s)}(m)}{ds} = \frac{d\varphi_{t(s)}(m)}{dt}\frac{dt}{ds} = f(\varphi_{t(s)}(m))\, X(\varphi_{t(s)}(m)),$$

using (14). Therefore $s \to \gamma(t(s))$ is an integral curve of $f \cdot X$ with $\gamma(t(0)) = \gamma(0) = m$, since $s(0) = 0$. We say that $f \cdot X$ is a *time rescaling* of X with *time rescaling function* f.

Consider the special case where the time rescaling f is a smooth integral of X. In this case we can integrate (14) to obtain $s = 1/f(m)\, t$. Hence the flow

$$\psi : \mathbf{R} \times M \to M : (s, m) \to \psi_s(m)$$

of the rescaled vector field $f \cdot X$ is $\varphi_{sf(m)}$. Specializing further, suppose that X_G is a Hamiltonian vector field on a symplectic manifold (M, ω) with flow φ_t^G. Let ϑ be the time one map of the flow of the rescaled vector field $f \cdot X_G$, where $f : M \to \mathbf{R}$ is a smooth integral of X_G. Then

$$\vartheta^*\omega = \omega + df \wedge dG. \tag{15}$$

(1.4) **Proof**: Let

$$\varphi^G : \mathbf{R} \times M \to M : (s, m) \to \varphi^G(s, m) = \varphi_s^G(m) = \varphi_m^G(s).$$

By definition $\vartheta(m) = \varphi^G(f(m), m)$. We compute the tangent to ϑ as follows. For $v_m, w_m \in T_m M$

$$\begin{aligned} T_m\vartheta\, v_m &= \left(\frac{d}{ds}\Big|_{s=0} \varphi_m^G\right)(f(m))v_m + T_m\varphi_{f(m)}^G v_m \\ &= (df(m)v_m)\, X_G(\varphi_{f(m)}^G(m)) + T_m\varphi_{f(m)}^G v_m. \end{aligned}$$

Consequently,

$$\begin{aligned} \vartheta^*\omega(m)(v_m, w_m) &= (df(m)v_m)\, \omega(\varphi_{f(m)}^G(m))\Big(X_G(\varphi_{f(m)}^G(m)), T_m\varphi_{f(m)}^G w_m \Big) \\ &\quad - (df(m)w_m)\, \omega(\varphi_{f(m)}^G(m))\Big(X_G(\varphi_{f(m)}^G(m)), T_m\varphi_{f(m)}^G v_m \Big) \\ &\quad + \omega(\varphi_{f(m)}^G(m))\Big(T_m\varphi_{f(m)}^G v_m, T_m\varphi_{f(m)}^G w_m \Big), \end{aligned}$$

using the definition of pull back and

$$\omega(\varphi^G_{f(m)}(m)\big(X_G(\varphi^G_{f(m)}(m)), X_G(\varphi^G_{f(m)}(m))\big) = 0.$$

Therefore

$$\begin{aligned}
\omega(\vartheta(m))\big(T_m\vartheta\, v_m, T_m\vartheta\, w_m\big) &= \\
&= (df(m)v_m)\, dG(\varphi^G_{f(m)}(m))T_m\varphi^G_{f(m)} w_m - (df(m)w_m)\, dG(\varphi^G_{f(m)}(m))T_m\varphi^G_{f(m)} v_m \\
&\qquad + \omega(m)(v_m, w_m), \\
&\qquad\qquad \text{since } X_G \lrcorner\, \omega = dG \text{ and } \varphi^G_{f(m)} \text{ is symplectic} \\
&= (df(m)v_m)\, dG(m)w_m - (df(m)w_m)\, dG(m)v_m + \omega(m)(v_m, w_m), \\
&\qquad\qquad \text{since } ((\varphi^G_{f(m)})^*G = G \\
&= (df \wedge dG)(m)(v_m, w_m) + \omega(m)(v_m, w_m). \quad \square
\end{aligned}$$

From (15) we see that ϑ is symplectic if and only if $df \wedge dG = 0$, that is, f is *locally* a smooth function of G. Thus the rescaled vector field $f \cdot X_G$ is a Hamiltonian vector field $X_{\widehat{G}}$ for some smooth function $\widehat{G}$ on M if there is a smooth function $\mathscr{G} : \mathbf{R} \to \mathbf{R}$ such that $\widehat{G} = \mathscr{G} \circ G$. When this is the case, the time rescaling function f is $\mathscr{G}' \circ G$.

We now apply these observations about time rescaling to the Delaunay vector field $X_{\widetilde{\mathscr{H}}}$. The Delaunay Hamiltonian

$$\widetilde{\mathscr{H}} : T^+S^3 \subseteq T\mathbf{R}^4 \to \mathbf{R} : (x, y) \to -\tfrac{1}{2}\,\frac{\mu^2}{\langle y, y\rangle}$$

is a smooth function of the geodesic Hamiltonian

$$H : T^+S^3 \subseteq T\mathbf{R}^4 \to \mathbf{R} : (x, y) \to \tfrac{1}{2}\,\langle y, y\rangle.$$

More precisely, $\widetilde{\mathscr{H}} = \mathscr{G} \circ H$, where $\mathscr{G} : \mathbf{R}_> \to \mathbf{R} : z \to -\frac{\mu^2}{4z}$. Thus $X_{\widetilde{\mathscr{H}}} = f \cdot X_H$ where

$$f = \mathscr{G}' \circ H : T^+S^3 \to \mathbf{R} : (x, y) \to \frac{\mu^2}{\langle y, y\rangle^2}.$$

Because the flow of the geodesic vector field X_H is

$$\varphi^H_t(x, y) = \begin{pmatrix} \cos \nu t & \frac{1}{\nu}\sin \nu t \\ -\nu \sin \nu t & \cos \nu t \end{pmatrix},$$

where $\nu^2 = \langle y, y\rangle$, it follows that the flow of the Delaunay vector field is

$$\varphi^{\widetilde{\mathscr{H}}}_t(x, y) = \begin{pmatrix} \cos \frac{\mu^2}{\nu^3}t & \frac{1}{\nu}\sin \frac{\mu^2}{\nu^3}t \\ -\nu \sin \frac{\mu^2}{\nu^3}t & \cos \frac{\mu^2}{\nu^3}t \end{pmatrix}. \tag{16}$$

Note that on the energy surface $\widetilde{\mathscr{H}}^{-1}\left(-\frac{1}{2}\,\mu^2/\nu^3\right)$ all the integral curves of the Delaunay vector field $X_{\widetilde{\mathscr{H}}}$ are periodic of period $2\pi\nu^3/\mu^2$.

2 The SO(4) momentum mapping

In this section we construct the momentum mapping associated to the SO(4) symmetry of the geodesic vector field on (TS^3, Ω_4) and study its geometric properties.

Recall that SO(4) is the Lie group of *orthogonal* linear mappings of $(\mathbf{R}^4, \langle\ ,\ \rangle)$ into itself with determinant 1. Consider the linear action of SO(4) on $\mathbf{R}^4$ defined by

$$\varphi : \mathrm{SO}(4) \times \mathbf{R}^4 \to \mathbf{R}^4 : (A, x) \to Ax.$$

This action lifts to an action of SO(4) on $(T\mathbf{R}^4, \omega_4)$ defined by

$$\Phi : \mathrm{SO}(4) \times T\mathbf{R}^4 \to T\mathbf{R}^4 : \big(A, (x, y)\big) \to (Ax, Ay).$$

▷ Φ preserves the 1-form $\theta = \langle y, dx\rangle$ on $T\mathbf{R}^4$.

(2.1) **Proof**: We compute

$$\Phi_A^*\theta = \langle Ay, dAx\rangle = \langle Ay, Adx\rangle = \langle A^t Ay, dx\rangle = \langle y, dx\rangle = \theta.$$

The second to last equality follows because $A \in \mathrm{SO}(4)$. □

Thus the action Φ is symplectic, for

$$\Phi_A^*\omega_4 = -\Phi_A^*(d\theta) = -d(\Phi_A^*\theta) = -d\theta = \omega_4.$$

▷ To show that Φ is a Hamiltonian action, we must verify that for every $a \in \mathrm{so}(4)$ (= the Lie algebra of SO(4)) the vector field

$$\begin{aligned} X^a(x, y) &= \frac{d}{dt}\Big|_{t=0} \Phi_{\exp ta}(x, y) = \frac{d}{dt}\Big|_{t=0} \big((\exp ta)x, (\exp ta)y\big) \\ &= (ax, ay) = (X_a(x), ay), \end{aligned}$$

which is the infinitesimal generator of Φ in the direction a, is a Hamiltonian vector field on $(T\mathbf{R}^4, \omega_4)$.

(2.2) **Proof**: From the momentum lemma (see appendix B ((3.6))) it follows that $X^a = X_{J^a}$ where

$$J^a : T\mathbf{R}^4 \to \mathbf{R} : (x, y) \to \theta(x, y)X_a(x) = \langle ax, y\rangle. \quad \square \tag{17}$$

Thus the action Φ has momentum mapping $J : T\mathbf{R}^4 \to \mathrm{so}(4)^*$ defined by

$$J(x, y)a = J^a(x, y) = \langle ax, y\rangle. \tag{18}$$

Choose a basis $\{e_{ij}\}_{1\le i<j\le 4}$ of so(4) where the $(k, \ell)^{th}$ entry of the 4×4 matrix e_{ij} is

$$\begin{cases} -1, & \text{if } (k, \ell) = (i, j) \\ 1, & \text{if } (k, \ell) = (j, i) \\ 0, & \text{otherwise.} \end{cases} \tag{19}$$

Then

$$J^{e_{ji}}(x, y) = \langle x, e_{ij} y\rangle = x_i y_j - x_j y_i = S_{ij}(x, y). \tag{20}$$

▷ The mapping J is coadjoint equivariant.

(2.3) **Proof**: We compute

$$\begin{aligned} J(\Phi_A(x,y))a &= J(Ax,Ay)a = \langle aAx, Ay\rangle = \langle A^{-1}aAx, y\rangle \\ &= J(x,y)(Ad_{A^{-1}}a) = Ad'_{A^{-1}}(J(x,y))a. \qquad \square \end{aligned}$$

Since Φ_A maps TS^3 into itself for every $A \in \mathrm{SO}(4)$, Φ restricts to an action $\widehat{\Phi}$ on TS^3 given by

$$\widehat{\Phi} : \mathrm{SO}(4) \times TS^3 \to TS^3 : \big(A, (x,y)\big) \to (Ax, Ay). \tag{21}$$

For every $a \in \mathrm{so}(4)$ the infinitesimal generator X^a of the SO(4) action Φ leaves TS^3 invariant because

$$\begin{aligned} \frac{d}{dt}\langle x,x\rangle &= 2\langle x,\dot{x}\rangle = 2\langle x, ax\rangle = 0, \\ \frac{d}{dt}\langle x,y\rangle &= \langle \dot{x},y\rangle + \langle x,\dot{y}\rangle = \langle x, ay\rangle + \langle ax, y\rangle = 0, \\ \frac{d}{dt}\langle y,y\rangle &= 2\langle y,\dot{y}\rangle = 2\langle y, ay\rangle = 0, \end{aligned}$$

since $a^t = -a$. Therefore $X^a|TS^3$ is a vector field on TS^3. $\widehat{\Phi}$ preserves the symplectic form Ω_4 on TS^3 because

$$\widehat{\Phi}_A^*\Omega_4 = \widehat{\Phi}_A^*(\omega_4|TS^3) = (\Phi_A^*\omega_4)|TS^3 = \omega_4|TS^3 = \Omega_4.$$

Claim: The action $\widehat{\Phi}$ on (TS^3, Ω_4) is Hamiltonian with momentum mapping

$$\widetilde{\mathcal{J}} = J|TS^3 : TS^3 \subseteq T\mathbf{R}^4 \to \mathrm{so}(4)^*. \tag{22}$$

(2.4) **Proof**: Because X^a leaves TS^3 invariant and $\Omega_4 = \omega_4|TS^3$, it follows that $X^a|TS^3 = X_{J^a}|TS^3$. Thus $X^a|TS^3$ is the infinitesimal generator of $\widehat{\Phi}$ on TS^3 in the direction a. □

▷ So far the SO(4) symmetry is not related to the geodesic flow on TS^3. But note, the Hamiltonian $\mathcal{H}^*$ is preserved by the action Φ, that is, for every $A \in \mathrm{SO}(4)$

$$\mathcal{H}^*(\Phi_A(x,y)) = \tfrac{1}{2}\big(\langle Ax,Ax\rangle\langle Ay,Ay\rangle - \langle Ax,Ay\rangle^2\big) = \mathcal{H}^*(x,y).$$

▷ Thus the function J^a is an integral of the vector field $X_{\mathcal{H}^*}$ for every $a \in \mathrm{so}(4)$.

(2.5) **Proof**: For every $a \in \mathrm{so}(4)$ we have $\Phi_{\exp ta}^*\mathcal{H}^* = \mathcal{H}^*$. Therefore

$$0 = \frac{d}{dt}\Big|_{t=0} \Phi_{\exp ta}^*\mathcal{H}^* = L_{X^a}\mathcal{H}^* = L_{X_{J^a}}\mathcal{H}^* = -L_{X_{\mathcal{H}^*}}J^a. \qquad \square$$

From the fact that Φ preserves both the Hamiltonian $\mathcal{H}^*$ and the manifold TS^3, it follows that $\widehat{\Phi}$ preserves the geodesic Hamiltonian $H = \mathcal{H}^*|TS^3$. Therefore for every $a \in \mathrm{so}(4)$ the function $J^a|TS^3$ is an integral of the geodesic vector field X_H. □

In order to study the geometry of the momentum mapping $\widetilde{\mathcal{J}}$ (22), we transform it into an easier to understand mapping, (see (27) below). We begin by recalling that the 4×4 skew symmetric matrices $\{e_{ij}\}_{1 \le i < j \le 4}$ (19) form a basis for the Lie algebra $(\mathrm{so}(4), [\, , \,])$. The vectors $\{e^*_{ij}\}_{1 \le i<j \le 4}$, where $e^*_{ij} = e^t_{ij}$, form the standard dual basis for $\mathrm{so}(4)^*$. The Lie bracket $\{\, , \,\}_{\mathrm{so}(4)^*}$ on $\mathrm{so}(4)^*$ is defined by

$$\{e^*_{ij}, e^*_{\ell k}\}_{\mathrm{so}(4)^*} = -\sum_{m,n} c^{mn}_{ij,\ell k}\, e^*_{mn},$$

where

$$[e_{ij}, e_{\ell k}] = \sum_{m,n} c^{mn}_{ij,\ell k}\, e_{mn}.$$

Let $\vartheta : \bigwedge^2 \mathbf{R}^4 \to \mathrm{so}(4)$ be the linear map defined by

$$\vartheta(u \wedge v)w = \langle w, v\rangle\, u - \langle w, u\rangle\, v, \tag{23}$$

for $u, v, w \in \mathbf{R}^4$. Using the basis $\{e_i \wedge e_j\}_{1 \le i<j \le 4}$ of $\bigwedge^2 \mathbf{R}^4$, we see that $\vartheta(e_i \wedge e_j) = e_{ij}$. Thus ϑ is an isomorphism. Consequently the mapping

$$\vartheta^t : \mathrm{so}(4)^* \to (\textstyle\bigwedge^2 \mathbf{R}^4)^* = \bigwedge^2 (\mathbf{R}^4)^*$$

sends e^*_{ij} to $e^*_i \wedge e^*_j$. Since

$$\begin{aligned} \vartheta^t(e^*_{ij})(x, y) &= (e^*_i \wedge e^*_j)(x, y) = e^*_i(x)\, e^*_j(y) - e^*_i(y) e^*_j(x) = x_i y_j - x_j y_i \\ &= S_{ij}(x, y) = \langle x, e_{ij}(y)\rangle, \quad \text{for every } x, y \in \mathbf{R}^4, \end{aligned} \tag{24}$$

$(\bigwedge^2 \mathbf{R}^4)^*$ is the space $\mathcal{S}$ of homogeneous quadratic functions on $T\mathbf{R}^4$ which is spanned by $\{S_{ij}\}_{1 \le i<j \le 4}$. As a subspace of $C^\infty(T\mathbf{R}^4)$, $\mathcal{S}$ has a Poisson bracket $\{\, , \,\}_{\mathcal{S}}$ which is induced from the standard Poisson bracket $\{\, , \,\}$ on the space of smooth functions on $(T\mathbf{R}^4, \omega_4)$. In other words, for every $(x, y) \in T\mathbf{R}^4$

$$\{S_{ij}, S_{\ell k}\}_{\mathcal{S}}(x, y) = \omega_4\big(X_{S_{ij}}(x, y), X_{S_{\ell k}}(x, y)\big), \tag{25}$$

where $X_{S_{rs}}$ is the Hamiltonian vector field on $(T\mathbf{R}^4, \omega_4)$ corresponding to the Hamiltonian function S_{rs}. A calculation using (25) gives table 2.1.

$\{A, B\}$	S_{12}	S_{13}	S_{14}	S_{23}	S_{24}	S_{34}	B
S_{12}	0	S_{23}	S_{24}	$-S_{13}$	$-S_{14}$	0	
S_{13}	$-S_{23}$	0	S_{34}	S_{12}	0	$-S_{14}$	
S_{14}	$-S_{24}$	$-S_{34}$	0	0	S_{12}	S_{13}	
S_{23}	S_{13}	$-S_{12}$	0	0	S_{34}	$-S_{24}$	
S_{24}	S_{14}	0	$-S_{12}$	$-S_{34}$	0	S_{23}	
S_{34}	0	S_{14}	$-S_{13}$	S_{24}	$-S_{23}$	0	
A							

Table 2.1 The Poisson bracket on $\mathcal{S}$.

Because the functions $f_1 = \frac{1}{2}\langle x, x\rangle$, $f_2 = \langle x, y\rangle$, and $f_3 = \frac{1}{2}\langle y, y\rangle$ are invariant under the SO(4) action Φ on $T\mathbf{R}^4$, the function J^a (17) is an integral of X_{f_i} for every $a \in \mathrm{so}(4)$.

In other words, $\{f_i, J^a\} = 0$ for $i = 1, 2, 3$ and $a \in \mathrm{so}(4)$. Thus the Lie algebra $(\mathrm{sl}(2, \mathbf{R}), \{\, , \,\})$ spanned by $\{f_i\}_{1 \le i \le 3}$ and the Lie algebra $(\mathcal{S}, \{\, , \,\}_{\mathcal{S}})$ are *dual pairs* in the Lie algebra of homogeneous quadratic functions on $T\mathbf{R}^4$ with Poisson bracket $\{\, , \,\}$. In other words, they have the following properties:

1. They centralize $\mathcal{H}^*$, that is, $\{\mathcal{H}^*, f_i\} = 0 = \{\mathcal{H}^*, S_{jk}\}$.
2. They centralize each other, that is, $\{f_i, S_{jk}\} = 0$.

▷ We now show that the Lie algebras $(\mathcal{S}, \{\, , \,\}_{\mathcal{S}})$ and $(\mathrm{so}(4)^*, \{\, , \,\}_{\mathrm{so}(4)^*})$ are isomorphic.

(2.6) **Proof**: From the definition of S_{ij} (24) we obtain

$$dS_{ij}(x, y) = \langle e_{ij}(y), dx\rangle - \langle e_{ij}(x), dy\rangle.$$

Since $\omega_4^\flat(dx) = -\frac{\partial}{\partial y}$ and $\omega_4^\flat(dy) = \frac{\partial}{\partial x}$, we find that

$$X_{S_{ij}}(x, y) = \omega_4^\flat(dS_{ij})(x, y) = -\langle e_{ij}(x), \frac{\partial}{\partial x}\rangle - \langle e_{ij}(y), \frac{\partial}{\partial y}\rangle.$$

Therefore

$$\begin{aligned}
\{S_{ij}, S_{\ell k}\}_{\mathcal{S}}(x, y) &= (X_{S_{\ell k}} \lrcorner\, dS_{ij})(x, y) = -\langle e_{\ell k}(x), e_{ij}(y)\rangle + \langle e_{\ell k}(y), e_{ij}(x)\rangle \\
&= -\langle x, (e_{ij}e_{\ell k} - e_{\ell k}e_{ij})y\rangle = -\langle x, [e_{ij}, e_{\ell k}]y\rangle \\
&= -\vartheta^t([e_{ij}, e_{\ell k}]^*)(x, y) = \vartheta^t(\{e_{ij}^*, e_{\ell k}^*\}_{\mathrm{so}(4)^*})(x, y). \qquad (26)
\end{aligned}$$

The last equality follows by definition of the Poisson bracket $\{\, , \,\}_{\mathrm{so}(4)^*}$. Hence ϑ^t is a Lie algebra isomorphism. □

On $\bigwedge^2\mathbf{R}^4$ define an inner product

$$K : \bigwedge^2\mathbf{R}^4 \times \bigwedge^2\mathbf{R}^4 \to \mathbf{R} : (u \wedge v, x \wedge y) \to \det\begin{pmatrix} \langle u, x\rangle & \langle u, y\rangle \\ \langle v, x\rangle & \langle v, y\rangle \end{pmatrix}.$$

Since $\{e_i \wedge e_j\}_{1 \le i < j \le 4}$ is an orthonormal basis of $(\bigwedge^2\mathbf{R}^4, K)$, we may identify $\bigwedge^2\mathbf{R}^4$ with $(\bigwedge^2\mathbf{R}^4)^*$.

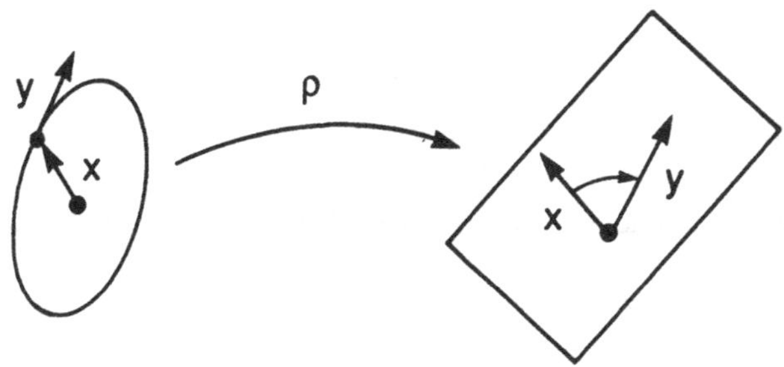

Figure 2.1. The mapping ρ.

Instead of studying the momentum mapping $\widetilde{\mathcal{J}}$ (22) we study the mapping

$$\rho : T^+S^3 \subseteq T\mathbf{R}^4 \to \bigwedge^2\mathbf{R}^4 : (x,y) \to x \wedge y = \sum_{1\le i<j\le 4} S_{ij}(x,y)\, e_i \wedge e_j, \qquad (27)$$

which is nothing but $K^\flat \circ \vartheta^t \circ \mathcal{J}$. The S_{ij} are the *Plücker coordinates* of the oriented 2-plane spanned by $\{x, y\}$ corresponding to the 2-vector $x \wedge y$. In other words, S_{ij} is the 2×2 minor formed from the ith and j^{th} columns of the 2×4 matrix with rows x and y, that is, $S_{ij} = \det\begin{pmatrix} x_i & x_j \\ y_i & y_j \end{pmatrix}$. Because

$$0 = (x \wedge y) \wedge (x \wedge y) = (S_{12}S_{34} - S_{13}S_{24} + S_{14}S_{23})\, e_1 \wedge e_2 \wedge e_3 \wedge e_4,$$

the Plücker coordinates of $x \wedge y$ satisfy *Plücker's equation*

$$S_{12}S_{34} - S_{13}S_{24} + S_{14}S_{23} = 0. \qquad (28)$$

Let C be the set of all nonzero 2-vectors on $\mathbf{R}^4$ whose Plücker coordinates satisfy (28). By ▷ definition $\rho(T^+S^3) \subseteq C$. Actually, C is the image of ρ.

(2.7) **Proof**: Suppose that $\theta \in C$. Then θ is decomposable, that is, there are vectors $u, v \in \mathbf{R}^4$ such that $\theta = u \wedge v$. To see this, let (S_{ij}) be the Plücker coordinates of θ. Since $\theta \neq 0$ not every S_{ij} is zero. Suppose that S_{12} is nonzero. Let $u = (1, 0, -S_{23}/S_{12}, -S_{24}/S_{12})$ and $v = (0, S_{12}, S_{13}, S_{14})$. Using Plücker's equation (28) it is easy to check that the Plücker coordinates of the 2-vector $u \wedge v$ are (S_{ij}). Therefore $\theta = u \wedge v$. A similar argument, which we omit, works in the other cases. Let $\{x, y\}$ be an orthonormal basis of the 2-plane spanned by $\{u, v\}$. Then $u \wedge v = \lambda\, x \wedge y$ for some nonzero λ. Therefore $\rho(x, \lambda y) = \theta$. □

For $h > 0$ let

$$H^{-1}(h) = \left\{ (x,y) \in T^+S^3 \subseteq T\mathbf{R}^4 \,\middle|\, \tfrac{1}{2}\langle y, y\rangle = h \right\}$$

be the h-level set of the geodesic Hamiltonian H (7). Consider the mapping

$$\rho_h : H^{-1}(h) \subseteq T^+S^3 \to C \subseteq \bigwedge^2\mathbf{R}^4 : (x,y) \to x \wedge y, \qquad (29)$$

which is the restriction of ρ (27) to $H^{-1}(h)$. From the identity

$$\sum_{1\le i<j\le 4} (x_i y_j - x_j y_i)^2 = \langle x, x\rangle\langle y, y\rangle - \langle x, y\rangle^2 \qquad (30)$$

we see that the image of ρ_h is contained in the submanifold C_h of C defined by

$$\sum_{1\le i<j\le 4} S_{ij}^2 = 2h. \qquad (31)$$

▷ C_h is diffeomorphic to $S^2_{\sqrt{h/2}} \times S^2_{\sqrt{h/2}}$.

(2.8) **Proof**: Adding and subtracting one half times (28) from one quarter times (31) and using the variables

$$\begin{aligned} \xi_1 &= \tfrac{1}{2}(S_{12} + S_{34}) & \eta_1 &= \tfrac{1}{2}(S_{12} - S_{34}) \\ \xi_2 &= \tfrac{1}{2}(S_{13} - S_{24}) & \eta_2 &= \tfrac{1}{2}(S_{13} + S_{24}) \\ \xi_3 &= \tfrac{1}{2}(S_{14} + S_{23}) & \eta_3 &= \tfrac{1}{2}(S_{14} - S_{23}), \end{aligned} \qquad (32)$$

we obtain

$$\xi_1^2 + \xi_2^2 + \xi_3^2 = h/2 \quad \text{and} \quad \eta_1^2 + \eta_2^2 + \eta_3^2 = h/2. \qquad \Box \tag{33}$$

We now investigate the geometry of the map ρ_h.

Claim: For every $h > 0$, the map $\rho_h : H^{-1}(h) \to C_h$ (29) is a surjective submersion each of whose fibers is a single oriented orbit of the geodesic vector field X_H of energy h.

(2.9) **Proof**: To show that ρ_h is surjective, suppose that $S = (S_{ij}) \in C_h$. Since C_h is contained in $C = \rho(T^+S^3)$, there is an $(x, y) \in T^+S^3$ such that $\rho(x, y) = S$. But $2h = \sum_{1 \le i < j \le 4} S_{ij}^2$ since $S \in C_h$. From (30), the definition of S_{ij} (24), and the fact that $(x, y) \in T^+S^3$, we find that $\frac{1}{2}\langle y, y\rangle = h$. Hence $(x, y) \in H^{-1}(h)$.

To show that ρ_h is a submersion, we must verify that the rank of $T_{(x,y)}\rho_h$ is 4 for every $(x, y) \in H^{-1}(h)$, because C_h is 4-dimensional. Towards this goal, let

$$V_{(x,y)} = \text{span}\,\{X_{S_{ij}}(x, y)\}_{1 \le i < j \le 4},$$

where $X_{S_{ij}}$ is the Hamiltonian vector field on $(T\mathbf{R}^4, \omega_4)$ corresponding to the Hamiltonian function

$$S_{ij} : T\mathbf{R}^4 \to \mathbf{R} : (x, y) \to x_i y_j - x_j y_i.$$

Since $S_{ij}|T^+S^3$ is an integral of of the geodesic vector field X_H on T^+S^3, we see that

$$V_{(x,y)} \subseteq \ker dH(x, y) = T_{(x,y)}H^{-1}(h)$$

for every $(x, y) \in T^+S^3$. Now

$$\left(T_{(x,y)}\rho_h\right)|V_{(x,y)} = \left(dS_{ij}(x, y)X_{S_{\ell k}}(x, y)\right) = (\{S_{ij}, S_{\ell k}\}_{\mathcal{G}}) = \widetilde{P}. \tag{34}$$

The 6×6 matrix $\widetilde{P}$ is conjugate to the matrix

$$P = \begin{pmatrix} (\{\xi_i, \xi_j\}_{\mathcal{G}}) & 0 \\ 0 & (\{\eta_i, \eta_j\}_{\mathcal{G}}) \end{pmatrix} = \begin{pmatrix} (\sum_k \varepsilon_{ijk}\xi_k) & 0 \\ 0 & (\sum_k \varepsilon_{ijk}\eta_k) \end{pmatrix},$$

using (33) and table 2.1. Here ε_{ijk} is 0 if $\{i, j, k\}$ is a proper subset of $\{1, 2, 3\}$, is $+1$ if $\{i, j, k\}$ is an even permutation of $1, 2, 3$, and is -1 otherwise. But $S = (S_{ij}) = \rho_h(x, y) \in C_h$. By (33) each of the 3×3 skew symmetric matrices $(\sum_k \varepsilon_{ijk}\xi_k)$ and $(\sum_k \varepsilon_{ijk}\eta_k)$ is nonzero and hence each has rank 2. Therefore, the rank of $T_{(x,y)}\rho_h$ is 4 for every $(x, y) \in H^{-1}(h)$. Thus ρ_h is a submersion.

Given $S = (S_{ij}) \in C_h$, the fiber $W = \rho_h^{-1}(S)$ is a union of orbits of the geodesic vector field X_H of energy h because $S_{ij}|T^+S^3$ are integrals of X_H. By definition of ρ_h (29), W is the set of all ordered pairs $\{x, y\}$ of orthogonal vectors in $\mathbf{R}^4$ such that $\langle x, x\rangle = 1$, $\langle y, y\rangle = 2h$ and the 2-plane Π spanned by $\{x, y\}$ has Plücker coordinates (S_{ij}). Since any two such bases of Π are related by a *counterclockwise* rotation in Π, we find that

$$W = \left\{(x\cos\theta - y\sin\theta, x\sin\theta + y\cos\theta) \in H^{-1}(h) \middle| \theta \in [0, 2\pi]\right\}.$$

Therefore W is a unique oriented orbit of X_H traced out by an integral curve of X_H. $\Box$

Corollary: C_h is the space of orbits of positive energy h of the geodesic vector field X_H on T^+S^3 with orbit mapping $\rho_h : H^{-1}(h) \to C_h$.

(2.10) **Proof**: The corollary follows immediately from the definition of orbit space (see appendix B section 2) and the claim. □

The goal of the following discussion is to construct a symplectic form on C_h. We begin by defining a Poisson bracket $\{\, ,\, \}$ on the space $C^\infty(\mathcal{S})$ of smooth functions on the Lie algebra $(\mathcal{S}, \{\, ,\, \}_{\mathcal{S}})$. For $f, g \in C^\infty(\mathcal{S})$ let

$$\{f, g\} = \sum_{\substack{1 \le i < j \le 4 \\ 1 \le \ell < k \le 4}} \frac{\partial f}{\partial S_{ij}} \frac{\partial g}{\partial S_{\ell k}} \{S_{ij}, S_{\ell k}\}_{\mathcal{S}}. \tag{35}$$

As is shown in example 1 of appendix A section 4, $(C^\infty(\mathcal{S}), \{\, ,\, \}_{\mathcal{S}})$ is a Lie algebra. On $C^\infty(\mathcal{S})$ define a multiplication $\cdot$ by $(f \cdot g)(S) = f(S)g(S)$ for every $S \in \mathcal{S}$. Then $(C^\infty(\mathcal{S}), \cdot)$ is a commutative ring with unit. Using (35) it is straightforward to check that Leibniz' rule holds, namely

$$\{f, g \cdot h\} = \{f, g\} \cdot h + \{f, h\} \cdot g,$$

for every $f, g, h \in C^\infty(\mathcal{S})$. Therefore $\mathcal{A} = (C^\infty(\mathcal{S}), \{\, ,\, \}_{\mathcal{S}}, \cdot)$ is a Poisson algebra. Consider the functions

$$C_1 = \sum_{1 \le i < j \le 4} S_{ij}^2 - 2h \quad \text{and} \quad C_2 = S_{12}S_{34} - S_{13}S_{24} + S_{14}S_{23}.$$

They are Casimirs for $\mathcal{A}$. In other words, $\{C_1, f\} = \{C_2, f\} = 0$ for every $f \in C^\infty(\mathcal{S})$. (From (35) it is enough to show that $\{C_1, S_{ij}\} = \{C_2, S_{ij}\} = 0$ for $1 \le i < j \le 4$. This is a direct verification using table 2.1.) Let $\mathcal{I}$ be the ideal in $(C^\infty(\mathcal{S}), \cdot)$ which is generated ▷ by C_1 and C_2. Then $\mathcal{I}$ is a Poisson ideal in $\mathcal{A}$, that is, if $f \in \mathcal{I}$, then $\{f, g\} \in \mathcal{I}$ for every $g \in C^\infty(\mathcal{S})$.

(2.11) **Proof**: Since $f \in \mathcal{I}$ there are $f_1, f_2 \in C^\infty(\mathcal{S})$ such that $f = f_1C_1 + f_2C_2$. Now

$$\begin{aligned} \{f, g\} &= \{f_1, g\} \cdot C_1 + f_1 \cdot \{C_1, g\} + \{f_2, g\} \cdot C_2 + f_2 \cdot \{C_2, g\}, \quad \text{by Leibniz' rule} \\ &= \{f_1, g\} \cdot C_1 + \{f_2, g\} \cdot C_2, \quad \text{because } C_1 \text{ and } C_2 \text{ are Casimirs} \\ &\in \mathcal{I}. \quad \square \end{aligned}$$

Therefore we can define a Poisson bracket $\{\, ,\, \}_{C_h}$ on $C^\infty(\mathcal{S}/\mathcal{I})$ by

$$\{f + \mathcal{I}, g + \mathcal{I}\}_{C_h} = \{f, g\}. \tag{36}$$

In order that we can identify the space $C^\infty(\mathcal{S}/\mathcal{I})$ with the space $C^\infty(C_h)$ of smooth functions on C_h, we need to know that $\mathcal{I}$ is the set of smooth functions vanishing identically on C_h. This is a consequence of the following general

Fact: Suppose that 0 is a regular value of the smooth map

$$F : \mathbf{R}^n \to \mathbf{R}^k : z \to \big(F_1(z), \ldots, F_k(z)\big).$$

Then $M = F^{-1}(0)$ is a smooth submanifold of $\mathbf{R}^n$ defined by

$$F_1(z) = \cdots = F_k(z) = 0.$$

If $G : \mathbf{R}^n \to \mathbf{R}$ is a smooth function, which vanishes identically on M, then there are smooth functions $g_i : \mathbf{R}^n \to \mathbf{R}$, $1 \le i \le k$, such that $G = \sum_{i=1}^k g_i F_i$.

(2.12) **Proof**: Since M is a smooth submanifold of $\mathbf{R}^n$, there is an atlas $\{U_i, \varphi_i\}_{i \in I}$ of $\mathbf{R}^n$ such that for every $i \in I$ for which $U_i \cap M \neq \emptyset$, we have

$$\varphi_i(U_i) = \widetilde{V}_i \times \widetilde{W}_i \subseteq \mathbf{R}^k \times \mathbf{R}^{n-k}$$

where $\widetilde{V}_i$ contains a ball of radius 1 about 0 in $\mathbf{R}^k$, $\widetilde{W}_i$ is an open subset of $\mathbf{R}^{n-k}$, and

$$\varphi_i(U_i \cap M) = \Big\{ (x, y) \in \varphi_i(U_i) \Big| x_1 = \cdots = x_k = 0 \Big\}.$$

Such an atlas of $\mathbf{R}^n$ is called a *submanifold atlas* for M. Let $\{\psi^\alpha, V^\alpha\}_{\alpha \in A}$ be a smooth *partition of unity* subordinate to the covering $\{U_i\}_{i \in I}$. For every $\alpha \in A$, there is an $i = i(\alpha) \in I$ such that $V^\alpha \subseteq U_{i(\alpha)}$. Let φ^α be the restriction of $\varphi_{i(\alpha)}$ to V^α multiplied by a suitable positive factor $r_{i(\alpha)}$. The factor $r_{i(\alpha)}$ is chosen so that $(r_{i(\alpha)}\varphi_{i(\alpha)})(V^\alpha) \cap \widetilde{V}_{i(\alpha)}$ contains a ball of radius 1 about 0. Then $\{V^\alpha, \varphi^\alpha\}_{\alpha \in A}$ is a submanifold atlas for M.

Let $G^\alpha = \psi^\alpha G$. Then G^α vanishes identically outside of V^α, since ψ^α does. Suppose that $V^\alpha \cap M = \emptyset$. Then $\sum_j F_j^2 \neq 0$. Let $g_j^\alpha = \frac{G^\alpha F_j}{\sum_j F_j^2}$. Then $g_j^\alpha : V^\alpha \to \mathbf{R}$ is smooth and on V^α we have

$$G^\alpha = \sum_{j=1}^k g_j^\alpha F_j. \tag{37}$$

Now suppose that $V^\alpha \cap M \neq \emptyset$. On

$$\varphi^\alpha(V^\alpha) = \big(\varphi_{i(\alpha)}(V^\alpha) \cap \widetilde{V}_{i(\alpha)}\big) \times \big(\varphi_{i(\alpha)}(V^\alpha) \cap \widetilde{W}_{i(\alpha)}\big) = \widetilde{V}^\alpha \times \widetilde{W}^\alpha,$$

the function $\widetilde{G}^\alpha = G^\alpha \circ (\varphi^\alpha)^{-1}$ is defined and vanishes identically on $\varphi^\alpha(V^\alpha \cap M)$. Therefore for every $(x, y) \in \widetilde{V}^\alpha \times \widetilde{W}^\alpha$,

$$\begin{aligned} \widetilde{G}^\alpha(x, y) &= \widetilde{G}^\alpha(0, y) + \int_0^1 \frac{d}{dt}\widetilde{G}^\alpha(tx, y)\,dt \\ &= 0 + \sum_{j=1}^k \Big(\int_0^1 \frac{\partial \widetilde{G}^\alpha}{\partial x_j}(tx, y)\,dt \Big) x_j = \sum_{j=1}^k \widetilde{g}_j^\alpha(x, y) x_j. \end{aligned}$$

Hence on V^α we have

$$G^\alpha = \sum_{j=1}^k g_j^\alpha F_j, \tag{38}$$

where $g_j^\alpha = \widetilde{g}_j^\alpha \circ \varphi^\alpha : V^\alpha \to \mathbf{R}$ is smooth. Pasting together the local results (37) and (38) using the smooth partition of unity (ψ^α, V^α), we obtain

$$G = \sum_\alpha \psi^\alpha G^\alpha = \sum_\alpha \psi^\alpha \Big(\sum_{j=1}^k g_j^\alpha F_j \Big) = \sum_{j=1}^k g_j F_j,$$

where $g_j = \sum_\alpha \psi^\alpha g_j^\alpha : \mathbf{R}^n \to \mathbf{R}$ is smooth. □

Consequently, we may define the quotient Poisson algebra

$$\mathcal{B} = \mathcal{A}/\mathcal{I} = (C^\infty(C_h), \{\,,\,\}_{C_h}, \cdot).$$

Because

$$\{S_{ij} + \mathcal{I}, S_{\ell k} + \mathcal{I}\}_{C_h} = \{S_{ij}, S_{\ell k}\}_{\mathcal{G}},$$

the matrix of Poisson brackets $(\{S_{ij} + \mathcal{I}, S_{\ell k} + \mathcal{I}\}_{C_h})$ has rank 4. Therefore C_h is a co- ▷ symplectic manifold. In other words, the Poisson bracket $\{\,,\,\}_{C_h}$ is *nondegenerate* and hence defines a symplectic form ω_h on C_h, (see appendix A section 4). Moreover, ω_h satisfies

$$\rho_h^* \omega_h = \widetilde{\Omega}_4 | H^{-1}(h). \tag{39}$$

(2.13) **Proof**: For every $(x, y) \in H^{-1}(h)$ we know that $T_{(x,y)}H^{-1}(h)$ is spanned by the vectors $\{X_{S_{ij}}(x, y)\}_{1\le i<j\le 4}$. Since $(T^+S^3, \widetilde{\Omega}_4)$ is a cosymplectic submanifold of $(T\mathbf{R}^4, \omega_4)$,

$$\begin{aligned}
\widetilde{\Omega}_4(x, y)\big(X_{S_{ij}}(x, y), X_{S_{\ell k}}(x, y)\big) &= \omega_4\big(X_{S_{ij}}(x, y), X_{S_{\ell k}}(x, y)\big) \\
&= \{S_{ij}, S_{\ell k}\}_{\mathcal{G}}(x, y) = \{S_{ij}, S_{\ell k}\}_{C_h}(\rho_h(x, y)) \\
&= \omega_h(\rho_h(x, y))\big(T_{(x,y)}\rho_h X_{S_{ij}}(x, y), T_{(x,y)}\rho_h X_{S_{\ell k}}(x, y)\big) \\
&= (\rho_h^*\omega_h)(x, y)\big(X_{S_{ij}}(x, y), X_{S_{\ell k}}(x, y)\big),
\end{aligned}$$

we obtain (39). □

We now prove the main result of this section, which describes the geometry of the mapping ρ (27). As a consequence, we know the geometry of the SO(4) momentum mapping $\widetilde{\mathcal{J}}$ (22) of the geodesic vector field X_H on (T^+S^3, Ω_4). Note that $\widetilde{\mathcal{J}}$ is also the momentum map of the Delaunay vector field $X_{\widetilde{\mathcal{H}}}$.

Claim: The mapping

$$\rho : T^+S^3 \subseteq T\mathbf{R}^4 \to C \subseteq \bigwedge^2 \mathbf{R}^4 : (x, y) \to x \wedge y$$

is a surjective submersion, each of whose fibers is a unique oriented orbit of the geodesic vector field X_H on $(T^+S^3, \widetilde{\Omega}_4)$.

(2.14) **Proof**: We have already shown that ρ is surjective ((2.6)). To show that each of its fibers is a unique oriented orbit of X_H we argue as follows. Suppose that $S = (S_{ij}) \in C$. Because S is nonzero, $\sum_{1\le i<j\le 4} S_{ij}^2 = 2h$ for some $h > 0$. Therefore $S \in C_h$. Since the fiber $\rho_h^{-1}(S)$ of ρ_h is a unique oriented orbit of X_H of energy h ((2.8)), so is the fiber $\rho^{-1}(S)$ of ρ because $\rho = \rho_h$ on $H^{-1}(h)$.

To show that ρ is a submersion, first note that by ((2.8)) the map $\rho_h : H^{-1}(h) \to C_h$ is a submersion. Note that $H^{-1}(h)$ and C_h are codimension 1 submanifolds of T^+S^3 and C, respectively. Since a normal direction to $H^{-1}(h)$ at $(x, y) \in T^+S^3$ and a normal direction to C_h at $\rho(x, y) \in C$ is spanned by grad $H(x, y)$ and grad $F(\rho(x, y))$ respectively, (where

$$F(S_{ij}) = \sum_{1\le i<j\le 4} S_{ij}^2 - 2h = 0$$

defines C_h as a submanifold of C), it suffices to show that

$$\left\langle T_{(x,y)}\rho \operatorname{grad} H(x,y), \operatorname{grad} F(\rho(x,y))\right\rangle$$

is nonzero. We compute. Clearly grad $H(x, y) = (0, y)$. Hence

$$T_{(x,y)}\rho\big(\operatorname{grad} H(x,y)\big) = x \wedge y = \big(S_{ij}(x,y)\big).$$

But grad $F(\rho(x, y)) = 2\big(S_{ij}(x, y)\big)$. Therefore

$$\left\langle T_{(x,y)}\rho\big(\operatorname{grad} H(x,y)\big), \operatorname{grad} F(\rho(x,y))\right\rangle = 2 \sum_{1\le i<j\le 4} S_{ij}^2(x,y) = 4h > 0. \quad \square$$

This claim has some interesting consequences.

Corollary 1: The *space of orbits* of the geodesic vector field with positive energy is the manifold C. The orbit map is $\rho : T^+S^3 \rightarrow C$.

(2.15) **Proof**: This follows immediately from the claim and the definition of orbit space, (see appendix B section 2). □

Observe that every smooth integral of the geodesic vector field on T^+S^3 is a smooth function of the integrals S_{ij}. More precisely we prove

Corollary 2: Suppose that $G : T^+S^3 \subseteq T\mathbf{R}^4 \rightarrow \mathbf{R}$ is a smooth integral of the geodesic vector field X_H on $(T^+S^3, \widetilde{\Omega}_4)$. Then there is a smooth function $\widehat{G} : C \subseteq \bigwedge^2\mathbf{R}^4 \rightarrow \mathbf{R}$ such that $G = \rho^*\widehat{G}$.

(2.16) **Proof**: Since G is a smooth integral of X_H on T^+S^3, it is constant on every orbit of X_H on T^+S^3. Because each fiber of ρ is a unique orbit of X_H on T^+S^3, G descends to a smooth function $\widehat{G}$ on the orbit space C. But $\rho : T^+S^3 \rightarrow C$ is the orbit map, so $G = \rho^*\widehat{G}$. □
The SO(4) momentum mapping

3 The Kepler problem

We investigate the bounded motion of a particle in $\mathbf{R}^3$ which is under the influence of a gravitational field of a second particle fixed at the origin. This is *Kepler's problem.*

3.1 The Kepler vector field

In this subsection we define the Kepler Hamiltonian system $(H, T_0\mathbf{R}^3, \omega_3)$. We then show that the Kepler Hamiltonian vector field X_H conserves energy H, angular momentum $\mathbf{J}$, and the eccentricity vector $\mathbf{e}$. On the set Σ_- of positions and momenta where the values of H are negative, the orbits of X_H are bounded, yet the flow of X_H is incomplete.

On the phase space $T_0\mathbf{R}^3 = (\mathbf{R}^3 - \{0\}) \times \mathbf{R}^3$ with coordinates (q, p) and symplectic form $\omega_3 = \sum_{i=1}^3 dq_i \wedge dp_i$, consider the *Kepler Hamiltonian*

$$H : T_0\mathbf{R}^3 \rightarrow \mathbf{R} : (q,p) \rightarrow \tfrac{1}{2}\langle p, p\rangle - \frac{\mu}{|q|}. \tag{40}$$

Here $\langle\, ,\, \rangle$ is the Euclidean inner product on $\mathbf{R}^3$ and $|q|$ is the length of the vector q. The integral curves of the Hamiltonian vector field X_H on $T_0\mathbf{R}^3$ satisfy the equations

$$\begin{cases} \dot{q} &= p \\ \dot{p} &= -\mu \dfrac{q}{|q|^3}, \end{cases} \tag{41}$$

which describe the motion of a particle of mass 1 about the origin under the influence of an inverse $|q|^2$ force — such as Newtonian gravity. We consider the case where the force is attractive, that is, $\mu > 0$. However, much of the following analysis can be carried out mutatis mutandis for $\mu < 0$. (See exercise 3 how to reduce the two body problem to the Kepler problem (41).)

The *Kepler vector field* X_H has some obvious integrals: the *total energy*

$$h = \tfrac{1}{2}\langle p, p\rangle - \frac{\mu}{|q|}, \tag{42}$$

which is nothing but the Hamiltonian H, and the *angular momentum*

$$\mathbf{J} = (J_1, J_2, J_3) = q \times p. \tag{43}$$

Here $\times$ is the *vector product* on $\mathbf{R}^3$.

(3.1) **Proof**: A direct way to see that $\mathbf{J}$ is an integral is to compute

$$\begin{aligned} \frac{d\mathbf{J}}{dt} &= \frac{dq}{dt} \times p + q \times \frac{dp}{dt} = p \times p - \frac{\mu}{|q|^3} q \times q, \quad \text{using (41)} \\ &= 0. \end{aligned}$$

A more sophisticated way to see this is to note that the SO(3) action

$$\mathrm{SO}(3) \times \mathbf{R}^3 \to \mathbf{R}^3 : (O, q) \to Oq$$

lifts to a Hamiltonian action

$$\mathrm{SO}(3) \times T_0\mathbf{R}^3 \to T_0\mathbf{R}^3 : \big(O, (q, p)\big) \to (Oq, Op).$$

This action has the momentum mapping

$$\widetilde{J} : T_0\mathbf{R}^3 \to \mathrm{so}(3)^* : (q, p) \to \begin{pmatrix} 0 & J_3 & -J_2 \\ -J_3 & 0 & J_1 \\ J_2 & -J_1 & 0 \end{pmatrix},$$

defined by $\widetilde{J}(q, p)X = \langle p, X(q)\rangle$ where $X \in \mathrm{so}(3)$. Now use the map $k^\flat$ associated to the Killing metric

$$k : \mathrm{so}(3) \times \mathrm{so}(3) \to \mathbf{R} : (X, Y) \to \tfrac{1}{2}\,\mathrm{tr}\, XY^t$$

to identify $\mathrm{so}(3)^*$ with $\mathrm{so}(3)$. This identification boils down to taking transposes. Follow this by the map

$$i : \mathrm{so}(3) \to \mathbf{R}^3 : X = \begin{pmatrix} 0 & -x_3 & x_2 \\ x_3 & 0 & -x_1 \\ -x_2 & x_1 & 0 \end{pmatrix} \to \mathbf{x} = (x_1, x_2, x_3),$$

which identifies so(3) with $\mathbf{R}^3$, (see chapter 3 section 1). Then $\widetilde{J}$ becomes the usual angular momentum $\mathbf{J} : T_0\mathbf{R}^3 \to \mathbf{R}^3 : (q, p) \to q \times p$. Since the SO(3) action on $(T_0\mathbf{R}^3, \omega_3)$ leaves the Kepler Hamiltonian H (40) invariant, every component of the angular momentum $\mathbf{J}$ is constant on the integral curves of X_H. □

▷ More interestingly, there is another integral of the Kepler vector field. This is the *eccentricity vector.*

$$\mathbf{e} = (e_1, e_2, e_3) = -\frac{q}{|q|} + \frac{1}{\mu} p \times (q \times p). \tag{44}$$

(3.2) **Proof**: To see this we calculate

$$\begin{aligned} \frac{d\mathbf{e}}{dt} &= -\frac{d}{dt}\frac{q}{|q|} + \frac{1}{\mu}\frac{dp}{dt} \times \mathbf{J} = \frac{1}{|q|^3}\langle \frac{dq}{dt}, q\rangle q - \frac{1}{|q|}\frac{dq}{dt} + \frac{1}{\mu}\frac{dp}{dt} \times \mathbf{J} \\ &= \frac{1}{|q|^3}\big(\langle q, p\rangle q - \langle q, q\rangle p\big) - \frac{1}{|q|^3} q \times \mathbf{J}, \quad \text{using (41)} \\ &= \frac{1}{|q|^3}\big(q \times (q \times p) - q \times (q \times p)\big) = 0. \qquad \square \end{aligned}$$

We now prove some properties of the flow of the Kepler vector field X_H.

Claim: If the energy h is negative, then the image of every integral curve of the Kepler vector field under the bundle projection $\tau : T_0\mathbf{R}^3 \to \mathbf{R}^3 : (q, p) \to q$ is bounded.

(3.3) **Proof**:
case 1. $\mathbf{J} = 0$. Since $\mathbf{e}$ is an integral of X_H and $\mathbf{J} = 0$, the direction $\mathbf{e} = -\frac{q}{|q|}$ of the motion is constant. Therefore the motion takes place on the line $q(t) = r(t)\,\mathbf{e}$. From conservation of energy we obtain $h + \frac{\mu}{r} = \frac{1}{2}\dot{r}^2 \geq 0$. Therefore $|q(t)| \leq \frac{\mu}{-h}$.

case 2. $\mathbf{J} \neq 0$. Since $J^2 = |q \times p|^2 = |q|^2|p|^2 - \langle q, p\rangle^2$,

$$h = \tfrac{1}{2}\langle p, p\rangle - \frac{\mu}{|q|} = \frac{1}{2}\frac{\langle q, p\rangle^2}{|q|^2} + \frac{1}{2}\frac{J^2}{|q|^2} - \frac{\mu}{|q|} \geq \frac{1}{2}\frac{J^2}{|q|^2} - \frac{\mu}{|q|}.$$

As is easily verified, the function $V_J(|q|) = \frac{1}{2}\frac{J^2}{|q|^2} - \frac{\mu}{|q|}$ has a unique nondegenerate minimum at $|q| = \frac{J^2}{\mu}$ corresponding to the critical value $-\frac{\mu^2}{2J^2}$. Since $\lim_{|q|\to 0^+} V_J(|q|) \nearrow \infty$ and $\lim_{|q|\to\infty} V_J(|q|) \nearrow 0^-$, the function V_J is proper on the set where it has negative values. Therefore $V_J^{-1}([-\frac{\mu^2}{2J^2}, h])$ is compact. Thus the length of $q(t)$ is bounded, when $h < 0$. □

Claim: The flow of the Kepler vector field X_H is *not complete*.

(3.4) **Proof**: Consider a bounded motion with $\mathbf{J} = 0$ and $h < 0$ which starts at $(r(0), \dot{r}(0)) = (\frac{\mu}{-h}, 0)$. The time it takes to reach the origin is

$$T = \int_0^{\frac{\mu}{-h}} \frac{dr}{\sqrt{\frac{2\mu}{r} + 2h}}.$$

This is obtained by separating variables in $\frac{1}{2}\dot{r}^2 = h + \frac{\mu}{r}$ and integrating. Performing the integral gives

$$T = \frac{\pi}{2} \frac{\mu}{(-2h)^{3/2}}, \tag{45}$$

which is finite. □

3.2 The so(4) momentum map

Recall that Σ_- is the open subset of $(T_0\mathbf{R}^3, \omega_3)$ where the energy H is negative. In this subsection we show that on Σ_- the components of the angular momentum $\mathbf{J}$ and the modified eccentricity vector $\widetilde{\mathbf{e}} = -\nu\, \mathbf{e}$ (where $\nu = \mu/\sqrt{-2H}$) form a Lie algebra under Poisson bracket which is isomorphic to so(4). This defines a representation of so(4) on the space of Hamiltonian vector fields on $(\Sigma_-, \widetilde{\omega}_3 = \omega_3|\Sigma_-)$ which has a momentum mapping $\mathcal{J}$. In fact $\mathcal{J}$ is a surjective submersion from Σ_- to

$$C = \left\{ (\mathbf{J}, \widetilde{\mathbf{e}}) \in \mathbf{R}^6 \,\middle|\, \langle \mathbf{J} + \widetilde{\mathbf{e}}, \mathbf{J} + \widetilde{\mathbf{e}} \rangle = \langle \mathbf{J} - \widetilde{\mathbf{e}}, \mathbf{J} - \widetilde{\mathbf{e}} \rangle > 0 \right\}$$

each of whose nonempty fibers is a unique oriented bounded orbit of X_H.

▷ First we show that on $(\Sigma_-, \widetilde{\omega}_3)$ the components of the angular momentum $\mathbf{J}$ and the modified eccentricity vector $\widetilde{\mathbf{e}}$ satisfy the Poisson bracket relations

$$\{J_i, J_j\} = \sum_k \varepsilon_{ijk} J_k, \quad \{J_i, \widetilde{e}_j\} = \sum_k \varepsilon_{ijk} \widetilde{e}_k, \quad \text{and} \quad \{\widetilde{e}_i, \widetilde{e}_j\} = \sum_k \varepsilon_{ijk} J_k. \tag{46}$$

(3.5) **Proof**: We verify only the third equality in (46). Let $\mathbf{A} = \mu\, \mathbf{e}$. Since $\{q_\ell, p_m\} = \delta_{\ell m}$, $\{q_i, q_j\} = 0$, and $\{p_i, p_j\} = 0$, we have

$$\{J_a, |q|\} = 0, \quad \{q_a, J_b\} = \sum_c \varepsilon_{abc} q_c, \quad \text{and} \quad \{p_a, J_b\} = \sum_c \varepsilon_{abc} p_c.$$

Using bilinearity and the derivation property of Poisson bracket, expand

$$\{A_i, A_j\} = \left\{ \sum_{j,k} \varepsilon_{ijk} p_j J_k - \frac{1}{|q|} q_i, \sum_{m,n} \varepsilon_{\ell mn} p_m J_n - \frac{1}{|q|} q_\ell \right\}$$

to obtain

$$\{A_i, A_j\} = -2H \sum_k \varepsilon_{ijk} J_k.$$

It helps to recall the identity

$$\sum_i \varepsilon_{ijk} \varepsilon_{i\ell m} = \delta_{mk}\delta_{\ell j} - \delta_{jm}\delta_{\ell k}. \quad \square$$

▷ The bracket relations (46) define a Lie algebra which is isomorphic to so(4).

(3.6) **Proof**: For $i = 1, 2, 3$ define

$$\xi_i = \tfrac{1}{2}(J_i + \widetilde{e}_i) \quad \text{and} \quad \eta_i = \tfrac{1}{2}(J_i - \widetilde{e}_i).$$

In terms of ξ_i and η_i the bracket relations (46) become

$$\{\xi_i, \xi_j\} = \sum_k \varepsilon_{ijk}\, \xi_k, \quad \{\eta_i, \eta_j\} = \sum_k \varepsilon_{ijk}\, \eta_k, \quad \text{and} \quad \{\xi_i, \eta_j\} = 0. \tag{47}$$

These relations define the Lie algebra so(3) × so(3), which is isomorphic to so(4). □

Thus the mappings

$$J_i \to \mathrm{ad}_{J_i} = -X_{J_i} \quad \text{and} \quad \widetilde{e}_i \to \mathrm{ad}_{\widetilde{e}_i} = -X_{\widetilde{e}_i}$$

define a representation of so(4) on the space of Hamiltonian vector fields on $(\Sigma_-, \widetilde{\omega}_3)$. In other words, we have a *Hamiltonian action* of the *Lie algebra* so(4) on $(\Sigma_-, \widetilde{\omega}_3)$. Associated to this Lie algebra action is the mapping

$$\mathcal{J} : \Sigma_- \subseteq T_0\mathbf{R}^3 \to \mathbf{R}^6 :$$
$$(q, p) \to (\mathbf{J}, \widetilde{\mathbf{e}}) = \left(q \times p, \nu\left(\frac{q}{|q|} - \frac{1}{\mu} p \times (q \times p)\right)\right). \tag{48}$$

Here we have chosen $\{\epsilon_i\}_{1 \le i \le 6} = \{J_1, J_2, J_3, \widetilde{e}_1, \widetilde{e}_2, \widetilde{e}_2\}$ as a basis for so(4) with Lie bracket { , }. Let $\mathcal{J}^{\epsilon_i}$ be the ith component of the mapping $\mathcal{J}$. Then the bracket relations (46) may be written as $\{\mathcal{J}^{\epsilon_i}, \mathcal{J}^{\epsilon_j}\} = \mathcal{J}^{\{\epsilon_i, \epsilon_j\}}$. Therefore we say that the map $\mathcal{J}$ is the *momentum map* of the so(4)-action on $(\Sigma_-, \widetilde{\omega}_3)$.

We now investigate the geometric properties of the mapping $\mathcal{J}$. We begin by noting that the vectors $\mathbf{J}$ and $\widetilde{\mathbf{e}}$ satisfy

$$\begin{cases} \langle \mathbf{J}, \widetilde{\mathbf{e}} \rangle &= 0 \\ \langle \mathbf{J}, \mathbf{J} \rangle + \langle \widetilde{\mathbf{e}}, \widetilde{\mathbf{e}} \rangle &= \nu^2 > 0. \end{cases} \tag{49}$$

The verification of the first equation in (49) is a straightforward. For the second, see (54). These relations define a smooth 4-dimensional manifold C_ν, which is diffeomorphic to $S^2_\nu \times S^2_\nu$ because (49) is equivalent to

$$\langle \mathbf{J} + \widetilde{\mathbf{e}}, \mathbf{J} + \widetilde{\mathbf{e}} \rangle = \langle \mathbf{J} - \widetilde{\mathbf{e}}, \mathbf{J} - \widetilde{\mathbf{e}} \rangle = \nu^2 > 0. \tag{50}$$

Write $\nu = \mu/\sqrt{-2h}$ for some $h < 0$ and consider the map

$$\mathcal{J}_h = \mathcal{J} | H^{-1}(h) : H^{-1}(h) \subseteq \Sigma_- \to C_\nu \subseteq \mathbf{R}^6. \tag{51}$$

Claim: $\mathcal{J}_h$ is a surjective submersion.

(3.7) **Proof**: Let $(q, p) \in H^{-1}(h)$ and let

$$V_{(q,p)} = \mathrm{span}\,\{X_{J_j}(q, p), X_{\widetilde{e}_j}(q, p)\}_{1 \le j \le 3}.$$

Since $\mathbf{J}$ and $\widetilde{\mathbf{e}}$ are integrals of X_H, it follows that $V_{(q,p)} \subseteq \ker dH(q, p)$ which is $T_{(q,p)}H^{-1}(h)$. Therefore

$$\begin{aligned} D\mathcal{J}_h(q, p) | V_{(q,p)} &= \begin{pmatrix} dJ_j(q, p) \\ d\widetilde{e}_j(q, p) \end{pmatrix} V_{(q,p)} \\ &= \begin{pmatrix} (\{J_i, J_j\}(q, p)) & (\{J_i, \widetilde{e}_j\}(q, p)) \\ (\{\widetilde{e}_i, J_j\}(q, p)) & (\{\widetilde{e}_i, \widetilde{e}_j\}(q, p)) \end{pmatrix} = P. \end{aligned}$$

On C_ν (49) the rank of P is 4, because P is conjugate to the matrix

$$\begin{pmatrix} (\{\xi_i, \xi_j\}(q,p)) & 0 \\ 0 & (\{\eta_i, \eta_j\}(q,p)) \end{pmatrix} = \begin{pmatrix} \left(\sum_k \varepsilon_{ijk} (J_k + \widetilde{e}_k)(q,p)\right) & 0 \\ 0 & \left(\sum_k \varepsilon_{ijk} (J_k - \widetilde{e}_k)(q,p)\right) \end{pmatrix}$$

(see (47) and (50)). Therefore $\mathcal{J}_h$ is a submersion.

To show that $\mathcal{J}_h$ is surjective, let $(\mathbf{J}, \widetilde{\mathbf{e}}) \in C_\nu$. Then $e = |\mathbf{e}| = \frac{1}{\nu} |\widetilde{\mathbf{e}}| \in [0,1]$ because $\nu^2 = \langle \mathbf{J}, \mathbf{J} \rangle + \langle \widetilde{\mathbf{e}}, \widetilde{\mathbf{e}} \rangle \ge \nu^2 \langle \mathbf{e}, \mathbf{e} \rangle$. Choose

$$(q,p) = \begin{cases} \left(-\frac{\nu}{\mu} \frac{(1-e)}{e} \widetilde{\mathbf{e}}, -\frac{\mu}{\nu^3} \frac{1}{e(1-e)} \mathbf{J} \times \widetilde{\mathbf{e}} \right), & \text{when } e \in (0,1) \text{ and } \mathbf{J} \ne 0 \\ \left(-\frac{\mu^2}{\nu^2} p \times \mathbf{J}, p \right), & \text{when } e = 0 \text{ and } \mathbf{J} \ne 0. \text{ Here } \langle p, \mathbf{J} \rangle = 0, |p| = \frac{\mu}{\nu} \\ \left(-\frac{\nu}{\mu} \widetilde{\mathbf{e}}, \frac{\nu}{\mu^2} \widetilde{\mathbf{e}} \right), & \text{when } \mathbf{J} = 0. \end{cases}$$

A straightforward calculation shows that $(q,p) \in H^{-1}(h)$ and $\mathcal{J}_h(q,p) = (\mathbf{J}, \widetilde{\mathbf{e}})$. □

Corollary: For every $c \in C_\nu$ the fiber $\mathcal{J}_h^{-1}(c)$ is a union of bounded Keplerian orbits.

(3.8) **Proof**: From the fact that $\mathcal{J}_h$ is a submersion, it follows that

$$\dim \ker D\mathcal{J}_h(q,p) = \dim T_{(q,p)} H^{-1}(h) - \dim \operatorname{im} D\mathcal{J}_h(q,p) = 5 - 4 = 1.$$

But $X_H(q,p) \in \ker D\mathcal{J}_h(q,p)$. Hence for every $c \in C_\nu$

$$T_{(q,p)} H^{-1}(h) = \ker D\mathcal{J}_h(q,p) = \operatorname{span}\{X_H(q,p)\}.$$

Therefore $\mathcal{J}_h^{-1}(c)$ is a union of bounded Keplerian orbits. □

The following claim is a substantial sharpening of the above corollary.

Claim: For every $c \in C_\nu$ the fiber $\mathcal{J}_h^{-1}(c)$ lies over

1. an oriented ellipse, when $c \notin C_\nu \cap \{\mathbf{J} = 0\}$;
2. a line which is the union of two half open line segments

$$\left\{ \left(\frac{\sigma}{\nu} \widetilde{\mathbf{e}}, \pm \frac{1}{\nu} \left(\sqrt{2h + \frac{2\mu}{\sigma}} \right) \widetilde{\mathbf{e}} \right) \in T_0 \mathbf{R}^3 \,\middle|\, \sigma \in (0, \frac{\mu}{-h}] \right\},$$

that join smoothly at $(\frac{\mu}{-\nu h} \widetilde{\mathbf{e}}, 0)$, when $c \in C_\nu \cap \{\mathbf{J} = 0\}$.

(3.9) **Proof**:
case 1. $c = (\mathbf{J}, \widetilde{\mathbf{e}}) \in C_\nu - C_\nu \cap \{\mathbf{J} = 0\}$. Let $h = -\mu^2/2\nu^2$. We have to show that the data $h < 0$, $\mathbf{J} \ne 0$, and $\mathbf{e} = -\frac{1}{\nu} \widetilde{\mathbf{e}}$ determine a unique oriented ellipse which is traced out by the projection $t \to q(t)$ of an integral curve $t \to (q(t), p(t))$ of X_H. Because $\mathbf{J} \ne 0$ and $\langle q(t), \mathbf{J} \rangle = \langle p(t), \mathbf{J} \rangle = 0$, the curves $t \to q(t)$ and $t \to p(t)$ lie in a plane $\Pi \subseteq \mathbf{R}^3$ which is perpendicular to $\mathbf{J}$. Since $\langle \mathbf{J}, \mathbf{e} \rangle = \mathbf{0}$, the eccentricity vector $\mathbf{e}$ also lies in Π. Therefore

we may write $\langle q, \mathbf{e}\rangle = |q| e \cos f$, where f is the *true anomaly*. From the definition of the eccentricity vector $\mathbf{e}$ (44) it follows that $\langle q, \mathbf{e}\rangle = -|q| + \frac{1}{\mu} J^2$. Therefore

$$|q| e \cos f = -|q| + \frac{1}{\mu} J^2. \tag{52}$$

Suppose that $e = 0$. Then (52) becomes $|q| = \frac{1}{\mu} J^2$, which defines a circle $\mathscr{C}$ in Π with center at the origin. Since

$$0 = \frac{d|q(t)|^2}{dt} = \langle q, \frac{dq}{dt}\rangle = \langle q, p\rangle,$$

the tangent vector $p(t)$ to $\mathscr{C}$ at $q(t)$ is perpendicular to $q(t)$. Because $\{q, p, p \times q\}$ is a positively oriented basis of $\mathbf{R}^3$, $\{q, p\}$ is a positively oriented basis for Π. Hence the circle traced out by $t \to q(t)$ is positively oriented.

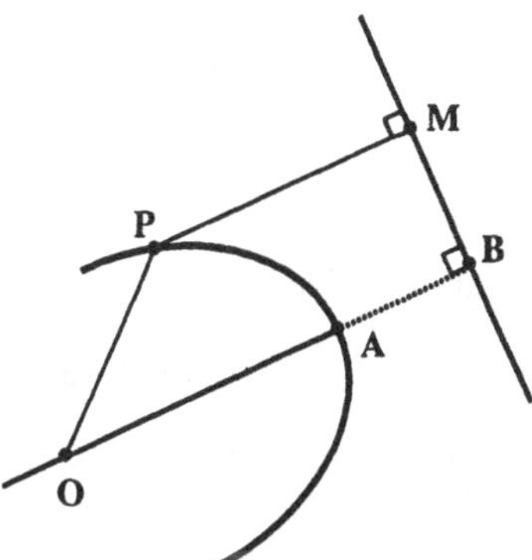

Figure 3.1. Ellipse in the plane Π.

Suppose that $e \neq 0$. Then equation (52) may be written as

$$e\left(\frac{J^2}{\mu e} - |q| \cos f\right) = |q|. \tag{53}$$

Equation (53) describes the locus of points P in the plane Π for which the ratio of the distance $\overline{OP}$ to the origin to the distance $\overline{PM}$ to the line MB is a constant e, (see figure 3.1). Thus the locus is a conic section. To see which conic it is, we calculate the size of e as follows.

$$\begin{aligned} e^2 &= |\mathbf{e}|^2 = 1 - \frac{2}{\mu|q|} |q \times p|^2 + \frac{1}{\mu^2} |p \times (q \times p)|^2, \quad \text{using (44)} \\ &= 1 - \frac{2}{\mu|q|} J^2 + \frac{1}{\mu^2}(|p|^2 J^2 - \langle p, \mathbf{J}\rangle^2), \quad \text{using } \mathbf{J} = q \times p \\ &= 1 + \frac{2}{\mu^2} J^2 h, \quad \text{since } \langle p, \mathbf{J}\rangle = 0 \text{ and } h = \tfrac{1}{2}\langle p, p\rangle - \tfrac{\mu}{|q|}. \end{aligned} \tag{54}$$

Since $h < 0$, it follows that $e \in [0, 1)$. Therefore the locus

$$|q| = \frac{J^2}{\mu} \frac{1}{1 + e \cos f} \tag{55}$$

is an ellipse in Π with *eccentricity* e and major semiaxis lying along $\mathbf{e}$ (which is directed from O to the periapse A) of length

$$a = \frac{J^2}{\mu}\frac{1}{1-e^2} = \frac{\mu}{-2h}.$$

$\triangleright$ When traced out by $t \to q(t)$, this ellipse is oriented in the direction of increasing true anomaly f.

(3.10) **Proof**: From the fact that $\{q, p\}$ is a positively oriented basis of the plane Π, we obtain

$$\begin{aligned} J &= |q \times p| = \text{area of the positively oriented parallogram spanned by } \{q, p\}. \\ &= \det\begin{pmatrix} \langle q, e^{-1}\mathbf{e}\rangle & \langle q, (Je)^{-1}\mathbf{J}\times\mathbf{e}\rangle \\ \langle p, e^{-1}\mathbf{e}\rangle & \langle p, (Je)^{-1}\mathbf{J}\times\mathbf{e}\rangle \end{pmatrix}, \\ &\quad \text{since } \{e^{-1}\mathbf{e}, (Je)^{-1}\mathbf{J}\times\mathbf{e}\} \text{ is a positively oriented orthonormal basis of } \Pi \\ &= |q|^2\frac{df}{dt}. \end{aligned} \tag{56}$$

Equation (56) follows by first differentiating

$$\langle q, e^{-1}\mathbf{e}\rangle = |q|\cos f \quad \text{and} \quad \langle q, (Je)^{-1}\mathbf{J}\times\mathbf{e}\rangle = |q|\sin f$$

along an integral curve of X_H and then using the fact that $p = \frac{dq}{dt}$ and $\dot{\mathbf{e}} = \dot{\mathbf{J}} = 0$ to obtain

$$\langle p, e^{-1}\mathbf{e}\rangle = \frac{d|q|}{dt}\cos f - |q|\sin f\frac{df}{dt}$$

and

$$\langle p, (Je)^{-1}\mathbf{J}\times\mathbf{e}\rangle = \frac{d|q|}{dt}\sin f + |q|\cos f\frac{df}{dt}.$$

From (56) we see that $\frac{df}{dt} > 0$. □

case 2. $c = (\mathbf{J}, \widetilde{\mathbf{e}}) \in C_\nu \cap \{\mathbf{J} = 0\}$. Since $\mathbf{J} = 0$, the modified eccentricity vector $\widetilde{\mathbf{e}} = \nu\frac{q}{|q|}$. Because $\widetilde{\mathbf{e}}$ is constant along integral curves $t \to (q(t), p(t))$ of X_H and $h < 0$, the image of $t \to q(t)$ lies along $\widetilde{\mathbf{e}}$ and is the half open line segment

$$\left\{ \frac{\sigma}{\nu}\widetilde{\mathbf{e}} \in \Pi \,\middle|\, \sigma \in \left(0, \frac{\mu}{-h}\right] \right\}.$$

From $\mathbf{J} = 0$ it follows that $p = \lambda\widetilde{\mathbf{e}}$ for some $\lambda \in \mathbf{R}$. In order that $(\frac{\sigma}{\nu}\widetilde{\mathbf{e}}, p) \in H^{-1}(h)$, where $h = -\mu^2/2\nu^2$, we must have $\lambda^2 = \langle p, p\rangle = 2h + 2\mu/\sigma$. Therefore $\mathcal{J}_h^{-1}(c)$ is the line which is the union of the two half open line segments

$$\left\{ \left(\frac{\sigma}{\nu}\widetilde{\mathbf{e}}, \pm\frac{1}{\nu}\left(\sqrt{2h + \frac{2\mu}{\sigma}}\right)\widetilde{\mathbf{e}}\right) \in T_O\mathbf{R}^3 \,\middle|\, \sigma \in (0, \frac{\mu}{-h}] \right\}$$

which join smoothly at $(\frac{\mu}{-h\nu}\widetilde{\mathbf{e}}, 0)$. □

It is not hard to show that on $\mathcal{J}_h^{-1}\big(C_\nu - (\{\mathbf{J}=0\} \cap C_\nu)\big)$ the mapping $\mathcal{J}_h$ is proper, whereas on $\mathcal{J}_h^{-1}(\{\mathbf{J}=0\} \cap C_\nu)$ it is not.

We now turn to examining the so(4) momentum mapping $\mathcal{J}$ (48). Let C be the submanifold of $\mathbf{R}^3 \times \mathbf{R}^3 - \{(0,0)\}$ defined by

$$\langle \mathbf{J}, \widetilde{\mathbf{e}} \rangle = 0. \tag{57}$$

Claim: The map

$$\mathcal{J} : \Sigma_- \subseteq T_0\mathbf{R}^3 \to C \subseteq \mathbf{R}^6 :$$
$$(q,p) \to \Big(q \times p, \nu\Big(\tfrac{q}{|q|} - \tfrac{1}{\mu}\, p \times (q \times p)\Big)\Big) = (\mathbf{J}, \widetilde{\mathbf{e}})$$

is a surjective submersion each of whose fibers is a unique bounded orbit of the Kepler vector field X_H.

(3.11) **Proof**: First we show that $\mathcal{J}$ is surjective. Supppose that $c = (\mathbf{J}, \widetilde{\mathbf{e}}) \in C$. Then

$$\|\mathbf{J} + \widetilde{\mathbf{e}}\|^2 = \|\mathbf{J} - \widetilde{\mathbf{e}}\|^2 = \nu^2$$

for some $\nu > 0$. Hence $(\mathbf{J}, \widetilde{\mathbf{e}}) \in C_\nu$. Let $h = -\mu^2/2\nu^2$. From ((3.7)) it follows that $\mathcal{J}_h^{-1}(c)$ is nonempty. Hence $\mathcal{J}^{-1}(c)$ is nonempty. Because $\mathcal{J}_h^{-1}(c)$ is a unique oriented bounded orbit of the Kepler vector field, $\mathcal{J}^{-1}(c)$ is as well.

Since C is a 5-dimensional smooth manifold, $\mathcal{J}$ is a submersion if for every $(q,p) \in \Sigma_-$ the rank of $D\mathcal{J}(q,p)$ is 5. Actually it suffices to show that for every $(q,p) \in H^{-1}(h)$ the vector $D\mathcal{J}(q,p)\,\mathrm{grad}H(q,p)$ is normal to C_ν at $\mathcal{J}(q,p)$, because

1. by ((3.7)), $D\mathcal{J}(q,p)T_{(q,p)}H^{-1}(h) = T_{\mathcal{J}(q,p)}C_\nu$;

2. a normal space to $H^{-1}(h)$ in Σ_- at (q,p) is spanned by $\mathrm{grad}H(q,p)$;

3. as a submanifold of C the manifold C_ν is defined by

$$F(\mathbf{J}, \widetilde{\mathbf{e}}) = \langle \mathbf{J}, \mathbf{J} \rangle + \langle \widetilde{\mathbf{e}}, \widetilde{\mathbf{e}} \rangle - \nu^2 = 0, \tag{58}$$

   where $\nu = \mu/\sqrt{-2H}$.

Since the normal space to C_ν at $\mathcal{J}(q,p) = (\mathbf{J}, \widetilde{\mathbf{e}}) \in C$ is spanned by $\mathrm{grad}F(\mathbf{J}, \widetilde{\mathbf{e}}) = 2(\mathbf{J}, \widetilde{\mathbf{e}})$, it suffices to check that

$$\Big\langle D\mathcal{J}(q,p)\,\mathrm{grad}H(q,p),\ \mathrm{grad}F(\mathcal{J}(q,p))\Big\rangle$$

is nonzero. The following calculation does this.

$$\begin{aligned} 0 &\neq \Big\langle \mathrm{grad}H(q,p), \mathrm{grad}H(q,p)\Big\rangle = DH(q,p)\,\mathrm{grad}H(q,p) \\ &= D\Big(-\frac{\mu^2}{2}\big(\langle \mathbf{J}, \mathbf{J}\rangle + \langle \widetilde{\mathbf{e}}, \widetilde{\mathbf{e}}\rangle\big)^{-1}\Big)(q,p)\,\mathrm{grad}H(q,p), \\ &\qquad \text{using } H = -\tfrac{\mu^2}{2\nu^2} \text{ and } (58) \end{aligned}$$

$$= \frac{\mu^2}{2}\big(\langle \mathbf{J},\mathbf{J}\rangle + \langle \widetilde{\mathbf{e}},\widetilde{\mathbf{e}}\rangle\big)^{-2}\big(\big\langle \mathbf{J}, D\mathbf{J}(q,p)\,\mathrm{grad}H(q,p)\big\rangle + \big\langle \widetilde{\mathbf{e}}, D\mathbf{J}(q,p)\,\mathrm{grad}H(q,p)\big\rangle\big)$$

$$= \frac{2H(q,p)^2}{\mu^2}\Big\langle D\mathcal{J}(q,p)\,\mathrm{grad}H(q,p), \mathrm{grad}F(\mathcal{J}(q,p))\Big\rangle. \qquad \square$$

The above result has several useful consequences.

Corollary 1. The smooth manifold C (57) is the *space of orbits* of negative energy of the Kepler vector field X_H and the momentum map $\mathcal{J} : \Sigma_- \to C$ is the orbit map.

(3.12) **Proof**: The corollary follows from ((3.11)) and the definition of orbit space. $\square$

The next corollary says that every smooth integral of the Kepler vector field on Σ_- is a smooth function of the components of angular momentum $\mathbf{J}$ and the modified eccentricity vector $\widetilde{\mathbf{e}}$. More precisely,

Corollary 2. Suppose that $G : \Sigma_- \subseteq T_0\mathbf{R}^3 \to \mathbf{R}$ is a smooth integral of the Kepler vector field X_H. Then there is a smooth function $\widehat{G} : C \subseteq \mathbf{R}^6 \to \mathbf{R}$ such that $G = \mathcal{J}^*\widehat{G}$.

(3.13) **Proof**: Since G is an integral of X_H on Σ_-, it is constant on each bounded orbit of X_H and hence is constant on the fibers of the momentum map $\mathcal{J}$. Because C is smooth and is the space of orbits of X_H on Σ_- with orbit mapping $\mathcal{J}$, G descends to a smooth function $\widehat{G} : C \subseteq \mathbf{R}^6 \to \mathbf{R}$. In other words, $G = \mathcal{J}^*\widehat{G}$. $\square$

3.3 Kepler's equation

So far we have only used the constants of motion to describe the orbits of the Kepler vector field X_H of negative energy. This means that we cannot tell where on the orbit the particle is at a given time.

In order to give a time parametrization of a bounded Keplerian orbit, we define a new time scale, the *eccentric anomaly* s, by

$$\frac{ds}{dt} = \frac{\sqrt{-2h}}{|q|}. \qquad (59)$$

Before finding a differential equation for $|q(s)|$, we use the integrals of energy and angular momentum to find a differential equation for $|q(t)|$. Multiplying the energy integral

$$h = \tfrac{1}{2}\langle p, p\rangle - \frac{\mu}{|q|}$$

by $2|q|^2$ gives

$$|q|^2|p|^2 = 2\mu\,|q| + 2h\,|q|^2.$$

But

$$|q|^2|p|^2 = |q\times p|^2 + \langle q,p\rangle^2 = J^2 + \langle p,q\rangle^2.$$

In other words,

$$|q|^2\left(\frac{d|q|}{dt}\right)^2 + J^2 = 2\mu\,|q| + 2h\,|q|^2. \qquad (60)$$

Using (59) to change to the time variable s and dividing by $-2h$ gives

$$\left(\frac{d|q|}{ds}\right)^2 + a^2(1-e^2) = 2a\,|q| - |q|^2, \tag{61}$$

since

$$a = \mu/(-2h) = \frac{J^2}{\mu}\frac{1}{1-e^2}.$$

Instead of separating variables and immediately integrating (61) we first change variables by $ea\,\rho = a - |q|$. Then (61) simplifies to

$$\left(\frac{d\rho}{ds}\right)^2 + \rho^2 = 1. \tag{62}$$

Since $|q(0)| = a(1-e)$, from the definition of ρ we obtain $\rho(0) = 1$. Therefore

$$|q(s)| = a - ae\cos s. \tag{63}$$

To find the relation between the eccentric anomaly time scale s and the physical time scale t, we substitute (63) into (59) and integrate to obtain

$$\sqrt{-2h}(t-\tau) = \sqrt{-2h}\int_\tau^t dt \;=\; \int_0^s (a - ae\cos s)ds \;=\; as - ae\sin s. \tag{64}$$

Here τ is a time related to the time of periapse passage. Its precise definition is given below. Dividing (64) by a and using $a = \mu/(-2h) = \nu^2/\mu$ gives *Kepler's equation*

$$s - e\sin s = \frac{\mu^2}{\nu^3}(t-\tau) \;=\; n\,\ell, \tag{65}$$

where ℓ is the *mean anomaly* and $n = \mu^2/\nu^3$ is the *mean motion*. Note that

$$\begin{aligned}\langle q, p\rangle &= \langle q, \frac{dq}{dt}\rangle = |q|\frac{d|q|}{dt} = |q|\frac{d|q|}{ds}\frac{ds}{dt} \\ &= \sqrt{-2h}\,ae\,\sin s, \quad \text{using (59) and (63)} \\ &= \nu\, e\sin s.\end{aligned} \tag{66}$$

When $t = \tau$ from Kepler's equation it follows that $s = 0$. Let τ' be the physical time corresponding to $s = 2\pi$ in (65). Then $\tau - \tau'$ is the period of elliptical motion, which according to Kepler's equation, is

$$2\pi/n = 2\pi\nu^3/\mu^2 \;=\; 2\pi a^{3/2}/\mu^{1/2}. \tag{67}$$

This is Kepler's third law of motion.

During elliptical motion the particle goes through the periapse periodically. Therefore the time τ in (65) is *not* uniquely determined by the initial condition $(q(0), p(0))$ which defines the integral curve of X_H. We will define τ as follows. In the interval $[-\pi, \pi]$ there are precisely *two* values $\varepsilon\widehat{s_0}$ (with $\varepsilon^2 = 1$) which satisfy $|q(0)| = a(1 - e\cos\varepsilon\widehat{s_0})$. To fix the choice of ε note that from (66) we have $\varepsilon = \langle q(0), p(0)\rangle/(\nu e \sin \widehat{s}_0)$, unless $\widehat{s}_0 = 0$ in which case ε is irrelevant. Set $s_0 = \varepsilon \widehat{s}_0$ and let $\tau = -\frac{1}{n}(s_0 - e \sin s_0)$. In words, we define τ as follows. If at $t = 0$ the particle is in the upper half of the ellipse, then τ is the first time *before* $t = 0$ when the particle passed through the periapse; otherwise it is the first time *on* or *after* $t = 0$ when the particle passes through the periapse.

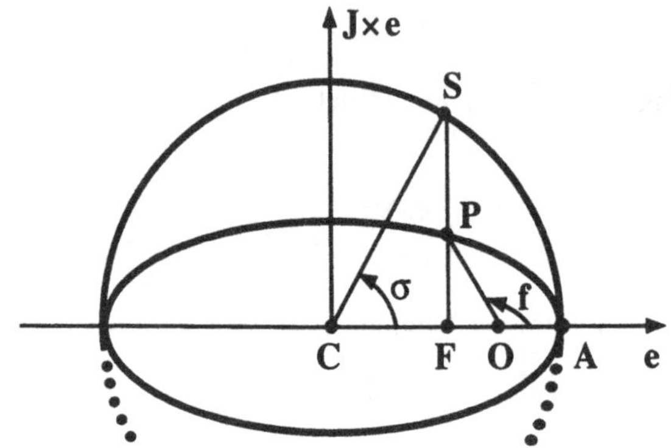

Figure 3.2. The eccentric anomaly.

To describe the geometric meaning of the eccentric anomaly s, consider the figure 3.2. Let O be the center of attraction, A the periapse and C the center of the ellipse of eccentricity e. The arrow on the ellipse indicates the direction of motion and P is the position of the particle on the ellipse with true anomaly f. Construct a line $\mathcal{L}$ through P which is perpendicular to the line CA. Project P parallel along $\mathcal{L}$ to the point S on the circle $\mathcal{C}$ with center C and radius equal to the distance $\overline{CA}$.

Claim: The eccentric anomaly s is the angle $\angle SCA$.

(3.14) **Proof**: Let $\sigma = \angle SCA$. From figure 3.2 we obtain $\overline{CS} = a$ and $\overline{CO} = ae$. Since $\overline{CF} = \overline{CO} + \overline{OF}$, we find that $a \cos \sigma = ae + |q| \cos f$. As the orbit is elliptical, $|q| = \frac{a(1-e^2)}{1+e\cos f}$. This may be rewritten as

$$|q| = a - e(ae + |q| \cos f) = a - ae \cos \sigma.$$

But

$$|q| = a - ae \cos s.$$

Hence $\cos s = \cos \sigma$. Since $s = 0$ when $\sigma = 0$, we obtain $\sigma = s$. □

As the point S traces out the circle $\mathcal{C}$ uniformly with speed n, the point P on the ellipse traces out the projection of an integral curve of the Kepler vector field in configuration space.

3.4 Regularization of the Kepler vector field

The incompleteness of the flow of the Kepler vector field may be removed by embedding it into a complete flow. This process is called regularization. On the subset of phase space where the Kepler Hamiltonian is negative, one can perform regularization in such a way that the embedding not only is symplectic and the resulting vector field is Hamiltonian but also that it linearizes and integrates the so(4)-action. That there is such a large hidden symmetry in the Kepler problem is quite remarkable because this symmetry does *not* arise from a lift of a symmetry on configuration space. We will *regularize all negative energy* Keplerian orbits at once using the *Ligon-Schaaf map* LS. We show that LS is the only symplectic mapping from the negative energy set $(\Sigma_- = \{(q,p) \in T_0\mathbf{R}^3 \mid H(q,p) < 0\}, \omega_3|\Sigma_-)$ to $(T^+S^3, \widetilde{\Omega}_4)$ which has the following properties:

1. It intertwines the Kepler and Delaunay vector fields, X_H and $X_{\widetilde{\mathcal{H}}}$, respectively;
2. It intertwines their so(4) momentum mappings $\mathcal{J}$ and $\widetilde{\mathcal{J}}$, respectively;
3. It maps Σ_- onto $T^+(S^3 - \mathrm{np})$, where $\mathrm{np} = (0,0,0,1)$.

We will show that the Ligon–Schaaf map is a diffeomorphism because it maps a one period of a parmetrized integral curve of the Kepler vector field of negative energy onto one period of a parmetrized integral curve of the Delaunay vector field on $T^+(S^3 - \mathrm{np})$.

We begin our search for the Ligon–Schaaf map by noting that the image of the so(4) momentum mapping $\mathcal{J}$ (48) of the Kepler vector field X_H on Σ_- is the *same* as the image of the so(4) momentum mapping $\widetilde{\mathcal{J}}$ (22) of the Delaunay vector field $X_{\widetilde{H}}$ on T^+S^3. Thus we suspect that the momentum maps are somehow related.

Claim: The smooth map

$$\Phi : \Sigma_- \subseteq T_0\mathbf{R}^3 \to T^+S^3 : (q,p) \to (x,y)$$

intertwines the so(4) momentum mappings $\mathcal{J}$ and $\widetilde{\mathcal{J}}$, that is, $\Phi^*\widetilde{\mathcal{J}} = \mathcal{J}$, if and only if

$$(x,y) = \Phi(q,p) = (A\sin\varphi + B\cos\varphi, -\nu A\cos\varphi + \nu B\sin\varphi), \tag{68}$$

where

$$A = (\widetilde{A}, A_4) = \Big(\frac{q}{|q|} - \frac{1}{\mu}\langle q,p\rangle p, \frac{1}{\nu}\langle q,p\rangle\Big) \tag{69}$$

$$B = (\widetilde{B}, B_4) = \Big(\frac{1}{\nu}|q|p, \frac{1}{\mu}\langle p,p\rangle|q| - 1\Big), \tag{70}$$

$\nu = \mu/\sqrt{-2H}$, and $\varphi = \varphi(q,p)$ is an arbitrary smooth real valued function on Σ_-.

.15) **Proof**: Suppose that Φ intertwines the momentum maps. Write

$$x = (\widetilde{x}, x_4),\ y = (\widetilde{y}, y_4) \in \mathbf{R}^3 \times \mathbf{R} = \mathbf{R}^4.$$

From the definitions of the momentum mappings $\mathcal{J}$ and $\widetilde{\mathcal{J}}$ it is straightforward to check that $\Phi^*\widetilde{\mathcal{J}} = \mathcal{J}$ is equivalent to

$$\widetilde{x} \times \widetilde{y} = q \times p \tag{71}$$

$$x_4\widetilde{y} - y_4\widetilde{x} = \nu\Big(\frac{q}{|q|} - \frac{1}{\mu}p \times (q \times p)\Big) = Mq + Np, \tag{72}$$

where

$$M = \nu \left(\frac{1}{|q|} - \frac{1}{\mu} \langle p, p \rangle \right) \quad \text{and} \quad N = \frac{\nu}{\mu} \langle q, p \rangle. \tag{73}$$

Suppose that $q \times p \neq 0$. From (71) it follows that $\widetilde{x}$ and $\widetilde{y}$ are orthogonal to $q \times p$. Since q and p are linearly independent, we obtain

$$\begin{cases} \widetilde{x} &= a\,q + b\,p \\ \widetilde{y} &= c\,q + d\,p \end{cases} \tag{74}$$

Because $q \times p = \widetilde{x} \times \widetilde{y} = (ad - bc)(q \times p)$, we find that

$$ad - bc = 1. \tag{75}$$

Substituting (74) into (72) and using the linear independence of q and p gives a set of linear equations for x_4 and y_4. Using (75) these equations may be solved to give

$$\begin{cases} x_4 &= a\,N - b\,M \\ y_4 &= c\,N - d\,M. \end{cases} \tag{76}$$

Since $(x, y) \in T^+S^3$,

$$\begin{cases} 1 &= \langle \widetilde{x}, \widetilde{x} \rangle + x_4^2 \\ 0 &= \langle \widetilde{x}, \widetilde{y} \rangle + x_4 y_4. \end{cases} \tag{77}$$

Substituting the expressions for $\widetilde{x}$ and $\widetilde{y}$ (74) and the expressions for x_4 and y_4 (76) into (77) gives

$$1 = a^2 \langle q, q \rangle + \left(\frac{\nu}{\mu} \langle q, p \rangle\, a + \frac{\nu}{|q|}\, b \right)^2 \tag{78}$$

$$0 = ac \left(\langle q, q \rangle + \frac{\nu^2}{\mu^2} \langle q, p \rangle^2 \right) + (ad + bc) \left(\frac{\nu^2}{\mu |q|} \langle q, p \rangle \right) + bd\, \frac{\nu^2}{\langle q, q \rangle}. \tag{79}$$

Here we have used the identities

$$\begin{cases} \langle q, q \rangle + N^2 &= \langle q, q \rangle + \frac{\nu^2}{\mu^2} \langle q, p \rangle^2 \\ \langle q, p \rangle - MN &= \frac{\nu^2}{\mu |q|} \langle q, p \rangle \\ \langle p, p \rangle + M^2 &= \dfrac{\nu^2}{\langle q, q \rangle}, \end{cases}$$

which follow from the definition of M and N (73) and the identity

$$\frac{1}{\mu} \langle p, p \rangle - \frac{1}{|q|} = \frac{1}{|q|} - \frac{\mu}{\nu^2},$$

which is a consequence of the definition of ν. Multiplying (78) by c and (79) by $-a$ and adding the resulting equations gives

$$c = -a \left(\frac{\nu^2}{\mu |q|} \langle q, p \rangle \right) - b\, \frac{\nu^2}{\langle q, q \rangle}, \tag{80}$$

after using (75). Similarly

$$d = a\left(\langle q, q\rangle + \frac{\nu^2}{\mu^2}\langle q, p\rangle^2\right) + b\left(\frac{\nu^2}{\mu|q|}\langle q, p\rangle\right). \tag{81}$$

All solutions of (78) are parametrized by

$$\begin{cases} a &= \dfrac{1}{|q|}\sin\varphi \\ b &= \dfrac{|q|}{\nu}\cos\varphi - \frac{1}{\mu}\langle q, p\rangle\sin\varphi, \end{cases} \tag{82}$$

where $\varphi = \varphi(q, p)$ is an arbitrary smooth function on Σ_-. Substituting (82) into (80) and (81) gives

$$\begin{cases} c &= -\dfrac{\nu}{|q|}\cos\varphi \\ d &= |q|\sin\varphi + \frac{\nu}{\mu}\langle q, p\rangle\cos\varphi. \end{cases} \tag{83}$$

Substituting (82) and (83) into (74) and (76) shows that the mapping Φ has the form (68) with the vectors A and B given by (69) and (70).

Since A and B are defined and continuous on all of Σ_-, the map Φ (68) is defined on *all* of Σ_- and not just on the open dense subset where $q \times p \neq 0$. A straightforward calculation shows that a map Φ of the form (68) intertwines the momentum mappings $\mathcal{J}$ and $\widetilde{\mathcal{J}}$. □

Corollary: Φ intertwines the Kepler and Delaunay Hamiltonians, that is, $\Phi^*\widetilde{\mathcal{H}} = H$ on Σ_-.

3.16) **Proof**: We compute.

$$\begin{aligned} \widetilde{\mathcal{H}}(q, p) &= -\frac{1}{2}\frac{\mu^2}{\langle y, y\rangle} = -\frac{1}{2}\frac{\mu^2}{\nu^2\,\| - A\cos\varphi + B\sin\varphi\|^2} = -\frac{1}{2}\frac{\mu^2}{\nu^2}, \\ &\qquad \text{since } \langle A, A\rangle = \langle B, B\rangle = 1 \text{ and } \langle A, B\rangle = 0, \text{ (see below)} \\ &= H(q, p), \quad \text{by definition of } \nu. \end{aligned}$$

Now

$$\begin{aligned} \langle A, A\rangle &= 1 - \frac{2}{\mu|q|}\langle q, p\rangle^2 + \frac{1}{\mu^2}\langle q, p\rangle^2\langle p, p\rangle + \frac{1}{\nu^2}\langle q, p\rangle^2 \\ &= 1 + \frac{2}{\mu^2}\langle q, p\rangle^2\left(\tfrac{1}{2}\langle p, p\rangle - \frac{\mu}{|q|}\right) + \frac{1}{\nu^2}\langle q, p\rangle^2 \\ &= 1, \quad \text{since } \nu^2 = \mu^2/(-2H), \end{aligned}$$

$$\begin{aligned} \langle B, B\rangle &= \frac{|q|^2}{\nu^2}\langle p, p\rangle\frac{|q|^2}{\mu^2}\langle p, p\rangle^2 - \frac{2|q|}{\mu}\langle p, p\rangle + 1 \\ &= \frac{2|q|^2}{\nu^2}\langle p, p\rangle\left(\tfrac{1}{2}\langle p, p\rangle - \frac{\mu}{|q|} + \frac{1}{2}\frac{\mu^2}{\nu^2}\right) + 1 \\ &= 1, \quad \text{since } H = -\mu^2/2\nu^2, \end{aligned}$$

and

$$\langle A, B\rangle = \frac{1}{\nu}|q|\left\langle p, \frac{q}{|q|} - \frac{1}{\mu}\langle q, p\rangle p\right\rangle + \frac{1}{\nu}\langle q, p\rangle\left(\frac{1}{\mu}\langle p, p\rangle|q| - 1\right) = 0. \qquad \square$$

From now on we will use the notation Φ_φ for the map Φ (68) when we wish to emphasize the dependence of Φ on the parameter φ.

We now show that the map Φ is determined up to the flow Ψ_t (16) of the Delaunay vector field $X_{\widetilde{\mathcal{H}}}$ (13). More precisely we prove

Claim: Let Φ_φ and $\Phi_{\varphi'}$ be maps of the form (68) with free parameter φ and φ', respectively. Then for every $(q, p) \in \Sigma_-$

$$\Phi_{\varphi'}(q, p) = \Psi_{\frac{\nu^3}{\mu^2}\left(\varphi(q,p) - \varphi'(q,p)\right)}(\Phi_\varphi(q, p)) \tag{84}$$

(3.17) **Proof**: To verify (84) we write

$$\Phi_\varphi = \begin{pmatrix} \cos\varphi & \sin\varphi \\ \nu\sin\varphi & -\nu\cos\varphi \end{pmatrix}\begin{pmatrix} A \\ B \end{pmatrix} \text{ and } \Phi_{\varphi'} = \begin{pmatrix} \cos\varphi' & \sin\varphi' \\ \nu\sin\varphi' & -\nu\cos\varphi' \end{pmatrix}\begin{pmatrix} A \\ B \end{pmatrix},$$

where $\varphi = \varphi(q, p)$, $\varphi' = \varphi'(q, p)$, and A and B are given by (69) and (70). Because the product $\begin{pmatrix} \cos(\varphi - \varphi') & \frac{1}{\nu}\sin(\varphi - \varphi') \\ -\nu\sin(\varphi - \varphi') & \cos(\varphi - \varphi') \end{pmatrix}$ and Φ_φ is equal to $\Phi_{\varphi'}$, we obtain (84) using the definition of Ψ_t. $\square$

Claim: The map Φ_φ (68) intertwines the Kepler and Delaunay vector fields, that is,

$$T\Phi_\varphi \circ X_H = X_{\widetilde{\mathcal{H}}} \circ \Phi_\varphi,$$

if and only if

$$\varphi = \varphi(q, p) = \frac{1}{\nu}\langle q, p\rangle - F(q, p), \tag{85}$$

where F is a smooth integral of X_H.

(3.18) **Proof**: Let $t \to (q(t), p(t))$ be an integral curve of the Kepler vector field X_H with negative energy. Then Φ_φ intertwines the Kepler and Delaunay vector fields if and only if $t \to (x(t), y(t)) = \Phi_\varphi(q(t), p(t))$ is an integral curve of the Delaunay vector field $X_{\widetilde{\mathcal{H}}}$. In other words, $t \to (x(t), y(t))$ satisfies

$$\frac{dx}{dt} = \frac{\mu^2}{\nu^4}y \tag{86}$$

$$\frac{dy}{dt} = -\frac{\mu^2}{\nu^2}x. \tag{87}$$

From (86) and (87) it follows that $L_{X_{\widetilde{\mathcal{H}}}}\langle y, y\rangle = 0$; while from

$$-\mu^2/\nu^2 = H(q, p) = \mathcal{H}(x, y) = -\mu^2/\langle y, y\rangle$$

it follows that $\langle y, y\rangle = \nu^2$. Differentiating

$$(x(t), y(t)) = (A\sin\varphi + B\cos\varphi, -\nu A\cos\varphi + \nu B\sin\varphi)$$

with respect to t and using

$$L_{X_H} A = \frac{\mu}{\nu|q|} B \quad \text{and} \quad L_{X_H} B = -\frac{\mu}{\nu|q|} A, \tag{88}$$

(which are proved in ((3.19)) below), gives

$$\begin{aligned}\frac{dx}{dt} &= \frac{dA}{dt}\sin\varphi + A\cos\varphi\frac{d\varphi}{dt} + \frac{dB}{dt}\cos\varphi - B\sin\varphi\frac{d\varphi}{dt} \\ &= \Big(\frac{\mu}{\nu|q|} - \frac{d\varphi}{dt}\Big)(B\sin\varphi - A\cos\varphi) = \frac{1}{\nu}\Big(\frac{\mu}{\nu|q|} - \frac{d\varphi}{dt}\Big)y\end{aligned} \tag{89}$$

and

$$\begin{aligned}\frac{dy}{dt} &= -\nu\frac{dA}{dt}\cos\varphi + \nu A\sin\varphi\frac{d\varphi}{dt} + \nu\frac{dB}{dt}\sin\varphi + \nu B\cos\varphi\frac{d\varphi}{dt} \\ &= -\nu\Big(\frac{\mu}{\nu|q|} - \frac{d\varphi}{dt}\Big)(A\sin\varphi + B\cos\varphi) = -\nu\Big(\frac{\mu}{\nu|q|} - \frac{d\varphi}{dt}\Big)x.\end{aligned} \tag{90}$$

Therefore $t \to (x(t), y(t))$ is an integral curve of $X_{\widetilde{\mathcal{H}}}$ if and only if

$$L_{X_H}\varphi - \frac{\mu}{\nu|q|} = -\frac{\mu^2}{\nu^3}. \tag{91}$$

The following calculation shows that $\varphi_{LS} = \frac{1}{\nu}\langle q, p\rangle$ satisfies (91):

$$\frac{d\varphi_{LS}}{dt} - \frac{\mu}{\nu|q|} = \frac{1}{\nu}\langle\frac{dq}{dt}, p\rangle + \frac{1}{\nu}\langle q, \frac{dp}{dt}\rangle - \frac{\mu}{\nu|q|} = \frac{2}{\nu}\Big(\tfrac{1}{2}\langle p, p\rangle - \frac{\mu}{|q|}\Big) = -\frac{\mu^2}{\nu^3},$$

since $-\frac{\mu^2}{2\nu^2} = \frac{1}{2}\langle p, p\rangle - \frac{\mu}{|q|}$. If φ is another smooth solution of (91), we see that

$$L_{X_H}(\varphi - \varphi_{LS}) = 0.$$

In other words, $\varphi(q, p) - \frac{1}{\nu}\langle q, p\rangle = -F(q, p)$ is a smooth integral of X_H on Σ_-. □

▷ The following argument proves (88).

(3.19) **Proof**: Let $t \to (q(t), p(t))$ be an integral curve of the Kepler vector field X_H. Let $A = A(q(t), p(t))$ and $B = B(q(t), p(t))$ where $A = (\widetilde{A}, A_4)$ and $B = (\widetilde{B}, B_4)$ are given by (69) and (70) respectively. To verify the first equation in (88) we compute.

$$\begin{aligned}\frac{d\widetilde{A}}{dt} &= -\frac{1}{|q|^3}\langle q, \frac{dq}{dt}\rangle q + \frac{1}{|q|}\frac{dq}{dt} - \frac{1}{\mu}\langle\frac{dq}{dt}, p\rangle p - \frac{1}{\mu}\langle q, \frac{dp}{dt}\rangle p - \frac{1}{\mu}\langle q, q\rangle\frac{dp}{dt} \\ &= -\mu\langle p, p\rangle p + \frac{2}{|q|}p \\ &= \frac{\mu}{\nu|q|}|q|p, \quad \text{since } -\tfrac{1}{2}\tfrac{\mu^2}{\nu^2} = \tfrac{1}{2}\langle p, p\rangle - \tfrac{\mu}{|q|} \\ &= \frac{\mu}{\nu|q|}\widetilde{B}\end{aligned}$$

and

$$\frac{dA_4}{dt} = \frac{1}{\nu}\Big[\langle \frac{dq}{dt}, p\rangle + \langle q, \frac{dp}{dt}\rangle\Big] = \frac{\mu}{\nu|q|}\Big(\frac{|q|}{\mu}\langle p,p\rangle - 1\Big) = \frac{\mu}{\nu|q|}B_4.$$

To verify the second equation in (88) we compute

$$\begin{aligned}\frac{d\widetilde{B}}{dt} &= \frac{1}{\nu}\Big[\frac{1}{|q|}\langle q, \frac{dq}{dt}\rangle p + |q|\frac{dp}{dt}\Big], \quad \text{since } L_{X_H}\nu = 0.\\ &= \frac{\mu}{\nu|q|}\Big(\langle q,p\rangle p - \frac{1}{|q|}q\Big) = -\frac{\mu}{\nu|q|}\widetilde{A}\end{aligned}$$

and

$$\begin{aligned}\frac{dB_4}{dt} &= \frac{1}{\mu}\Big[\frac{1}{|q|}\langle q, \frac{dq}{dt}\rangle + 2|q|\langle p, \frac{dp}{dt}\rangle\Big]\\ &= \frac{2\langle q,p\rangle}{\mu|q|}\Big(\tfrac{1}{2}\langle p,p\rangle - \frac{\mu}{|q|}\Big) = -\frac{\mu}{\nu|q|}\frac{1}{\nu}\langle q,p\rangle = -\frac{\mu}{\nu|q|}A_4. \quad \square\end{aligned}$$

If we choose the parameter φ in the mapping Φ_φ to be $\varphi_{LS} = \frac{1}{\nu}\langle q,p\rangle$, we call the resulting mapping LS the *Ligon-Schaaf map*.

Corollary: For every $(q,p) \in \Sigma_-$ we may write

$$\Phi_\varphi(q,p) = \Psi_{\frac{\nu^3}{\mu^2}F(q,p)}(LS(q,p)). \tag{92}$$

(3.20) **Proof**: Using ((3.17)) and $\varphi = \frac{1}{\nu}\langle q,p\rangle - F(q,p)$, we obtain (92). $\square$

Because $\frac{\nu^3}{\mu^2}F(q,p)$ is a smooth integral of the Kepler vector field X_H by ((3.13)) there is a smooth function $\check{F}$ on $C = \mathcal{J}(\Sigma_-) \subseteq \mathbf{R}^6$ such that $\frac{\nu^3}{\mu^2}F = \mathcal{J}^*\check{F}$. By ((3.15)), $(LS)^*\widetilde{\mathcal{J}} = \mathcal{J}$. Thus there is a smooth function $\widetilde{F} = \widetilde{\mathcal{J}}^*\check{F}$ on T^+S^3 such that $\frac{\nu^3}{\mu^2}F = (LS)^*\widetilde{F}$. Therefore we may rewrite (92) as

$$\Phi_\varphi = (LS)^*\widetilde{\Psi}, \tag{93}$$

where $\widetilde{\Psi}$ is the time one map of the flow of the rescaled Delaunay vector field $\widetilde{F}\cdot X_{\widetilde{H}}$. In other words, $\widetilde{\Psi}(x,y) = \Psi_{\widetilde{F}(x,y)}(x,y)$ for every $(x,y) \in \operatorname{im} LS(\Sigma_-)$. Note that the function $\widetilde{F}$ is a smooth integral of $X_{\widetilde{\mathcal{H}}}$. To see this we compute

$$\begin{aligned}L_{X_{\widetilde{\mathcal{H}}}}\widetilde{F}\big(LS(q,p)\big) &= d\widetilde{F}\big(LS(q,p)\big)X_{\widetilde{\mathcal{H}}}\big(LS(q,p)\big)\\ &= d\widetilde{F}\big(LS(q,p)\big)T_{(q,p)}LS\,X_H(q,p)\\ &\qquad \text{since } LS \text{ intertwines } X_H \text{ and } X_{\widetilde{\mathcal{H}}}\\ &= d\Big(\frac{\nu^3}{\mu^2}F(q,p)\Big)X_H(q,p),\\ &\qquad \text{since } \nu^3/\mu^2 = (LS)^*\widetilde{F}\\ &= 0, \qquad \text{since } \nu \text{ and } F \text{ are integrals of } X_H.\end{aligned}$$

We now prove

Claim: If $\Phi_\varphi : (\Sigma_-, \widetilde{\omega}_3) \to (T^+S^3, \widetilde{\Omega}_4)$ is symplectic, in addition to intertwining the momentum mappings $\mathcal{J}$ and $\widetilde{\mathcal{J}}$ and the Kepler and Delaunay vector fields, then the smooth integral $\widetilde{F}$ of the Delaunay vector field $X_{\widetilde{\mathcal{H}}}$ is a smooth function of $\widetilde{\mathcal{H}}$.

(3.21) **Proof**: Because $\widetilde{\Psi}$ is the time one map of the rescaled Delaunay vector field $\widetilde{F}\cdot X_{\widetilde{\mathcal{H}}}$, using ((3.18)) we obtain $\widetilde{\Psi}^*(\widetilde{\Omega}_4) = \widetilde{\Omega}_4 + d\widetilde{F}\wedge d\widetilde{\mathcal{H}}$. From (93) it follows that

$$\Phi_\varphi^*(\widetilde{\Omega}_4) = (LS)^*(\widetilde{\Psi}^*(\widetilde{\Omega}_4)) = (LS)^*(\Omega_4) + (LS)^*(d\widetilde{F}\wedge d\widetilde{\mathcal{H}}). \tag{94}$$

By hypothesis, Φ_φ is symplectic for every choice of smooth φ. Therefore LS is symplectic. Hence (94) becomes $\widetilde{\Omega}_4 = \widetilde{\Omega}_4 + (LS)^*(d\widetilde{F}\wedge d\widetilde{\mathcal{H}})$, that is, $(LS)^*(d\widetilde{F}\wedge d\widetilde{\mathcal{H}}) = 0$. Because LS is symplectic, $T_{(q,p)}LS$ is injective for every $(q,p)\in\Sigma_-$. Hence $d\widetilde{F}\wedge d\widetilde{\mathcal{H}} = 0$. This implies that the 1-form $\widetilde{F}d\widetilde{\mathcal{H}}$ is locally exact, and since T^+S^3 is simply connected, we may conclude that $\widetilde{F}d\widetilde{\mathcal{H}}$ is globally exact. Since $\widetilde{F}$ and $\widetilde{H}$ are both SO(4)-invariant, and SO(4) acts transitively on energy surfaces, we are reduced to understanding

$$\widetilde{F}d\widetilde{\mathcal{H}} = d\widetilde{K} \tag{95}$$

for some smooth function $\widetilde{K}$. Everything in (95) may be thought of as a smooth function of $|p|$, since the orbit space $T^+S^3/\mathrm{SO}(4)$ is $\{|p|>0\}$. Since $\frac{d\widetilde{\mathcal{H}}}{d|p|}\neq 0$, we may write (95) as $\widetilde{F} = \frac{d\widetilde{K}}{d\widetilde{\mathcal{H}}}$. Consequently, $\widetilde{F}$ is a function of $\widetilde{\mathcal{H}}$, that is,

$$\widetilde{F} = \widetilde{\mathcal{G}}\circ\widetilde{\mathcal{H}}, \tag{96}$$

where $\widetilde{\mathcal{G}} = \frac{d\widetilde{K}}{d\widetilde{\mathcal{H}}}$, on *all* of T^+S^3. □

We now prove the following characterization of the Ligon–Schaaf map.

Theorem: Suppose that $\Phi_\varphi(\Sigma_-) = T^+S^3_{\mathrm{np}}$, in addition to

$$\Phi_\varphi^*\widetilde{\mathcal{J}} = \mathcal{J},\quad T\Phi_\varphi\circ X_H = X_{\widetilde{\mathcal{H}}}\circ\Phi_\varphi,\quad\text{and}\quad \Phi_\varphi^*\widetilde{\Omega}_4 = \widetilde{\omega}_3.$$

Then $\Phi_\varphi = LS$.

(3.22) **Proof**: We now show that these hypotheses are not vacuous. From the facts that $\Phi_\varphi(\Sigma_-) = T^+S^3_{\mathrm{np}} = LS(\Sigma_-)$ (see ((3.23)) below) and $\Phi_\varphi = \widetilde{\Psi}\circ LS$ (93) it follows that $\widetilde{\Psi}$ maps $T^+S^3_{\mathrm{np}}$ onto itself. Hence $\widetilde{\Psi}$ maps $NP = T^+S^3 - T^+S^3_{\mathrm{np}}$ diffeomorphically onto itself. Note that NP is the fiber of the bundle

$$\tau : T^+S^3 \subseteq \mathbf{R}^4 \to S^3 \subseteq \mathbf{R}^4 : (x,y)\to x$$

over np $= (0,0,0,1)$. For some $\nu>0$ suppose that $g = \widetilde{\mathcal{G}}(-\mu^2/2\nu^2)$ is nonzero. Here $\widetilde{\mathcal{G}}$ is the smooth function constructed in the proof of ((3.21)). Then the flow Ψ_{gt} has no fixed points on $\widetilde{\mathcal{H}}^{-1}(-\mu^2/2\nu^2)$ because g is nonzero and the Delaunay vector field has no equilibrium points on $\widetilde{\mathcal{H}}^{-1}(-\mu^2/2\nu^2)$. This contradicts the fact that $\widetilde{\Psi}$ maps the set

$NP \cap \widetilde{\mathcal{H}}^{-1}(-\mu^2/2\nu^2)$ into itself. Therefore $g = 0$. Hence $\widetilde{\mathcal{G}}$ is identically zero. From (96) we see that the function $\widetilde{F}$ is identically zero. By definition $\widetilde{\Psi} = \text{id}$. From (93) it follows that $\Phi_\varphi = LS$. □

▷ To prove the converse to the above theorem we need only show that

1. $LS(\Sigma_-) = T^+S^3_{\text{np}}$
2. $(LS)^*(\widetilde{\Omega}_4) = \widetilde{\omega}_3$,

because by definition LS intertwines the momentum maps $\mathcal{J}$ and $\widetilde{\mathcal{J}}$ and by ((3.18)) it also intertwines the vector fields X_H and $X_{\widetilde{\mathcal{H}}}$.

(3.23) **Proof**:
1). Suppose that $(x, y) = LS(q, p)$ for some $(q, p) \in \Sigma_-$ and that $x = \text{np}$. Then

$$1 = x_4 = \frac{1}{\nu}\langle q, p\rangle \sin\varphi + \left(\frac{1}{\mu}\langle p, p\rangle |q| - 1\right)\cos\varphi. \tag{97}$$

Since $(x, y) \in T^+S^3$, it follows that $\langle x, y\rangle = 0$. Therefore

$$0 = \frac{1}{\nu}\, y_4 = -\frac{1}{\nu}\langle q, p\rangle \cos\varphi + \left(\frac{1}{\mu}\langle p, p\rangle |q| - 1\right)\sin\varphi. \tag{98}$$

Multiplying (97) by $\sin\varphi$ and adding the result to $-\cos\varphi$ times (98) gives

$$\sin\varphi = \frac{1}{\nu}\langle q, p\rangle = \varphi, \tag{99}$$

by definition of LS. Since $1 = \langle A, A\rangle$, we find that $\varphi \in [0, 1]$ using (69) and (99). Hence $\varphi = 0$ is the only solution of (99). Therefore (97) becomes $1 = \frac{1}{\mu}|q|\langle p, p\rangle - 1$, which is equivalent to $\frac{1}{2}\langle p, p\rangle - \frac{\mu}{|q|} = 0$. In other words, $(q, p) \notin \Sigma_-$. But this contradicts the hypothesis. Therefore $LS(\Sigma_-) \subseteq T^+S^3_{\text{np}}$. Actually equality holds. To see this look at the proof of ((3.15)). We see that when $\mathbf{J} \neq 0$, the Keplerian ellipse with data $h, \mathbf{J}, \widetilde{\mathbf{e}}$ is mapped by LS to an orbit of the Delaunay vector field which is a great circle which does not pass through NP. When $\mathbf{J} = 0$ a linear Keplerian orbit corresponds to a great circle which passes through np when $q = 0$, (see ((3.24))). □

2). This follows by taking the exterior derivative of

$$(LS)^*\langle y, dx\rangle = -\langle q, dp\rangle - d\langle q, p\rangle, \tag{100}$$

which we establish below, (see (104)). First we factor the map LS into the composition of the mapping

$$\begin{aligned} S : \Sigma_- \subseteq T_0\mathbf{R}^3 &\to \mathbf{R} \times T_1S^3 : (q, p) \to (\varphi, A, B) = \\ &= \left(\nu^{-1}\langle q, p\rangle, (\tfrac{q}{|q|} - \mu^{-1}\langle q, p\rangle\, p, \nu^{-1}\langle q, p\rangle), (\nu^{-1}|q|\, p, \mu^{-1}|q|\,\langle p, p\rangle - 1)\right) \end{aligned}$$

followed by the mapping

$$\begin{aligned} L : \mathbf{R} \times T_1S^3 &\to T^+S^3 : \\ (\varphi, A, B) &\to (x, y) = \big((\sin\varphi)A + (\cos\varphi)B, -\nu(\cos\varphi)A + \nu(\sin\varphi)B\big). \end{aligned}$$

Below we show that

$$L^*\langle y, dx\rangle = -\nu(d\varphi + \langle A, dB\rangle) \tag{101}$$

and

$$S^*(\nu\langle A, dB\rangle) = -\nu\langle q, p\rangle d(\nu^{-1}) + \langle q, dp\rangle \tag{102}$$

$$S^*(\nu\, d\varphi) = \nu\, d\left(\nu^{-1}\langle q, p\rangle\right). \tag{103}$$

Consequently,

$$\begin{aligned}(LS)^*\langle y, dx\rangle &= S^*(L^*(\langle y, dx\rangle)) = -S^*(\nu d\varphi + \nu\langle A, dB\rangle), \quad \text{by (101)}\\ &= -S^*(\nu d\varphi) + S^*(\nu\langle A, dB\rangle)\\ &= -\nu d\big(\nu^{-1}\langle q, p\rangle\big) + \nu\langle q, p\rangle\, d(\nu^{-1}) + \langle q, dp\rangle, \quad \text{by (102) and (103)}\\ &= -\langle q, dp\rangle - d\langle q, p\rangle. \end{aligned} \tag{104}$$

This proves (100).

We need only verify (101)–(103). Equation (103) follows immediately because $\varphi = \frac{1}{\nu}\langle q, p\rangle$. Equation (101) follows from the definition of L and

$$\langle A, dA\rangle = \langle B, dB\rangle = \langle A, dB\rangle + \langle B, dA\rangle = 0,$$

which is an immediate consequence of $\langle A, A\rangle = \langle B, B\rangle = 1$ and $\langle A, B\rangle = 0$. It remains to prove (102). From the definition of the mapping S we find that

$$S^*(\langle A, dB\rangle) =$$

$$\left[\mu^{-1}\nu^{-2}\langle q, p\rangle |q| \left(\langle p, p\rangle - \tfrac{\mu}{|q|}\right) d\nu + \mu^{-1}\nu^{-1}\langle q, p\rangle |q| \left(\langle p, dp\rangle + \tfrac{\mu}{|q|^3}\langle q, dq\rangle\right)\right] + \nu^{-1}\langle q, dp\rangle.$$

Differentiating $\nu = \mu(-2H)^{-1/2}$ gives $d\nu = \frac{\nu}{(-2H)^{3/2}}\, dH$. Also

$$dH = \langle p, dp\rangle + \frac{\mu}{|q|^3}\langle q, dq\rangle.$$

Hence the terms in the square brackets in the right hand side of the equation for $S^*(\langle A, dB\rangle)$ can be written as

$$\mu^{-1}\nu^{-1}\langle q, p\rangle \left[\frac{\mu - |q|\langle p, p\rangle + 2H|q|}{2H}\right] dH = \nu^{-2}\langle q, p\rangle d\nu,$$

because $\langle p, p\rangle = 2(H + \frac{\mu}{|q|})$ and $d\nu = \frac{\nu}{(-2H)^{3/2}}\, dH$. Therefore

$$S^*(\langle A, dB\rangle) = \nu^{-2}\langle q, p\rangle d\nu + \nu^{-1}\langle q, dp\rangle = -\langle q, p\rangle d(\nu^{-1}) + \nu^{-1}\langle q, dp\rangle.$$

This proves (102) and shows that LS is symplectic. □

▷ We now show that the Ligon–Schaaf map is a diffeomorphism.

(3.24) **Proof**: For every $c \in C$, the fibers $\mathcal{J}^{-1}(c)$ and $\widetilde{\mathcal{J}}^{-1}(c)$ are unique oriented orbits of the Kepler and Delaunay vector fields respectively. Since $(LS)^*\widetilde{\mathcal{J}} = \mathcal{J}$, LS maps a bounded

Kepler orbit onto an orbit of the Delaunay vector field. Thus we need only show that on orbits, LS is a diffeomorphism. Hence it suffices to show that LS preserves the periods of corresponding orbits. When $\mathbf{J} \neq 0$ and $H < 0$, the period of an elliptical Keplerian orbit is $2\pi \nu^3/\mu^2$. Since the period of the corresponding orbit of the Delaunay vector field is also $2\pi \nu^3/\mu^2$, LS preserves periods. When $\mathbf{J} = 0$ and $H < 0$, the Keplerian orbit projects to a linear orbit traversed twice. Hence the period of such a Keplerian orbit is *twice* the time it takes to reach the origin starting with momentum 0. Using (45) we see that again the period is given by $2\pi \nu^3/\mu^2$. The corresponding Delaunay orbit is periodic of period $2\pi \nu^3/\mu^2$. Therefore LS is a diffeomorphism. □

We now put the preceding arguments into perspective. Geometrically we think of T^+S^3 as a regularized model for the negative energy subset Σ_- of the phase space of the Kepler problem. On T^+S^3 the SO(4) symmetry $\widehat{\Phi}$ (21) is *globally* defined and its orbits

$$T_{\mu/\sqrt{-2h}}S^3 = \left\{ (x, y) \in T^+S^3 \middle| \langle y, y \rangle = -\frac{\mu^2}{2h} \right\}$$

are the regularization of the energy surface $H^{-1}(h) \subseteq \Sigma_-$ of the Kepler problem. Let

$$\mathscr{C} = \left\{ (x, y) \in T^+S^3 \middle| x_4 = 1 \right\}.$$

From ((3.24)) we see that $LS^{-1}(\mathscr{C})$ is the collision set

$$\left\{ (q, p) \in T\mathbf{R}^3 \middle| q = 0 \right\}$$

of the Kepler problem. Since $T_{\mu/\sqrt{-2h}}S^3 \cap \mathscr{C}$ is nonempty for every $h < 0$, we conclude that the SO(4) symmetry is *not globally* defined for the Kepler problem. On the other hand the flow of the Delaunay vector field $X_{\widetilde{\mathscr{H}}}$ is globally defined on T^+S^3.

Consider the function

$$\widetilde{J}(x, y) = \sum_{1 \le i < j \le 3} (x_i y_j - x_j y_i)^2 | T^+S^3 = \langle \widetilde{x} \times \widetilde{y}, \widetilde{x} \times \widetilde{y} \rangle.$$

Then $\widetilde{J}$ is an integral of the geodesic flow on T^+S^3 and hence is an integral of the Delaunay vector field $X_{\widetilde{\mathscr{H}}}$.

Claim: $\widetilde{J}^{-1}(0)$ is the set of all integral curves of $X_{\widetilde{\mathscr{H}}}$ which pass through $\mathscr{C}$.

(3.25) **Proof**: As all orbits of $X_{\widetilde{\mathscr{H}}}$ when projected onto S^3 are great circles, each projected orbit intersects the equator $\{x_4 = 0\} \cap S^3$. Hence every orbit of $X_{\widetilde{\mathscr{H}}}$ intersects the equator at some point $(\widetilde{x}, 0, \widetilde{y}, y_4)$. If $\widetilde{y} = 0$ then $\widetilde{J} = 0$. As seen from the formula (16) for the flow of $X_{\widetilde{\mathscr{H}}}$, the $X_{\widetilde{\mathscr{H}}}$ orbit through $(\widetilde{x}, 0, 0, y_4)$ passes through $\{x_4 = 1\}$. Now suppose that $\widetilde{y} \neq 0$. $\widetilde{J} = 0$ if and only if $\widetilde{x} \times \widetilde{y} = 0$, that is, $\widetilde{x}$ and $\widetilde{y}$ are linearly dependent. But this contradicts the fact that $0 = \langle x, y \rangle = \langle \widetilde{x}, \widetilde{y} \rangle$, and both $\widetilde{x}$ and $\widetilde{y}$ are nonzero. Therefore $\widetilde{y} = 0$. □

From ((3.15)) we see that $(LS)^{-1}(\widetilde{J}^{-1}(0)) = \mathscr{J}^{-1}(0) \cap \Sigma_-$. An integral curve γ of $X_{\widetilde{\mathscr{H}}}$ which lies on $\mathscr{J}^{-1}(0)$ is the image under LS of an integral curve of the Kepler vector field

X_H on Σ_- with zero angular momentum. In particular the projection of $(LS)^{-1}(\gamma - \text{np})$ is a ray from the origin. The particle moves on this ray being reflected at the origin. This is the regularized picture of collision in the Kepler problem. This regularization is geometrically faithful as the mapping LS is an open dense embedding of $(H, \Sigma_-, \widetilde{\omega}_3)$ into $(\widetilde{\mathcal{H}}, T^+S^3, \widetilde{\Omega}_4)$ with $(LS)^*\widetilde{\mathcal{H}} = H$.

4 Exercises

1. (The virial group.) For the Kepler problem show that the change of variables $(t, q, p) \to (a^3t, a^2q, a^{-1}p)$ defines an action of the multiplicative group $\mathbf{R}^*$ and leaves the equations of motion invariant. Deduce the period energy relation (Kepler's third law) from this symmetry.

2. (sl$(2, \mathbf{R})$ and the Kepler problem.) For the Kepler problem with rotationally symmetric Hamiltonian $H = \frac{1}{2} p \cdot p - \frac{1}{|q|}$ let $j = q \times p$, $x = q \cdot q$, $y = p \cdot p$, and $z = q \cdot p$.

   (a). Show that the functions x, y, z Poisson commute with the components of j. Moreover, show that the Poisson brackets of x,y, and z define a representation of sl$(2, \mathbf{R})$. Conclude that so(3) and sl$(2, \mathbf{R})$ form a dual pair in the Lie algebra of homogeneous quadratic polynomials.

   (b). In xyz-space draw that level sets of j^2 = const. for different values of the constant including zero. These are models for the SO(3) reduced space.

   (c). Draw the intersections of the h = const. surfaces with a given j = const. surface to see the integral curves of the reduced dynamics.

   (d). Show that the level sets j^2 = const. are symplectic leaves for the Poisson manifold sl$(2, \mathbf{R}) = \mathbf{R}^3$ with coordinates (x, y, z)

3. (Center of mass and the two body problem.) (a). For the two body problem in space show that regular reduction by the translation group can be interpreted as passing to a center of mass frame. Do the reduction of the translation and rotational symmetries in one step by using the Euclidean group $E(3)$.

   (b). Consider the spherical analogue of the planar two body problem. This is the motion of two particles connected by a spring constrained to move on the surface of a 2-sphere. The rotation group SO(3) is an obvious symmetry group of the problem, as compared to the Euclidean group $E(2)$ for the planar problem. Construct all the SO(3) reduced spaces. Show that there is *no notion* of a center of mass *frame*.

   (c)*. Is the spherical two body problem Liouville integrable?

4. (Kepler's equation.) Consider Kepler's equation

$$t = s - e \sin s = f(s). \tag{105}$$

   (a). For $|e| < 1$ show that f is a one to one real analytic diffeomorphism of $\mathbf{R}$. Expand $f(s)$ in a power series: $t = \sum_{k=0}^{\infty} a_k s^k$ and invert: $s = f^{-1}(t) =$

$\sum_{k=0}^{\infty} c_k t^k$. Estimate the radius of convergence of the series for f^{-1}. (Hint: see the discussion in Markushevich [145].)

(b). Another way to invert Kepler's equation goes as follows. Suppose we are given a function $t(s)$ in the form $t = F(s) = s - \varepsilon f(s)$, where $f(0) = 0$, $f(s+2\pi) = f(s)$, and $|\varepsilon f'(s)| < 1$. In other words, εf is a "small" periodic perturbation of the identity. Since $|\varepsilon f'(s)| < 1$, we have $\frac{df}{ds} = 1 - \varepsilon f'(s) > 0$. Thus by the inverse function theorem there is a well-defined inverse function $s(t)$. Furthermore, it is of the form $s = s(t) = t + g_\varepsilon(t)$, where $g_\varepsilon(t)$ is 2π-periodic with $g_\varepsilon(0) = 0$. We would like to expand g_ε in a Fourier series: $g_\varepsilon(t) = \frac{1}{2} a_0 + \sum_{k=1}^{\infty} (a_k \cos kt + b_k \sin kt)$, where

$$a_0 = \frac{1}{\pi} \int_0^{2\pi} g_\varepsilon(t)\, dt, \quad a_k = \frac{1}{\pi} \int_0^{2\pi} g_\varepsilon(t) \cos kt\, dt, \text{ and } b_k = \frac{1}{\pi} \int_0^{2\pi} g_\varepsilon(t) \sin kt\, dt.$$

The difficulty with this direct calculation is that we do not know the function $g_\varepsilon(t)$. We circumvent this by expressing $g_\varepsilon(t)$ in terms of s, namely, $g_\varepsilon(t) = s - F(s) = \varepsilon f(s)$, where s and t are related by $t = F(s)$. Now

$$\begin{aligned} a_0 &= \frac{1}{\pi} \int_0^{2\pi} g_\varepsilon(t)\, dt \\ &= \frac{1}{\pi} \int_0^{2\pi} g_\varepsilon(t(s)) \frac{dt}{ds}\, ds \\ &= \frac{1}{\pi} \int_0^{2\pi} \varepsilon f(s) \big(1 - \varepsilon f'(s)\big)\, ds, \end{aligned}$$

using the change of variable $f(s)$. A similar formula holds for a_k and b_k. Show that for the case of Kepler's equation, where $f(s) = e \sin s$, we have $a_k = 0$ for all $k \geq 0$. Change variable from t to s and use the fact that

$$J_k(x) = \int_0^{\pi} \cos(ks - x \sin s)\, ds$$

to conclude that $b_k = \frac{2}{k} J_k(ke)$.

(c). Show that $y(x) = J_k(x)$ satisfies

$$x^2 y'' + xy' + (x^2 - k^2) y = 0,$$

which is Bessel's equation.

5. (a). Let $G(2,4)$ be the space of all 2-planes in $\mathbf{R}^4$. Show that the orthogonal group $\mathrm{O}(4)$ acts transitively on $G(2,4)$ and that the subgroup of all elements of $\mathrm{O}(4)$, which leave the 2-plane span$\{e_1, e_2\}$ fixed, is isomorphic to $\mathrm{O}(2) \times \mathrm{O}(2)$. Deduce that $G(2,4)$ is diffeomorphic to the orbit space $\mathrm{O}(4)/(\mathrm{O}(2) \times \mathrm{O}(2))$.

   (b). Let $\widetilde{G}(2,4)$ be the space of all *oriented* 2-planes in $\mathbf{R}^4$. Show that the group $\mathrm{SO}(4)$ acts transitively on $\widetilde{G}(2,4)$ and that the subgroup of all elements of $\mathrm{SO}(4)$, which leave the oriented 2-plane span$\{e_1, e_2\}$ fixed, is isomorphic to $\mathrm{SO}(2) \times \mathrm{SO}(2)$. Deduce that $\widetilde{G}(2,4)$ is diffeomorphic to the orbit space $\mathrm{SO}(4)/(\mathrm{SO}(2) \times \mathrm{SO}(2))$.

   (c). Show that $\widetilde{G}(2,4)$ is diffeomorphic to $S^2 \times S^2$ and hence is simply connected. Deduce that $\widetilde{G}(2,4)$ is a 2-fold covering space of $G(2,4)$.

6. (a). Construct an isomorphism between the Lie algebra so(4) and so(3) $\times$ so(3).

   (b). Show that the corresponding Lie-Poisson algebras are isomorphic.

   (c). Write out Hamilton's equations on the Lie-Poisson algebra corresponding to so(3) $\times$ so(3).

7. (Time rescaling). Consider the Hamiltonian system (H, M, ω) with h a regular value of the Hamiltonian H. Let $f : M \to \mathbf{R}_>$ be a positive smooth function and define a new Hamiltonian $K : M \to \mathbf{R}$ by $K(m) = f(m)(H(m) - h)$. Show that $X_K|K^{-1}(0) = f\left(X_H|H^{-1}(h)\right)$. Show that the integral curves of X_K on $K^{-1}(0)$ are an orientation preserving time rescaling of the integral curves of X_H on $H^{-1}(h)$ obtained by defining a new time variable s by $\frac{ds}{dt} = \frac{1}{f}$.

8. (Souriau's linearization and regularization.) In the Kepler problem

$$\begin{aligned} \dot{q} &= p \\ \dot{p} &= -\tfrac{1}{r^3} q, \quad r = |q|, \end{aligned} \tag{106}$$

   let $H = \frac{1}{2}\langle p, p\rangle - \frac{1}{r}$ be the Hamiltonian and define a new time variable s by

$$s = \langle q, p\rangle - 2ht.$$

   (a). Show that $\frac{ds}{dt} = \frac{1}{r}$. Thus s is the eccentric anomaly.

   (b). Define a 4-vector ξ by $\xi = \begin{pmatrix} t \\ q \end{pmatrix}$. Let $\Xi = \mathrm{col}(\xi, \xi', \xi'', \xi''')$ be a 4×4 matrix, where $'$ is differentiation with respect to s. Show that Ξ satisfies the *linear* differential equation

$$\Xi' = A\,\Xi \tag{107}$$

   where $A = \begin{pmatrix} 0 & 1 & 0 & 0 \\ 0 & 0 & 1 & 0 \\ 0 & 0 & 0 & 1 \\ 0 & 2h & 0 & 0 \end{pmatrix}$.

   (c). Solve (107) and thus find $\xi(s)$. Note that because $\xi(s)$ is defined for *all* s and hence for all t by Kepler's equation, it follows that the Kepler problem has been regularized.

9. (Bacry-Györgyi variables and the conformal group.) Using the same notation in the Kepler problem as in exercise 8, set $\alpha = \sqrt{-2h}$, $P = \begin{pmatrix} t''' \\ \alpha^{-1} q''' \end{pmatrix}$ and $Q = \begin{pmatrix} \alpha t'' \\ q'' \end{pmatrix}$. Here we are confining ourselves to the case of bounded motions, namely, $h < 0$.

   (a). Show that $P^t P = Q^t Q = 1$ and $P^t Q = 0$.

   (b). Let ζ be the 6×6 matrix

$$\begin{pmatrix} QP^t - PQ^t & P & Q \\ P^t & 0 & 1 \\ Q^t & -1 & 0 \end{pmatrix}$$

(c). Show that $\zeta^2 = 0$.

(d). Show that the components of ζ satisfy the Poisson bracket relations for the Lie algebra so(4, 2).

(e). Show that the map from the regularized phase space of the negative energy orbits $(q, p, h) \to \zeta$ is a symplectic diffeomorphism if we equip the SO(4, 2)-coadjoint orbit through ζ with the symplectic structure given in appendix 1 section 1 example 2. The tricky part of this is deciding which component of the variety $\zeta^2 = 0, \zeta \neq 0$ in $\mathrm{so}(4, 2)^*$ you need to map to.

10. (Levi-Civita regularization). (a). Let $\mathbf{R}_0^2 = \mathbf{R}^2 - \{0\}$. On $T^*\mathbf{R}_0^2 = \mathbf{R}_0^2 \times \mathbf{R}^2$ with coordinates (x, y) and symplectic form $\omega = \sum_{i=1}^3 dx_i \wedge dy_i$ consider the Kepler Hamiltonian

$$H(x, y) = \tfrac{1}{2}(y_1^2 + y_2^2) - \frac{1}{\sqrt{x_1^2 + x_2^2}}. \tag{108}$$

Identify $\mathbf{R}_0^2$ with $\mathbf{C}_0 = \mathbf{C} - \{0\}$ and $T^*\mathbf{R}_0^2$ with $T^*\mathbf{C}_0 = \mathbf{C}_0 \times \mathbf{C}$. Introduce complex coordinates $q = x_1 + i\, x_2$ and $p = y_1 + i\, y_2$ on $T^*\mathbf{C}_0$. Show that $\omega = \mathrm{Re}(dq \wedge d\overline{p})$ and that the Kepler Hamiltonian becomes

$$H(q, p) = \tfrac{1}{2}|p|^2 - \frac{1}{|q|}. \tag{109}$$

(b). Using exercise 7 and the time rescaling $\frac{ds}{dt} = \frac{k}{2|q|}$ show that the integral curves of X_H on the level set $H^{-1}(-k^2/2)$ are a time reparametrization of the integral curves of the vector field $X_{\widetilde{K}}$ on the level set $\widetilde{K}^{-1}(0)$ where

$$\widetilde{K}(q, p) = \frac{2|q|}{k}\left(\tfrac{1}{2}|p|^2 - \frac{1}{|q|} + \frac{k^2}{2}\right) = \frac{1}{k}|q|\,|p|^2 + k\,|q| - \frac{2}{k}. \tag{110}$$

(c). Define the Levi-Civita map

$$\mathcal{L} : T^*\mathbf{C}_0 \to T^*\mathbf{C}_0 : (u, v) \to (q, p) = \left(\frac{1}{2k^2}u^2, k\frac{v}{\overline{u}}\right). \tag{111}$$

Show that $\mathcal{L}$ has the following properties:

1). $\mathcal{L}$ is a smooth two to one surjective submersion with $\mathcal{L}(-u, -v) = \mathcal{L}(u, v)$.

2). $\mathcal{L}^*(\mathrm{Re}\, q\, d\overline{p}) = \mathrm{Re}\,(u\, d\overline{v})$. Hence $\mathcal{L}$ is symplectic.

3). The Hamiltonian

$$K(u, v) = (\mathcal{L}^*\widetilde{K})(u, v) = \tfrac{1}{2}|v|^2 + \tfrac{1}{2}|u|^2 - \frac{2}{k} \tag{112}$$

is defined on $K^{-1}(0)$ which is a 3-sphere centered at the origin and having radius $2/\sqrt{k}$. Since $K^{-1}(0)$ is compact, all the integral curves of X_K on $K^{-1}(0)$ are defined for *all* time. Thus K is the Levi-Civita regularization of the Kepler Hamiltonian for negative energy orbits. Note that up to an additive constant, K is the harmonic oscillator Hamiltonian.

(d). The Levi-Civita map $\mathcal{L}$ is *not* an equivalence between the Hamiltonian systems $(K, T^*\mathbf{C}_0, \omega)$ and $(\widetilde{K}, T^*\mathbf{C}_0, \omega)$, because it is *not* a diffeomorphism. Show that that vector fields X_K on $K^{-1}(0)$ and $X_{\widetilde{K}}$ on $\widetilde{K}^{-1}(0)$ are $\mathcal{L}$-related, that is, $T\mathcal{L}\circ X_K = X_{\widetilde{K}}\circ\mathcal{L}$. Thus the image of an integral curve of X_K on $K^{-1}(0)$ under the Levi-Civita map $\mathcal{L}$ is an integral curve of $X_{\widetilde{K}}$ on $\widetilde{K}^{-1}(0)$.

(e). On $T^*\mathbf{C}_0$ define a $\mathbf{Z}_2$-action generated by $(u, v) \to (-u, -v)$. Show that this action is free, preserves the symplectic form ω, and preserves the Hamiltonian K. Thus there is an induced Hamiltonian $\mathcal{H}$ on $(T^*\mathbf{C}_0/\mathbf{Z}_2, \omega)$. Since the map $\mathcal{L}$ is invariant under the $\mathbf{Z}_2$-action, it induces an equivalence between the Hamiltonian systems $(\mathcal{H}, T^*\mathbf{C}_0/\mathbf{Z}_2, \omega)$ and $(K, T^*\mathbf{C}_0, \omega)$. Thus the regularized energy surface $H^{-1}(-k^2/2)$ of the Kepler Hamiltonian is $\mathcal{H}^{-1}(0) = S^3_{2/\sqrt{k}}/\mathbf{Z}_2$, which is real projective three space $\mathbf{RP}^3$.

11. (Moser regularization). (a). Let $\mathbf{R}^3_0 = \mathbf{R}^3 - \{0\}$. On $T^*\mathbf{R}^3_0 = \mathbf{R}^3_0 \times \mathbf{R}^3$ with coordinates (x, y) and symplectic form $\omega = \sum_{i=1}^3 dx_i \wedge dy_i$ consider the Kepler Hamiltonian

$$H(x, y) = \tfrac{1}{2}\,|y|^2 - \frac{\mu}{|x|}. \tag{113}$$

The Hamiltonian system $(H, T\mathbf{R}^3_0, \omega)$ describes the two body problem in 3-*space* with one body located at the origin.

(b). Introduce a new time scale s defined by $\frac{ds}{dt} = \frac{k}{|x|}$ and consider the new Hamiltonian

$$\widetilde{K}(x, y) = \frac{|x|}{k}\left(\tfrac{1}{2}\,|y|^2 - \frac{\mu}{|x|} + \frac{k^2}{2}\right) + \frac{\mu}{k} = \frac{1}{2k}\,|x|\,(|y|^2 + k^2). \tag{114}$$

Show that the integral curves of $X_{\widetilde{K}}$ on $\widetilde{K}^{-1}(\mu/k)$ are a time reparametrization of the integral curves of X_H on $H^{-1}(-k^2/2)$. Since the total energy of the Kepler Hamiltonian H is *negative*, we are only considering bounded Keplerian orbits.

(c). Define Moser's mapping by

$$\mathcal{M} : T^+S^3_{\text{np}} \subseteq T\mathbf{R}^4 \to T\mathbf{R}^3_0 : (q, p) \to (x, y)$$

where $S^3_{\text{np}} = S^3 - \{(0, 0, 0, 1)\}$ and for $i = 1, 2, 3$

$$\begin{aligned} x_i &= -\tfrac{1}{k}\,(p_i + q_i p_4 - p_i q_4) \\ y_i &= \frac{k q_i}{1 - q_4}. \end{aligned} \tag{115}$$

Note that the tangent bundle TS^3 of S^3 is defined by

$$q_1^2 + q_2^2 + q_3^2 + q_4^2 = 1 \quad \text{and} \quad q_1 p_1 + q_2 p_2 + q_3 p_3 + q_4 p_4 = 0.$$

1). Show that $\mathcal{M}$ is given by composing the tangent of stereographic projection of S^3 from the north pole $(0, 0, 0, 1)$ to the plane tangent to S^3 at $(0, 0, 0, -1)$ followed by the map $(x, y) \to (-y, x)$ which interchanges positions and velocities.

2). Show that

$$\begin{cases} |x|^2 &= \frac{1}{k^2}\,|p|^2(1-q_4)^2 \\ |y|^2+k^2 &= \dfrac{2k^2}{1-q_4} \\ \langle x, y\rangle &= -p_4. \end{cases} \tag{116}$$

Deduce that Moser's mapping $\mathcal{M}$ is a diffeomorphism with inverse given by

$$\begin{cases} q_i = \dfrac{2k\,y_i}{|y|^2+k^2}, \\ p_i = -\dfrac{1}{2k}\left[(|y|^2+k^2)x_i - 2\langle x, y\rangle\, y_i\right] \end{cases} \qquad i = 1, 2, 3$$

and

$$q_4 = \frac{|y|^2-k^2}{|y|^2+k^2} \qquad p_4 = -\langle x, y\rangle.$$

3). Show that

$$\mathcal{M}^*\Big(\sum_{i=1}^{3} x_i\, dy_i\Big) = -\left(\sum_{i=1}^{4} p_i\, dq_i\right)\Bigg|T^+S^3_{\mathrm{np}}. \tag{117}$$

Hence Moser's map is symplectic when we restrict the standard symplectic form $\sum_{i=1}^{4} dq_i \wedge dp_i$ on $T\mathbf{R}^3$ to $T^+S^3_{\mathrm{np}}$.

4). Show that

$$(\mathcal{M}^*\widetilde{K})(q, p) = |p| = \mathcal{H}(q, p). \tag{118}$$

$\mathcal{H}$ is defined and smooth on all of $T^+S^3 = TS^3 - \{(q, 0) \,|\, q \in S^3\}$. The flow of $X_{\mathcal{H}}$ is complete, because it is the geodesic flow on S^3. Thus we have regularized the flow of Kepler Hamiltonian on a negative energy level. The regularized negative energy level of the Kepler Hamiltonian is diffeomorphic to the unit tangent sphere bundle on S^3, that is, $S^2 \times S^3$.

12. (Kustaanheimo-Stiefel regularization). Let $x = (x_1, x_2, x_3)$ be a vector in $\mathbf{R}^3_0 = \mathbf{R}^3 - \{0\}$ and let $z = (z_1, z_2) \in \mathbf{C}^2_0 = \mathbf{C}^2 - \{0\} = \mathbf{R}^4 - \{0\}$. Define the 2×2 skew Hermitian matrices

$$\sigma_1 = \begin{pmatrix} 0 & 1 \\ 1 & 0 \end{pmatrix} \quad \sigma_2 = \begin{pmatrix} 0 & -i \\ i & 0 \end{pmatrix} \quad \sigma_3 = \begin{pmatrix} 1 & 0 \\ 0 & -1 \end{pmatrix}.$$

Let $\langle z, w\rangle = z_1\overline{w_1} + z_2\overline{w_2}$ be the standard Hermitian inner product on $\mathbf{C}^2$. Show that the mapping

$$\pi : \mathbf{C}^2_0 \to \mathbf{R}^3_0 : z \to \big(\langle z, \sigma_1(z)\rangle, \langle z, \sigma_2(z)\rangle, \langle z, \sigma_3(z)\rangle\big), \tag{119}$$

is the Hopf map.

(a). On $\mathbf{C}_0^2$ define an action

$$\varphi : U(1) \times \mathbf{C}_0^2 \to \mathbf{C}_0^2 : \left(e^{is}, (z_1, z_2)\right) \to (e^{is} z_1, e^{is} z_2).$$

Let $T^*\mathbf{C}_0^2 = (\mathbf{C}^2 - \{0\}) \times \mathbf{C}^2$. Lift φ to a $U(1)$-action

$$\Phi : S^1 \times T^*\mathbf{C}_0^2 \to T^*\mathbf{C}_0^2 : \left(e^{is}, z, w\right) \to \left(e^{is} z, e^{is} w\right).$$

Define a 1-form θ on $T^*\mathbf{C}_0^2$ by $\theta = -2i \operatorname{Im} \langle w, dz \rangle$. Show that $\Omega = -d\theta$ is a symplectic form on $T^*\mathbf{C}_0^2$ and that Φ is a Hamiltonian action with momentum map

$$\mathcal{J} : T^*\mathbf{C}_0^2 \to \mathbf{R} : (z, w) \to 2 \operatorname{Re} \langle w, z \rangle.$$

Let $\mathcal{J}_0 = \mathcal{J}^{-1}(0) - \{0\}$.

(b). The map π (119) lifts to the *Kustaanheimo-Stiefel* map

$$\mathcal{KS} : T^*\mathbf{C}_0^2 \to T^*\mathbf{R}_0^3 : (z, w) \to (x, y) =$$
$$\left(\langle\langle z, \sigma_j(z) \rangle\rangle, \frac{1}{\langle z, z \rangle} (\operatorname{Re} \langle w, \sigma_j(z) \rangle) \right), \qquad j = 1, 2, 3.$$

The following calculation shows that

$$(\mathcal{KS})^*(\vartheta | \mathcal{J}_0) = \theta | \mathcal{J}_0, \tag{120}$$

where $\vartheta = \langle y, dx \rangle$ is the canonical 1-form on $T^*\mathbf{R}^3$. For every $u, w, z \in \mathbf{C}^2$

$$\sum_{j=1}^{3} \operatorname{Re} \langle u, \sigma_j(z) \rangle \, \sigma_j(w) = 2 \langle w, z \rangle \, u - \langle u, z \rangle \, w. \tag{121}$$

Interchanging u with z in (121) and subtracting the result from (121) gives

$$\sum_{j=1}^{3} \operatorname{Re} \langle u, \sigma_j(z) \rangle \, \sigma_j(w) = \langle w, z \rangle \, u - \langle w, u \rangle \, z - i \operatorname{Im} \langle u, z \rangle \, w. \tag{122}$$

Taking the inner product of (122) with z and then adding the result to its complex conjugate gives

$$\sum_{j=1}^{3} \operatorname{Re} \langle u, \sigma_j(z) \rangle \operatorname{Re} \langle z, \sigma_j(w) \rangle = \operatorname{Re} \langle z, w \rangle \operatorname{Re} \langle u, z \rangle - \langle z, z \rangle \operatorname{Re} \langle u, w \rangle. \tag{123}$$

Replacing w in (123) with $-iw$ gives

$$\sum_{j=1}^{3} \operatorname{Re} \langle u, \sigma_j(z) \rangle \operatorname{Im} \langle z, \sigma_j(w) \rangle = \operatorname{Im} \langle z, w \rangle \operatorname{Re} \langle u, z \rangle - \langle z, z \rangle \operatorname{Im} \langle u, w \rangle. \tag{124}$$

Finally, replacing w by dz and u by w in (124) gives

$$\sum_{j=1}^{3} \operatorname{Re} \langle u, \sigma_j(z) \rangle \operatorname{Im} \langle z, \sigma_j(dz) \rangle = \operatorname{Im} \langle z, dz \rangle \operatorname{Re} \langle z, w \rangle - \langle z, z \rangle \operatorname{Im} \langle w, dz \rangle. \tag{125}$$

Consequently

$$(\mathcal{KS})^*\vartheta = 2i\left(\frac{\mathrm{Im}\langle z, dz\rangle \,\mathrm{Re}\,\langle z, w\rangle - \langle z, z\rangle \,\mathrm{Im}\,\langle w, dz\rangle}{\langle z, z\rangle}\right). \tag{126}$$

From (126) it follows that $(\mathcal{KS})^*(\vartheta|\mathcal{I}_0) = \theta|\mathcal{I}_0$.

(c). On $T^*\mathbf{R}_0^3$ with coordinates (x, y) and symplectic form $\omega = \sum_i dx_i \wedge dy_i$, consider the time rescaled Kepler Hamiltonian

$$\widetilde{K}(x, y) = \frac{1}{2k}\,\|x\|(\|y\|^2 + k^2)$$

whose $\frac{\mu}{k}$-level set corresponds to the $-k^2/2$-level set of the Kepler Hamiltonian (see exercise 11). Setting $u = w$ in (123) show that on $\mathcal{I}_0$ $\|(\mathcal{KS})^* y\|^2 = \frac{\langle w, w\rangle}{\langle z, z\rangle}$ and $\|(\mathcal{KS})^* x\|^2 = \langle z, z\rangle$. Therefore on $\mathcal{I}_0$ we obtain the regularized Hamiltonian

$$K = (\mathcal{KS})^*\widetilde{K} = \frac{1}{2k}\big(\langle w, w\rangle + k^2\langle z, z\rangle\big). \tag{127}$$

When $k = 1$ the regularized Hamiltonian is the harmonic oscillator Hamiltonian on $(T^*\mathbf{C}^2, \Omega)$ restricted to the open cone $\mathcal{I}_0$. Show that the regularized Hamiltonian K (127) is invariant under the $U(1)$-action Φ. Since the mapping $\mathcal{KS}$ is *not* a diffeomorphism, the harmonic oscillator vector field X_K is not equivalent to the Kepler vector field $X_{\widetilde{K}}$. Show that they are $\mathcal{KS}$-related on $\mathcal{I}_0$, that is, on $\mathcal{I}_0$ we have $T(\mathcal{KS}) \circ X_K = X_{\widetilde{K}} \circ (\mathcal{KS})$. Moreover, show that after dividing out the S^1 action Φ on $\mathcal{I}_0$ we obtain an equivalence of Hamiltonian systems. Show that the orbit space $\mathcal{I}_0/S^1$ is diffeomorphic to T^+S^3, the tangent bundle to S^3 less its zero section.

13. (Ligon-Schaaf map.) Consider the Ligon-Schaaf map

$$LS : \Sigma_- \subseteq T_0\mathbf{R}^3 \to T^+(S^3 - \mathrm{np}) \subseteq T\mathbf{R}^4 : (q, p) \to (x, y),$$

(see (68)) with $\varphi = \frac{1}{\nu}\langle q, p\rangle$. Show that its inverse is given by

$$\begin{aligned} q &= \mu^{-1}\langle y, y\rangle\big((\sin\varphi - \langle y, y\rangle^{-1/2} y_4)\widetilde{x} + \langle y, y\rangle^{-1/2}(x_4 - \cos\varphi)\widetilde{y}\big) \\ p &= \mu\langle y, y\rangle^{-1/2}\Big(\frac{\widetilde{x}\cos\varphi + \langle y, y\rangle^{-1/2}\widetilde{y}\sin\varphi}{1 - x_4\cos\varphi - \langle y, y\rangle^{-1/2} y_4 \sin\varphi}\Big), \end{aligned}$$

where $x = (\widetilde{x}, x_4)$, $y = (\widetilde{y}, y_4)$ and φ is a smooth solution of

$$\varphi - x_4\sin\varphi - y_4\langle y, y\rangle^{-1/2}\cos\varphi = 0.$$

Chapter III

The Euler Top

Mathematically, the motion of the Euler top is described by geodesics of a left invariant metric on the rotation group SO(3). Physically, the Euler top is a rigid body moving about its center of mass (which is fixed) without any forces acting on the body.

1 Facts about SO(3)

We begin by reviewing some basic facts about the group of rotations of $\mathbf{R}^3$.

1.1 The standard model

On $\mathbf{R}^3$ with Euclidean inner product $(\, ,\,)$, the orthogonal group O(3) is the group of linear maps which preserve the inner product, that is, $O \in \mathrm{O}(3)$ if and only if for every $x, y \in \mathbf{R}^3$, $(Ox, Oy) = (x, y)$. The *group of rotations* SO(3) of $(\mathbf{R}^3, (\, ,\,))$ is the identity component of O(3). Equivalently, $O \in \mathrm{SO}(3)$ if and only if $OO^t = I$ and $\det O = 1$. SO(3) is a connected compact Lie group with Lie algebra

$$T_e\,\mathrm{SO}(3) = \mathrm{so}(3) = \left\{ X \in \mathrm{gl}(3, \mathbf{R}) \middle| X + X^t = 0 \right\}.$$

$X \in \mathrm{so}(3)$ if and only if it is a 3×3 skew symmetric real matrix, that is, for every $x, y \in \mathbf{R}^3$, $(Xx, y) + (x, Xy) = 0$. The Lie algebra so(3) has a Lie bracket $[\, ,\,]$ defined by the relations

$$\begin{aligned} [E_1, E_2] &= E_1E_2 - E_2E_1 = E_3 \\ [E_2, E_3] &= E_1 \\ [E_3, E_1] &= E_2, \end{aligned}$$

where $\{E_1, E_2, E_3\}$ is the standard basis

$$E_1 = \begin{pmatrix} 0 & 0 & 0 \\ 0 & 0 & -1 \\ 0 & 1 & 0 \end{pmatrix}, \quad E_2 = \begin{pmatrix} 0 & 0 & 1 \\ 0 & 0 & 0 \\ -1 & 0 & 0 \end{pmatrix}, \quad E_3 = \begin{pmatrix} 0 & -1 & 0 \\ 1 & 0 & 0 \\ 0 & 0 & 0 \end{pmatrix}.$$

Rewritten, the bracket relations read

$$[E_i, E_j] = \sum_{k=1}^{3} \varepsilon_{ijk} E_k.$$

Here $\varepsilon_{ijk} = 0$, if i, j and k are not distinct. If i, j and k are distinct, ε_{ijk} is 1 if ijk is an even permutation of 123 and -1 otherwise.

On so(3) there is an inner product k : so(3) $\times$ so(3) $\rightarrow$ **R**, called the *Killing metric*. ▷ The Killing metric is defined by

$$k(X, Y) = -\tfrac{1}{2} \operatorname{tr} XY \tag{1}$$

and has the following properties.

1. k is positive definite.

2. For $O \in \mathrm{SO}(3)$ let $\mathrm{Ad}_O X = OXO^{-1} \in \mathrm{so}(3)$. k is Ad_O-invariant, that is, for $X, Y \in \mathrm{so}(3)$
$$k(\mathrm{Ad}_O X, \mathrm{Ad}_O Y) = k(X, Y).$$

3. For every $O \in \mathrm{SO}(3)$,
$$\mathrm{Ad}^t_{O^{-1}} = k^\sharp \circ \mathrm{Ad}_O \circ k^\flat. \tag{2}$$

4. For $X, Y, Z \in \mathrm{so}(3)$,
$$k([Z, X], Y) + k(X, [Z, Y]) = 0. \tag{3}$$

(1.1) **Proof**:

1). Let $X = \sum_{i=1}^{3} x_i E_i$. Then

$$|X|^2 = k(X, X) = -\tfrac{1}{2} \operatorname{tr} \begin{pmatrix} 0 & -x_3 & x_2 \\ x_3 & 0 & -x_1 \\ -x_2 & x_1 & 0 \end{pmatrix}^2 = x_1^2 + x_2^2 + x_3^2 \geq 0.$$

Equality holds if and only if $X = 0$.

2). For $O \in \mathrm{SO}(3)$, we have

$$k(\mathrm{Ad}_O X, \mathrm{Ad}_O Y) = -\tfrac{1}{2} \operatorname{tr} OXYO^{-1} = -\tfrac{1}{2} \operatorname{tr} XY = k(X, Y).$$

3). For every $X, Y \in \mathrm{so}(3)$ we have $k(\mathrm{Ad}_O X, Y) = k(X, \mathrm{Ad}_{O^{-1}} Y)$, because k is $\mathrm{Ad}_{O^{-1}}$-invariant. The preceding equation may be rewritten as

$$(k^\sharp(\mathrm{Ad}_O X))Y = k^\sharp(X)(\mathrm{Ad}_{O^{-1}} Y) = \left(\mathrm{Ad}^t_{O^{-1}} k^\sharp(X)\right)Y$$

for every $Y \in \mathrm{so}(3)$. Hence $(k^\sharp \circ \mathrm{Ad}_O)X = (\mathrm{Ad}^t_{O^{-1}} \circ k^\sharp)X$ for every $X \in \mathrm{so}(3)$, which proves (2).

4). Since

$$\left.\frac{d}{dt}\right|_{t=0} \mathrm{Ad}_{\exp tZ} W = \mathrm{ad}_Z W = [Z, W]$$

for every $Z, W \in \mathrm{so}(3)$, differentiating the equation

$$k(\mathrm{Ad}_{\exp tZ} X, \mathrm{Ad}_{\exp tZ} Y) = k(X, Y)$$

with respect to t and setting $t = 0$ gives

$$k(\mathrm{ad}_Z X, Y) + k(X, \mathrm{ad}_Z Y) = 0,$$

which is (3). □

Define the linear map

$$i : \mathrm{so}(3) \to \mathbf{R}^3 : X = \begin{pmatrix} 0 & -x_3 & x_2 \\ x_3 & 0 & -x_1 \\ -x_2 & x_1 & 0 \end{pmatrix} \to x = (x_1, x_2, x_3). \tag{4}$$

▷ The map i allows us to do calculations in $\mathbf{R}^3$ instead of in so(3). It has the following properties.

1. i is an *isometry* from $(\mathrm{so}(3), k)$ to $(\mathbf{R}^3, (\ ,\ ))$.
2. i is an isomorphism of the Lie algebra $(\mathrm{so}(3), [\ ,\ ])$ with the Lie algebra $(\mathbf{R}^3, \times)$, where $\times$ is the vector product on $\mathbf{R}^3$, (see exercise 3).
3. i *intertwines* the adjoint action of SO(3) on so(3) with the usual action of SO(3) on $\mathbf{R}^3$.

(1.2) **Proof:**
1). For $X = \sum_{i=1}^3 x_i E_i$ and $Y = \sum_{i=1}^3 y_i E_i$,

$$k(X, Y) = x_1 y_1 + x_2 y_2 + x_3 y_3 = (x, y) = (i(X), i(Y)).$$

2). For $X = \sum_{i=1}^3 x_i E_i$ and $Y = \sum_{i=1}^3 y_i E_i$,

$$\begin{aligned} i([X, Y]) &= i(\sum_{i,j,k=1}^{3} \varepsilon_{ijk}\, x_i y_j\, E_k) = \sum_{i,j,k=1}^{3} \varepsilon_{ijk}\, x_i y_j\, e_k \\ &= x \times y = i(X) \times i(Y) = Xy. \end{aligned}$$

3). For $O \in \mathrm{SO}(3)$ define the mapping $A_O : \mathbf{R}^3 \to \mathbf{R}^3 : x \to i(\mathrm{Ad}_O\, i^{-1}(x))$. Then $A_O \in \mathrm{O}(3)$, because

$$(A_O x, A_O y) = k(\mathrm{Ad}_O\, i^{-1}x, \mathrm{Ad}_O\, i^{-1}y) = k(i^{-1}x, i^{-1}y) = (x, y).$$

In fact $A_O \in \mathrm{SO}(3)$, because the map $\sigma : \mathrm{SO}(3) \to \mathrm{O}(3) : O \to A_O$ is continuous and sends the identity element into itself. Since σ is a group homomorphism and

$$T_e\sigma : T_e\,\mathrm{SO}(3) \to T_e O(3) = T_e\,\mathrm{SO}(3) : X \to i \circ \mathrm{ad}_X \circ i^{-1} = X,$$

it follows that σ is the inclusion map. Therefore $O = i \circ \mathrm{Ad}_O \circ i^{-1}$, that is, the map i intertwines the adjoint action of SO(3) on so(3) with the usual action of SO(3) on $\mathbf{R}^3$. □

▷ We now show that every element in so(3) has a normal form. More precisely, we show that for every element $X \in$ so(3) there is an $O \in$ SO(3) such that $\mathrm{Ad}_O X = rE_1$, where $r = |X|$.

(1.3) **Proof**: With respect to the *standard basis* $\{e_1, e_2, e_3\}$ of $\mathbf{R}^3$ the matrix of X is

$$\begin{pmatrix} 0 & -x_3 & x_2 \\ x_3 & 0 & -x_1 \\ -x_2 & x_1 & 0 \end{pmatrix}.$$

If $X = 0$ then X is already in normal form. So suppose that $X \neq 0$. Then $r = |X| > 0$ and the unit vector $x = \frac{1}{r}(x_1, x_2, x_3)$ is an eigenvector of X corresponding to the eigenvalue 0. Let Π be the plane orthogonal to the line spanned by the vector x. Π is invariant under X, for if $y \in \Pi$, then $Xy \in \Pi$ because $(Xy, x) = -(y, Xx) = 0$. On $(\Pi, (\ ,\ )|\Pi)$ the mapping $X|\Pi$ is skew symmetric and has characteristic polynomial $\lambda^2 + r^2$. Let $y \in \Pi$ be a vector of unit length. Then $\{y, \frac{1}{r}Xy\}$ is an *orthonormal basis* of Π because $(y, Xy) = -(Xy, y) = -(y, Xy)$ implies that $(y, \frac{1}{r}Xy) = 0$ and $(\frac{1}{r}Xy, \frac{1}{r}Xy) = (y, -\frac{1}{r^2}X^2y) = (y, y) = 1$. Thus the matrix $O^{-1} = \mathrm{col}(x, y, \frac{1}{r}Xy)$ of column vectors is orthogonal. In fact O^{-1} is in SO(3), because $\det \mathrm{col}(x, y, \frac{1}{r}Xy) = \frac{1}{r}(x \times y, Xy) = \frac{1}{r}(x \times y, x \times y) > 0$. The matrix of X with respect to the ordered orthonormal basis $\{x, y, \frac{1}{r}Xy\}$ is

$$Y = \begin{pmatrix} 0 & 0 & 0 \\ 0 & 0 & -r \\ 0 & r & 0 \end{pmatrix} = rE_1.$$

Clearly $r = |Y| = |X|$. □

▷ As a corollary of the above normal form, we find that for $X \in$ so(3) the linear map $\mathrm{ad}_X : \mathrm{so}(3) \to \mathrm{so}(3) : Y \to [X, Y]$ has eigenvalues $0, \pm ir$.

(1.4) **Proof**: For $O \in$ SO(3), it follows that $\mathrm{Ad}_O(\mathrm{ad}_X)(\mathrm{Ad}_O)^{-1} = \mathrm{ad}_{\mathrm{Ad}_O X}$, since

$$\mathrm{Ad}_O(\mathrm{ad}_X)(\mathrm{Ad}_O)^{-1}Y = \mathrm{Ad}_O\left[X, \mathrm{Ad}_{O^{-1}} Y\right] = [\mathrm{Ad}_O X, Y].$$

Therefore ad_X has the same eigenvalues as $\mathrm{ad}_{\mathrm{Ad}_O X}$. Choosing $O \in$ SO(3) so that $\mathrm{Ad}_O X = rE_1$, we see that ad_X has the same eigenvalues as $\mathrm{ad}_{rE_1} = r\,\mathrm{ad}_{E_1}$. But $\mathrm{ad}_{E_1} E_1 = 0$, $\mathrm{ad}_{E_1} E_2 = E_3$, $\mathrm{ad}_{E_1} E_3 = -E_2$. Therefore the matrix of ad_{rE_1} with respect to the basis $\{E_1, E_2, E_3\}$ is $\begin{pmatrix} 0 & 0 & 0 \\ 0 & 0 & -r \\ 0 & r & 0 \end{pmatrix}$. Consequently ad_X has eigenvalues $0, \pm ir$. □

1.2 The exponential map

We now derive some basic properties of the *exponential mapping*

$$\exp : \mathrm{so}(3) \to \mathrm{gl}(3, \mathbf{R}) : X \to \sum_{n=0}^{\infty} \frac{1}{n!} X^n.$$

▷ We start by showing that the image of exp is contained in SO(3).

(1.5) **Proof**: Since $X \in so(3)$ is skew symmetric,

$$(\exp X)^t = \sum_{n=0}^{\infty} \frac{1}{n!}(X^t)^n = \sum_{n=0}^{\infty} \frac{1}{n!}(-X)^n = \exp(-X) = (\exp X)^{-1}.$$

Therefore the image of exp is contained in $O(3)$. But exp is continuous and $\exp 0 = I$. Hence the image of exp is contained in SO(3). □

▷ Next we show that for a nonzero $X \in so(3)$

$$\exp X = I + \frac{\sin r}{r} X + \frac{1 - \cos r}{r^2} X^2, \tag{5}$$

where $r = |X|$. Since $\lim_{r\to 0} \frac{\sin r}{r} = 1$ and $\lim_{r \to 0} \frac{1-\cos r}{r^2} = \frac{1}{2}$, equation (5) is defined when $X = 0$ and gives $\exp 0 = I$.

(1.6) **Proof**: For $X \in so(3)$, a calculation shows that its characteristic polynomial is $\lambda^3 + r^2\lambda$. Therefore $X^3 + r^2 X = 0$, from which the formulæ

$$\begin{aligned} X^{2n+1} &= (-1)^n r^{2n} X, \\ X^{2n+2} &= (-1)^n r^{2n} X^2, \quad n \geq 0 \end{aligned}$$

follow by induction. Substituting these expressions into the power series for exp gives equation (5). □

When $|X| = 1$ we obtain the special case of (5):

$$(\exp sX)y = y + \sin s\,(x \times y) + (1 - \cos s)\,(x \times (x \times y)), \tag{6}$$

where $x = i(X)$. Because $(\exp sX)x = x$, equation (6) defines a *one parameter group* of rotations about the axis x. If $\{x, y\}$ are orthonormal vectors in $\mathbf{R}^3$, then (6) becomes

$$(\exp sX)y = y \cos s + (x \times y) \sin s, \tag{7}$$

since $x \times (x \times y) = -y$.

Because the function $r^2 : so(3) \to \mathbf{R} : X \to k(X, X)$ is differentiable, the functions $\sin r/r$ and $(1 - \cos r)/r^2$ are differentiable. From (5) it follows that exp is differentiable.

▷ Next we prove the following formula for the derivative of exp:

$$\exp(-X)(D\exp X) = \big(1 - \exp(-\operatorname{ad}_X)\big)/\operatorname{ad}_X. \tag{8}$$

The right hand side of (8) is to be thought of as a power series in ad_X.

(1.7) **Proof**: Consider the function

$$Z : \mathbf{R}^2 \to gl(3, \mathbf{R}) : (s, t) \to Z(s, t) = \exp(-sX - stY)\frac{\partial}{\partial t}\exp(sX + stY).$$

Then $Z(0,0)=0$, while

$$Z(1,0)=\exp(-X)\frac{d}{dt}\Big|_{t=0}\exp(X+tY) = \exp(-X)(D\exp X)Y.$$

Differentiating $Z(s,t)$ with respect to s gives

$$\begin{aligned}
\frac{\partial}{\partial s}Z(s,t) &= \frac{\partial}{\partial s}(\exp(-sX-stY))\frac{\partial}{\partial t}\exp(sX+stY) \\
&\qquad + \exp(-sX-stY)\frac{\partial^2}{\partial s\partial t}\exp(sX+stY) \\
&= -\exp(-sX-stY)(X+tY)\frac{\partial}{\partial t}\exp(sX+stY) \\
&\qquad + \exp(-sX-stY)\frac{\partial^2}{\partial t\partial s}\exp(sX+stY) \\
&= -\exp(-sX-stY)(X+tY)\frac{\partial}{\partial t}\exp(sX+stY) \\
&\qquad + \exp(-sX-stY)\frac{\partial}{\partial t}\big((X+tY)\exp(sX+stY)\big) \\
&= \exp(-sX-stY)Y\exp(sX+stY).
\end{aligned}$$

Consequently,

$$\begin{aligned}
Z(1,0) &= \int_0^1 \frac{d}{ds}Z(s,0)\,ds \\
&= \int_0^1 (\exp -sX)\,Y(\exp sX)\,ds = \int_0^1 \exp(-s\ \mathrm{ad}_X)Y\,ds.
\end{aligned}$$

The last equality follows because

1. For every $s\in\mathbf{R}$ the linear map

$$\gamma(s):\mathrm{so}(3)\to\mathrm{so}(3):Y\to(\exp -sX)Y(\exp sX)$$

   is invertible.

2. The map $\gamma:\mathbf{R}\to\mathrm{Gl}(\mathrm{so}(3),\mathbf{R}):s\to\gamma(s)$ is a one parameter subgroup.

3. $\gamma'(0)=-\mathrm{ad}_X$.

Therefore γ is the one parameter subgroup $s\to\exp(-s\ \mathrm{ad}_X)$. Expanding $\exp(-s\ \mathrm{ad}_X)$ in a power series in s and integrating with respect to s gives

$$\begin{aligned}
Z(1,0) &= \Big(1-\frac{1}{2!}\mathrm{ad}_X+\frac{1}{3!}(\mathrm{ad}_X)^2-\frac{1}{4!}(\mathrm{ad}_X)^3+\cdots\Big)Y \\
&= \left(\frac{1-\exp(-\mathrm{ad}_X)}{\mathrm{ad}_X}\right)Y.\ \square
\end{aligned}$$

1.3 The solid ball model

In this subsection we will show that the rotation group SO(3) is homeomorphic to a closed *solid ball* in $\mathbf{R}^3$ of radius π with antipodal points on its bounding 2-sphere identified. Such a homeomorphism can be constructed by sending the rotation O to the vector in the solid ball which is the axis of rotation of O normalized so that its length is the amount of rotation O makes in a right handed sense about its rotation axis. Our proof shows that this homeomorphism is defined by the exponential map.

Let $D^3_\pi = \{X \in \mathrm{so}(3) \,|\, k(X,X) < \pi^2\}$ be an open 3-ball in $(\mathrm{so}(3), k)$ of radius π and let $S^2_\pi = \{X \in \mathrm{so}(3) \,|\, k(X,X) = \pi^2\}$ be a 2-sphere of radius π, which is the boundary ∂D^3_π of D^3_π. The proof of the solid ball model of SO(3) goes in three steps.

1. The exponential map $\exp : \overline{D}^3_\pi \subseteq \mathrm{so}(3) \rightarrow \mathrm{SO}(3)$ is a diffeomorphism of D^3_π onto its image and is continuous on its closure $\overline{D}^3_\pi$.
2. The image of $\overline{D}^3_\pi$ under the exponential mapping is SO(3).
3. On $\partial \overline{D}^3_\pi = S^2_\pi$ the exponential map is two to one.

(1.8) **Proof**:
Step 1.
From (8) we see that the derivative of the exponential map is invertible if and only if the linear mapping $Z = (1 - \exp(-\mathrm{ad}_X))/\,\mathrm{ad}_X$ is invertible. If λ is an eigenvalue of ad_X, then

$$\begin{cases} (1 - e^{-\lambda})/\lambda, & \text{if } \lambda \neq 0 \\ 1, & \text{otherwise} \end{cases}$$

are eigenvalues of Z. If $X \neq 0$ then $\pm ir, 0$ are eigenvalues of ad_X. Consequently, $(1 - e^{-ir})/ir$, $(e^{ir} - 1)/ir$ and 1 are eigenvalues of Z. For $X \in D^3_\pi - \{0\}$, $0 < r < \pi$. Hence Z has no zero eigenvalues. If $X = 0$ then 1, 1, 1 are eigenvalues of Z. Thus $D\exp X$ is invertible for all $X \in D^3_\pi$. It is clearly continuous in D^3_π.

To show that exp is a diffeomorphism of D^3_π onto its image, we need only verify that $\exp | D^3_\pi$ is one to one. Toward this end, suppose that for some $X, Y \in D^3_\pi$ we have $\exp X = \exp Y$. Let $r = |X|$ and $s = |Y|$. Furthermore, suppose that r and s are greater than zero. The proofs of the other cases are easy and are omitted. Then $r, s \in (0, \pi)$. Using (5) we obtain

$$\frac{\sin r}{r} X + \frac{(1 - \cos r)}{r^2} X^2 = \frac{\sin s}{s} Y + \frac{(1 - \cos s)}{s^2} Y^2. \tag{9}$$

Subtracting (9) from its transpose and then dividing by 2 gives $\frac{\sin r}{r} X = \frac{\sin s}{s} Y$. Therefore

$$\sin^2 r = \frac{\sin^2 r}{r^2} k(X, X) = \frac{\sin^2 s}{s^2} k(Y, Y) = \sin^2 s.$$

Because $r, s \in (0, \pi)$, it follows that $\sin r = \sin s$ and therefore $\frac{1}{r} X = \frac{1}{s} Y$. Adding (9) to its transpose and using the preceding equation gives $\cos r = \cos s$, once we have noted that $X^2 \neq 0$. Therefore $r = s$, since $r, s \in (0, \pi)$. Consequently, $X = Y$.

From (5) we see that the exponential mapping is continuous on $\overline{D}^3_\pi$.

Step 2.
To show that the image of $\overline{D}^3_\pi$ under the exponential mapping is SO(3), we need the
▷ following normal form for rotations. Given $A \in SO(3)$, there is an $O \in SO(3)$ such that $\mathrm{Ad}_O A = \exp \theta E_1$ for some $\theta \in [0, \pi]$.

(1.9) **Proof**: Since $A \in SO(3)$, A has an eigenvalue $+1$. This follows from

$$\det(A - I) = \det\left((I - A)^t A\right) = \det(I - A) = -\det(A - I),$$

which implies $\det(A-I) = 0$. In other words, A leaves the line spanned by the eigenvector x corresponding to the eigenvalue 1 pointwise fixed. This line is called the *axis of rotation* of A. Normalize x so that its length is 1. Let Π be the plane in $\mathbf{R}^3$ orthogonal to x. Π is invariant under A, because if $y \in \Pi$ then $(Ay, x) = (Ay, Ax) = (y, x) = 0$, that is, $Ay \in \Pi$. Therefore $A|\Pi$ is an orthogonal linear map on $(\Pi, (\,,\,)|\Pi)$. Let y be a unit vector in Π. Then $\{y, x \times y\}$ is an orthonormal basis of Π, because $(y, x \times y) = \det \mathrm{col}(y, x, y) = 0$ and $(x \times y, x \times y) = (x, x)(y, y) - (x, y)^2 = 1$. With respect to the orthonormal basis $\{x, y, x \times y\}$ of $\mathbf{R}^3$ the matrix of A is $\begin{pmatrix} 1 & 0 \\ 0 & \widetilde{A} \end{pmatrix}$. Since $1 = \det A = \det \widetilde{A} = \det A|\Pi$, the map $A|\Pi$ lies in SO(2). Hence the matrix of $A|\Pi$ with respect to the orthonormal basis $\{y, x \times y\}$ is $\widetilde{A} = \begin{pmatrix} \cos\theta' & -\sin\theta' \\ \sin\theta' & \cos\theta' \end{pmatrix}$ for some $\theta' \in [0, 2\pi]$. If $\theta' \in [0, \pi]$, we are done. Otherwise, $\sin\theta' < 0$. Now use the orthonormal basis $\{-x, y, y \times x\}$. With respect to the orthonormal basis $\{y, y \times x\}$ the matrix of $A|\Pi$ is $\begin{pmatrix} \cos\theta'' & -\sin\theta'' \\ \sin\theta'' & \cos\theta'' \end{pmatrix}$ where $\theta'' = 2\pi - \theta' \in [0, \pi]$. Because either $\det \mathrm{col}(x, y, x \times y) = (x \times y, x \times y) > 0$, or $\det \mathrm{col}(-x, y, y \times x) = (-x \times y, y \times x) > 0$, the matrix O^{-1} formed by taking the vectors $\{x, y, x \times y\}$ or $\{-x, y, y \times x\}$ for its columns lies in SO(3). Moreover,

$$\mathrm{Ad}_O A = OAO^{-1} = \begin{pmatrix} 1 & 0 & 0 \\ 0 & \cos\theta & -\sin\theta \\ 0 & \sin\theta & \cos\theta \end{pmatrix} = \exp\theta E_1,$$

where $\theta = \theta'$ or θ''. □

The second step is proved by noting that $A = \mathrm{Ad}_{O^{-1}}(\exp\theta E_1) = \exp(\theta\, \mathrm{Ad}_{O^{-1}} E_1)$ and

$$k(\theta\, \mathrm{Ad}_{O^{-1}} E_1, \theta\, \mathrm{Ad}_{O^{-1}} E_1) = \theta^2 k(E_1, E_1) = \theta^2 \le \pi^2.$$

Step 3.
Suppose that $X, Y \in S^2_\pi$ and $\exp X = \exp Y$. Then using (5) we obtain

$$1 + \frac{2}{\pi^2} X^2 = \exp X = \exp Y = 1 + \frac{2}{\pi^2} Y^2$$

which implies that $X = \pm Y$. Conversely, if $X = \pm Y$ then $\exp X = \exp Y$. Therefore on S^2_π the exponential mapping is two to one with $\exp X = \exp(-X)$. In other words, exp maps antipodal points on S^2_π to the same element of SO(3).

This establishes the solid ball model of SO(3). □

1.4 The sphere bundle model

In this subsection we describe the *sphere bundle model* of the rotation group.

Let

$$T_1S^2 = \left\{ (x, y) \in T\mathbf{R}^3 = \mathbf{R}^3 \times \mathbf{R}^3 \middle| (x, x) = 1, \ \ (x, y) = 0 \ \& \ (y, y) = 1 \right\}$$

be the unit tangent sphere bundle to the unit 2-sphere S^2. In other words, T_1S^2 is the set of all *ordered pairs of orthonormal vectors* in $\mathbf{R}^3$. T_1S^2 is diffeomorphic to the rotation group SO(3) via the smooth map

$$\varphi : T_1S^2 \subseteq T\mathbf{R}^3 \to \mathrm{SO}(3) \subseteq \mathbf{R}^9 : (x, y) \to \mathrm{col}(x, y, x \times y), \tag{10}$$

whose smooth inverse is the restriction of the projection

$$\widehat{\pi} : \mathbf{R}^9 \to T\mathbf{R}^3 : A = \mathrm{col}(a_1, a_2, a_3) \to (a_1, a_2) \tag{11}$$

to SO(3).

T_1S^2 can be made into a Lie group by pushing forward the Lie group structure on SO(3) via the mapping $\widehat{\pi}$ (11). In more detail, define a multiplication $\cdot$ on T_1S^2 by

$$\begin{aligned} (x, y) \cdot (z, w) &= \widehat{\pi}\big(\mathrm{col}(x, y, x \times y) \cdot \mathrm{col}(z, w, z \times w) \big) \\ &= \big(z_1x + z_2y + z_3(x \times y), w_1x + w_2y + w_3(x \times y) \big). \end{aligned} \tag{12}$$

Observe that $e = (e_1, e_2)$ is the identity element of $(T_1S^2, \cdot)$.

Since

$$T_{(x,y)}(T_1S^2) = \left\{ (u, v) \in T\mathbf{R}^3 \middle| (x, u) = 0, \ \ (u, y) + (x, v) = 0 \ \& \ (y, v) = 0 \right\},$$

we find that

$$T_e(T_1S^2) = \left\{ \mathbf{u} = \big((0, u_3, -u_2), (-u_3, 0, u_1) \big) \in T\mathbf{R}^3 \middle| (u_1, u_2, u_3) \in \mathbf{R}^3 \right\}. \tag{13}$$

In other words,

$$T_e(T_1S^2) = \left\{ \mathbf{u} = \widehat{\pi}(U) \in T\mathbf{R}^3 \middle| U = \begin{pmatrix} 0 & -u_3 & u_2 \\ u_3 & 0 & -u_1 \\ -u_2 & u_1 & 0 \end{pmatrix} \in \mathrm{so}(3) \right\},$$

which has the standard basis $\{\boldsymbol{\varepsilon}_i = \widehat{\pi}(E_i)\}$. Define a Lie bracket $[\ ,\ ]$ on $T_e(T_1S^2)$ by

$$[\mathbf{u}, \mathbf{v}] = \widehat{\pi}([U, V]), \tag{14}$$

for $\mathbf{u}, \mathbf{v} \in T_e(T_1S^2)$. In terms of the standard basis $\{\boldsymbol{\varepsilon}_i\}$ we obtain the bracket relations $[\boldsymbol{\varepsilon}_i, \boldsymbol{\varepsilon}_j] = \sum_k \varepsilon_{ijk}\boldsymbol{\varepsilon}_k$. The following argument shows that $(T_e(T_1S^2), [\ ,\ ])$ is the Lie algebra of the Lie group $(T_1S^2, \cdot)$.

(1.10) **Proof**: For every $(x, y) \in T_1S^2$ define left translation by (x, y) by

$$L_{(x,y)} : T_1S^2 \to T_1S^2 : (z, w) \to (x, y) \cdot (z, w) \tag{15}$$

and right translation by (x, y) by

$$R_{(x,y)} : T_1S^2 \to T_1S^2 : (z, w) \to (z, w) \cdot (x, y).$$

Hence we may define the diffeomorphism

$$\mathrm{Int}_{(x,y)} : T_1S^2 \to T_1S^2 : (z, w) \to L_{(x,y)} \circ R_{(x,y)^{-1}}(z, w),$$

which induces the group homomorphism

$$\mathrm{Int} : T_1S^2 \to \mathrm{Diff}(T_1S^2) : (x, y) \to \mathrm{Int}_{(x,y)}.$$

Here $(\mathrm{Diff}(T_1S^2), \circ)$ is the group of diffeomorphisms of T_1S^2 with composition $\circ$ as multiplication. Differentiating $\mathrm{Int}_{(x,y)}$ at e, we obtain the linear map

$$\mathrm{Ad}_{(x,y)} : T_e(T_1S^2) \to T_e(T_1S^2) : \mathbf{v} \to T_e\mathrm{Int}_{(x,y)}\mathbf{v} = \widehat{\pi}(\mathrm{Ad}_{\mathrm{col}(x,y,x\times y)}V),$$

which gives rise to the group homomorphism

$$\mathrm{Ad} : T_1S^2 \to \mathrm{Gl}(T_e(T_1S^2), \mathbf{R}) : (x, y) \to \mathrm{Ad}_{(x,y)}.$$

Differentiating Ad at e gives the linear map

$$\mathrm{ad} : T_e(T_1S^2) \to \mathrm{gl}(T_e(T_1S^2), \mathbf{R}) : \mathbf{u} \to \mathrm{ad}_{\mathbf{u}},$$

where

$$\begin{aligned} \mathrm{ad}_{\mathbf{u}}\mathbf{v} &= T_e(\mathrm{Ad}_{(x,y)}\mathbf{v})\mathbf{u} = \widehat{\pi}\big(T_I(\mathrm{Ad}_{\mathrm{col}(x,y,x\times y)}V)U\big) \\ &= \widehat{\pi}(\mathrm{ad}_U V) = \widehat{\pi}([U, V]). \end{aligned}$$

The Lie bracket on $T_e(T_1S^2)$ is defined by $[\mathbf{u}, \mathbf{v}] = \mathrm{ad}_{\mathbf{u}}\mathbf{v}$, which agrees with (14). □

If we identify the $T_e(T_1S^2)$ with $\mathbf{R}^3$ using the mapping

$$i : T_e(T_1S^2) \to \mathbf{R}^3 : \mathbf{u} = \left(\begin{pmatrix} 0 \\ u_3 \\ -u_2 \end{pmatrix}, \begin{pmatrix} -u_3 \\ 0 \\ u_1 \end{pmatrix} \right) \to u = (u_1, u_2, u_3),$$

it follows that i is an isomorphism of the Lie algebra $(T_e(T_1S^2), [\,,\,])$ with the Lie algebra $(\mathbf{R}^3, \times)$, since $i(\boldsymbol{\varepsilon}_i) = e_i$. In other words,

$$i([\mathbf{u}, \mathbf{v}]) = i(\mathbf{u}) \times i(\mathbf{v})$$

for every $\mathbf{u}, \mathbf{v} \in T_e(T_1S^2)$.

Define the Killing metric k on $T_e(T_1S^2)$ as follows. For $\mathbf{u}, \mathbf{v} \in T_e(T_1S^2)$ let

$$k(\mathbf{u}, \mathbf{v}) = (i(\mathbf{u}), i(\mathbf{v})), \tag{16}$$

where (,) is the Euclidean inner product on $\mathbf{R}^3$. The Killing metric k is infinitesimally Ad-invariant because for $\mathbf{t}, \mathbf{u}, \mathbf{v} \in T_e(T_1 S^2)$

$$k(\mathbf{t}, [\mathbf{u}, \mathbf{v}]) = (i(\mathbf{t}), i(\mathbf{u}) \times i(\mathbf{v})) = (i(\mathbf{t}) \times i(\mathbf{u}), i(\mathbf{v})) = k([\mathbf{t}, \mathbf{u}], \mathbf{v}).$$

Since $T_1 S^2$ is connected, it follows that k is Ad-invariant, that is,

$$k(\mathrm{Ad}_{(x,y)}\mathbf{u}, \mathrm{Ad}_{(x,y)}\mathbf{v}) = k(\mathbf{u}, \mathbf{v}),$$

for every $(x, y) \in T_1 S^2$.

▷ We now turn to discussing the geometry of the sphere bundle model. One of the advantages of the sphere bundle model of SO(3) is that $T_1 S^2$ is the total space of an S^1-principal bundle over S^2 with bundle projection

$$\tau : T_1 S^2 \subseteq T\mathbf{R}^3 \to S^2 \subseteq \mathbf{R}^3 : (x, y) \to x. \tag{17}$$

(1.11) **Proof**: A calculation shows that for every $(v, w) \in T_{(x,y)}(T_1 S^2)$ we have $T_{(x,y)}\tau(v, w) = v$. Since $v \in T_x S^2$, it follows that τ is a submersion. In addition, τ is a surjective proper map. Therefore by the Ehresmann theorems of appendix C section 2, τ is a locally trivial bundle with fiber S^1.

To show that τ is a principal bundle (see appendix B section 2) we argue as follows. Consider the $S^1 = \mathbf{R}/2\pi\mathbf{Z}$ action

$$\begin{aligned}\Psi &: S^1 \times T_1 S^2 \to T_1 S^2 : \\ &\big(t, (x, y)\big) \to \big(x, \exp(-tX)y\big) = (x, y\cos t - (x \times y)\sin t),\end{aligned} \tag{18}$$

where X is the skew symmetric matrix $i^{-1}(x)$ and i is the map given by (4). The action Ψ has the following properties.

1. Ψ_t preserves the fibers of the bundle τ. In other words, Ψ_t maps the unit circle in $T_x S^2$ into itself.

2. The action Ψ is proper, since S^1 is compact.

3. The action Ψ is free, because if $(x, y) = \Psi_t(x, y) = (x, y\cos t - (x \times y)\sin t)$, then $y = y\cos t - (x \times y)\sin t$. Taking the inner product of both sides of this last equation with y and using the fact that $(y, y) = 1$, yields $1 = \cos t$, that is, $t = 2\pi n$ for some $n \in \mathbf{Z}$. Hence $t = e \in S^1$.

Therefore $T_1 S^2$ is the total space of an S^1 principal bundle over the smooth orbit space $V = T_1 S^2/S^1$ with bundle projection $\lambda : T_1 S^2 \to V$, (see appendix B ((2.9))). Since τ is invariant under Ψ, it induces a smooth map $\sigma : V \to S^2$ which makes the diagram 1.1 commute. The map σ is surjective because τ is. Also σ is injective, because the fiber of τ is a unique S^1 orbit of Ψ by property 1. Therefore σ is a homeomorphism, since V is a compact Hausdorff space. To verify that σ is a diffeomorphism, it suffices to show that for every $v \in V$, the tangent map $T_v\sigma : T_v V \to T_{\sigma(v)} S^2$ is injective, because $\dim T_v V =$

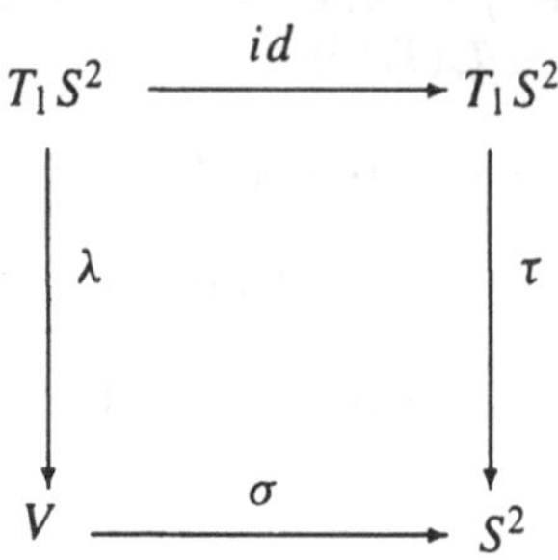

Diagram 1.1

dim $T_{\sigma(v)}S^2$. Towards this goal suppose that $0 = T_v\sigma(w_v)$ for some $w_v \in T_vV$. Since λ is a surjective submersion, there is a $w_{(x,y)} \in T_{\sigma(v)}(T_1S^2)$ such that $(T_{(x,y)}\lambda)w_{(x,y)} = w_v$ and $\lambda(x, y) = v$. Therefore $0 = (T_v\sigma \circ T_{(x,y)}\lambda)w_{(x,y)} = (T_{(x,y)}\tau)w_{(x,y)}$, since $\tau = \sigma \circ \lambda$. In other words

$$
\begin{aligned}
w_{(x,y)} \in \ker T_{(x,y)}\tau &= T_{(x,y)}\tau^{-1}(x), && \text{since } \tau \text{ is a smooth bundle} \\
&= T_{(x,y)}\lambda^{-1}(v), && \text{since } \tau^{-1}(x) \text{ and } \lambda^{-1}(v) \text{ are the same } S^1 \text{ orbit} \\
&= \ker T_{(x,y)}\lambda, && \text{since } \lambda \text{ is a smooth bundle.}
\end{aligned}
$$

Therefore $w_v = 0$. Thus τ and λ are isomorphic bundles. In fact they are isomorphic S^1 principal bundles, since the horizontal arrows in diagram 1.1 are diffeomorphisms and the map id interwines the S^1 action Ψ on T_1S^2 with itself. □

Now consider the diagram

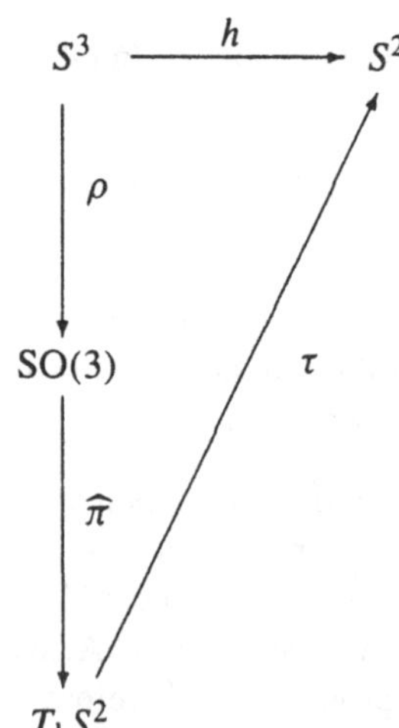

Diagram 1.2

Here h is the Hopf fibration (see chapter 1 ((4.2))), $\widehat{\pi}$ is the map (11) and $\rho : S^3 \rightarrow \mathrm{SO}(3)$ is a two to one covering map, (see (19) below).

Claim: The bundle projection $\tau : T_1 S^2 \rightarrow S^2$ is double covered by the Hopf fibration h.

(1.12) **Proof**: We start by defining the covering map $\rho : S^3 \rightarrow \mathrm{SO}(3)$. This involves an extensive excursion into *quaternions*.

As a real vector space the set of quaternions $\mathbf{H}$ has a basis $\{1, i, j, k\}$. Thus every $q \in \mathbf{H}$ can be written uniquely as $q = q_0\, 1 + q_1\, i + q_2\, j + q_3\, k$, where $(q_0, q_1, q_2, q_3) \in \mathbf{R}^4$. We make $\mathbf{H}$ into an associative noncommutative algebra by defining a multiplication $\cdot$ on its basis elements as follows.

1. 1 commutes with i, j and k.
2. $i^2 = j^2 = k^2 = -1$.
3. $i \cdot j = -j \cdot i$, $j \cdot k = -k \cdot j$ and $k \cdot i = -i \cdot k$.
4. $i \cdot j = k$, $j \cdot k = i$ and $k \cdot i = j$.

Using the above properties and the distributive law, it is easy to see that multiplication is defined for any two quaternions. For $q \in \mathbf{H}$ define the conjugate $\overline{q}$ of q by

$$\overline{q} = q_0\, 1 - q_1\, i - q_2\, j - q_3\, k.$$

It is straightforward to check that $\overline{q \cdot p} = \overline{p} \cdot \overline{q}$. The 3-sphere $S^3 \subseteq \mathbf{H}$ is the set of all quaternions q such that $\overline{q} \cdot q = 1$. It is easy to check that S^3 is a Lie group under quaternionic multiplication with identity element $e = 1$, (see exercise 8). Identify $\mathbf{R}^3$ with the vector subspace of $\mathbf{H}$ spanned by the vectors i, j and k. For every $q \in S^3 \subseteq \mathbf{H}$ consider the real linear map

$$L_q : \mathbf{R}^3 \subseteq \mathbf{H} \rightarrow \mathbf{R}^3 \subseteq \mathbf{H} : x \rightarrow q \cdot x \cdot \overline{q}.$$

▷ L_q is orthogonal.

(1.13) **Proof**: First observe that for $x = x_1\, i + x_2\, j + x_3\, k \in \mathbf{R}^3 \subseteq \mathbf{H}$ we have

$$x \cdot \overline{x} = x_1^2 + x_2^2 + x_3^2 = (x, x),$$

where (,) is the Euclidean inner product on $\mathbf{R}^3$. Consequently, for $x, y \in \mathbf{R}^3 \subseteq \mathbf{H}$

$$4(x, y) = (x + y) \cdot (\overline{x + y}) - (x - y) \cdot (\overline{x - y}).$$

Since L_q is a linear map, it suffices to show that it is length preserving. We verify this as follows:

$$\begin{aligned} L_q x \cdot \overline{L_q x} &= q \cdot x \cdot \overline{x} \cdot \overline{q}, \quad \text{since } \overline{q} \cdot q = 1. \\ &= q \cdot \overline{q} \cdot x \cdot \overline{x} = x \cdot \overline{x}. \end{aligned}$$ □

Thus we have a map

$$\rho : S^3 \rightarrow O(3) : q \rightarrow L_q. \tag{19}$$

ρ is a group homomorphism, since

$$L_{p\cdot q}x = p\cdot q\cdot x\cdot \overline{p\cdot q} = p\cdot(q\cdot x\cdot \overline{q})\cdot \overline{p} = (L_p \circ L_q)x.$$

Because $L_1 = id_{\mathbf{R}^3}$, ρ is continuous, and S^3 is connected, the image of ρ lies in ▷ SO(3). The following argument shows that ρ is a submersion and hence its image is *all* of SO(3).

(1.14) **Proof**: By its very definition, ρ is smooth. Because ρ is a group homomorphism, it suffices to show that $T_e\rho : T_eS^3 \to T_I\,\mathrm{SO}(3) = \mathrm{so}(3)$ is bijective. Since $(1,i) = (1,j) = (1,k) = 0$, we can identify $T_eS^3 = \{y \in \mathbf{H}|\,(y,1) = 0\}$ with $\mathbf{R}^3 \subseteq \mathbf{H}$. Using this identification we assert that

$$T_e\rho : \mathbf{R}^3 \to \mathrm{so}(3) : x = x_1\,i + x_2\,j + x_3\,k \to -2\begin{pmatrix} 0 & -x_3 & x_2 \\ x_3 & 0 & -x_1 \\ -x_2 & x_1 & 0 \end{pmatrix}. \tag{20}$$

To see this, observe that the curves $t \to e^{ti}$, $t \to e^{tj}$, and $t \to e^{tk}$ (which lie on S^3 and pass through 1) represent tangent vectors to S^3 at e in the direction i, j and k respectively, since $\frac{d}{dt}\Big|_{t=0} e^{ti} = i$, $\frac{d}{dt}\Big|_{t=0} e^{tj} = j$ and $\frac{d}{dt}\Big|_{t=0} e^{tk} = k$. Therefore $T_e\rho(i)$ is the linear map on $\mathbf{R}^3$ given by

$$x \to \frac{d}{dt}\Big|_{t=0} L_{e^{ti}}x = \frac{d}{dt}\Big|_{t=0} e^{ti}\cdot x\cdot e^{-ti} = i\cdot x - x\cdot i = -2(x_3 j - x_2 k).$$

Similarly, $T_e\rho(j)$ is the linear map $x \to -2(-x_3 i + x_1 k)$ and $T_e\rho(k)$ the linear map $x \to -2(x_2 i - x_1 j)$. This proves (20). Consequently the map $T_e\rho$ is bijective. Therefore ρ is a submersion. □

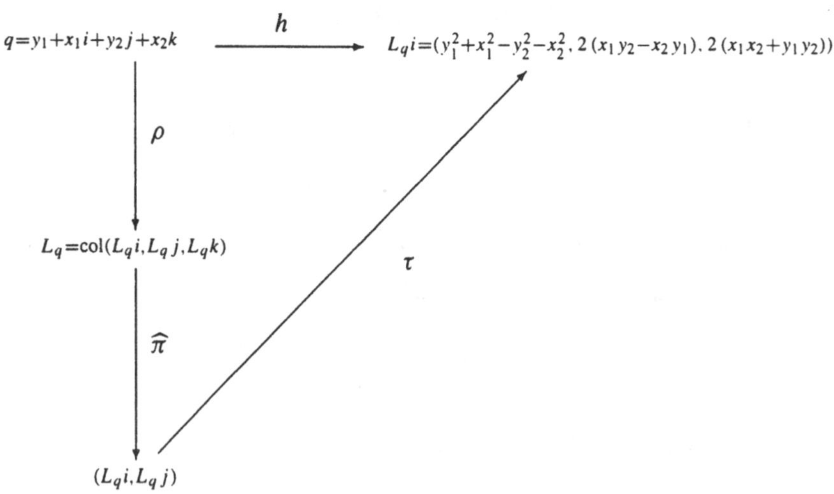

Diagram 1.3

▷ Next we show that the kernel of ρ is $\mathbf{Z}_2$.

1.15) **Proof**: Suppose that $L_q = id_{\mathbf{R}^3}$. Then for every $x \in \mathbf{R}^3 \subseteq \mathbf{H}$, we have $\overline{q} \cdot x \cdot q = x$, that is, $x \cdot q = q \cdot x$. Letting x be succesively i, j and k in the preceding equation, we see that $q = Q\,1$ where $Q \in \mathbf{R}$. Since $q \cdot \overline{q} = 1$, we have $Q^2 = 1$. Therefore $q = \pm 1$. Hence $\ker \rho = \mathbf{Z}_2$. □

Thus ρ is a two to one covering map. In other words, the twofold covering group of SO(3) is S^3. A geometric way of saying this is that if we act on S^3 by the fixed point free proper $\mathbf{Z}_2$ action generated by the map which sends the point q into $-q$, then the smooth orbit space (see appendix B ((2.9))) is SO(3). Thus SO(3) is diffeomorphic to the space formed by identifying antipodal points on S^3, that is, real projective three space $\mathbf{RP}^3$.

We now look at diagram 1.3. The top horizontal map h is the Hopf fibration, (see chapter 1 ((4.2))). The first vertical map ρ is the two to one submersion (19), while the second vertical map is the diffeomorphism $\widehat{\pi}$ (11). The diagonal map τ is the bundle projection (17). From diagram 1.3 it follows that diagram 1.2 commutes. This proves ((1.12)). □

▷ Next we determine the isomorphism type of the bundle τ.

1.16) **Proof**: For $y \in S^2$, we know that the fiber $h^{-1}(y)$ of the Hopf bundle is a great circle on S^3, (see chapter 1 ((4.2))). Let $\widetilde{\rho} = \widehat{\pi} \circ \rho$. From diagram 1.2 it follows that $\widetilde{\rho}(h^{-1}(y)) = \tau^{-1}(y)$. Moreover, the fiber of $\widetilde{\rho}$ is two antipodal points on S^3. Therefore, on fibers $\widetilde{\rho}$ is a two to one covering map. Let S^1_E be the equator of S^2 and let $\chi_h : S^1_E \rightarrow S^1 \subseteq S^3$ be the classifying map of the Hopf fibration, (see chapter 1 (38)). Then $\widetilde{\rho} \circ \chi_h$ is a classifying map $\chi_\tau : S^1_E \rightarrow S^1 \subseteq T_1 S^2$ of the bundle τ. We compute the degree of χ_τ as follows

$$\deg \chi_\tau = \deg \widetilde{\rho} \cdot \deg \chi_h = 2 \cdot 1 = 2.$$

This determines the homotopy class of χ_τ and hence the isomorphism type of τ. □

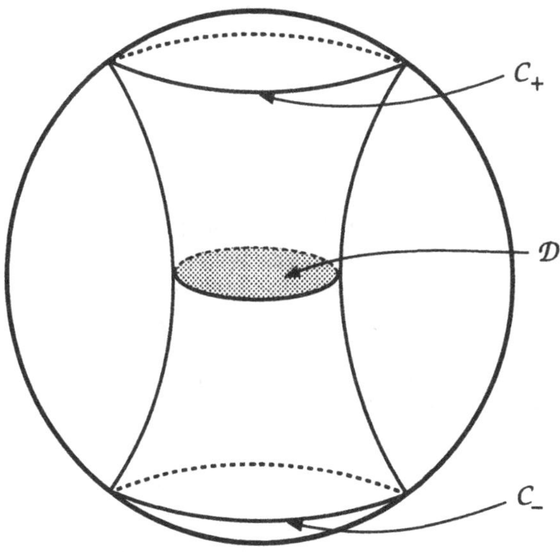

Figure 1.1.

We now give a way to visualize the bundle τ, see figure 1.1. Think of T_1S^2 as the closed solid ball $\overline{D}^3$ in $\mathbf{R}^3$: $x_1^2+x_2^2+x_3^2 \le 1$ with points on its boundary S^2: $x_1^2+x_2^2+x_3^2=1$ identified by the antipodal mapping $\mathbf{R}^3 \to \mathbf{R}^3 : x \to -x$. Look at the piece of the hyperboloid H in $\overline{D}^3$ defined by

$$x_1^2+x_2^2-x_3^2=\tfrac{1}{2} \qquad \text{and} \qquad x_1^2+x_2^2+x_3^2 \le 1.$$

This piece intersects S^2 in two disjoint circles

$$C_\pm : x_1^2+x_2^2=\tfrac{3}{4} \quad \& \quad x_3=\pm\tfrac{1}{2}.$$

Orienting S^2 positively induces a positive orientation on C_+ and C_-. Under the antipodal map, the closed 2-disc $\overline{D}_+$ bounded by the circle C_+ is identified with the closed 2-disc $\overline{D}_-$ bounded by the circle C_-. Because the oriented circles $C_\pm$ agree after identification by the antipodal map, H is a 2-torus T^2 in the solid ball model and *not* a Klein bottle. In fact, H bounds a *solid torus* ST_1 formed by identifying the oriented end 2-discs $\overline{D}_+$ and $\overline{D}_-$ of the solid cylinder $x_1^2+x_2^2-x_3^2 \le \frac{1}{2}$ and $x_1^2+x_2^2+x_3^2 \le 1$.

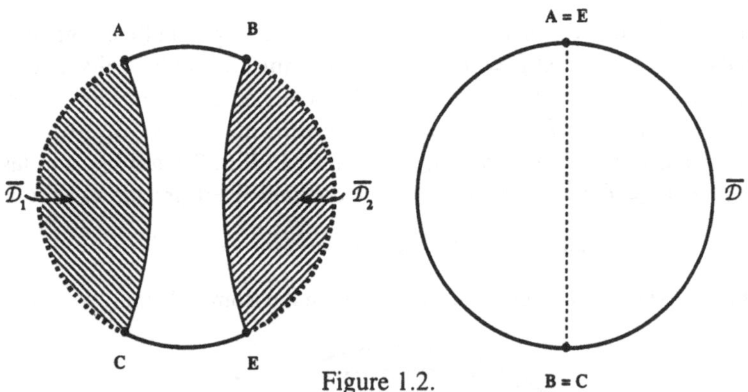

Figure 1.2.

Claim: $\overline{T_1S^2 - ST_1}$ is a solid torus ST_2.

(1.17) **Proof**: To see this, consider the slice $\{x_1=0\} \cap (\overline{T_1S^2 - ST_1})$ given by the shaded region of the left 2-disc in figure 1.2. Before identification this slice is two disjoint closed 2-discs $\overline{D}_1$ and $\overline{D}_2$. Under the antipodal map, the boundary piece AC of $\overline{D}_1$ is identified with the boundary piece EB of $\overline{D}_2$, thus forming a closed 2-disc $\overline{D}$, (see the right 2-disc in figure 1.2). The same argument holds for every slice of the solid ball model through the x_3-axis. Therefore $\overline{T_1S^2 - ST_1}$ is diffeomorphic to $\overline{D} \times S^1$, which is a solid torus ST_2. □

We have decomposed T_1S^2 into the *union of two solid tori* ST_1 and ST_2 which are identified along their common boundary T^2 by a map $\psi : T^2 \to T^2$. We now discuss the geometry of the *gluing map* ψ. Let $\overline{\mathfrak{D}}$ be the closed 2-disc defined by $x_1^2+x_2^2 \le \frac{1}{2}$ & $x_3=0$. Let S_1^m and S_2^m be the boundary $\partial\overline{\mathfrak{D}}$ of $\overline{\mathfrak{D}}$ and ∂D the boundary of $\overline{D}$, respectively, (see figure 1.3). Clearly $\partial\overline{\mathfrak{D}}$ and $\partial\overline{D}$ are closed curves on the 2-tori $\partial(ST_1)$ and $\partial(ST_2)$, respectively. Because $\overline{\mathfrak{D}}$ and $\overline{D}$ are contractible in ST_1 and ST_2, respectively, the curves S_1^m and S_2^m are *meridians* on $\partial(ST_1)$ and $\partial(ST_2)$, respectively. Consider the dashed curve S_1^ℓ on $\partial(ST_1)$

drawn in figure 1.3. It is the same as the curve S_2^ℓ on $\partial(ST_2)$. The curve S_1^ℓ is closed, since it joins the points P and P' which are antipodal on S^2 and hence are identified. S_1^ℓ intersects $\overline{D}$ once at P. Similarly, the curve S_2^ℓ intersects $\overline{\mathcal{D}}$ once at Q. Therefore, S_1^ℓ and S_2^ℓ are *not* contractible in ST_1 and ST_2, respectively. Hence S_1^ℓ and S_2^ℓ are *longitudes* on

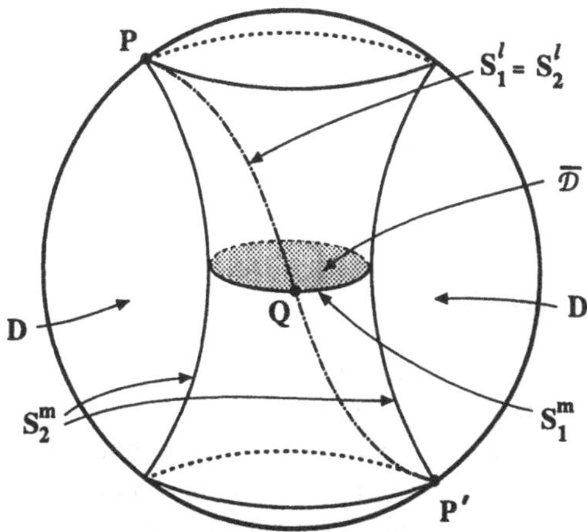

Figure 1.3.

$\partial(ST_1)$ and $\partial(ST_2)$, respectively. Thinking of S_2^m as a curve on $\partial(ST_1)$, we see that S_2^m intersects $\overline{\mathcal{D}}$ *twice* with intersection number of the *same* sign. Consequently, S_2^m is homotopic on $\partial(ST_1)$ to *two* times S_1^ℓ. Thus the gluing map

$$\psi : \partial(ST_2) = T^2 \to \partial(ST_1) = T^2$$

takes the longitude S_2^ℓ onto the longitude S_1^ℓ by the identity map and the meridian S_2^m onto a curve on $\partial(ST_1)$ which is homotopic to two times the longitude S_1^ℓ.

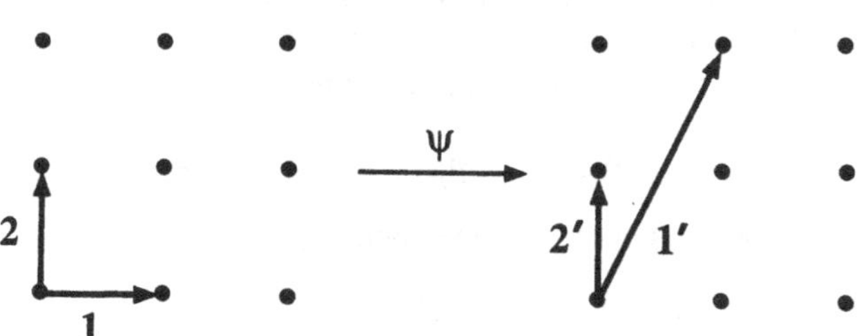

Figure 1.4.

We now give a map, which up to homotopy, is the gluing map ψ. Identify T^2 with $\mathbf{R}^2/\mathbf{Z}^2$

and consider the linear map

$$\widetilde{\Psi} : \mathbf{R}^2 \to \mathbf{R}^2 : \begin{pmatrix} \xi_1 \\ \xi_2 \end{pmatrix} \to A \begin{pmatrix} \xi_1 \\ \xi_2 \end{pmatrix} = \begin{pmatrix} 1 & 0 \\ 2 & 1 \end{pmatrix} \begin{pmatrix} \xi_1 \\ \xi_2 \end{pmatrix}.$$

Geometrically we visualize $\widetilde{\Psi}$ as in figure 1.4. Since $A \in \mathrm{Sl}(2, \mathbf{Z})$, the map $\widetilde{\Psi}$ preserves $\mathbf{Z}^2$ and hence induces a map $\Psi : T^2 \to T^2$. On T^2 the curves $\{\xi_1 = 0 \mod 1\}$ and $\{\xi_2 = 0 \mod 1\}$ are a longitude and a meridian, respectively. Under Ψ, the longitude $\{\xi_1 = 0 \mod 1\}$ is mapped bijectively onto itself; whereas the meridian $\{\xi_2 = 0 \mod 1\}$ is taken onto the curve $\{(\xi_1, 2\,\xi_1) \in T^2 | \, \xi_1 \mod 1\}$, which is homotopic to two times the longitude $\{\xi_1 = 0 \mod 1\}$. Thus homotopically Ψ has the same properties as the gluing map ψ.

2 Left invariant geodesics

In this section we present two models for the Euler top: one based on SO(3) and the other on $T_1 S^2$. Given a left invariant (dual) metric on the cotangent bundle, we consider the Hamiltonian formed by taking a cotangent vector to onehalf its length squared. The pull back by left trivialization of Hamilton's equations on the cotangent bundle gives the Euler-Arnol'd equations, whose solutions describe the motion of the top in space.

2.1 Euler-Arnol'd equations on SO(3) $\times$ $\mathbf{R}^3$

We begin by deriving the Euler-Arnol'd equations for the traditional SO(3) model of the Euler top.

Suppose that the initial position of the top is the *frame* $\{e_1, e_2, e_3\}$ in $\mathbf{R}^3$, which we identify with the identity element e of SO(3). An arbitrary position of the top is given by the frame $\{Be_1, Be_2, Be_3\}$, which is obtained by rotating the initial frame by B. We identify this new frame with the element B in SO(3). Thus the configuration space of the Euler top is the rotation group SO(3). The phase space is the cotangent bundle $T^*\,\mathrm{SO}(3)$ with its canonical symplectic form Ω, (see appendix A section 2).

To describe the motion of the Euler top we need a Hamiltonian function on phase space. Towards this end, let ρ be a left invariant (dual) metric on $T^*\,\mathrm{SO}(3)$. In other words, for every $A \in \mathrm{SO}(3)$, $\rho(A)$ is a nondegenerate inner product on $T^*_A\,\mathrm{SO}(3)$ such that for every $B \in \mathrm{SO}(3)$

$$\rho(BA)(\alpha_{BA}, \beta_{BA}) = \rho(A)(\alpha_A, \beta_A). \tag{21}$$

The cotangent vector α_A is defined by the mapping

$$\mathcal{L} : \mathrm{SO}(3) \times \mathrm{so}(3)^* \to T^*\,\mathrm{SO}(3) : (A, \alpha) \to (T_A L_{A^{-1}})^t \alpha = \alpha_A, \tag{22}$$

where

$$L_A : \mathrm{SO}(3) \to \mathrm{SO}(3) : B \to AB$$

is *left* translation. Thus $\mathcal{L}$ is a trivialization of the cotangent bundle

$$\tau : T^*\,\mathrm{SO}(3) \to \mathrm{SO}(3) : \alpha_A \to A.$$

By definition α is the momentum of the top at the position A. From left invariance it follows that ρ is completely determined by its value at e. Because ρ is nondegenerate, $\rho(e)$ may be written as

$$\rho(e)(\alpha, \beta) = \kappa((I^{-1})^t\alpha, \beta). \tag{23}$$

Here κ is the (dual) metric on so(3)* induced by the Killing metric k (1) on so(3). In particular, κ is defined by $\kappa(\alpha, \beta) = \beta(k^\flat(\alpha))$. The map I : so(3) → so(3) (which is uniquely determined by (23)) is k-symmetric and invertible. I is called the *moment of inertia tensor* of the Euler top, (see exercise 5). Let I_i be the eigenvalues of I. They are real and nonzero and are called the *principal moments of inertia* of the top. Below we will show that we may suppose that the matrix of $\rho(e)$ with respect to the dual basis $\{E_i^* = k^\sharp(E_i)\}$ of so(3)* is diag(I_1, I_2, I_3).

The motion of the Euler top on $(T^*\mathrm{SO}(3), \Omega)$ is given by the integral curves of the Hamiltonian vector field $X_{\mathcal{H}}$ corresponding to the Hamiltonian

$$\mathcal{H} : T^*\mathrm{SO}(3) \to \mathbf{R} : \alpha_A \to \tfrac{1}{2}\,\rho(A)(\alpha_A, \alpha_A). \tag{24}$$

Because $\mathcal{H}$ assigns to a cotangent vector one half its ρ-length squared, the vector field $X_{\mathcal{H}}$ is the geodesic vector field on $T^*\mathrm{SO}(3)$ associated to the *left* invariant metric ρ. The image of an integral curve of $X_{\mathcal{H}}$ under the bundle projection τ is a geodesic on SO(3) for the left invariant metric ρ, (see appendix A section 3).

To write out Hamilton's equations for $X_{\mathcal{H}}$ explicitly, we pull back the Hamiltonian system $(\mathcal{H}, T^*\mathrm{SO}(3), \Omega)$ by left trivialization $\mathcal{L}$ (22). We obtain the *equivalent Hamiltonian system* $(H, \mathrm{SO}(3) \times \mathrm{so}(3)^*, \omega)$, where the Hamiltonian $H = \mathcal{L}^*\mathcal{H}$ is

$$\begin{aligned} H &: \mathrm{SO}(3) \times \mathrm{so}(3)^* \to \mathbf{R} : \\ &(A, \alpha) \to \tfrac{1}{2}\,\rho(e)(\alpha, \alpha) = \tfrac{1}{2} k\big(k^\flat\big((I^{-1})^t\alpha\big), k^\flat(\alpha)\big) \end{aligned} \tag{25}$$

and the symplectic form $\omega = \mathcal{L}^*\Omega$ is

$$\omega(A, \alpha)\big((T_eL_AX, \beta), (T_eL_AY, \gamma)\big) = -\beta(Y) + \gamma(X) + \alpha([X, Y]), \tag{26}$$

for $X, Y \in$ so(3) and $\beta, \gamma \in$ so(3)*, (see appendix A section 2 equation (4)). Because

$$T_{(A,\alpha)}\big(\,\mathrm{SO}(3) \times \mathrm{so}(3)^*\big) = \Big\{(T_eL_AX = AX, \alpha)\Big|X \in \mathrm{so}(3) \text{ and } \alpha \in \mathrm{so}(3)^*\Big\},$$

we may write $X_H(A, \alpha) = (T_eL_AX_1, \alpha_1)$ for some $X_1 = X_1(A, \alpha) \in$ so(3) and some $\alpha_1 = \alpha_1(A, \alpha) \in$ so(3)*. By definition of Hamiltonian vector field,

$$dH(A, \alpha)(T_eL_AX_2, \alpha_2) = \omega(A, \alpha)\big(X_H(A, \alpha), (T_eL_AX_2, \alpha_2)\big) \tag{27}$$

for every $(T_eL_AX_2, \alpha_2) \in T_{(A,X)}\big(\,\mathrm{SO}(3) \times \mathrm{so}(3)^*\big)$. Differentiating H (25) and using the definition of ω (26), equation (27) becomes

$$\rho(e)(\alpha, \alpha_2) = -\alpha_1(X_2) + \alpha_2(X_1) + \alpha([X_1, X_2]) \tag{28}$$

for every $X_2 \in \mathrm{so}(3)$ and every $\alpha_2 \in \mathrm{so}(3)^*$. Setting $X_2 = 0$ in (28) gives $\alpha_2(X_1) = \alpha_2(\rho(e)^\sharp\alpha)$ for every $\alpha_2 \in \mathrm{so}(3)^*$. Hence $X_1 = \rho(e)^\sharp(\alpha)$. Similarly, setting $\alpha_2 = 0$ in (28) and using $X_1 = \rho(e)^\sharp(\alpha)$ gives

$$\alpha_1(X_2) = \alpha([\rho(e)^\sharp(\alpha), X_2]) = (\mathrm{ad}^t_{\rho(e)^\sharp(\alpha)}\alpha)(X_2)$$

for every $X_2 \in \mathrm{so}(3)$. Hence $\alpha_1 = \mathrm{ad}^t_{\rho(e)^\sharp(\alpha)}\alpha$. Consequently, the Hamiltonian vector field of the Euler top on $\mathrm{SO}(3) \times \mathrm{so}(3)^*$ is

$$X_H(A, \alpha) = \big(A\rho(e)^\sharp(\alpha), \mathrm{ad}^t_{\rho(e)^\sharp(\alpha)}\alpha\big). \tag{29}$$

▷ We now show how to bring the matrix of $\rho(e)$ into diagonal form.

(2.1) **Proof**: Using the map

$$j = k^\sharp \circ i^{-1} : \mathbf{R}^3 \to \mathrm{so}(3)^* : e_i \to E_i^*,$$

the inner product $\rho(e)$ on $\mathrm{so}(3)^*$ pulls back to an inner product $\overline{\rho}$ on $\mathbf{R}^3$. For $v, w \in \mathbf{R}^3$ we have

$$\begin{aligned}
\overline{\rho}(v, w) &= \rho(e)(j(v), j(w)) = k\big((k^\flat \circ (I^{-1})^t \circ k^\sharp)(i^{-1}v), i^{-1}w\big) \\
&= k\big(i^{-1}((I')^{-1}v), i^{-1}(w)\big), \quad \text{using the } k\text{-symmetry of } I. \\
&\qquad \text{Here } I' = i \circ I \circ i^{-1} \\
&= ((I')^{-1}v, w).
\end{aligned}$$

With respect to the standard basis $\{e_i\}$ of $(\mathbf{R}^3, (\,,\,))$ the matrix of $\overline{\rho}$ is invertible and symmetric. Hence there is a rotation O of $\mathbf{R}^3$ such that the matrix of $\overline{\rho}$ with respect to the basis $\{Oe_i\}$ is $\mathrm{diag}(I_1^{-1}, I_2^{-1}, I_3^{-1})$. Under the map j, the basis $\{Oe_i\}$ becomes the becomes the basis $\{\mathrm{Ad}^t_{O^{-1}}E_i^*\}$ of $\mathrm{so}(3)^*$. The matrix of $\rho(e)$ with respect to this basis is $\mathrm{diag}(I_1^{-1}, I_2^{-1}, I_3^{-1})$. Classically, this diagonalization procedure is called transforming the moment of inertia tensor to its *principal axes*. □

▷ This is not quite what we want because the principal axis transformation does *not* give an equivalent Hamiltonian system for the motion of the Euler top.

(2.2) **Proof**: Consider the diffeomorphism $R_O : \mathrm{SO}(3) \to \mathrm{SO}(3) : A \to AO$ of configuration space given by *right* translation by O, where O is the rotation constructed in ((2.1)). Physically, the initial position of the top is the new frame $\{Oe_1, Oe_2, Oe_3\}$ instead of the frame $\{e_1, e_2, e_3\}$ and its general position is $\{OBe_1, OBe_2, OBe_3\}$ instead of $\{Be_1, Be_2, Be_3\}$. The diffeomorphism R_O lifts to a symplectic diffeomorphism T^*R_O of $(T^*\mathrm{SO}(3), \Omega)$ given by

$$T^*R_O : T^*\mathrm{SO}(3) \to T^*\mathrm{SO}(3) : \alpha_A \to (T_A R_{O^{-1}})^t\alpha_A.$$

Pushing the left invariant metric ρ forward by T^*R_O gives a new *left* invariant metric ρ'

$$\rho'(A)(\alpha_A, \beta_A) = \rho(AO)\big((T_A R_{O^{-1}})^t\alpha_A, (T_A R_{O^{-1}})^t\alpha_A\big).$$

Pulling ρ' back by the left translation $L_{(AO)^{-1}}$ we obtain

$$\rho'(e)(\alpha, \beta) = \rho(e)(\mathrm{Ad}^t_{O^{-1}}\alpha, \mathrm{Ad}^t_{O^{-1}}\beta).$$

By construction of the rotation O in ((2.1)), the matrix of $\rho'(e)$ with respect to the basis $\{E_i^*\}$ of so(3)* is $\mathrm{diag}(I_1^{-1}, I_2^{-1}, I_3^{-1})$.

Let $(\mathcal{H}', T^*\mathrm{SO}(3), \Omega)$ be the Hamiltonian system obtained by pushing forward the original Hamiltonian system $(\mathcal{H}, T^*\mathrm{SO}(3), \Omega)$ of the Euler top by the symplectic diffeomorphism T^*R_O. Pulling back $(\mathcal{H}', T^*\mathrm{SO}(3), \Omega)$ by left trivialization $\mathcal{L}$ (22) we obtain a new Hamiltonian system $(H', \mathrm{SO}(3) \times \mathrm{so}(3)^*, \omega)$ with Hamiltonian

$$H' : \mathrm{SO}(3) \times \mathrm{so}(3)^* \to \mathbf{R} : (A, \alpha) \to \tfrac{1}{2}\rho'(e)(\alpha, \alpha)$$

and symplectic form ω (26). With respect to the basis $\{E_i^*\}$ of so(3)*, the matrix of ρ' is $\mathrm{diag}(I_1^{-1}, I_2^{-1}, I_3^{-1})$. □

From now on we will assume that $\rho(e)$ is $\mathrm{diag}(I_1^{-1}, I_2^{-1}, I_3^{-1})$. Writing $\alpha = \sum_i p_i E_i^*$ and using the relations

$$(I^{-1})^t E_i^* = I_i^{-1} E_i^*, \quad \mathrm{ad}_{E_i}^t E_j^* = -\sum_{k=1}^{3} \varepsilon_{ijk} E_k^*, \quad \text{and} \quad E_i^* = E_i^t,$$

it follows that the integral curves of X_H (29) on $(\mathrm{SO}(3) \times \mathrm{so}(3)^*, \omega)$ satisfy the *Euler-Arnol'd equations*:

$$\begin{aligned} \frac{dA}{dt} &= A \begin{pmatrix} 0 & -I_3^{-1}p_3 & I_2^{-1}p_2 \\ I_3^{-1}p_3 & 0 & -I_1^{-1}p_1 \\ -I_2^{-1}p_2 & I_1^{-1}p_1 & 0 \end{pmatrix} \\ \frac{d}{dt}\begin{pmatrix} 0 & p_3 & -p_2 \\ -p_3 & 0 & p_1 \\ p_2 & -p_1 & 0 \end{pmatrix} &= \\ &= \begin{pmatrix} 0 & -(I_2^{-1} - I_1^{-1})p_1p_2 & (I_1^{-1} - I_3^{-1})p_1p_3 \\ (I_2^{-1} - I_1^{-1})p_1p_2 & 0 & -(I_3^{-1} - I_2^{-1})p_2p_3 \\ -(I_1^{-1} - I_3^{-1})p_1p_3 & (I_3^{-1} - I_2^{-1})p_2p_3 & 0 \end{pmatrix}. \end{aligned}$$

Pulling the Euler-Arnol'd equations back by the diffeomorphism mapping $id \times j$ gives

$$\frac{dA}{dt} = A \begin{pmatrix} 0 & -I_3^{-1}p_3 & I_2^{-1}p_2 \\ I_3^{-1}p_3 & 0 & -I_1^{-1}p_1 \\ -I_2^{-1}p_2 & I_1^{-1}p_1 & 0 \end{pmatrix} \qquad (30)$$

$$\frac{dp}{dt} = p \times (I')^{-1} p \qquad (31)$$

on $(\mathrm{SO}(3) \times \mathbf{R}^3, \omega' = j^*\omega)$. Here $(I')^{-1}(p) = (I_1^{-1}p_1, I_2^{-1}p_2, I_3^{-1}p_3)$. (See exercise 4 for a more traditional derivation of the Euler-Arnol'd equations using a parametrization of SO(3) given by Euler angles.)

2.2 Euler-Arnol'd equations on $T_1 S^2 \times \mathbf{R}^3$

In this section we derive the Euler-Arnol'd equations for the Euler top in the nontraditional sphere bundle model.

Before launching into the details, we derive the nontraditional Euler-Arnol'd equations from the traditional ones, (30) and (31), as follows. Let $(x, y) \in T_1S^2$ and set $A = \mathrm{col}(x, y, x \times y)$ in (30). Taking the first two columns of the both sides of (30) gives

$$\begin{aligned} \dot{x} &= I_3^{-1} p_3 y - I_2^{-1} p_2 (x \times y) \\ \dot{y} &= -I_3^{-1} p_3 x + I_1^{-1} p_1 (x \times y). \end{aligned} \tag{32}$$

Clearly (31) remains unchanged:

$$\dot{p} = p \times (I')^{-1} p. \tag{33}$$

Of course this derviation does *not* show that the Euler-Arnol'd equations in the sphere ▷ bundle model are in Hamiltonian form. To do this we give an argument which parallels the derivation in the traditional SO(3) model.

(2.3) **Proof**: First, let ρ be a left invariant (dual) metric on the Lie group $(T_1S^2, \cdot)$. Because of left invariance, ρ is determined by its value at the identity element e of T_1S^2. In particular, for every $\boldsymbol{\alpha}, \boldsymbol{\beta} \in T_e^*(T_1S^2)$, the dual of the Lie algebra of T_1S^2,

$$\rho(e)(\boldsymbol{\alpha}, \boldsymbol{\beta}) = \kappa((I^{-1})^t \boldsymbol{\alpha}, \boldsymbol{\beta}). \tag{34}$$

Here κ is the (dual) metric on $T_e^*(T_1S^2)$ (induced by the Killing metric k (16) on $T_e(T_1S^2)$) is defined by $\kappa(\boldsymbol{\alpha}, \boldsymbol{\beta}) = \boldsymbol{\beta}(k^\flat(\boldsymbol{\alpha}))$. The linear map $\widehat{I} : T_e(T_1S^2) \to T_e(T_1S^2)$ (which is uniquely determined by (34)) is invertible and k-symmetric. $\widehat{I}$ is defined by

$$i\big(\widehat{I}(\mathbf{u})\big) = I'\big(i(\mathbf{u})\big) = (I_1u_1, I_2u_2, I_3u_3),$$

that is, $\widehat{I}(\mathbf{u}) = \big((0, I_3u_3, -I_2u_2), (-I_3u_3, 0, I_1u_1)\big)$.

Let

$$\begin{aligned} &\mathcal{L} : T_1S^2 \times T_e^*(T_1S^2) \to T^*(T_1S^2) : \\ &\big((x, y), \boldsymbol{\alpha}\big) \to (T_{(x,y)}L_{(x,y)^{-1}})^t \boldsymbol{\alpha} = \boldsymbol{\alpha}_{(x,y)}, \end{aligned}$$

where $L_{(x,y)}$ is left translation on $(T_1S^2, \cdot)$ by (x, y). $\mathcal{L}$ is a trivialization of the cotangent bundle

$$\tau : T^*(T_1S^2) \to T_1S^2 : \boldsymbol{\alpha}_{(x,y)} \to (x, y)$$

by *left* translation. On $T^*(T_1S^2)$ with its canonical symplectic form Ω, consider the Hamiltonian

$$\mathcal{H} : T^*(T_1S^2) \to \mathbf{R} : \boldsymbol{\alpha}_{(x,y)} \to \tfrac{1}{2}\, \rho(x, y)\big(\boldsymbol{\alpha}_{(x,y)}, \boldsymbol{\alpha}_{(x,y)}\big). \tag{35}$$

Pulling back the Hamiltonian system $(\mathcal{H}, T^*(T_1S^2), \Omega)$ by the left trivialization $\mathcal{L}$ gives the equivalent Hamiltonian system $(H, T_1S^2 \times T_e^*(T_1S^2), \omega)$, where the Hamiltonian $H = \mathcal{L}^*\mathcal{H}$ is

$$H : T_1S^2 \times T_e^*(T_1S^2) \to \mathbf{R} : \big((x, y), \boldsymbol{\alpha}\big) \to \tfrac{1}{2}\, \rho(e)(\boldsymbol{\alpha}, \boldsymbol{\alpha}) \tag{36}$$

and symplectic form $\omega = \mathcal{L}^*\Omega$ is

$$\omega(x, y, \boldsymbol{\alpha})\big((T_eL_{(x,y)}\mathbf{u}, \boldsymbol{\beta}), (T_eL_{(x,y)}\mathbf{v}, \boldsymbol{\gamma})\big) = -\boldsymbol{\beta}(\mathbf{v}) + \boldsymbol{\gamma}(\mathbf{u}) + \boldsymbol{\alpha}([\mathbf{u}, \mathbf{v}]), \tag{37}$$

for $\mathbf{u}, \mathbf{v} \in T_e(T_1S^2)$ and $\boldsymbol{\beta}, \boldsymbol{\gamma} \in T_e^*(T_1S^2)$.

To compute the Hamiltonian vector field X_H, note that

$$X_H(x, y, \boldsymbol{\alpha}) = \big(T_e L_{(x,y)}\mathbf{x}_1, \boldsymbol{\alpha}_1\big)$$

for some $\mathbf{x}_1 = \mathbf{x}_1(x, y, \boldsymbol{\alpha}) \in T_e(T_1S^2)$ and some $\boldsymbol{\alpha}_1 = \boldsymbol{\alpha}_1(x, y, \boldsymbol{\alpha}) \in T_e^*(T_1S^2)$. From the definition of X_H it follows that

$$dH(x, y, \boldsymbol{\alpha})\big(T_e L_{(x,y)}\mathbf{x}_2, \boldsymbol{\alpha}_2\big) = \omega(x, y, \boldsymbol{\alpha})\big(X_H(x, y, \boldsymbol{\alpha}), (T_e L_{(x,y)}\mathbf{x}_2, \boldsymbol{\alpha}_2)\big), \tag{38}$$

for every $\mathbf{x}_2 \in T_e(T_1S^2)$ and every $\boldsymbol{\alpha}_2 \in T_e^*(T_1S^2)$. Differentiating H and using the definition of ω (37), equation (38) becomes

$$\rho(e)(\boldsymbol{\alpha}, \boldsymbol{\alpha}_2) = -\boldsymbol{\alpha}_1(\mathbf{x}_2) + \boldsymbol{\alpha}_2(\mathbf{x}_2) + \boldsymbol{\alpha}([\mathbf{x}_1, \mathbf{x}_2]), \tag{39}$$

(compare with (28)). Arguing exactly as in section 2.1 we find that

$$X_H(x, y, \boldsymbol{\alpha}) \quad = \quad \big(T_e L_{(x,y)}(\rho(e)^\sharp(\boldsymbol{\alpha})), \mathrm{ad}^t_{\rho(e)^\sharp(\boldsymbol{\alpha})}\boldsymbol{\alpha}\big), \tag{40}$$

where $\rho(e)^\sharp(\boldsymbol{\alpha}) = k^\flat \circ (\widehat{I}^{-1})^t$. Hence on $(T_1S^2 \times T_e^*(T_1S^2), \omega)$ the Euler-Arnol'd equations are

$$\frac{d}{dt}(x, y) \quad = \quad T_e L_{(x,y)}(\rho(e)^\sharp\boldsymbol{\alpha}) \tag{41}$$

$$\frac{d\boldsymbol{\alpha}}{dt} \quad = \quad \mathrm{ad}^t_{\rho(e)^\sharp(\boldsymbol{\alpha})}\boldsymbol{\alpha}, \tag{42}$$

which are Hamilton's equations for X_H. □

To write (41) and (42) in a more convenient form, consider the isomorphism

$$j = k^\sharp \circ i^{-1} : \mathbf{R}^3 \to T_e^*(T_1S^2) : e_i \to \boldsymbol{\varepsilon}_i^*.$$

$\{\boldsymbol{\varepsilon}_i^*\}$ is a basis of $T_e^*(T_1S^2)$, which is dual basis to the *standard basis* $\{\boldsymbol{\varepsilon}_i\}$ of $T_e(T_1S^2)$, because

$$\boldsymbol{\varepsilon}_i^*(\boldsymbol{\varepsilon}_\ell) = k^\sharp(i^{-1}(e_i))\boldsymbol{\varepsilon}_\ell \ = \ k(i^{-1}(e_i), i^{-1}(e_\ell)) \ = \ (e_i, e_\ell) = \delta_i^\ell.$$

Pulling X_H back by the diffeomorphism

$$\psi = \mathrm{id} \times j : T_1S^2 \times \mathbf{R}^3 \to T_1S^2 \times T_e^*(T_1S^2) : (x, y, p) \to (x, y, \sum_i p_i\boldsymbol{\varepsilon}_i^*)$$

gives a Hamiltonian vector field $X_{H'}$ on $(T_1S^2 \times \mathbf{R}^3, \omega' = \psi^*\omega)$ with Hamiltonian

$$H' = \psi^* H : T_1S^2 \times \mathbf{R}^3 \to \mathbf{R} :$$
$$(x, y, p) \to \tfrac{1}{2}\,((I')^{-1}p, p) = \tfrac{1}{2}\left(I_1^{-1}p_1^2 + I_2^{-1}p_2^2 + I_3^{-1}p_3^2\right).$$

▷ Hamilton's equations for $X_{H'}$ are

$$\begin{cases} \dfrac{dx}{dt} = I_3^{-1} p_3 y - I_2^{-1} p_2 (x \times y) \\ \dfrac{dy}{dt} = -I_3^{-1} p_3 x + I_1^{-1} p_1 (x \times y) \end{cases} \tag{43}$$

$$\frac{dp}{dt} = p \times (I')^{-1} p, \tag{44}$$

where $(I')^{-1} p = (I_1^{-1} p_1, I_2^{-1} p_2, I_3^{-1} p_3)$.

(2.4) **Proof**: With respect to the bases $\{\boldsymbol{\varepsilon}_i^*\}$ of $T_e^*(T_1 S^2)$ and $\{\boldsymbol{\varepsilon}_i\}$ of $T_e(T_1 S^2)$ the matrix of $\rho(e)^\sharp : T_e^*(T_1 S^2) \to T_e(T_1 S^2)$ is $\mathrm{diag}(I_1^{-1}, I_2^{-1}, I_3^{-1})$, because

$$\begin{aligned} \boldsymbol{\varepsilon}_\ell^*\big(\rho(e)^\sharp(\boldsymbol{\varepsilon}_i^*)\big) &= \rho(e)\big(\boldsymbol{\varepsilon}_i^*, \boldsymbol{\varepsilon}_\ell^*\big) = \boldsymbol{\varepsilon}_\ell^*\big(k^\flat((\widehat{I}^{-1})^t \boldsymbol{\varepsilon}_i^*)\big) \\ &= \boldsymbol{\varepsilon}_\ell^*\big((k^\flat \circ (\widehat{I}^{-1})^t \circ k^\sharp)(\boldsymbol{\varepsilon}_i)\big) = \boldsymbol{\varepsilon}_\ell^*((I')^{-1} \boldsymbol{\varepsilon}_i) = \boldsymbol{\varepsilon}_\ell^*(I_i^{-1} \boldsymbol{\varepsilon}_i). \end{aligned}$$

Writing $\boldsymbol{\alpha} = \sum_i p_i \boldsymbol{\varepsilon}_\ell^*$, we see that $\rho(e)^\sharp(\boldsymbol{\alpha}) = \sum_i I_i^{-1} p_i \, \boldsymbol{\varepsilon}_i$. Differentiating the definition of left translation $L_{(x,y)}$ (15) gives

$$T_e L_{(x,y)} \mathbf{u} = \widehat{\pi}(\mathrm{col}(x, y, x \times y) \cdot U) = \big(u_3 y - u_2 (x \times y), -u_3 x + u_1 (x \times y)\big),$$

for $\mathbf{u} = \sum_i u_i \boldsymbol{\varepsilon}_i \in T_e(T_1 S^2)$. Setting $\mathbf{u} = \rho(e)^\sharp(\boldsymbol{\alpha})$ in the above equation and using (41) we obtain (43). To obtain (44) we first show that $\mathrm{ad}^t_{\boldsymbol{\varepsilon}_i} \boldsymbol{\varepsilon}_j^* = -\sum_k \varepsilon_{ijk} \, \boldsymbol{\varepsilon}_k^*$. We compute

$$(\mathrm{ad}^t_{\boldsymbol{\varepsilon}_i} \boldsymbol{\varepsilon}_j^*)(\boldsymbol{\varepsilon}_\ell) = \boldsymbol{\varepsilon}_j^*([\boldsymbol{\varepsilon}_i, \boldsymbol{\varepsilon}_\ell]) = \boldsymbol{\varepsilon}_j^*(\sum_k \varepsilon_{i\ell k} \, \boldsymbol{\varepsilon}_k) = \varepsilon_{i\ell j} = -\big(\sum_k \varepsilon_{ijk} \, \boldsymbol{\varepsilon}_k^*\big)(\boldsymbol{\varepsilon}_\ell).$$

From (42) we find that

$$\begin{aligned} \sum_i \left(\frac{dp}{dt}\right)_i \boldsymbol{\varepsilon}_i^* &= \sum_{ij} p_i (I_j^{-1} p_j) \, \mathrm{ad}^t_{\boldsymbol{\varepsilon}_j} \boldsymbol{\varepsilon}_i^* \\ &= \sum_{ijk} p_i (I_j^{-1} p_j) \, \varepsilon_{ijk} \, \boldsymbol{\varepsilon}_k^* = \sum_i (p \times (I')^{-1} p)_i \, \boldsymbol{\varepsilon}_i^*. \end{aligned}$$

Equating components gives (44). □

Note that the right hand side of (43) and (44) defines a vector field on $\mathbf{R}^9$ with coordinates (x, y, p). A calculation shows that $T_1 S^2 \times \mathbf{R}^3$ is an invariant manifold of this vector field.

3 Symmetry and reduction

In this section we discuss the SO(3) symmetry of the Euler top. Using the regular reduction theorem (see appendix B ((4.1))) to remove this symmetry, we obtain a Hamiltonian system on the 2-sphere S^2. We show that the integral curves of the reduced Hamiltonian vector field satisfy *Euler's equations*.

3.1 Construction of the reduced phase space

In this subsection we construct the reduced phase space.

Because the configuration space of the Euler top is the Lie group SO(3), it has a natural symmetry, namely, the action of SO(3) on itself by left translation

$$L : \mathrm{SO}(3) \times \mathrm{SO}(3) \to \mathrm{SO}(3) : (B, A) \to L_B A = BA. \tag{45}$$

The action L lifts to an action

$$\widehat{L} : \mathrm{SO}(3) \times T^* \mathrm{SO}(3) \to T^* \mathrm{SO}(3) : (B, \alpha_A) \to \alpha_{BA}, \tag{46}$$

▷ where $\alpha_A = (T_A L_{A^{-1}})^t \alpha$ for $\alpha \in \mathrm{so}(3)^*$. The action $\widehat{L}$ on $(T^* \mathrm{SO}(3), \Omega)$ is Hamiltonian with equivariant momentum mapping

$$\mathcal{J} : T^* \mathrm{SO}(3) \to \mathrm{so}(3)^* : \alpha_A \to (T_e R_A)^t \alpha, \tag{47}$$

where

$$R : \mathrm{SO}(3) \times \mathrm{SO}(3) \to \mathrm{SO}(3) : (B, A) \to R_A B = BA$$

is *right translation* by A.

(3.1) **Proof**: Let $\xi \in \mathrm{so}(3)$. The infinitesimal generator $X^\xi(\alpha_A) = \frac{d}{dt}\Big|_{t=0} \widehat{L}_{\exp t\xi}(\alpha_A)$ of $\widehat{L}$ in the direction ξ is the Hamiltonian vector field on $(T^* \mathrm{SO}(3), \Omega = -d\theta)$ corresponding to the Hamiltonian function

$$\mathcal{J}^\xi : T^* \mathrm{SO}(3) \to \mathbf{R} : \alpha_A \to (X^\xi \lrcorner\, \theta)(\alpha_A).$$

To check this observe that the canonical 1-form θ on $T^* \mathrm{SO}(3)$ is invariant under $\widehat{L}$. Therefore

$$0 = L_{X^\xi}\theta = d(X^\xi \lrcorner\, \theta) + X^\xi \lrcorner\, d\theta = d\mathcal{J}^\xi - X^\xi \lrcorner\, \Omega.$$

We now find another expression for $\mathcal{J}^\xi$. Let $X_\xi(A) = \frac{d}{dt}\Big|_{t=0} L_{\exp t\xi}(A)$ be the infinitesimal generator of L in the direction ξ. Then

$$X_\xi(A) = \frac{d}{dt}\Big|_{t=0} \exp t\xi\, A = \frac{d}{dt}\Big|_{t=0} R_A \exp t\xi = T_e R_A \xi. \tag{48}$$

From the momentum lemma (see appendix B ((3.6))), it follows that

$$\mathcal{J}^\xi(\alpha_A) = \alpha_A(X_\xi(A)) = \big((T_e R_A)^t(\alpha_A)\big)\xi,$$

for every $\xi \in \mathrm{so}(3)$. Since $\xi \to \mathcal{J}^\xi(\alpha_A)$ is a linear function on so(3), we may define the mapping $\mathcal{J} : T^* \mathrm{SO}(3) \to \mathrm{so}(3)^*$ by $\mathcal{J}(\alpha_A)\xi = \mathcal{J}^\xi(\alpha_A)$. $\mathcal{J}$ is coadjoint equivariant because for $B \in \mathrm{SO}(3)$

$$\begin{aligned}
\mathcal{J}(\widehat{L}_B \alpha_A) &= (T_e R_{BA})^t \alpha_{BA} = (T_e R_{BA})^t (T_{BA} L_{(BA)^{-1}})^t \alpha \\
&= (T_B L_{B^{-1}} \circ T_e R_B)^t (T_e R_A)^t (T_A L_{A^{-1}})^t \alpha \\
&= \mathrm{Ad}^t_{B^{-1}} \mathcal{J}(\alpha_A).
\end{aligned} \tag{49}$$

Infinitesimalizing (49) gives

$$\begin{aligned}\{\mathcal{J}^\xi, \mathcal{J}^\eta\}(\alpha_A) &= \left(L_{X_{\mathcal{J}^\eta}} \mathcal{J}^\xi\right)(\alpha_A) = \frac{d}{ds}\Big|_{s=0} \mathcal{J}^\xi\left(\widehat{L}_{\exp s\eta}(\alpha_A)\right) \\ &= \frac{d}{ds}\Big|_{s=0} \mathrm{Ad}^t_{\exp -s\eta}\left(\mathcal{J}^\xi(\alpha_A)\right) = \mathcal{J}^{\frac{d}{ds}\big|_{s=0} \mathrm{Ad}_{\exp -s\eta}\xi}(\alpha_A) \\ &= \mathcal{J}^{[\xi,\eta]}(\alpha_A),\end{aligned}$$

for every $\xi, \eta \in \mathrm{so}(3)$. Therefore $\mathcal{J}$ is a momentum mapping for the action $\widehat{L}$. □

Pulling back the SO(3)-action $\widehat{L}$ (46) by the left trivialization mapping

$$\mathcal{L} : \mathrm{SO}(3) \times \mathrm{so}(3)^* \to T^*\mathrm{SO}(3) : (A, \alpha) \to \alpha_A = (T_A L_{A^{-1}})^t \alpha$$

gives the action

$$\begin{aligned}\widehat{\ell} : \mathrm{SO}(3) \times \left(\mathrm{SO}(3) \times \mathrm{so}(3)^*\right) &\to \mathrm{SO}(3) \times \mathrm{so}(3)^* : \\ \left(B, (A, \alpha)\right) &\to (BA, \alpha).\end{aligned} \tag{50}$$

▷ $\widehat{\ell}$ is a Hamiltonian action on $(\mathrm{SO}(3) \times \mathrm{so}(3)^*, \omega = \mathcal{L}^*\Omega)$ with equivariant momentum mapping

$$J = \mathcal{L}^*\mathcal{J} : \mathrm{SO}(3) \times \mathrm{so}(3)^* \to \mathrm{so}(3)^* : (A, \alpha) \to \mathrm{Ad}^t_{A^{-1}}\alpha. \tag{51}$$

(3.2) **Proof**: To verify (51) we compute

$$\begin{aligned}J(A, \alpha) &= \mathcal{J}(\mathcal{L}(A, \alpha)) = \mathcal{J}(\alpha_A) \\ &= (T_e R_A)^t \alpha_A = (T_e R_A)^t (T_A L_{A^{-1}})^t \alpha = \mathrm{Ad}^t_{A^{-1}}\alpha.\end{aligned}$$

Coadjoint equivariance follows because for $B \in \mathrm{SO}(3)$

$$\begin{aligned}J\left(\widehat{\ell}_B(A, \alpha)\right) &= J(BA, \alpha) = \mathrm{Ad}^t_{(BA)^{-1}}\alpha \\ &= \mathrm{Ad}^t_{B^{-1}}(\mathrm{Ad}^t_{A^{-1}}\alpha) = \mathrm{Ad}^t_{B^{-1}}\left(J(A, \alpha)\right). \quad \square\end{aligned}$$

To start the process of reduction, let μ be a *nonzero* element of $\mathrm{so}(3)^*$. Then the μ-level set $J^{-1}(\mu)$ of the momentum map J is

$$\left\{ (A, \mathrm{Ad}^t_A \mu) \in \mathrm{SO}(3) \times \mathrm{so}(3)^* \,\middle|\, A \in \mathrm{SO}(3) \right\}.$$

$J^{-1}(\mu)$ is a smooth submanifold of $\mathrm{SO}(3) \times \mathrm{so}(3)^*$, which is diffeomorphic to SO(3), because it is the graph of the smooth mapping

$$\mathrm{SO}(3) \to \mathrm{so}(3)^* : A \to \mathrm{Ad}^t_A \mu.$$

We now want to find a subgroup of SO(3) which acts on $J^{-1}(\mu)$. Because J is coadjoint equivariant, we look at the isotropy group

$$\mathrm{SO}(3)_\mu = \{B \in \mathrm{SO}(3) \mid \mathrm{Ad}^t_{B^{-1}}\mu = \mu\}. \tag{52}$$

Since $\mathrm{Ad}'_{BA}\mu = \mathrm{Ad}'_A\mathrm{Ad}'_B\mu = \mathrm{Ad}'_A\mu$ for every $B \in \mathrm{SO}(3)_\mu$, it follows that

$$J(BA, \mathrm{Ad}'_{BA}\mu) = J(BA, \mathrm{Ad}'_A\mu).$$

Therefore restricting $\widehat{\ell}$ (50) to $\mathrm{SO}(3)_\mu \times J^{-1}(\mu)$ defines the action

$$\Phi : \mathrm{SO}(3)_\mu \times J^{-1}(\mu) \to J^{-1}(\mu) : \big(B, (A, \mathrm{Ad}'_A\mu)\big) \to (BA, \mathrm{Ad}'_A\mu). \tag{53}$$

To get a better idea of what the isotropy group $\mathrm{SO}(3)_\mu$ means, apply the map $\frac{1}{|\mu|}\, i \circ k^\flat$ to both sides of its defining equation (52). This gives

$$y = \frac{1}{|\mu|}\, i \circ k^\flat(\mu) = \frac{1}{|\mu|}\, i(k^\flat \circ \mathrm{Ad}'_{B^{-1}} \circ k^\sharp)(k^\flat(\mu)) = (i \circ \mathrm{Ad}_B \circ i^{-1})(y) = By.$$

Thus $\mathrm{SO}(3)_\mu$ is the set of all rotations which leave the unit vector $y = \frac{1}{|\mu|}\, i \circ k^\flat(\mu)$ fixed. Physically, $|\mu|\, y$ is the angular momentum *vector* of the Euler top. We now ▷ show that the isotropy group $\mathrm{SO}(3)_\mu$ is the image of the one parameter subgroup

$$\gamma : \mathbf{R} \to \mathrm{SO}(3) : t \to \exp tY,$$

where $Y = i^{-1}(y)$. Note that $|Y| = 1$.

(3.3) **Proof**: By definition, $\mathrm{SO}(3)_\mu$ is a closed subgroup of SO(3) and hence is a compact Lie group. Its Lie algebra

$$\mathrm{so}(3)_\mu = \Big\{ X \in \mathrm{so}(3) \,\Big|\, \mathrm{ad}'_X\mu = 0 \Big\}$$

is one dimensional. To see this apply the map $i \circ k^\flat$ to both sides of the defining equation of $\mathrm{so}(3)_\mu$ to obtain

$$0 = i\big(k^\flat \circ \mathrm{ad}'_X \circ k^\sharp\big)(k^\flat(\mu)) = -(i \circ \mathrm{ad}_X \circ i^{-1})(|\mu|\, y)) = -|\mu|\, i(X) \times i(Y),$$

where the second equality follow since ad_X is k-skew symmetric. Therefore X and Y are linearly dependent. Hence $\dim \mathrm{so}(3)_\mu = 1$.

We would be done if we knew that $\mathrm{SO}(3)_\mu$ was connected, because the image of the one parameter group γ is circle. Therefore we need to show that $\mathrm{SO}(3)_\mu$ is connected. The following argument shows that $\mathrm{SO}(3)_\mu$ is isomorphic to SO(2) and hence is connected. Write $\mathbf{R}^3 = \mathrm{span}\{y\} \oplus \Pi$, where Π is a plane in $\mathbf{R}^3$ orthogonal to the vector y. Π is invariant under every $B \in \mathrm{SO}(3)_\mu$, because $B \in \mathrm{SO}(3)$ and $By = y$. Moreover, $B|\Pi \in \mathrm{SO}(2)$, since $B \in \mathrm{SO}(3)$ implies that $B|\Pi$ preserves the length of every vector in Π and

$$1 = \det B = \det(B|\,\mathrm{span}\{y\}) \cdot \det(B|\Pi) = \det(B|\Pi).$$

The smooth map

$$\sigma : \mathrm{SO}(3)_\mu \subseteq \mathrm{SO}(3) \to \mathrm{SO}(2) : B \to B|\Pi$$

is a homomorphism of Lie groups. Actually, σ is an isomorphism. To see that σ is surjective, let $\widetilde{B} \in \mathrm{SO}(2)$ and choose $\{w, z\}$ to be a positively oriented orthonormal basis of Π such that $\{y, w, z\}$ is a positively oriented orthonormal basis of $\mathbf{R}^3$. Define the linear map $\mathcal{B} : \mathbf{R}^3 \to \mathbf{R}^3$ by requiring that $\mathcal{B}$ sends the positively oriented ordered orthonormal basis $\{y, w, z\}$ to the ordered basis $\{y, \widetilde{B}w, \widetilde{B}z\}$. Clearly $\mathcal{B}$ is an extension of $\widetilde{B}$. Since $\widetilde{B} \in \mathrm{SO}(2)$, $\{\widetilde{B}w, \widetilde{B}z\}$ is a positively oriented orthonormal basis of Π. Hence $\{y, \widetilde{B}w, \widetilde{B}z\}$ is a positively oriented orthonormal basis of $\mathbf{R}^3$. Therefore, $\mathcal{B} \in \mathrm{SO}(3)_\mu$. Hence σ is surjective. Because $\mathcal{B}$ is the unique rotation which extends $\widetilde{B}$, it follows that σ is injective. Thus σ is an isomorphism. □

We return to discussing the construction of the reduced space. The reduced space $P_\mu = J^{-1}(\mu)/\mathrm{SO}(3)_\mu$ is the space of $\mathrm{SO}(3)_\mu$-orbits on $J^{-1}(\mu)$. Since the action Φ (53) on $J^{-1}(\mu)$ is free and proper, P_μ is a smooth symplectic manifold, (see appendix B

▷ ((4.1))). The following argument shows that P_μ is the SO(3) coadjoint orbit $\mathcal{O}_\mu$ through μ.

(3.4) **Proof**: Consider the mapping

$$\pi_\mu : J^{-1}(\mu) \subseteq \mathrm{SO}(3) \times \mathrm{so}(3)^* \to \mathrm{so}(3)^* : (A, \mathrm{Ad}^t_A\mu) \to \nu = \mathrm{Ad}^t_A\mu. \tag{54}$$

Because the fiber $\pi_\mu^{-1}(\nu)$ is a single $\mathrm{SO}(3)_\mu$-orbit, the image of π_μ is the orbit space P_μ. By definition, $\pi_\mu(J^{-1}(\mu))$ is the coadjoint orbit

$$\mathcal{O}_\mu = \left\{ \nu = \mathrm{Ad}^t_A\mu \in \mathrm{so}(3)^* \middle| A \in \mathrm{SO}(3) \right\}.$$

From example 1 of appendix B section 4, we know that the symplectic form $\omega_{\mathcal{O}_\mu}$ on $\mathcal{O}_\mu$ is

$$\omega_{\mathcal{O}_\mu}(\nu)(\mathrm{ad}^t_\xi\nu, \mathrm{ad}^t_\eta\nu) = -\nu([\xi, \eta]), \tag{55}$$

where $\xi, \eta \in \mathrm{so}(3)$. □

3.2 Geometry of the reduction map

In this subsection we study the geometry of the reduction map π_μ (54).

▷ Our main result is that π_μ is double covered by the Hopf fibration.

(3.5) **Proof**: This follows from ((1.12)) once we prove the next claim. □

Claim: The $\mathrm{SO}(3)_\mu$ principal bundle $\pi_\mu : J^{-1}(\mu) \to \mathcal{O}_\mu$ is isomorphic to the S^1 principal bundle $\tau : T_1S^2 \to S^2$.

(3.6) **Proof**: We find a succession of principal bundle isomorphisms.

1. We begin by looking at the S^1 action

$$\psi : S^1 \times \mathrm{SO}(3) \to \mathrm{SO}(3) : (s, A) \to (\exp sY)A,$$

where $Y = \frac{1}{|\mu|}k^\flat(\mu)$. Since ψ is free and proper, SO(3) is the total space of an S^1-principal bundle over $\mathrm{SO}(3)/S^1 = S^2$ with bundle projection $\sigma^y : \mathrm{SO}(3) \to S^2 : A \to A^{-1}y$, where $y = i(Y)$. We now show that the principal bundles π_μ and σ^y are isomorphic. Consider diagram 3.1. Since $\mu = |\mu| k^\sharp \circ i^{-1}(y)$ and $A \in \mathrm{SO}(3)$, it follows that $\mathrm{Ad}^t_A\mu = \beta_1(A^{-1}y)$. Therefore diagram 3.1 commutes. In addition,

$$\Phi_{\exp sY}(\mathcal{B}_1(A)) = \left((\exp sY)A, \mathrm{Ad}^t_{(\exp sY)A}\mu\right) = \mathcal{B}_1(\psi_s(A)).$$

In other words, $\mathcal{B}_1$ intertwines the actions ψ and Φ. Because $\mathcal{B}_1$ and β_1 are diffeomorphisms, the principal bundles σ^y and π are isomorphic.

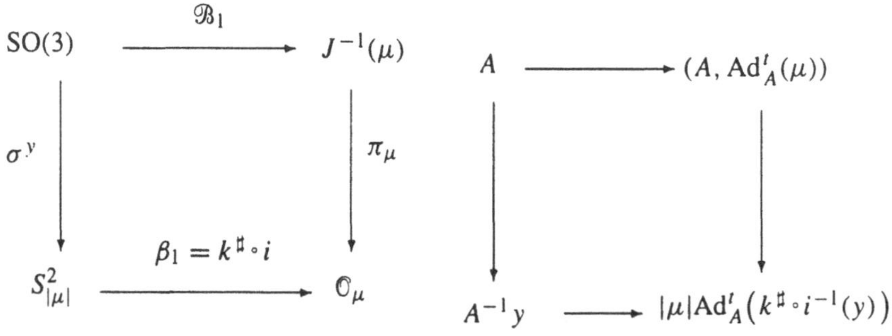

Diagram 3.1

2. Consider diagram 3.2. Because SO(3) acts transitively on the 2-sphere S^2, there is an $O \in$ SO(3) such that $Oy = e_1$. Since $(O^{-1}A)^{-1}y = A^{-1}Oy = A^{-1}e_1$, diagram 3.2 commutes.

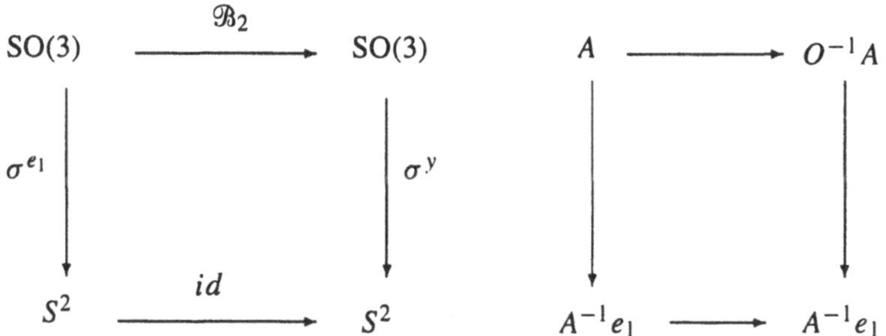

Diagram 3.2

Define the S^1 action

$$\theta : S^1 \times \mathrm{SO}(3) \to \mathrm{SO}(3) : (s, A) \to (\exp sE_1)A. \tag{56}$$

Since θ is free and proper, SO(3) is the total space of an S^1 principal bundle over $\mathrm{SO}(3)/S^1 = S^2$ with bundle projection $\sigma^{e_1} : \mathrm{SO}(3) \to S^2 : A \to A^{-1}e_1$. Because

$$\begin{aligned}\mathcal{B}_2(\theta_s(A)) &= (O^{-1}\exp sE_1 O)O^{-1}A = \big(\exp s(\mathrm{Ad}_{O^{-1}}E_1)\big)\mathcal{B}_2(A)\\ &= (\exp sY)(\mathcal{B}_2(A)), \quad \text{since } y = O^{-1}e_1\\ &= \psi_s(\mathcal{B}_2(A)),\end{aligned}$$

$\mathcal{B}_2$ intertwines the actions θ and ψ. Thus the bundles σ^{e_1} and σ^y are isomorphic principal bundles, since the maps $\mathcal{B}_2$ and id are diffeomorphisms.

3. To complete the proof, consider diagram 3.3. Clearly this diagram commutes. The S^1 action

$$\Psi : S^1 \times T_1S^2 \to T_1S^2 : \big(s, (x, y)\big) \to \big(x, y\cos s - (x \times y)\sin s\big) \qquad (57)$$

defining the principal bundle τ (17) extends to an S^1 action

$$\vartheta : S^1 \times \mathrm{SO}(3) \to \mathrm{SO}(3) :$$

$$\big(s, \mathrm{col}\,(x, y, x \times y)^t\big) \to \begin{pmatrix} x^t \\ y^t \cos s - (x \times y)^t \sin s \\ y^t \sin s + (x \times y)^t \cos s \end{pmatrix} = (\exp sE_1)\,\mathrm{col}\,(x, y, x \times y)^t,$$

ϑ is the *same* as the action θ (56). Therefore $\mathcal{B}_3(\Psi_s(x, y)) = \theta_s(\mathcal{B}_3(x, y))$, that is, $\mathcal{B}_3$ intertwines the actions θ and Ψ. Since $\mathcal{B}_3$ and id are diffeomorphisms, the principal bundles τ and σ^{e_1} are isomorphic.

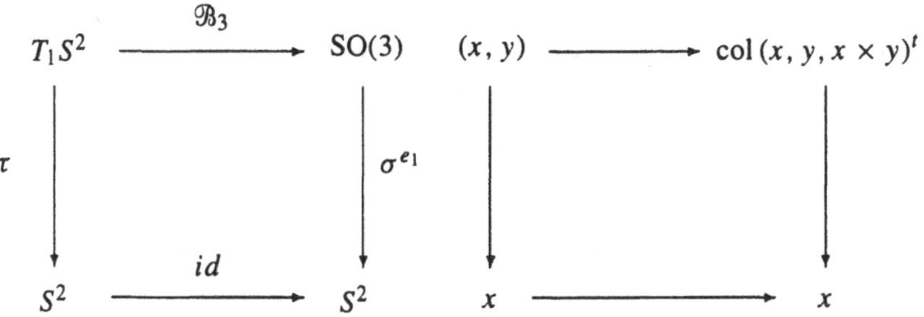

Diagram 3.3

Composing the bundle isomorphisms $(\mathcal{B}_3, id)$, $(\mathcal{B}_2, id)$ and $(\mathcal{B}_1, \beta_1)$ shows that the principal bundles τ and π_μ are isomorphic. □

3.3 Euler's equations

In this subsection we study the reduced Hamiltonian vector field of the Euler top. This vector field is obtained by removing the $\mathrm{SO}(3)_\mu$ symmetry of the Hamiltonian vector field X_H (29) restricted to the invariant manifold $J^{-1}(\mu)$. The integral curves of the reduced Hamiltonian vector field satisfy Euler's equations.

▷ We start by constructing the reduced Hamiltonian. Observe that the Hamiltonian of the Euler top

$$H : \mathrm{SO}(3) \times \mathrm{so}(3)^* \to \mathbf{R} : (A, \alpha) \to \tfrac{1}{2}\alpha\big((k^\flat(I^{-1})^t\alpha\big), \qquad (58)$$

when restricted to the μ-level set of the momentum J, is invariant under the left $\mathrm{SO}(3)_\mu$ action Φ (53).

(3.7) **Proof**: To see this recall that

$$J^{-1}(\mu) = \Big\{(A, \nu = \mathrm{Ad}^t_A\mu) \in \mathrm{SO}(3) \times \mathrm{so}(3)^* \big| A \in \mathrm{SO}(3)\Big\}.$$

Therefore

$$(H|J^{-1})(\mu)(A,\nu) = \nu\big(k^\flat (I^{-1})^t \nu\big).$$

Since $\Phi_B(A,\nu) = (BA,\nu)$ for every $B \in \mathrm{SO}(3)_\mu$, we see that $H|J^{-1}(\mu)$ is invariant under the action Φ. □

Thus $H|J^{-1}(\mu)$ induces a function H_μ on the $\mathrm{SO}(3)_\mu$-orbit space $J^{-1}(\mu)/\mathrm{SO}(3)_\mu = \mathcal{O}_\mu$ defined by $\pi_\mu^* H_\mu = H|J^{-1}(\mu)$. Here π_μ is the reduction mapping

$$\pi_\mu : J^{-1}(\mu) \subseteq \mathrm{SO}(3) \times \mathrm{so}(3)^* \to \mathcal{O}_\mu \subseteq \mathrm{so}(3)^* : (A,\nu) \to \nu. \tag{59}$$

Hence the reduced Hamiltonian is

$$H_\mu : \mathcal{O}_\mu \subseteq \mathrm{so}(3)^* \to \mathbf{R} : \nu \to \tfrac{1}{2}\,\nu\Big(k^\flat\big((I^{-1})^t(\nu)\big)\Big). \tag{60}$$

Since $\mathcal{O}_\mu$ is a symplectic manifold with symplectic form $\omega_{\mathcal{O}_\mu}$ (55), we obtain the reduced Hamiltonian system $(H_\mu, \mathcal{O}_\mu, \omega_{\mathcal{O}_\mu})$.

To see that the integral curves of X_{H_μ} on $\mathcal{O}_\mu$ satisfy Euler's equations, we need another model for the reduced system. Consider the diffeomorphism

$$\widetilde{J} = k^\sharp \circ i^{-1} : S_r^2 \subseteq \mathbf{R}^3 \to \mathcal{O}_\mu \subseteq \mathrm{so}(3)^* : p \to \nu \tag{61}$$

where S_r^2 is the 2-sphere of radius $r = |\mu|$. Pulling back the reduced Hamiltonian H_μ by $\widetilde{J}$ gives the Hamiltonian

$$H_r : S_r^2 \subseteq \mathbf{R}^3 \to \mathbf{R} : p = (p_1, p_2, p_3) \to \tfrac{1}{2}\,(I_1^{-1} p_1^2 + I_2^{-1} p_2^2 + I_3^{-1} p_3^2). \tag{62}$$

▷ The following calculation shows that $(\widetilde{J})^* \omega_{\mathcal{O}_\mu} = \omega_r$, where ω_r is the symplectic form on S_r^2 given by

$$\omega_r(p)(p \times x_1, p \times x_2) = -(p, x_1 \times x_2), \tag{63}$$

for $x_1, x_2 \in \mathbf{R}^3$.

(3.8) **Proof**: For some $A \in \mathrm{SO}(3)$, we may write $p = A(\widetilde{J}^{-1}(\mu))$. Then

$$i^{-1}(p) = (i^{-1} \circ A^{-1} \circ i)\big(k^\flat(\mu)\big) = \mathrm{Ad}_{A^{-1}}(k^\flat(\mu)) = k^\flat(\mathrm{Ad}_A^t \mu) = k^\flat(\nu).$$

Let $\xi = i^{-1}(x_1)$, then

$$k^\flat(\mathrm{ad}_\xi^t \nu) = -\mathrm{ad}_\xi(k^\flat(\nu)) = \Big[i^{-1}(p), i^{-1}(x_1)\Big] = i^{-1}(p \times x_1).$$

Similarly, if $\eta = i^{-1}(x_2)$, $k^\flat(\mathrm{ad}_\eta^t \nu) = i^{-1}(p \times x_2)$. We compute $(\widetilde{J})^* \omega_{\mathcal{O}_\mu}$ as follows:

$$\begin{aligned}
((\widetilde{J})^* \omega_{\mathcal{O}_\mu})(p)(p \times x_1, p \times x_2) &= \\
&= \omega_{\mathcal{O}_\mu}(\widetilde{J}(p))\big(T_p\widetilde{J}(p \times x_1), T_p\widetilde{J}(p \times x_2)\big) \\
&= \omega_{\mathcal{O}_\mu}(\widetilde{J}(p))\big(\widetilde{J}(p \times x_1), \widetilde{J}(p \times x_2)\big), \\
&\qquad \text{since } \widetilde{J} \text{ is the restriction of a linear map} \\
&= \omega_{\mathcal{O}_\mu}(\nu)(\mathrm{ad}_\xi^t \nu, \mathrm{ad}_\eta^t \nu) = -\nu([\xi, \eta]) \\
&= -\widetilde{J}(p)(i^{-1}(x_1 \times x_2)) = -k\big(i^{-1}(p), i^{-1}(x_1 \times x_2)\big) \\
&= -(p, x_1 \times x_2). \qquad \square
\end{aligned}$$

The integral curves of the reduced Hamiltonian vector field X_{H_μ} on $(\mathcal{O}_\mu, \omega_{\mathcal{O}_\mu})$ satisfy the
▷ equations $\dot{\nu} = \mathrm{ad}^t_{k^\flat((I^{-1})^t\nu)}\nu$, (see appendix B section 4 example 2). Pulling X_{H_μ} back by $\widetilde{J}$ gives the vector field X_{H_r} on (S_r^2, ω_r) whose integral curves satisfy Euler's equations

$$\dot{p} = p \times (I')^{-1}p \tag{64}$$

in angular momentum coordinates. Here $(I')^{-1}p = (I_1^{-1}p_1, I_2^{-1}p_2, I_3^{-1}p_3)$.

(3.9) **Proof**: Since X_{H_r} is a vector field on S_r^2, we may write $X_{H_r}(p) = p \times X$ for some $X = X(p) \in \mathbf{R}^3$. From the definition of Hamiltonian vector field and the symplectic form ω_r (63) we obtain

$$dH_r(p)(p \times x) = -(p, X \times x)$$

for every $x \in \mathbf{R}^3$. Differentiating (62) gives $dH_r(p)(p \times x) = \big((I')^{-1}(p), p \times x\big)$ for every $x \in \mathbf{R}^3$, that is, $((I')^{-1}(p) - X) \times p = 0$. Consequently, for some $\lambda \in \mathbf{R}$ we have $X = \lambda p + (I')^{-1}(p)$. In other words,

$$X_{H_r}(p) = p \times (\lambda p + (I')^{-1}(p)) = p \times (I')^{-1}(p).$$

Thus the integral curves of X_{H_r} satisfy Euler's equations (64). □

4 Qualitative behavior of the reduced system

In this section we give a global qualitative description of the solutions of Euler's equations (64) on the reduced space S_r^2. (For a quantitive treatment see exercise 2). This amounts to finding the topology of the *level sets* of the reduced Hamiltonian H_r (62).

▷ We begin by showing that H_r is a Morse function on S_r^2 with six critical points.

(4.1) **Proof**: Since S_r^2 is compact, H_r has a critical point q. Using Lagrange multipliers, we see that $q = (x, y, z)$ is a solution of the equations

$$\begin{aligned} 0 &= DH_r(q) - \lambda DG(q) = (I_1^{-1}x, I_2^{-1}y, I_3^{-1}z) - \lambda(x, y, z) \\ 0 &= G(q) = \tfrac{1}{2}(x^2 + y^2 + z^2 - r^2). \end{aligned}$$

In other words, q lies on the intersection of an eigenspace of the diagonal matrix $(I')^{-1} = \mathrm{diag}(I_1^{-1}, I_2^{-1}, I_3^{-1})$ with the 2-sphere S_r^2 of radius r. From now on we assume that

$$0 < I_3^{-1} < I_2^{-1} < I_1^{-1}. \tag{65}$$

Since the eigenvalues of $(I')^{-1}$ are distinct, the eigenspaces corresponding to the eigenvalues $I_1^{-1}, I_2^{-1}, I_3^{-1}$ are spanned by the vectors re_1, re_2, re_3 respectively, which lie on S_r^2. Therefore H_r has six critical points $\pm re_i$ $i = 1, 2, 3$ with corresponding Lagrange multiplier I_i^{-1}. At the critical point q with Lagrange multiplier λ the Hessian of H_r is

$$\begin{aligned} \mathrm{Hess}\, H_r(q) &= \big(D^2H_r(q) - \lambda D^2G(q)\big)\Big|_{T_qS_r^2} \\ &= \mathrm{diag}(I_1^{-1} - \lambda, I_2^{-1} - \lambda, I_3^{-1} - \lambda)\Big|_{T_qS_r^2}, \end{aligned}$$

(see appendix E section 2). Since

$$T_qS_r^2 = \ker DG(q) = \left\{ v \in \mathbf{R}^3 \middle| (v, q) = 0 \right\},$$

the tangent space $T_{\pm re_i}S_r^2$ is spanned by $\{e_j, e_k\}$ where $j \neq i, k \neq i$ and $k \neq j$. Therefore

$$\begin{aligned}
\text{Hess}\, H_r(\pm re_1) &= \text{diag}(I_2^{-1} - I_1^{-1}, I_3^{-1} - I_1^{-1}) \\
\text{Hess}\, H_r(\pm re_2) &= \text{diag}(I_1^{-1} - I_2^{-1}, I_3^{-1} - I_2^{-1}) \\
\text{Hess}\, H_r(\pm re_3) &= \text{diag}(I_1^{-1} - I_3^{-1}, I_2^{-1} - I_3^{-1}).
\end{aligned}$$

Because (65) holds, we see that q is a nondegenerate critical point. Thus H_r is a Morse function. In particular, the Morse index of H_r at the critical point $\pm re_i$ is 2 if $i = 1$, 1 if $i = 2$, and 0 if $i = 3$. Hence H_r has two maxima, two saddle points and two minima. □

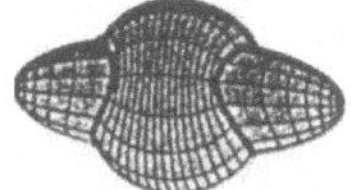
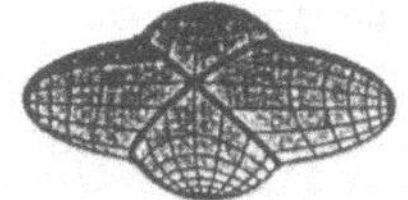
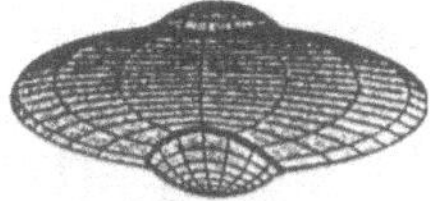

Figure 4.1. Intersection of $\mathscr{E}_h$ with S_r^2.

Geometrically the level set $H_r^{-1}(h)$ of the Hamiltonian is the intersection of a triaxial ellipsoid

$$\mathscr{E}_h : \tfrac{1}{2}(I_1^{-1}p_1^2 + I_2^{-1}p_2^2 + I_3^{-1}p_3^2) = h$$

with the 2-sphere $p_1^2 + p_2^2 + p_3^2 = r^2$, (see figure 4.1).

▷ To show the pictures in figure 4.1 are qualitatively correct we use Morse theory.

(4.2) **Proof**: First we show that the $\frac{1}{2}r^2I_3^{-1}$-level set of H_r on S_r^2 is two points $\{\pm re_3\}$. The point $q = (x, y, z)$ lies on $H_r^{-1}(\frac{1}{2}r^2I_3^{-1})$ if and only if

$$\tfrac{1}{2}(I_1^{-1}x^2 + I_2^{-1}y^2 + I_3^{-1}z^2) = \tfrac{1}{2}r^2I_3^{-1} \tag{66}$$

$$x^2 + y^2 + z^2 = r^2. \tag{67}$$

Multiplying (67) by I_3^{-1} and subtracting the result from two times (66) gives

$$(I_1^{-1} - I_3^{-1})x^2 + (I_2^{-1} - I_3^{-1})y^2 = 0. \tag{68}$$

Since (65) holds, $I_1^{-1} - I_3^{-1} > 0$, and $I_2^{-1} - I_3^{-1} > 0$. Thus (68) yields $x = y = 0$. Consequently, $z = \pm r$.

Since the critical points $\{\pm re_3\}$ are nondegenerate minima, we may apply the Morse lemma (see appendix E section 2) to conclude that for a value of h slightly greater than $\frac{1}{2}r^2I_3^{-1}$, the level set $H_r^{-1}(h)$ is diffeomorphic to two disjoint circles, one in the neighborhood of re_3 and the other in the neighborhood of $-re_3$. Thus $H_r^{-1}(h)$ is *not* connected. A similar argument shows that for h slightly less than $\frac{1}{2}r^2I_1^{-1}$, the level set $H_r^{-1}(h)$ is also the disjoint union of two circles near $\pm re_1$. Since H_r has no critical values in

$$\mathcal{I} = (\tfrac{1}{2}r^2I_3^{-1}, \tfrac{1}{2}r^2I_2^{-1}) \coprod (\tfrac{1}{2}r^2I_2^{-1}, \tfrac{1}{2}r^2I_1^{-1}), \tag{69}$$

using the Morse isotopy lemma (see appendix E section 3) we deduce that for $h \in \mathcal{I}$ the h-level set of H_r is diffeomorphic to the disjoint union of two circles.

To describe the remaining level set $H_r^{-1}(\frac{1}{2}r^2I_2^{-1})$ we note that $q = (x, y, z)$ lies in the $\frac{1}{2}r^2I_2^{-1}$-level set of H_r if and only if

$$\begin{aligned} \tfrac{1}{2}(I_1^{-1}x^2 + I_2^{-1}y^2 + I_3^{-1}z^2) &= \tfrac{1}{2}r^2I_2^{-1} &(70)\\ x^2 + y^2 + z^2 &= r^2. &(71) \end{aligned}$$

Multiplying (71) by I_2^{-1} and subtracting the result from two times (70) gives

$$0 = (I_2^{-1} - I_3^{-1})z^2 - (I_2^{-1} - I_1^{-1})x^2.$$

Thus the $\frac{1}{2}r^2I_2^{-1}$-level set of H_r is the intersection of the two 2-planes

$$\Pi_\pm = \left\{(x, y, z) \in \mathbf{R}^3 \,\middle|\, (\sqrt{I_2^{-1} - I_3^{-1}})z = \pm(\sqrt{I_2^{-1} - I_1^{-1}})x\right\}$$

with the 2-sphere S_r^2. Since $\Pi_+ \neq \Pi_-$ and $\Pi_+ \cap \Pi_-$ is the line spanned by the vector e_2, the level set $H_r^{-1}(\frac{1}{2}r^2I_2^{-1})$ is an *algebraic variety* $\widetilde{V}$, which is the union of two circles which intersect each other at $\{\pm re_2\}$. Because $\{\pm re_2\}$ are nondegenerate saddle points, using the Morse lemma we see that these circles intersect transversely. This completes the verification of figure 4.1. □

The information we have obtained about about the topology of the level sets of H_r is summarized in figure 4.2.

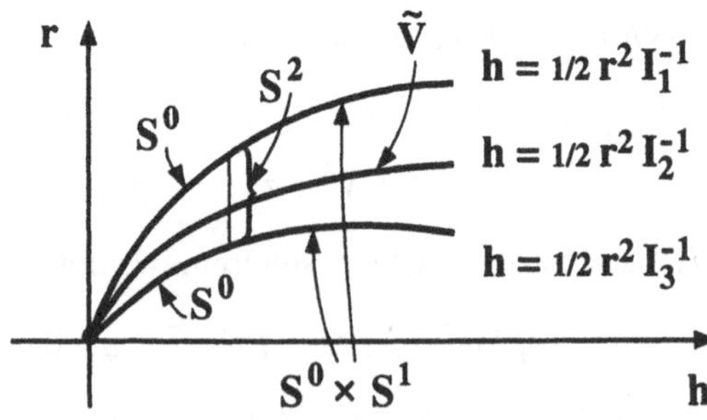

Figure 4.2. Bifurcation diagram for the level sets of H_r.

Putting all the pictures in figure 4.1 together gives figure 4.3. Since the connected components of the level sets of H_r on S_r^2 are orbits of the reduced vector field X_{H_r}, figure 4.3 gives a qualitative description of the solutions of Euler's equations of the Euler top.

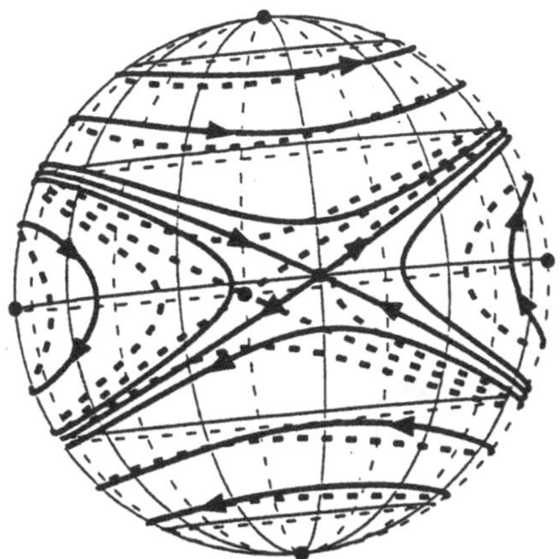

Figure 4.3. Level sets of H_r on S_r^2.

▷ We now verify that the orientations of the integral curves of the vector field X_{H_r} on S_r^2 in figure 4.3 are correct.

(4.3) **Proof**: By continuity, it suffices to show that the linearization of X_{H_r} at $(0,0,r)$ is an infinitesimal counterclockwise rotation around the positive p_3-axis. Differentiating (64) gives

$$DX_{H_r}(p_1, p_2, p_3) = \begin{pmatrix} 0 & (I_3^{-1} - I_2^{-1})p_3 & (I_3^{-1} - I_2^{-1})p_2 \\ -(I_3^{-1} - I_1^{-1})p_3 & 0 & -(I_3^{-1} - I_1^{-1})p_1 \\ (I_2^{-1} - I_1^{-1})p_2 & (I_2^{-1} - I_1^{-1})p_1 & 0 \end{pmatrix}.$$

Since $T_{(0,0,r)}S_r^2$ is spanned by $\{e_1, e_2\}$, the linearization of X_{H_r} at $(0,0,r)$ is

$$Y = DX_{H_r}(0,0,r)|T_{(0,0,r)}S_r^2 = \begin{pmatrix} 0 & -r(I_2^{-1} - I_3^{-1}) \\ r(I_1^{-1} - I_3^{-1}) & 0 \end{pmatrix}.$$

Let $P = \begin{pmatrix} u & 0 \\ 0 & u^{-1} \end{pmatrix}$ where $u = (I_2^{-1} - I_3^{-1})/(I_1^{-1} - I_3^{-1})$. A calculation shows that $Z = P^{-1}YP = \begin{pmatrix} 0 & -s \\ s & 0 \end{pmatrix}$, where $s = r\sqrt{(I_1^{-1} - I_3^{-1})(I_2^{-1} - I_3^{-1})}$. Since $s > 0$, Z is an infinitesimal counterclockwise rotation about the p_3-axis. Hence Y is also. □

5 Analysis of energy momentum mapping

In this section we study the energy momentum mapping $\mathcal{EM}$ of the Euler top:

$$\mathcal{EM} : \mathrm{SO}(3) \times \mathrm{so}(3)^* \to \mathbf{R} \times \mathrm{so}(3)^* :$$
$$(A, \alpha) \to \big(H(A,\alpha), J(A,\alpha)\big) = \big(\tfrac{1}{2}\,\alpha(k^\flat(I^{-1})^t\alpha), \mathrm{Ad}^t_{A^{-1}}\alpha\big).$$

Our main goal is to determine the topology of its fibers $\mathcal{EM}^{-1}(h, \mu)$, because these fibers are invariant under the flow of the Euler-Arnol'd vector field X_H, (see table 5.1). We also describe how the fibers $\mathcal{EM}^{-1}(h, \mu)$ foliate a level set $J^{-1}(\mu)$ of constant angular momentum μ, (see figure 5.2). This gives a complete qualitative description of the invariant sets of the Euler-Arnol'd vector field.

(h, μ)	topology of $H_r^{-1}(h)$	topology of $\mathcal{EM}^{-1}(h, \mu)$
$(0, 0)$	point	SO(3)
$h = \frac{1}{2} r^2 I_i^{-1}$ $i = 1, 3$	two points	$S^1 \coprod S^1$ whose double cover is once linked.
$h \in \mathcal{I}, r \neq 0$	$S^1 \coprod S^1$	$T^2 \coprod T^2$
$h = \frac{1}{2} r^2 I_2^{-1}$	$\widetilde{V}$, the union of two circles on S^2 which intersect transversely at two points.	$\widetilde{W}$, the union of two T^2 in SO(3) which intersect along two circles, whose double cover is once linked.

Table 5.1 Topology of the fibers of the energy momentum map $\mathcal{EM}$.

▷ To reconstruct the fiber $\mathcal{EM}^{-1}(h, \mu)$ from the h-level set of the reduced Hamiltonian H_r, observe that

$$\mathcal{EM}^{-1}(h, \mu) = \pi_r^{-1}(H_r^{-1}(h)), \tag{72}$$

where

$$\pi_r : J^{-1}(\mu) \subseteq \mathrm{SO}(3) \times \mathrm{so}(3)^* \to S_r^2 \subseteq \mathbf{R}^3 : (A, \mathrm{Ad}_A^t \mu) \to A^{-1}(i \circ k^\flat(\mu)) \tag{73}$$

is the reduction mapping.

(5.1) **Proof**: From the definition of the reduced Hamiltonian H_r (62) and the reduction mapping π_r (73), we find that

$$\pi_r^{-1}(H_r^{-1}(h)) = \left(H|J^{-1}(\mu)\right)^{-1}(h) = H^{-1}(h) \cap J^{-1}(\mu) = \mathcal{EM}^{-1}(h, \mu).$$

Using table 5.1 we obtain the bifurcation diagram figure 5.1. □

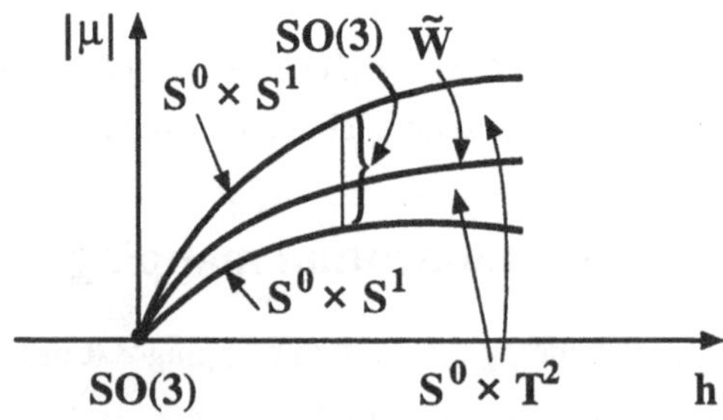

Figure 5.1. The Bifurcation diagram for the Euler top.

▷ The following argument verifies the entries in the third column of table 5.1.

(5.2) **Proof**: First we find the critical points and critical values of $\mathcal{EM}$.

Suppose that $\mu = 0$. Then

$$J^{-1}(0) = \Big\{ (A, 0) \in \mathrm{SO}(3) \times \mathrm{so}(3) \Big| \, A \in \mathrm{SO}(3) \Big\}$$

is diffeomorphic to SO(3). On $J^{-1}(0)$ the Hamiltonian H is the constant function 0. Therefore, every point of $J^{-1}(0)$ is a critical point of $H|J^{-1}(0)$ and hence is a critical point of $\mathcal{EM}$. The corresponding critical value of $\mathcal{EM}$ is $(0, 0)$.

From now on we suppose that $\mu \neq 0$. Then $J^{-1}(\mu)$ is diffeomorphic to SO(3).

Claim : The function

$$H|J^{-1}(\mu) : J^{-1}(\mu) \to \mathbf{R} : (A, \mathrm{Ad}^t_A \mu) \to \tfrac{1}{2}\, k\big(k^\flat(\mathrm{Ad}^t_A \mu), I^{-1}(k^\flat(\mathrm{Ad}^t_A \mu))\big)$$

is an $\mathrm{SO}(3)_\mu$–invariant Bott–Morse function with six nondegenerate critical $\mathrm{SO}(3)_\mu$-orbits

$$\gamma_i^\pm = \pi_r^{-1}(\pm r e_i)$$

for $i = 1, 2, 3$, two of Morse index 2, 1 and 0 respectively.

(5.3) **Proof**: By construction $H|J^{-1}(\mu) = (\pi_r)^* H_r$. Therefore p is a critical point of $H|J^{-1}(\mu)$ if and only if

$$D\big(H|J^{-1}(\mu)\big)(p) = DH_r(\pi_r(p)) D\pi_r(p)$$

is not surjective. Since π_r is a submersion, p is a critical point if and only if $DH_r(\pi_r(p))$ is not surjective, that is, if and only if $\pi_r(p)$ is a critical point of H_r. By ((4.1)) the set of critical points of of H_r is $\{\pm r\, e_i | \, i = 1, 2, 3\}$. Therefore the six $\mathrm{SO}(3)_\mu$-orbits

$$\gamma_i^\pm = \Big\{ \pi_r^{-1}(\pm r\, e_i) \Big| i = 1, 2, 3 \Big\}$$

form the set of critical points of $H|J^{-1}(\mu)$. Since an $\mathrm{SO}(3)_\mu$-orbit is diffeomorphic to a circle, the critical set of $H|J^{-1}(\mu)$ is the union of six circles $\gamma_i^\pm$ corresponding to the critical values $\frac{1}{2} r^2\, I_i^{-1}$.

We now show that each of the $\gamma_i^\pm$ is a nondegenerate critical manifold, (see appendix E section 2). Choose an open neighborhood $\mathcal{U}_i^\pm$ of $\pm r e_i$ in S_r^2 such that the bundle $\pi_r : J^{-1}(\mu) \to S_r^2$, when restricted to $\pi_r^{-1}(\mathcal{U}_i^\pm)$, is a trivial $\mathrm{SO}(3)_\mu$ principal bundle. Then there is a diffeomorphism

$$\varphi_i^\pm : \pi_r^{-1}(\mathcal{U}_i^\pm) \subseteq J^{-1}(\mu) \to \mathcal{U}_i^\pm \times \mathrm{SO}(3)_\mu$$

which intertwines the $\mathrm{SO}(3)_\mu$-action Φ (53) on $\pi_r^{-1}(\mathcal{U}_i^\pm)$ with the $\mathrm{SO}(3)_\mu$-action on $\mathcal{U}_i \times \mathrm{SO}(3)_\mu$ given by

$$\big(B, (q, A)\big) \to (q, BA).$$

If we parametrize the $SO(3)_\mu$-orbit $\gamma_i^\pm$ by $t \to \Phi_{\exp t\xi_i^\pm}(p_i^\pm)$ where $\xi_i^\pm \in so(3)_\mu$ and $p_i^\pm \in \gamma_i^\pm$, then $\varphi_i^\pm(\gamma_i^\pm)$ is parametrized by $t \to (\pm r\, e_i, \exp t\xi_i^\pm)$. Thus for fixed t_0 the set

$$\mathcal{S}_i^\pm = (\varphi_i^\pm)^{-1}(\mathcal{U}_i^\pm \times \{\exp t_0 \xi^\pm\})$$

is a slice to $\gamma_i^\pm$ at $\gamma_i^\pm(t_0)$. Since the mapping π_r restricted to $\mathcal{S}_i^\pm$ is a diffeomorphism onto $\mathcal{U}_i^\pm$, the Morse index of $H|J^{-1}(\mu)$ restricted to $\mathcal{S}_i^\pm$ at $\gamma_i^\pm(t_0)$ is the same as the Morse index of H_r at $\pm re_i$. Therefore $\gamma_i^\pm$ is a nondegenerate critical manifold of $H|J^{-1}(\mu)$. □

We return to verifying the third column of table 5.1. Suppose that h is a regular value of the Hamiltonian H_r. Then for $h \in \mathcal{I}$ (69) the level set $H_r^{-1}(h)$ is diffeomorphic to the disjoint union of two circles S_j^1. Since each S_j^1 is null homotopic in S_r^2, the circle S_j^1 bounds a disc $\overline{D}_j^2$, which is contractible in S_r^2. Therefore, the bundle $\pi_r : J^{-1}(\mu) \to S_r^2$ restricted to $\pi_r^{-1}(\overline{D}_j^2)$ is trivial, that is, $\pi_r^{-1}(\overline{D}_j^2)$ is diffeomorphic to $\overline{D}_j^2 \times S^1$, (see appendix C section 2). Hence, the manifold $\mathcal{EM}^{-1}(h,\mu) = \pi_r^{-1}(H_r^{-1}(h))$ is the disjoint union of two 2-tori $\partial\overline{D}_j^2 \times S^1$.

Now suppose that $h = \frac{1}{2}r^2 I_1^{-1}$ or $h = \frac{1}{2}r^2 I_3^{-1}$. Then the h-level set of H_r is the disjoint union of two points. Therefore, the manifold $\mathcal{EM}^{-1}(h,\mu)$ is the disjoint union of two circles, whose double covers are linked once. This follows because the reduction map π_r is double covered by the Hopf map and any two distinct fibers of the Hopf fibration are linked once.

Finally, suppose that $h = \frac{1}{2}r^2 I_2^{-1}$. Then $H_r^{-1}(h)$ is the union of two circles C_j which intersect each other transversely at $\pm re_2$. Since each of the circles bounds a contractible disc on S_r^2, each set $\pi_r^{-1}(C_j)$ is a 2-torus T_j^2. Using the Morse lemma (appendix E section 2), we see that these tori intersect each other transversely in $J^{-1}(\mu)$ along two ▷ circles $\gamma_2^\pm$. Hence $\gamma_2^\pm$ is a hyperbolic periodic orbit of the vector field $X_H|J^{-1}(\mu)$.

(5.4) **Proof**: Since $(\pi_r)^*\omega_r = \Omega|J^{-1}(\mu)$, the reduction mapping π_r is a symplectic diffeomorphism of the slice $(\mathcal{S}_2^\pm, \Omega|(\mathcal{S}_2^\pm)$ onto $(\mathcal{U}_2^\pm, \omega_r|\mathcal{U}_2^\pm)$. Therefore the vector field $X_H|\mathcal{S}_2^\pm$ pushes forward under π_r to the vector field $X_{H_r}|\mathcal{U}_2^\pm$, which clearly has a hyperbolic equilibrium point at $\pm re_2$. □

The union of the two 2-tori $\pi_r^{-1}(C_j)$ is the variety $\widetilde{W} = \pi_r^{-1}(\widetilde{V})$, which is the union of the closures of the stable and unstable manifolds of the hyperbolic periodic orbits γ_2^+ ▷ and γ_2^-. The local unstable manifold of $\gamma_2^\pm$ is *not* twisted.

(5.5) **Proof**: To see this, let $\mathcal{V}_2^\pm$ be an open neighborhood of $\pm re_2$ in S_r^2. Then the intersection of $\mathcal{V}_2^\pm$ with the closure of the local unstable manifold of $\pm re_2$ is a contractible open subset of S_r^2. Consequently, the closure of the local unstable manifold of $\gamma_2^\pm$ is a trivial bundle over S^1. Thus it is not twisted. □

This completes checking the third column of table 5.1. □

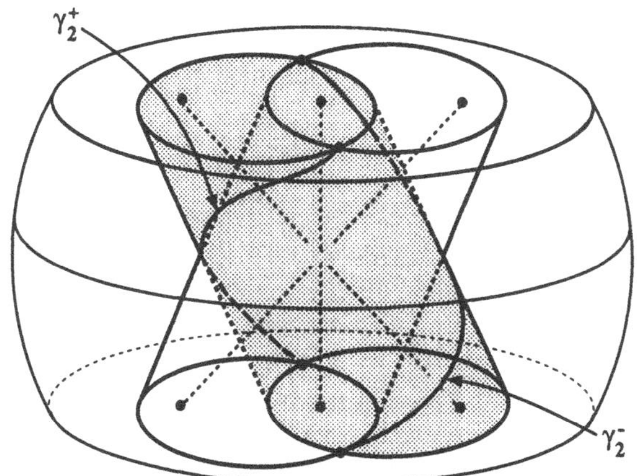

Figure 5.2. The fibration of $J^{-1}(\mu)$ by the level sets of H.

▷ We now show that figure 5.2 give a qualitatively correct picture of the level sets of H on $J^{-1}(\mu)$.

(5.6) **Proof**: Let $\overline{D} \subseteq S_r^2$ be a small closed 2-disc about the north pole $(0, 0, r)$ of the two sphere S_r^2. Using stereographic projection pr from the north pole, we find that the image of $S_r^2 - D$ is the closed 2-disc E in $\mathbf{R}^2$, (see figure 5.3). Since E is contractible, the bundle $\pi_r|\pi_r^{-1}(E)$ is trivial, that is, $\pi_r^{-1}(E)$ is diffeomorphic to the solid torus $S^1 \times E$. Thus we obtain a decomposition of $J^{-1}(\mu)$ (which is diffeomorphic to SO(3)) into the union of two solid tori $ST_1 = \pi_r^{-1}(\overline{D})$ and $ST_2 = \pi_r^{-1}(E)$.

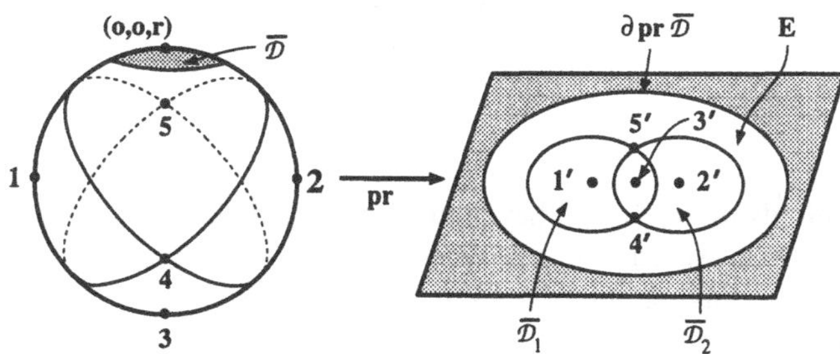

Figure 5.3. Stereographic projection of the level curves of H_r.

Let us investigate more carefully how the solid torus ST_2 fits into SO(3). First remove ST_2 from the solid ball D^3. Clearly, ST_2 is homeomorphic to $S^1 \times (\overline{D}_1 \cup \overline{D}_2)$, which is a solid torus on the two overlapping 2-discs $\overline{D}_1$ and $\overline{D}_2$. The solid torus ST_2 is formed by

1. Taking the cylinder $\mathscr{C} = [0, 1] \times (\overline{D}_1 \cup \overline{D}_2)$ and giving it a number of half twists.

2. Placing the result in the 3-disc D^3 less $\mathscr{C}$.

3. Identifying antipodal points on the two end 2-discs of $\mathcal{C}$ and also on $\partial(D^3 - \mathcal{C})$.

In figure 5.4 we illustrate this construction when the cylinder $\mathcal{C}$ has undergone *zero* half twists. The lines AA', BB', CC' in this figure are the center lines of the cylinders

$$[0,1] \times \overline{D}_1 \qquad [0,1] \times (\overline{D}_1 \cap \overline{D}_2) \qquad [0,1] \times \overline{D}_2.$$

After antipodal identification of the end 2-discs, we obtain *two* solid tori with center circles $AA'CC'$ and BB'. In figure 5.4 there are only *three* fibers which are center circles of solid tori. These fibers correspond to critical submanifolds of $H|J^{-1}(\mu)$ of index 0 or 2. According to ((5.3)) there are *four* such critical submanifolds. Hence figure 5.4 does *not* describe the foliation of $J^{-1}(\mu)$ by level sets of H. An obvious generalization of the above argument eliminates the possibility of replacing the cylinder $\mathcal{C}$ with an arbitary *even* number of half twists. Suppose that $\mathcal{C}$ is replaced in $D^3 - \mathcal{C}$ after an *odd* number of half twists *greater* than one. For the sake of argument say three. It is clear that the curves AA', BB', CC' and DD' are center circles of solid tori. However, the double cover of AA' and BB' in S^3 has linking number three, (see exercise 12). Thus AA' and BB' do *not* correspond to any critical submanifold of $H|J^{-1}(\mu)$ because the double covers of the

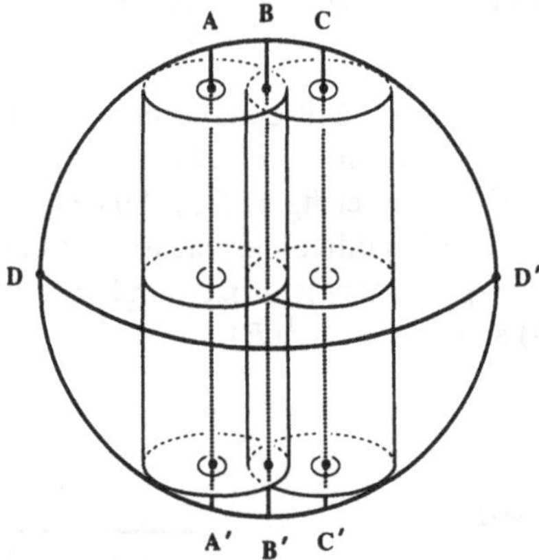

Figure 5.4. Replace the cylinder $\mathcal{C}$ with zero half twists.

critical manifolds of $H|J^{-1}(\mu)$ have linking number one. Thus we can only replace $\mathcal{C}$ with *one* half twist. Whether this is a clockwise or counterclockwise half twist depends on the *sign* of the linking number. We determine this sign as follows. Orient SO(3) so that its double cover S^3 is positively oriented. Give the solid tori ST_i in SO(3) the induced orientation. Orient a 2-disc in ST_i, which is transverse to the center circle, so that its image under the reduction map has the same orientation as a solution to Euler's equations which it contains. This, together with the orientation of ST_i, determines the orientation of the center circles $A'A$ and $B'B$. As integral curves of X_H on $J^{-1}(\mu)$, the curves $A'A$ and $B'B$ are positively oriented. Hence their double covers in S^3, which are integral curves of the harmonic oscillator on an energy surface, are also positively oriented. Thus their linking number in S^3 is $+1$. Therefore the cylinder $\mathcal{C}$ is given a *counterclockwise* half twist when looking in the direction of the positively oriented curve $B'B$.

This completes the verification of figure 5.2. □

6 Integration of the Euler-Arnol'd equations

In this section we integrate the equations

$$\begin{cases} \dfrac{dx}{dt} = I_3^{-1} p_3 y - I_2^{-1} p_2 (x \times y) \\ \dfrac{dy}{dt} = -I_3^{-1} p_3 x + I_1^{-1} p_1 (x \times y) \end{cases} \tag{74}$$

$$\frac{dp}{dt} = p \times (I')^{-1} p, \tag{75}$$

which are the Euler-Arnol'd equations in the sphere bundle model $(T_1 S^2 \times \mathbf{R}^3, \omega')$, (see section 2.2). Here $I'p = (I_1 p_1, I_2 p_2, I_3 p_3)$. Solutions of these equations describe motions of the Euler top in space.

We begin by looking at certain invariant manifolds of the Euler-Arnol'd equations. A straightforward calculation shows that the angular momentum

$$J' : T_1 S^2 \times \mathbf{R}^3 \to \mathbf{R}^3 : (x, y, p) \to \mathrm{col}(x, y, x \times y) p \tag{76}$$

and energy

$$H' : T_1 S^2 \times \mathbf{R}^3 \to \mathbf{R} : (x, y, p) \to \tfrac{1}{2} (I_1^{-1} p_1^2 + I_2^{-1} p_2^2 + I_3^{-1} p_3^2). \tag{77}$$

are integrals of these equations. Suppose that $(h, |\mu| e_3)$ is a *regular value* of the energy momentum mapping

$$\mathcal{EM}' : T_1 S^2 \times \mathbf{R}^3 \to \mathbf{R} \times \mathbf{R}^3 : (x, y, p) \to \big(H'(x, y, p), J'(x, y, p) \big).$$

Then $(\mathcal{EM}')^{-1}(h, |\mu|\, e_3)$ is diffeomorphic to the disjoint union of two 2-tori. Call one of them $T^2_{h,\mu}$. Then $T^2_{h,\mu}$ is an invariant manifold of the Hamiltonian vector field $X_{H'}$, whose
▷ integral curves satisfy the Euler-Arnol'd equations. We wish to describe $T^2_{h,\mu}$ as a bundle whose projection map is

$$\pi : T^2_{h,\mu} \subseteq T_1 S^2 \times \mathbf{R}^3 \to \mathbf{R}^3 : (x, y, p) \to p. \tag{78}$$

(6.1) **Proof**: Observe that $\pi(T^2_{h,\mu})$ is a connected component of the h-level set of the reduced Hamiltonian

$$H_r : S^2_{|\mu|} \to \mathbf{R} : p \to \tfrac{1}{2} \left((I')^{-1} p, p \right)$$

and hence is diffeomorphic to a circle $S^1_{h,\mu}$, since h is a regular value of H_r. $S^1_{h,\mu}$ is parametrized by a periodic solution of Euler's equations (75).

To find an explicit description of $T^2_{h,\mu}$ we begin by solving $J'(x, y, p) = |\mu| e_3$ for p. Using (76) we obtain $|\mu|^{-1} p = (\mathrm{col}(x, y, x \times y))^t e_3$, that is,

$$x_3 = |\mu|^{-1} p_1 \tag{79}$$

$$y_3 = |\mu|^{-1} p_2 \tag{80}$$

$$x_1 y_2 - x_2 y_1 = |\mu|^{-1} p_3. \tag{81}$$

Substituting (79) and (80) into the defining equations of $T_1 S^2$

$$\begin{aligned} x_1^2 + x_2^2 + x_3^2 &= 1 \\ x_1 y_1 + x_2 y_2 + x_3 y_3 &= 0 \\ y_1^2 + y_2^2 + y_3^2 &= 1 \end{aligned} \tag{82}$$

gives

$$x_1^2 + x_2^2 = 1 - |\mu|^{-2} p_1^2 \tag{83}$$

$$x_1 y_1 + x_2 y_2 = -|\mu|^{-2} p_1 p_2 \tag{84}$$

$$y_1^2 + y_2^2 = 1 - |\mu|^{-2} p_2^2. \tag{85}$$

Because h is a regular value of the reduced Hamiltonian H_r, $p_i \neq \pm|\mu|$ for $i = 1, 2, 3$. Therefore the right hand side of (83) is nonzero for every $p \in S^1_{h,\mu}$. Write (81) and (84) as the linear system

$$\begin{pmatrix} -x_2 & x_1 \\ x_1 & x_2 \end{pmatrix} \begin{pmatrix} y_1 \\ y_2 \end{pmatrix} = \begin{pmatrix} |\mu|^{-1} p_3 \\ -|\mu|^{-2} p_1 p_2 \end{pmatrix}. \tag{86}$$

Using (83), we obtain the solution

$$\begin{aligned} y_1 &= -\frac{1}{|\mu|^2 - p_1^2} (p_1 p_2 x_1 + |\mu| p_3 x_2) \\ y_2 &= \frac{1}{|\mu|^2 - p_1^2} (|\mu| p_3 x_1 - p_1 p_2 x_2). \end{aligned} \tag{87}$$

Therefore the 2-torus $T^2_{h,\mu}$ is the set of points $(x, y, p) \in T_1 S^2 \times \mathbf{R}^3$ where

$$\begin{aligned} & x_1^2 + x_2^2 = 1 - |\mu|^{-2} p_1^2, \ x_3 = |\mu|^{-1} p_1, \\ & y_1 = -\frac{1}{|\mu|^2 - p_1^2} (p_1 p_2 x_1 + |\mu| p_3 x_2), \\ & y_2 = \frac{1}{|\mu|^2 - p_1^2} (|\mu| p_3 x_1 - p_1 p_2 x_2), \ y_3 = |\mu|^{-1} p_2, \end{aligned} \tag{88}$$

and $p = (p_1, p_2, p_3)$ lies in the connected component of

$$\begin{aligned} I_1^{-1} p_1^2 + I_2^{-1} p_2^2 + I_3^{-1} p_3^2 &= 2h \\ p_1^2 + p_2^2 + p_3^2 &= |\mu|^2, \end{aligned}$$

which defines $S^1_{h,\mu}$. This gives the desired description of the bundle π. □

▷ Because $T^2_{h,\mu}$ is a smooth invariant manifold of $X_{H'}$, the restriction of $X_{H'}$ to $T^2_{h,\mu}$ is a vector field Y whose integral curves satisfy

$$\dot{x}_1 = \alpha(p) x_1 - \beta(p) x_2 \tag{89}$$

$$\dot{x}_2 = \beta(p) x_1 + \alpha(p) x_2 \tag{90}$$

$$\dot{p} = p \times (I')^{-1} p. \tag{91}$$

Here $p \in S^1_{h,\mu}$ and

$$\alpha(p) = -\frac{(I_3^{-1} - I_2^{-1})\, p_1 p_2 p_3}{|\mu|^2 - p_1^2} \quad \text{and} \quad \beta(p) = \frac{|\mu|(I_2^{-1} p_2^2 + I_3^{-1} p_3^2)}{|\mu|^2 - p_1^2}. \tag{92}$$

.2) **Proof**: Consider the Euler-Arnol'd equations, (74) and (75), subject to the constraints (82). Restrict these equations to $(J')^{-1}(|\mu| e_3)$. In other words, impose the additional constraints (79)–(81). Using (79) and (80) eliminate x_3 and y_3 from (74). We obtain

$$\dot{x}_1 = I_3^{-1} p_3 y_1 - |\mu|^{-1} I_2^{-1} p_2 (p_2 x_2 - p_1 y_2) \tag{93}$$

$$\dot{x}_2 = I_3^{-1} p_3 y_2 - |\mu|^{-1} I_2^{-1} p_2 (p_1 y_1 - p_2 x_1) \tag{94}$$

$$\dot{y}_1 = -I_3^{-1} p_3 x_1 + |\mu|^{-1} I^{-1} p_1 (p_2 x_2 - p_1 y_2) \tag{95}$$

$$\dot{y}_2 = -I_3^{-1} p_3 x_2 + |\mu|^{-1} I_1^{-1} p_1 (p_1 y_1 - p_2 x_1) \tag{96}$$

together with (91) and the constraints (81), (83–85). Restrict (93–96) to $(H')^{-1}(h)$. Since (87) holds on $T^2_{h,\mu}$ (a connected component of $(J')^{-1}(|\mu| e_3) \cap (H')^{-1}(h)$), equations (95) and (96) hold as soon as (93) and (94) do. Substituting (87) into (93) and (94) to eliminate y_1 and y_2 gives (89) and (90) where the functions $\alpha(p)$ and $\beta(p)$ are defined by (92). Clearly, (91) holds. Since $\pi(T^2_{h,\mu}) = S^1_{h,\mu}$, it follows that $p \in S^1_{h,\mu}$. □

Because of the hierarchical nature of the equations (89–91), we can solve them as follows. Let $t \to p(t)$ be a solution of Euler's equations (91), which parametrizes $S^1_{h,\mu}$, (see exercise 2). Substituting $t \to p(t)$ into (89) and (90) gives the time dependent linear system

$$\begin{aligned} \dot{x}_1 &= \alpha(t) x_1 - \beta(t) x_2 \\ \dot{x}_2 &= \beta(t) x_1 + \alpha(t) x_2. \end{aligned} \tag{97}$$

Introduce polar coordinates $r^2 = x_1^2 + x_2^2$ and $\theta = \tan^{-1} \frac{x_2}{x_1}$. Then (97) becomes

$$\frac{d\theta}{dt} = \beta(t) \quad \text{and} \quad \frac{dr}{dt} = \alpha(t). \tag{98}$$

Integrating (98) gives

$$r^2(t) = 1 - |\mu|^{-2} p_1^2(t) \quad \text{and} \quad \theta(t) = \int_0^t \beta(s)\, ds + \theta(0). \tag{99}$$

Therefore,

$$x_1(t) = \sqrt{1 - |\mu|^{-2} p_1^2(t)}\, \cos\theta(t) \quad \text{and} \quad x_2(t) = \sqrt{1 - |\mu|^{-2} p_1^2(t)}\, \sin\theta(t).$$

Substituting $x_1(t)$ and $x_2(t)$ into (87) and using (79) and (80) gives $t \to \big(x(t), y(t), p(t)\big)$, where

$$x(t) = \Big(\sqrt{1 - |\mu|^{-2} p_1^2(t)}\, \cos\theta(t), \sqrt{1 - |\mu|^{-2} p_1^2(t)}\, \sin\theta(t), |\mu|^{-1} p_1(t)\Big) \tag{100}$$

$$y(t) = \Big(-\frac{|\mu|^{-2}}{\sqrt{1-|\mu|^{-2}p_1^2(t)}} [p_1(t)p_2(t)\cos\theta(t) + |\mu|p_3(t)\sin\theta(t)],$$

$$\frac{|\mu|^{-2}}{\sqrt{1-|\mu|^{-2}p_1^2(t)}} [|\mu|p_3(t)\cos\theta(t) - p_1(t)p_2(t)\sin\theta(t)], |\mu|^{-1}p_2(t)\Big). \quad (101)$$

Note that $t \to \big(x(t), y(t), p(t)\big)$ is a solution of the Euler-Arnol'd equations which lies in $T^2_{h,\mu} \subseteq T_1S^2 \times \mathbf{R}^3$ and under π projects onto $t \to p(t)$ which is a solution of Euler's equation of energy h and angular momentum $|\mu|$. □

7 The rotation number

In this section we give two ways of finding the rotation number of the flow of the Euler-Arnol'd equations of the Euler top. The first one is an analytic formula based on the integration of the Euler-Arnol'd equations given in the preceding section. The second is a geometric interpretation of the classical Poinsot construction.

7.1 An analytic formula

We begin by deriving an analytic formula for the rotation number. Recall that $S^1_{h,\mu}$ is a connected component of the level set of the reduced Hamiltonian H_r on $S^2_{|\mu|}$ corresponding to the regular value h. Thus $S^1_{h,\mu}$ is parametrized by a periodic solution $t \to p(t)$ of Euler's equations of minimal period $T = T(h, \mu)$. Let $t \to \theta(t)$ be the function defined by (99). The quantity $\Delta\theta(T) = \theta(T) - \theta(0)$ is an important invariant of the flow of the Euler-Arnol'd equations.

Claim: $\frac{1}{2\pi}\Delta\theta(T)$ is the rotation number of the flow of $X_{H'}$ on $T^2_{h,\mu}$.

(7.1) **Proof**: Let γ be the one parameter subgroup of $(T_1S^2, \cdot)$ (12) defined by

$$\gamma : \mathbf{R} \to T_1S^2 : t \to \big((\cos t, \sin t, 0), (-\sin t, \cos t, 0)\big).$$

Observe that the image of γ is the isotropy group $\mathrm{SO}(3)_{|\mu|e_3}$. Considered as a subgroup of $(T_1S^2, \cdot)$, $\mathrm{SO}(3)_{|\mu|e_3}$ is diffeomorphic to S^1. Also γ induces an action

$$\Phi : \mathbf{R} \times (T_1S^2 \times \mathbf{R}^3) \to T_1S^2 \times \mathbf{R}^3 : \big(t, (x, y), p\big) \to \big(\gamma(t) \cdot (x, y), p\big). \quad (102)$$

Φ leaves the 2-torus $T^2_{h,\mu}$ invariant because it leaves the h-level set of the Hamiltonian H' (77) and the $|\mu|e_3$-level set of the momentum J' (76) invariant. This last assertion is proved as follows. Let $R_t = \begin{pmatrix} \cos t & -\sin t & 0 \\ \sin t & \cos t & 0 \\ 0 & 0 & 1 \end{pmatrix}$. For every $(x, y, p) \in (J')^{-1}(|\mu|e_3)$

$$\begin{aligned} J'(\Phi_t(x, y, p)) &= J'\big(\widehat{\pi}(R_t\,\mathrm{col}(x, y, x \times y)), p\big), \\ &= R_t\big(J'(x, y, p)\big) = |\mu|e_3. \end{aligned}$$

Thus the induced action $\Phi' = \Phi|(S^1 \times T^2_{h,\mu})$ is defined.

Since the action Φ does not change the p-coordinate of a point on $T^2_{h,\mu}$, every orbit of Φ' is a fiber of the bundle π (78). Because under π every integral curve of $X_{H'}|T^2_{h,\mu}$ projects to the periodic integral curve $S^1_{h,\mu}$ of Euler's equations, every fiber of π is a cross section for the flow $\varphi_t^{H'}$ of $X_{H'}$ on $T^2_{h,\mu}$. From the fact that along $S^1_{h,\mu}$ the function $t \to \beta(t)$ (92) is strictly positive and bounded away from 0, it follows that the function $t \to \theta(t)$ (99) is strictly increasing and is unbounded as $t \to \pm\infty$. Therefore for every $q \in T^2_{h,\mu}$ the image of the curve $t \to \Phi'_{\theta(t)-\theta(0)}(q)$ is the fiber of the bundle π over $p = \pi(q) \in S^1_{h,\mu}$. Since

$$t \to p(t) = \pi(\varphi_t^{H'}(q))$$

starts at p, parametrizes $S^1_{h,\mu}$, and has minimal period T, we see that $\varphi_T^{H'}(q) \in \pi^{-1}(p)$. Therefore for every $q \in T^2_{h,\mu}$ there is a minimal $\tau > 0$ such that

$$\Phi'_{\theta(\tau)-\theta(0)}(q) = \varphi_T^{H'}(q). \tag{103}$$

Below we show that $\tau = T$. For now on assume this. Using (103) and the definition of rotation number, it follows that $\frac{1}{2\pi}\Delta(T) = \frac{1}{2\pi}\big(\theta(T) - \theta(0)\big)$ is the rotation number of the flow of $X_{H'}$ on $T^2_{h,\mu}$. □

▷ We now show that $\tau = T$.

(7.2) **Proof**: Let $\big(\xi(t), \eta(t), \rho(t)\big) = \Phi'_{\theta(t)-\theta(0)}(q) = \Phi_\varphi(q)$ and let $\big(x(t), y(t), p(t)\big) = \varphi_t^{H'}(q)$. Calculate $\xi(t)$ as follows. From the definition of the action Φ (102), we find that

$$\begin{aligned}
\xi(t) &= \big(x_1(0)\cos\varphi - x_2(0)\sin\varphi,\, x_1(0)\sin\varphi + x_2(0)\cos\varphi,\, x_3(0)\big), \\
&\qquad \text{where } q = \big(x(0), y(0), p\big) \text{ and } p(0) = p \\
&= \Big(\sqrt{1-|\mu|^2p_1^2}\cos(\varphi+\theta(0)),\, \sqrt{1-|\mu|^2p_1^2}\sin(\varphi+\theta(0)),\, |\mu|^{-1}p_1\Big), \\
&\qquad \text{using (100) to compute } x(0) \\
&= \Big(\sqrt{1-|\mu|^2p_1^2}\cos\theta(t),\, \sqrt{1-|\mu|^2p_1^2}\sin\theta(t),\, |\mu|^{-1}p_1\Big).
\end{aligned}$$

Comparing the above expression for $\xi(t)$ with the expression for $x(t)$ given in (100) and looking at their third components, we see that $\xi(t) = x(t)$ if and only if $t = nT$, because $t \to p_1(t)$ is periodic with minimal period T. Since τ in (103) is minimal, we conclude that $\tau = T$. □

The rotation number has the following physical interpretation. Recall that the action of SO(3) on itself under left multiplication by A corresponds to changing the space frame $\{e_1, e_2, e_3\}$ to the frame $\{Ae_1, Ae_2, Ae_3\}$. When A is in the isotropy group $SO(3)_{|\mu|e_3}$, left multiplication by A leaves e_3 fixed. Thus the action of the isotropy group corresponds to the Euler top rotating around the e_3-axis. Hence 2π times the rotation number is the amount of rotation about the spatial e_3-axis the top makes in time T. Here the top has energy h and angular momentum $|\mu|e_3$.

7.2 Poinsot's construction

In this subsection we present a modern version of Poinsot's construction. (For a traditional treatment of this construction see exercise 6). Poinsot's construction gives another geometric interpretation of the rotation number of the Euler top.

Let $(h, \mu) \in \mathbf{R} \times \mathrm{so}(3)^*$ be a regular value of the energy momentum mapping

$$\begin{aligned} \mathscr{EM} : T\,\mathrm{SO}(3) \rightarrow \mathbf{R} \times \mathrm{so}(3)^* : \\ \alpha_A \rightarrow \big(\mathscr{H}(\alpha_A), \mathscr{J}(\alpha_A)\big) = \big(\tfrac{1}{2}\rho(A)(\alpha_A, \alpha_A), (T_e R_A)^t(\alpha_A)\big) \end{aligned} \tag{104}$$

of the Euler top. Here ρ is the left invariant metric on SO(3) defined by (23) and R_A is right translation by A.

Consider the mapping

$$\mathscr{R} : T^*\,\mathrm{SO}(3) \rightarrow \mathrm{so}(3) : \alpha_A \rightarrow T_A R_{A^{-1}}(\rho(A)^\sharp(\alpha_A)). \tag{105}$$

Claim: The image of $\mathscr{EM}^{-1}(h, \mu)$ under the map $\mathscr{R}$ is contained in the affine subspace $\mathscr{A}_\mu = \{\xi \in \mathrm{so}(3) | \, \mu(\xi) = 2h\}$.

(7.3) **Proof**: For $\alpha_A \in \mathscr{EM}^{-1}(h, \mu)$ we find that

$$\begin{aligned} \mu(\mathscr{R}(\alpha_A)) &= (T_e R_A)^t(\alpha_A)\big(T_A R_{A^{-1}}\rho(A)^\sharp(\alpha_A)\big), \quad \text{since } \alpha_A \in \mathscr{J}^{-1}(\mu) \\ &= \rho(A)(\alpha_A, \alpha_A) = 2h, \quad \text{since } \alpha_A \in \mathscr{H}^{-1}(h). \qquad \square \end{aligned}$$

Recall that the isotropy group $\mathrm{SO}(3)_\mu$ of the coadjoint action of SO(3) on so(3)* is the image of the one parameter subgroup

$$\gamma : \mathbf{R} \rightarrow \mathrm{SO}(3)_\mu \subseteq \mathrm{SO}(3) : t \rightarrow \exp tY,$$

where $Y = k^\flat(\mu)/|\mu|$. Since $|Y| = 1$, γ defines the $S^1 = \mathbf{R}/2\pi\mathbf{Z}$ action

$$\varphi : S^1 \times \mathrm{SO}(3) \rightarrow \mathrm{SO}(3) : (t, A) \rightarrow (\exp tY)A$$

which lifts to a Hamiltonian action

$$\widehat{\varphi} : S^1 \times T^*\,\mathrm{SO}(3) \rightarrow T^*\,\mathrm{SO}(3) : (t, \alpha_A) \rightarrow (T_A L_{\exp -tY})^t \alpha_A \tag{106}$$

on $(T^*\,\mathrm{SO}(3), \Omega)$. The momentum mapping of $\widehat{\varphi}$ is

$$L : T^*\,\mathrm{SO}(3) \rightarrow \mathbf{R} : \alpha_A \rightarrow \mathscr{J}(\alpha_A)Y, \tag{107}$$

which is the Y-component of the momentum map $\mathscr{J}$ (47). Because $\mathscr{J}^{-1}(\mu)$ and $\mathscr{H}^{-1}(h)$ are invariant manifolds of the infinitesimal generator X_L of $\widehat{\varphi}$, the 2-torus $T^2_{h,\mu}$ (which is a connected component of $\mathscr{EM}^{-1}(h, j)$) is an invariant manifold of X_L. Since every integral curve of X_L is periodic, the image of the map

$$\Gamma : \mathbf{R} \rightarrow T^2_{h,\mu} : t \rightarrow \widehat{\varphi}_t(\alpha_A)$$

is a smooth circle $\mathscr{C}$ through α_A for every $\alpha_A \in T^2_{h,\mu}$. Thus $\mathfrak{R}(\mathscr{C})$ is a smooth circle in $\mathscr{A}_\mu$. Actually

Claim: $\mathfrak{R}(\mathscr{C})$ is a circle in $(\mathscr{A}_\mu, k|\mathscr{A}_\mu)$ with center $V = \frac{2h}{|\mu|}Y$ and radius

$$\mathrm{rad}(\alpha_A) = |\mathrm{Ad}_A(\rho(e)^\sharp\alpha) - V|.$$

(7.4) **Proof**: For every $\alpha_A \in T^*_A\,\mathrm{SO}(3)$

$$\rho(A)^\sharp(\alpha_A) = T_e L_A(\rho(e)^\sharp\alpha),$$

since ρ is left invariant. Therefore

$$\mathfrak{R}(\alpha_A) = T_A R_{A^{-1}}(\rho(A)^\sharp(\alpha_A)) = \mathrm{Ad}_A(\rho(e)^\sharp\alpha).$$

Hence, for every $\alpha_A \in T^2_{h,\mu}$

$$|\mathfrak{R}(\widehat{\varphi}_t(\alpha_A)) - V| = |\mathrm{Ad}_{(\exp tY)A}(\rho(e)^\sharp\alpha) - V| = |\mathrm{Ad}_A(\rho(e)^\sharp\alpha) - V|,$$

because k is Ad_A-invariant and $\mathrm{Ad}_{\exp tY}Y = Y$. We only need to check that $V \in \mathscr{A}_\mu$: $\mu(V) = \frac{2h}{|\mu|}\,\mu(Y) = 2h$, by definition of Y. □

Corollary: The radius function

$$\mathrm{rad} : \mathscr{EM}^{-1}(h,\mu) \to \mathbf{R} : \alpha_A \to |\mathrm{Ad}_A(\rho(e)^\sharp\alpha) - V| \tag{108}$$

is invariant under the flow of $X_L|\mathscr{EM}^{-1}(h,\mu)$.

▷ We now want to show that the image of the 2-torus $T^2_{h,\mu}$ under the mapping $\mathfrak{R}$ is the annulus

$$\mathscr{A} = \left\{ \xi \in \mathscr{A}_\mu \,\middle|\, \min_{T^2_{h,\mu}} \mathrm{rad} \le |\xi| \le \max_{T^2_{h,\mu}} \mathrm{rad} \right\}. \tag{109}$$

This entails verifying that the radius function has a *strictly positive* minimum value. As preparation recall that the orbits of $X_L|\mathscr{EM}^{-1}(h,\mu)$ are fibers of the reduction mapping

$$\begin{aligned}\pi_\mu : \mathscr{J}^{-1}(\mu) \subseteq T^*\,\mathrm{SO}(3) &\to S^2_{|\mu|} : \\ \alpha_A &\to i\circ k^\flat(\mathrm{Ad}^t_A\mu) = |\mu|A^{-1}(i(Y)) = x.\end{aligned} \tag{110}$$

Since the radius function is invariant under the flow of $X_L|\mathscr{EM}^{-1}(h,\mu)$, it induces a function

$$\mathrm{rad}_\mu : \mathscr{S}_{h,\mu} = \pi_\mu\big(\mathscr{EM}^{-1}(h,\mu)\big) \subseteq S^2_{|\mu|} \to \mathbf{R},$$

defined by

$$\pi^*_\mu(\mathrm{rad}_\mu) = \mathrm{rad}|\mathscr{EM}^{-1}(h,\mu).$$

Claim: For every $x \in \mathscr{S}_{h,\mu}$

$$\mathrm{rad}_\mu(x) = |Dx - \frac{2h}{|\mu|^2}\,x|, \tag{111}$$

where $D = (I')^{-1}$.

(7.5) **Proof**: Let $\alpha_A \in \mathcal{EM}^{-1}(h,\mu)$ and set $x = \pi_\mu(\alpha_A)$. Calculate rad_μ as follows:

$$\begin{aligned}
\mathrm{rad}_\mu(x) &= \mathrm{rad}(\alpha_A) \\
&= |\mathrm{Ad}_A\big(\rho(e)^\sharp\alpha\big) - \frac{2h}{|\mu|}Y|, \quad \text{in } (\mathrm{so}(3), k) \\
&= |Ai\big((I^{-1})k^\flat(\alpha)\big) - \frac{2h}{|\mu|}i(Y)|, \quad \text{in } (\mathbf{R}^3, (\,,\,)) \\
&= |(I')^{-1}(i\circ k^\flat)(\alpha) - \frac{2h}{|\mu|^2}A^{-1}\big(|\mu|i(Y)\big)|, \\
&\qquad \text{since } A \in \mathrm{SO}(3),\ I \text{ is } k\text{-symmetric, and } I' = i\circ I\circ i^{-1} \\
&= ||\mu|(I')^{-1}A^{-1}\big(i(Y)\big) - \frac{2h}{|\mu|^2}A^{-1}\big(|\mu|i(Y)\big)|, \\
&\qquad \text{since } \alpha_A \in \mathcal{J}^{-1}(\mu) \text{ implies } \alpha = \mathrm{Ad}^t_A\mu. \\
&\qquad \text{Hence } i\circ k^\flat(\mathrm{Ad}^t_A\mu) = A^{-1}(i\circ k^\flat(\mu)). \text{ Moreover, } |\mu|Y = k^\flat(\mu) \\
&= |Dx - \frac{2h}{|\mu|^2}x|, \quad \text{using (110) and } D = (I')^{-1}. \qquad \square
\end{aligned}$$

▷ Next we show that rad_μ is strictly positive on $\mathcal{S}_{h.\mu}$.

(7.6) **Proof**: From (111) we see that $\mathrm{rad}_\mu(x) \ge 0$ for every $x \in \mathcal{S}_{h,\mu}$. Suppose that $\mathrm{rad}_\mu(x) = 0$ for some $x \in \mathcal{S}_{h,\mu}$. Then $Dx = \big(2h/|\mu|^2\big)x$. Therefore $2h/|\mu|^2$ is an eigenvalue of D and hence equals I_i^{-1} for some $i \in \{1,2,3\}$. In other words, $h = \frac{1}{2}I_i^{-1}|\mu|^2$. Thus (h,μ) is a critical value of the energy momentum map $\mathcal{EM}$ (104). This is contrary to our hypothesis. □

From the definition of rad_μ it follows that the radius function rad is strictly positive on $\mathcal{EM}^{-1}(h,\mu)$. Since $\mathcal{EM}^{-1}(h,\mu)$ is compact, rad has a positive minimum. Thus $\mathcal{A}$ (109) is an annulus. □

For latter purposes (see ((7.10))) we want to know more about the critical points and critical values of rad_μ. Towards this goal, consider the function

$$R : \mathcal{S}_{h,\mu} \subseteq S^2_{|\mu|} \to \mathbf{R} : x \to \big((D - 2h/|\mu|^2)^2x, x\big).$$

Claim: R and rad_μ have the same critical points with the same Morse index.

(7.7) **Proof**: Since R is the square of rad_μ, differentiating we obtain

$$DR(x)v = 2\,\mathrm{rad}_\mu(x)\,D\,\mathrm{rad}_\mu(x)v, \tag{112}$$

for every $x \in S^2_{|\mu|}$ and $v \in T_xS^2_{|\mu|}$. From (112) it follows that R and rad_μ have the same critical points, since $\mathrm{rad}_\mu(x)$ is strictly positive on $\mathcal{S}_{h,\mu}$ ((7.6)). Differentiating (112) gives

$$D^2R(x)(v,w) = 2\,D\,\mathrm{rad}_\mu(x)v\,D\,\mathrm{rad}_\mu(x)w + 2\,\mathrm{rad}_\mu(x)\,D^2\mathrm{rad}_\mu(x)(v,w), \tag{113}$$

for $x \in S^2_{|\mu|}$ and $v, w \in T_x S^2_{|\mu|}$. If x is a critical point of R on $\mathcal{S}_{h,\mu}$ then (113) becomes

$$D^2 R(x)(v, w) = 2\,\mathrm{rad}_\mu(x)\, D^2 \mathrm{rad}_\mu(x)(v, w). \tag{114}$$

Since rad_μ is strictly positive on $\mathcal{S}_{h,\mu}$, from (114) we see that x is a nondegenerate critical point of R if and only if it is a nondegenerate critical point of rad_μ. Moreover, they have the same Morse index. □

▷ We now show that R is a Morse function.

(7.8) **Proof**: Because $\mathcal{S}_{h,\mu}$ is compact and R is continuous, R has a critical point $p = (x_1^0, x_2^0, x_3^0)$. By Lagrange multipliers, p satisfies

$$\begin{aligned} \big((D - 2h/|\mu|^2)^2 + \lambda_1 + \lambda_2 D\big)p &= 0 &(115)\\ (p, p) &= |\mu|^2 &(116)\\ (Dp, p) &= 2h. &(117) \end{aligned}$$

Here $D = \mathrm{diag}(d_1, d_2, d_3) = (I_1^{-1}, I_2^{-1}, I_3^{-1})$ and $d_1 > d_2 > d_3 > 0$ (65).

Suppose that all of the components of p are nonzero. Then (115) becomes

$$\begin{aligned} (d_1 - 2h/|\mu|^2)^2 + \lambda_1 + \lambda_2 d_1 &= 0 &(118)\\ (d_2 - 2h/|\mu|^2)^2 + \lambda_1 + \lambda_2 d_2 &= 0 &(119)\\ (d_3 - 2h/|\mu|^2)^2 + \lambda_1 + \lambda_2 d_3 &= 0. &(120) \end{aligned}$$

Subtracting (118) from (119) and (119) from (120) gives

$$\lambda_2 = -(d_1 + d_2) + 4h/|\mu|^2 \quad \text{and} \quad \lambda_2 = -(d_2 + d_3) + 4h/|\mu|^2,$$

since $d_1 - d_2$ and $d_2 - d_3$ are nonzero. Consequently, $d_1 = d_3$, which is a contradiction. Therefore at least one of the components of p is zero.

Now suppose that two components of p are zero, say x_1^0 and x_2^0. (The other cases are similar and their proof is omitted). Then (116) and (117) give $h = \frac{1}{2} d_3 |\mu|^2$. Hence (h, μ) is a critical value of $\mathcal{EM}$. This is a contradiction. Therefore exactly one component of p is zero.

Suppose that $x_1^0 = 0$. (Again the other cases are handled similarly and the details are left to the reader.) Solving (116) and (117) we obtain

$$x_1^0 = 0 \quad x_2^0 = \pm\sqrt{\frac{2h - d_3|\mu|^2}{d_2 - d_3}}, \quad x_3^0 = \pm\sqrt{\frac{d_2|\mu|^2 - 2h}{d_2 - d_3}}, \tag{121}$$

where

$$h \in I_1 = (\tfrac{1}{2} d_3|\mu|^2, \tfrac{1}{2} d_2|\mu|^2).$$

Since x_2^0 and x_3^0 are nonzero, (115) becomes (119) and (120). Solving them for the Lagrange multipliers λ_1 and λ_2 we obtain

$$\lambda_1 = d_2 d_3 - 4h^2/|\mu|^4 \quad \text{and} \quad \lambda_2 = -(d_2 + d_3) + 4h/|\mu|^2. \tag{122}$$

Therefore (121) is a solution to (115) – (117). The set $\mathcal{S}_{h,\mu} = \pi_\mu(\mathcal{EM})$ has two connected components $\mathcal{S}^\varepsilon$ ($\varepsilon = \pm$) one of which is the circle $S^1_{h,\mu} = \pi_\mu(T^2_{h,\mu})$. A glance at figure 4.3 shows that for fixed ε the *two* critical points $p^\varepsilon_\pm$

$$x_1^0 = 0 \quad x_2^0 = \pm\sqrt{\frac{2h - d_3|\mu|^2}{d_2 - d_3}}, \quad x_3^0 = \varepsilon\sqrt{\frac{d_2|\mu|^2 - 2h}{d_2 - d_3}},$$

lie on $\mathcal{S}^\varepsilon$. The value of R at $p^\varepsilon_\pm$ is

$$4(\tfrac{1}{2}d_2|\mu|^2 - h)(h - \tfrac{1}{2}d_3|\mu|^2)/|\mu|^2.$$

At p_ε the Hessian of R is the diagonal 3×3 matrix with entries $(d_1 - 2h/|\mu|^2)^2 + \lambda_1 + \lambda_2 d_1$, $(d_2 - 2h/|\mu|^2)^2 + \lambda_1 + \lambda_2 d_2$, and $(d_3 - 2h/|\mu|^2)^2 + \lambda_1 + \lambda_2 d_3$ restricted to $T_{p_\varepsilon}\mathcal{S}^\varepsilon$. Since

$$T_{p^\varepsilon_\pm}\mathcal{S}^\varepsilon = \ker\begin{pmatrix} x_1 & x_2 & x_3 \\ d_1x_1 & d_2x_2 & d_3x_3 \end{pmatrix}\Bigg|_{(0,\, x_2^0,\, x_3^0)}$$

is spanned by the vector e_1, it follows that

$$\begin{aligned} \mathrm{Hess}_{p^\varepsilon_\pm} R &= \left(d_1 - 2h/|\mu|^2\right)^2 + \left(d_2d_3 - 4h^2/|\mu|^4\right) + d_1\left(-(d_2 + d_3) + 4h/|\mu|^2\right), \\ &\qquad \text{using (122)} \\ &= d_1^2 - d_1(d_2 + d_3) + d_2d_3 = \left(d_1 - \tfrac{1}{2}(d_2 + d_2)\right)^2 - \left(\tfrac{1}{2}(d_2 - d_3)\right)^2. \end{aligned}$$

But $0 < d_3 < d_2 < d_1$ which implies that

$$d_1 - \tfrac{1}{2}(d_2 + d_3) > d_2 - \tfrac{1}{2}(d_2 + d_3) = \tfrac{1}{2}(d_2 - d_3) > 0.$$

Hence $\mathrm{Hess}_{p^\varepsilon_\pm} R > 0$, that is, $p^\varepsilon_\pm$ is a nondegenerate minimum of R.

Let $I_2 = (\tfrac{1}{2}d_2|\mu|^2, \tfrac{1}{2}d_1|\mu|^2)$. In table 7.1 we summarize information about the critical points of the function R.

critical points & conditions	critical value	index
1. $\left(0, \pm\sqrt{\frac{2h-d_3\|\mu\|^2}{d_2-d_3}}, \varepsilon\sqrt{\frac{d_2\|\mu\|^2-2h}{d_2-d_3}}\right)$, if $h \in I_1$	$\frac{4}{\|\mu\|^2}\left(\frac{1}{2}d_2\|\mu\|^2-h\right)\left(h-\frac{1}{2}d_3\|\mu\|^2\right)$	0
2. $\left(\varepsilon\sqrt{\frac{2h-d_3\|\mu\|^2}{d_1-d_3}}, 0, \pm\sqrt{\frac{d_1\|\mu\|^2-2h}{d_1-d_3}}\right)$, if $h \in I_1$	$\frac{4}{\|\mu\|^2}\left(\frac{1}{2}d_1\|\mu\|^2-h\right)\left(h-\frac{1}{2}d_3\|\mu\|^2\right)$	1
3. $\left(\pm\sqrt{\frac{2h-d_3\|\mu\|^2}{d_1-d_3}}, 0, \varepsilon\sqrt{\frac{d_1\|\mu\|^2-2h}{d_1-d_3}}\right)$, if $h \in I_2$	$\frac{4}{\|\mu\|^2}\left(\frac{1}{2}d_1\|\mu\|^2-h\right)\left(h-\frac{1}{2}d_3\|\mu\|^2\right)$	1
4. $\left(\varepsilon\sqrt{\frac{2h-d_2\|\mu\|^2}{d_1-d_2}}, \pm\sqrt{\frac{d_1\|\mu\|^2-2h}{d_1-d_2}}, 0\right)$, if $h \in I_2$	$\frac{4}{\|\mu\|^2}\left(\frac{1}{2}d_1\|\mu\|^2-h\right)\left(h-\frac{1}{2}d_2\|\mu\|^2\right)$	0

Table 7.1 Critical points of R and their Morse index.

Thus R is a Morse function. □

Using ((7.7)) we deduce that $\mathrm{rad}_\mu : \mathcal{S}_{h,\mu} \subseteq S^2_{|\mu|} \rightarrow \mathbf{R}$ is a Morse function. Fixing ε and looking at table 7.1 we see that rad_μ has *two* nondegenerate minima and *two* nondegenerate maxima on $\mathcal{S}^\varepsilon$ with corresponding ciritical values $r_{\min}$ and $r_{\max}$.

From the above discussion we see that the image of the 2-torus $T^2_{h,\mu}$ under the mapping $\mathcal{R}$ (105) is a closed annulus $\mathcal{A}$ in the affine plane $\mathcal{A}_\mu$, which is bounded by two circles $C_{\min}$ and $C_{\max}$ with center at $(2h/|\mu|)Y$ and radius $r_{\min}$ and $r_{\max}$. Moreover, both $\mathcal{R}^{-1}(C_{\min})$ and $\mathcal{R}^{-1}(C_{\max})$ consist of *two* orbits of $X_L|T^2_{h,\mu}$ (107), because each orbit is the inverse image under the reduction map π_μ of a critical point of rad_μ on $S^1_{h,\mu}$. Since the Hessian of rad_μ is definite at these critical points, the mapping $\mathcal{R}$ has a fold singularity along each orbit in $\mathcal{R}^{-1}(C_{\min})$ and $\mathcal{R}^{-1}(C_{\max})$. We leave it to the reader to deduce from these geometric facts that $T^2_{h,\mu}$ cannot be embedded in $\mathbf{R}^3$, (see exercise 10).

Let $\mathcal{C}_1$ be an integral curve of $X_L|T^2_{h,\mu}$ whose image under the mapping $\mathcal{R}$ is the oriented circle $C_{\max} \subseteq A_\mu$. $\mathcal{C}_1$ is a cross section for the flow of $X_H|T^2_{h,\mu}$. Hence for $p \in \mathcal{C}_1$ there is a smallest positive time $T = T(h,\mu)$ such that $q = \varphi^H_T(p) \in \mathcal{C}_1$, where $t \rightarrow \varphi^H_t$ is the flow of X_H. Let $\theta_{h,\mu}$ be the rotation number of the flow of $X_H|T^2_{h,\mu}$.

Claim: $2\pi\theta_{h,\mu}$ of $X_H|T^2_{h,\mu}$ is equal to the angle $\vartheta_{h,\mu}$ on the oriented arc on $C_{\max}$ between $\mathcal{R}(p)$ and $\mathcal{R}(q)$, (see figure 7.1).

(7.9) **Proof**: Recall that $\widehat{\varphi}_t$ (106) is the flow of X_L. From $\widehat{\varphi}_{2\pi\theta_{h,\mu}}(p) = q$ it follows that

$$\mathcal{R}(q) = \mathcal{R}\big(\widehat{\varphi}_{2\pi\theta_{h,\mu}}(p)\big) = \widehat{\psi}_{2\pi\theta_{h,\mu}}\big(\mathcal{R}(p)\big). \tag{123}$$

Here $\widehat{\psi}_t$ is the pull back of $\widehat{\varphi}_t$ by $\mathcal{R}$, that is, $\widehat{\psi}_t(W) = \mathcal{R}\big(\widehat{\varphi}_t(\mathcal{R}^{-1}(W))\big)$. Since $\widehat{\varphi}_t|\mathcal{C}_1$ is a rotation through an angle t, from (123) we see that the angle $\vartheta_{h,\mu}$ swept out by the oriented arc on $C_{\max}$ between $\mathcal{R}(p)$ and $\mathcal{R}(q)$ is equal to $2\pi\theta_{h,\mu}$. □

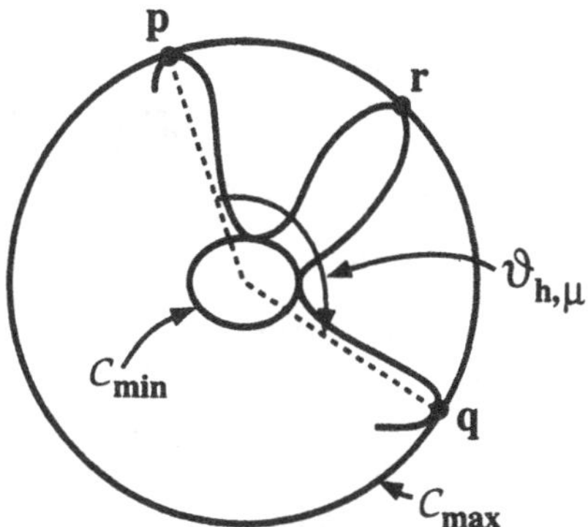

Figure 7.1. The Poinsot description of the rotation number.

▷ Note that q is *not the first* point after p where the curve $\Gamma : t \rightarrow \mathcal{R}(\varphi^H_t(p))$ meets $\mathcal{C}_{\max}$ again, but is the *second*.

(7.10) **Proof**: To see this, recall that $\mathcal{R}^{-1}(\mathcal{C}_{\max})$ is the disjoint union of *two* X_L orbits $\mathcal{C}_1$ and $\mathcal{C}_2$ on $T^2_{h,\mu}$. Before crossing $\mathcal{C}_1$, the curve $\gamma : t \to \varphi_t^H(p)$ must cross $\mathcal{C}_2$ transversely, say at the point P. This follows from the fact that every orbit of X_L on $T^2_{h,\mu}$ is a cross section for the flow of $X_H | T^2_{h,\mu}$. Thus $\mathcal{R}(P) = r \in \mathcal{C}_{\max}$ is the *first* point where Γ crosses $\mathcal{C}_{\max}$. A similar argument shows that starting at P the curve γ crosses $\mathcal{C}_1$ for the first time at the point $Q = \mathcal{R}^{-1}(q)$. □

8 A twisting phenomenon

In this subsection we describe and explain a twisting phenomenon which a tennis racket-like rigid body performs.

Consider a triaxial rigid body with principal axes $\{\widehat{e}_1, \widehat{e}_2, \widehat{e}_3\}$ fixed to the body at its center of mass. Suppose that the corresponding principal moments of inertia I_i satisfy $0 < I_1 < I_2 < I_3$. We say that the rigid body is *tennis racket-like* if it is nearly planar that is,

$$I_3 \approx I_1 + I_2, \tag{124}$$

(I_3 is approximately equal to $I_1 + I_2$) and

$$I_1 \ll I_2, \tag{125}$$

(I_1 is much less than I_2). Since an old fashioned wooden tennis racket fulfills all of these conditions nicely, we will talk of a tennis racket.

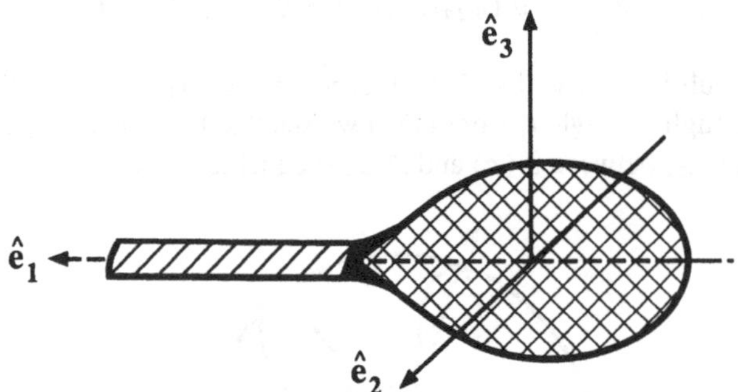

Figure 8.1. The principal axes of a tennis racket.

The following experiment demonstrates the twisting phenomenon quite dramatically. Take a tennis racket and mark its faces so that they can be distinguished. Call one rough and the other smooth. Hold the racket horizontally so that the *smooth* face is up. Toss the racket attempting to make it rotate about the intermediate principal axis $\widehat{e}_2$. After one rotation catch the racket by its handle. The *rough* face will almost always be up! Thus the racket has made a near half twist around its handle.

The problem, of course, is to explain this twisting phenomenon. In what follows we give a qualitative explanation. Since we are interested in the rotational motion of the racket, we can forget about the motion of its center of mass. In other words we suppose that the center of mass of the racket is fixed. The racket rotating around its intermediate principal axis $\widehat{e}_2$ corresponds to one of the hyperbolic period orbits $\gamma_2^{\pm}$ of the Euler-Arnol'd equations of the Euler top with energy $h = \frac{1}{2}|\mu|^2 I_2^{-1}$ and angular momentum $\mu = |\mu|e_3$, (see figure 5.2). In section 5 we showed that γ_2^+ with γ_2^- are heteroclinic, that is, the closure of the unstable manifold of γ_2^+ contains γ_2^-. Therefore every motion of the tennis racket which starts "near" γ_2^+ eventually comes "near" γ_2^-. (Warning: the racket spinning exactly about its $\widehat{e}_2$-axis does *not* twist at all. Thus we should quantify what we mean by "near". For an estimate of the size of the region about γ_2^+ where the racket does *not* twist, see exercise 13.) Since γ_1^+ and γ_2^+ are heteroclinic for *every* triaxial rigid body, this qualitative feature of the motion cannot be the whole story why the racket twists. What it does *not* explain is why the racket makes a *twist* when going from γ_2^+ to γ_2^-.

To say what we mean by a twist, we must have some reference plane *fixed* in space. The following discussion shows that the handle of the racket moves nearly in the plane. Suppose that the racket has energy h and angular momentum $\mu = |\mu|e_3$. Moreover, assume that the principal axis frame $\{\widehat{e}_1, \widehat{e}_2, \widehat{e}_3\}$ and the frame $\{e_1, e_2, e_3\}$ fixed in space are the same at time $t = 0$.

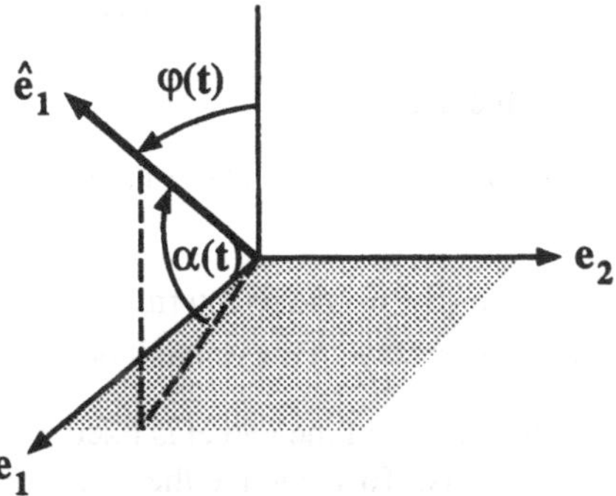

Figure 8.2. The angles $\alpha(t)$ and $\varphi(t)$.

Claim. Let the position of the racket be as given in figure 8.2 and let $\alpha(t)$ be the angle the handle $\widehat{e}_1$ makes with the e_1–e_2 plane at time t. Then

$$0 \le \tan\alpha(t) \le \sqrt{\big(I_1(2I_3h - |\mu|^2)\big)/\big(I_3(|\mu|^2 - 2I_1h)\big)}. \tag{126}$$

(8.11) **Proof**: Since the angular momentum of the rigid body is $\mu = |\mu|e_3$, its component along the $\widehat{e}_1$-axis at time t is $p_1(t)$, where $t \to p(t) = \big(p_1(t), p_2(t), p_3(t)\big)$ is a solution of Euler's equations (64) on $S^2_{|\mu|}$ of energy h. Thus

$$|\mu|^{-1}p_1(t) = |\cos\varphi(t)| = \sin\alpha(t). \tag{127}$$

Using the explicit solution of Euler's equations in terms of Jacobi elliptic functions (see exercise 2), we find that when $h \in [\frac{1}{2}|\mu|^2 I_3^{-1}, \frac{1}{2}|\mu|^2 I_2^{-1}]$,

$$p_1(t) = \sqrt{\big(2h - |\mu|^2 I_3^{-1}\big)/\big(I_1^{-1} - I_3^{-1}\big)}\,\mathrm{cn}(nt;k).$$

Here

$$n = \sqrt{(I_1^{-1} - I_2^{-1})(2h - |\mu|^2 I_3^{-1})} \quad \text{and} \quad k = \sqrt{\frac{(I_2^{-1} - I_3^{-1})(|\mu|^2 I_1^{-1} - 2h)}{(I_1^{-1} - I_3^{-1})(2h - |\mu|^2 I_3^{-1})}}.$$

Since $|\mathrm{cn}(nt; k)| \le 1$, (see exercise 1), we obtain

$$0 \le \sin\alpha(t) \le \frac{1}{|\mu|}\sqrt{(2h - |\mu|^2 I_3^{-1})/(I_1^{-1} - I_3^{-1})}, \tag{128}$$

which yields (126). Similarly when $h \in [\frac{1}{2}|\mu|^2 I_2^{-1}, \frac{1}{2}|\mu|^2 I_1^{-1}]$ we find that

$$p_1(t) = \sqrt{(2h - |\mu|^2 I_3^{-1})/(I_1^{-1} - I_3^{-1})}\,\mathrm{dn}(\widetilde{n}t; \widetilde{k}).$$

Here

$$\widetilde{n} = \sqrt{(I_2^{-1} - I_3^{-1})(|\mu|^2 I_1^{-1} - 2h)} \quad \text{and} \quad \widetilde{k} = \sqrt{\frac{(I_1^{-1} - I_2^{-1})(2h - |\mu|^2 I_1^{-1})}{(I_2^{-1} - I_3^{-1})(|\mu|^2 I_1^{-1} - 2h)}}.$$

Since $|\mathrm{dn}(\widetilde{n}t; \widetilde{k})| \le 1$, we again obtain (128). □

Corollary: For a tennis racket rotating almost about its intermediate axis the angle $\alpha(t)$ is small for all t.

(8.12) **Proof**: When the racket rotates nearly around its intermediate axis, $h \approx \frac{1}{2}|\mu|^2 I_2^{-1}$. Using (124) the right hand side of (126) is $\approx \frac{I_1}{I_2}$. This is small because of (125). □

Thus the plane swept out by the handle of the racket is fixed in space. With respect to this ▷ plane we can see if the racket twists. To show that the racket does indeed twist, consider its motion along the unstable manifold of γ_2^+ as it goes from a neighborhood of γ_2^+ to a neighborhood of γ_2^-.

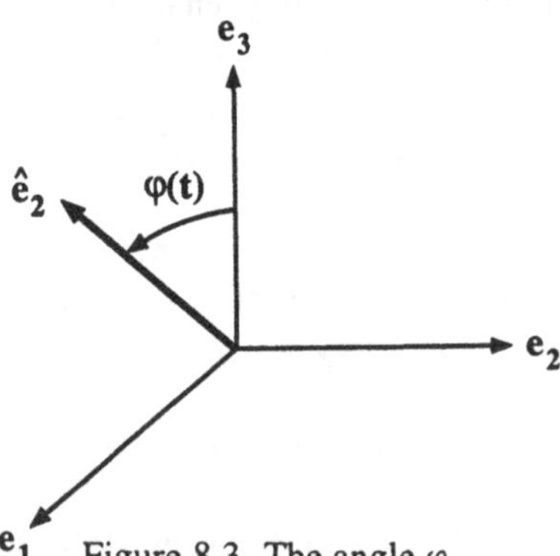

Figure 8.3. The angle φ.

(8.13) **Proof**: Let the position of the racket be as given in figure 8.3. At time t the $\widehat{e_2}$-component $p_2(t)$ of the angular momentum $\mu = |\mu| e_3$ of the racket is

$$p_2(t) = |\mu|\cos\varphi(t). \tag{129}$$

Using exercises 1 and 2, the solution $t \to p(t) = (p_1(t), p_2(t), p_3(t))$ of Euler's equation of energy $h = \frac{1}{2}|\mu|^2 I_2^{-1}$ which goes from the hyperbolic equilibrium point $|\mu| e_2$ along its unstable manifold to the hyperbolic equilibrium point $-|\mu| e_2$ is given by

$$p(t) = |\mu| \Big(\sqrt{\frac{I_2^{-1} - I_3^{-1}}{I_1^{-1} - I_3^{-1}}} \operatorname{sech} nt, -\tanh nt, \sqrt{\frac{I_1^{-1} - I_2^{-1}}{I_1^{-1} - I_3^{-1}}} \operatorname{sech} nt \Big). \tag{130}$$

Here $n = |\mu| \sqrt{(I_1^{-1} - I_2^{-1})(I_2^{-1} - I_3^{-1})}$. Therefore $\cos\varphi(t) = -\tanh nt$. From (130) we see that at $t = -\infty$, $p(-\infty) = (0, |\mu|, 0)$, while at $t = \infty$, $p(\infty) = (0, -|\mu|, 0)$. Therefore $\varphi(-\infty) = 0$, while $\varphi(\infty) = \pi$. Refering to figure 8.3, this says that the $\widehat{e}_2$-axis of the racket starts at e_3 *vertically above* the invariant e_1–e_2 plane and finishes at $-e_3$ *vertically below*. Thus the racket has made a near *half* twist in going from near γ_2^+ to near γ_2^- along the unstable manifold. □

To explain the experiment described at the beginning of this section in finer detail we should show that

1. The racket has enough time perform a twist, that is, the time it takes the handle to make one revolution is longer than the time it takes to perform a half twist.
2. The racket is likely to be caught after making a half twist.
3. The experiment is repeatable because the handle revolves nearly uniformly.

We refer the reader to exercise 13 for a treatment of these points.

9 Exercises

1. (Jacobi elliptic functions.) Consider the system of differential equations

$$\begin{cases} \dot{x} &= yz \\ \dot{y} &= -xz \\ \dot{z} &= -k^2 xy, \end{cases} \tag{131}$$

where $0 < k^2 < 1$. Define the *Jacobi elliptic functions* as the integral curve

$$t \to \big(x(t), y(t), z(t)\big) = \big(\operatorname{sn}(t;k), \operatorname{cn}(t;k), \operatorname{dn}(t;k)\big)$$

of (131) which passes through $(0, 1, 1)$ at $t = 0$.

(a). Show that the functions $x^2 + y^2$ and $k^2x^2 + z^2$ are integrals of (131) and deduce that

$$\begin{aligned} \operatorname{sn}^2(t;k) + \operatorname{cn}^2(t;k) &= 1 \\ k^2\operatorname{sn}^2(t;k) + \operatorname{dn}^2(t;k) &= 1. \end{aligned}$$

Consequently, for all t

$$|\operatorname{sn}(t;k)| \le 1, \quad |\operatorname{cn}(t;k)| \le 1 \text{ and } k' = \sqrt{1-k^2} \le \operatorname{dn}(t;k) \le 1.$$

(b). Let $x(t) = \mathrm{sn}(t;k)$. From (131) deduce that $\frac{dt}{dx} = \frac{1}{\sqrt{(1-x^2)(1-k^2x^2)}} > 0$ for $x \in (-1, 1)$. Hence

$$t(x) = \int_0^x \frac{1}{\sqrt{(1-x^2)(1-k^2x^2)}}\, dx$$

is a smooth inverse of $\mathrm{sn}(t;k)$ on $(-1,1)$. Since $t(\pm 1) = \pm K(k) = \pm K$ is finite, $t(x)$ is continuous on $[-1,1]$. Therefore, $\mathrm{sn}(K;k) = 1$. Hence $\mathrm{cn}(K;k) = 0$ and $\mathrm{dn}(K;k) = k'$.

(c). Show that $\mathrm{sn}(t;0) = \sin t$, $\mathrm{cn}(t;0) = \cos t$, $\mathrm{dn}(t;0) = 1$ and $\mathrm{sn}(t;1) = \tanh t$, $\mathrm{cn}(t;1) = \mathrm{sech}\, t$, $\mathrm{dn}(t;1) = \mathrm{sech}\, t$.

(d). Define

$$\xi(t) = \frac{\mathrm{cn}(t;k)}{\mathrm{dn}(t;k)}, \quad \eta(t) = -k'\frac{\mathrm{sn}(t;k)}{\mathrm{dn}(t;k)}, \quad \zeta(t) = k'\frac{1}{\mathrm{dn}(t;k)}.$$

Show that $t \to (\xi(t), \eta(t), \zeta(t))$ is an integral curve of (131) passing through $(1, 0, k')$ at $t = 0$. Since $t \to \big(\mathrm{sn}(t+K;k), \mathrm{cn}(t+K;k), \mathrm{dn}(t+K;k)\big)$ is also an integral curve of (131) passing through $(1, 0, k')$ at $t = 0$, deduce that

$$\begin{cases} \mathrm{sn}(t+K;k) &= \dfrac{\mathrm{cn}(t;k)}{\mathrm{dn}(t;k)} \\[2ex] \mathrm{cn}(t+K;k) &= -k'\dfrac{\mathrm{sn}(t;k)}{\mathrm{dn}(t;k)} \\[2ex] \mathrm{dn}(t+K;k) &= k'\dfrac{1}{\mathrm{dn}(t;k)}. \end{cases}$$

Conclude that $\mathrm{sn}(t;k)$ and $\mathrm{cn}(t;k)$ are periodic of period $4K$ while $\mathrm{dn}(t;k)$ is periodic of period $2K$.

(e). Use (131) to prove the addition formulæ

$$\begin{aligned} \mathrm{sn}(u+v;k) &= \frac{\mathrm{sn}(u;k)\mathrm{cn}(v;k)\mathrm{dn}(v;k) + \mathrm{sn}(v;k)\mathrm{cn}(u;k)\mathrm{dn}(u;k)}{1 - k^2\mathrm{sn}^2(u;k)\mathrm{sn}^2(v;k)} \\ \mathrm{cn}(u+v;k) &= \frac{\mathrm{cn}(u;k)\mathrm{cn}(v;k) - \mathrm{sn}(u;k)\mathrm{sn}(v;k)\mathrm{dn}(u;k)\mathrm{dn}(v;k)}{1 - k^2\mathrm{sn}^2(u;k)\mathrm{sn}^2(v;k)} \\ \mathrm{dn}(u+v;k) &= \frac{\mathrm{dn}(u;k)\mathrm{dn}(v;k) - k^2\mathrm{sn}(u;k)\mathrm{sn}(v;k)\mathrm{cn}(u;k)\mathrm{cn}(v;k)}{1 - k^2\mathrm{sn}^2(u;k)\mathrm{sn}^2(v;k)}. \end{aligned}$$

(f). Using the substitution $x = iy/\sqrt{1-y^2}$ the integral

$$u = \int_0^x \frac{1}{\sqrt{(1-x^2)(1-k^2x^2)}}\, dx$$

becomes

$$u = i\int_0^{iy/\sqrt{1-y^2}} \frac{1}{\sqrt{(1-y^2)(1-(k')^2y^2)}}\,dy.$$

Let $w = iu$. Then $y = sn(w; k')$. Therefore

$$\begin{cases} sn(u;k) &= i\,\dfrac{\mathrm{sn}(w;k')}{\mathrm{cn}(w;k')} \\[2ex] \mathrm{cn}(u;k) &= \dfrac{1}{\mathrm{cn}(w;k')} \\[2ex] \mathrm{dn}(u;k) &= \dfrac{\mathrm{dn}(w;k')}{\mathrm{cn}(w;k')}. \end{cases}$$

Show that $\mathrm{sn}(w;k')$, $\mathrm{cn}(w;k')$ and $\mathrm{dn}(w;k')$ are periodic of period $4K'$, $4K'$ and $2K'$ respectively where $K' = \int_0^1 \frac{1}{\sqrt{(1-y^2)(1-(k')^2y^2)}}\,dy$. Thus sn, cn and dn have a second purely imaginary period of $4iK'$, $4iK'$ and $2iK'$ respectively.

2. (Euler's equations on S^2_ℓ.) Let $\mathbf{x} \in \mathbf{R}^3$ and let $\langle\, ,\, \rangle$ be the Euclidean inner product on $\mathbf{R}^3$. On the 2-sphere $S^2_\ell \subseteq \mathbf{R}^3$ given by $\langle \mathbf{x}, \mathbf{x}\rangle = \ell^2$ show that every vector in $T_{(x,y,z)}S^2_\ell = \{\xi \in \mathbf{R}^3 |\, \langle \mathbf{x}, \xi\rangle = 0\}$ can be written as $\mathbf{x} \times \mathbf{p}$ for some $\mathbf{p} \in \mathbf{R}^3$.

(a). Define a 2-form ω_ℓ on S^2_ℓ by $\omega_\ell(\mathbf{x})\big(\mathbf{x}\times\mathbf{p}, \mathbf{x}\times\mathbf{s}\big) = \langle \mathbf{x}, \mathbf{p}\times\mathbf{s}\rangle$, where $\mathbf{p}, \mathbf{s} \in \mathbf{R}^3$. Show that ω_ℓ is the element of surface area of S^2_ℓ and that it is a symplectic form.

(b). On the symplectic manifold (S^2_ℓ, ω_ℓ) consider the Hamiltonian function

$$H : S^2_\ell \subseteq \mathbf{R}^3 \to \mathbf{R} : (x, y, z) \to \tfrac{1}{2}\,(ax^2 + by^2 + cz^2),$$

where $a > b > c > 0$. Show that the Hamiltonian vector field X_H has integral curves which satisfy Euler's equations

$$\begin{cases} \dot{x} &= -(b-c)yz \\ \dot{y} &= \;\;(a-c)xz \\ \dot{z} &= -(a-b)xy. \end{cases} \tag{132}$$

Note that (132) defines a vector field on $\mathbf{R}^3$ which has S^2_ℓ as an invariant manifold.

(c). Integrate Euler's equations using Jacobi elliptic functions, (see exercise 1). In particular show that

i). When $\ell^2 b \ge 2h \ge \ell^2 c$,

$$\begin{cases} x(t) &= A\,\mathrm{cn}\,(nt;k) \\ y(t) &= B\,\mathrm{sn}\,(nt;k) \\ z(t) &= C\,\mathrm{dn}\,(nt;k), \end{cases}$$

where

$$A^2 = \frac{2h - c\ell^2}{a-c}, \quad B^2 = \frac{2h - c\ell^2}{b-c}, \quad C^2 = \frac{a\ell^2 - 2h}{a-c}$$

$$n = \sqrt{(a\ell^2 - 2h)(b-c)} \quad \text{and} \quad k = \sqrt{\frac{(a-b)(2h-c\ell^2)}{(b-c)(a\ell^2-2h)}}.$$

ii). When $\ell^2 a \ge 2h \ge \ell^2 b$,

$$\begin{cases} x(t) &= A\,\mathrm{dn}\,(nt;k) \\ y(t) &= B\,\mathrm{sn}\,(nt;k) \\ z(t) &= C\,\mathrm{cn}\,(nt;k), \end{cases}$$

where

$$A^2 = \frac{2h - c\ell^2}{a-c}, \quad B^2 = \frac{a\ell^2 - 2h}{a-b}, \quad C^2 = \frac{a\ell^2 - 2h}{a-c}$$

$$n = \sqrt{(a-b)(2h-c\ell^2)} \quad \text{and} \quad k = \sqrt{\frac{(b-c)(a\ell^2-2h)}{(a-b)(2h-c\ell^2)}}.$$

The signs of A, B and C above are chosen so that $x(t)$, $y(t)$ and $z(t)$ lie in one of the connected components of

$$\begin{cases} x^2+y^2+z^2 &= \ell^2 \\ ax^2+by^2+cz^2 &= 2h, \end{cases}$$

when $2h \ne \ell^2 b$. When $2h = \ell^2 b$ all choices of sign are possible.

3. Let $\times$ be the vector product and $\langle\, , \,\rangle$ the Euclidean inner product on $\mathbf{R}^3$. For $x, y, z \in \mathbf{R}^3$ show that

(a). $x \times (y \times z) = \langle x, z\rangle\, y - \langle x, y\rangle\, z$.

(b). $x \times (y \times z) = (x \times y) \times z + y \times (x \times z)$.

(c). $\langle x, y \times z\rangle = \det(\mathrm{col}(x, y, z)) = \langle x \times y, z\rangle$.

(d). $\langle x \times y, x \times y\rangle + \langle x, y\rangle^2 = \langle x, x\rangle\langle y, y\rangle$.

From b) deduce that $(\mathbf{R}^3, \times)$ is a Lie algebra, which is isomorphic to $(\mathrm{so}(3), [\, , \,])$.

4. (Euler angles.) Represent a proper rotation in $(\mathbf{R}^3, \langle\, , \,\rangle)$ as the product $B = A_3 A_2 A_1$ of three right handed rotations where

$$A_1 = \begin{pmatrix} \cos\varphi & -\sin\varphi & 0 \\ \sin\varphi & \cos\varphi & 0 \\ 0 & 0 & 1 \end{pmatrix}, \quad A_2 = \begin{pmatrix} 1 & 0 & 0 \\ 0 & \cos\theta & -\sin\theta \\ 0 & \sin\theta & \cos\theta \end{pmatrix},$$

and

$$A_3 = \begin{pmatrix} \cos\psi & -\sin\psi & 0 \\ \sin\psi & \cos\psi & 0 \\ 0 & 0 & 1 \end{pmatrix}.$$

The angles $\{\varphi, \theta, \psi\}$ are called the *Euler angles* of the positively oriented frame $\{Be_1, Be_2, Be_3\}$ with respect to the frame $\{e_1, e_2, e_3\}$.

(a). Show that $B = B(\varphi, \theta, \psi) =$

$$\begin{pmatrix} \cos\varphi\cos\psi - \cos\theta\sin\varphi\sin\psi & -\sin\varphi\cos\psi - \cos\varphi\cos\theta\sin\psi & \sin\theta\sin\psi \\ \cos\varphi\sin\psi + \cos\theta\sin\varphi\cos\psi & -\sin\varphi\sin\psi + \cos\varphi\cos\theta\cos\psi & -\sin\theta\cos\psi \\ \sin\varphi\sin\theta & \cos\varphi\sin\theta & \cos\theta \end{pmatrix}.$$

Show that the map

$$B : U = (0, 2\pi) \times (0, \pi) \times (0, \pi) \to \mathrm{SO}(3) : (\varphi, \theta, \psi) \to B(\varphi, \theta, \psi)$$

is a chart for $SO(3)$. Describe the set of proper rotations which do not lie in the image of this chart. Using the solid ball model of SO(3) find three charts which define an atlas for SO(3). Find two other Euler angle like charts for SO(3). Do these charts form an atlas for SO(3)?

(b). Let

$$\mathcal{L} : \mathrm{SO}(3) \times \mathrm{SO}(3) \to T\,\mathrm{SO}(3) : (A, X) \to X_A = T_e L_A X = AX = \dot{A}$$

be the left trivialization of $T\,\mathrm{SO}(3)$. Since $X = A^t \dot{A} \in \mathrm{so}(3)$, we may write

$$X = \begin{pmatrix} 0 & -\omega_3 & \omega_2 \\ \omega_3 & 0 & -\omega_1 \\ -\omega_2 & \omega_1 & 0 \end{pmatrix}.$$

Give a physical argument to show that $\omega = (\omega_1, \omega_2, \omega_3) = i(X)$ is the *angular velocity vector* of a rigid body at the configuration A. In terms of Euler angles $\{\varphi, \theta, \psi\}$ show that

$$\omega = \left(\cos\varphi\,\dot{\theta} + \sin\varphi\sin\theta\,\dot{\psi},\, -\sin\varphi\,\dot{\theta} + \cos\varphi\sin\theta\,\dot{\psi},\, \dot{\varphi} + \cos\theta\,\dot{\psi}\right).$$

(c). Identify so(3) with so(3)* using the left invariant metric

$$\rho(e) : \mathrm{so}(3) \times \mathrm{so}(3) \to \mathbf{R} : (X, Y) \to k(I(X), Y).$$

Here k is the Killing metric on so(3) and $I : \mathrm{so}(3) \to \mathrm{so}(3)$ is the moment of inertia tensor. We assume that the matrix of I with respect to the k-orthonormal basis $\{E_1, E_2, E_3\}$ is $I = \mathrm{diag}(I_1, I_2, I_3)$ and is invertible. Show that the angular momentum $p = (p_1, p_2, p_3)$ corresponding to the angular velocity ω is $p = I(\omega)$, having identified so(3) with $\mathbf{R}^3$ using the map i. Define the momenta $(p_\theta, p_\psi, p_\varphi)$ so that

$$\begin{aligned} &T^*B : T^*U \to T^*\,\mathrm{SO}(3) : \\ &\quad (\theta, \psi, \varphi, p_\theta, p_\psi, p_\varphi) \to \Big(B(\theta, \psi, \varphi), \big((DB(\theta, \psi, \varphi))^{-1}\big)^t (p_\theta, p_\psi, p_\varphi)\Big) \end{aligned}$$

is a natural chart for $T^*\,\mathrm{SO}(3)$ for which the pullback of the standard symplectic form by T^*B is the symplectic form $d\theta \wedge p_\theta + d\psi \wedge dp_\psi + d\varphi \wedge p_\varphi$ on T^*U. Since

$$\begin{pmatrix} \omega_1 \\ \omega_2 \\ \omega_3 \end{pmatrix} = \begin{pmatrix} \cos\varphi & \sin\varphi\sin\theta & 0 \\ -\sin\varphi & \cos\varphi\sin\theta & 0 \\ 0 & \cos\theta & 1 \end{pmatrix} \begin{pmatrix} \dot{\theta} \\ \dot{\psi} \\ \dot{\varphi} \end{pmatrix},$$

it follows that

$$\begin{pmatrix} p_1 \\ p_2 \\ p_3 \end{pmatrix} = \begin{pmatrix} \cos\varphi & \sin\varphi\csc\theta & -\sin\varphi\cot\theta \\ -\sin\varphi & \cos\varphi\csc\theta & -\cos\varphi\cot\theta \\ 0 & 0 & 1 \end{pmatrix} \begin{pmatrix} p_\theta \\ p_\psi \\ p_\varphi \end{pmatrix}. \tag{133}$$

Note that the right hand side of (133) is the inverse transpose of the right hand side of the preceding matrix. The Hamiltonian of the Euler top on $\mathrm{SO}(3) \times \mathrm{so}(3)^*$ is $H(A,p) = \frac{1}{2}(I_1^{-1}p_1^2 + I_2^{-1}p_2^2 + I_3^{-1}p_3^2)$. Show that the Hamiltonian of the Euler top on T^*U is

$$\begin{aligned} H(\theta,\varphi,\psi,p_\theta,p_\varphi,p_\psi) &= \tfrac{1}{2}\Big[I_1^{-1}\big(\cos\varphi\, p_\theta + \sin\varphi \csc\theta\,(p_\psi - \cos\theta\, p_\varphi)\big)^2 \\ &+ I_2^{-1}\big(-\sin\varphi\, p_\theta + \sin\varphi\csc\theta\,(p_\psi - \cos\theta\, p_\varphi)\big)^2 + I_3^{-1}p_\varphi^2 \Big]. \end{aligned}$$

Write out Hamilton's equations. They are the Euler-Arnol'd equations for the Euler top in Euler angles.

5. (Moment of inertia tensor.) (a). A body is a set of points $B \subseteq \mathbf{R}^3$ with a mass distribution given by a positive measure $d\mu$ whose support is B. Assume that the center of mass of the body is at the origin in $\mathbf{R}^3$, that is, the first moments $\int x_i\, d\mu(x)$ of $d\mu$ are zero for $i = 1,2,3$. In addition assume that the body is not contained in any line through the origin. A body is *rigid* if the Euclidean distance between any two of its points does not change when the body is moving. Let $\xi(t,x)$ be the position of a point in the body at time t whose position at time 0 was x. Show that for a rigid body there is a unique $A(t) \in \mathrm{O}(3)$ such that $\xi(t,x) = A(t)x$. Suppose that the curve $\mathbf{R} \to \mathrm{O}(3) : t \to A(t)$ is smooth and that the initial position $A(0)$ of the body is id. Then $A(t) \in \mathrm{SO}(3)$.

   (b). The kinetic energy (of rotation) of a rigid body is

$$\begin{aligned} K &= \tfrac{1}{2}\int \langle \dot{\xi}(t,x), \dot{\xi}(t,x)\rangle\, d\mu(x) = \tfrac{1}{2}\int \langle \dot{A}x, \dot{A}x\rangle\, d\mu(x) \\ &= \tfrac{1}{2}\int \langle A^{-1}\dot{A}x, A^{-1}\dot{A}x\rangle d\mu(x), \end{aligned}$$

   where the last equality follows because the Euclidean inner product $\langle\, ,\, \rangle$ on $\mathbf{R}^3$ is invariant under $\mathrm{SO}(3)$. But $\Omega = A^{-1}\dot{A} \in \mathrm{so}(3)$ is the angular velocity of the body at A. In other words, $\omega = i(\Omega) \in \mathbf{R}^3$ is the angular velocity vector of the body with respect to the fixed frame $\{e_1, e_2, e_3\}$. Thus $K = \frac{1}{2}\int \langle -\Omega^2 x, x\rangle d\mu(x)$. Using the fact that $\Omega x = \omega \times x$, show that $K = \frac{1}{2}\int \Big(\langle\omega,\omega\rangle\langle x,x\rangle - \langle\omega,x\rangle^2\Big)\, d\mu(x)$. Let $M = (M_{ij}) = (\int x_i x_j\, d\mu(x))$ be the matrix of second moments of the mass distribution $d\mu$. M is symmetric and $M_{ii} \ge 0$ for $i = 1,2,3$. Show that $M = \int x \otimes x^*\, d\mu(x)$, where $x \otimes x^* \in \mathbf{R}^3 \otimes \mathbf{R}^{3^*} = \mathrm{gl}(3,\mathbf{R})$ is defined by $(x\otimes x^*)(y) = x^*(y)x = \langle x,y\rangle x$. Show that $K = \frac{1}{2}\Big(\langle\omega,\omega\rangle \operatorname{tr} M - \langle M\omega,\omega\rangle\Big)$.

   (c). Let the position of the body at time 0 be $O^{-1} \in \mathrm{SO}(3)$ instead of I. Using $O\Omega O^{-1}x = O\omega \times x$ show that

$$\begin{aligned} K &= \tfrac{1}{2}\int \langle \Omega O^{-1}x, \Omega O^{-1}x\rangle\, d\mu(x) \\ &= \tfrac{1}{2}\langle\omega,\omega\rangle \operatorname{tr}(O^t M O) - \tfrac{1}{2}\langle O^t M O\omega, \omega\rangle. \qquad (134) \end{aligned}$$

Show that $O^t M O = \int (O^{-1}x) \otimes (O^{-1}x)^* \, d\mu(x)$ for every $O \in SO(3)$. Choose $O \in SO(3)$ so that $O^t M O = \text{diag}(M_1, M_2, M_3)$. Such an O is called a *principal axis transformation*. Deduce that $M_i = \int (O^{-1}x)_i^2 \, d\mu(x) \geq 0$ for $i = 1, 2, 3$. Show that (134) can be written as

$$K = \tfrac{1}{2} \left(I_1 \omega_1^2 + I_2 \omega_1^2 + I_3 \omega_3^2 \right),$$

where $I_1 = M_2 + M_3$, $I_2 = M_1 + M_3$ and $I_3 = M_1 + M_2$. The I_i are called the *principal moments of inertia* of the body. Show that

$$0 \leq I_1 \leq I_2 + I_3, \quad 0 \leq I_2 \leq I_1 + I_3, \quad \text{and } 0 \leq I_3 \leq I_1 + I_2. \tag{135}$$

(d). The following argument exhibits the physical significance of the inequalities (135). Suppose that one wishes to spin a rigid body about the axis corresponding to the largest moment starting from rest. By alternately applying torques about the other two axes and using the Lie algebra property of infinitesimal rotations, show that if the inequalities did not hold one could violate conservation of energy.

6. (Poinsot's construction.) Here we give a derivation of kinematical description of the motion of the Euler top. Let $Ox_1x_2x_3$ be a set of rectangular coordinate axes moving with the top. Let $I_1 x_1^2 + I_2 x_2^2 + I_3 x_3^2 = 1$ be the inertia ellipsoid $\mathscr{E}$ of the top with the fixed point at the origin O. Suppose that Π is a tangent plane to $\mathscr{E}$ which is perpendicular to the fixed angular momentum vector j, (which spans the invariant line). Let E be the total energy of the top and let $p = \frac{1}{|j|}(I_1\omega_1, I_2\omega_2, I_3\omega_3)$, where ω_i is the component of j along the x_i-axis. Show that $(p, p) = 2E/|j|$. Since $|p|$ is constant and p lies on the line spanned by j, the plane Π is fixed in space (and hence is called the invariable plane). Thus the inertia ellipsoid $\mathscr{E}$ touches the plane Π at one point (x_1', x_2', x_3'). Using $I_1 x_1 x_1' + I_2 x_2 x_2' + I_3 x_3 x_3' = 1$ and $I_1 x_1 \omega_1 + I_2 x_2 \omega_2 + I_3 x_3 \omega_3 = |p||j|$, show that the point of contact is $(x_1', x_2', x_3') = \frac{1}{|p||j|}(\omega_1, \omega_2, \omega_3)$. Deduce that the Euler top moves as if it is rigidly attached to the inertia ellipsoid, which rolls on the invariable plane without slipping so that its center is a fixed distance from the invariable plane.

7. (Equation of herpolhode.) In the Poinsot description of the motion of the Euler top given the energy and angular momentum derive the equation of the curve traced out by the point of contact of the moment of inertia ellipsoid on the invariable plane. This curve is called a *herpolhode*. Use a computer to draw some typical herpolhodes.

8. Show that every $q \in \mathbf{H}$ can be written uniquely as $\alpha + \beta \cdot j$ for some $\alpha, \beta \in \mathbf{C} = \mathbf{R} + \mathbf{R} \cdot i \subseteq \mathbf{H}$. For every $\alpha \in \mathbf{C}$ show that $j \cdot \alpha \cdot j = -\overline{\alpha}$. Verify that the map $\theta : S^3 \subseteq \mathbf{H} \to SU(2) : \alpha + \beta \cdot j \to \begin{pmatrix} \alpha & \beta \\ -\overline{\beta} & \overline{\alpha} \end{pmatrix}$ is an isomorphism of Lie groups.

9. Recall that the algebra of polynomials invariant under the S^1 action

$$\Phi : S^1 \times T\mathbf{R}^3 \to T\mathbf{R}^3 : (t, x, y) \to (R_t x, R_t y)$$

where $R_t = \begin{pmatrix} \cos t & -\sin t & 0 \\ \sin t & \cos t & 0 \\ 0 & 0 & 1 \end{pmatrix}$ is generated by

$$\sigma_1 = x_1 y_2 - x_2 y_1, \quad \sigma_2 = x_3, \quad \sigma_3 = y_3,$$
$$\sigma_4 = x_1 y_1 + x_2 y_2, \quad \sigma_5 = x_1^2 + x_2^2, \quad \sigma_6 = y_1^2 + y_2^2$$

and satisfy the relation $\sigma_1^2 + \sigma_4^2 = \sigma_5 \sigma_6$, $\sigma_5 \geq 0$ & $\sigma_6 \geq 0$, (see chapter 4 ((2.1))).

(a). Show that the algebra of polynomials on $T_1 S^2$, which are invariant under the S^1 action $\Phi|(S^1 \times T_1 S^2)$ is generated by $\sigma_1, \ldots, \sigma_6$ and satisfy the addtional relations $\sigma_5 + \sigma_2^2 = 1$, $\sigma_4 + \sigma_2\sigma_3 = 0$, and $\sigma_6 + \sigma_3^2 = 1$. Show that these relations define an S^2 embedded in $\mathbf{R}^3$.

(b). Explicitly determine the fiber of the Hilbert map of the S^1 action $\Phi|(S^1 \times T_1 S^2)$ and deduce that it is a single orbit. Show that the orbit space $T_1 S^2 / S^1$ is S^2.

10. Let $T^2 \subseteq \mathbf{R}^4$ be a 2-dimensional torus. Suppose that $\pi : T^2 \to [0, 1] \times S^1$ is a smooth surjective mapping such that $\pi^{-1}(\{0\} \times S^1)$ and $\pi^{-1}(\{1\} \times S^1)$ are each the union of two disjoint circles which are the only singularities of π and are fold points. Show that T^2 cannot be embedded in $\mathbf{R}^3$.

11. Consider $T_1 S^2 \subseteq \mathbf{R}^6$ as a Lie group. Give a symplectic form on $T\mathbf{R}^6$ whose restriction to $T_1 S^2 \times \mathbf{R}^3$ is the canonical symplectic form on $T(T_1 S^2)$ pulled back by left trivialization.

12. Consider a vector field on X on a flat 2-torus $T^2 = \mathbf{R}^2/\mathbf{Z}^2$ in $\mathbf{R}^4$ whose flow is a projection of the straight line flow on $\mathbf{R}^4$ with slope p/q, where p, q are relatively prime positive integers. Let $j : T^2 \to S^3$ be an embedding. Give a vector field Y on S^3 such that $j^*(Y|j(T^2)) = X$.

    (a). Let $\pi : S^3 - \{(0, 0, 0, 1)\} \to \mathbf{R}^3$ be stereographic projection from the north pole. Show that $\pi(T^2)$ is a 2-torus in $\mathbf{R}^3$. An integral curve γ of Y is called a *p-q* torus knot. Draw a picture of $\pi(\gamma)$ when $(p, q) = (2, 3)$. This knot is called the *trefoil* knot. Draw a picture of the projection of $\pi(\gamma)$ on $\mathbf{R}^2$ using a convention which distinguishes over crossings from under crossings. Here we assume that $\pi(\gamma)$ is oriented. Construct an orientable surface bounded by $\pi(\gamma)$ from its planar projection by replacing every crossover with a rectangle which has been given a counterclockwise half twist and whose edges lie on $\pi(\gamma)$. Fill in the remaining pieces of $\pi(\gamma)$ with rectangles. Do this for the trefoil knot. Find a different filling procedure which produces a nonorientable surface bounded by $\pi(\gamma)$. Show that the 2-1 torus knot bounds a Möbius band.

    (b). Suppose that γ and γ' are two distinct integral curves of the vector field X on T^2. Then γ and γ' are called *parallel p-q torus knots*. Find a formula for the linking number of $\pi(\gamma)$ and $\pi(\gamma')$.

13. (The twisting phenomenon.) Let $\{\widehat{e}_1, \widehat{e}_2, \widehat{e}_3\}$ be the principal axes of a triaxial rigid body with principal moments of inertia $0 < I_1 < I_2 < I_3$. Let $\{e_1, e_2, e_3\}$ be a frame fixed in space which is the principal axis frame at time $t = 0$. Assume that the body is a tennis racket-like, that is, $I_3 \approx I_1 + I_2$ and $I_1 \ll I_2$.

(a). (Almost uniform rotation of the handle.) Because the handle $\widehat{e}_1$ moves nearly in the e_1–e_2 plane (see ((8.1))), the amount that its projection on the e_1–e_2 plane rotates around the e_3-axis is nearly the same as the amount that $\widehat{e}_1$ rotates around e_3. This latter rotation at time t is $\theta(t) = \int_0^t \beta(s)\, ds$, where

$$\beta(t) = |\mu| \left(I_2^{-1} p_2^2(t) + I_3^{-1} p_3^2(t) \right) \left(|\mu|^2 - p_1^2(t) \right)^{-1},$$

and $t \to p(t)$ is a periodic solution of Euler's equations on $S^2_{h,\mu}$ of energy h and period $T = T(h, \mu)$. Prove the following. There is an $M > 0$ such that for every $t \in \mathbf{R}$,

$$|\theta(t) - \overline{\beta}\, t| \le M, \tag{136}$$

where $\overline{\beta} = \frac{1}{T} \int_0^T \beta(s)\, ds$. First show that

$$\beta(t) = \frac{2h}{|\mu|} - \frac{1}{|\mu|} (|\mu|^2 I_1^{-1} - 2h) \frac{p_1^2(t)}{|\mu|^2 - p_1^2(t)}. \tag{137}$$

From the fact the $\beta(t) - \overline{\beta}$ is a periodic function of period T and average value 0 deduce that

$$\left| \int_0^t (\beta(s) - \overline{\beta})\, ds \right| \le \int_0^T |\beta(s) - \overline{\beta}|\, ds = M$$

for every $t \in \mathbf{R}$. For a tennis racket-like body M is small, when h is close to $\frac{1}{2} |\mu|^2 I_2^{-1}$. To see this use the triangle inequality to show that

$$M \le \frac{2}{|\mu|} (|\mu|^2 I_1^{-1} - 2h) \int_0^T \frac{p_1^2(s)}{|\mu|^2 - p_1^2(s)}\, ds. \tag{138}$$

Using (138) and the fact that $|p_1(t)| \le \sqrt{(2h - |\mu|^2 I_3^{-1})/(I_1^{-1} - I_3^{-1})}$ (see the proof of ((8.1))) deduce the estimate: $M \le 2T(2h - |\mu|^2 I_3^{-1})/|\mu|$. When $h \approx |\mu|^2/2I_2$ show that $|\mu| T \approx 2\pi/\sqrt{(I_1^{-1} - I_3^{-1})(I_2^{-1} - I_3^{-1})}$. Note that the right hand side of the preceding formula is the period of rotation of the body around its intermediate axis. Using the fact that the rigid body is tennis racket-like, it follows that $M \approx 4\pi I_1/I_2$ which is small.

Use the same kind argument as in estimating the right hand side of (138) to prove $\overline{\beta} \ge \frac{2h}{|\mu|} - \frac{1}{|\mu|}(2h - |\mu|^2 I_3^{-1}) = |\mu|/I_3$. Thus the time the projection of $\widehat{e}_1$ on the e_1–e_2 plane needs to make one revolution about e_3 is at *most* $2\pi I_3/|\mu|$.

(b). (Enough time to twist.) From (136) it follows immediately that $\overline{\beta} \le 2h/|\mu|$. When $h = \frac{1}{2} |\mu|^2/I_2$, we find that $\overline{\beta} \le |\mu|/I_2$. Therefore the time needed for the projection of $\widehat{e}_1$ on the e_1–e_2 plane to make one revolution about e_3 is at *least* $t_* = 2\pi I_2/|\mu|$. For a tennis racket-like body show that t_* is larger than the characteristic twisting time $2\pi/\sqrt{(I_1^{-1} - I_3^{-1})(I_2^{-1} - I_3^{-1})}$.

(c). (A long time near γ_2^+.) Show that the racket moving near the unstable (stable) manifold of γ_2^+ (γ_2^-) spends most of its time near γ_2^- (γ_2^+). The hyperbolic character of the periodic orbits $\gamma_2^\pm$ is essential here.

(d). (No twist region.) Consider the function

$$F(p_1, p_2, p_3) = (I_1^{-1} - I_2^{-1})p_1^2 + (I_2^{-1} - I_3^{-1})p_3^2 \tag{139}$$

on the energy surface $h = \frac{1}{2}\left(I_1^{-1}p_1^2 + I_2^{-1}p_2^2 + I_3^{-1}p_3^2\right)$. The following argument shows that a tennis racket will *not* make a half twist if

$$F(0) < 2I_3 h(I_2^{-1} - I_3^{-1})\,\mathrm{e}^{2nt_0}, \tag{140}$$

where $t_0 = \frac{2\pi I_3}{|\mu|}$ and $n = |\mu|\sqrt{(I_2^{-1} - I_3^{-1})(I_1^{-1} - I_3^{-1})}$. First show that

$$\dot{F} = L_X F \;=\; -4(I_1^{-1} - I_3^{-1})(I_2^{-1} - I_3^{-1})\, p_1 p_2 p_3,$$

where X is the Euler vector field $X(p) = p \times (I')^{-1} p$ on $S^2_{|\mu|}$. Using the inequalities $|p_2| \le |\mu|$ and $|p_1 p_3| \le \frac{1}{2}(\sigma^{-1}p_1^2 + \sigma p_3^2)$ with $\sigma = \sqrt{(I_2^{-1} - I_3^{-1})/(I_1^{-1} - I_3^{-1})}$ deduce that $|\dot{F}| \le 2|\mu| n F$. This inequality integrates to $F(t) \le F(0)\,\mathrm{e}^{2nt}$. By (a) the largest time required for the projection of the handle on the e_1–e_2 plane to is $t_0 = 2\pi I_3/|\mu|$. Therefore if $F(0)$ is sufficiently small then $F(t)$ is not very large for $t \in [0, t_0]$. Thus we need a bound on $F(t)$ which excludes the occurence of a half twist. From its definition $\lambda = F(p(t)) = F(t)$ determines a family of ellipses

$$\mathscr{E}_\lambda : (I_1^{-1} - I_2^{-1})p_1^2 + (I_2^{-1} - I_3^{-1})p_3^2 = \lambda.$$

Show that if for every $t \in [0, t_0]$ the curve $\mathscr{E}_\lambda$ lies in the interior of the ellipse

$$\mathscr{E} : 2h = I_1^{-1}p_1^2 + I_3^{-1}p_3^2$$

then the integral curve $t \to p(t)$ of the vector field X on the energy surface does *not* cross the $\{p_2 = 0\}$ plane. From $I_3 - I_2 \le I_2 - I_1$ deduce the estimate

$$F(t) \ge I_3(I_2^{-1} - I_3^{-1})(I_1^{-1}p_1^2 + I_3^{-1}p^3).$$

Show that $\mathscr{E}_\lambda$ lies in the interior of $\mathscr{E}$ if

$$F(t) < 2hI_3(I_2^{-1} - I_3^{-1}).$$

Show that if (140) holds on the energy surface then *no* twist occurs.

Chapter IV

The spherical pendulum

In this chapter we treat the spherical pendulum as a constrained Hamiltonian system. We derive Hamilton's equations and show that there is an axial symmetry which gives rise to a conserved angular momentum. Thus the spherical pendulum is a Liouville integrable Hamiltonian system. Using the technique of singular reduction (see appendix B section 5) we remove the axial symmetry to obtain a Hamiltonian systems with one degree of freedom which we analyze. From the qualitative description of the reduced system we obtain a complete qualitative picture of the motion of the spherical pendulum. Because of monodromy, the Liouville tori fit together in a nontrivial way. This precludes the existence of global action coordinates, (see appendix D section 2).

1 Liouville integrability

Recall some standard facts about Hamiltonian systems on $T\mathbf{R}^3$, (see appendix A section 3). Let $\langle\, , \,\rangle$ be the Euclidean inner product and $\times$ the usual vector product on $\mathbf{R}^3$. Let $\zeta = (x, y)$ be canonical coordinates on $T\mathbf{R}^3$, that is, the canonical symplectic 2-form is $\omega = \sum_i dx_i \wedge dy_i$. Corresponding to the Hamiltonian function $H : T\mathbf{R}^3 \to \mathbf{R}$ is the Hamiltonian vector field X_H whose integral curves satisfy Hamilton's equations

$$\begin{aligned} \frac{dx}{dt} &= \frac{\partial H}{\partial y} \\ \frac{dy}{dt} &= -\frac{\partial H}{\partial x}. \end{aligned} \tag{1}$$

Using the symplectic form ω, we define a Poisson bracket $\{\, , \,\}_{T\mathbf{R}^3}$ on $C^\infty(T\mathbf{R}^3)$ by

$$\{f, g\}_{T\mathbf{R}^3} = \omega(X_f, X_g) = \sum_{i,j} \frac{\partial f}{\partial \zeta_i}\frac{\partial g}{\partial \zeta_j}\{\zeta_i, \zeta_j\}_{T\mathbf{R}^3},$$

whose structure matrix is

$$(\{\zeta_i, \zeta_j\}_{T\mathbf{R}^3}) = \left(\begin{array}{ccc|ccc} 0 & 0 & 0 & 1 & 0 & 0 \\ 0 & 0 & 0 & 0 & 1 & 0 \\ 0 & 0 & 0 & 0 & 0 & 1 \\ \hline -1 & 0 & 0 & 0 & 0 & 0 \\ 0 & -1 & 0 & 0 & 0 & 0 \\ 0 & 0 & -1 & 0 & 0 & 0 \end{array}\right).$$

In terms of Poisson brackets Hamilton's equations for X_H read

$$\begin{aligned} \dot{x} &= \{x, H\}_{T\mathbf{R}^3} \\ \dot{y} &= \{y, H\}_{T\mathbf{R}^3}. \end{aligned} \tag{2}$$

We now describe the spherical pendulum as a constrained Hamiltonian system. First consider the unconstrained Hamiltonian system $(H, T\mathbf{R}^3, \omega)$ with Hamiltonian

$$H : T\mathbf{R}^3 \to \mathbf{R} : (x, y) \to \tfrac{1}{2}\langle y, y\rangle + \gamma\langle x, e_3\rangle. \tag{3}$$

The integral curves of X_H give the motion of a particle in $\mathbf{R}^3$ under a constant vertical gravitational force. Choosing an appropriate unit of length, we may assume that $\gamma = 1$. ▷ Now *constrain* the particle to move on the 2-sphere $S^2 \subseteq \mathbf{R}^3$. The following argument shows that the 2-form $\omega|TS^2$ is symplectic.

(1.1) **Proof**: Let $T_0\mathbf{R}^3 = (\mathbf{R}^3 - \{0\}) \times \mathbf{R}^3$. Consider the constraint functions

$$c_1 : T_0\mathbf{R}^3 \to \mathbf{R} : (x, y) \to \langle x, x\rangle - 1 \quad \text{and} \quad c_2 : T_0\mathbf{R}^3 \to \mathbf{R} : (x, y) \to \langle x, y\rangle.$$

Observe that the constraint manifold TS^2 is $\mathscr{C}^{-1}(0)$, where

$$\mathscr{C} : T_0\mathbf{R}^3 \to \mathbf{R}^2 : m \to \big(c_1(m), c_2(m)\big).$$

Since 0 is a regular value of $\mathscr{C}$, TS^2 is a smooth manifold. Because the matrix

$$\big(\{c_i, c_j\}_{T\mathbf{R}^3}(x, y)\big)$$

is invertible for every $(x, y) \in T_0\mathbf{R}^3$ with inverse

$$C = (C_{ij}) = \frac{1}{2\langle x, x\rangle}\begin{pmatrix} 0 & -1 \\ 1 & 0 \end{pmatrix},$$

TS^2 is cosymplectic. Therefore $\omega|TS^2$ is a symplectic form, (see appendix A ((4.10))). □

The *spherical pendulum* is the constrained system $(H|TS^2, TS^2, \omega|TS^2)$.

Using the modified Dirac brackets (see appendix A section 4), we compute Hamilton's equations for the constrained Hamiltonian $H|TS^2$ as follows. Let

$$H^* = H - \sum_{i,j=1}^{2} (\{H, c_i\}_{T\mathbf{R}^3} + H_i)\, C_{ij}\, c_j, \tag{4}$$

where H_j is in the ideal generated by the constraints c_j. Then

$$\begin{aligned} H^*(x,y) &= \tfrac{1}{2}\langle y,y\rangle + \langle x,e_3\rangle + \frac{1}{2\langle x,x\rangle}\big(-2\langle x,y\rangle + H_1\big)\langle x,y\rangle \\ &\quad - \frac{1}{2\langle x,x\rangle}\big(-\langle y,y\rangle + \langle x,e_3\rangle + H_2\big)\big(\langle x,x\rangle - 1\big). \end{aligned}$$

Choose

$$H_1 = -\langle x,y\rangle\big(\langle x,x\rangle - 2\big) \quad \text{and} \quad H_2 = -\big(\langle x,x\rangle - 1\big)\big(\langle y,y\rangle - \langle x,e_3\rangle\big).$$

Then

$$H^*(x,y) = \tfrac{1}{2}\langle y,y\rangle + \langle x,e_3\rangle + \tfrac{1}{2}\big(\langle y,y\rangle - \langle x,e_3\rangle\big)\big(\langle x,x\rangle - 1\big) - \tfrac{1}{2}\langle x,y\rangle^2,$$

Using (2), a calculation shows that on $T_0\mathbf{R}^3$ the Hamiltonian vector field X_{H^*} is

$$\begin{aligned} \frac{dx}{dt} &= y + y\big(\langle x,x\rangle - 1\big) - \langle x,y\rangle x \\ \frac{dy}{dt} &= -e_3 + \tfrac{1}{2}e_3\big(\langle x,x\rangle - 1\big) - \big(\langle y,y\rangle - \langle x,e_3\rangle\big)x + \langle x,y\rangle y. \end{aligned}$$

We have $X_{H|TS^2} = X_{H^*|TS^2} = X_{H^*}|TS^2$. Therefore Hamilton's equations for the spherical pendulum are

$$\begin{aligned} \frac{dx}{dt} &= y \\ \frac{dy}{dt} &= -e_3 + \big(\langle x,e_3\rangle - \langle y,y\rangle\big)x. \end{aligned} \tag{5}$$

Note that (5) is a system of differential equations on $T\mathbf{R}^3$ for which TS^2 is an invariant manifold.

We now look at the symmetries of the spherical pendulum. As a physical system in $\mathbf{R}^3$, the spherical pendulum is invariant under rotation about the x_3-axis. Lifting this to $T\mathbf{R}^3$ gives the S^1 action

$$\Phi : S^1 \times T\mathbf{R}^3 \to T\mathbf{R}^3 : (x,y) \to (R_t x, R_t y) \tag{6}$$

where $R_t = \begin{pmatrix} \cos t & -\sin t & 0 \\ \sin t & \cos t & 0 \\ 0 & 0 & 1 \end{pmatrix}$. The infinitesimal generator of Φ is the vector field Y whose integral curves satisfy

$$\begin{aligned} \frac{dx}{dt} &= -x \times e_3 \\ \frac{dy}{dt} &= -y \times e_3. \end{aligned} \tag{7}$$

Using (1) we see that Y is the Hamiltonian vector field X_J corresponding to the Hamiltonian function

$$J : T\mathbf{R}^3 \to \mathbf{R} : (x,y) \to \langle x \times y, e_3\rangle = x_1 y_2 - x_2 y_1. \tag{8}$$

In other words, J is the momentum mapping of the S^1 action Φ. Since H and TS^2 are invariant under Φ,

$$\{J|TS^2, H|TS^2\}_{TS^2} = -L_{X_{J|TS^2}}\big(H|TS^2\big) = -\{H^*, J^*\}_{T\mathbf{R}^3}|TS^2 = 0. \tag{9}$$

Here J^* is obtained by replacing H in (4) with J. Therefore $J|TS^2$ is an integral of the spherical pendulum. From (9) we see that the spherical pendulum is Liouville integrable.

2 Reduction of the S^1 symmetry

We remove the S^1 symmetry $\Phi|(S^1 \times TS^2)$ of the spherical pendulum using singular reduction, (see appendix B section 5). The regular reduction theorem does *not* apply, because the action $\Phi|(S^1 \times TS^2)$ leaves the points $\pm(0,0,1,0,0,0)$ fixed. After reduction we obtain a Hamiltonian system with *one degree of freedom* on a *singular* reduced phase space.

The first step in the reduction process is to find the invariants of the action Φ (6).

Claim: The algebra $\mathbf{R}\,[x, y]^{S^1}$ of polynomials invariant under Φ is generated by

$$\begin{array}{lll} \tau_1 = x_3 & \tau_3 = y_1^2 + y_2^2 & \tau_5 = x_1^2 + x_2^2 \\ \tau_2 = y_3 & \tau_4 = x_1y_1 + x_2y_2 & \tau_6 = x_1y_2 - x_2y_1. \end{array} \tag{10}$$

(2.1) **Proof**: We begin by noting that the action Φ fixes every point on the 2-plane

$$\Pi = \Big\{(0,0,x_3,0,0,y_3) \in T\mathbf{R}^3 \big| (x_3, y_3) \in \mathbf{R}^2\Big\}.$$

Therefore the algebra of polynomials invariant under $\Phi|(S^1 \times \Pi)$ is the algebra of polynomials $\mathbf{R}[x_3, y_3]$ in the variables x_3 and y_3. On $T(\mathbf{R}^2 \times \{0\})$ the action $\Phi|\big(S^1 \times T(\mathbf{R}^2 \times \{0\})\big)$ is

$$\big(t, (\widetilde{x}, \widetilde{y})\big) \to (S_t\widetilde{x}, S_t\widetilde{y}) \tag{11}$$

where

$$\widetilde{x} = (x_1, x_2), \;\; \widetilde{y} = (y_1, y_2) \;\; \text{and} \;\; S_t = \begin{pmatrix} \cos t & -\sin t \\ \sin t & \cos t \end{pmatrix}.$$

In other words, (11) is the diagonal action of $\mathrm{SO}(2, \mathbf{R})$ on $\mathbf{R}^2 \times \mathbf{R}^2$. Therefore $\mathbf{R}[x, y]^{S^1}$ is equal to $\mathbf{R}[x_3, y_3] \otimes \mathbf{R}[\widetilde{x}, \widetilde{y}]^{S^1}$, where $\mathbf{R}[\widetilde{x}, \widetilde{y}]^{S^1}$ is the algebra of polynomials invariant under $\Phi|T(\mathbf{R}^2 \times \{0\})$. Thus we need only show that $\mathbf{R}[\widetilde{x}, \widetilde{y}]^{S^1}$ is generated by τ_i for $i = 3, 4, 5, 6$. This follows from ((5.2)) of chapter 1. (Watch out for a change in notation.) □

▷ The generators τ_i of $\mathbf{R}[x, y]^{S^1}$ satisfy a single polynomial relation

$$\tau_4^2 + \tau_6^2 = \tau_3\tau_5 \qquad \tau_3 \geq 0 \;\&\; \tau_5 \geq 0. \tag{12}$$

(2.2) **Proof**: This follows from ((5.3)) of chapter 1. □

Claim: Let

$$\pi : T\mathbf{R}^3 \to \mathbf{R}^6 : (x, y) \to \big(\tau_1(x, y), \ldots, \tau_6(x, y)\big) \tag{13}$$

be the *Hilbert map* of the S^1 action Φ (6). As a set, the space of S^1 orbits $T\mathbf{R}^3/S^1$ is the *semialgebraic variety* W defined by (12).

(2.3) **Proof**: We want to show that π is onto W. Let $p = (\tau_1, \ldots, \tau_6)$ be a given point on W. We now find necessary conditions that the point $(x, y) \in T\mathbf{R}^3$ satisfies $\pi(x, y) = p$. Suppose that $\tau_3 = y_1^2 + y_2^2 > 0$. Then solving

$$\begin{aligned}\tau_4 &= x_1y_1 + x_2y_2\\ \tau_6 &= x_2y_1 - x_1y_2\end{aligned}$$

for (x_1, x_2) gives

$$\begin{aligned}x_1 &= (y_1\tau_4 - y_2\tau_6)/\tau_3\\ x_2 &= (y_2\tau_4 + y_1\tau_6)/\tau_3.\end{aligned} \tag{14}$$

Thus $\pi^{-1}(p) \subseteq \mathcal{V}$ where $\mathcal{V}$ is

$$\Big\{\big((y_1\tau_4 - y_2\tau_6)/\tau_3, (y_2\tau_4 + y_1\tau_6)/\tau_3, \tau_1, y_1, y_2, \tau_2\big) \in T\mathbf{R}^3 \big| \tau_3 = y_1^2 + y_2^2 > 0\Big\}.$$

Using the fact that $\tau_5 = (\tau_4^2 + \tau_6^2)/\tau_3$ on $\mathcal{V}$, a calculation shows that for every $q \in \mathcal{V}$, we have $\pi(q) = p$, that is, $\mathcal{V} \subseteq \pi^{-1}(p)$. Thus $\mathcal{V} = \pi^{-1}(p)$. Clearly $\mathcal{V}$ is a single orbit of the S^1 action Φ (6). Under the hypothesis that $\tau_5 = x_1^2 + x_2^2 > 0$, a similar argument shows that $\pi^{-1}(p) = \mathcal{V}'$ where $\mathcal{V}'$ is

$$\Big\{(x_1, x_2, \tau_1, (x_1\tau_4 + x_2\tau_6)/\tau_5, (x_2\tau_4 - x_1\tau_6)/\tau_5, \tau_2) \in T\mathbf{R}^3 \big| \tau_5 = x_1^2 + x_2^2 > 0\Big\}.$$

Hence the variety W is the image of the Hilbert map π. Note that when $\tau_3 = 0$ or $\tau_5 = 0$, the fiber $\pi^{-1}(\tau_1, \tau_2, 0, 0, 0, 0)$ is a single point $(0, 0, \tau_1, 0, 0, \tau_2)$, which is a fixed point of Φ. □

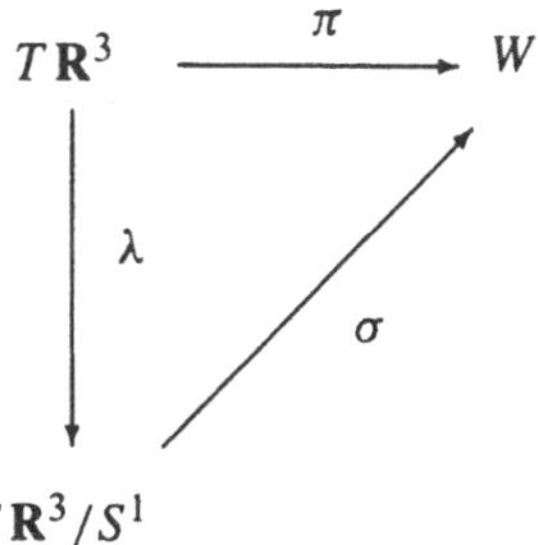

Diagram 2.1

Claim: W is homeomorphic to the orbit space $T\mathbf{R}^3/S^1$.

(2.4) **Proof**: In diagram 2.1 the map λ is the orbit mapping of the action Φ. Since the components of π are invariant under Φ, the Hilbert mapping π induces a mapping σ, which is continuous because π is. Because every fiber of π is a single S^1 orbit, the mapping σ is bijective.

To show that the inverse of σ is continuous, it is enough to show that the map π has a continuous local cross section. In particular, this shows that the fibers of π are connected. We do this as follows. Define new generators of $\mathbf{R}[x, y]^{S^1}$ by

$$\begin{aligned} \xi_1 &= 2\tau_4 & \xi_3 &= \tau_3 - \tau_5 & \eta_1 &= \tau_1 \\ \xi_2 &= 2\tau_6 & \xi_4 &= \tau_3 + \tau_5 & \eta_2 &= \tau_2. \end{aligned} \tag{15}$$

Then the variety $W \subseteq \mathbf{R}^6$ (12) becomes

$$\xi_1^2 + \xi_2^2 + \xi_3^2 = \xi_4^2, \qquad \xi_4 \ge 0. \tag{16}$$

Here (η_1, η_2) is free to range over $\mathbf{R}^2$. Hence W is the Cartesian product of $\mathbf{R}^2$ and a cone on S^2 with vertex 0. Thus W is homeomorphic to $\mathbf{R}^5$. Using the generators (ξ, η), rewrite the Hilbert map π as

$$\pi : T\mathbf{R}^3 \to W \subseteq \mathbf{R}^6 : (x, y) \to (\xi, \eta). \tag{17}$$

Notice that π is just the Cartesian product of the Hopf mapping and the identity mapping on $\mathbf{R}^2$, (see chapter 1 ((5.3))). Now consider the mappings

$$\begin{aligned} &\psi_1 : W_1 = W - \{\xi_3 = \xi_4\} \to T\mathbf{R}^3 : \\ &\qquad (\xi, \eta) \to \Big(0, \sqrt{\tfrac{1}{2}(\xi_4 - \xi_3)}, \eta_1, \frac{\xi_2}{\sqrt{2(\xi_4 - \xi_3)}}, \frac{\xi_1}{\sqrt{2(\xi_4 - \xi_3)}}, \eta_2\Big) \end{aligned}$$

and

$$\begin{aligned} &\psi_2 : W_2 = W - \{-\xi_3 = \xi_4\} \to T\mathbf{R}^3 : \\ &\qquad (\xi, \eta) \to \Big(-\frac{\xi_2}{\sqrt{2(\xi_4 + \xi_3)}}, \frac{\xi_1}{\sqrt{2(\xi_4 + \xi_3)}}, \eta_1, 0, \sqrt{\tfrac{1}{2}(\xi_4 + \xi_3)}, \eta_2\Big). \end{aligned}$$

Clearly ψ_1 and ψ_2 are continuous, and $\pi \circ \psi_1 = id_{W_1}$ and $\pi \circ \psi_2 = id_{W_2}$. Thus ψ_1 and ψ_2 are continuous local sections of π. However, W_1 and W_2 do *not* cover W, since

$$W_1 \cup W_2 = W - \Big\{(0, 0, \eta_1, 0, 0, \eta_2) \Big| (\eta_1, \eta_2) \in \mathbf{R}^2\Big\}.$$

To finish the proof we will show that ψ_1 has a continuous extension $\overline{\psi}_1$ to

$$W_1 \cup \Big\{(0, 0, \eta_1, 0, 0, \eta_2) \Big| (\eta_1, \eta_2) \in \mathbf{R}^2\Big\},$$

namely, $\overline{\psi}_1(0, 0, \eta_1, 0, 0, \eta_2) = (0, 0, \eta_1, 0, 0, \eta_2)$ and $\overline{\psi}_1 | W_1 = \psi_1$. This we do as follows. From (16) we see that $|\xi_j| \le \sqrt{\xi_4^2 - \xi_3^2}$, for $j = 1, 2$. Consequently,

$$\frac{|\xi_j|}{\sqrt{\tfrac{1}{2}(\xi_4 - \xi_3)}} \le \sqrt{2}\sqrt{\xi_4 + \xi_3} \le 2\sqrt{\xi_4}, \quad \text{for } j = 1, 2. \tag{18}$$

From (18) and the definition of ψ_1 we obtain

$$\overline{\psi}_1(0, 0, \eta_1, 0, 0, \eta_2) = \lim_{\substack{\xi_4 \to 0^+ \\ (\xi, \eta) \in W_1}} \psi_1(\xi, \eta) = (0, 0, \eta_1, 0, 0, \eta_2).$$

Thus the function $\overline{\psi}_1$ is a continuous extension of ψ_1, as desired. □

We continue with the construction of the singular reduced space. For later purposes it is convenient to employ yet another set of generators for $\mathbf{R}[x, y]^{S^1}$, namely

$$\begin{array}{lll} \sigma_1 = \tau_1 & \sigma_3 = \tau_3 + \tau_2^2 & \sigma_5 = \tau_5 \\ \sigma_2 = \tau_2 & \sigma_4 = \tau_4 & \sigma_6 = \tau_6. \end{array} \tag{19}$$

Since $TS^2 \subseteq T\mathbf{R}^3$ is invariant under the action $\Phi|(S^1 \times TS^2)$, the orbit space TS^2/S^1 is the semialgebraic variety V defined by

$$\begin{aligned} \sigma_4^2 + \sigma_6^2 &= \sigma_5(\sigma_3 - \sigma_2^2), \quad \sigma_3 - \sigma_2^2 \geq 0 \ \& \ \sigma_5 \geq 0 \\ \sigma_5 + \sigma_1^2 &= 1, \\ \sigma_4 + \sigma_1\sigma_2 &= 0. \end{aligned} \tag{20}$$

Because $J|TS^2$ (8) is the momentum mapping of the action $\Phi|(S^1 \times TS^2)$, the singular reduced phase space $V_j = (J|TS^2)^{-1}(j)/S^1$ is the semialgebraic subvariety of V (20) defined by $\sigma_6 = j$. Eliminating σ_4 and σ_5 from (20) and setting $\sigma_6 = j$ gives the semialgebraic variety M_j

$$\sigma_3(1 - \sigma_1^2) - \sigma_2^2 - j^2 = 0, \quad \text{where } |\sigma_1| \leq 1 \ \& \ \sigma_3 - \sigma_2^2 \geq 0. \tag{21}$$

In other words, the image of the projection

$$\mu : \mathbf{R}^6 \to \mathbf{R}^3 : (\sigma_1, \ldots, \sigma_6) \to (\sigma_1, \sigma_2, \sigma_3) \tag{22}$$

restricted to V_j is the semialgebraic variety M_j. As is easily checked, the inverse of $\mu|V_j$ is the map

$$\nu : M_j \subseteq \mathbf{R}^3 \to \mathbf{R}^6 : (\sigma_1, \sigma_2, \sigma_3) \to (\sigma_1, \sigma_2, \sigma_3, -\sigma_1\sigma_2, 1 - \sigma_1^2, j). \tag{23}$$

Thus M_j is isomorphic as a semialgebraic variety to V_j and hence serves as a model of the reduced space.

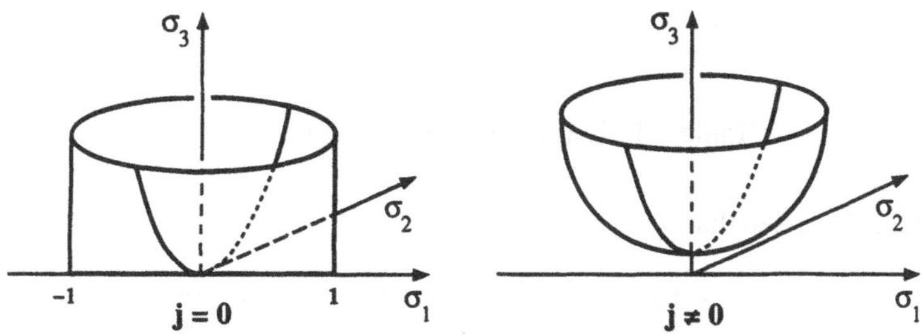

Figure 2.1. The reduced phase space M_j.

When $j \neq 0$, M_j is diffeomorphic to $\mathbf{R}^2$, being the graph of the function

$$\sigma_3 = G(\sigma_1, \sigma_2) = \frac{j^2 + \sigma_2^2}{1 - \sigma_1^2}, \quad |\sigma_1| < 1.$$

When $j = 0$, M_0 is *not* the graph of a function, because it contains the vertical lines $\{(\pm 1, 0, \sigma_3) \in \mathbf{R}^3 | \sigma_3 \geq 0\}$, (see figure 2.1). M_0 is *not* smooth because $p_\pm = (\pm 1, 0, 0)$ are singular points. However, M_0 is homeomorphic to $\mathbf{R}^2$. In fact, M_0 can be obtained as an orbit space of a $\mathbf{Z}_2$-action on TS^1 defined by

$$\varphi : \mathbf{Z}_2 \times TS^1 \rightarrow TS^1 : (x, y) \rightarrow (x, -y).$$

Geometrically the orbit space $TS^1/\mathbf{Z}_2$ of the action φ is obtained by first pinching the zero section of the cylinder TS^1 to a closed interval which forms a crease and then folding this pinched cylinder along the crease until the ends match up. Thus M_0 is the phase space TS^1 of the mathematical pendulum with points identified by the $\mathbf{Z}_2$ action. This identification is necessary because we can not distinguish positive and negative velocities in the spherical pendulum.

From this discussion we see that singular reduction of the spherical pendulum not only produces an accurate symplectic model of the reduced phase spaces, but also a geometrically faithful representation of the $j \rightarrow 0$ limit.

Since the Hamiltonain H (3) is invariant under the action $\Phi|(S^1 \times TS^2)$, it induces the function

$$H_j : \mathbf{R}^3 \rightarrow \mathbf{R} : (\sigma_1, \sigma_2, \sigma_3) \rightarrow \tfrac{1}{2}\sigma_3 + \sigma_1. \tag{24}$$

To find Hamilton's derivation for the reduced Hamiltonian $H_j|M_j$ (see appendix B section 5) we must determine the Poisson structure on $C^\infty(M_j)$. Let $(\sigma_1, \ldots, \sigma_6)$ be coordinates on $\mathbf{R}^6$. $C^\infty(\mathbf{R}^6)$ has a Poisson bracket $\{\, , \,\}_{\mathbf{R}^6}$, see table 2.1.

$\{A, B\}$	σ_1	σ_2	σ_3	σ_4	σ_5	σ_6	B
σ_1	0	1	$2\sigma_2$	0	0	0	
σ_2		0	0	0	0	0	
σ_3			0	$2(\sigma_2^2 - \sigma_3)$	$-4\sigma_4$	0	
σ_4				0	$-2\sigma_5$	0	
σ_5					0	0	
σ_6						0	
A							

Table 2.1. Half the structure matrix for $\{\, , \,\}_{\mathbf{R}^6}$.

From table 2.1 we see that

$$C_1 = \sigma_4^2 + \sigma_6^2 - \sigma_5(\sigma_3 - \sigma_2^2) \quad \text{and} \quad C_2 = \sigma_6$$

are Casimir elements of the Poisson algebra $(C^\infty(\mathbf{R}^6), \{\, , \,\}_{\mathbf{R}^6}, \cdot)$, that is, $\{C_1, f\}_{\mathbf{R}^6} = \{C_2, f\}_{\mathbf{R}^6} = 0$ for every $f \in C^\infty(\mathbf{R}^6)$. Since the orbit space W (12) is defined by $C_1 = 0$, the Poisson bracket on $C^\infty(W)$ has the same structure matrix as the Poisson bracket on $C^\infty(\mathbf{R}^6)$. Poisson bracket $\{\, , \,\}_V$ on V (20) may be computed using the Dirac prescription, because

$$\{\sigma_5 + \sigma_1^2, \sigma_4 + \sigma_1\sigma_2\}_{\mathbf{R}^6}|V = 2.$$

In particular, for every $f, g \in C^\infty(\mathbf{R}^6)$

$$\{f|V, g|V\}_V = \{f^*, g^*\}_{\mathbf{R}^6}|V,$$

where

$$f^* = f - \tfrac{1}{2}(\sigma_5 + \sigma_1{}^2 - 1)\{f, \sigma_4 + \sigma_1\sigma_2\}_{\mathbf{R}^6} + \tfrac{1}{2}(\sigma_4 + \sigma_1\sigma_2)\{f, \sigma_5 + \sigma_1{}^2\}_{\mathbf{R}^6}$$

and similarly for g^*. Half the skew symmetric structure matrix for $\{\,,\,\}_V$ is given in table 2.2.

$\{A, B\}$	σ_1	σ_2	σ_3	σ_6	B
σ_1	0	$1-\sigma_1^2$	$2\sigma_2$	0	
σ_2		0	$-2\sigma_1\sigma_3$	0	
σ_3			0	0	
σ_6				0	
A					

Table 2.2. Half the structure matrix of $\{\,,\,\}_V$

The reduced space V_j is defined by $C_2 = j$. As C_2 is a Casimir in $(C^\infty(V), \{\,,\,\}_V)$, the Poisson bracket $\{\,,\,\}_{V_j}$ on $C^\infty(V_j)$ has the same structure matrix as the Poisson bracket $\{\,,\,\}_V$ on $C^\infty(V)$. Because V_j is diffeomorphic to M_j by the projection map μ (22), the Poisson bracket $\{\,,\,\}_{M_j}$ on $C^\infty(M_j)$ has the same structure matrix as that given in table 2.2 with the last row and column deleted.

It is easier to calculate in the ambient space $\mathbf{R}^3$ with coordinates $(\sigma_1, \sigma_2, \sigma_3)$ than on the reduced space M_j. On $C^\infty(\mathbf{R}^3)$ define a Poisson bracket $\{\,,\,\}_{\mathbf{R}^3}$ half of whose structure matrix is given in table 2.3.

$\{A, B\}$	σ_1	σ_2	σ_3	B
σ_1	0	$1-\sigma_1^2$	$2\sigma_2$	
σ_2		0	$-2\sigma_1\sigma_3$	
σ_3			0	
A				

Table 2.3. Half the structure matrix for $\{\,,\,\}_{\mathbf{R}^3}$.

An inspection of table 2.3 shows that

$$\{\sigma_i, \sigma_j\}_{\mathbf{R}^3} = \sum_{k=1}^{3} \varepsilon_{ijk} \frac{\partial\psi}{\partial\sigma_k} \tag{25}$$

where

$$\psi(\sigma_1, \sigma_2, \sigma_3) = \sigma_3(1-\sigma_1^2) - \sigma_2^2 - j^2. \tag{26}$$

Note that $\psi = 0$ is the defining equation of M_j. Since ψ is a Casimir element of $(C^\infty(\mathbf{R}^3), \{\,,\,\}_{\mathbf{R}^3})$, the Poisson bracket $\{\,,\,\}_{M_j}$ on $C^\infty(M_j)$ is obtained by restricting $\{\,,\,\}_{\mathbf{R}^3}$. From (25) it follows that for every $f, g \in C^\infty(\mathbf{R}^3)$

$$\{f, g\}_{\mathbf{R}^3} = \langle \nabla f \times \nabla g, \nabla \psi \rangle.$$

Consequently, Hamilton's derivation $-\mathrm{ad}_{H_j|M_j}$ on $C^\infty(M_j)$ for the reduced Hamiltonian $H_j|M_j$ (24) is

$$\sigma_2 \frac{\partial}{\partial \sigma_1} - \big((1-\sigma_1^2) + \sigma_1\sigma_3\big) \frac{\partial}{\partial \sigma_2} - 2\sigma_2 \frac{\partial}{\partial \sigma_3}. \tag{27}$$

Note that ψ is an integral of (27), because it is a Casimir.

3 The energy momentum mapping

In this section we study the qualitative properties of the *energy momentum mapping*

$$\begin{aligned} \mathcal{EM} : TS^2 \subseteq T\mathbf{R}^3 \to \mathbf{R}^2 : \\ (x, y) \to (h, j) = \big(\tfrac{1}{2}(y_1^2 + y_2^2 + y_3^2) + x_3,\; x_1y_2 - x_2y_1\big) \end{aligned} \tag{28}$$

of the spherical pendulum. In particular, we will determine its set of critical values, its range and the topology of its fibers. The results are summarized in figure 3.1.

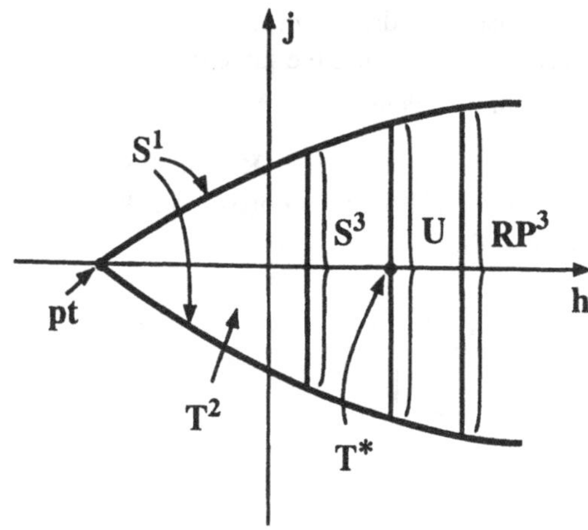

Figure 3.1. The bifurcation diagram of the energy momentum mapping $\mathcal{EM}$ of the spherical pendulum.

▷ We begin by finding the set Σ of critical values of $\mathcal{EM}$, that is, the image under $\mathcal{EM}$ of the set of points where its derivative has less than full rank.

(3.1) **Proof**: The energy momentum mapping has a critical value (h, j) if and only if h is a critical value of the reduced Hamiltonian $H_j|M_j$. Since M_j is possibly singular, we must define what we mean by a critical value. In geometric terms, $(h, j) \in \Sigma$ if and only if the 2-plane

$$\Pi_h : \tfrac{1}{2}\sigma_3 + \sigma_1 = h \tag{29}$$

intersects the semialgebraic variety

$$M_j : \sigma_2^2 + j^2 = \sigma_3(1 - \sigma_1^2), \qquad |\sigma_1| \le 1 \ \& \ \sigma_3 \ge 0 \tag{30}$$

at a point $\sigma^0 = \left(\sigma_1^0, \sigma_2^0, \sigma_3^0 = 2(h - \sigma_1^0)\right)$ with multiplicity greater than one. To explain this last phrase, consider the polynomial

$$Q(\sigma_1, \sigma_2) = \sigma_2^2 + j^2 - 2(h - \sigma_1)(1 - \sigma_1^2), \tag{31}$$

which is obtained by eliminating σ_3 from (29) and (30). σ^0 has multiplicity greater than one if and only if the *Taylor polynomial* of Q at $(\sigma_1^0, \sigma_2^0) \in [-1, 1] \times \mathbf{R}$ has no constant or linear terms. This condition is satisfied if and only if

$$\sigma_2 = \sigma_2^0 = 0 \tag{32}$$

and the polynomial

$$\widetilde{P}(\sigma_1) = (h - \sigma_1)(1 - \sigma_1^2) - \tfrac{1}{2} j^2 = \sigma_1^3 - h\sigma_1^2 - \sigma_1 + h - \tfrac{1}{2} j^2 \tag{33}$$

has a multiple root $\sigma_1^0 \in [-1, 1]$. □

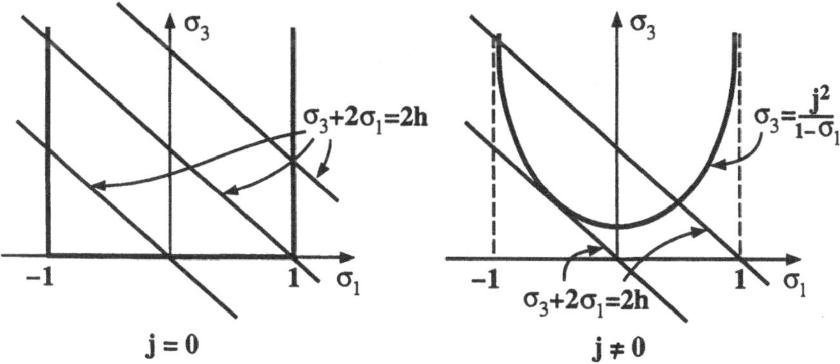

Figure 3.2. The critical points and critical values of the reduced Hamiltonian H_j on $M_j \cap \{\sigma_2 = 0\}$.

Another way to formulate the above discussion goes as follows. The point (h, j) is a critical value of $\mathcal{EM}$ if and only if the line

$$\ell_h : \tfrac{1}{2}\sigma_3 + \sigma_1 = h \ \& \ \sigma_2 = 0 \tag{34}$$

intersects the curve

$$\mathcal{F} : \sigma_3(1 - \sigma_1^2) = j^2, \quad |\sigma_1| \le 1 \ \& \ \sigma_3 \ge 0 \tag{35}$$

at a point $(\sigma_1^0, 0, \sigma_3^0)$ with multiplicity greater than one, (see figure 3.2). Note that $\mathcal{F}$ is the boundary of the image of the *fold curve* of the projection

$$\rho : M_j \subseteq \mathbf{R}^3 \to \mathbf{R}^2 : (\sigma_1, \sigma_2, \sigma_3) \to (\sigma_1, 0, \sigma_3), \tag{36}$$

namely, $\mathcal{F}$ is the boundary of

$$\sigma_3(1 - \sigma_1^2) \geq j^2 \qquad |\sigma_1| \leq 1 \ \& \ \sigma_3 \geq 0.$$

Geometrically this means that over every point p in the interior of $\rho(M_j)$, the fiber $\rho^{-1}(p)$ consists of two distinct points; while over every point p on $\mathcal{F}$, the fiber $\rho^{-1}(p)$ consists of one point.

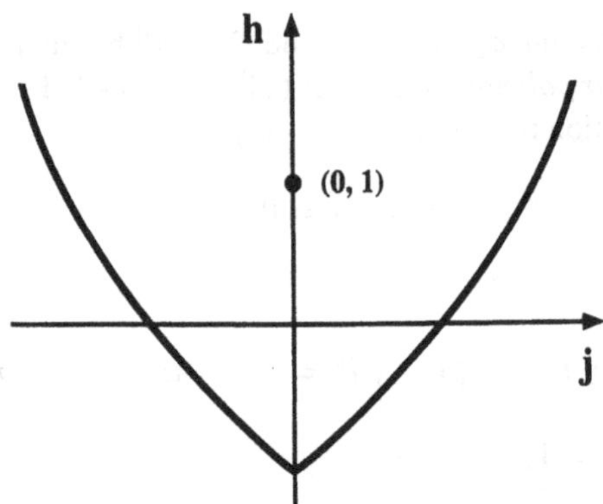

Figure 3.3. The set $\Delta_{\widetilde{P}}$ of critical values of $\mathcal{EM}$. The set $\mathcal{R}$ of regular values of $\mathcal{EM}$ is the region lying above the darkened curves of $\Delta_{\widetilde{P}}$ excluding the point (0, 1).

Let

$$\Delta_{\widetilde{P}} = \left\{ (h, j) \in \mathbf{R}^2 \middle| \widetilde{P} \text{ has a multiple root in } [-1, 1] \right\}.$$

Note that $\Delta_{\widetilde{P}}$ is the set of critical values of $\mathcal{EM}$. We now prove

Claim: $\Delta_{\widetilde{P}}$ is parametrized by

$$\begin{cases} h &= \frac{3}{2}s - \dfrac{1}{2s} \\ j &= \pm \dfrac{1}{\sqrt{-s}}(1 - s^2), \end{cases} \qquad \text{for } s \in [-1, 0) \cup \{1\}. \tag{37}$$

(3.2) **Proof**: For every $(h, j) \in \Delta_{\widetilde{P}}$, the polynomial $\widetilde{P}$ (33) factors as

$$(\sigma_1 - s)^2(\sigma_1 - t) = \sigma_1^3 - (2s + t)\sigma_1^2 + (2st + s^2)\sigma_1 - ts^2 \tag{38}$$

for some $s \in [-1, 1]$ and $t \in \mathbf{R}$. Equating coefficients of (33) and (38) gives

$$\begin{cases} 2s + t &= h \\ 2st + s^2 &= -1 \\ ts^2 &= \frac{1}{2} j^2 - h. \end{cases} \tag{39}$$

If $s = 0$, then the second equation in (39) becomes $0 = -1$, which is a contradiction. Thus $s \neq 0$. Eliminating t from (39) gives

$$\begin{cases} h &= \frac{3}{2}s - \dfrac{1}{2s} \\ j^2 &= -\dfrac{1}{s}(1-s^2)^2. \end{cases} \tag{40}$$

For j to be real we must have $s \in [-1, 0) \cup \{1\}$. □

A *branch* $\mathcal{B}_\pm$ of the discriminant locus $\Delta_{\widetilde{P}}$ is defined by (37) with parameter $s \in [-1, 0)$.

Claim: $\mathcal{B}_\pm$ is the curve

$$j = \pm \frac{2}{9}\left(3 - h^2 + h\sqrt{h^2+3}\right)\sqrt{h + \sqrt{h^2+3}} = \pm B(h) \tag{41}$$

where $h \geq -1$. Note that (41) defines the function $\pm B$.

(3.3) **Proof**: This follows by eliminating s from (37). □

We now verify that figure 3.3 is correct. Since $j = 0$ when $h = -1$, the branches $\mathcal{B}_\pm$ join continuously at $(-1, 0)$. Because $\left.\frac{dj}{dh}\right|_{h=-1} = \pm 1$, the branches do *not* join smoothly at $(-1, 0)$, but make an angle of $\frac{\pi}{2}$ with each other.

The point $(1, 0) \in \Delta_{\widetilde{P}}$, which corresponds to the parameter s being equal to 1, does not lie on either of the branches $\mathcal{B}_\pm$. Hence $(1, 0)$ is an *isolated* point of $\Delta_{\widetilde{P}}$ and therefore is an *isolated critical value* of $\mathcal{EM}$. □

The critical values $(\pm 1, 0)$ of $\mathcal{EM}$ are special because the corresponding critical points $p_\pm = (\pm 1, 0, 0)$ of the reduced Hamiltonian H_0 are *singular* points of the reduced space M_0. These critical points do *not* depend on the reduced Hamiltonian, but are a consequence of symmetry *alone*.

We now describe the qualitative features of the reduced system. From figure 3.2 we can read off the topology of the h-level set of the reduced Hamiltonian H_j. The results are given in table 3.1.

Conditions on (h, j)	Topology of $H_j^{-1}(h)$
1. $j = \pm B(h)$, $h \geq -1$	a point
2. $\lvert j\rvert < B(h)$, $h > -1$ and $(h, j) \neq (1, 0)$	a smooth S^1
3. $(1, 0)$	a topological S^1 with a conical singular point

Table 3.1 The topology of $H_j^{-1}(h)$.

▷ We now verify the entries in the second column of table 3.1.

(3.4) **Proof:**

1). When $|j| = B(h)$ and $h > -1$, the line ℓ_h (34) intersects the curve $\mathcal{F}$ (35) at a nonsingular point $p_{h,j}$. Thus $\rho(H_j^{-1}(h)) = p_{h,j}$. Since $p_{h,j} \in \mathcal{F}$, we infer that $H_j^{-1}(h)$ is a point. Thus h is the minimum value of H_j on M_j. This shows that the set of (h, j) bounded by the branches $\mathcal{B}_\pm$ and containing the point $(1, 0)$ is the range of the energy momentum mapping $\mathcal{EM}$. When $h = -1$, we have $j = \pm B(-1) = 0$. The line ℓ_0 meets $\mathcal{F}$ at the point $p_{-1,0} = (-1, 0)$. Thus $H_0^{-1}(1)$ is the singular point $(-1, 0, 0)$ of M_0.

2). When $|j| < B(h)$, $-1 < h$ and $(h, j) \neq (1, 0)$, the line ℓ_h intersects $\rho(M_j)$ in a closed line segment $L_{h,j}$ whose end points lie on $\mathcal{F}$. Over every point in the interior of $L_{h,j}$ the fiber of ρ consists of two distinct points; while over the end points the fiber of ρ is just a single point. Thus $H_j^{-1}(h) = \rho^{-1}(L_{h,j})$ is a topological circle. Since (h, j) is a regular value of $\mathcal{EM}$, the value h is a regular value for the reduced Hamiltonian H_j. Hence $H_j^{-1}(h)$ is a smooth circle.

3). When $h = 1$, the line ℓ_1 intersects $\rho(M_0)$ in a closed line segment $L_{1,0}$. Hence $H_0^{-1}(1) = \rho^{-1}(L_{1,0})$ is a topological circle. As a semialgebraic variety in $\mathbf{R}^3$, the 1-level set of H_0 is defined by

$$\begin{cases} \sigma_2^2 = \sigma_3(1 - \sigma_1^2), & |\sigma_1| \le 1 \ \& \ \sigma_3 \ge 0 \\ 1 = \frac{1}{2}\sigma_3 + \sigma_1. \end{cases} \tag{42}$$

Eliminating σ_3 from these equations yields

$$\sigma_2^2 = 4(1 - \sigma_1)^2 - 2(1 - \sigma_1)^3, \qquad |\sigma_1| \le 1.$$

Hence $H_0^{-1}(1)$ has a nondegenerate tangent cone at the singular point $(1, 0, 0)$. At other points $H_0^{-1}(1)$ is smooth. This completes the verification of table 3.1. □

Conditions on (h, j)	Topology of $\mathcal{EM}^{-1}(h, j)$
1. $|j| < B(h)$, $-1 < h$ and $(h, j) \neq (1, 0)$	T^2, a smooth 2-torus
2. $j = \pm B(h)$, $h > -1$	S^1, a smooth circle
3. $(-1, 0)$	a point
4. $(1, 0)$	T^*, a 2-torus with a circle of longitude pinched to a point.

Table 3.2 The topology of $\mathcal{EM}^{-1}(h)$.

We are now in position to describe the topology of the fibers of the energy momentum mapping of the spherical pendulum, (see figure 3.1). The results are given in table 3.2. The verification of the entries in the second column of table 3.2 is based on the

Facts.

1. $\mathcal{EM}^{-1}(h, j) = \pi_j^{-1}(H_j^{-1}(h))$.

2. The reduction mapping

$$\pi_j : J^{-1}(j) \to M_j \subseteq \mathbf{R}^3 : (x,y) \to \big(\sigma_1(x,y), \sigma_2(x,y), \sigma_3(x,y)\big) \qquad (43)$$

is smooth and has fibers $\pi_j^{-1}(m) = \begin{cases} S^1, & \text{if } M_j \text{ is smooth at } m \\ \text{point}, & \text{otherwise.} \end{cases}$

(3.5) **Proof**:

1). Because the reduction mapping π_j is surjective and $\pi_j{}^* H_j = H|J^{-1}(j)$, we have

$$\mathscr{EM}^{-1}(h,j) = H^{-1}(h) \cap J^{-1}(j) = \pi_j^{-1}(H_j^{-1}(h)).$$

2). Let f be a *smooth function* on the reduced space M_j, (see appendix B section 5). By definition, there is a smooth function F on $\mathbf{R}^3$ such that $F|M_j = f$. Let π be the Hilbert map of the action Φ (6) and μ the projection

$$\mu : \mathbf{R}^6 \to \mathbf{R}^3 : (\sigma_1, \ldots, \sigma_6) \to (\sigma_1, \sigma_2, \sigma_3).$$

Then $(\mu \circ \pi)^* F$ is a smooth function on $T\mathbf{R}^3$. Hence $((\mu \circ \pi)^* F)|J^{-1}(j)$ is a smooth function on $J^{-1}(j)$. Suppose that F' is a smooth function such that $F'|M_j = f$. Then $\mu^*(F - F')$ vanishes on $V_j = \pi(J^{-1}(j))$. Consequently, $\pi^*\big(\mu^*(F - F')\big)$ vanishes on $J^{-1}(j)$ and therefore $\big((\mu \circ \pi)^* F'\big)|J^{-1}(j) = \big((\mu \circ \pi)^* F\big)|J^{-1}(j)$. Hence the mapping

$$\xi : C^\infty(M_j) \to C^\infty(J^{-1}(j)) : f \to \big((\mu \circ \pi)^* F\big)|J^{-1}(j)$$

is well defined. As $\pi_j = (\mu \circ \pi)|J^{-1}(j)$, it follows that $\xi = \pi_j^*$. By definition π_j is *smooth*, (see appendix B section 5).

If m is a point where the reduced space M_j is nonsingular, then $\mu^{-1}(m)$ is a nonsingular point of V_j and hence is a nonsingular point of the orbit space W (12). From ((2.3)) it follows that $\pi^{-1}(\mu^{-1}(m)) = \pi_j^{-1}(m)$ is a smooth S^1. If m is a singular point of M_j, then $\pi^{-1}(\mu^{-1}(m))$ is a point of $J^{-1}(j)$ where the isotropy group of the action $\Phi|J^{-1}(j)$ is *nontrivial*. This can only happen when $j = 0$ and $\pi_0^{-1}(m)$ is a *fixed point* of $\Phi|J^{-1}(0)$. □

▷ The verification of the second column of table 3.2 proceeds as follows:

(3.6) **Proof**:

1). The conditions on (h,j) in the first entry of the first column of table 3.2 are equivalent to requiring that (h,j) is a regular value of the energy momentum mapping. Hence h is a regular value of the reduced Hamiltonian H_j on M_j. By table 3.1 the level set $H_j^{-1}(h)$ is diffeomorphic to S^1. From ((3.5)) it follows that

$$\widehat{\pi}_j = \pi_j|\mathscr{EM}^{-1}(h,j) : \mathscr{EM}^{-1}(h,j) \to H_j^{-1}(h)$$

is a smooth bundle with fiber S^1. Since the reduced space M_j is homeomorphic to $\mathbf{R}^2$, the level set $H_j^{-1}(h)$ bounds a 2-disc which is contractible in M_j to a point. Thus the bundle $\widehat{\pi}_j$ is trivial, (see appendix C section 2). In other words $\mathscr{EM}^{-1}(h,j)$ is a 2-torus $T^2_{h,\mu}$. By construction $T^2_{h,j}$ is invariant under the flow of the vector fields $X_{H|TS^2}$ and $X_{J|TS^2}$.

A good way to picture the 2-torus $T^2_{h,j}$ is to view it as some sort of bundle over its image under the projection

$$\pi_{TS^2} : TS^2 \subseteq T\mathbf{R}^3 \rightarrow S^2 \subseteq \mathbf{R}^3 : (x, y) \rightarrow x,$$

(see figure 3.4).

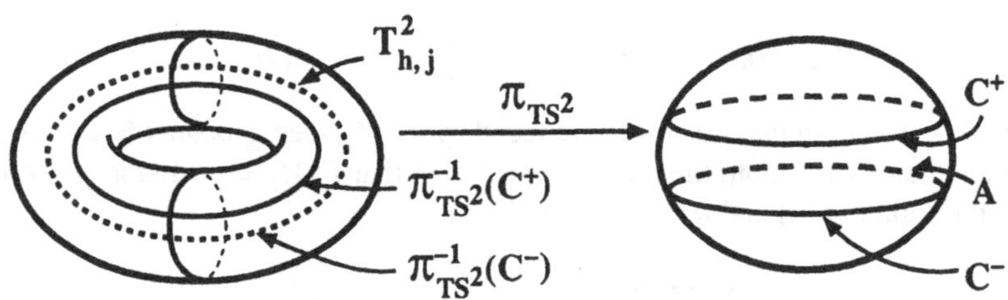

Figure 3.4. Image of $T^2_{h,j}$ under π_{TS^2}.

$T^2_{h,j} \subseteq T\mathbf{R}^3$ is defined by

$$\left\{ \begin{array}{rl} x_1^2 + x_2^2 + x_3^2 & = 1 \\ x_1y_1 + x_2y_2 + x_3y_3 & = 0 \\ \frac{1}{2}(y_1^2 + y_2^2 + y_3^2) + x_3 & = h \\ x_2y_1 - x_1y_2 & = j. \end{array} \right. \tag{44}$$

Substituting (44) into the identity

$$(x_1y_2 - x_2y_1)^2 + (x_1y_1 + x_2y_2)^2 = (x_1^2 + x_2^2)(y_1^2 + y_2^2)$$

and simplifying gives

$$0 \leq y_3^2 = P(x_3) = 2(h - x_3)(1 - x_3^2) - j^2 \tag{45}$$

where $|x_3| \leq 1$. Consequently, $x_3 \in [x_3^-, x_3^+]$ where $P(x_3^{\pm}) = 0$. Because (h, j) is a regular value of $\mathscr{EM}$, it follows that $-1 < x_3^- < x_3^+ < 1$. Therefore the image of $T^2_{h,j}$ under π_{TS^2} is contained in the annulus

$$\mathscr{A} = \left\{ x \in S^2 \subseteq \mathbf{R}^3 \middle| x_3^- \leq x_3 \leq x_3^+ \right\}.$$

Given $x \in \mathscr{A}$, from (45) we find that $y_3 = \varepsilon\sqrt{P(x_3)}$, where $\varepsilon^2 = 1$. Solving

$$\begin{array}{rcl} x_1y_2 - x_2y_1 & = & j \\ x_1y_1 + x_2y_2 & = & -x_3y_3 = -\varepsilon x_3\sqrt{P(x_3)} \end{array}$$

for (y_1, y_2) gives

$$y_1 = -\frac{1}{1-x_3^2}\left[jx_2 + \varepsilon x_1 x_3 \sqrt{P(x_3)}\right]$$
$$y_2 = -\frac{1}{1-x_3^2}\left[-jx_1 + \varepsilon x_2 x_3 \sqrt{P(x_3)}\right].$$

Therefore for every x in the interior of $\mathscr{A}$ there are exactly two points $p_\pm \in T^2_{h,j}$ such that $\pi_{TS^2}(p_\pm) = x$; while for every x on the boundary $\partial\mathscr{A}$ of $\mathscr{A}$, there is exactly one point p_0 such that $\pi_{TS^2}(p_0) = x$. In other words, $\pi_{TS^2}|T^2_{h,j}$ has a *fold singularity* at every point on the two curves

$$\mathscr{C}^\pm = \pi^{-1}_{TS^2}(\{x \in S^2 | x_3 = x_3^\pm\}). \tag{46}$$

Thus $T^2_{h,j}$ is a *pinched* S^0 bundle over $\mathscr{A}$ with bundle projection $\pi_{TS^2}|T^2_{h,j}$. Observe that $\mathscr{C}^\pm$ is an orbit of $X_{J|TS^2}$ on $T^2_{h,j}$.

2). To verify the second entry in the second column of table 3.2, note that the conditions on (h, j) in the first column are precisely those for which the h-level set of the reduced Hamiltonian H_j is a nonsingular point of M_j. From ((3.5)) it follows that $\mathscr{EM}^{-1}(h, j)$ is a smooth S^1.

3). When $(h, j) = (-1, 0)$ the line ℓ_{-1} (34) intersects the curve $\mathscr{F}$ (35) at the point $(\sigma_1^0, \sigma_3^0) = (-1, 0)$. In other words, the 2-plane Π_{-1} (29) intersects the reduced space M_0 at the singular point $(\sigma_1^0, \sigma_2^0, \sigma_3^0) = (-1, 0, 0)$. From ((3.5)) it follows that $\mathscr{EM}^{-1}(1, 0)$ is the point $(0, 0, -1, 0, 0, 0) \in T\mathbf{R}^3$.

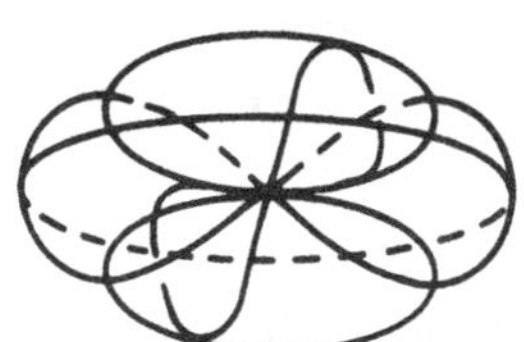

Figure 3.5. The level set $\mathscr{EM}^{-1}(1, 0)$.

4). When $(h, j) = (1, 0)$ the line ℓ_1 intersects $\rho(M_0)$ in a closed line segment L_{10}, whose end points lie on the curve $\mathscr{F}$. One end point is $\sigma^0 = (\sigma_1^0, \sigma_3^0) = (1, 0)$. Thus the 2-plane Π_1 intersects M_0 in a topological circle $H_0^{-1}(1)$ with singular point $p^0 = \rho^{-1}(\sigma^0) = (1, 0, 0)$. From ((3.5)) it follows that

$$\pi_0|(H_0^{-1}(1) - \{p^0\}) : \pi_0^{-1}(H_0^{-1}(1) - \{p^0\}) \to H_0^{-1}(1) - \{p^0\}$$

is a bundle with fiber S^1. Since $H_0^{-1}(1) - \{p^0\}$ is contractible in M_0 to a point, the bundle $\pi_0|(H_0^{-1}(1) - \{p^0\})$ is trivial. Thus $\pi_0^{-1}(H_0^{-1}(1) - \{p^0\})$ is topologically a cylinder $S^1 \times \mathbf{R}$. Because p^0 is a singular point of M_0, the fiber $\pi_0^{-1}(p^0)$ is a point. Therefore $\mathscr{EM}^{-1}(1, 0)$ is a one point compactification of a cylinder, that is, a cylinder with its ends identified to a point. In other words, it is a torus T^* with a *longitudinal* circle pinched to a point. In ((3.7)) below we show that $\pi_0^{-1}(p^0) = (0,0,1,0,0,0)$ is a critical point of $H|TS^2$ of Morse index 2. Thus $\pi_0^{-1}(p^0)$ is a conical singular point of $\mathcal{EM}^{-1}(0,1)$. Another way of describing this compactification is to say that $\mathcal{EM}^{-1}(1,0)$ is a 2-torus with a *meridial* circle pinched to a point. To see this project $\mathcal{EM}^{-1}(1,0)$ onto a 2-sphere S^2 using the bundle projection π_{TS^2}. If $|h| < 1$, then the annulus $\mathcal{A}$ in part 1) of this proof becomes the 2-disc $\{x \in S^2 | -1 \le x_3 \le h\}$. In other words, $\pi_{TS^2}(\mathcal{EM}^{-1}(h,0))$ is a 2-disc on S^2 with boundary $\{x = h\}$. As $h \to 1^-$, $x_3^+ \to 1$. Hence $\pi_{TS^2}(\mathcal{EM}^{-1}(1,0))$ is S^2. Moreover, for $p \in S^2 - \{(0,0,1)\}$ the fiber $\pi_{TS^2}^{-1}(p)$ consists of two distinct points; while at $(0,0,1)$ the fiber $\pi_{TS^2}^{-1}(p)$ is a single point. From this information we can reconstruct $\mathcal{EM}^{-1}(1,0)$, (see figure 3.5). We find that it is a surface of revolution formed by rotating a figure eight about an axis through the crossing point. Topologically this surface is just a 2-torus with a merdial circle pinched to a point. The topological equivalence between a torus with a meridial circle pinched to a point and a torus with a longitudinal circle pinched to a point cannot be realized by a homeomorphism of $\mathbf{R}^3$ but can be by a homeomorphism of S^3, (see exercise 2).

This completes the verification of table 3.2. □

▷ Next we determine the topology of the energy surfaces $H^{-1}(h)$ of the spherical pendulum. The results are summarized in table 3.3.

Conditions	Topology of $H^{-1}(h)$
1. $h = -1$	point
2. $-1 < h < 1$	a three sphere, S^3
3. $h = 1$	U, the unit tangent S^1 bundle over S^2 with one fiber *pinched* to a point
4. $h > 1$	$\mathbf{RP}^3$, real projective 3-space.

Table 3.3 Topology of the level sets $H^{-1}(h)$.

To verify the entries in the second column of table 3.3, we use Morse theory. As a submanifold of $T\mathbf{R}^3$ with coordinates (z, y), the space TS^2 is defined by

$$F_1(z,y) = z_1^2 + z_2^2 + z_3^2 - 1 = 0 \tag{47}$$
$$F_2(z,y) = z_1y_1 + z_2y_2 + z_3y_3 = 0. \tag{48}$$

On $T\mathbf{R}^3$ consider the function

$$H : T\mathbf{R}^3 \to \mathbf{R} : (z,y) \to \tfrac{1}{2}(y_1^2 + y_2^2 + y_3^2) + z_3.$$

▷ We now show that $H|TS^2$ is a Morse function.

(3.7) **Proof**: Because $H|TS^2$ is proper and is bounded below, it has a critical point $p = (z, y)$. By Lagrange multipliers the critical point p satisfies

$$DH(z,y) + \lambda_1\, DF_1(z,y) + \lambda_2\, DF_2(z,y) = 0, \quad F_1(z,y) = 0, \quad F_2(z,y) = 0. \tag{49}$$

Writing out (49) gives

$$\begin{aligned} 2\lambda_1 z_1 + \lambda_2 y_1 &= 0, & y_1 + \lambda_2 z_1 &= 0, \\ 2\lambda_1 z_2 + \lambda_2 y_2 &= 0, & y_2 + \lambda_2 z_2 &= 0, \\ 1 + 2\lambda_1 z_3 + \lambda_3 y_3 &= 0, & y_3 + \lambda_2 z_3 &= 0, \end{aligned}$$

in addition to (47) and (48). Therefore

$$\lambda_2 = \lambda_2(z_1^2 + z_2^2 + z_3^2) = -(z_1 y_1 + z_2 y_2 + z_3 y_3) = 0. \tag{50}$$

Hence $y_1 = y_2 = y_3 = 0$. Consequently,

$$2\lambda_1 z_1 = 0 \tag{51}$$

$$2\lambda_1 z_2 = 0 \tag{52}$$

$$2\lambda_1 z_3 = -1. \tag{53}$$

Suppose that $2\lambda_1 = 0$. Then (53) gives $0 = -1$, a contradiction. Therefore $\lambda_1 \neq 0$. Hence (51) and (52) become

$$z_1 = z_2 = 0. \tag{54}$$

From (47) we obtain $z_3 = \varepsilon$ where $\varepsilon^2 = 1$, while from (53) we obtain $\lambda_1 = -\frac{1}{2}\varepsilon$. Thus $H|TS^2$ has two critical points $p_\varepsilon = (\varepsilon e_3, 0)$ with Lagrange multipliers $\lambda_1 = \frac{1}{2}\varepsilon$ and $\lambda_2 = 0$.

To show that p_ε is nondegenerate critical point of $H|TS^2$, first note that the tangent space $T_{p_\varepsilon}(TS^2)$ to TS^2 at p_ε is

$$\ker\begin{pmatrix} DF_1(p_\varepsilon) \\ DF_2(p_\varepsilon) \end{pmatrix} = \ker\begin{pmatrix} 0 & 0 & 2\varepsilon & 0 & 0 & 0 \\ 0 & 0 & 0 & 0 & 0 & \varepsilon \end{pmatrix},$$

which is spanned by the vectors $\{e_1, e_2, e_4, e_5\}$. Therefore

$$Q = D^2(H|TS^2)(p_\varepsilon) = \left(D^2H + \lambda_1 D^2F_1 + \lambda_2 D^2F_2\right)\Big|_{T_{p_\varepsilon}TS^2}(p_\varepsilon) = \mathrm{diag}(-\varepsilon, -\varepsilon, 1, 1),$$

whereupon the critical points p_ε are nondegenerate. Thus the Morse index of Q is 2 if $\varepsilon = 1$ or 0 if $\varepsilon = -1$. Hence, $p_{-1} = (-e_3, 0)$ is a nondegenerate minimum of H with corresponding minimum value -1, and $p_{+1} = (e_3, 0)$ is a nondegenerate saddle point of index 2 corresponding to the critical value 1. □

▷ We now verify the entries in the second column of table 3.3.

(3.8) **Proof**:

1). At the critical value $h = -1$, the level set $(H|TS^2)^{-1}(h)$ is a point.

2). By the Morse lemma (see appendix E section 2) for values of h slightly greater than the minimum value -1, the h-level set of $H|TS^2$ is diffeomorphic to a three sphere S^3. Using the Morse isotopy lemma (see appendix E section 3) it follows that the level set $(H|TS^2)^{-1}(h)$ is diffeomorphic to S^3 for every $h \in (-1, 1)$.

4). For $h > 1$ we claim that $(H|TS^2)^{-1}(h)$ is diffeomorphic to the unit tangent S^1 bundle $T_1S^2 \subseteq \mathbf{R}^3 \times \mathbf{R}^3$ of the 2-sphere S^2. Consider the smooth mapping

$$\varphi : (H|TS^2)^{-1}(h) \to T_1S^2 : (z, w) \to \left(z, w/\sqrt{2h - 2z_3}\right) = (\xi, \eta).$$

The level set $(H|TS^2)^{-1}(h) \subseteq \mathbf{R}^3 \times \mathbf{R}^3$ is defined by

$$\langle z, z\rangle = 1, \ \langle z, w\rangle = 0, \ \text{and } h = \tfrac{1}{2}\langle w, w\rangle + \langle z, e_3\rangle.$$

A computation shows that $(\xi, \eta) \in \varphi\big((H|TS^2)^{-1}(h)\big)$ satisfy

$$\langle \xi, \xi\rangle = 1, \ \langle \xi, \eta\rangle = 0 \ \text{and} \ \langle \eta, \eta\rangle = (2h - 2z_3)^{-1}\langle w, w\rangle = 1. \tag{55}$$

Therefore the image of φ is contained in T_1S^2. Since $(\xi, \eta) \to \left(\xi, \eta\sqrt{2h - 2\xi_3}\right)$ is a smooth inverse of φ, the mapping φ is a diffeomorphism.

From (55) it follows that T_1S^2 is the set of ordered pairs of orthonormal vectors $\xi, \eta \in \mathbf{R}^3$. Extend the ordered pair $\{\xi, \eta\}$ to the positively oriented ordered orthonormal basis $\{\xi, \eta, \xi \times \eta\}$ of $\mathbf{R}^3$. Every such basis may be identified with a proper rotation of $\mathbf{R}^3$, whose matrix $\mathrm{col}(\xi, \eta, \xi \times \eta)$ has columns ξ, η and $\xi \times \eta$. Thus

$$\psi : T_1S^2 \to \mathrm{SO}(3) : (\xi, \eta) \to \mathrm{col}(\xi, \eta, \xi \times \eta)$$

is a smooth map with smooth inverse

$$\sigma : \mathrm{SO}(3) \to T_1S^2 : \mathrm{col}(\xi, \eta, \zeta) \to (\xi, \eta).$$

Hence T_1S^2 is diffeomorphic to SO(3), which in turn is diffeomorphic to real *projective 3-space* $\mathbf{RP}^3$ by ((1.15)) of chapter 3.

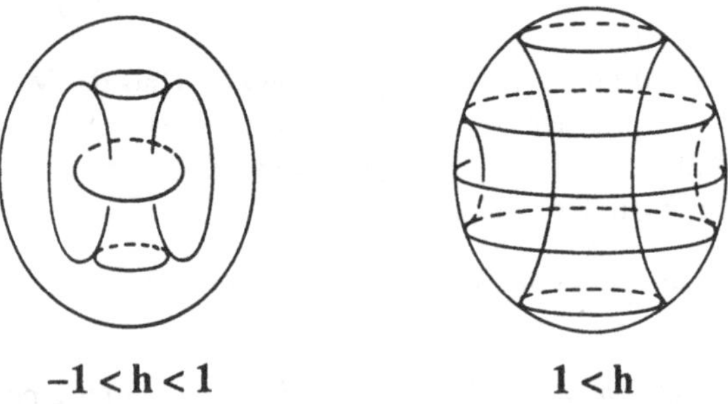

Figure 3.6. Foliation of an energy level set by level sets of angular momentum.

3). At the critical value $h = 1$, the 1-level set of $H|TS^2$ is an algebraic variety U with one singular point $p_1 = (e_3, 0)$. Considered as a "bundle" over S^2 with bundle projection $\widehat{\pi} = \pi_{TS^2}|U$, it is straightforward to check that the fiber of $\widehat{\pi}$ over every point of $S^2 - \{(0,0,1)\}$ is an S^1; while the fiber over $(0,0,1)$ is the point p_1. Therefore U is T_1S^2, the unit tangent sphere bundle over S^2, with the tangent circle $\widehat{\pi}^{-1}(0,0,1)$ pinched to a point. Because p_1 is a nondegenerate critical point of $H|TS^2$ with Morse index 2 ((3.7)), it follows from the Morse lemma that a neighborhood of p_1 in U is diffeomorphic to a neighborhood of the vertex of the cone

$$C = \left\{(x_1, x_2, y_1, y_2) \in \mathbf{R}^4 \middle| x_1^2 + x_2^2 = y_1^2 + y_2^2\right\}.$$

Note that C is a cone on a 2-torus and that C has two 2-planes

$$\left\{(x_1, x_2, \varepsilon x_1, -\varepsilon x_2) \middle| (x_1, x_2) \in \mathbf{R}^2\right\}$$

where $\varepsilon^2 = 1$ which intersect transversely at 0. This describes the singularity of U and also shows why it is impossible to draw it in $\mathbf{R}^3$ in a geometrically faithful way.

This completes the verification of table 3.3. □

In summary, we have verified the bifurcation diagram (see figure 3.1) for the energy momentum mapping $\mathcal{EM}$ of the spherical pendulum. From figure 3.1 it is clear that for regular values of h the energy level $H^{-1}(h)$ is foliated by 2-tori with two singular S^1 fibers, (see figure 3.6).

4 Rotation number and first return time

In this section we derive formulæ for the rotation number and time of first return to a cross section for the flow of $X_{H|TS^2}$ on the 2-torus $T^2_{h,j}$. We show that the rotation number is a *multivalued* real analytic function on the set of regular values of the energy momentum map, while the first return time is a *single valued* real analytic function.

The *rotation number* is defined as follows. When (h, j) is a regular value of $\mathcal{EM}$, the differentials $d(H|TS^2)$, $d(J|TS^2)$ are linearly independent at every point of $T^2_{h,j} = \mathcal{EM}^{-1}(h, j)$ and therefore so are the vector fields $X_{H|TS^2}$ and $X_{J|TS^2}$. Consider the curve $\mathcal{C}^+$ (46) on $T^2_{h,j}$, which is the image of the integral curve $t \rightarrow \varphi_t^{J|TS^2}(p)$ of $X_{J|TS^2}$ through $p \in \pi_{TS^2}^{-1}(\{x_3 = x_3^+\})$. $\mathcal{C}^+$ is a cross section for the flow $\varphi_t^{H|TS^2}$ of $X_{H|TS^2}$ on $T^2_{h,j}$. To see this, first note that $X_{H|TS^2}(q)$ is transverse to $\mathcal{C}^+$ at every $q \in \mathcal{C}^+$. Second, note that the image of the integral curve $\Gamma : t \rightarrow \varphi_t^{H|TS^2}(p)$ of $X_{H|TS^2}$ through p under the reduction mapping π_j (43) is a periodic integral curve $t \rightarrow \varphi_t^{H_j}(\pi_j(p))$ of the reduced vector field X_{H_j}. Therefore there is a smallest $T = T(p) > 0$ such that Γ intersects $\mathcal{C}^+$. From the fact that Γ intersects $\mathcal{C}^+$ transversely, it follows that T is a smooth (indeed, real analytic) function of p. Let $\widetilde{\theta}$ be the smallest positive number such that

$$\varphi_{2\pi\widetilde{\theta}}^{J|TS^2}(p) = \varphi_T^{H|TS^2}(p),$$

(see figure 3.4). By definition $\widetilde{\theta}$ is the rotation number of the flow of $X_{H|TS^2}$ on $T^2_{h,j}$. Because $X_{H|TS^2}$ is invariant under the S^1-action generated by the flow of $X_{J|TS^2}$, the rotation number and the time of first return do not depend on the choice of the point p on $\mathcal{C}^+$. Since $\mathcal{C}^+$ can be an arbitrary integral curve of $X_{J|TS^2}$ on $T^2_{h,j}$, the rotation number and the time of first return depend only on (h, j).

We now derive a formula for the rotation number $\widetilde{\theta} = \theta_{h,j}$. First use the bundle map π_{TS^2} to project an integral curve Γ of $X_{H|TS^2}$ on $T^2_{h,j}$ onto a curve γ in the annular region $\mathcal{A} = \pi_{TS^2}(\mathcal{EM}^{-1}(h, j))$ of S^2. Let $x_i|\mathcal{A}$ be coordinates on $\mathcal{A}$ with (x_1, x_2, x_3) being coordinates on $\mathbf{R}^3$. Furthermore, let $\theta = \tan^{-1}\frac{x_2}{x_1}$ and x_3 be coordinates on the *universal covering space* $\widetilde{\mathcal{A}}$ of $\mathcal{A}$.

▷ The following argument shows that a lift $\widetilde{\gamma}$ of γ to $\widetilde{\mathcal{A}}$ satisfies

$$\frac{d\theta}{dt} = \frac{j}{1 - x_3^2} \tag{56}$$

$$\frac{dx_3}{dt} = \varepsilon\sqrt{P(x_3)}, \tag{57}$$

with $\varepsilon^2 = 1$. How the sign of ε is determined is discussed below.

(4.1) **Proof**: By definition of Lie derivative, we find that

$$\begin{aligned} \frac{d\theta}{dt} &= L_{X_{H|TS^2}}\theta|TS^2 = \frac{1}{x_1^2 + x_2^2}\Big(x_1\frac{dx_2}{dt} - x_2\frac{dx_1}{dt}\Big) \\ &= \frac{x_1y_2 - x_2y_1}{x_1^2 + x_2^2} = \frac{j}{1 - x_3^2}. \end{aligned}$$

Also,

$$\begin{aligned} \frac{dx_3}{dt} &= L_{X_{H|TS^2}}(x_3|TS^2) = \pi_j^*(L_{X_{H_j|M_j}}\sigma_1) = \pi_j^*(\sigma_2), \quad \text{using (27)} \\ &= \pm\pi_j^*\left(\sqrt{\sigma_3(1 - \sigma_1^2) - j^2}\right), \quad \text{since } \psi \text{ (26) is an integral of } X_{H_j} \\ &= \pm\pi_j^*\left(\sqrt{2(h - \sigma_1)(1 - \sigma_1^2) - j^2}\right), \quad \text{since } \widetilde{\gamma}(t) \subseteq (H|TS^2)^{-1}(h) \\ &= \varepsilon\sqrt{P(x_3)}, \end{aligned}$$

where $\varepsilon^2 = 1$ and $P(x_3)$ is given by (45). The sign ambiguity in (57) is handled by the following *sign convention*. Suppose that $\gamma(t_0) = x_3^{\pm}$ and that at $t_0 - \delta > 0$ for $\delta > 0$ and small the value of ε is known. Then at time $t_0 + \delta$ the value of ε is the negative of ε at $t_0 - \delta$. Since $\dot{\theta} \neq 0$ and $\dot{x}_3(t_0) = 0$ when $\widetilde{\gamma}(t_0) \in \partial\widetilde{\mathcal{A}}$, the curve $\widetilde{\gamma}$ has at least first order contact with $\partial\widetilde{\mathcal{A}}$ at $\widetilde{\gamma}(t_0)$. Because Γ crosses $\mathcal{C}^{\pm} = \pi_{TS^2}^{-1}(\{x_3 = x_3^{\pm}\})$ transversely at $\Gamma(t_0)$ and the mapping $\pi_{TS^2}|T^2_{h,j}$ has a fold singularity at $\Gamma(t_0)$, the curve $\widetilde{\gamma}$ has second order contact with $\partial\widetilde{\mathcal{A}}$ at $\widetilde{\gamma}(t_0)$. The sign convention ensures that the solutions of (56) and (57) in $\widetilde{\mathcal{A}}$ are real analytic. Since these solutions have no self intersections (see exercise 3), equations (56) and (57) together with the sign convention define a vector field on $\widetilde{\mathcal{A}}$.□

Now consider the curve $s \to \varphi_s^{J|TS^2}(p)$ for $s \in [0, 2\pi\theta_{h,j})$ where $p \in \mathscr{C}^+$. The projected curve

$$s \to \pi_{TS^2}(\varphi_s^{J|TS^2}(p))$$

is an arc of the small circle $\{x_3 = x_3^+\}$ on S^2 which joins two successive points of intersection $\gamma(0)$ and $\gamma(T)$ with $\{x_3 = x_3^+\}$. Let $2\pi\vartheta_{h,j}$ be the angle between $\gamma(0)$ and $\gamma(T)$ as ▷ measured from the center of the small circle $\{x_3 = x_3^+\}$. Then $\vartheta_{h,j}$ is equal to the rotation number $\theta_{h,j}$.

(4.2) **Proof**: This follows because

$$\begin{aligned} 2\pi\vartheta_{h,j} &= \int_0^{2\pi\vartheta_{h,j}} d\theta = \int_0^{2\pi\theta_{h,j}} L_{X_{J|TS^2}}\theta \, ds, \quad \text{by definition of rotation number} \\ &= \int_0^{2\pi\theta_{h,j}} ds, \quad \text{using (7) and the definition of } \theta \\ &= 2\pi\theta_{h,j}. \quad \square \end{aligned}$$

▷ We now find a formula for $\theta_{h,j}$.

(4.3) **Proof**: Since $P(x_3) > 0$ for $x = (x_1, x_2, x_3)$ in the interior of $\mathscr{A}$, from (57) we see that $\frac{dx_3}{dt} \neq 0$. Hence for $t \in (0, T/2) \cup (T/2, T)$ we may parametrize γ by x_3. Chasing down the minus signs, we find

$$2\pi\theta_{h,j} = \int_0^T L_{X_{H|TS^2}}\theta \, dt = -\int_{x_3^+}^{x_3^-} \frac{d\theta}{dx_3} dx_3 + \int_{x_3^-}^{x_3^+} \frac{d\theta}{dx_3} dx_3 = 2\int_{x_3^-}^{x_3^+} \frac{\frac{d\theta}{dt}}{\frac{dx_3}{dt}} dx_3.$$

Using (56) and (57) we get

$$2\pi\theta_{h,j} = 2\pi\theta(h,j) = 2j\int_{x_3^-}^{x_3^+} \frac{1}{(1-x_3^2)\sqrt{P(x_3)}} dx_3, \tag{58}$$

which is the desired formula for the rotation number. Some care needs to be taken in interpreting (58) when $j = 0$, for then the integral is infinite. Using techniques from complex analysis, we will show that the limit of the right hand side of (58) exists as $j \to 0$ for every h greater than -1 but not equal to 1. □

Next we investigate the analytic properties of the rotation number.

Claim: Let $\mathfrak{R}$ be the set of regular values of the energy momentum mapping, (see figure 3.3). Let $x_\pm$ be *roots of the polynomial* $\widetilde{P}$ (33) in $(-1, 1)$. Then the rotation number

$$\theta : \mathfrak{R} \to \mathbf{R} : (h,j) \to \frac{j}{2\sqrt{2}\pi}\int_{x_-}^{x_+} \frac{1}{(1-x^2)\sqrt{\widetilde{P}(x)}} dx \tag{59}$$

is *locally* a single valued real analytic function. Because $\mathfrak{R}$ is not simply connected, θ need *not* be single valued on all of $\mathfrak{R}$.

Before proving the claim we recall some facts from complex analysis. Let

$$Q(z) = (z - x_-)(z - x_+)(z - x_0),$$

where $x_{\pm,0}$ are three distinct real numbers such that $-1 < x_- < x_+ < 1 < x_0$. Cut the extended complex plane $\widetilde{\mathbf{C}}$ along the real axis twice: once between x_- and x_+ and again between x_0 and ∞. Consider the complex 1-form

$$\omega = \frac{1}{(1-z^2)\sqrt{Q(z)}}, \tag{60}$$

▷ where z is a coordinate on $\widetilde{\mathbf{C}}$. Then ω is meromorphic with simple poles at $z = \pm$.

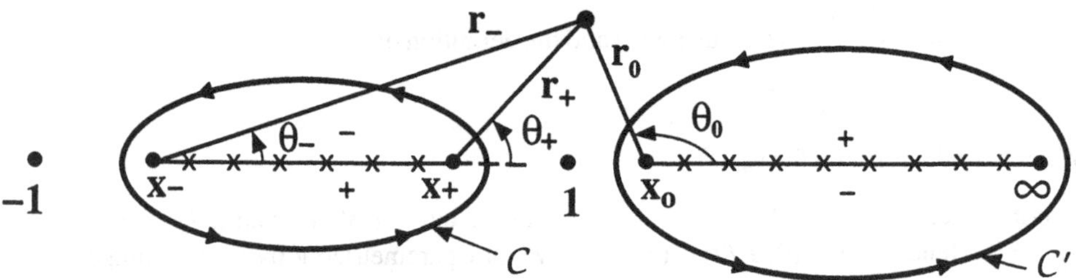

Figure 4.1. The choice of square root.

(4.4) **Proof**: Fix a choice of square root by setting $z - x_{\pm,0} = r_{\pm,0} \exp i\theta_{\pm,0}$, where $\theta_{\pm,0} \in [0, 2\pi]$ are as given in figure 4.1, and defining

$$\sqrt{Q(z)} = \sqrt{r_- r_0 r_+}\, \mathrm{e}^{i(\theta_- + \theta_0 + \theta_+)/2}.$$

Let $\mathscr{C}$ be the positively oriented path encircling the cut between x_- and x_+ as indicated in figure 4.1. As z makes a circuit around $\mathscr{C}$, the 1-form ω returns to its original value. This follows because $\theta_\pm$ increases by 2π, while θ_0 does not change. Therefore $\sqrt{z - x_0}$ keeps the same sign, whereas $\sqrt{z - x_\pm}$ changes sign. Hence $\sqrt{Q(z)}$ does not change sign as z runs around $\mathscr{C}$ once. A similar argument shows that ω is single valued as z makes a circuit around the path $\mathscr{C}'$ which encircles the cut between x_0 and ∞. Hence ω is single valued on the cut extended complex plane. It is meromorphic with simple poles at $z = \pm 1$ because $Q(\pm 1) \neq 0$ by hypothesis. □

Let the positively oriented path $\mathscr{C}$ shrink to the cut between x_- and x_+. Then on the upper part of the cut the square root of $Q(z)$ is negative because $\sqrt{z - x_+} = i\sqrt{x_+ - x}$, $\sqrt{z - x_0} = i\sqrt{x_0 - x}$, and $\sqrt{z - x_-} = \sqrt{x - x_-}$. On the lower part of the cut the square root of $Q(z)$ is positive because $\sqrt{z - x_+} = i\sqrt{x_+ - x}$, $\sqrt{z - x_-} = -\sqrt{x - x_-}$, and $\sqrt{z - x_0} = i\sqrt{x - x_0}$. Using Cauchy's theorem in the domain bounded by $\mathscr{C}$ and both sides of the cut, we find that

$$\int_{\mathscr{C}} \omega = \left(-\int_{x_+}^{x_-} + \int_{x_-}^{x_+} \right) \frac{1}{(1-x^2)\sqrt{Q(x)}}\, dx = 2\int_{x_-}^{x_+} \frac{1}{(1-x^2)\sqrt{Q(x)}}\, dx. \tag{61}$$

We now start proving the claim about the analytic properties of the function θ.

(4.5) **Proof**: Let (h_0, j_0) be a fixed element of $\mathcal{R}^* = \mathcal{R} - \{j = 0\}$. Then the polynomial $\widetilde{P}_0(x) = (h_0 - x)(1 - x^2) - \frac{1}{2} j_0^2$ has three distinct real roots $x^0_{\pm,0}$ where $x^0_\pm \in (-1, 1)$ and $x^0_0 > 1$. Since $x^0_{\pm,0}$ are simple roots, applying the implicit function theorem shows that there is an open neighborhood $\mathcal{U}_0$ of (h_0, j_0) in $\mathcal{R}^*$ such that for every $(h, j) \in \mathcal{U}_0$ the polynomial $\widetilde{P}(x) = (h - x)(1 - x^2) - \frac{1}{2} j^2$ has distinct real roots $x_{\pm,0} = x_{\pm,0}(h, j)$ where $x_\pm$ lie in a fixed open interval in $(-1, 1)$ and x_0 lies in a fixed bounded open interval in $(1, \infty)$. Thus for every $(h, j) \in \mathcal{U}_0$ there is a positively oriented path $\mathcal{C}$ encircling the cut between x_- and x_+ with $z = \pm 1$ lying outside of the domain bounded by $\mathcal{C}$. Because the hypotheses of ((4.3)) hold for all $(h, j) \in \mathcal{U}_0$, we may take $Q(x) = \widetilde{P}(x)$. After multiplying both sides by $\frac{j}{2\pi\sqrt{2}}$, we obtain

$$\frac{j}{2\pi\sqrt{2}} \int_{\mathcal{C}} \omega = \frac{j}{\pi\sqrt{2}} \int_{x_-}^{x_+} \frac{1}{(1 - x^2)\sqrt{\widetilde{P}(x)}} dx = \theta(h, j). \tag{62}$$

Now *complexify* the variables $(h, j) \in \mathcal{U}^{\mathbf{C}} \subseteq \mathbf{C}^2$. For a fixed j_0 consider the function

$$\theta^{\mathbf{C}}_{j_0} : \mathcal{U}^{\mathbf{C}} \cap (\mathbf{C} \times \{j_0\}) \to \mathbf{C} : h \to \theta(h, j_0).$$

$\theta^{\mathbf{C}}_{j_0}$ is complex analytic because

$$\frac{d\theta^{\mathbf{C}}_{j_0}}{d\overline{h}} = \frac{j}{2\pi\sqrt{2}} \int_{\mathcal{C}} \frac{d\omega}{d\overline{h}} = 0,$$

since ω does not depend on $\overline{h}$. Similarly $\theta^{\mathbf{C}}_{h_0}$ is complex analytic in $\mathcal{U}^{\mathbf{C}} \cap \big(\{h_0\} \times \mathbf{C}\big)$. Therefore by *Hartog's theorem*, $\theta^{\mathbf{C}}$ is complex analytic in an open neighborhood of (h_0, j_0) in $\mathbf{C}^2$. Hence $\theta = \theta^{\mathbf{C}} | \mathbf{R}^2$ is real analytic in an open neighborhood of (h_0, j_0). □

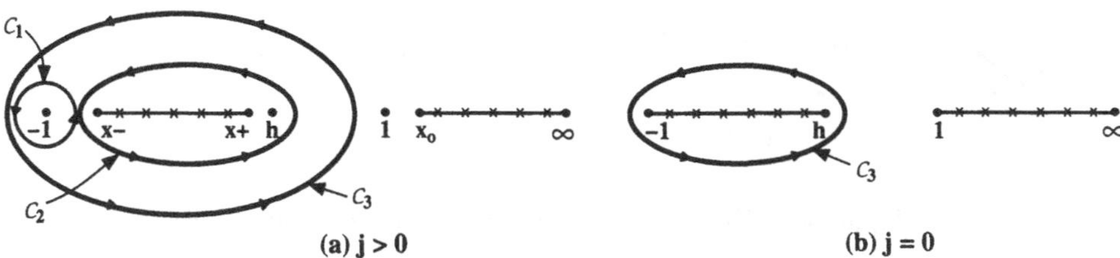

Figure 4.2. The curves $\mathcal{C}_i$ on the cut extended complex plane.

To show that θ can be extended to a *locally* single valued real analytic function on *all* of $\mathcal{R}$, it suffices to show that it remains bounded as $j \to 0$. Thus it suffices to prove

Claim:

$$\lim_{j \to 0} \theta(h, j) = \begin{cases} \frac{1}{2}, & \text{if } -1 < h < 1 \\ 1, & \text{if } h > 1. \end{cases} \tag{63}$$

(4.6) **Proof**:

Case 1. $-1 < h < 1$. For $i = 1, 2, 3$ let $\mathscr{C}_i$ be oriented closed curves as indicated in figure 4.2(a). By *Cauchy's theorem*

$$\frac{j}{2\pi\sqrt{2}}\int_{\mathscr{C}_3}\omega = \frac{j}{2\pi\sqrt{2}}\left(\int_{\mathscr{C}_1}+\int_{\mathscr{C}_2}\right)\omega = \frac{j}{2\pi\sqrt{2}}\int_{\mathscr{C}_1}\omega + \theta(h,j). \tag{64}$$

ω has a first order pole at -1 whose residue is

$$\operatorname*{Res}_{z=-1}\omega = \lim_{z\to -1}\frac{z+1}{(1-z^2)\sqrt{\widetilde{P}(z)}} = \frac{1}{2\sqrt{\widetilde{P}(-1)}} = \frac{i}{j\sqrt{2}}.$$

The plus sign in the last equality above comes from the choice of square root indicated in figure 4.1. Thus

$$\frac{j}{2\pi\sqrt{2}}\int_{\mathscr{C}_1}\omega = \frac{j}{2\pi\sqrt{2}}\,2\pi i \operatorname*{Res}_{z=-1}\omega = -\tfrac{1}{2}. \tag{65}$$

Letting $j \to 0$, we see that $x_- \to -1$, $x_+ \to h$, and $x_0 \to 1$. Cut $\widetilde{\mathbf{C}}$ as given in figure 4.2(b). As the 1-form ω is analytic and single valued in the domain bounded by $\mathscr{C}_3$, we find that

$$\lim_{j\to 0}\int_{\mathscr{C}_3}\omega = \int_{\mathscr{C}_3}\lim_{j\to 0}\omega = \int_{\mathscr{C}_3}\frac{1}{(1-z^2)\sqrt{(h-z)(1-z^2)}}\,dz,$$

which is finite. Hence letting $j \to 0$ in (64) gives $0 = -\frac{1}{2} + \lim_{j\to 0}\theta(h,j)$. This establishes the first case.

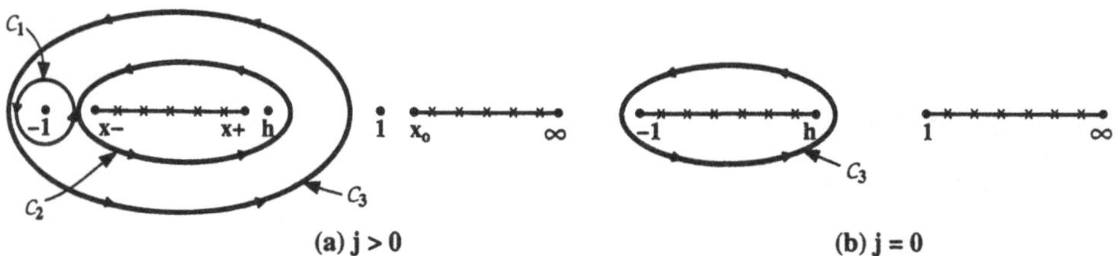

Figure 4.3. The curves $\mathscr{C}_i$ in the cut extended complex plane.

Case 2. $h > 1$. Consider the closed oriented paths $\mathscr{C}_i$ for $i = 1, \ldots, 4$ given in figure 4.3. Using Cauchy's theorem and the *residue theorem*, we see that

$$\frac{j}{2\pi\sqrt{2}}\int_{\mathscr{C}_3}\omega = -\tfrac{1}{2}.$$

Thus $\lim_{j\to 0}\theta(h,j) = 1$. The reader can easily supply the missing details. □

This completes the proof of ((4.4)). □

We now turn to describing the *global* real analytic function θ on $\mathcal{R}$ which is defined by *analytically continuing* the locally defined real analytic functions $\theta|\mathcal{U}_i$. Here $\{\mathcal{U}_i\}$ is a suitable open covering of $\mathcal{R}$. Let $\widetilde{\mathcal{R}}$ be the universal covering space of $\mathcal{R}$ with covering mapping $\pi : \widetilde{\mathcal{R}} \rightarrow \mathcal{R}$. Then $\widetilde{\mathcal{R}}$ is a real analytic manifold, which is diffeomorphic to a 2-disc $\mathcal{D}$ in $\mathbf{R}^2$ because $\mathcal{R}$ is diffeomorphic to an annulus. The covering projection π is real analytic. Each function element $\theta|\mathcal{U}_i$ lifts to a locally defined real analytic function element $\widetilde{\theta}|\widetilde{\mathcal{U}}_{i,j}$ where $\pi(\widetilde{\mathcal{U}}_{i,j}) = \mathcal{U}_i$ and $\bigcup \widetilde{\mathcal{U}}_{i,j} = \pi^{-1}(\mathcal{U}_i)$. Here $\widetilde{\theta}|\widetilde{\mathcal{U}}_{i,j} = (\theta \circ \pi)|\widetilde{\mathcal{U}}_{i,j}$.
▷ Since $\widetilde{\mathcal{R}}$ is simply connected, there is a single valued real analytic function $\widetilde{\theta}$ whose local function elements are $\widetilde{\theta}|\widetilde{\mathcal{U}}_{i,j}$.

(4.7) **Proof**: Let Γ be a simple closed curve in $\widetilde{\mathcal{R}}$. Then Γ bounds a 2-disc D. Since $\widetilde{\mathcal{R}}$ is simply connected, there is a homotopy Γ_s such that $\Gamma_1 = \Gamma$ and Γ_0 is a point $p \in D$. Let

$$S = \left\{ s \in [0,1] \,\middle|\, \widetilde{\theta} \text{ is a single valued real analytic function in the disc } D_s \text{ bounded by } \Gamma_s \right\}.$$

S is nonempty, because for some (i,j) the point $p \in \widetilde{\mathcal{U}}_{i,j}$. Then there is an $s_0 > 0$ such that $\Gamma_{s_0} \subseteq \widetilde{\mathcal{U}}_{i,j}$. Let $\sigma = \sup\{s \in S\}$. Then $\sigma > s_0 > 0$. Now suppose that $\sigma < 1$. Cover Γ_σ with open discs $\Delta_k \subseteq \widetilde{\mathcal{U}}_{i(k),j(k)}$ of radius $\frac{3}{4}\delta$ with center at the point $p_k \in \Gamma_\sigma$ where for $k \ge 1$ the distance between p_{k-1} and p_k is less than δ. Then the discs Δ_k pairwise overlap and a finite number of them cover Γ_σ. There is an $s_1 < \sigma$ such that $\Gamma_{s_1} \subseteq D_\sigma \cup \bigcup_k \Delta_k$. By definition of σ, the function $\widetilde{\theta}$ is single valued and real analytic in D_{s_1}. By analytic continuation $\widetilde{\theta}$ is single valued and real analytic on $\bigcup_k \Delta_k$. Therefore, $\widetilde{\theta}$ is single valued and real analytic on $D_{s_1} \cup \bigcup_k \Delta_k$, which contains D_σ as a proper subset. There is an $s_2 > \sigma$ such that $\Gamma_{s_2} \subseteq D_\sigma \cup \bigcup_k \Delta_k$. On D_{s_2} the function $\widetilde{\theta}$ is single valued and real analytic. But this contradicts the definition of σ. Therefore $\sigma = 1$. In other words, the function $\widetilde{\theta}$ is real analytic and single valued on Γ and hence on all of $\widetilde{\mathcal{R}}$. □

The above result just states that θ *could* be a multivalued real analytic function on $\mathcal{R}$. The following discussion shows that θ *is* multivalued. We begin by proving

Fact: The variation of θ along an oriented closed curve Γ in $\mathcal{R}$ depends only on the homotopy class of Γ.

(4.8) **Proof**: Suppose that Γ_1 and Γ_2 are homotopic positively oriented closed curves in $\mathcal{R}$. Then the curve $\gamma = \Gamma_1 - \Gamma_2$ is null homotopic in $\mathcal{R}$. Since the fundamental group of $\mathcal{R}$ is isomorphic to $\mathbf{Z}$, it is abelian and hence is isomorphic to the first homology group of $\mathcal{R}$. Therefore γ is the boundary of some domain $\mathcal{E}$ in $\mathcal{R}$. Hence

$$\int_{\Gamma_1} d\theta - \int_{\Gamma_2} d\theta = \int_{\gamma = \partial \mathcal{E}} d\theta = \int_{\mathcal{E}} d^2\theta = 0,$$

where the second to last equality follows by Stokes' thoerem. □

Next we prove

Claim: Let Γ be a positively oriented curve which generates the fundamental group of $\mathcal{R}$. As (h, j) makes a circuit around Γ the value of θ *decreases* by 1.

(4.9) **Proof**: Using the above fact we may choose Γ to be the positively oriented non null-homotopic curve in $\mathcal{R} \cup \{(1,0)\}$ made up of line segments joining the points (h_1, j_0), (h_0, j_0), $(h_0, -j_0)$, and $(h_1, -j_0)$ for some $j_0 > 0$. Let Γ_1^j be the oriented line segment joining the points (h_1, j) and (h_0, j) and Γ_2^j the oriented line segment joining $(h_0, -j)$ and $(h_1, -j)$ (see figure 4.4). Here $0 < j < j_0$. The curve Γ is homotopic to the curve $\Gamma^j = \Gamma_1^j \cup S_0^j \cup \Gamma_2^j \cup S_1^j$, where S_1^j is the line segment joint $(h_1, -j)$ to (h_1, j) and S_0^j is the line segment joining (h_0, j) to $(h_0, -j)$. Therefore

$$\begin{aligned}
\int_\Gamma d\theta &= \lim_{j\to 0^+} \int_{\Gamma^j} d\theta = \lim_{j\to 0^+} \left[2\int_{\Gamma_1^j} d\theta + \int_{S_1^j} d\theta + \int_{S_0^j} d\theta \right], \\
&\qquad \text{because } \theta(h,-j) = -\theta(h,j) \text{ when } j \neq 0 \\
&= \lim_{j\to 0^+} \Big[2\big(\theta(h_0,j) - \theta(h_1,j)\big) + \big(\theta(h_1,j) - \theta(h_1,-j)\big) \\
&\qquad + \big(\theta(h_0,j) - \theta(h_0,-j)\big)\Big] \\
&= -1, \quad \text{using (63) and the fact that } \theta \text{ is locally real analytic and hence locally continuous.}
\end{aligned}$$

Hence the variation of θ along Γ in $\mathcal{R}$ is -1. □

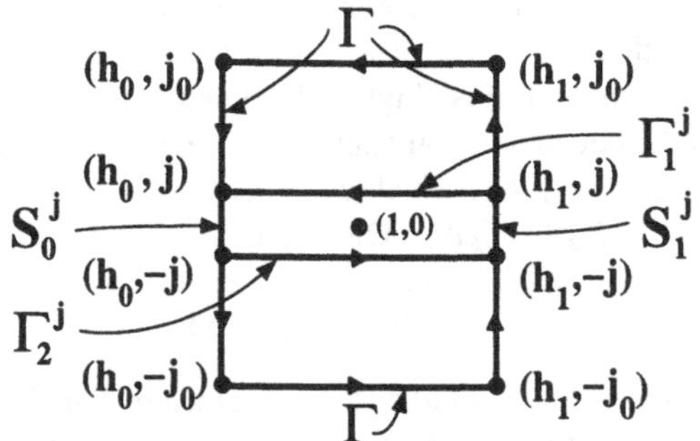

Figure 4.4. Variation of θ on the curve Γ.

We end this section by investigating the analytic properties of the time T of first return. From the definition of the cross section $\mathcal{C}^+$ (46) and the sign convention for η in (57), it follows that the time of first return is

$$T = \int_{x_3^-}^{x_3^+} \frac{dx_3}{dt}\, dt - \int_{x_3^+}^{x_3^-} \frac{dx_3}{dt}\, dt = \sqrt{2} \int_{x^-}^{x^+} \frac{dx}{\sqrt{\widetilde{P}(x)}} dx, \tag{66}$$

where $x_\pm$ are roots of the polynomial $\widetilde{P}$ (33) in $(-1, 1)$. Our main result is

Claim: The function $T : \mathcal{R} \to \mathbf{R} : (h, j) \to T(h, j)$ is a *single valued* real analytic function.

(4.10) **Proof**: We will give only an outline, leaving the reader to fill in the details. An argument similar to the one used to prove ((4.4)) shows that T is *locally* a single valued real analytic function. The same argument used in ((4.8)) shows that the variation of T along an oriented closed curve Γ in $\mathcal{R}$ depends only on the homotopy class of Γ. Choose the curves Γ and Γ^j as in figure 4.4. Then

$$\begin{aligned}
\int_\Gamma dT &= \lim_{j\to 0^+} \int_{\Gamma^j} dT \\
&= \lim_{j\to 0^+} \left[\int_{\Gamma_1^j} dT + \int_{S_0^j} dT + \int_{\Gamma_2^j} dT + \int_{S_1^j} dT \right] \\
&= \lim_{j\to 0^+} \Big[\big((T(h_0,j) - T(h_1,j)\big) + \big(T(h_1,-j) - T(h_0,-j)\big) \\
&\qquad + \big((T(h_0,-j) - T(h_0,j)\big) + \big((T(h_1,j) - T(h_1,-j)\big)\Big] \\
&= 0, \qquad \text{because } T \text{ is locally real analytic and hence is locally continuous}
\end{aligned}$$

Thus T is a single valued real analytic function in $\mathcal{R}$. □

5 Monodromy

In this section we show that the regular fibers of the energy momentum map $\mathcal{EM}$ of the spherical pendulum fit together in a nontrivial way. More precisely, let Γ be a closed non-null homotopic curve in the set $\mathcal{R}$ of regular values of $\mathcal{EM}$, which bounds a disc containing the point $(1,0)$ in its interior. We will show that the 2-torus bundle $\mathcal{EM}^{-1}(\Gamma) \to \Gamma$ is *nontrivial*, that is, $\mathcal{EM}^{-1}(\Gamma) \to \Gamma$ is not isomorphic to $S^1 \times T^2$. In other words, the classifying map χ of $\mathcal{EM}^{-1}(\Gamma) \to \Gamma$, which *glues* together the end 2-tori of the trivial bundle $\mathcal{EM}^{-1}(\Gamma - \{pt\}) \to \Gamma - \{pt\}$, is not homotopic to the identity map. In fact, the map

$$\chi_* : H_1(\mathcal{EM}^{-1}(pt), \mathbf{Z}) \to H_1(\mathcal{EM}^{-1}(pt), \mathbf{Z}) \tag{67}$$

induced on the first *homology group* of the end 2-torus $\mathcal{EM}^{-1}(pt)$ by the classifying map χ is $\begin{pmatrix} 1 & 0 \\ 1 & 1 \end{pmatrix}$ with respect to a suitably chosen basis. The map χ_* is called the *monodromy map* of the bundle $\mathcal{EM}^{-1}(\Gamma) \to \Gamma$ and depends only on the homotopy class of the curve Γ in $\mathcal{R}$. To see this suppose that $\widetilde{\Gamma}$ is a closed curve in $\mathcal{R}$ which is homotopic to Γ, then

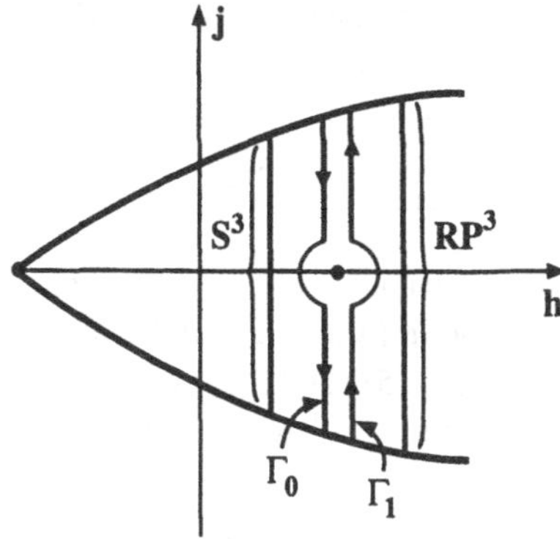

Figure 5.1. The geometric situation.

the bundles $\mathscr{EM}^{-1}(\Gamma)$ and $\mathscr{EM}^{-1}(\widetilde{\Gamma})$ are isomorphic. Therefore their classifying maps are
▷ homotopic, which implies that their monodromy maps are equal. The following geometric argument shows that the bundle $\mathscr{EM}^{-1}(\Gamma) \rightarrow \Gamma$ is *not* trivial.

(5.1) **Proof**: From table 3.3 we see that for $h_1 > 1$ the energy surface $H^{-1}(h_1)$ is diffeomorphic to $\mathbf{RP}^3$; while for $-1 < h_0 < 1$ the energy surface $H^{-1}(h_0)$ is diffeomorphic to S^3. Since $H_1(\mathbf{RP}^3) = \mathbf{Z}_2$ and $H_1(S^3) = 0$, it follows that $\mathbf{RP}^3$ is not even homeomorphic to S^3 let alone diffeomorphic to S^3. Suppose that the bundle $\mathscr{EM}^{-1}(\Gamma)$ is trivial. Then the manifolds $\mathscr{EM}^{-1}(\Gamma_0)$ and $\mathscr{EM}^{-1}(\Gamma_1)$ are diffeomorphic, (see figure 5.1). But $\mathscr{EM}^{-1}(\Gamma_0)$ is isotopic to $H^{-1}(h_0)$; while $\mathscr{EM}^{-1}(\Gamma_1)$ is isotopic to $H^{-1}(h_1)$. Therefore $H^{-1}(h_0)$ is diffeomorphic to $H^{-1}(h_1)$, which is false. Consequently, the bundle $\mathscr{EM}^{-1}(\Gamma) \rightarrow \Gamma$ is *not* trivial. □

To compute the monodromy map of the bundle $\mathscr{EM}^{-1}(\Gamma)$ we construct a bundle $\mathscr{P}$ of lattices over Γ whose transition maps with respect to a suitably chosen collection of trivializations are fixed elements of $\mathrm{Sl}(2, \mathbf{Z})$. For each $(h, j) \in \Gamma$ the fiber $\mathscr{P}_{(h,j)}$ of $\mathscr{P}$ is the period lattice of the 2-torus $\mathscr{EM}^{-1}(h, j)$. The period lattice is obtained in the following way. Let $\varphi_t^{H|TS^2}$ and $\varphi_t^{J|TS^2}$ be the flows of the Hamiltonian vector fields $X_{H|TS^2}$ and $X_{J|TS^2}$ on $(TS^2, \omega|TS^2)$. Since $\mathscr{EM}^{-1}(h, j)$ is a compact manifold which is invariant under the commuting flows $\varphi_t^{H|TS^2}$ and $\varphi_t^{J|TS^2}$, there is an action of $\mathbf{R}^2$ on $\mathscr{EM}^{-1}(h, j)$ defined by

$$\Phi : \mathbf{R}^2 \times \mathscr{EM}^{-1}(h, j) \rightarrow \mathscr{EM}^{-1}(h, j) : \big((s, t), p\big) \rightarrow \varphi_s^{J|TS^2} \circ \varphi_t^{H|TS^2}(p).$$

For $p_0 \in \mathscr{EM}^{-1}(h, j)$, let

$$L = \Big\{(T_1, T_2) \in \mathbf{R}^2 \big| \Phi_{(T_1, T_2)}(p_0) = p_0\Big\}$$

be the isotropy group of the action Φ at p_0. Because (h, j) is a regular value of $\mathscr{EM}$, the vector fields $X_{H|TS^2}$ and $X_{J|TS^2}$ span each tangent space to $\mathscr{EM}^{-1}(h, j)$. Thus Φ is a locally transitive action. Since $\mathscr{EM}^{-1}(h, j)$ is connected, the action Φ is transitive. Thus the isotropy group L does *not* depend on the point p_0. Because $\mathscr{EM}^{-1}(h, j)$ is compact, L is a discrete subgroup of $(\mathbf{R}^2, +)$, that is, L is a lattice. The lattice L depends only on (h, j) and is called the *period lattice* $\mathscr{P}_{(h,j)}$ of $\mathscr{EM}^{-1}(h, j)$. From transitivity it follows that $\mathscr{EM}^{-1}(h, j)$ is diffeomorphic to the space $\mathbf{R}^2/\mathbf{Z}^2$ of orbits of the affine action of L on $\mathbf{R}^2$, which is a 2-torus. We now give an explicit description of $\mathscr{P}_{(h,j)}$. For fixed $(h, j) \in \Gamma$ consider the Hamiltonian functions

$$F_1 = 2\pi J \quad \text{and} \quad F_2 = -2\pi\theta(h, j)J + T(h, j)H.$$

Note that F_2 is well defined because $(h, j) \in \Gamma$ is fixed and the value of the rotation number $\theta(h, j)$ is chosen to lie in $[\,0, 1)$. From the definition of the rotation number θ and the time of first return T (see (58) and (66)) we see that the flows $\varphi_t^{F_1}$ and $\varphi_t^{F_2}$ of the Hamiltonian vector fields X_{F_1} and X_{F_2} have the following properties:

1. They commute.

2. They leave $\mathscr{EM}^{-1}(h, j)$ invariant.

3. They are periodic of period 1.

Therefore, the period lattice $\mathcal{P}_{(h,j)}$ is generated by the vectors $X_{F_1}(p_0)$ and $X_{F_2}(p_0)$, that is, the vectors

$$(2\pi, 0) \quad \text{and} \quad \big(-2\pi\theta(h,j), T(h,j)\big)$$

with respect to the basis $X_{J|TS^2}(p_0)$ and $X_{H|TS^2}(p_0)$ of $T_{p_0}\mathcal{EM}^{-1}(h,j)$.

We are now in position to construct a bundle of lattices with total space $\mathcal{P}$. From ((4.4)) it follows that the rotation number θ is *locally* smooth on $\mathcal{R}$. Note that the time T of first return is globally smooth on $\mathcal{R}$. Thus there is a *good covering* of $\mathcal{R}$ (in the sense of Leray) which restricts to a good covering of Γ. This means that there is a collection of open subsets $\{U^\alpha\}$ of $\mathcal{R}$ such that

1. U^α, $U^\alpha \cap U^\beta$, $U^\alpha \cap \Gamma$, $U^\alpha \cap U^\beta \cap \Gamma$ are connected and contractible.
2. $U^\alpha \cap U^\beta \cap U^\gamma = \varnothing$.
3. The functions $T^\alpha = T|U^\alpha$ and $\theta^\alpha = \theta|U^\alpha$ are smooth and single valued.

A parametrization of $\mathcal{P}$ over $U^\alpha \cap \Gamma$ is given by the smooth family of lattices

$$\begin{aligned} \sigma_\alpha : \big(U^\alpha \cap \Gamma\big) \times \mathbf{Z}^2 \to \mathcal{P}_{(h,j)} \subseteq \mathbf{R}^2 : \\ \big((h,j),(n,m)\big) \to n(2\pi, 0) + m\big(-2\pi\theta^\alpha(h,j), T^\alpha(h,j)\big). \end{aligned} \tag{68}$$

▷ On the overlap the transition function for the bundle $\mathcal{P}$ is

$$\begin{aligned} \sigma^{\alpha\beta} = \sigma_\beta \circ \sigma_\alpha^{-1} : (U^\alpha \cap U^\beta \cap \Gamma) \times \mathbf{Z}^2 \to (U^\alpha \cap U^\beta \cap \Gamma) \times \mathbf{Z}^2 : \\ \big((h,j),(n,m)\big) \to \big((h,j), g^{\alpha\beta}(n,m)\big), \end{aligned} \tag{69}$$

where $g^{\alpha\beta} \in \mathrm{Sl}(2, \mathbf{Z})$.

(5.2) **Proof**: Let $p \in \mathcal{EM}^{-1}(U^\alpha)$. Consider the functions

$$F_1^\alpha(p) = J(p)$$

and

$$F_2^\alpha(p) = -2\pi\theta^\alpha(\mathcal{EM}(p)) \cdot J(p) + T^\alpha(\mathcal{EM}(p)) \cdot H(p).$$

The flows $\varphi_t^{F_1^\alpha}$ and $\varphi_t^{F_2^\alpha}$ of the Hamiltonian vector fields $X_{F_1^\alpha}$ and $X_{F_2^\alpha}$ commute, leave $\mathcal{EM}^{-1}(h,j)$ invariant for every $(h,j) \in U^\alpha$ and are periodic of period 1. (Note that F_1^α and F_2^α are action variables, (see appendix D).) Thus they generate a $T^2 = \mathbf{R}^2/\mathbf{Z}^2$ action

$$\Phi^\alpha : T^2 \times \mathcal{EM}^{-1}(U^\alpha) \to \mathcal{EM}^{-1}(U^\alpha) : \big((s,t), p\big) \to \varphi_s^{F_1^\alpha} \circ \varphi_t^{F_2^\alpha}(p)$$

on $\mathcal{EM}^{-1}(U^\alpha)$ which is free. Therefore $\mathcal{EM}^{-1}(U^\alpha)$ is a *principal* T^2 bundle with trivialization $\tau_\alpha : \mathcal{EM}^{-1}(U^\alpha) \to U^\alpha \times T^2$, (see appendix B ((2.12))). τ_α is a diffeomorphism intertwining the T^2 action Φ^α with the affine action of T^2 on itself. From this intertwining property it follows that for every $(h,j) \in U^\alpha$ the tangent of the mapping

$$\tau_{\alpha,(h,j)}^{-1} : T^2 \to \mathcal{EM}^{-1}(h,j) : (s,t) \to \tau_\alpha^{-1}\big((h,j),(s,t)\big) = q$$

maps the lattice $\mathbf{Z}^2 \subseteq T_{(s,t)}T^2$ onto the lattice in $T_q\mathcal{EM}^{-1}(h,j)$ spanned by the vectors $\{X_{F_1^\alpha}(q), X_{F_2^\alpha}(q)\}$. But this is just the period lattice $\mathcal{P}_{(h,j)}$ of $\mathcal{EM}^{-1}(h,j)$. Since $\mathcal{P}_{(h,j)}$ does *not* depend on the point $q \in \mathcal{EM}^{-1}(h,j)$, the mapping $T_{(s,t)}\tau^{-1}_{\alpha,(h,j)}$ does not depend on $(s,t) \in T^2$. From this it follows that the inverse of σ_α (68) is given by

$$\sigma_\alpha^{-1} : \coprod_{(h,j)\in U^\alpha} \mathcal{P}_{(h,j)} \to (U^\alpha \cap \Gamma) \times \mathbf{Z}^2 : p \to \left(\mathcal{EM}(p), T\tau_\alpha \mathcal{P}_{(h,j)}\right).$$

For every $(h,j) \in U^\alpha \cap U^\beta$ the partial transition map

$$\tau^{\alpha\beta}_{(h,j)} = \tau_{\beta,(h,j)} \circ \tau^{-1}_{\alpha,(h,j)} : T^2 \to T^2$$

has a tangent map which does not depend on the point $(s,t) \in T^2$ and is a linear isomorphism of $\mathbf{R}^2$ which preserves the lattice $\mathbf{Z}^2$. The set of all such isomorphisms forms the group $\mathrm{Sl}(2,\mathbf{Z})$, which is a discrete subgroup of the Lie group $\mathrm{Sl}(2,\mathbf{R})$. Since $U^\alpha \cap U^\beta$ is connected, it follows that the map

$$U^\alpha \cap U^\beta \cap \Gamma \to \mathrm{Sl}(2,\mathbf{Z}) : (h,j) \to T\tau^{\alpha\beta}_{(h,j)}$$

is constant, namely, $g^{\alpha\beta}$. Thus (69) is precisely the transition map for the bundle $\mathcal{P}$ of period lattices. □

▷ The period lattice bundle is isomorphic to the bundle of homology groups

$$\coprod_{(h,j)\in\Gamma} H_1(\mathcal{EM}^{-1}(h,j),\mathbf{Z}) = H_1(\mathcal{EM}^{-1}(\Gamma),\mathbf{Z}) \to \Gamma$$

of the fibers of the energy momentum mapping over Γ.

(5.3) **Proof**: For every $(n,m) \in \mathbf{Z}^2$ and $p \in \mathcal{EM}^{-1}(h,j)$ where $(h,j) \in U^\alpha \cap \Gamma$ consider the mapping

$$\begin{aligned} \psi^\alpha : \mathcal{P}_{(h,j)} &\to H_1(\mathcal{EM}^{-1}(h,j),\mathbf{Z}) : \\ (nX_{F_1^\alpha} + mX_{F_2^\alpha})(p) &\to \left[t \to \varphi^{F_1^\alpha}_{nt} \circ \varphi^{F_2^\alpha}_{mt}(p)\right] = \left[\gamma_p^{nm}\right]. \end{aligned} \tag{70}$$

The homology class $\left[\gamma_p^{nm}\right]$ of the curve γ_p^{nm} does not depend on the point p, because of the transitivity of the $\mathbf{R}^2$ action. Since the period lattice $\mathcal{P}_{(h,j)}$ does not depend on the choice of the point p, the mapping ψ^α is well defined. The curve γ_p^{nm} is homotopic to n times the curve γ^1 followed by m times the curve γ^2 where

$$\gamma^1 : [0,1] \to \mathcal{EM}^{-1}(h,j) : t \to \varphi_t^{F_1^\alpha}(p)$$

and

$$\gamma^2 : [0,1] \to \mathcal{EM}^{-1}(h,j) : t \to \varphi_t^{F_2^\alpha}(p).$$

Thus

$$\left[\gamma_p^{nm}\right] = n\left[\gamma^1\right] + m\left[\gamma^2\right].$$

In other words, the mapping ψ^α (70) is linear. Since the homology class of γ^1 and γ^2 generate the lattice $H_1(\mathcal{EM}^{-1}(h, j), \mathbf{Z})$, the mapping ψ^α is an isomorphism of lattices. As a bundle map covering the identity map on U^α, the map ψ^α is a bundle isomorphism. Thus the bundle $\mathcal{P}$ of period lattices over Γ is isomorphic to the bundle of homology groups $H_1(\mathcal{EM}^{-1}(\Gamma), \mathbf{Z})$ over Γ. □

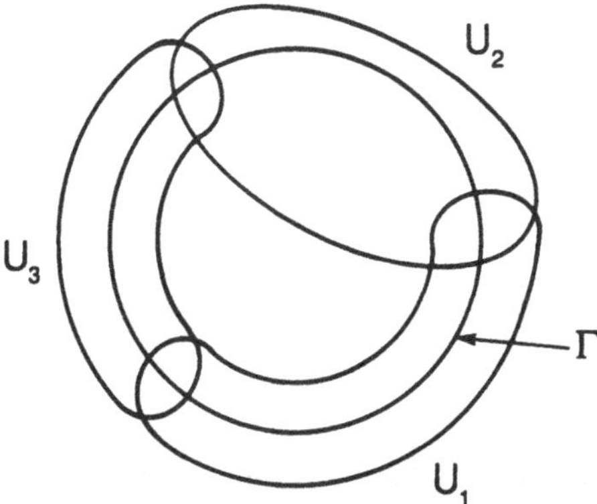

Figure 5.2. The circle Γ and the covering $\{U^i \cap \Gamma\}_{i=1}^3$ of Γ.

▷ After all this preparation we compute the monodromy map of the bundle $\mathcal{EM}^{-1}(\Gamma) \to \Gamma$ as the *variation* of the period lattice along Γ.

(5.4) **Proof**: We choose Γ and the covering $\{U^\alpha\}_{\alpha=1}^\ell$ as indicated in figure 5.2. For $(h, j) \in U^\alpha$ the period lattice $\mathcal{P}_{(h,j)}$ is generated by the vectors

$$(2\pi, 0) \qquad \text{and} \qquad \big(-2\pi\theta^\alpha(h, j), T^\alpha(h, j)\big). \tag{71}$$

When $\alpha = 1$, let $\theta^1(h_1, 0) = 0$. This fixes the global multivalued function θ. Then $\theta^\ell(h_1, 0) = -1$ by ((4.5)). On the *overlaps* $U^\alpha \cap U^{\alpha+1}$, where $\alpha = 1, \ldots, \ell - 1$, from (71) it follows that $g^{\alpha,\alpha+1} = \begin{pmatrix} 1 & 0 \\ 0 & 1 \end{pmatrix}$, that is, there is *no* variation in the period lattice when going from U^α to $U^{\alpha+1}$, because $\theta^\alpha = \theta^{\alpha+1}$ and $T^\alpha = T^{\alpha+1}$ on $U^\alpha \cap U^{\alpha+1}$. Of course the period lattice varies in U^α. On the overlap $U^1 \cap U^\ell$ the period lattice generated by

$$\Big\{ (2\pi, 0), \big(-2\pi\theta^1(h_1, 0), T^1(h_1, 0)\big) \Big\}$$

is transformed into the period lattice generated by

$$\Big\{ (2\pi, 0), \big(-2\pi\theta^\ell(h_1, 0), T^\ell(h_1, 0)\big) \Big\} =$$

$$= \Big\{ (2\pi, 0), \big(-2\pi(\theta^1(h_1, 0) - 1), T^1(h_1, 0)\big) \Big\},$$

recall that T is single valued on all of $\mathcal{R}$

$$= \Big\{ (2\pi, 0), (2\pi, 0) + \big(-2\pi\theta^1(h_1, 0), T^1(h_1, 0)\big) \Big\}$$

Thus $g^{1,\ell} = \begin{pmatrix} 1 & 0 \\ 1 & 1 \end{pmatrix}$. Hence the monodromy map

$$\chi_* : H_1(\mathcal{EM}^{-1}(h_1, 0), \mathbf{Z}) \to H_1(\mathcal{EM}^{-1}(h_1, 0), \mathbf{Z})$$

of the bundle $\mathcal{EM}^{-1}(\Gamma) \to \Gamma$ is the variation of the period lattice around Γ. This is just the product of the variations on the overlaps, that is,

$$g^{1,\ell} \prod_{1 \le \alpha \le \ell - 1} g^{\alpha,\alpha+1} = \begin{pmatrix} 1 & 0 \\ 1 & 1 \end{pmatrix}.$$

6 Exercises

1. Show that there is no homeomorphism of $\mathbf{R}^3$ which maps a 2-torus with a meridial circle pinched to a point to a 2-torus in $\mathbf{R}^3$ with a longitudinal circle pinched to a point.

2. If (h, ℓ) is a regular value of the energy momentum mapping $\mathcal{EM}$ of the spherical pendulum, show that the image of an integral curve of $X_H | \mathcal{EM}^{-1}(h, \ell)$ under the bundle projection $\pi : TS^2 \subseteq T\mathbf{R}^3 \to S^2 \subseteq \mathbf{R}^3$, after being lifted to the universal covering space of $\pi(\mathcal{EM}^{-1}(h, \ell))$, has no self-intersections.

3. (Estimate for the rotation number). If (h, ℓ) is a regular values of the energy momentum map $\mathcal{EM}$ of the spherical pendulum, then 2π times the rotation number of the flow on the 2-torus $\mathcal{EM}^{-1}(h, \ell)$ is

$$\theta(h, \ell) = 2\ell \int_{x_3^-}^{x_3^+} \frac{dx_3}{(1 - x_3^2)\sqrt{2(h - x_3)(1 - x_3^2) - \ell^2}}. \tag{1}$$

Note that $\theta(h, -\ell) = -\theta(h, \ell)$, when $\ell \neq 0$. The goal of this exercise is to show that

$$\pi < \theta(h, \ell) < 2\pi, \tag{2}$$

when $\ell > 0$. We will use complex analysis.

(a). Let

$$\omega = \frac{\ell \, dz}{(1 - z^2)\sqrt{2(h - z)(1 - z^2) - \ell^2}}.$$

Since $-1 < x_3^- < x_3^+ < 1 < x_3^0$ are real roots of $2(h - z)(1 - z^2) - \ell^2$, we cut the extended complex plane $\widetilde{\mathbf{C}}$ along the real axis between x_3^- and x_3^+ and again between x_3^0 and ∞. Choosing the square root as in figure 4.1, we see that ω is a meromorphic 1-form on $\mathbf{C}' = \widetilde{\mathbf{C}} - ([x_3^-, x_3^+] \cup [x_3^0, \infty])$ with first order poles at ± 1. Show that $\operatorname{Res}_{z=\pm 1} \omega = \frac{1}{2} i$. Let $\mathcal{C}_1$, $\mathcal{C}_3$, and $\mathcal{C}_4$ be positively oriented closed curves in $\mathbf{C}'$ which enclose $[x_3^-, x_3^+]$ but not ± 1, $[x_3^-, x_3^+]$ and -1 but not $+1$ and

$[x_3^-, x_3^+]$ and ± 1, respectively. Show that

$$\begin{aligned} \theta(h,\ell) &= \int_{\mathscr{C}_1} \omega = \int_{\mathscr{C}_3} \omega - 2\pi i \operatorname*{Res}_{z=-1} \omega \\ &= \int_{\mathscr{C}_4} \omega - 2\pi i \left(\operatorname*{Res}_{z=-1} \omega + \operatorname*{Res}_{z=1} \omega \right) = \int_{\mathscr{C}_4} \omega + 2\pi. \end{aligned} \tag{3}$$

Let $\mathscr{C}_2$ be the *positively* oriented curve in $\mathbf{C}'$ which encloses the cut $[x_3^0, \infty]$ but not 1. Show that $\mathscr{C}_2$ goes from near ∞ to x_3^- along the underside of the cut and then back to the original starting point. Show that the curves $\mathscr{C}_4$ and $\mathscr{C}_2$ are homotopic in $\mathbf{C}' - \{\pm 1\}$. Deduce that

$$\int_{\mathscr{C}_4} \omega = \int_{\mathscr{C}_2} \omega = -2 \int_{x_3^0}^{\infty} \frac{1}{(x_3^2 - 1)\sqrt{2(h - x_3)(1 - x_3^2) - \ell^2}} \, dx_3. \tag{4}$$

Since $x_3^0 > 1$, the integrand in (4) is positive. Therefore from (3) we obtain $\theta(h,\ell) < 2\pi$.

(b). Consider the positively oriented vertical line $L : z = \xi + i\,\eta$ where $\eta \in [-\infty, \infty]$, $\xi \in (x_3^+, 1)$ and L is closer to x_3^- than x_3^+, that is,

$$x_3^- + \xi < x_3^+ - \xi. \tag{5}$$

Show that L is homotopic to $\mathscr{C}_3$ in $\mathbf{C}' - \{\pm 1\}$. Deduce that

$$\theta(h,\ell) = \int_{\mathscr{C}_1} \omega = \int_L \omega + \pi. \tag{6}$$

Let $z = \xi + i\,\eta$ with $\eta \geq 0$. Define $\theta^{0,\pm}$ and $r^{0,\pm}$ by $z - x_3^{0,\pm} = r^{0,\pm} \mathrm{e}^{i\,\theta^{0,\pm}}$, where $0 \leq \theta^{0,\pm} < 2\pi$. Show that

$$\theta^- + \theta^+ > \pi. \tag{7}$$

Hence

$$\pi > \alpha(\eta) = \tfrac{1}{2}\left(\theta^0 + \theta^- + \theta^+\right) > \pi/2. \tag{8}$$

If $z = \xi + i\,\eta$ with $\eta < 0$, show that $\alpha(\eta) = \alpha(-\eta)$. Justify each step of the following calculation.

$$\begin{aligned} \int_L \omega &= \int_{-\infty}^{\infty} \frac{i\ell\, d\eta}{\left(1 - (\xi + i\,\eta)^2\right) \mathrm{e}^{i\alpha(\eta)} \sqrt{2}\sqrt{r^0 r^- r^+}} \\ &= 2\ell \int_0^{\infty} \left[\frac{(1 - \xi^2 + \eta^2)\sin\alpha(\eta) - 2\xi\eta\cos\alpha(\eta)}{(1 - \xi^2 + \eta^2)^2 + 4\xi^2\eta^2} \right] \frac{d\eta}{\sqrt{2}\sqrt{r^0 r^- r^+}} \\ &> 0. \end{aligned}$$

Therefore $\theta(h,\ell) > \pi$.

4. (Horozov's theorem). We know that the regular values of the energy momentum map $\mathcal{EM}$ of the spherical pendulum is the set $\mathcal{R}$ of $(h, \ell) \in \mathbf{R}^2$ such that

$$h > -1, \; \ell^2 < \frac{4(3+h^2)^{3/2} + 4h(9-h^2)}{27}, \text{ and } (h, \ell) \neq (1, 0). \tag{9}$$

For $(h, \ell) \in \mathcal{R}$, 2π times the rotation number of the flow of the spherical pendulum on the 2-torus $\mathcal{EM}^{-1}(h, \ell)$ is

$$\theta(h, \ell) = 2\ell \int_{x_3^-}^{x_3^+} \frac{dx_3}{(1-x_3^2)\sqrt{2(h-x_3)(1-x_3^2) - \ell^2}},$$

where $-1 < x_3^- < x_3^+ < 1 < x_3^0$ where $x^{0,\pm}$ are roots of $2(h-x_3)(1-x_3^2) - \ell^2$. The time $T(h, \ell)$ of first return of the flow of the spherical pendulum to a cross section on $\mathcal{EM}^{-1}(h, \ell)$ (given by an orbit of the angular momentum vector field) is

$$T(h, \ell) = 2 \int_{x_3^-}^{x_3^+} \frac{dx_3}{\sqrt{2(h-x_3)(1-x_3^2) - \ell^2}}.$$

We know that locally T and θ are real analytic functions in $\mathcal{R}$. The goal of this exercise is to show that they are *coordinates* on $\mathcal{R}$, that is, $dT \wedge d\theta \neq 0$. In other words, for every $(h, \ell) \in \mathcal{R}$

$$D = \det \begin{pmatrix} \frac{\partial T}{\partial h} & \frac{\partial T}{\partial \ell} \\ \frac{\partial \theta}{\partial h} & \frac{\partial \theta}{\partial \ell} \end{pmatrix} \neq 0. \tag{10}$$

(a). Use a computer to draw the level curves of θ and T. Notice that these curves are like polar coordinates centered at $(1, 0)$ with θ being the radial coordinate and T the angular coordinate.

(b). Consider the elliptic curve

$$\Gamma_{h,\ell} : y^2 = 2(h-z)(1-z^2) - \ell^2$$

with $(h, \ell) \in \mathcal{R}$. Let $\gamma_{h,\ell}$ be a positively oriented curve in the cut extended complex plane $\mathbf{C}' = \mathbf{C} - ([x_3^-, x_3^+] \cup [x_3^0, \infty])$, which encloses the cut $[x_3^-, x_3^+]$ but not ± 1. Note that

$$T(h, \ell) = \int_{\gamma_{h,\ell}} \frac{1}{y} \, dz \quad \text{and} \quad \theta(h, \ell) = \int_{\gamma_{h,\ell}} \frac{\ell \, dz}{(1-z^2)\, y}.$$

Since we can homotope the curve $\gamma_{h,\ell}$ in $\mathbf{C}' - \{\pm 1\}$ to another without changing T or θ, it follows that we can compute the partial derivative of T and θ by differentiating under the integral sign. Let

$$w_0 = \int_{\gamma_{h,\ell}} \frac{1}{y^3} \, dz \quad \text{and} \quad w_1 = \int_{\gamma_{h,\ell}} \frac{z}{y^3} \, dz.$$

Show that

$$D = \tfrac{4}{3}\,(h\,w_0 - w_1)(w_0 - h\,w_1) - \ell^2 w_0^2 \tag{11}$$

by checking the following

i). $\frac{\partial T}{\partial h} = \int_{\gamma_{h,\ell}} \frac{1-z^2}{y^3}\,dz \;=\; w_0 - \int_{\gamma_{h,\ell}} \frac{z^2}{y^3}\,dz$. But

$$\int_{\gamma_{h,\ell}} \frac{z^2}{y^3}\,dz = \frac{1}{6}\int_{\gamma_{h,\ell}} \frac{d(y^2+2hz^2+2z+\ell^2-2h)}{y^3}.$$

Therefore

$$\frac{\partial T}{\partial h} = \tfrac{2}{3}\,h\,w_1 + \tfrac{1}{3}\,w_0.$$

ii). $\frac{\partial \theta}{\partial \ell} = \int_{\gamma_{h,\ell}} \frac{\ell^2+y^2}{(1-z^2)\,y^3}\,dz \;=\; \int_{\gamma_{h,\ell}} \frac{2(h-z)(1-z^2)}{(1-z^2)\,y^3}\,dz$. Therefore

$$\frac{\partial \theta}{\partial \ell} = 2(h\,w_0 - w_1).$$

iii).

$$\frac{\partial \theta}{\partial h} = \frac{\partial T}{\partial \ell} = \ell\,w_0.$$

From (11), we see that we may assume that $\ell \geq 0$.

(c). Using the translation $z = x + \frac{1}{3}h$ and the scaling $y = \alpha v$ and $x = \beta u$, where $3\beta = (3+h^2)^{1/2}$ and $\alpha = \beta^{3/2}$, show that the elliptic curve $\Gamma_{h,\ell}$ becomes the elliptic curve

$$\Gamma_p : \tfrac{1}{2}\,v^2 = u^3 - 3u + p,$$

where

$$p = p(h,\ell) = \frac{2h(9-h^2)}{(3+h^2)^{3/2}} - \frac{27\ell^2}{2(3+h^2)^{3/2}}.$$

When $(h,\ell) \in \mathfrak{R}$ show that $-2 < p(h,\ell) < 2$. Also check that

$$w_0 = \frac{\beta}{\alpha^3}\int_{\gamma_p} \frac{du}{v^3} \quad \text{and} \quad w_1 = \frac{\beta}{\alpha^3}\int_{\gamma_p} \frac{\beta u + h/3}{v^3}\,du.$$

We now explain how the curve γ_p is chosen. Let $\mathbf{C}'$ be the extended complex plane $\widetilde{\mathbf{C}}$ cut between u_- and u_+ and again between u_0 and ∞. Here $u_{0,\pm}$ are real roots of $u^3 - 3u + p$, $-2 < p < 2$, and $u_- < u_+ < u_0$. Let γ_p be a closed positively oriented curve in $\mathbf{C}'$ which encloses the cut $[u_-, u_+]$ but not ± 1.

(d). Let

$$\theta_0(p) = \int_{\gamma_p} \frac{du}{v^3} \quad \text{and} \quad \theta_1(p) = \int_{\gamma_p} \frac{u\,du}{v^3}.$$

Show that (θ_0, θ_1) satisfy the Picard Fuchs equation

$$6(4-p^2)\,\frac{d}{dp}\begin{pmatrix}\theta_0\\ \theta_1\end{pmatrix} = \begin{pmatrix}7p & 10\\ 14 & 5p\end{pmatrix}\begin{pmatrix}\theta_0\\ \theta_1\end{pmatrix}.$$

Let $r(p) = \theta_0(p)/\theta_1(p)$. For $p \in (-2,2)$ show that r satisfies the Ricatti equation

$$3(4-p^2)\,\frac{dr}{dp} = 7 - pr - 5r^2 \tag{12}$$

and has the following properties.

i). $r(-2) = 7/5$ and $r(2) = 1$.

ii). $r'(p) < 0$ for every $p \in (-2,2)$.

To see that ii) holds we argue as follows. First we show that $r(p) > 0$ for every $p \in (-2,2)$. Suppose not. Let $p_0 \in (-2,2)$ be the smallest zero of r. Then $r'(p_0) \leq 0$. Using (12), we obtain $r'(p_0) = \frac{7}{3(4-{p_0}^2)} > 0$, which is a contradiction. Suppose that $r'(p_0) = 0$, for some $p_0 \in (-2,2)$. Differentiating (12) gives

$$3(4-p_0^2)\,\frac{d^2r}{dp^2}(p_0) = -r(p_0) < 0.$$

Thus every critical point of r is a nondegenerate local maximum. Suppose that for some $p' \in (-2,2)$, $r(p') < 1$. Since $r(-2) = 7/5 > 1 = r(2)$, it follows that r has a minimum in $(-2,2)$. This is a contradiction. Hence $r \geq 1$ on $(-2,2)$. From (12) it follows that $r'(p) = 0$ for $p \in (-2,2)$ if and only if $(p, r(p))$ lie on

$$0 = 7 - xy - 5y^2 \qquad -2 \leq x \leq 2, \ \&\ y \geq 1. \tag{13}$$

Equation (13) defines a smooth function $x \rightarrow y(x)$, which is strictly decreasing on $[-2,2]$ because

$$y' = \frac{-y}{10y + x} < 0.$$

Note that $y(-2) = 7/5$ and $y(2) = 1$. Suppose that $r'(p_0) = 0$ for some $p_0 \in (-2,2)$. Then $(p_0, r(p_0))$ satisfies (13). Hence $r(p_0) < 7/5$. But $r(-2) = 7/5$. Because $r(p_0)$ is a nondegenerate local maximum, we see that r has a local minimum in $(-2, p_0)$. This is a contradiction. Hence r has *no* critical points in $(-2,2)$. Since $r(-2) = 7/5$ and $r(2) = 1$, it follows that $r'(p) < 0$ for every $p \in (-2,2)$.

(e). Show that $w_0(p)w_1(p) > 0$ for every $p \in (-2,2)$. Argue as follows. Since

$$w_0(p) = 2\int_{u_-}^{u_+} \frac{du}{(u^3 - 3u + p)^{3/2}},$$

we see that $w_0(p) \neq 0$ for every $p \in (-2,2)$. Suppose that for some $(h^0, \ell^0) \in \mathcal{R}$ with $p^0 = p(h^0, \ell^0)$ we have $w_1(p^0) = 0$. From the definition of w_1 we obtain

$$r(p^0) = -\frac{h^0}{3\beta} = -h^0\big(3 + (h^0)^2\big)^{-1/3}.$$

Therefore $r(p^0) \leq 0$ when $h^0 > 0$ and $0 \leq r(p^0) \leq \frac{1}{2}$ when $-1 < h \leq 0$. This contradicts the fact that $r(p) \in (1, 7/5)$ when $p \in (-2,2)$. Therefore

$w_1(p(h,\ell)) \neq 0$ for every $(h,\ell) \in \mathcal{R}$. Setting $h = 0$ in the definition of w_1 gives

$$\frac{w_1(\widetilde{p})}{w_0(\widetilde{p})} = \frac{1}{\beta}\frac{\theta_1(\widetilde{p})}{\theta_0(\widetilde{p})} = \frac{1}{\beta}\, r(\widetilde{p}) > 0,$$

where $\widetilde{p} = p(0,\ell)$. Therefore $w_0(p)w_1(p) > 0$ for every $p \in (-2,2)$.

(f). We show that D (11) is nonzero when $(h,\ell) \in \mathcal{R}$ by considering three cases.

<u>case 1</u>. $h \le 0$, $\ell > 0$. From (e) and (11) it follows that $D < 0$.

<u>case 2</u>. $\ell = 0$. When $\ell = 0$ the spherical pendulum moves in a plane as if it were a mathematical pendulum. From the geometric interpretation of $\theta(h,0)$ ((4.2)) we see that

$$\theta(h,0) = \begin{cases} \pi, & \text{if } -1 < h < 1 \\ 2\pi, & \text{if } h > 1. \end{cases} \tag{14}$$

Therefore $\dfrac{\partial\theta}{\partial h}(h,0) = 0$. Now

$$T(h,0) = 2\int_{x_3^-}^{x_3^+} \frac{dx_3}{\sqrt{2(h-x_3)(1-x_3^2)}}$$

is the period of the mathematical pendulum as a function of energy. In exercise 3 of the introduction we show that $\dfrac{\partial T}{\partial h}(h,0) > 0$. Because θ is a real analytic function on $\mathcal{R}$, we have

$$\theta(h,\ell) = \theta(h,0) + \ell\,\frac{\partial\theta}{\partial\ell}(h,0) + O(\ell^2).$$

From (14) and exercise 4 it follows that

$$\pi < \pi + \ell\,\frac{\partial\theta}{\partial\ell}(h,0) + O(\ell^2)$$

for $(h,\ell) \in \mathcal{R}$ and $-1 < h < 1$, $\ell \neq 0$. Therefore $\dfrac{\partial\theta}{\partial\ell}(h,0) \neq 0$. Similarly, when $h > 1$ we find that $\dfrac{\partial\theta}{\partial\ell}(h,0) \neq 0$. Therefore $D \neq 0$ when $\ell = 0$. Setting $h = \ell = 0$ in (11) we see that $D < 0$. Hence $D < 0$ when $\ell = 0$.

<u>case 3</u>. $h > 0$ and $\ell > 0$. Using (11) show that

$$D = \tfrac{4}{3}\,\beta h w_0^2\, F(p,\nu), \tag{15}$$

where $\nu = \nu(h) = \frac{3\beta}{h}$ and $F(p,\nu) = r(p)\big(r(p) - 2\nu\big) + \nu p - 1$. Consider the mapping

$$\Psi : \mathcal{R} \cap (\mathbf{R}_>)^2 \to \mathcal{S} = \Big\{(p,\nu) \in \mathbf{R}^2 \,\big|\, \nu \in (\tfrac{1}{2},\infty)\ \&\ p \in \Big(-2, \frac{3\nu^2-1}{\nu^3}\Big)\Big\} :$$
$$(h,\ell) \to \big(p(h,\ell), \nu(h)\big).$$

Show that Ψ is a diffeomorphism which maps the half-line $\Psi(\{(h,0)\,|\, h > 0\})$ bijectively onto the curve $p = \big(3\nu^2-1\big)/\nu^3$, $\nu \in (\frac{1}{2},\infty)$. To show that $F(p,\nu) < 0$

we argue as follows. For every fixed $p_0 \in (-2,2)$, the function $F_{p_0}(\nu) = F(p_0, \nu)$, where $\nu \in \mathcal{S} \cap \{p = p_0\}$ is strictly decreasing. To see this differentiate the definition of F and obtain

$$\frac{\partial F_p}{\partial \nu} = -2r(p) + p.$$

At $p = 2$, we know that $r(2) = 1$. Hence $\dfrac{\partial F_2}{\partial \nu} = 0$. However,

$$\frac{\partial}{\partial p}\left(\frac{\partial F_p}{\partial \nu}\right) = -2r'(p) + 1 > 0,$$

since $r'(p) < 0$ for every $p \in (-2,2)$. Hence $\dfrac{\partial F_{p_0}}{\partial \nu} < 0$ for every $p_0 \in (-2,2)$. For any $(p_0, \nu_0) \in \mathcal{S}$ there is a $\nu_1 \in (\frac{1}{2}, \nu_0)$ such that $p_0 = \left(3\nu_1^2 - 1\right)/\nu_1^3$. Since F_{p_0} is strictly decreasing, $F(p_0, \nu_0) < F(p_0, \nu_1)$. Using (15) and the result of case 2 that $D < 0$ when $\ell = 0$, we see that $F(p, \nu_1) < 0$. Therefore $F(p, \nu) < 0$.

Chapter V

The Lagrange top

1 The basic model

Physically, the Lagrange top is a symmetric rigid body spinning about its figure axis whose base point is fixed. A constant vertical gravitational force acts on the center of mass of the top, which lies on its axis.

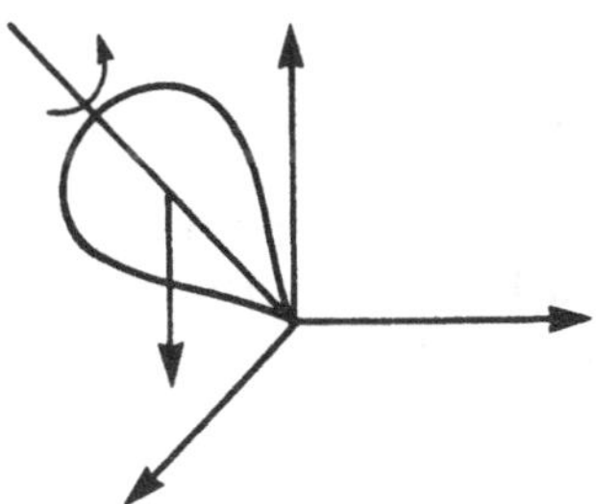

Figure 1.1. The Lagrange top.

Mathematically, the top is a Hamiltonian system on the phase space $(T\mathrm{SO}(3), \Omega_\rho)$. The symplectic form Ω_ρ on $T\mathrm{SO}(3)$ is the pull back of the canonical symplectic form Ω on $T^*\mathrm{SO}(3)$ by the map $\rho^\sharp$ associated to a left invariant metric ρ on SO(3). The metric ρ is uniquely determined by its value at the identity element e, which is given by

$$\begin{aligned} \rho(e) : T_e\,\mathrm{SO}(3) \times T_e\,\mathrm{SO}(3) = \mathrm{so}(3) \times \mathrm{so}(3) \longrightarrow \mathbf{R} : \\ (X, Y) \to k(I(X), Y) = I_1 k(X, Y) + (I_3 - I_1) k(X, E_3)\, k(Y, E_3). \end{aligned} \tag{1}$$

Here k is the Killing metric on so(3) (see chapter 3 section 1) and the moment of inertia tensor $I : \mathrm{so}(3) \to \mathrm{so}(3)$ is a k-symmetric linear map. The matrix of I with respect to the standard k-orthonormal basis $\{E_1, E_2, E_3\}$ is $\mathrm{diag}(I_1, I_1, I_3)$, where I_i are the *principal moments of inertia* of the top. Because ρ is a metric, I is invertible. We will assume that

$$0 < I_1 < I_3. \tag{2}$$

To make sure that I is the moment of inertia tensor of a physically realizable top, we require that

$$I_3 \leq 2 I_1, \tag{3}$$

(see chapter 3 exercise 5).

The Hamiltonian

$$\mathcal{H} : TSO(3) \to \mathbf{R} : X_A \to \mathcal{K}(X_A) + \tau \circ \mathcal{V}(X_A), \tag{4}$$

which determines the dynamics of the Lagrange top, is the sum of kinetic and potential energy. The kinetic energy

$$\mathcal{K} : TSO(3) \to \mathbf{R} : X_A \to \tfrac{1}{2}\, \rho(A)(X_A, X_A)$$

is one half the ρ-length squared of a tangent vector to SO(3). The potential energy

$$\mathcal{V} : \mathrm{SO}(3) \to \mathbf{R} : A \to \chi k(\mathrm{Ad}_A E_3, E_3)$$

measures the height of the center of mass of the top. To be able to define the Hamiltonian, we must pull back $\mathcal{V}$ by the bundle projection

$$\tau : TSO(3) \to \mathbf{R} : X_A \to A$$

so that it is a function on $TSO(3)$.

2 Liouville integrability

In this section we show that the Lagrange top is Liouville integrable, (see appendix D). To do this we need two additional integrals of motion other than the Hamiltonian. These two extra integrals arise from two rotational symmetries of the top, namely, one about the *vertical axis* e_3, which is fixed in space, and the other about the *figure axis* fixed in the top.

We now investigate these symmetries more carefully. Let

$$S^1 = \left\{ B \in \mathrm{SO}(3) \,\middle|\, \mathrm{Ad}_B E_3 = E_3 \right\}.$$

Then S^1 acts on the *left* on $TSO(3)$ by

$$\Phi^\ell : S^1 \times TSO(3) \to TSO(3) : \left(B, X_A\right) \to T_A L_B X_A = X_{BA}. \tag{5}$$

▷ Physically this action corresponds to rotating the top about the vertical axis.

(2.1) **Proof**: To see this let $A(t) \in \mathrm{SO}(3)$ be the configuration of the top at time t with respect to the *fixed frame* $\{e_1, e_2, e_3\}$ in $\mathbf{R}^3$. The figure axis of the top at time t is $A(t)e_3$. Acting on the top on the left by $B \in S^1$, we obtain the new configuration $BA(t)$ Applying the mapping $i : \mathrm{so}(3) \to \mathbf{R}^3$ (see chapter 3 ((1.2))) to the condition $E_3 = \mathrm{Ad}_B E_3$ defining S^1 gives $e_3 = Be_3$. Since $(BA(t)e_3, e_3) = (A(t)e_3, B^{-1}e_3) = (A(t)e_3, e_3)$, the angle between the figure axis and e_3 remains invariant under the left S^1 action. Thus the left S^1 action corresponds to rotating the top about the vertical axis e_3. □

▷ The action Φ^ℓ is Hamiltonian with momentum mapping

$$\mathcal{J}_\ell : TSO(3) \to \mathbf{R} : X_A \to \rho(A)(T_e R_A E_3, X_A). \tag{6}$$

Physically, $\mathcal{J}_\ell$ is the angular momentum of the top about the e_3-axis.

(2.2) **Proof**: The formula for the momentum mapping follows from the momentum lemma, (see appendix B ((3.6))). We provide some details. Let θ be the canonical 1-form on $T^* \mathrm{SO}(3)$. First we show that the 1-form $\theta_\rho = (\rho^\sharp)^*\theta$ is invariant under the action Φ^ℓ. For $W_{X_A} \in T_{X_A}(T\mathrm{SO}(3))$ and $B \in S^1$ we have

$$\begin{aligned}
\big((\Phi^\ell_B)^*\theta_\rho\big)(X_A)W_{X_A} &= \rho(BA)\big(T\tau(T_{X_A}\Phi^\ell_B)W_{X_A}, X_{AB}\big) \\
&= \rho(BA)\big(T_A L_B(T\tau W_{X_A}), T_A L_B X_A\big), \qquad \text{since } \tau\circ\Phi^\ell_B = L_B\circ\tau \\
&= \rho(A)(T\tau W_{X_A}, X_A), \qquad \text{by left invariance of } \rho \\
&= \theta_\rho(X_A)W_{X_A}.
\end{aligned}$$

Second, note that the Lie algebra $T_e S^1 \subseteq \mathrm{so}(3)$ of S^1 is spanned by E_3. Therefore the infinitesimal generator of the action

$$S^1 \times \mathrm{SO}(3) \to \mathrm{SO}(3) : (B, A) \to BA$$

is the vector field

$$X_{\mathrm{SO}(3)}(A) = \frac{d}{dt}\bigg|_{t=0} (\exp tE_3)A = \frac{d}{dt}\bigg|_{t=0} R_A \exp tE_3 = T_e R_A E_3$$

and the infinitesimal generator of Φ^ℓ is

$$X_{T\,\mathrm{SO}(3)}(X_A) = \frac{d}{dt}\bigg|_{t=0} \Phi^\ell_{\exp tE_3}(X_A).$$

Since $L_{\exp tE_3}\circ\tau = \tau\circ\Phi^\ell_{\exp tE_3}$, it follows that $T\tau X_{T\,\mathrm{SO}(3)}(X_A) = X_{\mathrm{SO}(3)}(A)$. Therefore

$$\begin{aligned}
\mathcal{J}_\ell(X_A) &= \theta_\rho(X_A)(X_{T\,\mathrm{SO}(3)}(X_A)) \\
&= \rho(A)(T\tau X_{T\,\mathrm{SO}(3)}(X_A), X_A) = \rho(A)(X_{\mathrm{SO}(3)}(A), X_A). \qquad \square
\end{aligned}$$

▷ To show that S^1 is a (left) symmetry of the Lagrange top we need only verify that the action Φ^ℓ preserves the Hamiltonian $\mathcal{H}$.

(2.3) **Proof**: For $B \in S^1$,

$$\begin{aligned}
(\Phi^\ell_B)^*\mathcal{H}(X_A) &= \mathcal{H}(X_{BA}) = \tfrac{1}{2}\rho(BA)(X_{BA}, X_{BA}) + \chi k(\mathrm{Ad}_{BA} E_3, E_3) \\
&= \tfrac{1}{2}\rho(A)(X_A, X_A) + \chi k(\mathrm{Ad}_A E_3, \mathrm{Ad}_{B^{-1}} E_3) \\
&= \mathcal{H}(X_A). \qquad \square
\end{aligned}$$

Therefore $\mathcal{J}_\ell$ is constant on the integral curves of the Hamiltonian vector field $X_{\mathcal{H}}$, that is, $\mathcal{J}_\ell$ is an integral of the Lagrange top.

The group $S^1 = \{B \in \mathrm{SO}(3) | \, \mathrm{Ad}_B E_3 = E_3\}$ also acts on the *right* on $T\mathrm{SO}(3)$ by

$$\Phi^r : T\mathrm{SO}(3) \times S^1 \to T\mathrm{SO}(3) : (X_A, B) \to X_{AB} = T_A R_B X_A. \tag{7}$$

▷ Physically, this action corresponds to a rotation about the figure axis of the top.

(2.4) **Proof**: To see this, let $A(t) \in \mathrm{SO}(3)$ be the configuration of the top at time t with respect to the fixed frame $\{e_1, e_2, e_3\}$ in $\mathbf{R}^3$. Acting on the right by $B \in S^1$ gives the new configuration $A(t)B$. Since $Be_3 = e_3$, we find that $A(t)Be_3 = A(t)e_3$. Thus the figure axis of the top in the new configuration is the same as in the original configuration. Hence the right S^1 action corresponds to rotating the top about its figure axis. □

▷ The right S^1 action Φ^r is Hamiltonian with momentum mapping

$$\mathcal{J}_r : T\mathrm{SO}(3) \to \mathbf{R} : X_A \to \rho(A)(T_e L_A E_3, X_A). \tag{8}$$

Physically, $\mathcal{J}_r$ is the angular momentum of the top about its figure axis.

(2.5) **Proof**: The argument follows along the same lines as the proof of ((2.2)). We only show that the 1-form θ_ρ is invariant under Φ^r. First we verify that the metric ρ (1) is *invariant* under the right S^1 action Φ^r, that is,

$$I \circ \mathrm{Ad}_B = \mathrm{Ad}_B \circ I, \tag{9}$$

for every $B \in S^1$. By left invariance of ρ

$$\rho(AB)(T_A R_B X_A, T_A R_B Y_A) = \rho(e)(\mathrm{Ad}_{B^{-1}} X, \mathrm{Ad}_{B^{-1}} Y).$$

From the definition of $\rho(e)$ and the fact that $B \in S^1$, it follows that

$$\begin{aligned} \rho(e)(\mathrm{Ad}_{B^{-1}} X, \mathrm{Ad}_{B^{-1}} Y) &= \\ &= I_1 k(\mathrm{Ad}_{B^{-1}} X, \mathrm{Ad}_{B^{-1}} Y) + (I_3 - I_1) k(\mathrm{Ad}_{B^{-1}} X, E_3)(\mathrm{Ad}_{B^{-1}} Y, E_3) \\ &= \rho(e)(X, Y), \qquad \text{since } k \text{ is Ad-invariant} \\ &= \rho(A)(X_A, Y_A). \end{aligned}$$

Thus ρ is Φ^r-invariant. The 1-form θ_ρ is Φ^r-invariant, because

$$\begin{aligned} (\Phi^r)^* \theta_\rho(X_A) W_{X_A} &= \rho(AB)\big(T\tau(T\Phi^r_B W_{X_A}), X_{AB}\big) \\ &= \rho(AB)\big(T_A R_B (T\tau W_{X_A}), T_A R_B X_A\big) \\ &= \theta_\rho(X_A)(W_{X_A}) \end{aligned}$$ □

▷ To show that S^1 is a (right) symmetry of the Lagrange top it suffices to verify that the Hamiltonian $\mathcal{H}$ is invariant under Φ^r.

(2.6) **Proof**: For $B \in S^1$ we have

$$\begin{aligned} (\Phi^r_B)^* \mathcal{H}(X_A) &= \mathcal{H}(X_{AB}) = \tfrac{1}{2}\rho(AB)(X_{AB}, X_{AB}) + \chi k(\mathrm{Ad}_{AB} E_3, E_3) \\ &= \tfrac{1}{2}\rho(A)(X_A, X_A) + \chi k(\mathrm{Ad}_A \mathrm{Ad}_B E_3, E_3), \\ &\qquad \text{since } \rho \text{ is } \Phi^r\text{-invariant} \\ &= \mathcal{H}(X_A), \qquad \text{since } B \in S^1. \end{aligned}$$ □

Therefore $\mathcal{J}_r$ is an integral of the vector field $X_{\mathcal{H}}$.

▷ To complete the argument that the Lagrange top is Liouville integrable, we need only show that the Poisson bracket of any two of the integrals $\{\mathcal{H}, \mathcal{J}_\ell, \mathcal{J}_r\}$ vanishes identically. Since $\mathcal{J}_\ell$ and $\mathcal{J}_r$ are integrals of $X_{\mathcal{H}}$, their Poisson bracket with $\mathcal{H}$ vanishes identically. We now show that the Poisson bracket of $\mathcal{J}_\ell$ and $\mathcal{J}_r$ vanishes identically.

(2.7) **Proof**: To see this it suffices to show that $\mathcal{J}_\ell$ is constant on the orbits of Φ^r. For $B \in S^1$ we have

$$\begin{aligned}
((\Phi^r_B)^*\mathcal{J}_\ell)(X_A) &= \mathcal{J}_\ell(X_{AB}) = \rho(AB)(T_eR_{AB}E_3, X_{AB}) \\
&= \rho(AB)(T_AR_B\, T_eR_AE_3, T_AR_BX_A) \\
&= \rho(A)(T_eR_AE_3, X_A), \qquad \text{since } \rho \text{ is } \Phi^r\text{-invariant} \\
&= \mathcal{J}_\ell(X_A). \qquad \square
\end{aligned}$$

3 Reduction of right S^1 action

In this section we remove the symmetry about the figure axis of the Lagrange top using the technique of regular reduction, (see appendix B ((4.1))). After reduction we obtain a Hamiltonian system with two degrees of freedom which is equivalent to the magnetic spherical pendulum.

3.1 Reduction to Euler-Poisson equations

In this subsection we reduce the symmetry given by the *right* S^1 action and obtain a Hamiltonian system for which Hamilton's equations are the Euler-Poisson equations.

To find an initial model for the reduced phase space, we follow the proof of the regular reduction theorem. This constructs the orbit space of the right S^1 action Φ^r (7) on the ▷ a-level set of the right angular momentum mapping $\mathcal{J}_r$. We start by checking that $\mathcal{J}_r^{-1}(a)$ is a smooth manifold for every $a \in \mathbf{R}$.

(3.1) **Proof**: Pulling back the right S^1 Hamiltonian action Φ^r by the left trivialization

$$\mathcal{L} : \mathrm{SO}(3) \times \mathrm{so}(3) \to T\mathrm{SO}(3) : (A, X) \to T_eL_AX = X_A \tag{10}$$

gives the right S^1 action

$$\begin{aligned}
\varphi^r : \big(\, \mathrm{SO}(3) \times \mathrm{so}(3)\big) \times S^1 &\to \mathrm{SO}(3) \times \mathrm{so}(3) : \\
\big((A, X), B\,\big) &\to (AB, \mathrm{Ad}_{B^{-1}} X),
\end{aligned} \tag{11}$$

because

$$\begin{aligned}
(\mathrm{Ad}_{B^{-1}} X)_{AB} &= T_eL_{AB}(T_BL_{B^{-1}}\, T_eR_BX) = T_BL_A\, T_eR_BX \\
&= T_eL_{AB}X = X_{AB}.
\end{aligned}$$

The action φ^r has a momentum mapping

$$J_r = \mathcal{L}^* \mathcal{J}_r : \mathrm{SO}(3) \times \mathrm{so}(3) \to \mathbf{R} : (A, X) \to \mathcal{J}_r(X_A)E_3 = \rho(e)(X, E_3). \tag{12}$$

Because the derivative of J_r at (A, X)

$$DJ_r(A,X)(V_A, Y) = \frac{d}{dt}\bigg|_{t=0} J_r(A \exp tV, X + tY) = \rho(e)(E_3, Y)$$

is a surjective linear mapping from $T_{(A,X)}\big(\mathrm{SO}(3) \times \mathrm{so}(3)\big)$ to $\mathbf{R}$, the level set $J_r^{-1}(a)$ is a smooth manifold. □

The level set $J_r^{-1}(a)$ is invariant under the action φ^r, because for every $B \in S^1$

$$\begin{aligned} J_r(AB, \mathrm{Ad}_{B^{-1}} X) &= k\big(I(\mathrm{Ad}_{B^{-1}} X), E_3\big) = k\big(\mathrm{Ad}_{B^{-1}}(I(X)), E_3\big), \quad \text{using (9)} \\ &= J_r(A, X). \end{aligned}$$

Therefore φ^r restricts to an S^1 action $\varphi^r|\big(S^1 \times J_r^{-1}(a)\big)$ on $J_r^{-1}(a)$. This induced action is a free, because if $(A, X) = (AB, \mathrm{Ad}_{B^{-1}} X)$, then $A = AB$, that is, B is the identity element e. Hence the space $J_r^{-1}(a)/S^1$ of S^1 orbits of $\varphi^r|(S^1 \times J_r^{-1}(a))$ is a smooth manifold called the reduced phase space, (see appendix B ((2.9))). This description of the reduced phase space is somewhat abstract. We now provide a more concrete model:

$$\mathcal{P}^a = \Big\{(Z, W) \in \mathrm{so}(3) \times \mathrm{so}(3) \Big| k(Z, Z) = 1 \ \&\ k(Z, W) = a\Big\}. \tag{13}$$

Claim: $\mathcal{P}^a$ is diffeomorphic to $J_r^{-1}(a)/S^1$.

(3.2) **Proof**: Consider the mapping

$$\begin{aligned} \pi^a : J_r^{-1}(a) \subseteq \mathrm{SO}(3) \times \mathrm{so}(3) \to \mathcal{P}^a \subseteq \mathrm{so}(3) \times \mathrm{so}(3) : \\ (A, X) \to \big(\mathrm{Ad}_A E_3, \mathrm{Ad}_A I(X)\big) = (Z, W). \end{aligned} \tag{14}$$

π^a is surjective. To see this suppose that $(Z, W) \in \mathcal{P}^a$. Then there is an $A \in \mathrm{SO}(3)$ such that $\mathrm{Ad}_A E_3 = Z$. Let $X = I^{-1}(\mathrm{Ad}_{A^{-1}} W)$. Using (14) we see that $\pi^a(A, X) = (Z, W)$. We are done once we can show that $(A, X) \in J_r^{-1}(a)$. This follows because

$$J_r(A, X) = k(I(X), E_3) = k(\mathrm{Ad}_{A^{-1}} W, \mathrm{Ad}_{A^{-1}} Z) = k(Z, W) = a,$$

since $(Z, W) \in \mathcal{P}^a$. Now π^a maps an orbit of $\varphi^r|\big(S^1 \times J_r^{-1}(a)\big)$ onto a point of $\mathcal{P}^a$ because for every $B \in S^1$,

$$\begin{aligned} \pi^a(AB, \mathrm{Ad}_{B^{-1}} X) &= \big(\mathrm{Ad}_{AB} E_3, \mathrm{Ad}_{AB} I(\mathrm{Ad}_{B^{-1}} X)\big) \\ &= \big(\mathrm{Ad}_A(\mathrm{Ad}_B E_3), \mathrm{Ad}_A \mathrm{Ad}_B I(\mathrm{Ad}_{B^{-1}} X)\big) \\ &= \big(\mathrm{Ad}_A E_3, \mathrm{Ad}_A I(X)\big), \quad \text{since } B \in S^1 \text{ and (9)} \\ &= (Z, W). \end{aligned}$$

Hence $J_r^{-1}(a)/S^1$ is a smooth manifold and π^a induces a smooth mapping $\sigma^a : J_r^{-1}(a)/S^1 \to \mathcal{P}^a$ such that the diagram 3.1 commutes, (see appendix B section 2). In diagram 3.1 the mapping $\rho^a : J_r^{-1}(a) \rightarrow J_r^{-1}(a)/S^1$ is called the orbit map, because it assigns to each point in $J_r^{-1}(a)$ the orbit of $\varphi^r | \big(S^1 \times J_r^{-1}(a)\big)$ through the given point.

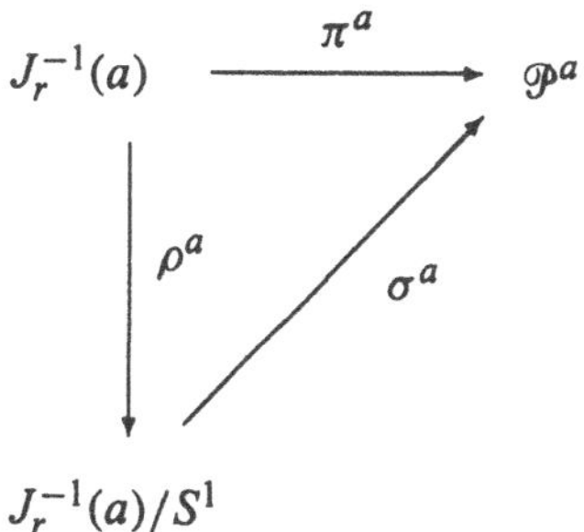

Diagram 3.1

We now show that σ^a is a diffeomorphism. Because π^a is surjective, it follows that σ^a is surjective. To show that σ^a is injective, it suffices to verify that the fiber $(\pi^a)^{-1}(Z, W)$ is a single orbit of φ^r. To see this suppose that (A, X) and (C, Y) lie in the fiber $(\pi^a)^{-1}(Z, W)$. Then $\mathrm{Ad}_A E_3 = \mathrm{Ad}_C E_3$, that is, $B = A^{-1}C \in S^1$. Since $\mathrm{Ad}_A I(X) = \mathrm{Ad}_C I(Y)$, it follows that $\mathrm{Ad}_A I(X) = \mathrm{Ad}_A \mathrm{Ad}_B I(Y) = \mathrm{Ad}_A I(\mathrm{Ad}_B Y)$. Consequently, $Y = \mathrm{Ad}_{B^{-1}} X$ because Ad_A and I are invertible. Therefore (C, Y) lies in the same $\varphi^r | \big(S^1 \times J_r^{-1}(a)\big)$ orbit as (A, X). Next we show that the inverse of σ^a is smooth. Because π^a is a proper submersion, the bundle $\pi^a : J_r^{-1}(a) \rightarrow \mathcal{P}^a$ is locally trivial, (see appendix C ((2.1))). Thus for a suitable open subset $\mathcal{U} \subseteq \mathcal{P}^a$ about (Z, W), the bundle π^a restricted to $(\pi^a)^{-1}(\mathcal{U})$ has a smooth cross section $\tau : \mathcal{U} \rightarrow (\pi^a)^{-1}(\mathcal{U})$. Therefore on $\mathcal{U}$, we have $\sigma^a \circ (\rho^a \circ \tau) = \pi^a \circ \tau = id_{\mathcal{U}}$. To verify that $(\rho^a \circ \tau) \circ \sigma^a = id_{(\sigma^a)^{-1}(\mathcal{U})}$, let $r \in (\sigma^a)^{-1}(\mathcal{U})$ and set $s = (\rho^a \circ \tau) \circ \sigma^a(r)$. Then $\sigma^a(s) = (\sigma^a \circ \rho^a \circ \tau)\big(\sigma^a(r)\big) = \sigma^a(r)$. But σ^a is injective, so $s = r$ and we are done. Therefore $(\sigma^a)^{-1} = \rho^a \circ \tau$ is smooth. Consequently σ^a is a diffeomorphism. □

Claim: The 2-form Ω^a on $\mathcal{P}^a$ defined by

$$\Omega_\rho | J_r^{-1}(a) = (\pi^a)^* \Omega^a \tag{15}$$

is symplectic.

(3.3) **Proof**: We find an explicit expression for the 2-form Ω^a. We begin by calculating the tangent of the mapping π^a (14). For $(U_A, T) \in T_{(A,X)}\big(\mathrm{SO}(3) \times \mathrm{so}(3)\big)$,

$$\begin{aligned}
T_{(A,X)}\pi^a(U_A, T) &= \frac{d}{dt}\Big|_{t=0} \pi^a(A \exp tU, X + tT) \\
&= \Big(\frac{d}{dt}\Big|_{t=0} \mathrm{Ad}_{A \exp tU} E_3, \frac{d}{dt}\Big|_{t=0} \mathrm{Ad}_{A \exp tU}(I(X) + tI(T))\Big) \\
&= \big(\mathrm{Ad}_A(\mathrm{ad}_U E_3), \mathrm{Ad}_A(\mathrm{ad}_U I(X) + I(T))\big) \\
&= \big([\mathrm{Ad}_A U, \mathrm{Ad}_A E_3], [\mathrm{Ad}_A U, \mathrm{Ad}_A I(X)] + \mathrm{Ad}_A I(T)\big) \\
&= \xi(A, X, U, T).
\end{aligned}$$

We compute the 2-form Ω^a as follows:

$$\begin{aligned}
\Omega^a(\mathrm{Ad}_A E_3, \mathrm{Ad}_A I(X))\big(\xi(A,X,U,T), \xi(A,X,R,S)\big) &= \\
&= \Omega^a(\pi^a(A,X))\big(T_{(A,X)}\pi^a(U_A,T), T_{(A,X)}\pi^a(R_A,S)\big) \\
&= \Omega_\rho(A,X)\big((U_A,T),(R_A,S)\big), \qquad \text{by (15)} \\
&= -k(I(T),R) + k(I(S),U) + k(I(X),[U,R]),
\end{aligned}$$

using (4) in section 2 of appendix A and $\rho(e)^b$ to identify so(3)* with so(3).

$$\begin{aligned}
&= -k(\mathrm{Ad}_A I(T), \mathrm{Ad}_A R) + k(\mathrm{Ad}_A I(S), \mathrm{Ad}_A I(U)) \\
&\qquad + k(\mathrm{Ad}_A I(X), [\mathrm{Ad}_A U, \mathrm{Ad}_A R]).
\end{aligned}$$

Ω^a is nondegenerate because

$$\pi^a \Omega^a = \ker \Omega_\rho | J_r^{-1}(a) = \ker T\pi^a.$$

Moreover, Ω^a is closed, because using (15) and the fact that Ω_ρ is closed, we find that

$$0 = (d\Omega_\rho | J_r^{-1}(a) = d(\Omega_\rho | J_r^{-1}(a)) = (\pi^a)^* d\Omega^a.$$

Since π^a is a submersion, it follows that $d\Omega^a = 0$. □

By the regular reduction theorem, the orbit space $J_a^{-1}(r)/S^1$ has a symplectic form $\widetilde{\Omega}^a$ ▷ defined by $\Omega_\rho | J_r^{-1}(a) = (\rho^a)^* \widetilde{\Omega}^a$. From ((3.3)) it follows that

$$\sigma^a : (J_a^{-1}(r)/S^1, \widetilde{\Omega}^a) \to (\mathcal{P}^a, \Omega^a)$$

is a symplectic diffeomorphism. In other words, $(\mathcal{P}^a, \Omega^a)$ is a model for the reduced space $(J_a^{-1}(r)/S^1, \widetilde{\Omega}^a)$.

(3.4) **Proof**: To see this let $\tau : \mathcal{U} \to (\pi^a)^{-1}(\mathcal{U})$ be a local cross section for the trivial bundle $\pi^a | (\pi^a)^{-1}(\mathcal{U})$. From the definition of the symplectic forms $\widetilde{\Omega}^a$ and Ω^a, we have $(\rho^a)^* \widetilde{\Omega}^a = (\pi^a)^* \Omega^a$. Therefore on $\mathcal{U}$

$$\Omega^a = (\pi^a \circ \tau)^* \Omega^a = (\rho^a \circ \tau)^* \widetilde{\Omega}^a = ((\sigma^a)^{-1})^* \widetilde{\Omega}^a. \qquad \square$$

We are now ready to compute the reduced Hamiltonian $\mathcal{H}^a$ on the reduced space $(\mathcal{P}^a, \Omega^a)$. First we treat the kinetic energy. From the definition of the left trivialization $\mathcal{L}$ (10) it follows that the pull back of the kinetic energy $\mathcal{K}$ by $\mathcal{L}$ is

$$\begin{aligned}
&K : \mathrm{SO}(3) \times \mathrm{so}(3) \to \mathbf{R} : \\
&(A,X) \to \tfrac{1}{2} k(I(X),X) = \tfrac{1}{2} I_1 k(X,X) + \tfrac{1}{2}(I_3 - I_1)(k(X,E_3))^2.
\end{aligned} \tag{16}$$

From (16) we find that

$$\tfrac{1}{2} k(I^{-1}(X),X) = \tfrac{1}{2} I_1^{-1} k(X,X) - \tfrac{1}{2}(I_1^{-1} - I_3^{-1})(k(X,E_3))^2. \tag{17}$$

Replacing X with $\mathrm{Ad}_{A^{-1}} W$ in (17) gives

$$k\big(I^{-1}(\mathrm{Ad}_{A^{-1}} W), \mathrm{Ad}_{A^{-1}} W\big) = \tfrac{1}{2} I_1^{-1} k(W,W) - \tfrac{1}{2}(I_1^{-1} - I_3^{-1})\big(k(W, \mathrm{Ad}_A E_3)\big)^2,$$

since k is Ad-invariant. By definition of π^a (14), the reduced kinetic energy is

$$\begin{aligned} \mathcal{K}^a : \mathcal{P}^a \subseteq \mathrm{so}(3) \times \mathrm{so}(3) \to \mathbf{R} : \\ (Z, W) \to \tfrac{1}{2} I_1^{-1} k(W, W) - \tfrac{1}{2} (I_1^{-1} - I_3^{-1})(k(W, Z))^2. \end{aligned} \tag{18}$$

Pulling back the potential energy $\mathcal{V}$ by the left trivialization $\mathcal{L}$, give the reduced potential energy

$$\mathcal{V}^a : \mathcal{P}^a \subseteq \mathrm{so}(3) \times \mathrm{so}(3) \to \mathbf{R} : (Z, W) \to \chi\,(Z, e_3) = \chi\, Z_3. \tag{19}$$

Thus the reduced Hamiltonian $\mathcal{H}^a$ is

$$\mathcal{H}^a : \mathcal{P}^a \subseteq \mathrm{so}(3) \times \mathrm{so}(3) \to \mathbf{R} : (Z, W) \to \tfrac{1}{2} I_1^{-1} k(W, W) + \chi\, Z_3. \tag{20}$$

In (20) we have omitted the additive constant $-\frac{1}{2}(I_1^{-1} - I_3^{-1})a^2$. In other words, if h is the value of the Hamiltonian $\mathcal{H}$ and h^a is the value of the reduced Hamiltonian $\mathcal{H}^a$, then

$$h^a = h + \tfrac{1}{2}(I_1^{-1} - I_3^{-1})a^2. \tag{21}$$

In order to compute the Hamiltonian vector field of the reduced Hamiltonian $\mathcal{H}^a$ on $(\mathcal{P}^a, \Omega^a)$, it is necessary to use another model for the reduced Hamiltonian system $(\mathcal{H}^a, \mathcal{P}^a, \Omega^a)$. The new model (H^a, P^a, ω^a) is obtained by pulling the old model back by the map $i : \mathrm{so}(3) \to \mathbf{R}^3$, (see chapter 3 ((1.2))). The reduced phase space P^a is

$$\left\{ (z, w) \in \mathbf{R}^3 \times \mathbf{R}^3 \middle| (z, z) = 1 \ \ \& \ \ (z, w) = a \right\}, \tag{22}$$

where $(\ ,\ )$ is the Euclidean inner product on $\mathbf{R}^3$. Here $z = i(Z) = i(\mathrm{Ad}_A E_3)$ and $w = i(W) = i(\mathrm{Ad}_A I(X))$. The symplectic form $\omega^a = i^*\Omega^a$ is

$$\omega^a(z, w)\big((u \times z, u \times w + t), (r \times z, r \times w + s)\big) = -(t, r) + (u, s) + (w, u \times r), \tag{23}$$

where $u = i(\mathrm{Ad}_A U) = Ai(U)$, $t = i(\mathrm{Ad}_A I(T)) = Ai(I(T))$, $r = i(\mathrm{Ad}_A R) = Ai(R)$ and $s = i(\mathrm{Ad}_A I(S)) = Ai(I(S))$. Here we have used the fact that $i([\mathrm{Ad}_A U,\ \mathrm{Ad}_A E_3])$ $= i(\mathrm{Ad}_A U) \times i(\mathrm{Ad}_A E_3) = u \times z$ and similarly $i([\mathrm{Ad}_A U, \mathrm{Ad}_A I(X)]) = u \times w$. Note that $(z, t) = 0$ and $(z, s) = 0$, since $(u \times z, u \times w + t)$ and $(r \times z, r \times w + s)$ lie in $T_{(z,w)}P^a$. From (18) we see that the reduced kinetic energy is

$$\begin{aligned} K^a = i^*\mathcal{K}^a : P^a \subseteq \mathbf{R}^3 \times \mathbf{R}^3 \to \mathbf{R} : \\ (z, w) \to \tfrac{1}{2} I_1^{-1}(w, w) - \tfrac{1}{2}(I_1^{-1} - I_3^{-1})(w, z)^2, \end{aligned} \tag{24}$$

while from (19) it follows that the reduced potential energy is

$$V^a = i^*\mathcal{V}^a : P^a \subseteq \mathbf{R}^3 \times \mathbf{R}^3 \to \mathbf{R} : (z, w) \to \chi\,(z, e_3) = \chi\, z_3. \tag{25}$$

Therefore, up to an additive constant the reduced Hamiltonian is

$$H^a = i^*\mathcal{H}^a : P^a \subseteq \mathbf{R}^3 \times \mathbf{R}^3 \to \mathbf{R} : (z, w) \to \tfrac{1}{2} I_1{}^{-1}(w, w) + \chi\, z_3. \tag{26}$$

▷ Now we are in position to show that for any smooth Hamiltonian $H : P^a \to \mathbf{R}$ the integral curves of the Hamiltonian vector field X_H on (P^a, ω^a) satisfy

$$\begin{cases} \dot{z} = \dfrac{\partial H}{\partial w} \times z \\ \dot{w} = \dfrac{\partial H}{\partial z} \times z + \dfrac{\partial H}{\partial w} \times w. \end{cases} \tag{27}$$

(3.5) **Proof**: Write

$$X_H(z, w) = (X_1 \times z, X_1 \times w + X_2),$$

where $(z, X_2) = 0$ because $X_H(z, w) \in T_{(z,w)}P^a$. From the definition of Hamiltonian vector field we find that

$$\begin{aligned} dH(z, w)\big((u \times z, u \times w + v)\big) &= \\ &= \omega^a(z, w)\big((X_1 \times z, X_1 \times w + X_2), (u \times z, u \times w + v)\big), \end{aligned} \tag{28}$$

for every $(u \times z, u \times w + v) \in T_{(z,w)}P^a$. Using $dH = (\frac{\partial H}{\partial z}, \frac{\partial H}{\partial w})$ and the definition (23) of the symplectic form ω^a, we see that (28) is the same as

$$(\frac{\partial H}{\partial z}, u \times z) + (\frac{\partial H}{\partial w}, u \times w + v) = -(X_2, u) + (X_1, v) + (w, X_1 \times u). \tag{29}$$

Set $u = 0$. Then (29) becomes $(\frac{\partial H}{\partial w}, v) = (X_1, v)$ for every $v \in \mathbf{R}^3$ such that $(z, v) = 0$. This last condition must hold in order that $(u \times z, u \times w + v) \in T_{(z,w)}P^a$. Therefore $X_1 = \frac{\partial H}{\partial w} + \lambda_0 z$ for some $\lambda_0 \in \mathbf{R}$. Set $v = 0$. Then (29) becomes

$$-(\frac{\partial H}{\partial z} \times z, u) - (\frac{\partial H}{\partial w} \times w, u) = -(X_2, u) + (w \times (\frac{\partial H}{\partial w} + \lambda_0 z), u)$$

for every $u \in \mathbf{R}^3$. Hence $X_2 = \frac{\partial H}{\partial z} \times z + \lambda_0(w \times z)$. Note that $(X_2, z) = 0$ holds automatically. Consequently, the integral curves X_H satisfy

$$\begin{aligned} \dot{z} &= (\frac{\partial H}{\partial w} + \lambda_0 z) \times z = \frac{\partial H}{\partial w} \times z \\ \dot{w} &= \frac{\partial H}{\partial w} \times w + \lambda_0(z \times w) + \frac{\partial H}{\partial z} \times z + \lambda_0(w \times z) \\ &= \frac{\partial H}{\partial z} \times z + \frac{\partial H}{\partial w} \times w. \quad \square \end{aligned} \tag{30}$$

Actually (30) defines a system of differential equations on $\mathbf{R}^3 \times \mathbf{R}^3$, called the *Euler-Poisson equations* of H. Note that the reduced space P^a is an invariant manifold of the Euler-Poisson equations.

We now show that if we put a nonstandard Poisson structure on $\mathbf{R}^3 \times \mathbf{R}^3$, then the Euler-Poisson equations are in Hamiltonian form. Explicitly, define a Poisson bracket $\{\, ,\, \}$ on $C^\infty(\mathbf{R}^3 \times \mathbf{R}^3)$ whose structure matrix W is given in table 3.1.

$\{A,B\}$	z_1	z_2	z_3	w_1	w_2	w_3
z_1	0	0	0	0	z_3	$-z_2$
z_2	0	0	0	$-z_3$	0	z_1
z_3	0	0	0	z_2	$-z_1$	0
w_1	0	z_3	$-z_2$	0	w_3	$-w_2$
w_2	$-z_3$	0	z_1	$-w_3$	0	w_1
w_3	z_2	$-z_1$	0	w_2	$-w_1$	0

(Rows: A; columns: B)

Table 3.1. Structure matrix for $C^\infty(\mathbf{R}^3 \times \mathbf{R}^3)$

Setting $\zeta = (\zeta_1, \cdots, \zeta_6) = (z, w)$ define the Poisson bracket of $f, g \in C^\infty(\mathbf{R}^3 \times \mathbf{R}^3)$ by

$$\{f, g\} = \sum_{i,j} \frac{\partial f}{\partial \zeta_i} \frac{\partial g}{\partial \zeta_j} \{\zeta_i, \zeta_j\}.$$

Hence $\left(C^\infty(\mathbf{R}^3 \times \mathbf{R}^3), \{\,,\,\}, \cdot\right)$ is a Poisson algebra, (see appendix A section 4). For $H \in C^\infty(\mathbf{R}^3 \times \mathbf{R}^3)$ the integral curves of the Hamiltonian vector field $-\mathrm{ad}_H$ satisfy

$$\begin{cases} \dot{z} = -\{H, z\} = \dfrac{\partial H}{\partial w} \times z \\ \dot{w} = -\{H, w\} = \dfrac{\partial H}{\partial w} \times w + \dfrac{\partial H}{\partial z} \times z. \end{cases} \tag{31}$$

These are the Euler-Poisson equations (30). Specializing (31) to the case where the Hamiltonian is the reduced Hamiltonian H^a (26) of the Lagrange top gives

$$\begin{cases} \dot{z} = I_1^{-1} w \times z \\ \dot{w} = \chi e_3 \times z. \end{cases} \tag{32}$$

The solutions of (32) on P^a describe the motion of the Lagrange top after rotation about its figure axis has been removed, which is a model for the motion of the tip of the figure axis of the top with a given body angular momentum a.

3.2 The magnetic spherical pendulum

In this subsection we show that after reduction of the S^1 symmetry about its figure axis, the Lagrange top is equivalent to the *magnetic spherical pendulum* up to a time rescaling.

Physically, the magnetic spherical pendulum is the motion of a massive electrically charged particle which moves on a 2-sphere S^2 under the combined influence of a constant vertical gravitational force and a radial magnetic field of strength a due to a *monopole* placed at the center of S^2.

Mathematically, the magnetic spherical pendulum is a Hamiltonian system on the phase space (TS^2, Ω_a), where TS^2 is the tangent bundle of S^2

$$\left\{(x, y) \in T\mathbf{R}^3 \,\middle|\, (x, x) = 1 \;\&\; (x, y) = 0\right\},$$

and Ω_a is the symplectic form

$$\Omega_a(x,y)\big((u,r),(v,s)\big) = -(r,v) + (u,s) + a(x, u \times v), \tag{33}$$

with $(u,r), (v,s) \in T_{(x,y)}(TS^2)$. Note that Ω_a is the sum of the standard symplectic form on TS^2 plus a *magnetic term* $a(x, u \times v)$. Integrating this magnetic term over a domain on S^2 gives a magnetic flux which is proportional to the surface area of the domain. The Hamiltonian of the magnetic spherical pendulum is

$$\widetilde{F} : TS^2 \to \mathbf{R} : (x,y) \to \tfrac{1}{2} I_1^{-1}(y,y) + \chi\,(x, e_3) + \tfrac{1}{2} I_1{}^{-1} a^2. \tag{34}$$

Claim: The mapping

$$\varphi : TS^2 \to P^a : (x,y) \to (x, x \times y + ax) = (z,w) \tag{35}$$

is an equivalence between the time rescaled magnetic spherical pendulum Hamiltonian system $(\widetilde{F}, TS^2, \Omega_a)$ and the Hamiltonian system (H^a, P^a, ω^a) of the Lagrange top after reduction of rotation about its figure axis.

(3.6) **Proof**: It is straightforward to check that the inverse of φ is the smooth mapping

$$P^a \to TS^2 : (z,w) \to (z, w \times z).$$

Hence φ is a diffeomorphism. We now compute the tangent of φ. From the definition of TS^2 it follows that $(u,r) \in T_{(x,y)}(T\mathbf{R}^3)$ is in $T_{(x,y)}(TS^2)$ if and only if in addition to $(x,x) = 1$ and $(x,y) = 0$, the conditions $(x,u) = 0$ and $(u,y) + (x,r) = 0$ hold. Differentiating (35) gives

$$T_{(x,y)}\varphi : T_{(x,y)}(TS^2) \to T_{\varphi(x,y)}P^a : (u,r) \to (u, u \times y + x \times r + au)$$

▷ This expression is not useful because the tangent vector $(u, u \times y + x \times r + au)$ at $T_{\varphi(x,y)}P^a$ is not in the form $(\widetilde{u} \times x, \widetilde{u} \times (x \times y + ax) + \widetilde{r})$, where $(x, \widetilde{r}) = 0$. The following argument remedies this by showing that

$$(u, u \times y + x \times r + au) = (\widetilde{u} \times x, \widetilde{u} \times (x \times y + ax) + x \times r) \tag{36}$$

where $\widetilde{u} = x \times u$. Note that $(x, x \times r) = 0$ in (36).

(3.7) **Proof**: Since $(x,x) = 1$ and $(x,u) = 0$, we find that $u = (x \times u) \times x = \widetilde{u} \times x$. Therefore

$$\begin{aligned} u \times y + x \times r + au &= (\widetilde{u} \times x) \times y + x \times r + a\widetilde{u} \times x \\ &= -\widetilde{u}(x,y) + x(\widetilde{u},y) + a\widetilde{u} \times x + x \times r \\ &= x(\widetilde{u},y) - y(x,\widetilde{u}) + a\widetilde{u} \times x + x \times r, \\ &\qquad \text{since } (x,y) = 0 \text{ and } (x,\widetilde{u}) = 0 \\ &= \widetilde{u} \times (x \times y + ax) + x \times r. \end{aligned}$$

□

Next we show that $\varphi^*\omega^a = \Omega_a$:

$$\begin{aligned} (\varphi^*\omega^a)(x,y)\big((u,r),(v,s)\big) &= \omega^a(\varphi(x,y))\big(T_{(x,y)}\varphi(u,r), T_{(x,y)}\varphi(v,s)\big) \\ &= \omega^a(x, x \times y + ax)\big((u, u \times y + x \times r + au), (v, v \times y + x \times s + av)\big) \\ &= \omega^a(z,w)\big((\widetilde{u} \times z, \widetilde{u} \times w + \widetilde{r}), (\widetilde{v} \times z, \widetilde{v} \times w + \widetilde{s})\big), \end{aligned}$$

using (36). Here $z = x$, $w = x \times y + ax$, $\widetilde{u} = x \times u$, $\widetilde{v} = x \times v$, $\widetilde{r} = x \times r$ and $\widetilde{s} = x \times s$. Thus

$$\begin{aligned}(\varphi^*\omega^a)(x,y)\big((u,r),(v,s)\big) &= \\ &= -(\widetilde{r},\widetilde{v}) + (\widetilde{u},\widetilde{s}) + (w, \widetilde{u}\times\widetilde{v}), \qquad \text{by definition of } \omega^a \text{ (23)} \\ &= -(x\times r, x\times v) + (x\times u, x\times s) + (x\times y + ax, (x\times u)\times(x\times v)) \\ &= -(r,v) + (u,s) + a(x, u\times v),\end{aligned}$$

since $(x,u) = (x,v) = 0$ and $(x,x) = 1$. Therefore, $\varphi^*\omega^a = \Omega_a$.

To finish proving the equivalence we compute $\varphi^* H^a$:

$$\begin{aligned}(\varphi^* H^a)(x,y) &= \tfrac{1}{2} I_1^{-1}(x\times y + ax, x\times y + ax) + \chi\,(x, e_3) \\ &= \tfrac{1}{2} I_1^{-1}(y,y) + \chi\,(x,e_3) + \tfrac{1}{2} I_1^{-1} a^2 \\ &= \widetilde{F}. \qquad \square\end{aligned}$$

Introduce a new time scale s by setting $s = \frac{1}{I_1} t$. Then the Hamiltonian system $(\widetilde{F}, TS^2, \Omega_a)$ becomes the magnetic spherical pendulum (F, TS^2, Ω_a) with Hamiltonian

$$F(x,y) = I_1(\widetilde{F}(x,y) - \frac{1}{2I_1}a^2) = \tfrac{1}{2}(y,y) + \lambda x_3, \tag{37}$$

where $\lambda = I_1 \chi$.

To find Hamilton's equations for the integral curves of X_F, we consider the magnetic spherical pendulum to be a *constrained* Hamiltonian system. Give the manifold $T_0\mathbf{R}^3 = (\mathbf{R}^3 - \{0\}) \times \mathbf{R}^3$ the *nonstandard* symplectic structure defined by the 2-form

$$\widehat{\Omega}_a(x,y)\big((u,r),(v,s)\big) = -(v,r) + (s,u) + a(\frac{x}{|x|^3}, u\times v). \tag{38}$$

For any smooth function $H : T_0\mathbf{R}^3 \to \mathbf{R}$ it is straightforward to check that the Hamiltonian vector field X_H on $(T_0\mathbf{R}^3, \widehat{\Omega}_a)$ has integral curves which satisfy

$$\begin{aligned}\dot{x} &= \frac{\partial H}{\partial y} \\ \dot{y} &= -\frac{\partial H}{\partial x} + a\,\frac{x}{|x|^3} \times \frac{\partial H}{\partial y}.\end{aligned} \tag{39}$$

Let $M = T_0\mathbf{R}^3$. On $(M, \widehat{\Omega}^a)$ define the constraint functions

$$c_1 : M \to \mathbf{R} : (x,y) \to (x,x) - 1 \quad \text{and} \quad c_2 : M \to \mathbf{R} : (x,y) \to (x,y).$$

Since 0 is a regular value of the map

$$\mathscr{C} : M \to \mathbf{R}^2 : m \to \big(c_1(m), c_2(m)\big),$$

the constraint set $\mathcal{C}^{-1}(0)$ is the smooth manifold TS^2. Because the matrix $(\{c_i, c_j\})$ of Poisson brackets is invertible on M with inverse

$$C = (C_{ij}) = \frac{1}{2\langle x, x\rangle} \begin{pmatrix} 0 & -1 \\ 1 & 0 \end{pmatrix},$$

the symplectic form $\widehat{\Omega}_a$ restricted to TS^2 is the symplectic form Ω_a (33) on TS^2. To compute the Hamiltonian vector field of the Hamiltonian

$$\widehat{F} : M \subseteq T\mathbf{R}^3 \to \mathbf{R} : (x, y) \to \tfrac{1}{2}\langle y, y\rangle + \lambda \langle \frac{x}{|x|}, e_3\rangle \tag{40}$$

constrained to TS^2, we use the modified Dirac procedure, (see appendix A section 4). Note that $\widehat{F}|TS^2$ is the Hamiltonian F of the magnetic spherical pendulum. To start the Dirac procedure let

$$\widehat{F}^* = \widehat{F} - \sum_{i,j=1}^{3} (\{\widehat{F}, c_i\} + \widehat{F}_i) C_{ij}\, c_j,$$

where $\widehat{F}_i$ are smooth functions in the ideal generated by the constraint functions c_i. The Poisson bracket on $C^\infty(M)$ is computed with respect to the symplectic form $\widehat{\Omega}_a$. Because $\{\widehat{F}, c_1\} = -\langle x, y\rangle$ and $\{\widehat{F}, c_2\} = -\langle y, y\rangle$, we may choose

$$\widehat{F}_1(x, y) = -\langle x, y\rangle\big(\langle x, x\rangle - 1\big) \quad \text{and} \quad \widehat{F}_2(x, y) = -\langle y, y\rangle\big(\langle x, x\rangle - 1\big).$$

Then

$$\widehat{F}^*(x, y) = \tfrac{1}{2}\langle y, y\rangle + \lambda\langle \frac{x}{|x|}, e_3\rangle + \langle y, y\rangle\big(\langle x, x\rangle - 1\big) - \tfrac{1}{2}\langle x, y\rangle^2. \tag{41}$$

Using (39), it follows that the integral curves of the Hamiltonian vector field $X_{\widehat{F}^*}$ on $(M, \widehat{\Omega}_a)$ satisfy

$$\begin{aligned} \dot{x} &= y + \big(\langle x, x\rangle - 1\big)y - \langle x, y\rangle x \\ \dot{y} &= \frac{\lambda}{|x|^3} x \times (x \times e_3) + \langle x, y\rangle y - \langle y, y\rangle x \\ &\quad + + a \frac{x}{|x|^3} \times \big(y + (\langle x, x\rangle - 1)y - \langle x, y\rangle x\big). \end{aligned}$$

Since $\{\widehat{F}^*, F_1\}|TS^2 = \{\widehat{F}^*, F_2\}|TS^2 = 0$, we see that TS^2 is an invariant manifold of $X_{\widehat{F}^*}$. Thus the integral curves of the constrained Hamiltonian vector field $X_F = X_{\widehat{F}^*}|TS^2$ satisfy the equations

$$\begin{aligned} \dot{x} &= y \\ \dot{y} &= -\lambda e_3 + \big(\lambda\langle x, e_3\rangle - \langle y, y\rangle\big)x + a\, x \times y, \end{aligned} \tag{42}$$

which are Hamilton's equations for the magnetic spherical pendulum. Note that when $a = 0$ and $\lambda = 1$ (42) reduce to Hamilton's equations for the spherical pendulum, (see chapter 4 equation (5)).

4 Reduction of the left S^1 action

In this section we complete the reduction of the Lagrange top to a Hamiltonian system with one degree of freedom by removing the symmetry of the left S^1 action on the reduced space P^a. Because this S^1 action on P^a has fixed points, the regular reduction theorem does *not* hold. We use invariant theory to carry out singular reduction, (see appendix B section 5).

We begin by finding the left S^1 action on the reduced space P^a induced by the left S^1 action Φ^ℓ (5) on $TSO(3)$. Pulling back Φ^ℓ by the left trivialization $\mathcal{L}$ gives the left S^1 action

$$\varphi^\ell : S^1 \times \big(\,\mathrm{SO}(3) \times \mathrm{so}(3)\big) \to \mathrm{SO}(3) \times \mathrm{so}(3) : \big(B, (A, X)\big) \to (BA, X). \tag{43}$$

Because

$$J_r^{-1}(a) = \Big\{(A, X) \in \mathrm{SO}(3) \times \mathrm{so}(3) \big| k(E_3, I(X)) = a\Big\}$$

is invariant under φ^ℓ, the induced action $\varphi^\ell|\big(S^1 \times J_r^{-1}(a)\big)$ is defined. Consider the S^1 action

$$\begin{aligned} \delta^\ell : S^1 \times \big(\,\mathrm{so}(3) \times \mathrm{so}(3)\big) &\to \mathrm{so}(3) \times \mathrm{so}(3) : \\ \big(B, (Z, W)\big) &\to (\mathrm{Ad}_B\, Z, \mathrm{Ad}_B\, W). \end{aligned} \tag{44}$$

δ^ℓ leaves the reduced space $\mathcal{P}^a$ (13) invariant, for if $(Z, W) \in \mathcal{P}^a$, then $k(\mathrm{Ad}_B\, Z, \mathrm{Ad}_B\, Z) = k(Z, Z) = 1$ and $k(\mathrm{Ad}_B\, Z, \mathrm{Ad}_B\, W) = k(Z, W) = a$. Therefore the induced action $\delta^\ell|(S^1 \times \mathcal{P}^a)$ is defined.

Claim: The map

$$\begin{aligned} \pi^a : J_r^{-1}(a) \subseteq \mathrm{SO}(3) \times \mathrm{so}(3) &\to \mathcal{P}^a \subseteq \mathrm{so}(3) \times \mathrm{so}(3) : \\ (A, X) &\to \big(\,\mathrm{Ad}_A\, E_3, \mathrm{Ad}_A\, I(X)\big) = (Z, W), \end{aligned} \tag{45}$$

intertwines the actions $\varphi^\ell|(S^1 \times J_r^{-1}(a))$ and $\delta^\ell|(S^1 \times \mathcal{P}^a)$, that is, for every $B \in S^1$

$$(\pi^a \circ \varphi^\ell_B)|J_r^{-1}(a) = \big(\delta^\ell_B|\mathcal{P}^a\big) \circ \pi^a.$$

(4.1) **Proof**: For $(A, X) \in J_r^{-1}(a)$, we compute

$$\begin{aligned} \pi^a\big(\varphi^\ell_B(A, X)\big) &= \pi^a(BA, X) \\ &= \big(\,\mathrm{Ad}_B(\mathrm{Ad}_A)E_3, \mathrm{Ad}_B(\mathrm{Ad}_A\, I(X))\big) = \delta^\ell_B\big(\pi^a(A, X)\big). \quad \square \end{aligned}$$

Using the identification mapping $i : \mathrm{so}(3) \to \mathbf{R}^3$ (see chapter 3 ((1.2))), the action δ^ℓ becomes

$$\Delta : S^1 \times (\mathbf{R}^3 \times \mathbf{R}^3) \to \mathbf{R}^3 \times \mathbf{R}^3 : \big(t, (z, w)\big) \to (R_t z, R_t w), \tag{46}$$

where

$$S^1 = \left\{ R_t = \begin{pmatrix} \cos t & -\sin t & 0 \\ \sin t & \cos t & 0 \\ 0 & 0 & 1 \end{pmatrix} \in \mathrm{SO}(3) \,\middle|\, t \in \mathbf{R} \right\}.$$

Since the reduced space P^a (22) is invariant under Δ, the induced action $\Delta|(S^1 \times P^a)$ is ▷ defined. It is Hamiltonian with momentum mapping

$$J_\ell^a : P^a \subseteq \mathbf{R}^3 \times \mathbf{R}^3 \to \mathbf{R} : (z, w) \to (e_3, w) = w_3. \tag{47}$$

(4.1) **Proof**: The map i intertwines the actions $\delta^\ell|(S^1 \times \mathcal{P}^a)$ and $\Delta|(S^1 \times P^a)$. To find the momentum mapping of $\delta^\ell|(S^1 \times \mathcal{P}^a)$ we pull back the momentum mapping

$$\mathcal{J}_\ell : TSO(3) \to \mathbf{R} : X_A \to \rho(A)(T_e R_A E_3, X_A)$$

of the left S^1 action Φ^ℓ (5) by the left trivialization $\mathcal{L}$ obtaining the momentum map

$$J_\ell : SO(3) \times so(3) \to \mathbf{R} : (A, X) \to k(E_3, \mathrm{Ad}_A I(X)) \tag{48}$$

of the left action φ^ℓ. Because

$$J_\ell(AB, \mathrm{Ad}_{B^{-1}} X) = k\big(E_3, \mathrm{Ad}_A(\mathrm{Ad}_B I(\mathrm{Ad}_{B^{-1}}(X)))\big) = J_\ell(A, X)$$

for every $B \in S^1$, $J_\ell|J_r^{-1}(a)$ induces a function on the orbit space $J_r^{-1}(a)/S^1 = \mathcal{P}^a$:

$$\widetilde{J}_\ell^a : \mathcal{P}^a \subseteq so(3) \times so(3) \to \mathbf{R} : (Z, W) \to k(E_3, W),$$

that is, $(\pi^a)^* \widetilde{J}_\ell^a = J_\ell|J_r^{-1}(a)$. Since π^a intertwines the actions $\varphi^\ell|(S^1 \times J_r^{-1}(a))$ and $\delta^\ell|(S^1 \times \mathcal{P}^a)$ and in addition $(\pi^a)^*\omega^a = \Omega^a|\mathcal{P}^a$, it follows that $\widetilde{J}_\ell^a$ is the momentum map of $\delta^a|(S^1 \times \mathcal{P}^a)$. Pulling back $\widetilde{J}_\ell^a$ by the map i gives

$$J_\ell^a : P^a \subseteq T\mathbf{R}^3 \to \mathbf{R} : (z, w) \to (e_3, w) = w_3, \tag{49}$$

which is the momentum map of $\Delta|(S^1 \times P^a)$ because i intertwines the actions $\delta^\ell|(S^1 \times \mathcal{P}^a)$ and $\Delta|(S^1 \times P^a)$ and is also a symplectic diffeomorphism of $(\mathcal{P}^a, \Omega_a)$ onto (P^a, ω^a). □

It is interesting to see what all this means in the magnetic spherical pendulum model. The unconstrained Hamiltonian $\widehat{F}$ (40) is invariant under the S^1 action

$$S^1 \times T\mathbf{R}^3 \to T\mathbf{R}^3 : \big(t, (x, y)\big) \to (R_t x, R_t y), \tag{50}$$

where $t \to R_t$ is a one parameter group of rotations about the e_3-axis. It is easy to check that the infinitesimal generator of this S^1 action is the vector field

$$X = (e_3 \times x, \frac{\partial}{\partial x}) + (e_3 \times y, \frac{\partial}{\partial y})$$

and that $M = T_0\mathbf{R}^3$ is an invariant manifold of X. Using (39) and the fact that the partial derivative of $x \to \langle \frac{x}{|x|}, e_3 \rangle$ is $\frac{1}{|x|^3}\big(x \times (e_3 \times x)\big)$, we see that X is a Hamiltonian vector field on $(T_0\mathbf{R}^3, \widehat{\Omega}_a)$ corresponding to the Hamiltonian function

$$\widehat{J}_a : T_0\mathbf{R}^3 \to \mathbf{R} : (x, y) \to (x \times y, e_3) + a(\frac{x}{|x|}, e_3).$$

In other words, the S^1 action (50) is Hamiltonian with momentum mapping $\widehat{J}_a$. Since TS^2 is an invariant manifold of $X_{\widehat{J}_a}$, it follows that the S^1 action (50) restricted to $(TS^2, \widehat{\Omega}_a|TS^2)$

is Hamiltonian with momentum mapping $J_a = \widehat{J}_a|TS^2$. Because the unconstrained Hamiltonian $\widehat{F}$ (40) and the modified Hamiltonian $\widehat{F}^*$ (41) are invariant under the S^1 action (50), it follows that $\{F, J_a\}_{TS^2} = \{\widehat{F}^*, \widehat{J}_a\}|TS^2 = L_X F^* = 0$. Thus

$$J_a : TS^2 \subseteq T\mathbf{R}^3 \to \mathbf{R} : (x, y) \to (x \times y, e_3) + a(x, e_3)$$

is an integral of the constrained Hamiltonian vector field X_F on $(TS^2, \widehat{\Omega}_a|TS^2)$. Here $F = \widehat{F}|TS^2$ is the Hamiltonian of the magnetic spherical pendulum.

We return to finding the reduced space of the left S^1 action $\Delta|(S^1 \times P^a)$ on P^a (22). This action has fixed points $p_\varepsilon = \varepsilon(0, 0, 1, 0, 0, a)$ where $\varepsilon^2 = 1$, because the action Δ (46) fixes every point of the 2-plane in $\mathbf{R}^3 \times \mathbf{R}^3$ spanned by $\{e_3, e_6\}$ and this 2-plane intersects P^a at p_ε. Since p_ε lies in the εa-level set of J^a_ℓ, we can not apply the regular reduction theorem to find the reduced phase space of the left S^1 action $\Delta|(S^1 \times P^a)$ on $(J^a_\ell)^{-1}(\varepsilon a)$. The reason for this is that the orbit space $(J^a_\ell)^{-1}(\varepsilon a)/S^1$ is *not* a smooth manifold. To construct a model for the orbit space $(J^a_\ell)^{-1}(b)/S^1$ for *every* value of the momentum map J^a_ℓ (49) we use invariant theory.

Claim: The algebra $\mathbf{R}[z, w]^{S^1}$ of polynomials on $\mathbf{R}^3 \times \mathbf{R}^3$ which are invariant under the S^1 action Δ (46) is generated by the monomials

$$\begin{aligned} \pi_1 &= z_3, & \pi_3 &= z_1 w_1 + z_2 w_2, & \pi_5 &= z_1^2 + z_2^2, \\ \pi_2 &= w_3, & \pi_4 &= z_2 w_1 - z_1 w_2, & \pi_6 &= w_1^2 + w_2^2. \end{aligned} \tag{51}$$

(4.2) **Proof**: See chapter 4 ((2.1)). □

The first step is to find a model for the orbit space $(\mathbf{R}^3 \times \mathbf{R}^3)/S^1$ of the action Δ. Let

$$\pi : \mathbf{R}^3 \times \mathbf{R}^3 \to \mathbf{R}^4 \times (\mathbf{R}_\geq)^2 : (z, w) \to \big(\pi_1(z, w), \ldots, \pi_6(z, w)\big). \tag{52}$$

▷ π is the *Hilbert map* of the S^1 action Δ. The image of π is the real semialgebraic variety V in $\mathbf{R}^6$ with coordinates $(\pi_1, \ldots, \pi_6)$ defined by

$$\pi_3^2 + \pi_4^2 = \pi_5 \pi_6, \quad \text{where } \pi_5 \geq 0 \ \& \ \pi_6 \geq 0. \tag{53}$$

(4.3) **Proof**: See chapter 4 ((2.2)). □

▷ V is homeomorphic to the orbit space $(\mathbf{R}^3 \times \mathbf{R}^3)/S^1$.

(4.4) **Proof**: See chapter 4 ((2.3)). □

The next step is to find the orbit space $(J^a_\ell)^{-1}(b)/S^1$. The level set $(J^a_\ell)^{-1}(b)$ is defined by

$$\begin{cases} z_1^2 + z_2^2 + z_3^2 &= 1 \\ z_1 w_1 + z_2 w_2 + z_3 w_3 &= a \\ w_3 &= b. \end{cases} \tag{54}$$

From ((4.3)) it follows that $\pi\big((J_\ell^a)^{-1}(b)\big) = W$ is the space of S^1 orbits of the action $\Delta|\big(S^1 \times (J_\ell^a)^{-1}(b)\big)$. W is the semialgebraic variety of $\mathbf{R}^6$ defined by

$$\begin{cases} \pi_3^2 + \pi_4^2 &= \pi_5\pi_6, \quad \pi_5 \geq 0 \;\&\; \pi_6 \geq 0 \\ \pi_5 + \pi_1^2 &= 1, \\ \pi_3 + \pi_1\pi_2 &= 0, \\ \pi_2 &= b. \end{cases} \tag{55}$$

These equations and inequalities are obtained from the conditions (53) defining the orbit space V and then expressing the conditions (54) defining $(J_\ell^a)^{-1}(b)$ in terms of invariants. To find a simpler description of W, consider the mapping

$$\begin{aligned} &\lambda : \mathbf{R}^3 \to \mathbf{R}^6 : \\ &(\sigma_1, \sigma_2, \sigma_3) \to (\pi_1, \pi_2, \pi_3, \pi_4, \pi_5, \pi_6) = (\sigma_1, b, a - b\sigma_1, \sigma_2, 1 - \sigma_1^2, \sigma_3). \end{aligned} \tag{56}$$

λ is a diffeomorphism onto its image. A straightforward calculation shows that $\lambda^{-1}(W)$ is the semialgebraic variety P_b^a in $\mathbf{R}^3$ defined by

$$F(\sigma_1, \sigma_2, \sigma_3) = \sigma_2^2 + (a - b\sigma_1)^2 - (1 - \sigma_1^2)\sigma_3 = 0, \quad |\sigma_1| \leq 1 \;\&\; \sigma_3 \geq 0. \tag{57}$$

▷ The variety P_b^a is homeomorphic to $(J_\ell^a)^{-1}(b)/S^1$.

(4.5) **Proof**: Consider diagram 4.1, where $\widetilde{\pi}$ is the map $(\lambda^{-1} \circ \pi)|(J_\ell^a)^{-1}(b)$ and $\widetilde{\rho}$ is the orbit map which assigns to each S^1 orbit of $\Delta|\big(S^1 \times (J_\ell^a)^{-1}(b)\big)$ a point in the orbit space $(J_\ell^a)^{-1}(b)/S^1$. Since the fiber of the bundle $\widetilde{\pi}$ is a single S^1 orbit on $(J_\ell^a)^{-1}(b)$ and $\widetilde{\pi}$ has a continuous local cross section, it follows that $\widetilde{\sigma}$ is invertible and hence is a homeomorphism. □

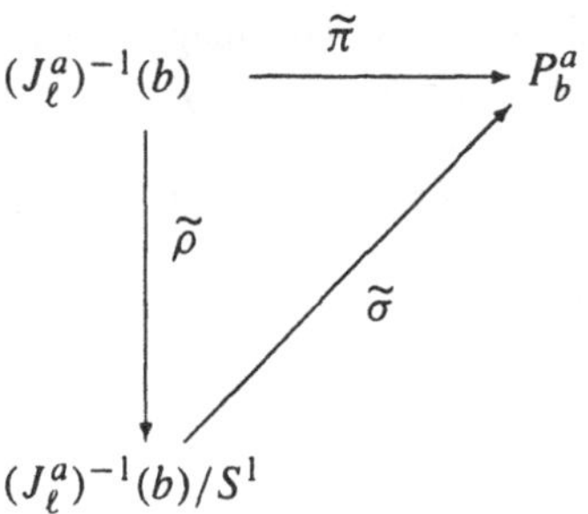

Diagram 4.1

Therefore P_b^a is a model for the singular reduced space $(J_\ell^a)^{-1}(b)/S^1$. In figures 4.1 and ▷ 4.2 we have sketched P_b^a. We now verify that these figures are qualitatively correct.

(4.6) **Proof**: When $b \neq \varepsilon a$, P_b^a is a smooth manifold diffeomorphic to $\mathbf{R}^2$. To see this suppose

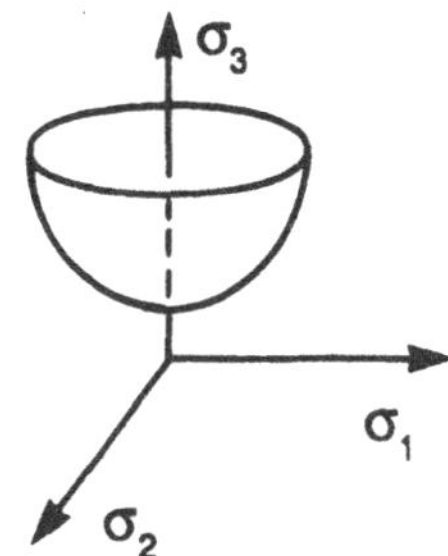

Figure 4.1. The reduced space P_b^a, when $b \neq \varepsilon a$.

that $\widetilde{\sigma} = (\widetilde{\sigma}_1, \widetilde{\sigma}_2, \widetilde{\sigma}_3)$ is a singular point of P_b^a. Then

$$(0,0,0) = \nabla F(\widetilde{\sigma}) = \big(-2b(a - b\widetilde{\sigma}_1) + 2\widetilde{\sigma}_1\widetilde{\sigma}_3, 2\widetilde{\sigma}_2, \widetilde{\sigma}_1^2 - 1 \big), \tag{58}$$

from which we obtain $\widetilde{\sigma_1} = \varepsilon$, $\widetilde{\sigma_2} = 0$, $\widetilde{\sigma_3} = \varepsilon b(a - \varepsilon b)$, where $\varepsilon^2 = 1$. But $F(\widetilde{\sigma}) = 0$, which gives $b = \varepsilon a$. Therefore $\sigma = (\varepsilon, 0, 0)$ is the only singular point of $P_{\varepsilon a}^a$, if $a \neq 0$; while $\sigma = \pm(1,0,0)$ are singular points of P_0^0 if $a = 0$. Hence when $b \neq \varepsilon a$, P_b^a is smooth and is diffeomorphic to $\mathbf{R}^2$ because it is the graph of the smooth function

$$G : (-1,1) \times \mathbf{R} \to \mathbf{R} : (\sigma_1, \sigma_2) \to \frac{\sigma_2^2 + (a - b\sigma_1)^2}{1 - \sigma_1^2}, \quad |\sigma_1| < 1.$$

To see this, solve the defining equation (57) of P_b^a for σ_3. We check that $|\sigma_1| < 1$. Suppose that $\sigma_1 = \varepsilon$. Then (57) becomes $0 = \sigma_2^2 + (a - \varepsilon b)^2$, which implies that $b = \varepsilon a$. But this contradicts our assumption that $b \neq \varepsilon a$. When $b = \varepsilon a$, $a \neq 0$ or when $b = a = 0$, each singular point $P_{\varepsilon a}^a$ has a nondegenerate *tangent cone*. Consequently, at each singular point $P_{\varepsilon a}^a$ has a conical singularity. Therefore $P_{\varepsilon a}^a$ is homeomorphic, but *not* diffeomorphic, to $\mathbf{R}^2$. □

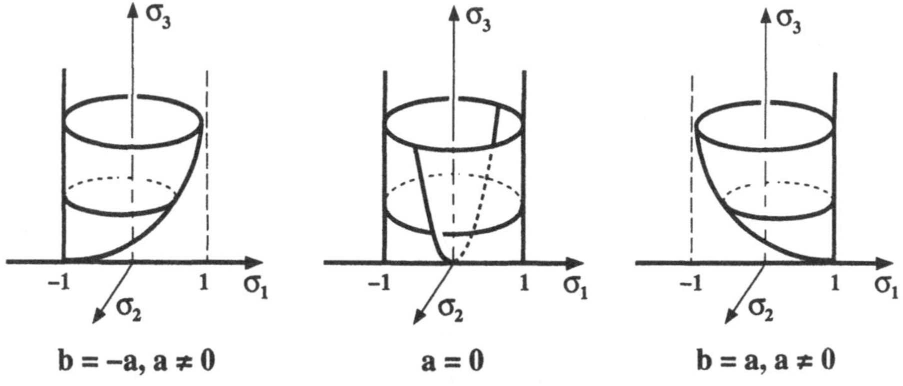

Figure 4.2. The reduced space $P_{\varepsilon a}^a$.

Having constructed the reduced space P_b^a, we turn to finding the reduced Hamiltonian.

Since the Hamiltonian

$$H^a : P^a \subseteq \mathbf{R}^3 \times \mathbf{R}^3 \to \mathbf{R} : (z, w) \to \tfrac{1}{2} I_1{}^{-1}(w, w) + \chi z_3$$

is invariant under the left S^1 action $\Delta|\big(S^1 \times (J_\ell^a)^{-1}(b)\big)$, there is an induced Hamiltonian on P_b^a given by

$$H_b^a : P_b^a \subseteq \mathbf{R}^3 \to \mathbf{R} : (\sigma_1, \sigma_2, \sigma_3) \to \tfrac{1}{2} I_1^{-1}\sigma_3 + \chi\sigma_1. \tag{59}$$

More precisely, $(\pi_b^a)^* H_b^a = H^a|(J_\ell^a)^{-1}(b)$, where

$$\begin{aligned} &\pi_b^a : (J_\ell^a)^{-1}(b) \subseteq P^a \to P_b^a : \\ &(z, w) \to \big(\sigma_1(z, w), \sigma_2(z, w), \sigma_3(z, w)\big) = (z_3, z_2 w_1 - z_1 w_2, w_1^2 + w_2^2) \end{aligned} \tag{60}$$

is the $\Delta|\big(S^1 \times (J_\ell^a)^{-1}(b)\big)$ orbit mapping $(\lambda^{-1} \circ \pi)|(J_\ell^a)^{-1}(b)$. Note that the $h^a + \frac{1}{2} I_1^{-1} b^2$-level set of H^a is mapped by π_b^a to the h_b^a-level set of H_b^a.

5 The Poisson structure

In this section we find a Poisson structure $\{\,,\,\}_{P_b^a}$ on the reduced space P_b^a so that the Poisson variety $(P_b^a, \{\,,\,\}_{P_b^a})$ is isomorphic to the singular reduced space $((J_\ell^a)^{-1}(b)/S^1, \{\,,\,\})$. Here $\{\,,\,\}$ is the Poisson structure given by the singular reduction theorem, (see appendix B section 5).

We begin by constructing the Poisson bracket $\{\,,\,\}_{P_b^a}$. We will think of the invariants σ_i for $i = 1, 2, 3$ defined in (56) as coordinates on $\mathbf{R}^3$. On $C^\infty(\mathbf{R}^3)$ define a Poisson bracket $\{\,,\,\}_{\mathbf{R}^3}$ by the skew symmetric structure matrix given in table 5.1.

$\{A, B\}$	σ_1	σ_2	σ_3	B
σ_1	0	$-(1-\sigma_1^2)$	$-2\sigma_2$	
σ_2		0	$-2b(a - b\sigma_1) + 2\sigma_1\sigma_3$	
σ_3			0	
A				

Table 5.1 Half of the structure matrix for $\{\,,\,\}_{\mathbf{R}^3}$.

From table 5.1 it follows that

$$\{\sigma_i, \sigma_j\}_{\mathbf{R}^3} = \sum_{k=1}^{3} \varepsilon_{ijk} \frac{\partial F}{\partial \sigma_k}, \tag{61}$$

where

$$F(\sigma) = \sigma_2^2 + (a - b\sigma_1)^2 - \sigma_3(1 - \sigma_1^2) \tag{62}$$

is the defining equation for P_b^a. Using (61) we find that

$$\{f, g\}_{\mathbf{R}^3} = \sum_{i,j=1}^{3} \frac{\partial f}{\partial \sigma_i} \frac{\partial g}{\partial \sigma_j} \{\sigma_i, \sigma_j\}_{\mathbf{R}^3} = (\nabla f \times \nabla g, \nabla F), \tag{63}$$

for $f, g \in C^{\infty}(\mathbf{R}^3)$. Therefore, the Hamiltonian derivation $-\mathrm{ad}_h$ corresponding to $h \in C^{\infty}(\mathbf{R}^3)$ is

$$\begin{aligned}&\Big((1-\sigma_1^2)\frac{\partial h}{\partial \sigma_1} - 2\sigma_2\frac{\partial h}{\partial \sigma_2}\Big)\frac{\partial}{\partial \sigma_1}\\ &\quad + \Big((1-\sigma_1^2)\frac{\partial h}{\partial \sigma_1} - (2b(a-b\sigma_1) - 2\sigma_1\sigma_3)\frac{\partial h}{\partial \sigma_3}\Big)\frac{\partial}{\partial \sigma_2}\\ &\qquad + 2\Big(\sigma_2\frac{\partial h}{\partial \sigma_1} + (b(a-b\sigma_1) - \sigma_1\sigma_3)\frac{\partial h}{\partial \sigma_2}\Big)\frac{\partial}{\partial \sigma_3}.\end{aligned}$$

From (63) it follows that F is a Casimir element of the Poisson algebra $(C^{\infty}(\mathbf{R}^3), \{\,,\,\}_{\mathbf{R}^3})$, that is, $\{F, h\}_{\mathbf{R}^3} = 0$ for every $h \in C^{\infty}(\mathbf{R}^3)$. Therefore the algebraic variety $\mathcal{F}$ defined by $F(\sigma) = 0$ is invariant under the flow of $-\mathrm{ad}_h$ for every $h \in C^{\infty}(\mathbf{R}^3)$. $\mathcal{F}$ is singular when $b = \pm a$ and $a \neq 0$ and then its only singular point is $(\pm 1, 0, 0)$ or when $b = a = 0$ and then its only singular points are $\pm(1, 0, 0)$. From the formula for $-\mathrm{ad}_h$ we see that each singular point of $\mathcal{F}$ is an equilibrium point of $-\mathrm{ad}_h$ for every $h \in C^{\infty}(\mathbf{R}^3)$.

Claim: The variety P_b^a is invariant under the flow of $-\mathrm{ad}_h$ for every $h \in C^{\infty}(\mathbf{R}^3)$.

(5.1) **Proof**: To prove the claim requires a detailed understanding of the topology of the variety $\mathcal{F}$, which is sketched in figure 5.1. We leave the verification of this figure as an exercise.

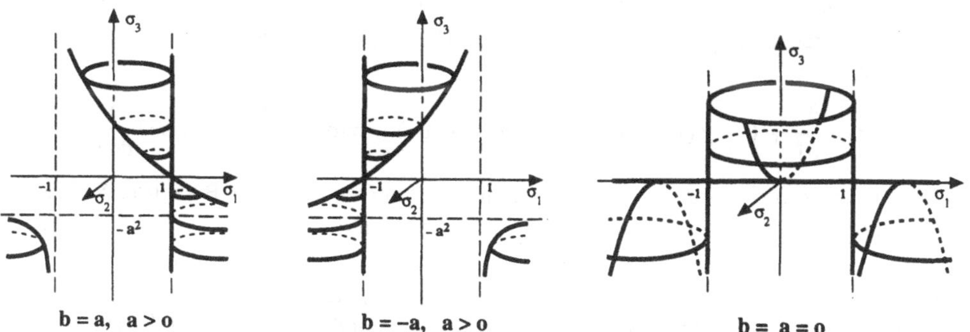

Figure 5.1 The variety $\mathcal{F}$.

Suppose that p is a singular point of P_b^a. Then it is an equilibrium point of $-\mathrm{ad}_h$ and hence remains fixed under its flow. Now suppose that p is a nonsingular point of P_b^a and hence of $\mathcal{F}$. Let γ be the integral curve of $-\mathrm{ad}_h$ starting at p. Then γ lies in the connected component $\widehat{\mathcal{F}}$ of $\mathcal{F}$ containing p. If $b \neq \pm a$, then $\widehat{\mathcal{F}} = P_b^a$. Therefore P_b^a is $-\mathrm{ad}_h$ invariant. If $b = \pm a$, then removing the singular points of $\mathcal{F}$ which lie in $\widehat{\mathcal{F}}$ disconnects $\widehat{\mathcal{F}}$ into $\widetilde{\mathcal{F}}$, which consists of at least two connected components. Suppose that for some $t_0 > 0$ we know that $\gamma(t_0)$ lies in a component of $\widetilde{\mathcal{F}}$ which does not contain p. Then at some $t_1 \in (0, t_0)$ the curve γ passes through a singular point of $\widehat{\mathcal{F}}$. But this is impossible, since every singular point of $\mathcal{F}$ is an equilibrium point of $-\mathrm{ad}_h$. Therefore $\gamma(t) \in P_b^a$ for every t in the domain of the flow of $-\mathrm{ad}_h$, that is, P_b^a is invariant under $-\mathrm{ad}_h$. □

Before we can define a Poisson structure $\{\, , \,\}_{P^a_b}$, we must first define the space of smooth functions on P^a_b. We say that a function $g : P^a_b \to \mathbf{R}$ is *smooth*, if there is a smooth function $G : \mathbf{R}^3 \to \mathbf{R}$ such that $G|P^a_b = g$. Let $\widehat{C}^\infty(P^a_b)$ be the space of smooth functions on P^a_b. From its very definition it follows that $\widehat{C}^\infty(P^a_b) = C^\infty(\mathbf{R}^3)/I$, where I is the ideal of smooth functions on $\mathbf{R}^3$ which vanish identically on P^a_b.

Claim: I is a Poisson ideal of $(C^\infty(\mathbf{R}^3), \{\, , \,\}_{\mathbf{R}^3}, \cdot)$, that is, if $G \in I$, then $\{G, H\}_{\mathbf{R}^3} \in I$ for every $H \in C^\infty(\mathbf{R}^3)$.

(5.2) **Proof**: Let $p \in P^a_b$. Since P^a_b is invariant under $-\mathrm{ad}_H$, the integral curve $t \to \varphi^H_t(p)$ lies in P^a_b. Therefore,

$$\{G, H\}_{\mathbf{R}^3}(p) = \left.\frac{d}{dt}\right|_{t=0} G(\varphi^H_t(p)) = 0,$$

since $G|P^a_b = 0$. Thus $\{G, H\}_{\mathbf{R}^3} \in I$. □

Since I is a Poisson ideal, the Poisson bracket $\{\, , \,\}_{\mathbf{R}^3}$ on $C^\infty(\mathbf{R}^3)$ induces a Poisson bracket $\{\, , \,\}_{P^a_b}$ on $\widehat{C}^\infty(P^a_b)$. In other words, for every $g, h \in \widehat{C}^\infty(P^a_b)$,

$$\{g, h\}_{P^a_b} = \{G, H\}_{\mathbf{R}^3}|P^a_b, \tag{64}$$

where $G, H \in C^\infty(\mathbf{R}^3)$ are chosen so that $G|P^a_b = g$ and $H|P^a_b = h$. This definition can be rewritten as $-\mathrm{ad}_{H|P^a_b} = -\mathrm{ad}_H|P^a_b$. Using (63) we see that the Hamiltonian derivation $-\mathrm{ad}_{H^a_b|P^a_b}$ corresponding to the Hamiltonian H^a_b (59) is given by the restriction to P^a_b of the derivation

$$-(I_1\sigma_2)\frac{\partial}{\partial\sigma_1} + \big(-\chi(1-\sigma_1{}^2) + I_1^{-1}(b(a-b\sigma_1) - \sigma_1\sigma_3)\big)\frac{\partial}{\partial\sigma_2} - (2\chi\sigma_2)\frac{\partial}{\partial\sigma_3}.$$

Following the proof of the singular reduction theorem (see appendix B section 5), we define another Poisson structure $\{\, , \,\}$ on P^a_b. We say that the function $f : P^a_b \to \mathbf{R}$ is *smooth* if and only if there is a smooth function $F : \mathbf{R}^3 \times \mathbf{R}^3 \to \mathbf{R}$, which is invariant under the action Δ (46), such that $(\pi^a_b)^* f = F|(J^a_\ell)^{-1}(b)$. Here π^a_b is the reduction map defined by (60). Let $C^\infty(P^a_b)$ be the vector space of smooth functions on P^a_b. The two spaces $\widehat{C}^\infty(P^a_b)$ and $C^\infty(P^a_b)$ are the same. More precisely,

Claim: Let

$$\sigma : P^a_b \to \mathbf{R}^3 : p \to \big(\sigma_1(p), \sigma_2(p), \sigma_3(p)\big),$$

where σ_i are defined in (56). In other words σ is the inclusion mapping. Then the induced mapping

$$\sigma^* : C^\infty(\mathbf{R}^3) \to C^\infty(P^a_b) : f \to f \circ \sigma$$

is surjective with kernel I. Therefore, σ^* induces an isomorphism

$$\widehat{\sigma}^* : \widehat{C}^\infty(P^a_b) \to C^\infty(P^a_b). \tag{65}$$

(5.3) **Proof**: If $h \in C^\infty(\mathbf{R}^3)$, then the function

$$H : \mathbf{R}^3 \times \mathbf{R}^3 \to \mathbf{R} : (z, w) \to h\big(\sigma_1(z, w), \sigma_2(z, w), \sigma_3(z, w)\big)$$

is smooth and invariant under the action Δ. Therefore $H|(J^a_\ell)^{-1}(b)$ induces the smooth function $\sigma^* h$ on P^a_b. This defines the linear map σ^*. The kernel of σ^* is the ideal I. To see that σ^* is surjective, let $f \in C^\infty(P^a_b)$. Then there is a Δ-invariant smooth function F on $\mathbf{R}^3 \times \mathbf{R}^3$ such that $F|(J^a_\ell)^{-1}(b) = (\pi^a_b)^* f$. Therefore, F is a smooth function of the generators $\pi_1, \ldots, \pi_6$ of the algebra of Δ-invariant polynomials on $\mathbf{R}^3 \times \mathbf{R}^3$. In other words, there is a smooth function h on $\mathbf{R}^6$ such that $F = h \circ \pi$, where π is the Hilbert map (52) of the Δ action. For $p \in (J^a_\ell)^{-1}(b)$ we have

$$\begin{aligned}((\pi^a_b)^* f)(p) &= F|(J^a_\ell)^{-1}(b)(p) = (\pi^* h)(p) \\ &= (h \circ \lambda \circ \sigma \circ \pi^a_b)(p) = ((\pi^a_b)^*(\sigma^*(h \circ \lambda))(p).\end{aligned}$$

In the second to last equality above, λ is the map defined in (56). Because the mapping $\pi^a_b : (J^\ell_a)^{-1}(b) \to P^a_b$ is surjective, it follows that $f = \sigma^*(h \circ \lambda)$. Therefore σ^* is surjective. Hence σ^* induces the linear isomorphism $\widehat{\sigma}^*$ between $C^\infty(\mathbf{R}^3)/I = \widehat{C}^\infty(P^a_b)$ and $C^\infty(P^a_b)$. □

To define a Poisson bracket $\{\,,\,\}$ on $C^\infty(P^a_b)$ we begin by observing that the space of smooth functions on $\mathbf{R}^3 \times \mathbf{R}^3$ has a Poisson bracket $\{\,,\,\}_{\mathbf{R}^3 \times \mathbf{R}^3}$ whose structure matrix is given in table 3.1. The space $C^\infty(\mathbf{R}^3 \times \mathbf{R}^3)^{S^1}$ of smooth Δ-invariant functions on $\mathbf{R}^3 \times \mathbf{R}^3$ is a Lie subalgebra of $(C^\infty(\mathbf{R}^3 \times \mathbf{R}^3), \{\,,\,\}_{\mathbf{R}^3 \times \mathbf{R}^3})$. Let $\mathcal{J}$ be the ideal of smooth Δ-invariant functions on $\mathbf{R}^3 \times \mathbf{R}^3$ which vanish on $(J^a_\ell)^{-1}(b)$.

Claim: $\mathcal{J}$ is a Poisson ideal in $(C^\infty(\mathbf{R}^3 \times \mathbf{R}^3), \{\,,\,\}_{\mathbf{R}^3 \times \mathbf{R}^3}, \cdot)$.

(5.4) **Proof**: Let Y be the infinitesimal generator of the action $\Delta|P^a_b$. Since $\Delta|P^a_b$ has a momentum mapping J^a_ℓ (49), the vector field $Y = X_{J^a_\ell}|P^a_b$. For $p \in (J^a_\ell)^{-1}(b)$ and $F \in C^\infty(\mathbf{R}^3 \times \mathbf{R}^3)^{S^1}$ we find that

$$\{F, J^a_\ell\}_{P^a}(p) = (L_{X_{J^a_\ell}})(p) = (L_Y F)(p) = 0,$$

since F is Δ invariant. Therefore, $(J^a_\ell)^{-1}(b)$ is an invariant manifold of the Hamiltonian vector field $-\mathrm{ad}_F$. In other words, if $p \in (J^a_\ell)^{-1}(b)$, the integral curve $t \to \varphi^F_t(p)$ of $-\mathrm{ad}_F$ starting at p lies in $(J^a_\ell)^{-1}(b)$. For any $H \in \mathcal{J}$

$$\{H, F\}_{\mathbf{R}^3 \times \mathbf{R}^3}(p) = -(\mathrm{ad}_F H)(p) = \left.\frac{d}{dt}\right|_{t=0} H(\varphi^F_t(p)) = 0,$$

since $\varphi^F_t(p)$ lies in $(J^a_\ell)^{-1}(b)$ and $H|(J^a_\ell)^{-1}(b) = 0$. Thus $\{H, F\}_{\mathbf{R}^3 \times \mathbf{R}^3} \in \mathcal{J}$. □

We are ready to define a Poisson bracket $\{\,,\,\}$ on $C^\infty(P^a_b)$. For $f, g \in C^\infty(P^a_b)$ there are $F, G \in C^\infty(\mathbf{R}^3 \times \mathbf{R}^3)^{S^1}$ such that $(\pi^a_b)^* f = F|(J^a_\ell)^{-1}(b)$ and $(\pi^a_b)^* g = G|(J^a_\ell)^{-1}(b)$. Define

$$\{f, g\} = \{F, G\}_{\mathbf{R}^3 \times \mathbf{R}^3}|(J^a_\ell)^{-1}(b).$$

Because $\mathcal{J}$ is a Poisson ideal in $C^\infty(\mathbf{R}^3 \times \mathbf{R}^3)^{S^1}$, the bracket $\{\,,\,\}$ does not depend on the choice of either F or G.

The goal of the rest of this section is to prove

Claim: The Poisson algebras $(\widehat{C}^{\infty}(P^a_b), \{\, , \,\}_{P^a_b}, \cdot)$ and $(C^{\infty}(P^a_b), \{\, , \,\}, \cdot)$ are isomorphic via the map $\widehat{\sigma}^*$ (65).

(5.5) **Proof**: In the course of proving this claim we construct isomorphisms of several intermediate Poisson algebras.

Let $M = (J^a_\ell)^{-1}(b)$. Since $\mathcal{I}$ is a Poisson ideal, we may define a Poisson bracket $\{\, , \,\}_M$ on $C^{\infty}(M)^{S^1}$ by restriction, that is, for $G, H \in C^{\infty}(\mathbf{R}^3 \times \mathbf{R}^3)^{S^1}$

$$\{G|M, H|M\}_M = \{G, H\}_{\mathbf{R}^3 \times \mathbf{R}^3}|M.$$

Consider the generators $\pi_1, \ldots, \pi_6$ of the algebra of Δ-invariant polynomials as coordinates on $\mathbf{R}^6$. Define a Poisson bracket $\{\, , \,\}_{\mathbf{R}^6}$ on $C^{\infty}(\mathbf{R}^6)$ by the skew symmetric structure matrix in table 5.2.

$\{A, B\}$	π_1	π_2	π_3	π_4	π_5	π_6	B
π_1	0	0	0	π_5	0	$2\pi_4$	
π_2		0	0	0	0	0	
π_3			0	$-\pi_2\pi_5$	0	$-2\pi_2\pi_4$	
π_4				0	$2\pi_1\pi_5$	$2(\pi_2\pi_3 - \pi_1\pi_6)$	
π_5					0	$-4\pi_1\pi_4$	
π_6						0	
A							

Table 5.2 Half the structure matrix for $\{\, , \,\}_{\mathbf{R}^6}$.

Using table 5.2 a straightforward calculation shows that the Hilbert map $\pi : \mathbf{R}^3 \times \mathbf{R}^3 \to \mathbf{R}^6$ of the Δ action induces a Poisson map

$$\pi^* : (C^{\infty}(\mathbf{R}^6), \{\, , \,\}_{\mathbf{R}^6}, \cdot) \to (C^{\infty}(\mathbf{R}^3 \times \mathbf{R}^3)^{S^1}, \{\, , \,\}_{\mathbf{R}^3 \times \mathbf{R}^3}, \cdot),$$

that is, $\pi^*(\{\pi_i, \pi_j\}_{\mathbf{R}^6}) = \{\pi^*\pi_i, \pi^*\pi_j\}_{\mathbf{R}^3 \times \mathbf{R}^3}$. Note that π^* is surjective. The image of M under π is the semialgebraic variety W in $\mathbf{R}^6$ defined by

$$\begin{aligned} C_1 &= \pi_3^2 + \pi_4^2 - \pi_5\pi_6 = 0, \qquad \pi_5 \geq 0 \ \&\ \pi_6 \geq 0 \\ C_2 &= \pi_5 + \pi_1^2 - 1 = 0 \\ C_3 &= \pi_3 + \pi_1\pi_2 - a = 0 \\ C_4 &= \pi_2 - b = 0. \end{aligned}$$

Observe that the functions C_i are Casimirs in $(C^{\infty}(\mathbf{R}^6), \{\, , \,\}_{\mathbf{R}^6}, \cdot)$. Since

$$\lambda : \mathbf{R}^3 \to \mathbf{R}^6 : (\sigma_1, \sigma_2, \sigma_3) \to (\sigma_1, b, a - b\sigma_1, \sigma_2, 1 - {\sigma_1}^2, \sigma_3)$$

restricted to P^a_b is an isomorphism of the semialgebraic variety P^a_b onto W, an argument similar to the one used to prove ((5.2)) shows that the ideal $\mathcal{J}$ of smooth functions on $\mathbf{R}^6$

which vanish identically on W is a Poisson ideal in $(C^{\infty}(\mathbf{R}^6), \{\,,\,\}_{\mathbf{R}^6})$. Therefore we may define a Poisson bracket $\{\,,\,\}_W$ on $C^{\infty}(W)$ by restriction, that is, if $G, H \in C^{\infty}(\mathbf{R}^6)$ then

$$\{G|W, H|W\}_W = \{G, H\}_{\mathbf{R}^6}|W.$$

▷ Let $\widehat{\pi} = \pi|M$. Then $\widehat{\pi}$ induces a map

$$\widehat{\pi}^* : (C^{\infty}(W), \{\,,\,\}_W, \cdot) \rightarrow (C^{\infty}(M)^{S^1}, \{\,,\,\}_M, \cdot),$$

which is an isomorphism of Poisson algebras.

(5.6) **Proof**: Because the inclusion map $\nu : M \rightarrow \mathbf{R}^3 \times \mathbf{R}^3$ is injective, the induced map

$$\nu^* : C^{\infty}(\mathbf{R}^3 \times \mathbf{R}^3)^{S^1} \rightarrow C^{\infty}(M)^{S^1}$$

is surjective. Since π^* is surjective, the linear map

$$(\pi \circ \nu)^* : C^{\infty}(\mathbf{R}^6) \rightarrow C^{\infty}(M)^{S^1}$$

is surjective. The kernel of $(\pi \circ \nu)^*$ is $\mathcal{J}$, since $\pi(M) = W$. Therefore $(\pi \circ \nu)^*$ induces the bijective linear map $\widehat{\pi}^*$. From the definition of the Poisson brackets $\{\,,\,\}_W$ and $\{\,,\,\}_M$ and the fact that π^* is a Poisson map, it follows that $\widehat{\pi}^*$ is a Poisson map. In more detail, for $m \in M$

$$\begin{aligned} \widehat{\pi}^*(\{\pi_i|W, \pi_j|W\}_W)(m) &= \{\pi_i|W, \pi_j|W\}_W(\pi(m)) = \pi^*(\{\pi_i, \pi_j\}_{\mathbf{R}^6})(m) \\ &= \{\pi^*\pi_i, \pi^*\pi_j\}_{\mathbf{R}^3 \times \mathbf{R}^3}(m) = \{\pi^*\pi_i, \pi^*\pi_j\}_M(m). \quad \square \end{aligned}$$

Because $\widehat{\lambda} = \lambda|P^a_b$ is an isomorphism of the semialgebraic variety P^a_b onto the semialgebraic variety W, it follows from the definition of the Poisson brackets $\{\,,\,\}_W$ and $\{\,,\,\}_{P^a_b}$ that the induced mapping

$$\widehat{\lambda}^* : (C^{\infty}(W), \{\,,\,\}_W, \cdot) \rightarrow (\widehat{C}^{\infty}(P^a_b), \{\,,\,\}_{P^a_b}, \cdot)$$

is an isomorphism of Poisson algebras.

▷ The map π^a_b induces the linear map

$$(\pi^a_b)^* : (C^{\infty}(P^a_b), \{\,,\,\}, \cdot) \rightarrow (C^{\infty}(M)^{S^1}, \{\,,\,\}_M, \cdot),$$

which is an isomorphism of Poisson algebras.

(5.7) **Proof**: From the definition of the Poisson bracket $\{\,,\,\}$ and $\{\,,\,\}_M$ it follows that $\{f, g\}(m) = \{(\pi^a_b)^* f, (\pi^a_b)^* g\}_M(m)$, for $m \in M$ that is, $(\pi^a_b)^*$ is a Poisson map. Also from the definition of $C^{\infty}(P^a_b)$ we find that $(\pi^a_b)^*$ is surjective; for if $G \in C^{\infty}(M)^{S^1}$ then G is $\Delta|(S^1 \times M)$-invariant and hence induces a function $g \in C^{\infty}(P^a_b)$ such that $(\pi^a_b)^* g = G$. The map $(\pi^a_b)^*$ is injective; for if $(\pi^a_b)^* g = 0$, then g vanishes identically on P^a_b. $\square$

We are now ready to prove ((5.5)). By construction, the map ρ, which is defined to be $((\pi^a_b)^{-1})^* \circ \widehat{\pi}^* \circ ((\widehat{\lambda})^*)^{-1}$, is an isomorphism of $(\widehat{C}^{\infty}(P^a_b), \{\,,\,\}_{P^a_b}, \cdot)$ onto the Poisson algebra $(C^{\infty}(P^a_b), \{\,,\,\}, \cdot)$. Since $\pi = \lambda \circ \sigma \circ \pi^a_b$, we find that $\rho = \widehat{\sigma}^*$. $\square$

6 The Euler-Poisson equations

In this section we describe the invariant manifolds of the Euler-Poisson vector field X_{H^a} (32) on the reduced space P^a (22) by studying the geometry of the energy momentum mapping:

$$\begin{aligned} EM^a &: P^a \to \mathbf{R}^2 : \\ (z, w) &\to \left(H^a(z, w), J^a_\ell(z, w)\right) = \left(\tfrac{1}{2} I_1^{-1}(w, w) + \chi z_3, w_3\right) \end{aligned} \tag{66}$$

Because the reduced energy H^a and and angular momentum J^a_ℓ are integrals of X_{H^a}, the fiber $(EM^a)^{-1}(h^a, b)$ is invariant under the flow of X_{H^a}. To understand these invariant sets, we need to know the following about the energy momentum mapping EM^a.

1. What its critical points, critical values, and its range are.
2. What the topology of every fiber $(EM^a)^{-1}(h^a, b)$ is.
3. How these fibers foliate an energy level set.

We also study the qualitative properties of the image of the integral curves of X_{H^a} restricted to $(EM^a)^{-1}(h^a, b)$ under the bundle projection

$$\tau^a : P^a \subseteq \mathbf{R}^3 \times \mathbf{R}^3 \to S^2 \subseteq \mathbf{R}^3 : (z, w) \to z$$

when (h^a, b) is a regular value of EM^a. These curves on S^2 describe the motion of the tip of the figure axis of the top.

6.1 The twice reduced system

In this subsection we study the qualitative behavior of the (twice) reduced Hamiltonian system $(H^a_b, P^a_b, \{,\}_{P^a_b})$.

Recall that the (twice) reduced space P^a_b is the semialgebraic variety in $\mathbf{R}^3$ defined by

$$F(\sigma) = \sigma_2^2 + (a - b\sigma_1)^2 - \sigma_3(1 - \sigma_1^2) = 0, \quad \text{where } |\sigma_1| \leq 1 \ \&\ \sigma_3 \geq 0 \tag{67}$$

and the (twice) reduced Hamiltonian is

$$H^a_b : P^a_b \subseteq \mathbf{R}^3 \to \mathbf{R} : (\sigma_1, \sigma_2, \sigma_3) \to \tfrac{1}{2} I_1^{-1} \sigma_3 + \chi \sigma_1. \tag{68}$$

In order to determine the topology of the level set $(H^a_b)^{-1}(h^a_b)$, we first find the critical points and critical values of H^a_b. Because P^a_b is not necessarily smooth, we use an algebraic definition of critical point. We say that $\widetilde{\sigma} = (\widetilde{\sigma}_1, \widetilde{\sigma}_2, \widetilde{\sigma}_3)$ is a critical point of H^a_b corresponding to the critical value h^a_b if and only if the 2-plane

$$\Pi : \tfrac{1}{2} I_1{}^{-1} \sigma_3 + \chi\, \sigma_1 = h^a_b$$

intersects the reduced space P^a_b at $\widetilde{\sigma}$ with *multiplicity* greater than one. Let

$$\alpha = I_1 h^a_b \text{ and } \beta = I_1 \chi$$

and form the polynomial

$$P(\sigma_1, \sigma_2) = \sigma_2^2 - 2(\alpha - \beta\sigma_1)(1 - \sigma_1^2) + (a - b\sigma_1)^2,$$

by eliminating σ_3 from (67) using the definition of Π. The point $\widetilde{\sigma}$ has multiplicity greater than one if and only if the *Taylor polynomial* of P at $(\widetilde{\sigma}_1, \widetilde{\sigma}_2)$ has no constant or linear terms. Therefore, $\widetilde{\sigma}$ is a critical point of H_b^a corresponding to the critical value h_b^a if and only if $\widetilde{\sigma}_2 = 0$ and $\widetilde{\sigma}_1$ is a multiple root of

$$V(\sigma_1) = 2(\alpha - \beta\sigma_1)(1 - \sigma_1^2) - (a - b\sigma_1)^2 \tag{69}$$

in $[-1, 1]$. Note that $\widetilde{\sigma}_3 = 2(\alpha - \beta\widetilde{\sigma}_1)$.

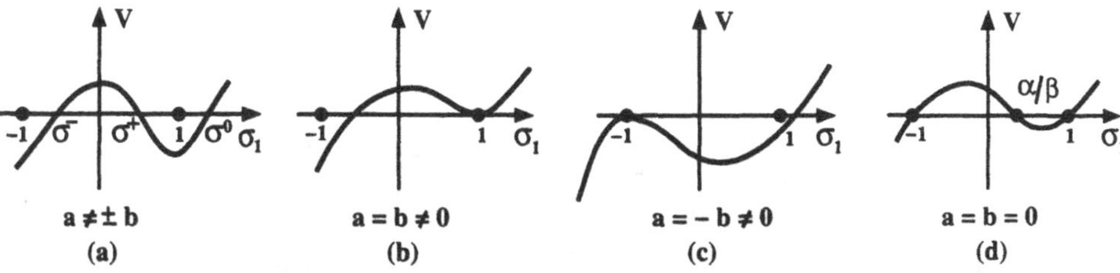

Figure 6.1. The graph of V.

▷ Every critical point of H_b^a lies on the curve $\mathscr{C} = \{\sigma_2 = 0\} \cap P_b^a$, which is the fold curve of the projection

$$\rho : \mathbf{R}^3 \to \mathbf{R}^2 : \sigma = (\sigma_1, \sigma_2, \sigma_3) \to (\sigma_1, \sigma_3)$$

restricted to P_b^a.

(6.1) **Proof**: The image of P_b^a under the mapping ρ is the set of points (σ_1, σ_3) in $\mathbf{R}^2$ which satisfy

$$\sigma_3(1 - \sigma_1^2) - (a - b\sigma_1)^2 \geq 0, \quad |\sigma_1| \leq 1 \ \& \ \sigma_3 \geq 0.$$

Over every point in the interior of $\rho(P_b^a)$ the fiber of $\rho|P_b^a$ consists of two distinct nonsingular points on P_b^a, while over every point on the boundary $\rho(\mathscr{C})$ of $\rho(P_b^a)$, the fiber is a single point. As is easily checked, table 6.1 gives all the possibilities for $\rho(\mathscr{C})$. □

Conditions	$\rho(\mathscr{C})$
1. $b \neq \pm a, \ a \neq 0$	$\sigma_3 = \frac{(a-b\sigma_1)^2}{1-\sigma_1^2}, \quad \lvert\sigma_1\rvert < 1$
2. $b = a, \ a \neq 0$	$\{\sigma_3 = \frac{a^2(1-\sigma_1)}{1+\sigma_1}, \ -1 < \sigma_1 \leq 1\} \cup \{(1, 0, \sigma_3) \mid \sigma_3 \geq 0\}$
3. $b = -a, \ a \neq 0$	$\{\sigma_3 = \frac{a^2(1+\sigma_1)}{1-\sigma_1}, \ -1 < \sigma_1 \leq 1\} \cup \{(-1, 0, \sigma_3) \mid \sigma_3 \geq 0\}$
4. $b = a = 0$	$\{(\pm 1, 0, \sigma_3) \mid \sigma_3 \geq 0\} \cup \{(\sigma_1, 0, 0) \mid \lvert\sigma_1\rvert \leq 1\}$

Table 6.1.

From (67) and the definition of the 2-plane Π, the point $\widetilde{\sigma} = (\widetilde{\sigma}_1, 0, \widetilde{\sigma}_3)$ is a critical point of $H^a_b|\mathcal{C}$ if and only if the line $\ell_\alpha : \sigma_3 + 2\beta\sigma_1 = 2\alpha$ & $\sigma_2 = 0$ is tangent to $\mathcal{C}$ at a nonsingular point or passes through a singular point of $\mathcal{C}$. Applying the projection mapping ρ does not change this geometric situation. In figure 6.2 we give the various geometric possibilities.

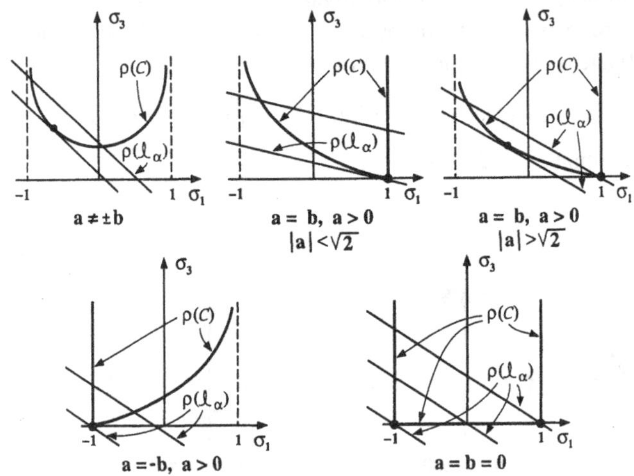

Figure 6.2. Intersection of the line $\rho(\ell_\alpha)$ with fold curve $\rho(\mathcal{C})$. The points $(\widetilde{\sigma}_1, \widetilde{\sigma}_3)$ are large dots. The corresponding critical point of H^a_b is $(\widetilde{\sigma}_1, 0, \widetilde{\sigma}_3)$ with critical value α.

Next we give a geometric description of the set of critical values of H^a_b, (see figure 6.3). In the preceding we showed that $h^a_b = I_1^{-1}\alpha$ is a critical value of H^a_b if and only if

$$W(\sigma) = \sigma^3 - (2\alpha + b^2)\sigma^2 + (2ab - 1)\sigma + 2\alpha - a^2. \tag{70}$$

Here $W(\sigma) = \frac{1}{\beta}V(\sigma)$ and $2\beta = 1$. (This can be arranged by an appropriate choice of physical units.) The set of $(a, b, \alpha) \in \mathbf{R}^3$ where W has a multiple root in $[-1, 1]$ is the *discriminant locus* Δ_W of W. A good way to present Δ_W is to give a parametrization

$$\mathcal{P} : U \subseteq [-1, 1] \times \mathbf{R} \to \Delta_W : (s, a) \to \big(a, b(s, a), \alpha(s, a)\big). \tag{71}$$

Claim: Let U be the set of all $(s, a) \in [-1, 1] \times \mathbf{R}$ such that

$$s \in I_a = \begin{cases} [-1, 0) \cup \{1\}, & \text{if } a = 0 \\ [-1, 0) \cup (0, \frac{1}{2}a^2\,] \cup \{1\}, & \text{if } 0 < |a| < \sqrt{2} \\ [-1, 0) \cup (0, 1\,], & \text{if } |a| \geq \sqrt{2}. \end{cases}$$

Then the parametrization $\mathcal{P}$ (71) consists of pieces

$$b_\pm(s, a) = \begin{cases} \mp \frac{1}{\sqrt{-2s}}(1 - s^2), & \text{if } a = 0 \\ \frac{1}{2s}\left[a(1 + s^2) \pm (1 - s^2)\sqrt{a^2 - 2s}\right], & \text{if } a \neq 0 \end{cases} \tag{72}$$

and

$$\alpha_{\pm}(s,a) = \begin{cases} \frac{1}{4}s(1+s^2), \text{ if } a=0 \\ \frac{1}{4}\left[a^2(1-s^2) \mp a(1-s^2)\sqrt{a^2-2s} + s(1+s^2)\right], \text{ if } a \neq 0. \end{cases} \tag{73}$$

(6.2) **Proof**: Note that $(s,t) \in [-1,1] \times \mathbf{R}$ lies in Δ_W if and only if W can be factored as

$$(\sigma - s)^2(\sigma - t) = \sigma^3 - (2s+t)\sigma^2 + (s^2+2st)\sigma - s^2 t. \tag{74}$$

Equating coefficients in (70) and (74) gives

$$\begin{cases} 2s+t &= 2\alpha + b^2 \\ s^2+2st &= 2ab - 1 \\ s^2 t &= a^2 - 2\alpha. \end{cases} \tag{75}$$

To find $b(s,a)$, add the first and third equations in (75) to obtain

$$t = \frac{1}{1+s^2}(a^2+b^2-2s). \tag{76}$$

Substituting (76) into the second equation in (75) gives

$$Q(b) = 2sb^2 - 2(1+s^2)ab + 2sa^2 + (1-s^2)^2 = 0. \tag{77}$$

Suppose that $s \neq 0$, then Q has real roots if and only if $4(1-s^2)^2(a^2-2s) \geq 0$, that is, if and only if

$$s \in \left(\left[-1, \min\{\tfrac{1}{2}a^2, 1\}\right] \cup \{1\}\right) - \{0\} = I_a.$$

Solving (77) for b gives (72). When $s = 0$, (77) becomes $2ab = 1$. Therefore when $a \neq 0$, $b(0,a) = \frac{1}{2a}$. Since $\lim_{s\to 0} b_-(s,a) = \frac{1}{2a}$ when $a \neq 0$, we can extend the domain of definition of b_- to $\overline{U}$.

To find $\alpha(s,a)$ we use (76) to eliminate t from the first equation in (75). We obtain

$$\alpha = -\frac{s^2}{2(1+s^2)}b^2 + \frac{1}{2(1+s^2)}a^2 + \frac{s^3}{1+s^2}. \tag{78}$$

Substituting (72) into (78) gives (73). □

For a fixed, the curves

$$\gamma_{\pm}(s,a) = \big(b_{\pm}(s,a), \alpha_{\pm}(s,a)\big)$$

are pieces of a parametrization of an a-slice of Δ_W. We now determine how the the curves

s	$\gamma_{\pm}(s,a)$
1. -1	$(-a, -\frac{1}{2})$
2. $\frac{1}{2}a^2$, $0 < \lvert a\rvert < \sqrt{2}$	$\left(\frac{1}{4a}(4+a^4), \frac{a^2}{32}(12-a^4)\right)$
3. 1	$(a, \frac{1}{2})$

Table 6.2. Points where γ_+ joins γ_-.

$\gamma_\pm$ fit together. A straightforward calculation gives table 6.2, which lists the points where γ_+ and γ_- join. We make the following observations. When $a \neq 0$ and is fixed, γ_- is defined and continuous on $\overline{I}_a$. For every $a \in \mathbf{R}$, $\lim_{s\to 0^\pm} b_\pm = \mp\infty$ and $\lim_{s\to 0^\pm} \alpha_\pm = 0$. When $|a| < \sqrt{2}$, $\{1\}$ is an isolated point of I_a. Hence $(a, \frac{1}{2})$ is an *isolated point* of $\gamma_\pm$. From these facts and table 6.2 we see that each of the curves γ_i defined below is continuous.

1. When $a = 0$, let γ_1 be the curve formed by joining $\gamma_+|[-1, 0)$ and $\gamma_-|[-1, 0)$ at $s = -1$.
2. When $0 < |a| < \sqrt{2}$, let γ_2 be the curve formed by joining $\gamma_+|[-1, 0)$ and $\gamma_-|(-1, \frac{1}{2}a^2]$ at $s = -1$ and $\gamma_-|(-1, \frac{1}{2}a^2]$ and $\gamma_+|(0, \frac{1}{2}a^2]$ at $s = \frac{1}{2}a^2$.
3. When $|a| \geq \sqrt{2}$, let γ_3 be the curve formed by joining $\gamma_+|[-1, 0)$ and $\gamma_-|[-1, 1]$ at $s = -1$ and $\gamma_-|[-1, 1]$ and $\gamma_+|(0, 1]$ at $s = 1$.

Gathering the above information together we have proved

Claim: An a-slice of the discriminant locus Δ_W of W (70) is

$$\begin{cases} \gamma_1 \cup \{(0, \frac{1}{2})\}, & \text{if } a = 0 \\ \gamma_2 \cup \{(a, \frac{1}{2})\}, & \text{if } 0 < |a| < \sqrt{2} \\ \gamma_3, & \text{if } |a| \geq \sqrt{2}. \end{cases}$$

Recall that $\alpha = I_1 h^a_b$. Applying the map

$$\mathbf{R}^2 \to \mathbf{R}^2 : (b, \alpha) \to (b, I_1{}^{-1}\alpha) = (b, h^a_b)$$

to a fixed a-slice of Δ_W gives the set of critical values of the (twice) reduced Hamiltonian H^a_b for a fixed value of a, (see figure 6.3). From the algebraic definition of critical point, it is clear that a point of the a-slice of Δ_W which lies on the curve $\gamma_\pm$ is a minimum of $I_1 h^a_b$. Thus for a fixed value of a, the range of H^a_b is the set of (b, h^a_b) values which lie on or above the curves given in figure 6.3.

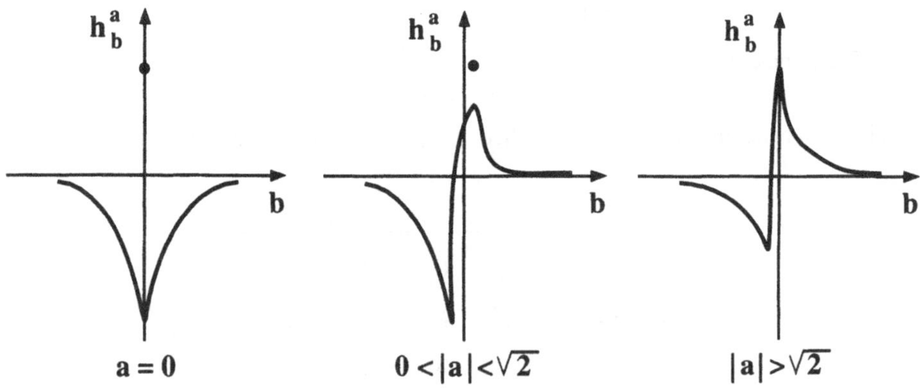

Figure 6.3. The set of critical values of the (twice) reduced Hamiltonian H^a_b for a fixed. The figures for $a < 0$ are obtained by reflection in the h^a_b-axis.

▷ Next we determine the topology of the level sets of the (twice) reduced Hamiltonian H^a_b. The results are given in table 6.3.

$[p_1, p_2]$	Topology of $(H^a_b)^{-1}(h^a_b)$	Conditions on roots of V, where $\beta = \frac{1}{2}$
1. $p_1 = p_2$ is nonsingular point of $\rho(\mathscr{C})$	point	A double real root in $(-1, 1)$ and no other root in $[-1, 1]$
2. $p_1 = p_2$ is singular point of $\rho(\mathscr{C})$	point	A triple real root at $+1$ or a double root at ± 1 and no other root in $[-1, 1]$
3. $p_1 \neq p_2$ and p_2 is singular point of $\rho(\mathscr{C})$	A topological S^1 which is smooth except for one conical singular point	A double root at 1 with $|a| < \sqrt{2}$, and one other simple real root in $[-1, 1]$
4. $p_1 \neq p_2$, p_1 and p_2 are nonsingular points of $\rho(\mathscr{C})$	A smooth S^1	Two simple real roots in $[-1, 1]$

Table 6.3. Topology of $H^a_b | P^a_b$.

(6.3) **Proof**: To find the topology of the level sets of H^a_b, we return to the geometric situation sketched in figure 6.2. Because the image of the fold curve $\mathscr{C}$ under the projection ρ bounds a convex subset C of $\mathbf{R}^2$, the line $\rho(\ell_\alpha)$ intersects $\rho(\mathscr{C})$ in at most two points p_1 and p_2, where p_2 lies to the right of p_1. Let $[p_1, p_2]$ be the segment of $\rho(\ell_\alpha)$ lying in C. Then $[p_1, p_2] = \rho\big((H^a_b)^{-1}(h^a_b)\big)$.

We now verify the entries in the second column to table 6.3. The first and second entries follow from the fact that the fiber of ρ over a singular point of $\rho(\mathscr{C})$ is a single point of $\mathscr{C}$. For the third and fourth entry we observe that $\rho^{-1}([p_1, p_2])$ is a *pinched* S^0 bundle, that is, a two point bundle over the interior of $[p_1, p_2]$ with fiber over p_1 and p_2 pinched to a point. Thus the h^a_b-level set of H^a_b is a topological circle. If h^a_b is a regular value of H^a_b, then $(H^a_b)^{-1}(h^a_b)$ is smooth. Otherwise, $p_1 = (-1, 0)$ or $p_2 = (1, 0)$ is the only singular point of $\rho(\mathscr{C})$ because the line $\rho(\ell_1)$ has negative slope (β is positive). When $p_1 = (-1, 0)$ is a singular point, then $b = -a$ and $h^a_{-a} = -\frac{1}{2} I_1^{-1}$. Hence the h^a_{-a}-level set of H^a_{-a} is a point. When $p_2 = (1, 0)$ is a singular point, then $b = a$ with $|a| < \sqrt{2}$ and $h^a_a = \frac{1}{2} I_1^{-1}$. Hence the h^a_a-level set of H^a_a is the semialgebraic variety

$$\begin{aligned} \sigma_1 + \sigma_3 &= 1 \\ \sigma_2^2 + a^2(1-\sigma_1)^2 - \sigma_3(1-\sigma_1^2) &= 0, \qquad |\sigma_1| \le 1 \ \&\ \sigma_3 \ge 0. \end{aligned}$$

Projecting $(H^a_a)^{-1}(h^a_a)$ onto the σ_1–σ_2 plane by eliminating σ_3 from the above equations gives

$$0 = \sigma_2^2 - (2 - a^2)(1-\sigma_1)^2 + (1-\sigma_1)^3, \qquad |\sigma_1| \le 1,$$

which has a conical singular point at $(1,0)$. Hence $\rho^{-1}(p_2)$ is a conical singularity of $(H_a^a)^{-1}(h_a^a)$. □

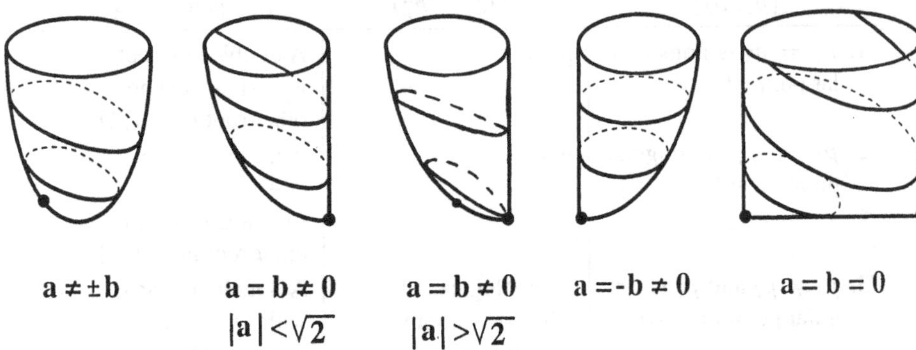

Figure 6.4. Orbits of $-\mathrm{ad}_{H_b^a|P_b^a}$.

A connected component of a level set of H_b^a on P_b^a is an orbit of the Hamiltonian derivation $-\mathrm{ad}_{H_b^a|P_b^a}$. This derivation is a vector field $X_{H_b^a}$ on the manifold of nonsingular points of P_b^a. If p is a singular point of $(H_b^a)^{-1}(h_b^a)$, which is a *nonsingular* point of P_b^a, then p is an equilibrium point of $X_{H_b^a}$. When this is the case, the usual definitions of elliptic and hyperbolic equilibrium point apply. On the other hand, if p is a singular point of $(H_b^a)^{-1}(h_b^a)$, which is a *singular* point of P_b^a, we use the following definitions. We say that p is an *elliptic equilibrium point* of $-\mathrm{ad}_{H_b^a|P_b^a}$ if for every $\widetilde{h}$ slightly larger than h_b^a, the $\widetilde{h}$-level set of H_b^a is a smooth circle, which shrinks to p as $\widetilde{h} \to (h_b^a)^+$. We say that p is a *hyperbolic equilibrium point* of $-\mathrm{ad}_{H_b^a|P_b^a}$ if the connected component of the h_b^a-level set of H_b^a containing p has points which do not lie in any sufficiently small neighborhood of p. In figure 6.4 we sketch the orbits of $-\mathrm{ad}_{H_b^a|P_b^a}$.

6.2 The energy momentum mapping

In this subsection we study the geometry of the energy momentum mapping EM^a (66) of the Euler-Poisson vector field X_{H^a} (32). First we reconstruct the topology of the (h^a, b)-level set of EM^a using table 6.3 and some basic facts about the geometry of the reduction mapping π_b^a (60). The results are summarized in table 6.4. Second we determine the topology of the level set $(H^a)^{-1}(h^a)$ (see table 6.5) and show how it is foliated by level sets of J_ℓ^a.

Knowing the geometry of the reduction mapping $\pi_b^a : (J_\ell^a)^{-1}(b) \to P_b^a$ and the topology of the h_b^a-level set of the (twice) reduced Hamiltonian H_b^a, we can reconstruct the topology of the (h^a, b)-level set of EM^a because

$$(EM^a)^{-1}(h^a, b) = (H^a)^{-1}(h^a) \cap (J_\ell^a)^{-1}(b) = (\pi_b^a)^{-1}\big((H_b^a)^{-1}(h_b^a)\big).$$

Here $h_b^a = h^a - \frac{1}{2} I_1^{-1} b^2$.

Topology of $(H_b^a)^{-1}(h_b^a)$	Topology of $(EM^a)^{-1}(h^a, b)$	Conditions
1. a nonsingular point of P_b^a	smooth S^1	$(a, b, \alpha) \in \Delta_W$, $b \neq \pm a, a \neq 0$
2. a singular point of P_b^a	point	$b = a$, $\lvert a\rvert \geq \sqrt{2}$, $\alpha = \frac{1}{2} + \frac{1}{2}a^2$, or $b = -a$, $\alpha = -\frac{1}{2} + \frac{1}{2}a^2$
3. a topological circle with a conical singular point	a 2-torus with a fiber pinched to a point	$b = a, \alpha = \frac{1}{2} + \frac{1}{2}a^2$, $\lvert a\rvert < \sqrt{2}$
4. a smooth S^1	a smooth 2-torus	(h^a, b) a regular value

Table 6.4. Topology of the level sets of $(EM^a)^{-1}(h^a, b)$. Here $h^a = h_b^a + \frac{1}{2} I_1^{-1} b^2$, $\alpha = I_1 h_b^a$, and $\beta = I_1 \chi = \frac{1}{2}$.

▷ We now verify table 6.4.

(6.4) **Proof**: The first and third columns of table 6.4 are the same as the second and third columns of table 6.3, respectively,.

Next we check the entries in the second column. Suppose that the h_b^a-level set of H_b^a is a point p. If p is a nonsingular point of P_b^a, then the fiber $(\pi_b^a)^{-1}(p)$ is a smooth S^1; otherwise it is a point, because it is a fixed point of the left S^1 action δ^ℓ (44).

To verify the fourth entry suppose that $(h^a, b) = (h_b^a + \frac{1}{2} I_1^{-1} b^2, b)$ is a regular value of EM^a which lies in its range. Then h^a is a regular value of H^a and hence h_b^a is a regular value of H_b^a. Therefore the h_b^a-level set of H_b^a is a smooth S^1. Let $N = (EM^a)^{-1}(h^a, b)$. Since π_b^a is a proper smooth submersion of N onto $(H_b^a)^{-1}(h_b^a)$, N is a smooth S^1 bundle over S^1, (see appendix C section 2). The total space of this bundle is connected, because the base and fiber are connected. Moreover, N is orientable because it is the preimage of 0 under the mapping

$$\mathbf{R}^3 \times \mathbf{R}^3 \to \mathbf{R}^4 : (z, w) \to \big(G_1(z, w), G_2(z, w), G_3, (z, w), G_4(z, w)\big),$$

where

$$G_1(z, w) = z_1^2 + z_2^2 + z_3^2 - 1, \quad G_2(z, w) = z_1 w_1 + z_2 w_2 + z_3 w_3 - a,$$
$$G_3(z, w) = \tfrac{1}{2} I_1^{-1}(w_1^2 + w_2^2 + w_3^2) + \chi z_3 - h^a, \quad G_4(z, w) = w_3 - b.$$

Therefore N is a smooth two dimensional torus. An alternative way to see this, which mimics part of the proof of the existence of action angle coordinates (see appendix D), goes as follows. Because (h^a, b) is a regular value of EM^a, at each $n \in N$ the differentials $dH^a(n)$ and $dJ_\ell^a(n)$ are linearly independent. Therefore, the vector fields $X_{H^a}|N$ and $X_{J_\ell^a}|N$ are tangent to N and are linearly independent at each $n \in N$. Since N is compact and $\{H^a, J_\ell^a\}_{P^a} = 0$, the flows of $X_{H^a}|N$ and $X_{J_\ell^a}|N$ define an $\mathbf{R}^2$ action on N which is transitive. Again because N is compact, the isotropy group L of the $\mathbf{R}^2$ action is a rank 2 lattice. Therefore N is diffeomorphic to $\mathbf{R}^2/L$ which is a two dimensional torus.

We verify the third entry as follows. Suppose that $(H^a_b)^{-1}(h^a_b)$ is a topological circle with one singular point $p = (1, 0, 0)$. Then $q = (\pi^a_b)^{-1}(p)$ is the only singular point of the variety N. The map π^a_b is a proper smooth submersion of $N - \{q\}$ onto $(H^a_b)^{-1}(h^a_b) - \{p\}$. Since $(H^a_b)^{-1}(h^a_b) - \{p\}$ is diffeomorphic to $\mathbf{R}$, it follows that $N - \{q\}$ is diffeomorphic to a cylinder $S^1 \times \mathbf{R}$, (see appendix C section 2). An alternative argument which establishes this goes as follows. Consider the vector fields $X_{H^a}|(N - \{q\})$ and $X_{J^a_\ell}|(N - \{q\})$. Since the integral curves of $X_{J^a_\ell}$ on P^a are periodic and q is an equilibrium point, the flow of $X_{J^a_\ell}|(N - \{q\})$ is periodic and hence is complete. On the other hand, the only way that the flow of the vector field $X_{H^a}|(N - \{q\})$ could be incomplete is for one of its integral curves to reach q in finite time. But this is impossible, because q is an equilibrium point of X_{H^a}. Consequently, the flows of $X_{H^a}|(N - \{q\})$ and $X_{J^a_\ell}|(N - \{q\})$ define a transitive $\mathbf{R}^2$ action on $N - \{q\}$, which has a rank 1 lattice L as isotropy group. Therefore $N - \{q\}$ is diffeomorphic to $\mathbf{R}^2/L$ which is a cylinder $S^1 \times \mathbf{R}$. Hence N is the one point compactification of a cylinder, that is, a two dimensional torus with a meridial circle pinched to a point. A more detailed argument shows that q is a hyperbolic equilibrium point for X_{H^a}. Thus q is a conical singularity of N being the transverse intersection of the stable and unstable manifolds of q. This completes the verification of table 6.4. □

Topology of $(H^a)^{-1}(h^a)$	Conditions
1. point	$h^a = -\chi + \frac{1}{2} I_1^{-1} a^2$
2. a three sphere, S^3	$-\chi + \frac{1}{2} I_1^{-1} a^2 < h^a < \chi + \frac{1}{2} I_1^{-1} a^2$
3. unit tangent S^1 bundle over S^2 with one fiber pinched to a point	$h^a = \chi + \frac{1}{2} I_1^{-1} a^2$
4. unit tangent S^1 bundle over S^2	$h^a > \chi + \frac{1}{2} I_1^{-1} a^2$

Table 6.5 Topology of the level sets $(H^a)^{-1}(h^a)$.

We now determine the topology of the level sets of H^a. The results are given in table 6.5.

To verify the entries in table 6.5 we use Morse theory. The phase space P^a is the preimage of the regular value $(0, 0)$ of the mapping

$$\mathbf{R}^3 \times \mathbf{R}^3 \rightarrow \mathbf{R}^2 : (z, w) \rightarrow \big(F_1(z, w), F_2(z, w) \big),$$

where

$$F_1(z, w) = z_1^2 + z_2^2 + z_3^2 - 1 \quad \text{and} \quad F_2(z, w) = z_1 w_1 + z_2 w_2 + z_3 w_3 - a.$$

Consider the function

$$H : \mathbf{R}^3 \times \mathbf{R}^3 \rightarrow \mathbf{R} : (z, w) \rightarrow \tfrac{1}{2} I_1^{-1}(w_1^2 + w_2^2 + w_3^2) + \chi z_3.$$

▷ The following argument shows that $H^a = H|P^a$ is a Morse function.

(6.5) **Proof**: H^a is bounded below and is proper. Therefore, H^a has a critical point (z, w), which we find using Lagrange multipliers. At (z, w)

$$DH(z, w) + \lambda_1 DF_1(z, w) + \lambda_2 DF_2(z, w) = 0$$

and $F_1(z, w) = F_2(z, w) = 0$. In other words,

$$\begin{aligned} 2\lambda_1 z_1 + \lambda_2 w_1 &= 0, & \lambda_2 z_1 + I_1^{-1} w_1 &= 0, \\ 2\lambda_1 z_2 + \lambda_2 w_2 &= 0, & \lambda_2 z_2 + I_1^{-1} w_2 &= 0, \\ 2\lambda_1 z_3 + \lambda_2 w_3 + \chi &= 0, & \lambda_2 z_3 + I_1^{-1} w_3 &= 0, \end{aligned}$$

$z_1^2 + z_2^2 + z_3^2 = 1$, and $z_1 w_1 + z_2 w_2 + z_3 w_3 = a$. Therefore

$$\lambda_2 = \lambda_2(z_1^2 + z_2^2 + z_3^2) = -I_1^{-1}(z_1 w_1 + z_2 w_2 + z_3 w_3) = -I_1^{-1} a,$$

which gives $w_i = a z_i$ for $i = 1, 2, 3$. Hence

$$\begin{aligned} (2\lambda_1 - I_1^{-1} a^2) z_1 &= 0 \\ (2\lambda_1 - I_1^{-1} a^2) z_2 &= 0 \\ (2\lambda_1 - I_1^{-1} a^2) z_3 + \chi &= 0. \end{aligned}$$

If $2\lambda_1 - I_1^{-1} a^2 = 0$, then $\chi = 0$, which is a contradiction. Therefore $2\lambda_1 - I_1^{-1} a^2 \neq 0$, which gives $z_1 = z_2 = 0$. Consequently, $z_3 = \varepsilon$ with $\varepsilon^2 = 1$ and $w_1 = w_2 = 0$ and $w_3 = \varepsilon a$. Thus we have shown that H^a has two critical points $p_\varepsilon = \varepsilon(e_3, a e_3)$ with Lagrange multipliers $\lambda_1 = \frac{1}{2}(-\varepsilon\chi + I_1^{-1} a^2)$ and $\lambda_2 = -I_1^{-1} a$.

We now show that the critical points p_ε are nondegenerate. The tangent space $T_{p_\varepsilon} P^a$ is $\ker\begin{pmatrix} DF_1(p_\varepsilon) \\ DF_2(p_\varepsilon) \end{pmatrix} = \ker\begin{pmatrix} 0 & 0 & 2\varepsilon & 0 & 0 & 0 \\ 0 & 0 & \varepsilon a & 0 & 0 & \varepsilon \end{pmatrix}$, which is spanned by $\{e_1, e_2, e_4, e_5\}$. Therefore

$$\begin{aligned} D^2 H^a(p_\varepsilon) &= \left(D^2 H + \lambda_1 D^2 F_1 + \lambda_2 D^2 F_2\right)\Big|_{T_{p_\varepsilon} P^a} (p_\varepsilon) \\ &= \begin{pmatrix} I_1^{-1} a^2 - \varepsilon\chi & 0 & -a I_1^{-1} & 0 \\ 0 & I_1^{-1} a^2 - \varepsilon\chi & 0 & -a I_1^{-1} \\ -a I_1^{-1} & 0 & I_1^{-1} & 0 \\ 0 & -a I_1^{-1} & 0 & I_1^{-1} \end{pmatrix}. \end{aligned}$$

Since $\det D^2 H^a(p_\varepsilon) = \chi^2 I_1^{-2} \neq 0$, the critical points p_ε are nondegenerate. Hence H^a is a Morse function. The characteristic polynomial of $D^2 H^a(p_\varepsilon)$ is the square of the polynomial $\lambda^2 - ((a^2+1) I_1^{-1} - \varepsilon\chi)\lambda - \varepsilon\chi I_1^{-1}$. Thus $D^2 H^a(p_\varepsilon)$ has two negative and two positive eigenvalues when $\varepsilon = 1$ and four positive eigenvalues when $\varepsilon = -1$. Therefore $p_1 = (e_3, a e_3)$ is a nondegenerate saddle point of H^a of Morse index 2 corresponding to the critical value $\chi + \frac{1}{2} I_1^{-1} a^2$; whereas $p_{-1} = (-e_3, -a e_3)$ is a nondegenerate minimum corresponding to the critical value $-\chi + \frac{1}{2} I_1^{-1} a^2$. □

▷ We are now ready to determine the topology of the level sets of H^a using Morse theory.

(6.6) **Proof**: When $h^a = -\chi + \frac{1}{2}I_1^{-1}a^2$, the level set $(H^a)^{-1}(h^a)$ is the point p_{-1}. By the Morse lemma (see appendix E section 2) in a neighborhood of p_{-1}, the function H^a is equal to its second derivative at p_{-1} up to a smooth change of coordinates. Therefore for h^a values slightly greater than $-\chi + \frac{1}{2}I_1^{-1}a^2$ the h^a-level set of H^a is diffeomorphic to a three dimensional sphere S^3. By the Morse isotopy lemma (see appendix E section 3) for every $h^a \in (-\chi + \frac{1}{2}I_1^{-1}a^2, \chi + \frac{1}{2}I_1^{-1}a^2)$ the level set $(H^a)^{-1}(h^a)$ is diffeomorphic to S^3.

For $h^a > \chi + \frac{1}{2}I_1^{-1}a^2$ the h^a-level set of H^a is diffeomorphic to the unit tangent S^1 bundle T_1S^2 over S^2. (We have seen in chapter 3 section 1 that the unit tangent circle bundle T_1S^2 of S^2 is diffeomorphic to real projective 3-space $\mathbf{RP}^3$.) To check this we view $(H^a)^{-1}(h^a)$ as a bundle over S^2 by applying the bundle isomorphism

$$\varphi : P^a \to TS^2 : (z, w) \to (x, y) = (z, w - az).$$

The total space $\varphi\big((H^a)^{-1}(h^a)\big)$ of this new bundle over S^2 is defined by

$$\begin{aligned} x_1^2 + x_2^2 + x_3^2 &= 1 \\ x_1y_1 + x_2y_2 + x_3y_3 &= 0 \qquad (79) \\ \tfrac{1}{2}I_1^{-1}(y_1^2 + y_2^2 + y_3^2) + \chi x_3 &= h^a - \tfrac{1}{2}I_1^{-1}a^2. \qquad (80) \end{aligned}$$

Fix a point x on S^2. Then (80) defines a 2-sphere

$$y_1^2 + y_2^2 + y_3^2 = 2I_1\big(h^a - (\chi x_3 + \tfrac{1}{2}I_1^{-1}a^2)\big),$$

which has a positive radius since $h^a > \chi + \frac{1}{2}I_1^{-1}a^2 \geq \chi x_3 + \frac{1}{2}I_1^{-1}a^2$. Intersecting this 2-sphere with the 2-plane (79) defines a circle which is tangent to S^2 at x. This circle is the fiber over x of the projection $TS^2 \to S^2 : (x, y) \to x$. Hence $\varphi\big((H^a)^{-1}(h^a)\big)$ is diffeomorphic to T_1S^2. Consequently $(H^a)^{-1}(h^a)$ is diffeomorphic to T_1S^2.

At the critical value $h^a = \chi + \frac{1}{2}I_1^{-1}a^2$, the h^a-level set of H^a is the algebraic variety

$$\begin{aligned} z_1^2 + z_2^2 + z_3^2 &= 1 \\ z_1w_1 + z_2w_2 + z_3w_3 &= a \\ \tfrac{1}{2}I_1^{-1}(w_1^2 + w_2^2 + w_3^2) + \chi z_3 &= \chi + \tfrac{1}{2}I_1^{-1}a^2, \end{aligned}$$

which is singular only at $p_1 = (e_3, ae_3)$. Because p_1 is a nondegenerate critical point of Morse index 2, $(H^a)^{-1}(h^a)$ is locally diffeomorphic to the zero level set of $D^2H^a(p_1)$, that is, to a neighborhood of the vertex of the cone defined by $z_1^2 - z_2^2 + w_1^2 - w_2^2 = 0$. Thus p_1 is a conical singularity of $(H^a)^{-1}(h^a)$. Because this cone over a 2-torus contains the two 2-planes $\{(z_1, z_2, \pm z_2, \pm z_1) | (z_1, z_2) \in \mathbf{R}^2\}$, which intersect transversely, it is impossible to draw a geometrically correct picture of this singularity in $\mathbf{R}^3$. Arguing as in the case when $h^a > \chi + \frac{1}{2}I_1^{-1}a^2$, we see that $(H^a)^{-1}(h^a) - \{p_1\}$ is diffeomorphic to $T_1S^2 - \{(e_3, e_3)\}$. Thus $(H^a)^{-1}(h^a)$ is homeomorphic to T_1S^2 with one fiber *pinched* to a point. □

This completes the verification of table 6.5. □

From table 6.5 we see that as the value of h^a increases through the critical value $\chi + \frac{1}{2} I_1^{-1} a^2$, the topology of the h^a-level set of H^a changes. This bifurcation is due to monodromy about the isolated critical value $(\chi + \frac{1}{2} I_1^{-1} a^2, a)$ of the energy momentum mapping EM^a when $|a| < \sqrt{4\chi I_1}$, (see figure 6.3). The same argument used to compute the monodromy in the spherical pendulum (see chapter 4 ((5.1))) shows that for a small circle S^1 in the set of regular values of EM^a with center at the isolated critical value the 2-torus bundle $(EM^a)^{-1}(S^1)$ is *nontrivial.*

We now discuss how the level sets of angular momentum foliate an energy level set. For $h^a \in (-\chi + \frac{1}{2} I_1^{-1} a^2, \chi + \frac{1}{2} I_1^{-1} a^2)$ the energy surface $(H^a)^{-1}(h^a) = S^3$ is foliated by 2-tori $(J_\ell^a)^{-1}(b) \cap S^3$ as in the harmonic oscillator. When $h^a > \chi + \frac{1}{2} I_1^{-1} a^2$ the energy surface $(H^a)^{-1}(h^a) = \mathbf{RP}^3$ is foliated by 2-tori so that its twofold covering is the same as the foliation of S^3 by 2-tori of the harmonic oscillator, (see chapter 3 section 5 figure 5.2).

This completes our analysis of the geometry of the energy momentum mapping of the Euler-Poisson equations.

6.3 Motion of the tip of the figure axis

In this subsection we discuss the qualitative behavior of the integral curves of the Euler-Poisson vector field X_{H^a} (32) on the smooth 2-torus $T^2_{H^a,b} = (EM^a)^{-1}(h^a, b)$. Here (h^a, b) is a regular value of the energy momentum mapping EM^a (66). Thinking of the reduced space

$$P^a : z_1^2 + z_2^2 + z_3^2 = 1 \quad \text{and} \quad z_1 w_1 + z_2 w_2 + z_3 w_3 = a$$

as a bundle over S^2 with bundle projection

$$\tau : P^a \subseteq \mathbf{R}^3 \times \mathbf{R}^3 \to S^2 \subseteq \mathbf{R}^3 : (z, w) \to z,$$

we can view the 2-torus $T^2_{h^a,b} \subseteq P^a$ as a bundle over $R = \tau(T^2_{h^a,b}) \subseteq S^2$. Physically, the image of an integral curve of $X_{H^a}|T^2_{h^a,b}$ under the projection τ is the curve in space traced out by the tip of the figure axis of the Lagrange top. We classify the possible motions of the tip.

We will begin by describing the various subsets R of S^2 which are the image of $\tau|T^2_{h^a,b}$.

Claim: Let $\sigma^\pm$ be roots of the polynomial V (see (82) below) which lie in $[-1, 1]$. Then R is one of the following sets.

1. When $b \neq \pm a$ & $a \neq 0$, R is a band

$$B = \left\{ z \in S^2 \subseteq \mathbf{R}^3 \middle| -1 < \sigma^- \leq z_3 \leq \sigma^+ < 1 \right\}$$

with boundary $\partial B = \left\{ z_3 = \sigma^\pm \right\}$.

2. When $b = a$ & $a \neq 0$, R is a cap

$$D_+ = \left\{ z \in S^2 \subseteq \mathbf{R}^3 \middle| -1 < \sigma^- \leq z_3 \leq 1 \right\}$$

which contains the north pole $np = (0, 0, 1)$ of S^2 and has boundary $\partial D_+ = \left\{ z_3 = \sigma^- \right\}$.

3. When $b = -a$ & $a \neq 0$, R is a cap

$$D_- = \left\{ z \in S^2 \subseteq \mathbf{R}^3 \mid -1 \leq z_3 \leq \sigma^+ < 1 \right\}$$

which contains the south pole $sp = (0, 0, -1)$ of S^2 and has boundary $\partial D_- = \left\{ z_3 = \sigma^+ \right\}$.

4. When $b = a = 0$, R is all of S^2.

(6.7) **Proof**: By construction, the 2-torus $T^2_{h^a,b} \subseteq \mathbf{R}^3 \times \mathbf{R}^3$ is given by

$$\begin{cases} z_1^2 + z_2^2 + z_3^2 &= 1 \\ z_1 w_1 + z_2 w_2 + z_3 w_3 &= a \\ \frac{1}{2} I_1^{-1}(w_1^2 + w_2^2 + w_3^2) + \chi z_3 &= h^a \\ w_3 &= b. \end{cases} \tag{81}$$

Substituting the above equations into the identity

$$(z_1 w_2 - z_2 w_1)^2 = (z_1^2 + z_2^2)(w_1^2 + w_2^2) - (z_1 w_1 + z_2 w_2)^2$$

gives

$$0 \leq (z_1 w_2 - z_2 w_1)^2 = 2(\alpha - \beta z_3)(1 - z_3^2) - (a - b z_3)^2 = V(z_3) \tag{82}$$

with $|z_3| \leq 1$. Here

$$\beta = I_1 \chi \quad \text{and} \quad \alpha = I_1(h^a - \tfrac{1}{2} b^2) = I_1 h^a_b.$$

Since (h^a, b) is a regular value in the image of EM^a, V has two simple roots $\sigma^\pm$ in $[-1, 1]$, (see table 6.3). Therefore,

$$R = \tau(T^2_{h^a,b}) \subseteq R' = \left\{ z \in S^2 \mid -1 \leq \sigma^- \leq z_3 \leq \sigma^+ \leq 1 \right\}.$$

Looking at table 6.3, we see that R' is the band B when $\sigma^\pm \in (-1, 1)$; the disc D_+, when $\sigma^+ = 1$; the disc D_-, when $\sigma^- = -1$; and the 2-sphere S^2 when $\sigma^- = -1$ and $\sigma^+ = 1$.

To finish the argument it suffices to show that $R' = \tau(T^2_{h^a,b})$. A nice way to do this, which also explains the geometry of the projection $\tau | T^2_{h^a,b}$, is to apply the diffeomorphism

$$\begin{aligned} &\varphi : T^2_{h^a,b} \subseteq R' \times \mathbf{R}^3 \to R' \times \mathbf{R}^3 : \\ &(z, w) \to (\xi, \eta) = \big(z_1, z_2, z_3, -w_2 + \mu(z_3) z_2, w_1 - \mu(z_3) z_1, b\big) \end{aligned} \tag{83}$$

where

$$\mu(z_3) = \begin{cases} (a - bz_3)(1 - z_3^2)^{-1}, \text{ if } b \neq \pm a \text{ and } a \neq 0 \\ a(1 + z_3)^{-1}, \text{ if } b = a \text{ and } a \neq 0 \\ a(1 - z_3)^{-1}, \text{ if } b = -a \text{ and } a \neq 0 \\ 0, \text{ if } b = a = 0. \end{cases}$$

A calculation shows that $\varphi(T^2_{h^a,b})$ is the 2-torus T^2 in $\mathbf{R}^3 \times \mathbf{R}^3$ defined by

$$\begin{cases} \xi_1^2 + \xi_2^2 &= 1 - \xi_3^2, \quad \xi = (\xi_1, \xi_2, \xi_3) \in R' \\ \xi_1\eta_2 - \xi_2\eta_1 &= 0 \\ \eta_1^2 + \eta_2^2 &= \nu(\xi_3) \\ \eta_3 &= b, \end{cases} \tag{84}$$

where

$$\nu(\xi_3) = \begin{cases} V(\xi_3)(1 - \xi_3^2)^{-1}, \text{ if } b \neq \pm a \text{ and } a \neq 0 \\ [2(\alpha - \beta\xi_3)(1 + \xi_3) - a(1 - \xi_3)]\,(1 + \xi_3)^{-1}, \text{ if } b = a \text{ and } a \neq 0 \\ [2(\alpha - \beta\xi_3)(1 - \xi_3) - a(1 + \xi_3)]\,(1 - \xi_3)^{-1}, \text{ if } b = -a \text{ and } a \neq 0 \\ 2(\alpha - \beta\xi_3), \text{ if } b = a = 0. \end{cases}$$

Let λ be the projection $T^2 \subseteq R' \times \mathbf{R}^3 \to R' : (\xi, \eta) \to \xi$. For $\xi \in R'$, the fiber $\lambda^{-1}(\xi)$ is diffeomorphic to the fiber $\tau^{-1}(\varphi^{-1}(\xi))$. Therefore it suffices to show that the fibers of λ are nonempty. If $\xi \in R' - \{np, sp\}$, then $|\xi_3| < 1$. From the first equation in (84) it follows that ξ_1 and ξ_2 are not both zero. Therefore for fixed ξ_1 and ξ_2 the second equation in (84) defines a line in the η_1-η_2 plane which passes through the origin and intersects the circle defined by the third equation in (84) in two diametrically opposite points. Consequently $\lambda^{-1}(\xi)$ is a point in real *projective 1-space* $\mathbf{RP}^1$. If $\xi = np$ or sp, then using the first equation in (84) we obtain $\xi_1 = \xi_2 = 0$. When this is the case, the second equation in (84) puts no restrictions on η_1 or η_2. Therefore $\lambda^{-1}(np)$ is the circle

$$\left\{(0, 0, 1, \eta_1, \eta_2, b) \in T^2 \middle| \eta_1^2 + \eta_2^2 = 2(\alpha - \beta) > 0\right\}$$

and $\lambda^{-1}(sp)$ is the circle

$$\left\{(0, 0, -1, \eta_1, \eta_2, b) \in T^2 \middle| \eta_1^2 + \eta_2^2 = 2(\alpha + \beta) > 0\right\}.$$

Since (h^a, b) is a regular value of EM^a, it follows that $\alpha > \pm\beta$. Therefore, for every $\xi \in R'$ the fiber $\lambda^{-1}(\xi)$ is a nonempty. Hence $R' = \tau(T^2_{h^a,b}) = R$. □

We now analyze the geometric behavior of the mapping $\widehat{\tau} = \tau | T^2_{h^a,b}$. When the image of $\widehat{\tau}$ is the band B, $\widehat{\tau}$ has fold singularities along the curves $\tau^{-1}(\{z_3 = \sigma^{\pm}\})$, (see figure 6.5). Thus $T^2_{h^a,b}$ is an S^0 bundle over B with the fiber over a point in ∂B pinched to a point. When the image of $\widehat{\tau}$ is the disc D_+, the map $\widehat{\tau}$ has a fold singularity over ∂D_+. After changing coordinates by the diffeomorphism φ (83), the projection $\widehat{\tau}$ becomes the mapping $\widehat{\lambda} = \lambda|\varphi(T^2_{h^a,b})$, which is the real analogue of the *blow up map* of algebraic geometry at np. The blow up map assigns to every point in the unit disc in $\mathbf{R}^2$ punctured at the origin, a point in $\mathbf{R}^2$ and inflates the origin into a circle $\mathbf{RP}^1$ by assigning to every line through the origin a point on this circle. Similarly, when the image of $\widehat{\tau}$ is the disc D_-, the mapping $\widehat{\tau}$ has a fold singularity over ∂D_- and blows up sp. When the image is S^2, $\widehat{\tau}$ blows up the north and south poles of S^2.

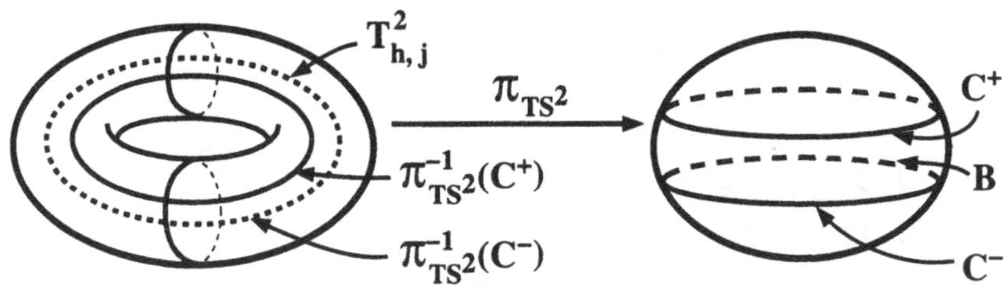

Figure 6.5. Projection map $\tau : T^2_{h^a,b} \to B$.

Next project the vector field $X_{H^a}|T^2_{h^a,b}$ on R. Consider the strip $\widetilde{\mathscr{A}} = [\sigma^-, \sigma^+] \times \mathbf{R}$ with coordinates (σ, φ) where $\sigma = z_3$, $\varphi = \tan^{-1}\frac{z_1}{z_2}$, and $z = (z_1, z_2, z_3) \in R \subseteq S^2$. $\widetilde{\mathscr{A}}$ is the universal covering space of $\mathscr{A}$ which is B, the blow up of $D_\pm$, or the blow up of S^2. Let $p \in T^2_{h^a,b}$ and let $\Gamma : t \to \varphi_t^{H^a}(p)$ be the integral curve of $X_{H^a}|T^2_{h^a,b}$ through p. Set $\gamma : \mathbf{R} \to \mathscr{A} : t \to \tau(\Gamma(t))$ and let $\widetilde{\gamma} : \mathbf{R} \to \widetilde{\mathscr{A}}$ be the lift of γ to $\widetilde{\mathscr{A}}$.

Claim: The curve $\widetilde{\gamma} : \mathbf{R} \to \widetilde{\mathscr{A}} : t \to \big(\sigma(t), \varphi(t)\big)$ satisfies

$$\begin{cases} \dot{\sigma} &= \eta I_1^{-1}\sqrt{V(\sigma)}, \quad \eta^2 = 1 \\ \dot{\varphi} &= \dfrac{I_1^{-1}(b - a\sigma)}{1-\sigma^2} \end{cases} \tag{85}$$

where V is given by (82).

(6.8) **Proof**: We compute. The first equation in (85) is obtained as follows:

$$\begin{aligned} \dot{\sigma} &= L_{X_{H^a}|T^2_{h^a,b}} z_3 = (L_{X_{H^a}} z_3)|T^2_{h^a,b} \\ &= \big(I_1^{-1}(z_2 w_1 - z_1 w_2)\big)|T^2_{h^a,b}, \\ &\qquad \text{using the Euler-Poisson equations:} \\ &\qquad \dot{z} = I_1^{-1} w \times z,\ \dot{w} = \chi e_3 \times z \text{ for } H^a \\ &= \eta I_1^{-1}\sqrt{V(z_3|T^2_{h^a,b})}, \quad \text{using (82)} \\ &= \eta I_1^{-1}\sqrt{V(\sigma)}. \end{aligned}$$

The second equation is obtained as follows:

$$\begin{aligned}\dot{\varphi} &= L_{X_{H^a}|T^2_{h^a,b}}\varphi = \frac{1}{z_1^2+z_2^2}(z_1\dot{z}_2 - z_2\dot{z}_1), \quad \text{by definition of } \varphi \\ &= \frac{I_1^{-1}}{z_1^2+z_2^2}\big(w_3(z_1^2+z_2^2) - z_3(z_1w_1+z_2w_2)\big), \\ &\qquad\qquad \text{using the Euler-Poisson equations for } H^a \\ &= \frac{I_1^{-1}(b-a\sigma)}{1-\sigma^2}, \quad \text{since } \Gamma \text{ lies on } T^2_{h^a,b} \text{ which is defined by (81).} \quad \square\end{aligned}$$

The sign ambiguity in (85) is handled in the following way. The integral curve Γ crosses each of the curves

$$\mathscr{C}^{\pm} = \tau^{-1}(\{z_3 = \sigma^{\pm}\}) \tag{86}$$

transversely, because $\mathscr{C}^{\pm}$ is an integral curve of $X_{J^a_\ell}|T^2_{h^a,b}$ and the vector fields X_{H^a} and $X_{J^a_\ell}$ are linearly independent on $T^2_{h^a,b}$. Thus the curve $t \to \widetilde{\gamma}(t) = \big(\sigma(t), \varphi(t)\big)$ does not stop when it reaches the boundary of $\widetilde{\mathscr{A}}$, say at time t_0. In other words, $\frac{d\widetilde{\gamma}}{dt}(t_0) \neq 0$.

Claim: Suppose that $\frac{d\widetilde{\gamma}}{dt}(t_0) \neq 0$ and $q_0 = \widetilde{\gamma}(t_0) \in \partial\widetilde{\mathscr{A}}$, while $\widetilde{\gamma}(t)$ lies in the interior int $\widetilde{\mathscr{A}}$ of $\widetilde{\mathscr{A}}$ for all t slightly smaller than t_0. Then one of the following holds.

1. $\dot{\varphi}(t_0) = 0$. Then $\frac{d\widetilde{\gamma}}{dt}(t_0)$ is an outward normal to $\partial\widetilde{\mathscr{A}}$ at $\widetilde{\gamma}(t_0)$.

2. $\dot{\varphi}(t_0) \neq 0$. Then $\frac{d\widetilde{\gamma}}{dt}(t_0)$ is tangent to $\partial\widetilde{\mathscr{A}}$ at $\widetilde{\gamma}(t_0)$.

(6.9) **Proof**: 1) Since $\frac{d\widetilde{\gamma}}{dt} = (\dot{\varphi}, \dot{\sigma})$ is nonzero at t_0 and $\dot{\varphi}(t_0) = 0$ by hypothesis, it follows that $\dot{\sigma}(t_0) \neq 0$.

2) Observe that continuity implies that $\dot{\varphi} \neq 0$ for points in $\widetilde{\mathscr{A}}$ near q_0. Hence near q_0 we may parametrize $\widetilde{\gamma}$ by φ instead of t. Using (85) we obtain

$$\frac{d\sigma}{d\varphi} = \frac{\dot{\sigma}}{\dot{\varphi}} = \eta\,\frac{\sqrt{V(\sigma)}}{I_1\dot{\varphi}}.$$

Therefore $\frac{d\sigma}{d\varphi} = 0$ at q_0. $\square$

Suppose that $\widetilde{\gamma}(t_0) = \sigma^-$ or σ^+ and for some sufficiently small $\varepsilon > 0$ the value of η at time $t_0 - \varepsilon$ is known. Then at time $t_0 + \varepsilon$ the value of η is defined to be the negative of its value at $t_0 - \varepsilon$. From ((6.9)) it follows that this *sign convention* insures that the image of $\widetilde{\gamma}$ lies in $\widetilde{\mathscr{A}}$ for all time and is continuous. With this convention in force, equation (85) does *not* define a vector field on $\widetilde{\mathscr{A}}$, because $\widetilde{\gamma}$ may have self intersections. Even worse, $\widetilde{\gamma}$ may *not* be smooth, (see figure 6.11). When $\widetilde{\mathscr{A}}$ is the universal covering space of the band B, the fiber $\tau^{-1}(\partial B)$ is the union of two fold curves $\mathscr{C}^{\pm}$ (86) of $\tau|T^2_{h^a,b}$. In this case, $\widetilde{\gamma}$ has second order contact with $\partial\widetilde{\mathscr{A}}$ at $\widetilde{\gamma}(t_0)$ and is real analytic, although it still may have self intersections, (see figure 6.7).

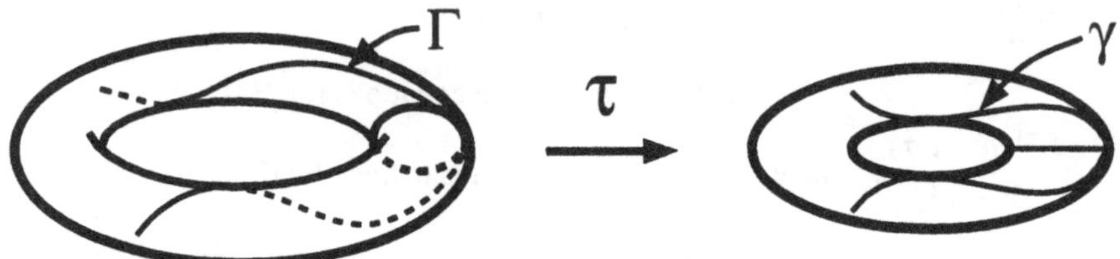

Figure 6.6. Tangential singularity.

We now determine the rotation number Θ of the flow of X_{H^a} restricted to the 2-torus $T^2_{h^a,b}$. Because every integral curve of $X_{J^a_\ell}$ on $T^2_{h^a,b}$ is periodic and projects under the bundle projection τ onto a periodic integral curve of $X_{H^a_b}$ of energy h^a_b, it follows that $\mathcal{C} = \mathcal{C}^-$ (86) is a cross section for the flow $\varphi^{H^a}_t$ of $X_{H^a}|T^2_{h^a,b}$. Thus for any $p \in \mathcal{C}$ there is a smallest $T = T(h^a, b) > 0$ such that $\varphi^{H^a}_T(p) \in \mathcal{C}$. Let φ_s be the flow of $X_{J^a_\ell}$. By definition of the rotation number Θ we have $\varphi_{2\pi\Theta}(p) = \varphi^{H^a}_T(p)$. From the choice of cross section $\mathcal{C}$ it follows that $\tau(\mathcal{C})$ lies on the boundary component of the annulus $\mathcal{A}$ corresponding to the lower boundary $\{\sigma = \sigma^-\}$ of the strip $\widetilde{\mathcal{A}}$. Let

$$\widehat{\Gamma} : [0, 2\pi] \to T^2_{h^a_b,b} : s \to \varphi_s(p).$$

The lift of $\tau \circ \widehat{\Gamma}$ to $\widetilde{\mathcal{A}}$ is the line segment on $\{\sigma = \sigma^-\}$ which joins $\widetilde{\gamma}(0)$ to $\widetilde{\gamma}(T)$. Let ϑ be the difference in the φ-coordinates of $\widetilde{\gamma}(T)$ and $\widetilde{\gamma}(0)$.

Claim: $\vartheta/2\pi$ is equal to the rotation number Θ.

(6.10) **Proof**: We compute.

$$\vartheta/2\pi = \int_0^{\vartheta/2\pi} d\varphi = \int_0^{\Theta} L_{X_{J^a_\ell}} \varphi \, ds = \int_0^{\Theta} ds = \Theta.$$

The second to last equality above follows from the Euler-Poisson equations

$$\dot{z} = e_3 \times z \quad \text{and} \quad \dot{w} = e_3 \times w$$

for $X_{J^a_\ell}$ and the definition of the coordinate φ. □

▷ The following is an explicit formula for the rotation number Θ:

$$2\pi\Theta = 2 \int_{\sigma^-}^{\sigma^+} \frac{b - a\sigma}{(1-\sigma^2)\sqrt{V(\sigma)}} \, d\sigma. \tag{87}$$

(6.11) **Proof**: If σ lies in the interior int $\widetilde{\mathcal{A}}$ of $\widetilde{\mathcal{A}}$, then $V(\sigma) > 0$. Therefore $\dot{\sigma} \neq 0$ for every $\sigma \in \text{int}\,\widetilde{\mathcal{A}}$. Suppose that $\sigma(0) = \sigma^-$. Then there is a $t' > 0$ such that $\sigma(t') = \sigma^+$.

Moreover, we may parametrize $\widetilde{\gamma}$ by σ in (σ^-, σ^+) instead of $t \in (0, t') \cup (t', T)$. In (85) choose $\eta = -1$ for $t = \varepsilon$. Then

$$2\pi\Theta = \int_0^T L_{X_{H^a}|T^2_{h^a,b}} \varphi \, dt = -\int_{\sigma^+}^{\sigma^-} \frac{d\varphi}{d\sigma} d\sigma + \int_{\sigma^-}^{\sigma^+} \frac{d\varphi}{d\sigma} d\sigma = 2\int_{\sigma^-}^{\sigma^+} \frac{\dot{\varphi}}{\dot{\sigma}} d\sigma,$$

which using (85) gives (87). □

We now describe the qualitative behavior of a solution $\widetilde{\gamma}$ of (85). We assume that the sign convention is in force. Then $\widetilde{\gamma}$ is defined for all time and lies in $\widetilde{\mathscr{A}}$. If at $q_0 \in \partial\widetilde{\mathscr{A}}$ we have $\dot{\varphi} \neq 0$, then we say that $\widetilde{\gamma}$ has a *tangential singularity* at q_0. Until further notice we shall assume that $\widetilde{\gamma}$ has only tangential singularities.

To classify the possible qualitative behavior of $\widetilde{\gamma}$, it suffices to look at a time interval when $\widetilde{\gamma}$ has three successive tangential singularities, say $q_0 = (\sigma^-, \varphi^0) = \gamma(0)$, $q_1 = (\sigma^+, \varphi^1) = \gamma(t')$, and $q_2 = (\sigma^-, \varphi^2) = \gamma(T)$, (see figure 6.7). The reason why this suffices is that the image of $\widetilde{\gamma}$ in $\widetilde{\mathscr{A}}$ is invariant under the translation mapping

$$\text{trans} : (\sigma, \varphi) \to (\sigma, \varphi + 2\pi\Theta).$$

In more detail: during an elapsed time T the φ-coordinate of $\widetilde{\gamma}$ increases by $2\pi\Theta$. Hence the image of $\widetilde{\gamma}$ is invariant under the translation map trans .

Claim: The possible qualitative behavior of $\widetilde{\gamma}$ in $\widetilde{\mathscr{A}}$ with only tangential singularities is given in figure 6.7.

(6.12) **Proof**: We assume that $\dot{\varphi} > 0$ at q_0. The argument when $\dot{\varphi} < 0$ is similar and is omitted. Since $\dot{\sigma} \neq 0$ in int $\widetilde{\mathscr{A}}$, the curves $\widetilde{\gamma}_1 = \widetilde{\gamma}|(0, t')$ and $\widetilde{\gamma}_2 = \widetilde{\gamma}|(t', T)$ may be parametrized by σ instead of t. As curves parametrized by σ, $\widetilde{\gamma}_1$ and $\widetilde{\gamma}_2$ are each a graph of a smooth function F_1 and F_2, respectively, which are defined on (σ^-, σ^+).

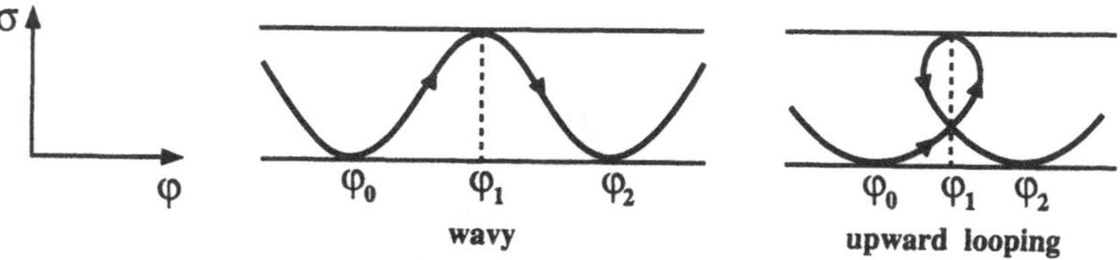

Figure 6.7. The possible qualitative behaviors of $\widetilde{\gamma}$ having only tangential singularities.

Case I. Suppose that $\dot{\varphi} > 0$ at q_1. Then (85) confirms that $\dot{\varphi} > 0$ throughout $\widetilde{\mathscr{A}}$. Because the sign of $\dot{\varphi}$ does not change as $\widetilde{\gamma}$ passes through q_1, whereas the sign of $\dot{\sigma}$ does by our sign convention, the function F_1 is strictly increasing; whereas the function F_2 is strictly decreasing. Thus the closures of the graphs of F_1 and F_2 intersect only at q_1. Since $\widetilde{\gamma}$ has a tangential singularity at q_1, $\widetilde{\gamma}$ is convex downward at q_1. A similar argument shows that $\widetilde{\gamma}$ is convex upward at q_0 and q_2. Therefore $\gamma|[0, T]$ is a *wave*. Since $\varphi^2 > \varphi^0$, the rotation number $\Theta = (\varphi^2 - \varphi^0)/2\pi$ is positive.

Case II. Suppose that $\dot{\varphi} < 0$ at q_1. From (85) it follows that $\frac{d\varphi}{d\sigma}$ vanishes exactly once in (σ^-, σ^+). Suppose that $a = 0$. From (85) we see that $\dot{\varphi}$ has the same sign as b throughout $\widetilde{\mathcal{A}}$. But this is contrary to the hypothesis that $\dot{\varphi}$ changes sign in $\widetilde{\mathcal{A}}$. Therefore, $a \neq 0$. Suppose that $a < 0$. From the fact that $\dot{\varphi} > 0$ at q_0 it follows that $b - a\sigma^- > 0$. Therefore $b/a < \sigma^-$. From the fact that $\dot{\varphi} < 0$ at q_1, it follows that $b - a\sigma^+ < 0$. Therefore $b/a > \sigma^+$. This is a contradiction because $\sigma^+ > \sigma^-$. Consequently $a > 0$. Since $\frac{d\varphi}{d\sigma}$ has only one zero at $\sigma^* = b/a$, each of the functions F_1 and F_2 has only one critical point at σ^* in (σ^-, σ^+). Now $\frac{dF_1}{d\sigma}$ and $\frac{dF_2}{d\sigma}$ have opposite signs near q_1. To be specific suppose that $\frac{dF_1}{d\sigma} < 0$ near q_1. (The argument for the other case is similar to the one given below and is omitted.) Then $\frac{dF_2}{d\sigma} > 0$ near q_1. It follows that F_1 has a strict maximum at σ^*, whereas F_2 has a strict minimum there. Because $\frac{dF_1}{d\sigma} < 0$ and $\frac{dF_2}{d\sigma} > 0$ in $(\sigma^*, \sigma^+]$, we find that

$$\begin{aligned} \varphi^1 - F_1(\sigma^*) &= F_1(\sigma^+) - F_1(\sigma^*) = \int_{\sigma^*}^{\sigma^+} \frac{dF_1}{d\sigma}\, d\sigma < \int_{\sigma^*}^{\sigma^+} \frac{dF_2}{d\sigma}\, d\sigma \\ &= F_2(\sigma^+) - F_2(\sigma^*) = \varphi^1 - F_2(\sigma^*), \end{aligned}$$

that is, $F_1(\sigma^*) > F_2(\sigma^*)$. Therefore the point $q^* = (\sigma^*, F_2(\sigma^*))$ does *not* lie on the graph of F_1. There are three possible locations for q_2 relative to the graph of F_1. Each of the three locations gives rise to a different possible qualitative behavior.

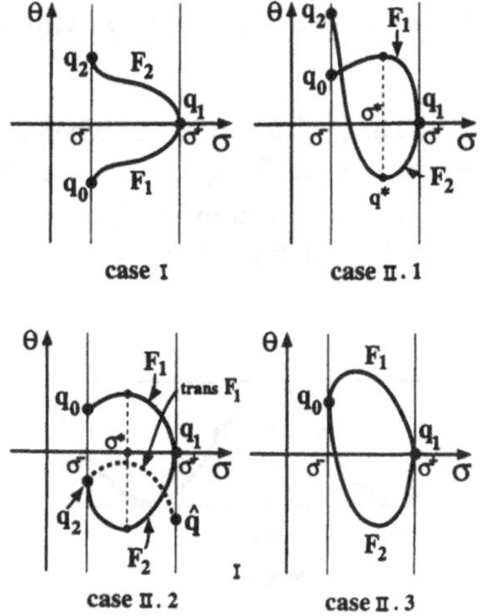

Figure 6.8. The geometric situation.

1. Suppose that q_2 lies above the graph of F_1. Then q_2 and q^* lie on opposite sides of the graph of F_1. Thus the graph of $F_2|(\sigma^-, \sigma^*]$ crosses the graph of F_1 at least once. It crosses exactly once because $F_2|(\sigma^-, \sigma^*]$ is strictly monotonic. Therefore in $\widetilde{\mathcal{A}}$ the curve $\widetilde{\gamma}|[0, T]$ makes an *upward pointing loop*. Note that the rotation number $\Theta = (\varphi^2 - \varphi^0)/2\pi$ is positive.

2. Suppose that q_2 lies below the graph of F_1. Then q_2 and q^* lie on the same side of the graph of F_1. Note that $F_1(\sigma^*) > F_1(\sigma^-) = \varphi^0$, while $F_2(\sigma^*) < F_2(\sigma^-) = \varphi^2$. Applying the translation mapping trans to the graph of F_1 gives trans $(q_0) = q_2$ and trans $(q_1) = \widehat{q}$.

Since $F_1|(\sigma^-, \sigma^*]$ is strictly increasing, so is trans $(F_1|(\sigma^-, \sigma^*])$. Because $F_2|(\sigma^-, \sigma^*]$ is strictly decreasing and

$$\begin{aligned} \text{trans}\,(F_1(\sigma^*)) &= F_1(\sigma^*) + \big(F_2(\sigma^-) - F_1(\sigma^-)\big) \\ &= \big(F_1(\sigma^*) - F_1(\sigma^-)\big) + F_2(\sigma^-) > F_2(\sigma^-), \end{aligned}$$

it follows that trans $(F_1(\sigma^*))$ lies above the graph of F_2. Since

$$F_2(\sigma^+) = q_1 > \text{trans}\,(q_1) = \widehat{q} = \text{trans}\,(F_1(\sigma^+)),$$

the point $\widehat{q}$ lies below the graph of F_2. Thus we find that the graph of trans $(F_1|[\sigma^*, \sigma^+))$ crosses the graph of F_2 at least once. It crosses exactly once because $F_1|[\sigma^*, \sigma^+)$ is strictly monotonic and hence trans $(F_1|[\sigma^*, \sigma^+))$ is strictly monotonic. Therefore in $\widetilde{\mathscr{A}}$ the curve $\widetilde{\gamma}|[0, T]$ makes a *downward pointing loop*. Note that the rotation number Θ is *negative*.

3. Suppose that q_2 lies on the graph of F_1. Then $q_2 = q_0$. Thus the closures of the graphs of F_1 and F_2 form a smooth closed curve, which does not have any self intersections. Thus $\widetilde{\gamma}|[0, T]$ forms a *wheel*. The rotation number Θ is zero.

▷ We now show that downward looping and wheeling motion do *not* occur, because the rotation number Θ is positive when $b/a \in (-1, 1)$ and $a > 0$.

(6.13) **Proof**: Consider the oriented curves $\mathscr{C}_i$ and $\mathscr{D}_i$ in the extended complex plane $\widetilde{\mathbf{C}}$ cut as given in figure 6.9. Note that the real root σ^0 of V, which does not lie in $[-1, 1]$, is strictly greater than 1. Let $\omega = \frac{b - a\sigma}{(1-\sigma^2)\sqrt{V(\sigma)}}$. Then ω is a meromorphic 1-form on $\widetilde{\mathbf{C}}$ with first

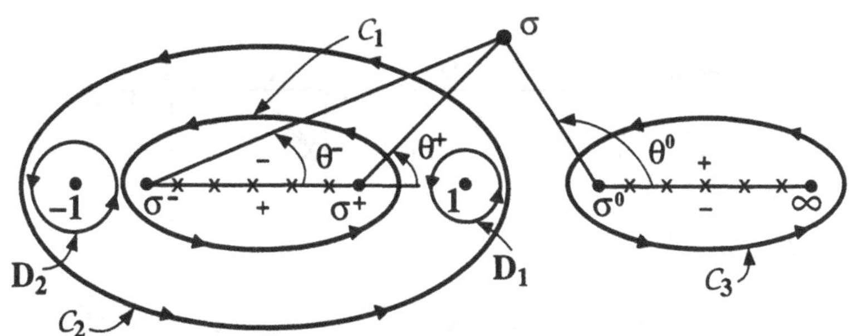

Figure 6.9 The contours $\mathscr{C}_i$ and $\mathscr{D}_i$ in $\widetilde{\mathbf{C}}$ and the choice of square root.

order poles at $\varepsilon = \pm 1$ which have residue $\operatorname{Res}_{z=\varepsilon} \omega = \frac{1}{2}i$. This follows from the choice of square root indicated in figure 6.9 and the fact that $b/a \in [-1, 1]$. Hence by the residue theorem

$$\int_{\mathscr{C}_2} \omega = 2\pi i \operatorname*{Res}_{z=-1} \omega + 2\pi i \operatorname*{Res}_{z=1} \omega + \int_{\mathscr{C}_1} \omega = -2\pi + \int_{\mathscr{C}_1} \omega.$$

By Cauchy's theorem

$$\int_{\mathscr{C}_2} \omega = \int_{\mathscr{C}_3} \omega = 2a \int_{\sigma_0}^{\infty} \frac{\sigma - b/a}{(\sigma^2 - 1)\sqrt{V(\sigma)}}\, d\sigma > 0,$$

since $b/a \in [-1, 1]$, $a > 0$ and $\sigma^0 > 1$. This proves the assertion because $2\pi\Theta = \int_{\mathscr{C}_1} \omega$. □

This completes the verification of figure 6.7. □

Figure 6.10. Motion of the tip of the figure axis of the top in the band B on S^2.

We now translate the motions of $\widetilde{\gamma}$ on $\widetilde{\mathscr{A}}$ into motions γ of the tip of the figure axis of the top on $\mathscr{A}$. There are two cases: when $\widetilde{\mathscr{A}}$ is the universal covering space of a band B or the universal covering space of the disc $D_\pm$ blown up at its center. In the case of a band, the translation is straightforward and the results are given in figure 6.10. In the case of a disc, the translation is a bit tricky. Consider the curve γ in D_+ given in figure 6.11(a),

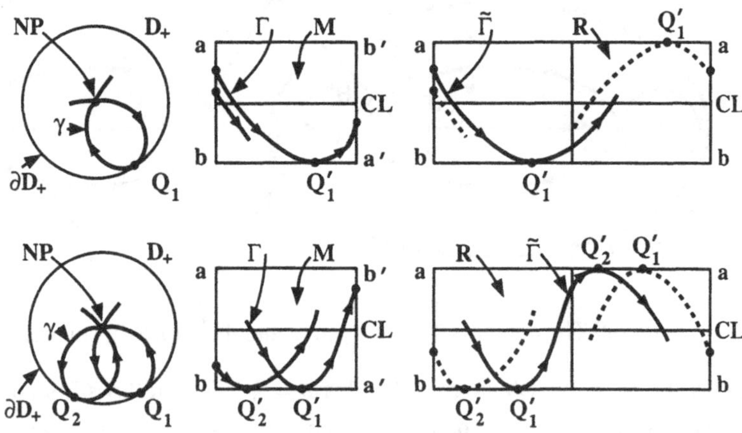

Figure 6.11. (a) and (a)′. Motion of the tip of the figure axis in the disc through north pole. (b) and (b)′. Blown up motion in the Möbius band M. (c) and (c)′. Motion in the strip R or the annulus.

which represents a motion of the tip of the figure axis of the top which passes through the north pole NP. Under the blow up map the north pole becomes the center line CL of a Möbius band M, (see figure 6.11(b), where the edge $\overline{\mathrm{ab}}$ is identified with the edge $\overline{\mathrm{a'b'}}$). The blown up curve Γ crosses CL transversely at α and β (which correspond to the two distinct tangent lines to γ at NP). Note that at Q_1' (which corresponds to Q_1) the curve Γ has second order contact with the boundary of M. This follows because γ has second order contact with ∂D_+ at Q_1 and the blow up map is a diffeomorphism of $D_+ - \mathrm{NP}$ onto $\mathrm{M} - \mathrm{CL}$. The double cover of Γ given in figure 6.11(b) is the curve $\widetilde{\Gamma}$ given in figure 6.11(c), where the edge $\overline{\mathrm{ab}}$ is identified with the edge $\overline{\mathrm{a'b'}}$. Now consider the curve γ in D_+ given in figure 6.11(a)'. Under the blow up map γ becomes the curve Γ in the Möbius band given in figure 6.11(b)'. The double cover of Γ is the curve $\widetilde{\Gamma}$ in the annulus given in figure 6.11(c)'. Clearly $\widetilde{\Gamma}$ is a *wavy* motion. From this discussion we see that motion of the tip of the axis through the north pole is *not* a limit of cuspy motion in a band whose upper boundary shrinks to the north pole. For more details see after ((6.15)) below.

We now turn to discussing singularities of $\widetilde{\gamma}$ which are *not* tangential. If $\dot{\varphi} = 0$ at $q_0 \in \partial\widetilde{\mathscr{A}}$, then we say that $\widetilde{\gamma}$ has a *nontangential singularity* at q_0. From ((6.9)) we see that $\frac{d\widetilde{\gamma}}{dt}(t_0)$ is an outward pointing normal to $\partial\widetilde{\mathscr{A}}$ at $q_0 = \widetilde{\gamma}(t_0)$. If in addition $\dot{\varphi}$ is not identically zero on $\mathscr{A}$, we say that q is *nondegenerate*; otherwise we say that it is *degenerate*.

Claim: A nondegenerate nontangential singularity of $\widetilde{\gamma}$ can only occur on the upper boundary $\{\sigma = \sigma^+\}$ of $\widetilde{\mathscr{A}}$.

Before we begin the proof, here is a physical argument using the magnetic spherical pendulum model. When $\dot{\varphi} = 0$, the figure axis of the top is not moving. Hence the *Lorentz force* on the electrically charged particle from the magnetic monopole field is zero. Thus only downward gravity is acting. Since the particle is moving in a band, this can only happen at the upper boundary.

(6.14) **Proof**: Suppose that $\widetilde{\gamma}$ has a nondegenerate nontangential singularity at $q_0 = (\sigma^-, \varphi_0)$. From the definition of this kind of singularity and using (85) it follows that $a \neq 0$, $\sigma^- = b/a$ and $|b/a| < 1$. Because σ^- is a root of V, from (82) we obtain $\sigma_2 = 0$. But $\tau^{-1}(q_0)$ lies on an integral curve of $X_{H^a}|T^2_{h^a,b}$. Hence its image under the reduction map π^a_b (60) lies on an integral curve of $-\mathrm{ad}_{H^a_b}$ which intersects $P^a_b \cap \{\sigma_2 = 0\}$. Since (h^a, b) is a regular value of EM^a, from figure 6.2 we see that the line

$$\rho(\ell_\alpha) : \ \sigma_3 + 2\beta\sigma_1 = 2\alpha$$

intersects the image of the fold curve

$$\rho(\mathscr{C}) : \sigma_3 = (b - a\sigma_1)(1 - \sigma_1^2)^{-1}, \quad |\sigma_1| < 1$$

in two distinct points (σ^+, σ_3^+) and (σ^-, σ_3^-) with $\sigma^- < \sigma^+$ because $\sigma^\pm$ are simple roots of V in $(-1, 1)$. Since $\sigma^- = b/a$, it follows that $\sigma_3^- = 0$. However, the slope of the line $\rho(\ell_\alpha)$ is negative and $\rho(\mathscr{C})$ lies in the half plane $\sigma_3 \geq 0$. Therefore $(\sigma^-, \sigma_3^-) = (b/a, 0)$ cannot be the left most point of intersection of $\rho(\ell_\alpha)$ with $\rho(\mathscr{C})$. This contradicts the definition of σ^-. □

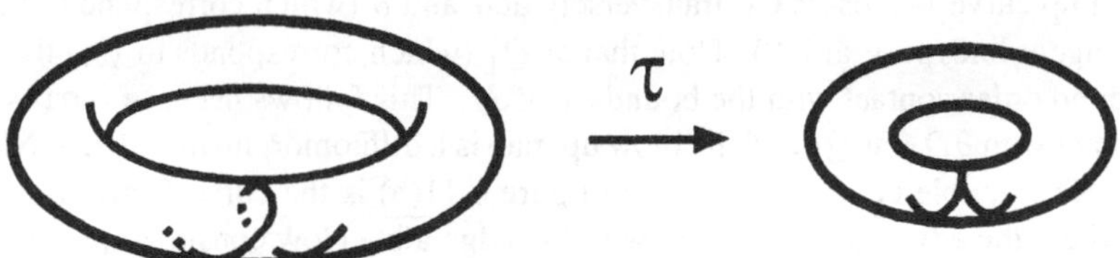

Figure 6.12. Appearance of a cusp.

Claim: Suppose that $q_1 = (\sigma^+, \varphi^1) \in \partial\widetilde{A}$ is a nondegenerate nontangential singularity of $\widetilde{\gamma}$. Then $\widetilde{\gamma}$ has a cusp at q_1.

(6.15) **Proof**: From the definition of nondegenerate nontangential singularity and (85) we find that $a \neq 0$, $\sigma^+ = b/a$ and $|b/a| < 1$. Parametrizing $\widetilde{\gamma}$ near q_1 by σ we obtain

$$\frac{d\varphi}{d\sigma} = \frac{a(\sigma^+ - \sigma)}{(1-\sigma^2)\sqrt{V(\sigma)}}. \tag{88}$$

Since $T^2_{h^a,b}$ is a smooth 2-torus, V has two simple real roots $\sigma^\pm$ in $(-1, 1)$ and one real root $\sigma_0 > 1$. Therefore $V'(\sigma^+) < 0$. Expanding $(1-\sigma^2)^{-1}$ and $V^{-1/2}$ in a Taylor series about σ^+, equation (88) becomes

$$\frac{d\varphi}{d\sigma} = \frac{a}{(1-(\sigma^+)^2)\sqrt{-V'(\sigma^+)}}(\sigma^+ - \sigma)^{1/2} + \mathrm{O}\big((\sigma^+ - \sigma)\big).$$

Integrating the above equation gives

$$\varphi(\sigma) = -\frac{2a}{3(1-(\sigma^+)^2)\sqrt{-V'(\sigma^+)}}(\sigma^+ - \sigma)^{3/2} + \mathrm{O}\big((\sigma^+ - \sigma)^2\big).$$

Therefore γ has a cusp at q_1. □

Figure 6.13. Transition between wavy and looping motion without becoming cuspy.

From the proof of ((6.15)) we see that $\widetilde{\gamma}$ has a *cusp singularity* if and only if V has two simple roots σ^- and $\sigma^+ = b/a$ in $(-1, 1)$ with $0 < |b| < |a|$. Consequently the tip of the figure axis of the top makes a *cuspy motion* only when $\widetilde{\mathscr{A}}$ is the universal covering space of the band B and only at those (a, b, h^a) values where $\alpha = \frac{b}{a}\beta$, since $(b/a, 0)$ lies on $\rho(\ell_\alpha)$. (Recall that $\alpha = I_1 h^a - \frac{1}{2}b^2$ and $\beta = I_1\chi$). Thus the north pole of S^2 is *never* a cusp point of the motion of the tip of the figure axis.

In figure 6.13 we have sketched how the tip of the figure axis can pass from a wavy to an upward looping motion without passing through a cuspy motion by a continuous variation in the parameter (a, b, h^a) and the initial condition. The motion of the tip in the middle drawing in figure 6.13 is *not* smooth as parametrized by arclength, even though the motion disregarding the time parametrization *is*. This lack of smoothness is due to the geometric behavior of the projection mapping τ.

Type of motion	Conditions	The roots $\sigma^\pm$
1. wavy	$b \neq \pm a,\ a \neq 0,\ b/a > 0$	$\sigma^\pm$ are simple roots of V in $(-1, 1)$. $\lvert\sigma^\pm\rvert \neq 1$. $\frac{d\varphi}{d\sigma}$ has same sign at $\sigma^\pm$.
2. upward looping	$b \neq \pm a,\ a \neq 0, b/a < 0$	$\sigma^\pm$ are simple roots of V in $(-1, 1)$. $\sigma^- < b/a$ & $\sigma^+ > 0$ $\frac{d\varphi}{d\sigma}$ has opposite signs at $\sigma^\pm$.
3. cuspy	$0 < \lvert b\rvert < \lvert a\rvert,\ \alpha = \beta\dfrac{b}{a}$	$\sigma^\pm$ are simple roots of V in $(-1, 1)$. $\sigma^+ = a/b$. $\frac{d\varphi}{d\sigma} = 0$ at σ^+.
4. through north pole	$b = a,\ a \neq 0,\ \alpha \neq \beta$	σ^- and $\sigma^+ = 1$ are simple roots of V in $(-1, 1]$. $\frac{d\varphi}{d\sigma} \neq 0$ at $\sigma^\pm$.
5. through south pole	$b = -a,\ a \neq 0,\ \alpha \neq -\beta$	$\sigma^- = -1$ and σ^+ are simple roots of V in $[-1, 1)$. $\frac{d\varphi}{d\sigma} \neq 0$ at $\sigma^\pm$.
6. arc of great circle	$b = a = 0,\ \lvert\alpha\rvert < \beta$	$\sigma^- = -1$ and $\sigma^+ = \alpha/\beta$ are simple roots of V in $[-1, 1)$.
7. great circle	$b = a = 0,\ \alpha > \beta$	$\sigma^\pm = \pm 1$ are simple roots of V.

Table 6.6 Types of motion of the tip of the figure axis. Here (h^a, b) is a regular value of the energy momentum mapping EM^a.

Suppose that $\widetilde{\gamma}$ has a *degenerate* nontangential singularity in $\widetilde{\mathscr{A}}$. Then $\widetilde{\gamma}$ moves periodically with period T along a vertical line joining the two boundary components of $\widetilde{\mathscr{A}}$. Since $\dot{\varphi} = 0$ on all of $\widetilde{\mathscr{A}}$, from (85) it follows that $a = b = 0$. Hence $V(\sigma) = 2(\alpha - \beta\sigma)(1 - \sigma^2)$. Suppose that $\alpha/\beta < 1$. Then $V(\sigma) \geq 0$ for $[-1, \alpha/\beta]$. Hence $\sigma^- = -1$ and $\sigma^+ = \alpha/\beta$.

Thus $\widetilde{\mathcal{A}}$ is the universal covering space of the disc D_- blown up at the south pole. So the tip of the figure axis periodically traces out an arc of a great circle on S^2 which passes through the south pole of S^2. Its rotation number is 0. If $\alpha/\beta > 1$, then $\sigma^{\pm} = \pm 1$. In this case, $\widetilde{\mathcal{A}}$ is the universal covering space of S^2 blown up at the north and south poles. The tip of the figure axis traverses a great circle through the north and south poles of S^2. Its rotation number is ± 1.

In table 6.6 we summarize the classification of motions of the tip of the figure axis of the Lagrange top.

7 The energy momentum mapping

In this section we investigate the geometry of the *energy momentum mapping*

$$\widetilde{\mathcal{EM}} : T\mathrm{SO}(3) \to \mathbf{R}^3 : X_A \to \big(\mathcal{H}(X_A), \mathcal{J}_r(X_A)\mathcal{J}_\ell(X_A)\big)$$

of the Lagrange top. Here $\mathcal{H}$ is the Hamiltonian (4), $\mathcal{J}_r$ the angular momentum (8) of the right S^1 action, and $\mathcal{J}_\ell$ the angular momentum (6) of the left S^1 action. Because the left trivialization mapping $\mathcal{L}$ (10) of $T\mathrm{SO}(3)$ is a diffeomorphism, it suffices to look at the geometry of

$$\begin{aligned} &\mathcal{EM} : \mathrm{SO}(3) \times \mathrm{so}(3) \to \mathbf{R}^3 : \\ &\qquad (A, X) \to \big(H(A,X), J_r(A,X), J_\ell(A,X)\big). \end{aligned} \tag{89}$$

Here $H = \mathcal{L}^*\mathcal{H}$ is given by (90), $J_r = \mathcal{L}^*\mathcal{J}_r$ by (12), and $J_\ell = \mathcal{L}^*\mathcal{J}_\ell$ by (48). If we understand the topology of the fibers $\mathcal{EM}^{-1}(h,a,b)$ and how they fit together to form $H^{-1}(h)$, then we have a complete qualitative picture of the invariant manifolds of the Hamiltonian vector field X_H, whose integral curves give the motions of the Lagrange top.

7.1 Topology of $\mathcal{EM}^{-1}(h,a,b)$ and $H^{-1}(h)$

In this subsection we reconstruct the topology of $\mathcal{EM}^{-1}(h,a,b)$ from the topology of the level set $(H_b^a)^{-1}(h_b^a)$ and the geometry of the reduction mapping $\pi : J_r^{-1}(a) \to P_b^a$. The map π is the composition of the reduction map

$$\pi^a : J_r^{-1}(a) \to P^a$$

(14) of the right S^1 action (11) and the reduction map

$$\pi_b^a : (J_\ell^a)^{-1}(b) \subseteq P^a \to P_b^a$$

(60) of the induced left S^1 action (43) on P^a. Reconstruction is possible because

$$\mathcal{EM}^{-1}(h,a,b) = \pi^{-1}\big((H_b^a)^{-1}(h_b^a)\big),$$

where

$$h_b^a = h - \tfrac{1}{2}I_1^{-1}(b^2 - a^2) - \tfrac{1}{2}I_3^{-1}a^2.$$

The results are given in table 7.1.

Conditions	Topology of $(H^a_b)^{-1}(h^a_b)$	Topology of $\mathcal{EM}^{-1}(h, a, b)$
1. h^a_b is a regular value of H^a_b	a smooth S^1	a smooth 3-torus T^3
2. V has a multiple root in $(-1, 1)$	a nonsingular point of P^a_b	a smooth 2-torus T^2
3. $a = b, h^a_b = \chi + \frac{1}{2I_1}a^2$, $\lvert a\rvert < \sqrt{4\beta}$	a topological S^1 with a conical singular point	a 3-torus with a normal crossing along an S^1
4. $a = b, h^a_b = \chi + \frac{1}{2I_1}a^2$, $\lvert a\rvert \geq \sqrt{4\beta}$	a singular point of P^a_a	a smooth S^1
5. $a = -b$, $h^a_b = -\chi + \frac{1}{2I_1}a^2$	a smooth S^1	a smooth 3-torus T^3

Table 7.1. Topology of $\mathcal{EM}^{-1}(h, a, b)$

▷ We now verify the entries in the third column of table 7.1.

(7.1) **Proof**: Over every nonsingular point of P^a_b the fiber of the reduction mapping π is a smooth 2-torus. From this the second entry follows.

Since h^a_b is a regular value of H^a_b, the level set $(H^a_b)^{-1}(h^a_b)$ bounds a contractible 2-disc in P^a_b. Therefore $\pi^{-1}\big((H^a_b)^{-1}(h^a_b)\big)$ is a trivial 2-torus bundle over S^1, that is, $\mathcal{EM}^{-1}(h, a, b)$ is a smooth 3-torus. This verifies the first entry.

Over a singular point $\widetilde{\sigma}$ of $P^a_{\varepsilon a}$ with $\varepsilon^2 = 1$ the fiber of π is a circle. This follows because the fiber $(\pi^a_b)^{-1}(\widetilde{\sigma})$ is a point p, being a fixed point of the left S^1 action on P^a; while the fiber of $(\pi^a)^{-1}(p)$ is the smooth circle $\pi^{-1}(\widetilde{\sigma})$. This verifies the fourth and fifth entries.

To verify the third entry suppose that $(H^a_a)^{-1}(h^a_a)$ is a topological circle $\mathcal{C}$ with a conical singular point $\widetilde{\sigma} = (1, 0, 0)$, which is the singular point of P^a_a. Since $\mathcal{C} - \{\widetilde{\sigma}\}$ is contractible, $\pi^{-1}(\mathcal{C} - \{\widetilde{\sigma}\})$ is diffeomorphic to $T^2 \times \mathbf{R}$. The fiber $\pi^{-1}(\widetilde{\sigma})$ is a nondegenerate critical circle of $H|\big(J_r^{-1}(a) \cap J_\ell^{-1}(b)\big)$ of Morse index 2 because p is a nondegenerate critical point of $H^a|(J^a_\ell)^{-1}(b)$ of Morse index 2 (see ((6.5)). By the Morse lemma the local stable and unstable manifolds of $H^a|(J^a_\ell)^{-1}(b)$ at p are diffeomorphic to $x_1^2 - x_2^2 - x_3^2 = 0$, which is a cone C on S^1. The local stable and unstable manifolds of $\pi^{-1}(\widetilde{\sigma})$ are untwisted, because the local stable and unstable manifolds of p are contractible and a fiber of π^a is an S^1. Therefore a neighborhood of $\pi^{-1}(\widetilde{\sigma})$ in the fiber $\pi^{-1}(\mathcal{C})$ is diffeomorphic to $S^1 \times C$, that is, is a cone on T^2. Consequently, $\pi^{-1}(\mathcal{C})$ is a 3-torus with a normal crossing along the circle $\pi^{-1}(\widetilde{\sigma})$. In other words, $\pi^{-1}(\mathcal{C})$ is a product of an S^1 with a 2-torus with a meridial circle pinched to a point. This completes the verification of the third entry. □

For later purposes (see figure 7.4) we note that the singular point $\sigma_\varepsilon = \varepsilon(1, 0, 0) \in P^a_{\varepsilon a}$ is a local minimum of $H^a_{\varepsilon a}$, when $\varepsilon = 1$, $b = a$ and $|a| > \sqrt{4\beta}$, or when $\varepsilon = -1$ and $b = -a$, because σ_ε is an elliptic equilibrium point of $-\mathrm{ad}_{H^a_b}$. Under these restrictions it follows that $\pi^{-1}(\sigma_\varepsilon)$ is a local minimum for H restricted to $J_r^{-1}(a) \cap J_\ell^{-1}(\varepsilon a)$.

Next we determine the topology of the level sets of the Hamiltonian

$$H : \mathrm{SO}(3) \times \mathrm{so}(3) \to \mathbf{R} : (A, X) \to \tfrac{1}{2}k(I(X), X) + \chi k(\mathrm{Ad}_{A^{-1}} E_3, E_3). \tag{90}$$

The results are given in table 7.2.

h	Topology of $H^{-1}(h)$
1. $-\chi$	a smooth S^1
2. $-\chi < h < \chi$	$S^1 \times S^4$
3. χ	an S^2-bundle over SO(3) with fibers over a non-null homotopic S^1 pinched to a point
4. $h > \chi$	$\mathrm{SO}(3) \times S^2$

Table 7.2. Topology of $H^{-1}(h)$.

First we prove

Claim: H is a Bott-Morse function.

(7.2) **Proof**: The group

$$S^1 = \left\{ B \in \mathrm{SO}(3) \,\middle|\, \mathrm{Ad}_{B^{-1}} E_3 = E_3 \right\}$$

acts freely and properly on SO(3) × so(3) via

$$\varphi^\ell : S^1 \times \big(\mathrm{SO}(3) \times \mathrm{so}(3) \big) \to \mathrm{SO}(3) \times \mathrm{so}(3) : \big(B, (A, X)\big) \to (BA, X).$$

Moreover, the orbit map of the action φ^ℓ is

$$\rho : \mathrm{SO}(3) \times \mathrm{so}(3) \to S^2 \times \mathbf{R}^3 : (A, X) \to \big(i(\mathrm{Ad}_{A^{-1}} E_3), i(X)\big) = (z, w).$$

Since H is invariant under φ^ℓ, it induces a smooth function

$$\widehat{H} : S^2 \times \mathbf{R}^3 \to \mathbf{R} : (z, w) \to \tfrac{1}{2} I_1^{-1}(w_1^2 + w_2^2) + \tfrac{1}{2} I_3^{-1} w_3^2 + \chi z_3.$$

To prove the claim it suffices to show that $\widehat{H}$ is a Morse function. Since $\widehat{H}$ is proper and bounded below, it has a critical point (z, w). By Lagrange multipliers

$$\begin{gathered} 2\lambda z_1 = 0, \quad I_1^{-1} w_1 = 0, \quad 2\lambda z_2 = 0, \\ I_1^{-1} w_2 = 0, \quad \chi + 2\lambda z_3 = 0, \quad I_3^{-1} w_3 = 0, \end{gathered}$$

and $z_1^2 + z_2^2 + z_3^2 = 1$. If $\lambda = 0$, then $\chi = 0$, which is a contradiction. Therefore $\lambda \neq 0$. Consequently, the points $p_\varepsilon = (0, 0, \varepsilon, 0, 0, 0)$ with $\varepsilon^2 = 1$ are critical points of $\widehat{H}$ with Lagrange multiplier $\lambda = -\frac{1}{2}\varepsilon\chi$. Since the tangent space

$$T_{p_\varepsilon}(S^2 \times \mathbf{R}^3) = \ker(0, 0, 2\varepsilon, 0, 0, 0)$$

is spanned by $\{e_1, e_2, e_4, e_5, e_6\}$, the Hessian of $\widehat{H}$ at p_ε is

$$\mathrm{diag}(-\varepsilon\chi, -\varepsilon\chi, I_1^{-1}, I_1^{-1}, I_3^{-1}).$$

This is nondegenerate (and has Morse index 2 if $\varepsilon = 1$ and index 0 if $\varepsilon = -1$). Therefore $\widehat{H}$ is a Morse function. □

▷ We now verify the entries in the second column of table 7.2.

(7.3) **Proof**: For the first entry note that $-\chi$ is the minimum value of $\widehat{H}$. This minimum is assumed at p_{-1}. Therefore $-\chi$ is the minimum value of H and it is assumed on the circle $\rho^{-1}(p_{-1})$.

To verify the second entry we argue as follows. By an equivariant version of the Morse lemma (see exercise 1 of appendix E), $\rho^{-1}(p_{-1})$ has a tubular neighborhood of the form $S^1 \times N$ where N is a neighborhood of 0 in $\mathbf{R}^5$ in a normal slice to $\rho^{-1}(p_{-1})$. Restricting H to the normal slice gives

$$H|N = z_1^2 + z_2^2 + w_1^2 + w_2^2 + w_3^2.$$

Therefore for h values slightly greater than $-\chi$, the h-level set of H is diffeomorphic to $S^1 \times S^4$.

To verify the fourth entry let

$$\tau : \mathrm{SO}(3) \times \mathrm{so}(3) \to \mathrm{SO}(3) : (A, X) \to A$$

be the bundle projection. For every $A \in \mathrm{SO}(3)$ we have $|k(\mathrm{Ad}_{A^{-1}} E_3, E_3)| \le 1$. This follows from the Cauchy-Schwarz inequality since

$$(k(\mathrm{Ad}_{A^{-1}} E_3, E_3))^2 \le k(\mathrm{Ad}_{A^{-1}} E_3, \mathrm{Ad}_{A^{-1}} E_3)\, k(E_3, E_3) = 1.$$

We now show that $H^{-1}(h)$ is a bundle over SO(3) with fiber S^2. Fix $A \in \mathrm{SO}(3)$ and suppose that $(A, X) \in H^{-1}(h)$. Then

$$\tfrac{1}{2} k(I(X), X) = h - \chi k(\mathrm{Ad}_{A^{-1}} E_3, E_3) \ge h - \chi > 0.$$

Thus the fiber $(\tau|H^{-1}(h))^{-1}(A)$ is a 2-sphere contained in $\{A\} \times \mathrm{so}(3)$. Since SO(3) is a Lie group, $H^{-1}(h)$ is a trivial S^2 bundle over SO(3).

We now look at the third entry. From a close examination of the argument proving the fourth entry, one sees that $\tau(H^{-1}(\chi)) = \mathrm{SO}(3)$. However, the fiber of τ over each point of the set

$$E = \left\{ A \in \mathrm{SO}(3) \middle| k(\mathrm{Ad}_{A^{-1}} E_3, E_3) = 1 \right\}$$

degenerates to a point. E is the image of the one parameter group $t \to \exp t E_3$. To see this consider the function

$$f : \mathrm{SO}(3) \to \mathbf{R} : A \to k(\mathrm{Ad}_{A^{-1}} E_3, E_3).$$

f is invariant under the S^1 action

$$S^1 \times \mathrm{SO}(3) \to \mathrm{SO}(3) : (t, A) \to (\exp t E_3) A.$$

This action is free and proper and has orbit map

$$\mathrm{SO}(3) \to S^2 : A \to i(\mathrm{Ad}_{A^{-1}} E_3) = z.$$

Hence f induces the function $\widetilde{f} : S^2 \subseteq \mathbf{R}^3 \to \mathbf{R} : z \to z_3$, which is the height function on S^2. Therefore $\widetilde{f}$ is a Morse function with two nondegenerate critical points e_3 and $-e_3$ of index 2 and 0, respectively. Therefore f is a Bott-Morse function with two nondegenerate critical circles, namely, $t \to \exp t E_3$ of index 2 and $t \to (\exp t E_3)\,\mathrm{diag}(-1, 1, -1)$ of index 0. Since $f(\exp t E_3) = 1$, E is the image of the circle $t \to \exp t E_3$, which is non-null homotopic in SO(3).

This completes the verification of table 7.2. □

We now investigate the structure of the singularity of $H^{-1}(\chi)$ a little more carefully. Because $\rho^{-1}(p_1) = E \times \{0\}$, it can be identified with E. Therefore the double cover of $\rho^{-1}(p_1)$ is a great circle Γ on S^3. Since Γ bounds a contractible disc in S^3, its normal bundle is trivial. Because the *antipodal map* on S^3 is orientation preserving, the normal bundle to $\rho^{-1}(p_1)$ is trivial. Let N be a normal slice to the $\rho^{-1}(p_1)$ at q_1. By the Morse lemma, the χ level set of $H|N$ is locally diffeomorphic to the zero level set of $D^2H^a(q_1)$, which is

$$-z_1^2 - z_2^2 + w_1^2 + w_2^2 + w_3^2 = 0.$$

Intersecting this cone with the 4-sphere

$$z_1^2 + z_2^2 + w_1^2 + w_2^2 + w_3^2 = r^2,$$

we see that near p_1 the level set $(H|N)^{-1}(\chi)$ is a cone on $S^1 \times S^2$ with vertex at 0. Since the local stable and unstable manifolds of p_1 are contractible and a fiber of ρ is S^1, the local stable and unstable manifolds of $\rho^{-1}(p_1)$ are untwisted. Therefore $\rho^{-1}(p_1)$ has a neighborhood which is diffeomorphic to a product of S^1 and a cone on $S^1 \times S^2$.

7.2 The discriminant locus

In this section we examine the set of critical values of the energy momentum mapping $\mathcal{EM}$ (89) of the Lagrange top.

From the reduction to a one degree of freedom Hamiltonian system (see section 4.4), it follows that the set of critical values of $\mathcal{EM}$ is very closely related to the set of critical values of the twice reduced Hamiltonian H_b^a (68). More precisely, (h, a, b) is a critical value of $\mathcal{EM}$ if and only if

$$h_b^a = h - \tfrac{1}{2} I_1^{-1}(b^2 - a^2) - \tfrac{1}{2} I_3^{-1} a^2$$

is a critical value of H_b^a. In other words, the polynomial

$$V(\sigma) = 2(\alpha - \beta\sigma)(1 - \sigma^2) - (a - b\sigma)^2$$

with

$$\alpha = I_1 h_b^a \quad \text{and} \quad \beta = I_1 \chi$$

has a multiple root in $[-1, 1]$, (see section 6.1). Thus (a, b, α) lies in the *discriminant locus* Δ_V of V. In the following we describe the singularities of the discriminant locus of ▷ V. We start by finding a parametrization of Δ_V.

(7.4) **Proof**: To simplify the discussion we choose physical units so that $\beta = \frac{1}{2}$. Then V becomes

$$W(\sigma) = \sigma^3 - (2\alpha + b^2)\sigma^2 + (2ab - 1)\sigma + 2\alpha - a^2. \tag{91}$$

Note that (a, b, α) lies in the discriminant locus Δ_W of W if and only if for $(s, t) \in [-1, 1] \times \mathbf{R}$ the polynomial W can be factored as

$$(\sigma - s)^2(\sigma - t) = \sigma^3 - (2s + t)\sigma^2 + (s^2 + 2st)\sigma - s^2 t.$$

Therefore

$$\begin{cases} 2\alpha + b^2 &= 2s + t \\ 2ab - 1 &= s^2 + 2st \\ a^2 - 2\alpha &= s^2 t. \end{cases} \tag{92}$$

Thinking of (s, t) as parameters, we solve (92) for a, b, α as follows. Adding the first and third equations in (92) to eliminate α gives

$$a^2 + b^2 = 2s + t + s^2 t \tag{93}$$

together with

$$2ab = s^2 + 2st + 1. \tag{94}$$

Adding and subtracting (93) and (94) gives

$$0 \le (a + b)^2 = (s + 1)^2(t + 1) \quad \text{and} \quad 0 \le (a - b)^2 = (1 - s)^2(t - 1). \tag{95}$$

For the inequalities in (95) to hold, (s, t) must lie in the set $\mathcal{S}$, (see figure 7.1. More

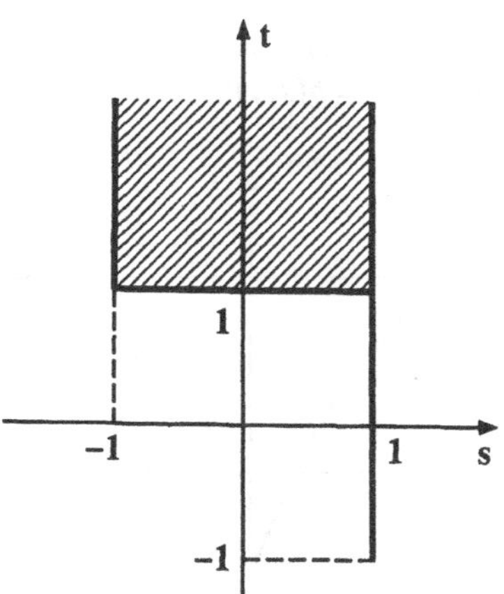

Figure 7.1. The set $\mathcal{S}$ of allowable values of (s, t).

precisely one of the following holds: $(s,t) \in (-1,1)\times(1,\infty)$, or $(1,t) \in \{1\}\times[-1,\infty)$, or $(-1,t) \in \{-1\}\times[1,\infty)$, or $(s,1) \in [-1,1]\times\{1\}$. Extracting the square root of the right hand sides of the equations in (95) gives

$$a+b = \varepsilon_1(s+1)\sqrt{t+1} \text{ and } a-b = \varepsilon_2(1-s)\sqrt{t-1},$$

where $\varepsilon_1^2 = \varepsilon_2^2 = 1$. Therefore

$$\begin{cases} a(s,t) &= \frac{1}{2}\varepsilon_1(s+1)\sqrt{t+1} + \frac{1}{2}\varepsilon_2(1-s)\sqrt{t-1} \\ b(s,t) &= \frac{1}{2}\varepsilon_1(s+1)\sqrt{t+1} - \frac{1}{2}\varepsilon_2(1-s)\sqrt{t-1}. \end{cases} \tag{96}$$

Substituting (96) into the third equation of (92) we obtain

$$\alpha(s,t) = \frac{1}{4}\big(2s + (1-s^2)t + \varepsilon_1\varepsilon_2(1-s^2)\sqrt{t^2-1}\big). \tag{97}$$

Therefore for each choice of ε_1 and ε_2 the map

$$\mathcal{P} : \mathbf{R}^2 \to \mathbf{R}^3 : (s,t) \to \big(a(s,t), b(s,t), \alpha(s,t)\big) \tag{98}$$

parametrizes a piece of Δ_W. □

Claim: The parametrization (98) is smooth except possibly when $(s,t) \in \partial\mathcal{S}$.

(7.5) **Proof**: Define

$$f : \mathbf{R}^5 \to \mathbf{R}^3 :$$
$$(s,t,a,b,\alpha) \to \big(2\alpha + b^2 - 2s - t, 2ab - 1 - s^2 - 2st, a^2 - 2\alpha - s^2 t\big).$$

Then

$$Df(s,t,a,b,\alpha) = \begin{pmatrix} -2 & -1 & 0 & 2b & 2 \\ -2s-2t & -2s & 2b & 2a & 0 \\ -2st & -s^2 & 2a & 0 & -2 \end{pmatrix}.$$

Since the [345]-minor of Df is $8(b^2-a^2)$, Df has rank 3 when $a^2 \neq b^2$. By the implicit function theorem, the level set $f^{-1}(0)$ is the graph of the mapping (98), except possibly for those values of (s,t) where $a(s,t)^2 = b(s,t)^2$. Using (95) it is straightforward to check that $a(s,t) = b(s,t)$ if and only if $s = 1$ & $t \in [-1,\infty)$ or $s \in [-1,1)$ & $t = 1$. Similarly, we see that $a(s,t) = -b(s,t)$ if and only if $s = -1$ & $t \in [1,\infty)$ or $s = 1$ & $t = -1$. In other words, $a(s,t)^2 = b(s,t)^2$ if and only if $(s,t) \in \partial\mathcal{S}$. □

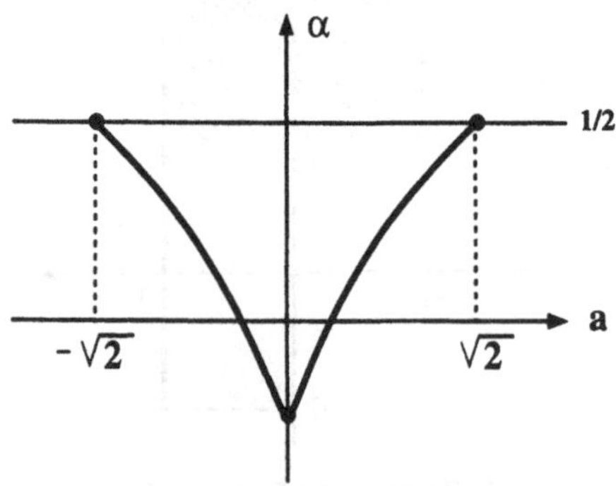

Figure 7.2. The $\{a = b\}$ slice of Δ_W.

We now investigate the $\{a = \pm b\}$ slices of Δ_W. We begin with the $\{a = b\}$ slice. Let $s = 1$. Then (96) and (97) become

$$a = \varepsilon\sqrt{1+t},\ b = \varepsilon\sqrt{1+t},\ \alpha = \tfrac{1}{2} \tag{99}$$

for $\varepsilon^2 = 1$ and $t \geq -1$. Let $t = 1$. Then (96) and (97) become

$$a = \tfrac{1}{2}\varepsilon\sqrt{2}(1+s),\ b = \tfrac{1}{2}\varepsilon\sqrt{2}(1+s),\ \alpha = \tfrac{1}{4} + \tfrac{1}{2}s - \tfrac{1}{4}s^2 \tag{100}$$

for $\varepsilon^2 = 1$ and $s \in [-1, 1]$. Thus $\{a = b\} \cap \Delta_W$ is parametrized by (99) and (100), (see figure 7.2). For the $\{a = -b\}$ slice let $s = -1$. Then (96) and (97) become

$$a = \varepsilon\sqrt{t-1},\ b = -\varepsilon\sqrt{t-1},\ \alpha = -\tfrac{1}{2} \tag{101}$$

with $\varepsilon^2 = 1$ and $t \geq 1$. For $s = 1$ and $t = 1$ we obtain the point

$$a = 0,\ b = 0,\ \alpha = \tfrac{1}{2}. \tag{102}$$

Thus $\{a = -b\} \cap \Delta_W$ is parametrized by (101) and (102), (see figure 7.3).

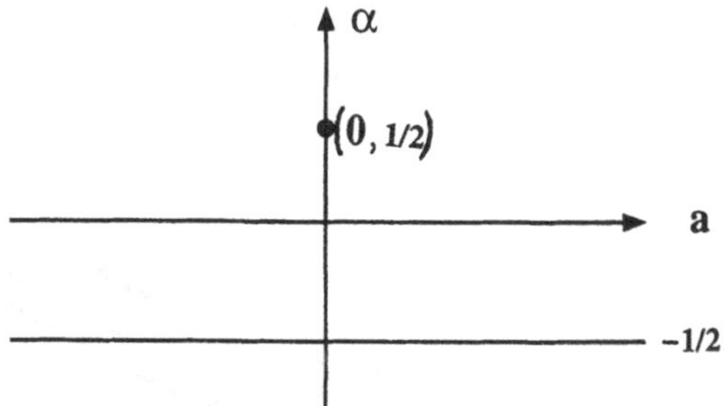

Figure 7.3. The $\{a = -b\}$ slice of Δ_W.

▷ Next we analyze the geometry of Δ_W near the slices $\{a = \pm b\}$.

(7.6) **Proof**: First suppose that (a, b, α) lies on the line ℓ_ε parametrized by

$$a \to (a, \varepsilon a, \tfrac{1}{2}\varepsilon)$$

with $\varepsilon^2 = 1$. Set $\sigma = v + \varepsilon$ and introduce new variables $x = \varepsilon - 2\alpha,\ y = a - \varepsilon b,\ z = b$, which turns the line ℓ_ε into the z-axis. Then $W(\sigma)$ (91) becomes

$$U(v) = v^3 + (2\varepsilon + x - z^2)v^2 + 2(\varepsilon x + yz)v - y^2.$$

Recall that a cubic polynomial $p(u) = au^3 + bu^2 + cu + d$ with $a \neq 0$ has a multiple zero if and only if its *discriminant* D_p vanishes, that is, if and only if

$$D_p = b^2c^2 - 4ac^3 - 4b^3d - 27a^2d^2 + 18abcd = 0. \tag{103}$$

Therefore U has a multiple zero if and only if

$$\begin{aligned} D_U &= 4(2\varepsilon + x - z^2)^2(\varepsilon x + yz)^2 - 32(\varepsilon x + yz)^3 - 27y^4 \\ &\quad + 4y^2(2\varepsilon + x - z^2)^3 - 36y^2(2\varepsilon + x - z^2)(\varepsilon x + yz) = 0. \end{aligned} \tag{104}$$

Since the constant and linear terms of the Taylor polynomial of D_U at $(0, 0, z)$ vanish, the z-axis is in the locus of singular points of $\{D_U = 0\}$. To find the tangent cone to $\{D_U = 0\}$ at $(0, 0, z)$ in a plane parallel to the x–y plane (which is normal to the z-axis at $(0, 0, z)$), we need to know the terms of degree two in x and y in the Taylor polynomial of D_U. A calculation gives

$$(2\varepsilon - z^2)^2(x^2 + 2zxy + 2\varepsilon y^2). \tag{105}$$

Let $\varepsilon = 1$. At each point of the segment $I_< = \{|z| < \sqrt{2}\}$ of the z-axis the tangent cone to $\{D_U = 0\}$ in a normal slice consists of a point. Since $(a, b, \alpha) = (0, 0, \frac{1}{2})$ is an isolated point of the $\{a = -b\} \cap \Delta_W$, it follows that $(x, y, z) = (0, 0, 0)$ is an isolated point of the intersection of $\{D_U = 0\}$ and the normal slice. Therefore the line segment $I_<$ is *isolated* in $\{D_U = 0\}$. In other words, the line segment parametrized by $a \to (a, a, \frac{1}{2})$ for $|a| < \sqrt{2}$ forms an isolated one dimensional piece of Δ_W called the *thread*. This thread is a perhaps the most remarkable feature of the set of critical values of the energy momentum mapping of the Lagrange top. At each point of the segment $I_> = \{|z| > \sqrt{2}\}$ of the z-axis when $\varepsilon = -1$ or on the z-axis when $\varepsilon = 1$ the tangent cone to $\{D_U = 0\}$ in the normal slice consists of two intersecting lines. Hence the tangent cone to $\{D_U = 0\}$ at each point in $I_>$ is two transversely intersecting 2-planes. Recall that for (a, b, α) on the line ℓ_{-1} or on the line segments ℓ_1 when $|a| > \sqrt{2}$, the Hamiltonian $H|(J_r^{-1}(a) \cap J_\ell^{-1}(\varepsilon a))$ assumes its minimum value. Therefore the tangent cone to Δ_W forms a *crease*, (see figure 7.4).

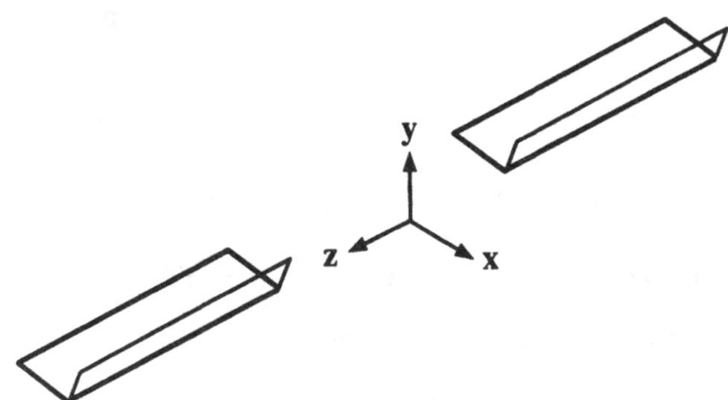

Figure 7.4. The crease singularity of Δ_W.

Second suppose that (a, b, α) lies along the curves $\mathscr{C}_\varepsilon$ parametrized by

$$a \to \left(a, a, \tfrac{1}{2}(-a^2 + 2\sqrt{2}\varepsilon a - 1)\right),$$

where $\varepsilon^2 = 1$ and $a \in I$. Here I is the interval $(0, \sqrt{2})$ if $\varepsilon = 1$ or the interval $(-\sqrt{2}, 0)$ if $\varepsilon = -1$. Set $\sigma = \upsilon + 1$ and introduce new variables $x = 2\alpha + (a^2 - 2\sqrt{2}\varepsilon a + 1)$, $y = a - b$, and $z = b$, which turns $\mathscr{C}_\varepsilon$ into the segment I of the z-axis. Then $W(\sigma)$ (91) becomes

$$\begin{aligned} U(\upsilon) = \upsilon^3 &+ (4 - x + y^2 + 2yz - 2\varepsilon\sqrt{2}y - 2\varepsilon\sqrt{2}z)\upsilon^2 \\ &+ 2(2 - x + y^2 + 3yz + z^2 - 2\varepsilon\sqrt{2}y - 2\varepsilon\sqrt{2}z)\upsilon - y^2. \end{aligned} \tag{106}$$

A calculation shows that the constant term of the Taylor polynomial of D_U vanishes at every z in the segment I of the z-axis, whereas the linear terms in the normal slice do not. Hence $\{D_U = 0\}$ is smooth along I, that is, Δ_W is smooth along $\mathscr{C}_\varepsilon$.

Finally, suppose that $(a, b, \alpha) = (\varepsilon\sqrt{2}, \varepsilon\sqrt{2}, \frac{1}{2})$. Set $\sigma = v + 1$ and introduce new variables $x = 1 - 2\alpha$, $y = a - b$ and $z = b - \varepsilon\sqrt{2}$. Then $(\varepsilon\sqrt{2}, \varepsilon\sqrt{2}, \frac{1}{2})$ becomes the point $(x, y, z) = (0, 0, 0)$ and $W(\sigma)$ (91) becomes

$$U(v) = v^3 + (x - \varepsilon\sqrt{2}z - z^2)v^2 + 2(x + \sqrt{2}\varepsilon y + yz)v - y^2. \qquad (107)$$

We want to describe the discriminant locus Δ_U, that is, the set of (x, y, z) where U has a multiple root in $[0, 2]$.

Claim: Let D_F be the discriminant of the cubic polynomial

$$F(u) = au^3 - bu^2 + cu - d^2, \qquad (108)$$

where $a > 0$ and the constant term is a square. Let D_G be the discriminant of the special quartic polynomial

$$G(u) = u^4 - \frac{b}{2a}u^2 + \frac{d}{\sqrt{a}}u + (\frac{b^2}{16a^2} - \frac{c}{4a}). \qquad (109)$$

Then $\{D_F = 0\} = \{D_G = 0\}$.

(7.7) **Proof**: Using (103) we see that the discriminant of F is $a^2 D_F = 4s^3 - 27t^2$, where

$$s = \frac{1}{3}b^2 - ac \quad \text{and} \quad t = \frac{2}{27}b^3 - \frac{1}{3}abc + a^2d^2.$$

For a special quartic polynomial

$$Q(u) = u^4 + Au^2 + Bu + C \qquad (110)$$

the *discriminant* is $D_Q = 4S^3 - 27T^2$, where

$$S = A^2 + 12C \quad \text{and} \quad T = 2A^3 + 27B^2 - 72AC.$$

For G we find that

$$S = \frac{3}{a^2}(\frac{1}{3}b^2 - ac) \quad \text{and} \quad T = \frac{3^3}{a^3}(\frac{2}{27}b^3 - \frac{1}{3}abc + a^2d^2).$$

Hence

$$\frac{a^6}{27}D_G = a^2 D_F. \qquad \square$$

From ((7.7)) it follows that the locus $\{D_U = 0\}$ of multiple zeroes of the cubic polynomial U (107) is the same as the multiple zero locus of the special quartic polynomial

$$Y(v) = v^4 + \tfrac{1}{2}(x - 2\varepsilon\sqrt{2}z - z^2)v^2 + yv + \tfrac{1}{16}(x - 2\varepsilon\sqrt{2}z - z^2)^2 - \tfrac{1}{2}(x + \varepsilon\sqrt{2}y + yz),$$

which is the well known *swallowtail surface*, (see figure 7.5).

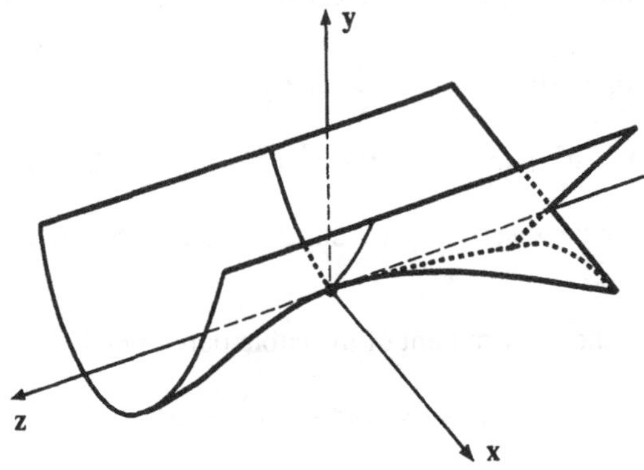

Figure 7.5. When $\varepsilon = 1$ the swallowtail surface $\{D_Y = 0\}$ near $x = y = z = 0$. When $\varepsilon = -1$ the surface must be reflected in the x–y plane.

The *double line* of the special quartic Q (110) is the union of the *line of self intersection* given by

$$B = 0 \ \& \ A^2 - 4C = 0 \ \& \ A \leq 0$$

and the *whisker* given by

$$B = 0 \ \& \ A^2 - 4C = 0 \ \& \ A \geq 0.$$

Therefore the double line of $\{D_Y = 0\}$ is the z–axis, on which

$$Y(v) = \left(v^2 - \frac{(2\varepsilon\sqrt{2}z + z^2)}{4}\right)^2.$$

On the z-axis Y has a zero of multiplicity 4 at $z = 0$ and $z = -2\varepsilon\sqrt{2}$, which are *swallowtail points*. Y has a double real root on the z-axis when $z \in I_1$ where I_1 is either $(-\infty, -2\sqrt{2}) \cup (0, \infty)$ when $\varepsilon = 1$ or $(-\infty, 0) \cup (0, 2\sqrt{2})$ when $\varepsilon = -1$. Y has a double purely imaginary root on the z-axis when $z \in I_2$ where I_2 is either $(-2\sqrt{2}, 0)$ when $\varepsilon = 1$ or $(0, 2\sqrt{2})$ when $\varepsilon = -1$. Therefore the line of self intersection of $\{D_U = 0\}$ is the union of the segments of the z-axis where $z \in I_1$; whereas the whisker of $\{D_U = 0\}$ is the segment of the z-axis where $z \in I_2$.

For the discriminant locus Δ_W the above discussion translates into the following. The points $(a, b, \alpha) = (\varepsilon\sqrt{2}, \varepsilon\sqrt{2}, \frac{1}{2})$ are swallowtail points. The whisker is the thread, which is parametrized by $a \rightarrow (a, a, \frac{1}{2})$ where $|a| < \sqrt{2}$. The line of self intersection is parametrized by $a \rightarrow (a, a, \frac{1}{2})$ where $|a| > \sqrt{2}$. Since the line of self intersection of $\{D_Y = 0\}$ is a crease singularity, the discriminant locus Δ_W is missing the *tail* of the swallowtail surface, (see figure 7.6). Another way of saying this is that the tail of the swallowtail surface of $\{D_Y = 0\}$ corresponds to multiple roots of Y which do not lie in $[0, 2]$. In figure 7.6 we have sketched the discriminant locus near a swallowtail point. □

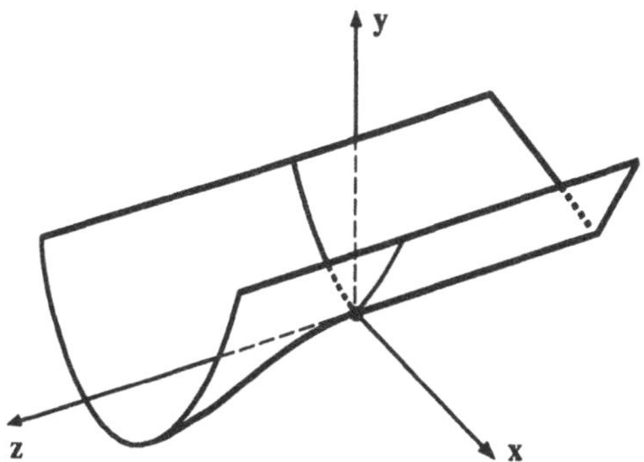

Figure 7.6. The discriminant locus Δ_W near the swallowtail point $(\sqrt{2}, \sqrt{2}, \frac{1}{2})$ when $\varepsilon = 1$. When $\varepsilon = -1$ the surface must be reflected in the x–y plane.

Under the mapping

$$(a, b, \alpha) \to (a, b, h) = \left(a, b, I_1^{-1}\alpha + \tfrac{1}{2} I_1^{-1}(b^2 - a^2) + \tfrac{1}{2} I_3^{-1}a^2\right)$$

the image of Δ_W is the set of critical values of $\mathcal{EM}$, (see figure 7.7).

We end this subsection by describing the motion of the top which corresponds to a fixed critical value of the energy momentum mapping $\mathcal{EM}$. Suppose that the critical value (h, a, b) lies on the smooth two dimensional piece of the set of critical values. In phase space, the motion takes place on a two dimensional torus, (see table 7.1). In physical space the top is spinning at a constant speed about its figure axis (which makes a fixed angle with the vertical axis) while turning uniformly about the vertical axis. In other words, the top is undergoing *regular precession*. If the critical value lies on a crease in the $\{a = -b\}$ plane, then the top is spinning at a constant speed about its figure axis which is pointing vertically downward. If the critical value lies on the crease in the $\{a = b\}$ plane, then the top is spinning at a constant speed about a figure axis, which is pointing vertically upward. Here the top is said to be *sleeping*. This motion is stable because the total energy of the top on the intersection of the appropriate level sets of the angular momenta is at a minimum. If the critical value lies on the thread, then the top is either spinning at a constant speed about its figure axis which is vertical or moving so that it is asymptotic to this motion as time goes to $\pm\infty$. Here the top is said to be *waking*. This motion is unstable except at the points where the thread attaches itself to the two dimensional piece of the set of critical values. At the points of attachment the motion is stable because it is surrounded by *bounded* motions which either lie on tori or on the stable or unstable manifolds of hyperbolic periodic orbits.

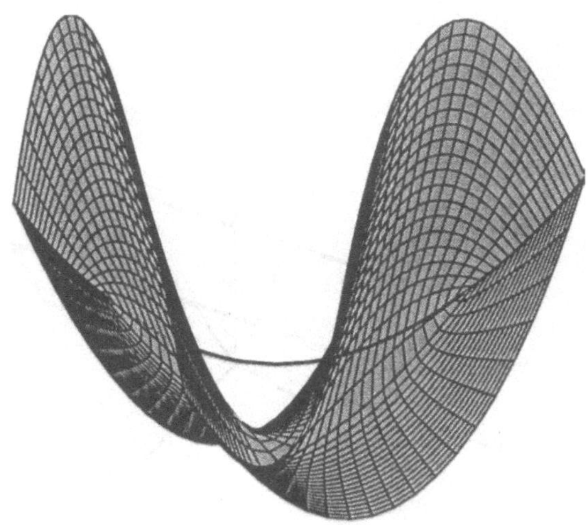

Figure 7.7. The set of critical values of the energy momentum mapping $\mathcal{EM}$ of the Lagrange top.

7.3 The period lattice and monodromy

Let $\mathfrak{R}$ be the set of regular values which lie in the image of $\mathcal{EM}$. We have shown that if $(h, a, b) \in \mathfrak{R}$, then $\mathcal{EM}^{-1}(h, a, b)$ is a smooth 3-torus $T^3_{h,a,b}$. On $T^3_{h,a,b}$ the motion of the top is the superposition of three circular motions, namely, a constant *spin* about its figure axis; a constant *precession* of its figure axis about the vertical axis; and a variable up and down *nutating* motion of its figure axis. In this subsection we examine how the 3-torus fibers of $\mathcal{EM}$ fit together. Let $\widetilde{\Gamma}$ be a loop in $\mathfrak{R}$ which bounds a disc in $\mathbf{R}^3$ which intersects the thread at one point. $\widetilde{\Gamma}$ is not null homotopic in $\mathfrak{R}$. As $\widetilde{\Gamma}$ is traced out once, we will show that the *variation* of the period lattice associated to the 3-torus $T^3_{h,a,b}$, $(h, a, b) \in \widetilde{\Gamma}$, is nonzero and gives rise to monodromy. Another way to say this is that the level sets of $\mathcal{EM}$ over points in $\widetilde{\Gamma}$ fit together to form a *nontrivial* smooth 3-torus bundle over $\widetilde{\Gamma}$.

We start our discussion of monodromy by defining the concept of period lattice $\mathcal{P}(h, a, b)$ associated to the smooth 3-torus $T^3_{h,a,b}$ where $(h, a, b) \in \mathfrak{R}$. The *period lattice* $\mathcal{P}(h, a, b)$ is a lattice over $\mathbf{Z}$ generated by certain Hamiltonian vector fields X_F. Specifically, for some open neighborhood $\mathcal{U}$ of (h, a, b) in $\mathfrak{R}$, F lies in the free $C^\infty(\mathcal{U})$-module generated by the angular momenta J_r, J_ℓ and the energy H. Moreover, the flow of the vector field $X_F|T^3_{h',a',b'}$ is periodic of period 1 for every $(h', a', b') \in \mathcal{U}$.

▷ The following argument constructs a basis for $\mathcal{P}(h, a, b)$.

(7.8) **Proof**: Clearly, the vector fields $2\pi X_{J_r}$ and $2\pi X_{J_\ell}$ on $T^3_{h,a,b}$ are elements of a basis for

$\mathcal{P}(h, a, b)$. To construct the third basis element, note that

$$T^3_{h.a.b} = (\pi^a_b \circ \pi^a)^{-1}\big((H^a_b)^{-1}(h^a_b)\big),$$

where $\pi^a : J_r^{-1}(a) \to P^a$ is the reduction map of the action $\varphi^r|\big(S^1 \times J_r^{-1}(a)\big)$ (11) and $\pi^a_b : (J^a_\ell)^{-1}(b) \subseteq P^a \to P^a_b$ is the reduction map of the induced action δ^ℓ (44) on P^a. On the smooth 2-torus $T^2_{h^a.b} = (\pi^a_b)^{-1}\big((H^a_b)^{-1}(h^a_b)\big)$ the vector field $2\pi X_{J^a_\ell}$ has a periodic flow of period 1. Note that $\pi^a_b(T^3_{h.a.b}) = T^2_{h^a.b}$. Here $h^a = h^a_b + \frac{1}{2} I_1^{-1} b^2$ and $\pi^a_b((H^a)^{-1}(h^a)) = (H^a_b)^{-1}(h^a_b)$. Let Θ_ℓ be the rotation number of the flow of $X_{H^a}|T^2_{h^a,b}$ and let T be the period of the flow of $-\mathrm{ad}_{H^a_b}$ on the circle $(H^a_b)^{-1}(h^a_b)$. From the definition of rotation number it follows that the vector field $T X_{H^a} - 2\pi\Theta_\ell X_{J^a_\ell}$ on $T^2_{h^a,b}$ has a periodic flow of period 1. Choose a point p on $T^3_{h,a,b}$ and let $\mathcal{C}$ be the image of an integral curve of X_{J_r} through p. Since X_{J_r} has a periodic flow on $T^3_{h.a.b}$ of period 1, $\mathcal{C}$ is a closed curve. Consider an integral curve $t \to \gamma(t)$ of $Y = X_H - \frac{2\pi\Theta_\ell}{T} X_{J_\ell}$ which starts at p. Since $t \to \pi^a_b \circ \pi^a(\gamma(t))$ is a periodic integral curve of $-\mathrm{ad}_{H^a_b}$ of period T, it follows that T is the least positive time such that $\gamma(T) \in \mathcal{C}$. Thus $\mathcal{C}$ is a cross section for the flow of Y on $T^3_{h.a.b}$. During the time T the integral curve of X_{J_r}, which started at p, has travelled through an angle $2\pi\Theta_r$. By construction, the vector field

$$T X_H - 2\pi\Theta_\ell X_{J_\ell} - 2\pi\Theta_r X_{J_r}$$

has periodic flow on $\mathcal{EM}^{-1}(h, a, b)$ of period 1. Thus the period lattice $\mathcal{P}(h, a, b)$ of $\mathcal{EM}^{-1}(h, a, b)$ has a basis

$$\Big\{ X_{2\pi J_r},\ X_{2\pi J_\ell},\ X_{TH - 2\pi\Theta_\ell J_\ell - 2\pi\Theta_r J_r} \Big\}. \qquad \square$$

To give a geometric interpretation of the rotation number Θ_r, we reduce the Lagrange top to the Hamiltonian system $(H^a_b, P^a_b, \{\,,\,\}_{P^a_b})$ with one degree of freedom. This time we first reduce by the *left* S^1 action and then by the induced *right* S^1 action. Because the argument follows along the same lines as that given in sections 3.1 and 4, we give only the high points, leaving the details to the reader. Recall that $S^1 = \{B \in SO(3) | \mathrm{Ad}_B E_3 = E_3\}$. The left S^1 action on $TSO(3)$

$$\Phi_\ell : S^1 \times TSO(3) \to TSO(3) : \big(B, X_A\big) \to X_{BA}$$

becomes the action

$$\varphi_\ell : S^1 \times \big(\mathrm{SO}(3) \times \mathrm{so}(3)\big) \to \mathrm{SO}(3) \times \mathrm{so}(3) : \big(B, (A, X)\big) \to (BA, X)$$

after pulling back by the left trivialization $\mathcal{L}$ (10). The action Φ_ℓ is Hamiltonian with momentum mapping

$$\mathcal{J}_\ell : TSO(3) \to \mathbf{R} : X_A \to \rho(e)(\mathrm{Ad}_{A^{-1}} E_3, X).$$

Under $\mathcal{L}$, the map $\mathcal{J}_\ell$ pulls back to the momentum mapping

$$J_\ell : \mathrm{SO}(3) \times \mathrm{so}(3) \to \mathbf{R} : (A, X) \to k\big(\mathrm{Ad}_{A^{-1}} E_3, I(X)\big)$$

of the action φ_ℓ. Since every value b is a regular value of J_ℓ, the level set $J_\ell^{-1}(b)$ is a smooth manifold. Because φ_ℓ acts freely and properly on $J_\ell^{-1}(b)$, the orbit space $J_\ell^{-1}(b)/S^1$ is a smooth manifold. Let

$$\mathcal{P}_b = \left\{ (Z, W) \in \mathrm{SO}(3) \times \mathrm{so}(3) \middle| k(Z, Z) = 1 \ \& \ k(Z, W) = b \right\}.$$

From the fact that the orbit map

$$\pi_b : J_\ell^{-1}(b) \to \mathcal{P}_b : (A, X) \to \left(\mathrm{Ad}_{A^{-1}} E_3, I(X) \right) = (Z, W)$$

is a submersion and every fiber $\pi_b^{-1}(Z, W)$ is a single φ_ℓ orbit, it follows that the orbit space $J_\ell^{-1}(b)/S^1$ is diffeomorphic to $\mathcal{P}_b$. By the regular reduction theorem, the reduced space $\mathcal{P}_b$ has a symplectic form

$$\begin{aligned}\Omega_b(Z, W)\big(T\pi_b(-\left[U, \mathrm{Ad}_{A^{-1}} E_3\right], I(R)), T\pi_b(-\left[V, \mathrm{Ad}_{A^{-1}} E_3\right], I(S))\big) &= \\ = -k(I(R), V) + k(I(S), U) + k(I(Z), [U, V]),&\end{aligned}$$

where the arguments of $\Omega_b(Z, W)$ lie in $T_{(z.w)}P^a$. Under $\mathcal{L}$ the Hamiltonian

$$\mathcal{H} : T\mathrm{SO}(3) \to \mathbf{R} : X_A \to \tfrac{1}{2}\rho(A)(X_A, X_A) + \chi k(\mathrm{Ad}_{A^{-1}} E_3, E_3)$$

pulls back to the Hamiltonian

$$H : \mathrm{SO}(3) \times \mathrm{so}(3) \to \mathbf{R} : (X, A) \to \tfrac{1}{2} k(I(X), X) + \chi k(\mathrm{Ad}_{A^{-1}} E_3, E_3).$$

Since H is invariant under the action φ_ℓ, it induces a function

$$\widetilde{H}_b : \mathcal{P}_b \to \mathbf{R} : (Z, W) \to \tfrac{1}{2} k(I^{-1}(W), W) + \chi k(Z, E_3).$$

If h is the value of the Hamiltonian H, then $h_b = h$ is the value of the reduced Hamiltonian $\widetilde{H}_b$.

Using the identification map $i : \mathrm{so}(3) \to \mathbf{R}^3$ (see chapter 3 ((1.2))), we obtain a second model (H_b, P_b, ω_b) of the reduced Hamiltonian system $(\widetilde{H}_b, \mathcal{P}_b, \Omega_b)$. Here the reduced phase space is

$$P_b = \left\{ (z, w) \in \mathbf{R}^3 \times \mathbf{R}^3 \middle| (z, z) = 1 \ \& \ (z, w) = b \right\} \tag{111}$$

with reduction map

$$\pi_b : J_\ell^{-1}(b) \to P_b : (A, X) \to (z, w) = \big(i(Z), i(W)\big). \tag{112}$$

The symplectic form is

$$\omega_b(z, w)\big((-u \times z, \xi), (-v \times z, \eta)\big) = -(\xi, v) + (\eta, u) + (w, u \times v).$$

The arguments of $\omega_b(z, w)$ lie in $T_{(z,w)}P_b$. The reduced Hamiltonian is

$$H_b : P_b \subseteq \mathbf{R}^3 \times \mathbf{R}^3 \to \mathbf{R} : (z, w) \to \tfrac{1}{2} I_1^{-1}(w_1^2 + w_2^2) + \tfrac{1}{2} I_3^{-1} w_3^2 + \chi z_3.$$

On $\mathbf{R}^3 \times \mathbf{R}^3$ the equations

$$\begin{cases} \dot{z} &= z \times I^{-1}(w) \\ \dot{w} &= z \times \chi e_3 + w \times I^{-1}(w), \end{cases} \tag{113}$$

where $I^{-1}(w) = (I_1^{-1} w_1, I_2^{-1} w_2, I_3^{-1} w_3)$, are the *left* Euler-Poisson equations for H_b. They are the negative of the *right* Euler-Poisson equations (27) of section 3.1. Restricting the left Euler-Poisson equations to P_b gives the Hamiltonian vector field X_{H_b} of the reduced Hamiltonian H_b.

The right Hamiltonian action

$$\Phi^r : TSO(3) \times S^1 \to TSO(3) : (X_A, B) \to X_{AB}$$

with momentum mapping

$$\mathcal{J}_r : TSO(3) \to \mathbf{R} : X_A \to \rho(A)(X_A, T_e L_A E_3)$$

when pulled back by the left trivialization $\mathcal{L}$ gives the right Hamiltonian action

$$\varphi^r : \big(\mathrm{SO}(3) \times \mathrm{so}(3) \big) \times S^1 \to \mathrm{SO}(3) \times \mathrm{so}(3) : \big((A, X), B \big) \to (AB, \mathrm{Ad}_{B^{-1}} X)$$

with momentum mapping

$$J_r : \mathrm{SO}(3) \times \mathrm{so}(3) \to \mathbf{R} : (A, X) \to k(I(X), E_3).$$

Note that the induced right S^1 action on P_a is the restriction of the diagonal action

$$\Delta : S^1 \times (\mathbf{R}^3 \times \mathbf{R}^3) \to \mathbf{R}^3 \times \mathbf{R}^3 : \big(t, (z, w)\big) \to (R_t z, R_t w), \tag{114}$$

where $R_t = \begin{pmatrix} \cos t & -\sin t & 0 \\ \sin t & \cos t & 0 \\ 0 & 0 & 1 \end{pmatrix}$. Since $\mathcal{J}_r$ is invariant under the left S^1 action Φ_ℓ, it induces the function

$$J_r^b : P_b \to \mathbf{R} : (z, w) \to w_3,$$

which is an integral of the reduced vector field X_{H_b}. Let

$$EM_b : P_b \to \mathbf{R}^2 : (z, w) \to \big(H_b(z, w), J_r(z, w) \big)$$

be the energy momentum mapping of X_{H_b}.

After this preparation we are ready to give a geometric interpretation of the angle $2\pi \Theta_r$.

Claim: Let (h_b, a) be a regular value of EM_b. On the smooth 2-torus $T^2_{h_b,a}$, which is the (h_b, a)-level set of EM_b, the rotation number of the flow of X_{H_b} is Θ_r.

(7.9) **Proof**: Note that the image under the reduction map π_b (112) of the 3-torus $T^3_{h,a,b}$ is a smooth 2-torus $T^2_{h_b,a}$. Since $T\pi_b X_H = X_{H_b} \circ \pi_b$, $T\pi_b X_{J_r} = X_{J_r^b} \circ \pi_b$, and $T\pi_b X_{J_\ell} = 0$, we see that the image of the vector field

$$Y = TX_H - 2\pi \Theta_\ell X_{J_\ell} - 2\pi \Theta_r X_{J_r}$$

on $T^3_{h.a.b}$ under $T\pi_b$ is the vector field

$$Z = (TX_{H_b} - 2\pi\Theta_r X_{J^b_r})\circ\pi_b$$

on $T^2_{h_b,a}$. The vector field Z has a periodic flow of period 1 on $T^2_{h_b,a}$ because the vector field Y belongs to the period lattice $\mathcal{P}(h,a,b)$ associated to the 3-torus $T^3_{h,a,b}$. Therefore Θ_r is the rotation number of X_{H_b} on $T^2_{h_b,a}$. □

▷ In order to compute the monodromy we need an explicit formula for the rotation number Θ_r, (see (117)).

(7.10) **Proof**: We find a formula for Θ_r using the same technique as we used to find the rotation number Θ_ℓ, (see section 6.3). Look at the reduced space P_b as a bundle over S^2 with bundle projection $\tau : P_b \to S^2 : (z,w) \to z$. On $\tau(T^2_{h_b,a})$ introduce coordinates $\varphi = \tan^{-1}\frac{z_2}{z_1}$ and z_3. The equations on $\tau(T^2_{h_b,a})$, which are satisfied by the image of an integral curve of $X_{H_b}|T^2_{h_b,a}$ under τ, are obtained as follows. From the third component of the first Euler-Poisson equation (113) we find that

$$\begin{aligned} \dot{z_3} &= I_1^{-1}(z_1w_2 - z_2w_1) \\ &= \eta I_1^{-1}\sqrt{(z_1^2+z_2^2)(w_1^2+w_2^2)-(z_1w_1+z_2w_2)^2}, \qquad \text{where } \eta^2 = 1 \\ &= \eta I_1^{-1}\sqrt{(1-z_3^2)\big(2I_1(h-\tfrac{1}{2}I_3^{-1}a^2-\chi z_3)\big)-(b-az_3)^2} \end{aligned} \tag{115}$$

since

$$\left\{ \begin{aligned} z_1^2+z_2^2+z_3^2 &= 1, \\ z_1w_1+z_2w_2+z_3w_3 &= b, \\ \tfrac{1}{2}I_1^{-1}(w_1^2+w_2^2)+\tfrac{1}{2}I_3^{-1}w_3^2+\chi z_3 &= h_b = h \\ w_3 &= a \end{aligned} \right.$$

are the defining equations for $T^2_{h_b,a}$. From the definition of the angle φ it follows that

$$\dot{\varphi} = L_{X_{H_b}}\varphi = (z_1\dot{z}_2 - z_2\dot{z}_1)(z_1^2+z_2^2)^{-1}.$$

Using (113) we obtain

$$\begin{aligned} \dot{\varphi} &= \left[I_1^{-1}z_3(z_1w_1+z_2w_2) - I_3^{-1}w_3(z_1^2+z_2^2)\right](z_1^2+z_2^2)^{-1} \\ &= \left[I_1^{-1}z_3(b-az_3) - I_3^{-1}a(1-z_3^2)\right](1-z_3^2)^{-1}. \end{aligned} \tag{116}$$

Choosing $\eta = -1$ and using (115) and (116) it follows that the rotation number Θ_r is given by

$$2\pi\,\Theta_r = 2\int_{z_3^-}^{z_3^+} \frac{\dot{\varphi}}{\dot{z}_3}\,dz_3$$

$$= 2 \int_{z_3^-}^{z_3^+} \frac{z_3(b - az_3)\, dz_3}{(1 - z_3^2)\sqrt{2(\widetilde{\alpha} - \beta z_3)(1 - z_3^2) - (b - az_3)^2}}$$
$$-2a \frac{I_1}{I_3} \int_{z_3^-}^{z_3^+} \frac{dz_3}{\sqrt{2(\widetilde{\alpha} - \beta z_3)(1 - z_3^2) - (b - az_3)^2}}, \tag{117}$$

where

$$\widetilde{\alpha} = I_1(h - \tfrac{1}{2} I_3^{-1} a^2) \quad \text{and} \quad \beta = I_1 \chi.$$

In addition, $z_3^{\pm} \in [-1, 1]$ are roots of the polynomial $2(\widetilde{\alpha} - \beta z_3)(1 - z_3^2) - (b - az_3)^2$. □

We now turn to discussing monodromy. Let $\widetilde{\Gamma}$ be a smooth loop in the set of regular values $\mathfrak{R}$ of $\mathscr{EM}$, which bounds a disc intersecting the thread at one point. Since $\mathscr{EM}$ restricted to $\mathscr{EM}^{-1}(\widetilde{\Gamma})$ is a proper submersion, it follows that $\mathscr{EM}^{-1}(\widetilde{\Gamma})$ is a smooth 3-torus bundle over $\widetilde{\Gamma}$. Because $I = \widetilde{\Gamma} - \{\text{pt}\}$ is contractible, the bundle $\mathscr{EM}^{-1}(I)$ is trivial. Therefore the bundle $\mathscr{EM}^{-1}(\widetilde{\Gamma})$ is determined by an orientation preserving diffeomorphism χ of the fiber $T^3 = \mathscr{EM}^{-1}(\text{pt})$ into itself, called the monodromy map, (see appendix D section 2 or chapter 4 section 5). The map χ induces the map $\chi_* : H^1(T^3, \mathbf{Z}) \to H^1(T^3, \mathbf{Z})$, which is the monodromy of the bundle $\mathscr{EM}^{-1}(\widetilde{\Gamma})$. The goal of the rest of this subsection is to show that $\chi_* = \begin{pmatrix} 1 & 0 & 0 \\ 0 & 1 & 0 \\ 0 & 1 & 1 \end{pmatrix}$. Because χ_* is not the identity matrix, the map χ is not homotopic to the identity map on $\mathscr{EM}^{-1}(\text{pt})$. Consequently, the bundle is *nontrivial*. This fact was classically unknown and represents a new qualitative feature of the Lagrange top.

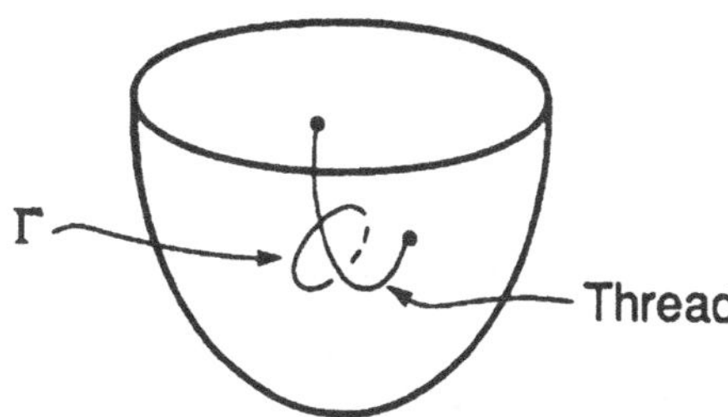

Figure 7.8. The loop Γ.

To compute the monodromy, we need to find the variation in the period lattice as (h, a, b) runs over the closed curve $\widetilde{\Gamma}$. The loop $\widetilde{\Gamma}$ may be smoothly homotoped within $\mathfrak{R}$ to a loop Γ in the $\{b = 0\}$ plane which bounds a disc which intersects the thread at $(0, 0, \frac{1}{2})$. This homotopy does not change the diffeomorphism type of the three torus bundle. In other words, the monodromy maps of the bundles $\mathscr{EM}^{-1}(\widetilde{\Gamma})$ and $\mathscr{EM}^{-1}(\Gamma)$ are homotopic and hence the ▷ bundles have the same monodromy. We now find the variation in the period lattice $\mathscr{P}(h, a, 0)$ associated to the 3-torus fiber $T^3_{h,a,0}$ as $(h, a, 0)$ runs once around the curve Γ, (see figure 7.8).

(7.11) **Proof**: The period lattice $\mathcal{P}(h, a, 0)$ has a basis

$$\left\{ 2\pi X_{J_r},\ 2\pi X_{J_\ell},\ \text{and } TX_H - 2\pi\Theta_r^0 X_{J_r} - 2\pi\Theta_\ell^0 X_{J_\ell} \right\},$$

where $\Theta_\ell^0 = \Theta_\ell \Big|_{b=0}$ and $\Theta_r^0 = \Theta_r \Big|_{b=0}$. From (87) we find that

$$2\pi\Theta_\ell^0 = -2a \int_{\sigma^-}^{\sigma^+} \frac{\sigma\, d\sigma}{(1-\sigma^2)\sqrt{2(\alpha-\beta\sigma)(1-\sigma^2)-a^2}},$$

where

$$\alpha = I_1 h^a = I_1\left(h + \tfrac{1}{2}(I_1^{-1} - I_3^{-1})a^2\right) \quad \text{and} \quad \beta = I_1\chi.$$

From (117) we get

$$\begin{aligned} 2\pi\Theta_r^0 &= -2a \int_{z_3^-}^{z_3^+} \frac{z_3^2\, dz_3}{(1-z_3^2)\sqrt{2(\widetilde{\alpha}-\beta z_3)(1-z_3^2)-a^2 z_3^2}} \\ &\quad -2aI_1 I_3^{-1} \int_{z_3^-}^{z_3^+} \frac{dz_3}{\sqrt{2(\widetilde{\alpha}-\beta z_3)(1-z_3^2)-a^2 z_3^2}}, \end{aligned}$$

where $\widetilde{\alpha} = I_1(h - \frac{1}{2} I_3^{-1} a^2)$. We calculate the variation in the period lattice along Γ by taking the limit of Γ as $a \to 0$, (see figure 7.9).

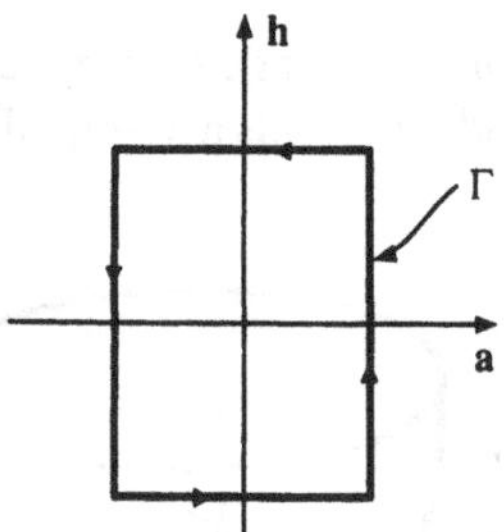

Figure 7.9. Γ as $a \to 0$.

Using a similar argument as in the spherical pendulum (see chapter 4 section 4), we find that $\widetilde{\Theta_\ell} = \lim_{a\to 0} \Theta_\ell^0$ and $\widetilde{\Theta_r} = \lim_{a\to 0} \Theta_r^0$ are equal to $-\frac{1}{2}$ if $h \in (-\chi, \chi)$ or -1 if $h \in (\chi, \infty)$. Thus the variation in Θ_ℓ^0 and Θ_r^0 after one loop around Γ is -1. Hence the initial basis

$$\left\{ 2\pi X_{J_r},\ 2\pi X_{J_\ell},\ TX_H - 2\pi\Theta_r^0 X_{J_r} - 2\pi\Theta_\ell^0 X_{J_\ell} \right\}.$$

of the period lattice $\mathcal{P}(h, a, 0)$ becomes the basis

$$\left\{ 2\pi X_{J_r},\ 2\pi X_{J_\ell},\ TX_H - 2\pi(\Theta_r^0 - 1)X_{J_r} - 2\pi(\Theta_\ell^0 - 1)X_{J_\ell} \right\}$$

after running around Γ once. Thus the *variation* in the bases of $\mathcal{P}(h, a, 0)$ is given by the matrix $\begin{pmatrix} 0 & 0 & 0 \\ 0 & 0 & 0 \\ 1 & 1 & 0 \end{pmatrix}$. Therefore the monodromy of $\mathcal{EM}^{-1}(\Gamma)$, which is the matrix taking the initial basis of $\mathcal{P}(h, a, 0)$ into the final basis, is $\chi_* = \begin{pmatrix} 1 & 0 & 0 \\ 0 & 1 & 0 \\ 1 & 1 & 1 \end{pmatrix}$. Conjugating χ_* by $P = \begin{pmatrix} 1 & 0 & 0 \\ 1 & 1 & 0 \\ 0 & 0 & 1 \end{pmatrix} \in \mathrm{Sl}(3, \mathbf{Z})$ gives $P\chi_* P^{-1} = \begin{pmatrix} 1 & 0 & 0 \\ 0 & 1 & 0 \\ 0 & 1 & 1 \end{pmatrix}$, which is a standard form for the monodromy mapping of the 3-torus bundle $\mathcal{E}\mathcal{M}^{-1}(\Gamma)$.

8 Hamiltonian Hopf bifurcation

Consider the family of motions of the Lagrange top where the figure axis of the top is vertical and the spin $|a|$ about the figure axis is increased through $|a| = \sqrt{4\beta}$. These motions are interesting because the top goes from unstable to stable motion. In physical terms, the top has become *gyroscopically stabilized*. Mathematically, this family of motions corresponds to the set of critical values of the energy momentum mapping where the thread attaches itself to the two dimensional piece of the discriminant locus and then becomes a crease. The goal of this section is to show that the top has undergone a Hamiltonian Hopf bifurcation, which is the mathematical explanation of gyroscopic stabilization.

8.1 The linear case

In this subsection we carry out a linear analysis of the motion of the top when its figure axis is vertical. We show that the Euler-Poisson equations undergo a linear Hamiltonian Hopf bifurcation.

After reducing the left S^1 action on SO(3) × so(3), the motion of the Lagrange top is governed by the *left* Euler-Poisson equations on $\mathbf{R}^3 \times \mathbf{R}^3$

$$\begin{aligned} \dot{z} &= z \times I^{-1}(z) \\ \dot{w} &= z \times \chi e_3 + w \times I^{-1}(w). \end{aligned} \tag{118}$$

Here $I = \mathrm{diag}(I_1, I_1, I_3)$ with $0 < I_3 \le 2I_1$, $0 < I_1 < I_3$ and $\chi > 0$. The Euler-Poisson equations are in Hamiltonian form on $(\mathbf{R}^3 \times \mathbf{R}^3, \{\,,\,\}_{\mathbf{R}^3 \times \mathbf{R}^3})$ with Hamiltonian

$$F_b : \mathbf{R}^3 \times \mathbf{R}^3 \to \mathbf{R} : (z, w) \to \tfrac{1}{2} I_1^{-1}(w_1^2 + w_2^2) + \tfrac{1}{2} I_3^{-1} w_3^2 + \chi z_3.$$

The structure matrix W of the Poisson bracket $\{\,,\,\}_{\mathbf{R}^3 \times \mathbf{R}^3}$ is given in table 8.1.

$\{A, B\}$	z_1	z_2	z_3	w_1	w_2	w_3	B
z_1	0	0	0	0	z_3	$-z_2$	
z_2	0	0	0	$-z_3$	0	z_1	
z_3	0	0	0	z_2	$-z_1$	0	
w_1	0	z_3	$-z_2$	0	w_3	$-w_2$	
w_2	$-z_3$	0	z_1	$-w_3$	0	w_1	
w_3	z_2	$-z_1$	0	w_2	$-w_1$	0	
A							

Table 8.1. The structure matrix W of the Poisson bracket $\{\,,\,\}_{\mathbf{R}^3 \times \mathbf{R}^3}$.

To simplify matters we rescale the time by setting $s = I_1 t$. Then (118) becomes

$$\begin{aligned} \dot{z} &= z \times \widetilde{I}^{-1}(w) \\ \dot{w} &= z \times \beta e_3 + w \times \widetilde{I}^{-1}(w), \end{aligned} \tag{119}$$

where $\widetilde{I} = \mathrm{diag}(1, 1, \gamma)$, $1 > \gamma = I_1/I_3 \geq \frac{1}{2}$ and $\beta = I_1 \chi > 0$. The rescaled Euler-Poisson equations (119) are in Hamiltonian form on $(\mathbf{R}^3 \times \mathbf{R}^3, \{\,,\,\}_{\mathbf{R}^3 \times \mathbf{R}^3})$ with Hamiltonian

$$\widetilde{F}_b : \mathbf{R}^3 \times \mathbf{R}^3 \to \mathbf{R} : (z, w) \to \tfrac{1}{2}(w_1^2 + w_2^2) + \tfrac{1}{2}\gamma w_3^2 + \beta z_3. \tag{120}$$

After reduction of the left S^1 action (see section 7.3), the motion of the top takes place on the reduced phase space P_b, which is defined by the 0-level sets of $C_1 = z_1^2 + z_2^2 + z_3^2 - 1$ and $C_2 = z_1 w_1 + z_2 w_2 + z_3 w_3 - b$. The sleeping or waking motion of the top occurs when the left and right S^1 angular momenta are equal, that is, $b = a$. After reduction, these motions correspond to the point $p_a = (e_3, a e_3)$ on P_a, which is an equilibrium point of $X_{\widetilde{F}_a} | P_a$. Linearizing $X_{\widetilde{F}_a}$ about p_a and then restricting to

$$T_{p_a} P_a = \ker \begin{pmatrix} 0 & 0 & 2 & 0 & 0 & 0 \\ 0 & 0 & a & 0 & 0 & 1 \end{pmatrix} = \mathrm{span}\{e_1, e_2, e_4, e_5\}$$

gives the smooth family of 4×4 matrices

$$a \to R_a = \begin{pmatrix} 0 & a\gamma & 0 & -1 \\ -a\gamma & 0 & 1 & 0 \\ 0 & \beta & 0 & a(\gamma - 1) \\ -\beta & 0 & -a(\gamma - 1) & 0 \end{pmatrix}. \tag{121}$$

Since $\widetilde{W}_a = W(p_a) | T_{p_a} P_a$ is invertible,

$$\Omega_a = (\widetilde{W}_a^{-1})^t = \begin{pmatrix} 0 & -a & 0 & 1 \\ a & 0 & -1 & 0 \\ 0 & 1 & 0 & 0 \\ -1 & 0 & 0 & 0 \end{pmatrix}$$

is a symplectic form on $T_{p_a} P_a$. The matrix R_a is infinitesimally symplectic with respect to Ω_a, that is, $R_a^t \Omega_a + \Omega_a R_a = 0$ or equivalently $R_a \in \mathrm{sp}(\Omega_a, \mathbf{R})$.

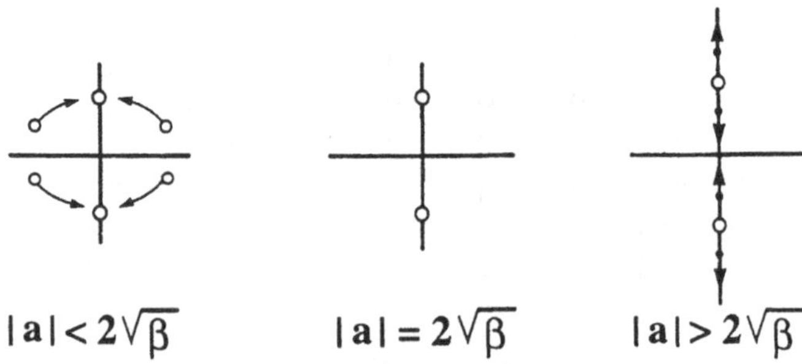

Figure 8.1. Movement of the eigenvalues of R_a for a near a_0.

We now look more closely at the smooth family $a \to R_a$. A calculation shows that the characteristic polynomial of R_a is $\lambda^4 + 2\alpha\lambda^2 + \delta^2$, where

$$\alpha = \tfrac{1}{2}a^2(\gamma^2 + (1-\gamma)^2) - \beta \quad \text{and} \quad \delta = -a^2\gamma(1-\gamma) + \beta.$$

Note that $\alpha + \delta = \frac{1}{2}a^2(2\gamma - 1)^2 \geq 0$ and $\alpha - \delta = \frac{1}{2}(a^2 - 4\beta)$. Thus the eigenvalues of R_a are

$$\begin{cases} \pm i\sqrt{\alpha \pm \sqrt{\alpha^2 - \delta^2}}, \text{ when } |a| > 2\sqrt{\beta} \\ \pm \frac{1}{2}\sqrt{2}(\sqrt{\delta - \alpha} \pm i\sqrt{\delta + \alpha}), \text{ when } |a| < 2\sqrt{\beta} \\ \pm i(2\gamma - 1)\sqrt{\beta}, \text{ when } |a| = 2\sqrt{\beta}. \end{cases}$$

The roots $\pm i(2\gamma - 1)\sqrt{\beta}$ are of multiplicity two. As a varies near $a_0 = \sqrt{4\beta}$ the movement of the eigenvalues of R_a is given in figure 8.1.

To understand the fusing of the eigenvalues at $a = a_0$, we examine

$$R = R_{a_0} = \begin{pmatrix} 0 & 2\gamma\sqrt{\beta} & 0 & -1 \\ -2\gamma\sqrt{\beta} & 0 & 1 & 0 \\ 0 & \beta & 0 & 2(\gamma - 1)\sqrt{\beta} \\ -\beta & 0 & -2(\gamma - 1)\sqrt{\beta} & 0 \end{pmatrix}.$$

The matrix R can be written uniquely as a commuting sum of a semisimple matrix

$$\mathcal{S} = \begin{pmatrix} 0 & r & 0 & 0 \\ -r & 0 & 0 & 0 \\ 0 & 0 & 0 & r \\ 0 & 0 & -r & 0 \end{pmatrix}$$

with $r = (2\gamma - 1)\sqrt{\beta}$ and a nilpotent matrix

$$\mathcal{N} = \begin{pmatrix} 0 & \sqrt{\beta} & 0 & -1 \\ -\sqrt{\beta} & 0 & 1 & 0 \\ 0 & \beta & 0 & -\sqrt{\beta} \\ -\beta & 0 & \sqrt{\beta} & 0 \end{pmatrix}.$$

Because R is infinitesimally symplectic with respect to the symplectic form

$$\Omega = \Omega_{a_0} = \begin{pmatrix} 0 & -2\sqrt{\beta} & 0 & 1 \\ 2\sqrt{\beta} & 0 & -1 & 0 \\ 0 & 1 & 0 & 0 \\ -1 & 0 & 0 & 0 \end{pmatrix},$$

it follows that $\mathcal{S}$ and $\mathcal{N}$ are also infinitesimally symplectic. To find the infinitesimally symplectic normal form of R, let $e = (0, 0, 1, 0)$ and set $f = e - \frac{1}{2r^2}\Omega(e, \mathcal{S}e)\mathcal{N}\mathcal{S}e$. From $\mathcal{S}^2 + r^2 = 0$, $\mathcal{N}^2 = 0$ and $\Omega(e, \mathcal{N}e) = -1$, it follows that $\Omega(f, \mathcal{N}f) = -1$ and $\Omega(f, \mathcal{S}f) = \Omega(\mathcal{N}f, \mathcal{S}f) = 0$. Consequently $\{f, \frac{1}{r}\mathcal{S}f, -\mathcal{N}f, -\frac{1}{r}\mathcal{S}\mathcal{N}f\}$ is a symplectic basis of $\mathbf{R}^4$ with respect to the standard symplectic form ω. The infinitesimally symplectic normal form of R is

$$\mathfrak{R} = P_0^{-1} R P_0 = \begin{pmatrix} 0 & -r & 0 & 0 \\ r & 0 & 0 & 0 \\ -1 & 0 & 0 & -r \\ 0 & -1 & r & 0 \end{pmatrix}.$$

where $P_0 = \mathrm{col}(f, \frac{1}{r}\mathcal{S}f, -\mathcal{N}f, -\frac{1}{r}\mathcal{S}\mathcal{N}f)$.

To see that the smooth family $a \to R_a$ undergoes a linear Hamiltonian Hopf bifurcation at $a = a_0$, we first transform it using the smooth coordinate change

$$a \to Q_a = \begin{pmatrix} 1 & 0 & 0 & 0 \\ 0 & 1 & 0 & 0 \\ \frac{1}{2}a & 0 & 0 & 1 \\ 0 & \frac{1}{2}a & -1 & 0 \end{pmatrix}. \tag{122}$$

The transformed smooth family is $a \to U_a = Q_a^{-1} R_a Q_a$, where

$$U_a = \begin{pmatrix} 0 & \frac{1}{2}a(2\gamma-1) & 1 & 0 \\ -\frac{1}{2}a(2\gamma-1) & 0 & 0 & 1 \\ \frac{1}{4}(4\beta - a^2) & 0 & 0 & \frac{1}{2}a(2\gamma-1) \\ 0 & \frac{1}{4}(4\beta - a^2) & -\frac{1}{2}a(2\gamma-1) & 0 \end{pmatrix} \tag{123}$$

and the transformed symplectic form $Q_a^t \Omega_a Q_a$ is ω. At $a = a_0$ the real symplectic linear map $Q = Q_{a_0}^{-1} P_0 \in \mathrm{Sp}(\omega, \mathbf{R})$ conjugates U_{a_0} into the normal form $\mathfrak{R}$, that is, $Q^{-1} U_{a_0} Q = \mathfrak{R}$. Using Q we transform $a \to U_a$ into the smooth family $a \to V_a = Q^{-1} U_a Q$, where $V_{a_0} = \mathfrak{R}$.

The following result gives a *smooth normal form* for the family $a \to V_a$ of infinitesimally symplectic linear maps on $(\mathbf{R}^4, \omega)$.

Claim: Suppose that $\gamma : \mathbf{R} \to \mathrm{sp}(\omega, \mathbf{R}) : \mu \to B_\mu$ is a smooth family of real infinitesimally symplectic matrices on $(\mathbf{R}^4, \omega)$ with

$$B_0 = \begin{pmatrix} 0 & -b & 0 & 0 \\ b & 0 & 0 & 0 \\ -1 & 0 & 0 & -b \\ 0 & -1 & b & 0 \end{pmatrix}.$$

For every μ in some open interval I containing 0 there is a smooth family

$$I \to \mathrm{Sp}(\omega, \mathbf{R}) : \mu \to P_\mu^{-1}$$

of real linear symplectic mappings which transforms the smooth family γ into the smooth normal form

$$\Gamma : I \to \mathrm{sp}(\omega, \mathbf{R}) :$$

$$\mu \to P_\mu B_\mu P_\mu^{-1} = \begin{pmatrix} 0 & -(b+\nu_1(\mu)) & \nu_2(\mu) & 0 \\ b+\nu_1(\mu) & 0 & 0 & \nu_2(\mu) \\ -1 & 0 & 0 & -(b+\nu_1(\mu)) \\ 0 & -1 & b+\nu_1(\mu) & 0 \end{pmatrix}.$$

Here $\nu_i : I \to \mathbf{R} : \mu \to \nu_i(\mu)$ is a smooth function with $\nu_i(0) = 0$.

(8.12) **Proof**: Consider the mapping

$$\varphi : \mathbf{R}^2 \times \mathrm{Sp}(\omega, \mathbf{R}) \to \mathrm{sp}(\omega, \mathbf{R}) : \big((\nu_1, \nu_2), P\big) \to P^{-1}(B_0 + \nu_1 X_S + \nu_2 X_M) P.$$

Here

$$X_S = \begin{pmatrix} 0 & -1 & 0 & 0 \\ 1 & 0 & 0 & 0 \\ 0 & 0 & 0 & -1 \\ 0 & 0 & 1 & 0 \end{pmatrix} \quad \text{and} \quad X_M = \begin{pmatrix} 0 & 0 & 1 & 0 \\ 0 & 0 & 0 & 1 \\ 0 & 0 & 0 & 0 \\ 0 & 0 & 0 & 0 \end{pmatrix},$$

which are both infinitesimally symplectic. Partially differentiating φ with respect to (ν_1, ν_2) we obtain $D_1\varphi(0, I)\begin{pmatrix} s \\ t \end{pmatrix} = sX_S + tX_M$. Moreover, the partial derivative of φ at $(0, I)$ with respect to P is $D_2\varphi(0, I)X = \mathrm{ad}_{B_0} X$ where $X \in T_I\,\mathrm{Sp}(\omega, \mathbf{R}) = \mathrm{sp}(\omega, \mathbf{R})$. Using the fact that

$$\mathrm{sp}(\omega, \mathbf{R}) = \mathrm{im}\,\mathrm{ad}_{B_0} \oplus \mathrm{span}\{X_S, X_M\}, \tag{124}$$

which is proved in ((8.13)) below, we see that $D\varphi(0, I)$ is surjective. From the implicit function theorem, it follows that the image of φ contains an open neighborhood $\mathcal{U}$ of B_0. Choose an open interval I containing 0 such that for every $\mu \in I$, the matrix $B(\mu) \in \mathcal{U}$. Again from the implicit function theorem, it follows that there are smooth functions

$$\nu_i : I \to \mathbf{R} : \mu \to \nu_i(\mu)$$

with $\nu_i(0) = 0$ and a smooth function

$$P : I \to \mathrm{Sp}(\omega, \mathbf{R}) : \mu \to P(\mu)$$

such that

$$B_\mu = \varphi\big(\nu_1(\mu), \nu_2(\mu), P_\mu\big) = P_\mu^{-1}\big(B_0 + \nu_1(\mu)X_S + \nu_2(\mu)X_M\big)P_\mu$$

for every $\mu \in I$. This proves the claim except for (124). □

(8.13) **Proof**: To prove (124) we show that $\{X_S, X_M\}$ spans $\ker \mathrm{ad}_{X_S} \cap \ker \mathrm{ad}_{X_M}$ and that a complement to $\mathrm{im}\,\mathrm{ad}_{B_0}$ is given by $\ker \mathrm{ad}_{X_S} \cap \ker \mathrm{ad}_{X_M}$. First observe that $\ker \mathrm{ad}_{X_S}$ is a Lie subalgebra of $\mathrm{sp}(\omega, \mathbf{R})$ with basis $\{X_M, X_N, X_T, X_S\}$ where

$$X_N = \begin{pmatrix} 0 & 0 & 0 & 0 \\ 0 & 0 & 0 & 0 \\ 1 & 0 & 0 & 0 \\ 0 & 1 & 0 & 0 \end{pmatrix} \quad \text{and} \quad X_T = \begin{pmatrix} 1 & 0 & 0 & 0 \\ 0 & 1 & 0 & 0 \\ 0 & 0 & -1 & 0 \\ 0 & 0 & 0 & -1 \end{pmatrix}.$$

This follows because

$$\begin{aligned} [X_T, X_M] &= X_T X_M - X_M X_T = 2X_M \\ [X_T, X_N] &= -2\,X_N \\ [X_M, X_N] &= X_T, \end{aligned}$$

and

$$[X_S, X_M] = [X_S, X_N] = [X_S, X_T] = 0.$$

From these bracket relations we see that the matrix of $\mathrm{ad}_{X_M} \mid \ker \mathrm{ad}_{X_S}$ with respect to the basis $\{X_M, X_N, X_T, X_S\}$ is

$$\begin{pmatrix} 0 & 0 & -2 & 0 \\ 0 & 0 & 0 & 0 \\ 0 & 1 & 0 & 0 \\ 0 & 0 & 0 & 0 \end{pmatrix}.$$

Therefore the intersection of $\ker \mathrm{ad}_{X_M}$ and $\ker \mathrm{ad}_{X_S}$ is spanned by $\{X_M, X_S\}$. Similarly,

$$\mathrm{ad}_{X_N} \mid \ker \mathrm{ad}_{X_S} = \begin{pmatrix} 0 & 0 & 0 & 0 \\ 0 & 0 & 2 & 0 \\ -1 & 0 & 0 & 0 \\ 0 & 0 & 0 & 0 \end{pmatrix}.$$

Therefore

$$\ker \mathrm{ad}_S = \left(\ker \mathrm{ad}_{X_M} \cap \ker \mathrm{ad}_{X_S} \right) \oplus \left(\mathrm{im}\, \mathrm{ad}_{X_N} \cap \ker \mathrm{ad}_{X_S} \right).$$

Since X_S is semisimple, $\mathrm{sp}(\omega, \mathbf{R}) = \ker \mathrm{ad}_{X_S} \oplus \mathrm{im}\, \mathrm{ad}_{X_S}$. Note that in ((8.12)) $B_0 = bX_S - X_N$. From $X_S X_N = X_N X_S$ and the fact that X_N is nilpotent we obtain

$$\mathrm{im}\, \mathrm{ad}_{B_0} = \mathrm{im}\, \mathrm{ad}_{X_S} \oplus \left(\mathrm{im}\, \mathrm{ad}_{X_N} \cap \ker \mathrm{ad}_{X_S} \right).$$

Consequently,

$$\mathrm{sp}(\omega, \mathbf{R}) = \mathrm{im}\, \mathrm{ad}_{B_0} \oplus \left(\ker \mathrm{ad}_{X_M} \cap \ker \mathrm{ad}_{X_S} \right).$$

This proves (124). □

In the course of the above proof we have shown that the tangent space at B_0 to the $\mathrm{Sp}(\omega, \mathbf{R})$-orbit $\mathcal{O} = \{P^{-1} B_0 P \mid P \in \mathrm{Sp}(\omega, \mathbf{R})\}$ through B_0 is $\mathrm{im}\, \mathrm{ad}_{B_0}$. Thus the plane spanned by X_S and X_M is *transverse* to $\mathcal{O}$ at B_0.

Since $V_{a_0} = \mathfrak{R}$, the smooth normal form of the family $a \to V_a$ near a_0 is the family $a \to Y_a$ where

$$Y_a = \begin{pmatrix} 0 & -(r + \nu_1(a)) & \nu_2(a) & 0 \\ r + \nu_1(a) & 0 & 0 & \nu_2(a) \\ -1 & 0 & 0 & -(r + \nu_1(a)) \\ 0 & -1 & r + \nu_1(a) & 0 \end{pmatrix}. \tag{125}$$

To compute the smooth functions ν_1 and ν_2 in (125) observe that the families $a \to Y_a$ and $a \to R_a$ are smoothly conjugate. Therefore the characteristic polynomial of Y_a

$$\lambda^4 + 2(r + \nu_1)^2 \lambda^2 + ((r + \nu_1)^2 + \nu_2)^2$$

is equal to the characteristic polynomial of R_a

$$\lambda^4 + 2\left(\tfrac{1}{2} a^2 (\gamma^2 + (\gamma - 1)^2) - \beta\right)\lambda^2 + \left(\beta - a^2 \gamma (1 - \gamma)\right)^2.$$

Equating coefficients and solving gives

$$\begin{cases} \nu_1(a) &= \sqrt{\tfrac{1}{2} a^2 (\gamma^2 + (\gamma - 1)^2) - \beta} - (2\gamma - 1)\sqrt{\beta} \\ \nu_2(a) &= -\tfrac{1}{2}\,(a^2 - 4\beta). \end{cases}$$

Note that $\nu_1(a_0) = \nu_2(a_0) = 0$. Hence $Y_{a_0} = \mathfrak{R}$. Since

$$\left.\frac{dY_a}{da}\right|_{a=a_0} = \left.\frac{d\nu_1}{da}\right|_{a=a_0} X_S + \left.\frac{d\nu_2}{da}\right|_{a=a_0} X_M \quad \text{and} \quad \left.\frac{d\nu_2}{da}\right|_{a=a_0} = -a_0 \neq 0,$$

the curve $a \to Y_a$ crosses the orbit $\mathcal{O} = \{P^{-1}Y_{a_0}P \mid P \in \mathrm{Sp}(\omega, \mathbf{R})\}$ transversely at Y_{a_0}. Thus the curve $a \to V_a$ crosses $\mathcal{O}$ transversely at V_{a_0}. Hence the eigenvalues of a small smooth perturbation of the curve $a \to R_a$ (such that the new curve lies in $\mathrm{sp}(\omega, \mathbf{R})$ and passes through R_{a_0}) has the same behavior as those of $a \to R_a$, (see figure 8.1). We say that the curve $a \to R_a$ undergoes a linear Hamiltonian Hopf bifurcation at $a = a_0$.

8.2 The nonlinear case

In this subsection we show that, after reduction of the left S^1 action, the Lagrange top undergoes a *nonlinear* Hamiltonian Hopf bifurcation when the left and right angular momenta are equal and the spin $|a|$ about the figure axis increases through $\sqrt{4\beta}$. This entails finding a smooth family $a \to \psi_a$ of local symplectic diffeomorphisms, which for every a near $\sqrt{4\beta}$ transforms the reduced Hamiltonian $\widetilde{F}_a$ (120) into a Hamiltonian H_a whose 4-jet at the origin is in normal form to second order. Specifically, we show that $H_a = H_a^2 + \varepsilon^2 H_a^4 + O(\varepsilon^4)$, where

$$H_a^2 = -\tfrac{1}{2}a(2\gamma - 1)(\xi_1\eta_2 - \xi_2\eta_1) + \tfrac{1}{2}(\eta_1^2 + \eta_2^2) + \tfrac{1}{8}(a^2 - 4\beta)(\xi_1^2 + \xi_2^2)$$

and

$$\begin{aligned} H_a^4 &= \tfrac{1}{96}(a^2 + 8\beta)(\xi_1^2 + \xi_2^2)^2 - \tfrac{1}{8}(\xi_1\eta_2 - \xi_2\eta_1)(\xi_1^2 + \xi_2^2) \\ &\quad + \tfrac{1}{6}(3\gamma - 2)(\xi_1\eta_2 - \xi_2\eta_1)^2. \end{aligned}$$

Since $\frac{1}{96}(a^2 + 8\beta) > 0$, the Hamiltonian H_a undergoes a nonlinear Hamiltonian Hopf bifurcation as a increases through $\sqrt{4\beta}$. The construction of the symplectic diffeomorphism ψ_a uses techniques from normal form theory. The argument is not straightforward as we need a constructive version of Darboux's theorem.

To start our analysis we choose a special parametrization of the reduced space P_a near $p_a = (e_3, ae_3)$ given by

$$\begin{aligned} \varphi_a : U = D_1^2 \times \mathbf{R}^2 \subseteq \mathbf{R}^4 \to P_a : (x_1, x_2, y_1, y_2) \to (z_1, z_2, z_3, w_1, w_2, w_3) = \\ = \big(x_1, x_2, (1 - x_1^2 - x_2^2)^{1/2}, y_1, y_2, (a - x_1y_1 - x_2y_2)(1 - x_1^2 - x_2^2)^{-1/2}\big). \quad (126) \end{aligned}$$

Here $\varphi_a(0) = p_a$ and D_1^2 is the 2-disc $\{x_1^2 + x_2^2 < 1\}$. The reason why φ_a is special is that it intertwines the right S^1 action on P_a (see (114)) with the S^1 action

$$S^1 \times \mathbf{R}^4 \to \mathbf{R}^4 : \big(s, (x, y)\big) \to (S_s x, S_s y), \tag{127}$$

where $S_s = \begin{pmatrix} \cos s & -\sin s \\ \sin s & \cos s \end{pmatrix}$. Because the reduced Hamiltonian $\widetilde{F}_a$ (120) is invariant under the right S^1 action on P_a, the Hamiltonian

$$\begin{aligned} \widehat{H}_a &= (\varphi_a)^* \widetilde{F}_a : U \subseteq \mathbf{R}^4 \to \mathbf{R} : (x, y) \to \tfrac{1}{2}(y_1^2 + y_2^2) \\ &\quad + \tfrac{1}{2}\gamma\big(a - x_1y_1 - x_2y_2\big)^2(1 - x_1^2 - x_2^2)^{-1} + \beta(1 - x_1^2 - x_2^2)^{1/2} \end{aligned} \tag{128}$$

is invariant under the S^1 action on $\mathbf{R}^4$.

We now find the induced symplectic structure on an open neighborhood of 0 in U. Recall that the reduced space P_a is defined by the 0-level sets of the Casimirs $z_1^2+z_2^2+z_3^2-1$ and $z_1w_1+z_2w_2+z_3w_3-a$ of the Poisson algebra $(C^\infty(\mathbf{R}^3\times\mathbf{R}^3), \{\,,\,\}_{\mathbf{R}^3\times\mathbf{R}^3}, \cdot)$. Thus the Poisson bracket $\{\,,\,\}_{P_a}$ on $C^\infty(P_a)$ is the restriction of $\{\,,\,\}_{\mathbf{R}^3\times\mathbf{R}^3}$ on $C^\infty(\mathbf{R}^3\times\mathbf{R}^3)$ to $C^\infty(P_a)$. On $C^\infty(U)$ define a Poisson bracket $\{\,,\,\}_U$ by pulling back $\{\,,\,\}_{P_a}$ by φ_a, that is, for $f, g\in C^\infty(U)$

$$\{f,g\}_U=\varphi_a^*\big(\{(\varphi_a^{-1})^*f,(\varphi_a^{-1})^*g\}_{P_a}\big).$$

The structure matrix $\widehat{W}_a$ of $\{\,,\,\}_U$ is given in table 8.2.

$\{A,B\}$	x_1	x_2	y_1	y_2	B
x_1	0	0	0	$\sqrt{1-x_1^2-x_2^2}$	
x_2		0	$-\sqrt{1-x_1^2-x_2^2}$	0	
y_1			0	$(a-x_1y_1-x_2y_2)(1-x_1^2-x_2^2)^{-1/2}$	
y_2				0	
A					

Table 8.2. Half of the skew symmetric structure matrix $\widehat{W}_a$.

Since $\widehat{W}_a(0)$ is invertible, the Poisson bracket $\{\,,\,\}_U$ defines a symplectic structure

$$\begin{aligned}\widehat{\Omega}_a(x,y) &= (\widehat{W}_a^{-1})^t = (a-x_1y_1-x_2y_2)(1-x_1^2-x_2^2)^{-3/2}dx_1\wedge dx_2\\ &\quad -(1-x_1^2-x_2^2)^{-1/2}(dx_1\wedge dy_2-dx_2\wedge dy_1)\end{aligned}\tag{129}$$

on an open neighborhood of 0 in $\mathbf{R}^4$. Note that $\widehat{\Omega}_a$ is invariant under the S^1 action on $\mathbf{R}^4$. Introduce new variables (ξ,η) by

$$\begin{pmatrix} x\\ y\end{pmatrix}=Q_a\begin{pmatrix}\xi\\ \eta\end{pmatrix},$$

where Q_a is given by (122). The pulled back symplectic form $Q_a^*\widehat{\Omega}_a=\widetilde{\Omega}_a$ is

$$\begin{aligned}\widetilde{\Omega}_a(\xi,\eta) &= \big(a-\tfrac12 a(\xi_1^2+\xi_2^2)-(\xi_1\eta_2-\xi_2\eta_1)\big)(1-\xi_1^2-\xi_2^2)^{-3/2}d\xi_1\wedge d\xi_2\\ &\quad +(1-\xi_1^2-\xi_2^2)^{-1/2}(d\xi_1\wedge d\eta_1+d\xi_2\wedge d\eta_2-a\,d\xi_1\wedge d\xi_2).\end{aligned}\tag{130}$$

Thus $\widetilde{\Omega}_a(0)=Q_a^t\widehat{\Omega}_a(0)Q_a$ is the standard symplectic form $\omega=d\xi_1\wedge d\eta_1+d\xi_2\wedge d\eta_2$ on $\mathbf{R}^4$. Since Q_a commutes with $\mathrm{diag}(S_s,S_s)$, the 2-form $\widetilde{\Omega}_a$ is invariant under the S^1 action on $\mathbf{R}^4$. Pulling back the Hamiltonian $\widehat{H}_a$ by Q_a gives the S^1 invariant Hamiltonian

$$\begin{aligned}\widetilde{H}_a(\xi,\eta) &= \tfrac18 a^2(\xi_1^2+\xi_2^2)+\tfrac12 a(\xi_1\eta_2-\xi_2\eta_1)+\tfrac12(\eta_1^2+\eta_2^2)\\ &\quad +\tfrac12\gamma\big(a-\tfrac12 a^2(\xi_1^2+\xi_2^2)-(\xi_1\eta_2-\xi_2\eta_1)\big)^2(1-\xi_1^2-\xi_2^2)^{-1}\\ &\quad +\beta(1-\xi_1^2-\xi_2^2)^{1/2}.\end{aligned}\tag{131}$$

Introduce a formal small parameter ε and replace the variables ξ and η by $\varepsilon\xi$ and $\varepsilon\eta$. Then the blown up Hamiltonian $\widetilde{H}_a'(\xi,\eta)=\frac{1}{\varepsilon^2}\widetilde{H}_a(\varepsilon\xi,\varepsilon\eta)$ and the blown up symplectic form

$\widetilde{\Omega}'_a(\xi,\eta) = \frac{1}{\varepsilon^2}\widetilde{\Omega}_a(\varepsilon\xi,\varepsilon\eta)$ have 4-jet at the origin given by

$$\widetilde{H}_a = \widetilde{H}_a^2 + \varepsilon^2\widetilde{H}_a^4 + O(\varepsilon^4),$$

where

$$\begin{aligned}\widetilde{H}_a^2 &= \tfrac{1}{2}(\eta_1^2+\eta_2^2) + \tfrac{1}{8}(a^2-4\beta)(\xi_1^2+\xi_2^2)\\ &\quad + \tfrac{1}{2}a(2\gamma-1)(\xi_1\eta_2-\xi_2\eta_1) \qquad (132)\\ \widetilde{H}_a^4 &= \tfrac{1}{8}(a\gamma^2-\beta)(\xi_1^2+\xi_2^2)^2 - \tfrac{1}{2}a\gamma(\xi_1\eta_2-\xi_2\eta_1)(\xi_1^2+\xi_2^2)\\ &\quad + \tfrac{1}{2}\gamma(\xi_1\eta_2-\xi_2\eta_1)^2 \qquad (133)\end{aligned}$$

and $\widetilde{\Omega}_a = \widetilde{\Omega}_a^0 + \varepsilon^2\widetilde{\Omega}_a^2 + O(\varepsilon^4)$, where

$$\begin{aligned}\widetilde{\Omega}_a^0 &= d\xi_1\wedge d\eta_1 + d\xi_2\wedge d\eta_2 = \omega \qquad (134)\\ \widetilde{\Omega}_a^2 &= \big(\tfrac{1}{2}a(\xi_1^2+\xi_2^2) - (\xi_1\eta_2-\xi_2\eta_1)\big)d\xi_1\wedge d\xi_2\\ &\quad + \tfrac{1}{2}(\xi_1^2+\xi_2^2)(d\xi_1\wedge d\eta_1 + d\xi_2\wedge d\eta_2), \qquad (135)\end{aligned}$$

after dropping the prime.

From the above equation we see that to zeroth order the symplectic form $\widetilde{\Omega}_a$ is the standard constant symplectic form ω. In other words, $\widetilde{\Omega}_a$ is flat to zeroth order. Darboux's theorem (see appendix A ((4.7))) states that there is a coordinate change which makes $\widetilde{\Omega}_a$ flat to all orders. However, the usual proof (see appendix B ((3.12))) does not give a constructive way to find this coordinate change. The goal of the following discussion is to show how to construct a coordinate change which removes the second order terms in $\widetilde{\Omega}_a$. We begin by giving a constructive proof of the Poincaré lemma.

Claim: Let A be a diagonalizable linear vector field on $\mathbf{R}^n$ with all of its eigenvalues strictly negative. Let φ_t be the flow of A. Suppose that β is a closed p-form with $p \geq 1$ in a closed ball $\overline{B}_r$ of radius r about 0. Then the $(p-1)$-form $\alpha = -\int_0^\infty \varphi_t^*(A \lrcorner \beta)dt$ satisfies $\beta = d\alpha$ in $\overline{B}_r$.

(8.14) **Proof**: Note that $\varphi_\infty = \lim_{t\to\infty}\varphi_t = 0$. The integral defining α exists because every coefficient of $A \lrcorner \beta$ is bounded in $\overline{B}_r$, while the pull back of the differentials by φ_t decay exponentially. To finish the argument we compute

$$\begin{aligned}\beta &= -(\varphi_\infty^*\beta - \varphi_0^*\beta) = -\int_0^\infty \frac{d}{dt}(\varphi_t^*\beta)dt\\ &= -\int_0^\infty \frac{d}{ds}\Big|_{s=0}(\varphi_{s+t}^*\beta)dt = -\int_0^\infty \varphi_t^*(L_A\beta)\,dt\\ &= -\int_0^\infty \varphi_t^*\big(A\lrcorner d\beta + d(A\lrcorner\beta)\big)dt\\ &= -d\Big(\int_0^\infty \varphi_t^*(A\lrcorner\beta)\,dt\Big), \quad \text{since } d\beta = 0 \text{ and } \varphi_t^*d = d\varphi_t^*\\ &= d\alpha. \quad \square\end{aligned}$$

Suppose that β is a p-form with coefficients which are homogeneous polynomials of degree ℓ. Choose A to be the vector field $-x_1\frac{\partial}{\partial x_1} - \cdots - x_n\frac{\partial}{\partial x_n}$. The flow of A is $\varphi_t(x) = e^{-t}x$. The above proof of the Poincaré lemma shows that $\alpha = -\frac{1}{\ell+p}(A \lrcorner \beta)$.

Before proceeding to state and prove the formal power series version of the Darboux theorem, we prove the following claim which gives the basic computational tools of normal form theory.

Claim: 1. Let X be a vector field on $\mathbf{R}^n$ on which we have coordinates $x = (x_1, \ldots, x_n)$. Then the formal power series in ε given by

$$(\exp \varepsilon L_X)x = \sum_{n\geq 0} \frac{\varepsilon^n}{n!} L_X^n x$$

is the *formal flow* of X. Here $L_X x$ is the vector whose i^{th} component is the Lie derivative $L_X x_i$ of the function x_i with respect to X.

2. Suppose that Q is a smooth geometric quantity on $\mathbf{R}^n$ such as a function, a vector field, or a differential form. Then the formal pull back of Q by the time ε map of the flow of X is given by the formal power series

$$(\exp \varepsilon L_X)^* Q = \sum_{n\geq 0} \frac{\varepsilon^n}{n!} L_X^n Q.$$

(8.15) **Proof**: 1) From the power series for exp it is easy to see that $\varepsilon \to \exp \varepsilon L_X$ is a one parameter group. Therefore $\varepsilon \to (\exp \varepsilon L_X)x$ is a one parameter group of invertible formal power series mappings on $\mathbf{R}^n$. Since $\left.\frac{d}{d\varepsilon}\right|_{\varepsilon=0} (\exp \varepsilon L_X)x = L_X x = X(x)$, the one parameter group $\varepsilon \to \exp \varepsilon L_X$ is the formal flow of X.

2) We compute

$$\begin{aligned}
\frac{d}{d\varepsilon}(\exp \varepsilon L_X)^* Q &= \left.\frac{d}{d\eta}\right|_{\eta=0} \big(\exp(\varepsilon+\eta)L_X\big)^* Q \\
&= (\exp \varepsilon L_X)^* \Big(\left.\frac{d}{d\eta}\right|_{\eta=0} (\exp \eta L_X)^* Q\Big) \\
&= (\exp \varepsilon L_X)^* (L_X Q).
\end{aligned}$$

Therefore by induction

$$\frac{d^n}{d\varepsilon^n}(\exp \varepsilon L_X)^* Q = (\exp \varepsilon L_X)^* (L_X^n Q).$$

Hence

$$(\exp \varepsilon L_X)^* Q = \sum_{n\geq 0} \frac{\varepsilon^n}{n!} \left.\frac{d^n}{d\varepsilon^n}\right|_{\varepsilon=0} (\exp \varepsilon L_X)^* Q = \sum_{n\geq 0} \frac{\varepsilon^n}{n!} L_X^n Q. \qquad \square$$

The following claim shows how to flatten a formal power series closed 2-form to second order.

Claim: Let $\Omega = \Omega_0 + \varepsilon\Omega_1 + \cdots$ be a formal power series closed 2-form on $\mathbf{R}^n$ with Ω_0 a constant symplectic form. By the Poincaré lemma, there is a formal power series 1-form $\alpha = \alpha_0 + \varepsilon\alpha_1 + \cdots$ such that $\Omega = d\alpha$. Define a vector field X by $X \lrcorner\, \Omega_0 = -\alpha_1$. Then changing coordinates by the time ε map of the flow of X flattens Ω to second order.

(8.16) **Proof**: Applying the coordinate change $\exp \varepsilon L_X$ to the 1-form α gives

$$\begin{aligned}
\widehat{\alpha} &= (\exp \varepsilon L_X)^* \alpha = \alpha + \varepsilon L_X \alpha + O(\varepsilon^2) \\
&= \alpha_0 + \varepsilon(\alpha_1 + L_X \alpha_0) + O(\varepsilon^2) \\
&= \alpha_0 + \varepsilon d(X \lrcorner\, \alpha_0) + \varepsilon(\alpha_1 + X \lrcorner\, d\alpha_0) + O(\varepsilon^2) \\
&= \alpha_0 + \varepsilon d(X \lrcorner\, \alpha_0) + O(\varepsilon^2),
\end{aligned}$$

since $\Omega_0 = d\alpha_0$ and $X \lrcorner\, \Omega_0 = -\alpha_1$. Hence

$$\begin{aligned}
(\exp \varepsilon L_X)^* \Omega &= (\exp \varepsilon L_X)^*(d\alpha) = d((\exp \varepsilon L_X)^* \alpha) \\
&= d\widehat{\alpha} = d\alpha_0 + O(\varepsilon^2) \\
&= \Omega_0 + O(\varepsilon^2). \quad \square
\end{aligned}$$

With these tools at hand we flatten $\widetilde{\Omega}_a$ to second order as follows. Clearly $\widetilde{\Omega}_a^0 = d\alpha^0$ where $\alpha^0 = \xi_1 d\eta_1 + \xi_2 d\eta_2$. Using the Poincaré lemma and the homogeneity of $\widetilde{\Omega}_a^2$, we find that $\widetilde{\Omega}_a^2 = d\alpha_a^2$, where

$$\alpha_a^2 = -\tfrac{1}{4}(A \lrcorner\, \widetilde{\Omega}_a^2) \quad \text{and} \quad A = -\xi_1 \frac{\partial}{\partial \xi_1} - \xi_2 \frac{\partial}{\partial \xi_2} - \eta_1 \frac{\partial}{\partial \eta_1} - \eta_2 \frac{\partial}{\partial \eta_2}.$$

A calculation gives

$$\begin{aligned}
\alpha_a^2 &= \Big[-\tfrac{1}{8}\eta_1(\xi_1^2 + \xi_2^2) - \tfrac{1}{8} a\xi_2(\xi_1^2 + \xi_2^2) + \tfrac{1}{4}\xi_2(\xi_1\eta_2 - \xi_2\eta_1)\Big] d\xi_1 \\
&\quad + \Big[-\tfrac{1}{8}\eta_2(\xi_1^2 + \xi_2^2) - \tfrac{1}{8} a\xi_1(\xi_1^2 + \xi_2^2) + \tfrac{1}{4}\xi_1(\xi_1\eta_2 - \xi_2\eta_1)\Big] d\xi_2 \\
&\quad + \tfrac{1}{8}\xi_1(\xi_1^2 + \xi_2^2)\, d\eta_1 + \tfrac{1}{8}\xi_2(\xi_1^2 + \xi_2^2)\, d\eta_2.
\end{aligned}$$

Clearly α^0 is S^1 invariant; while α_a^2 is S^1 invariant since A and $\widetilde{\Omega}_a^2$ are. Put $X_a^2 = -(\widetilde{\Omega}_a^0)^\flat(\alpha_a^2)$. Because

$$-d\xi_i = (\widetilde{\Omega}_a^0)^\sharp\Big(\frac{\partial}{\partial \eta_i}\Big) \quad \text{and} \quad d\eta_i = (\widetilde{\Omega}_a^0)^\sharp\Big(\frac{\partial}{\partial \xi_i}\Big),$$

we find that

$$\begin{aligned}
X_a^2 &= \Big[-\tfrac{1}{8}\eta_1(\xi_1^2 + \xi_2^2) - \tfrac{1}{8} a\xi_2(\xi_1^2 + \xi_2^2) + \tfrac{1}{4}\xi_2(\xi_1\eta_2 - \xi_2\eta_1)\Big] \frac{\partial}{\partial \eta_1} \\
&\quad + \Big[-\tfrac{1}{8}\eta_2(\xi_1^2 + \xi_2^2) - \tfrac{1}{8} a\xi_1(\xi_1^2 + \xi_2^2) + \tfrac{1}{4}\xi_1(\xi_1\eta_2 - \xi_2\eta_1)\Big] \frac{\partial}{\partial \eta_2} \\
&\quad - \tfrac{1}{8}\xi_1(\xi_1^2 + \xi_2^2) \frac{\partial}{\partial \xi_1} - \tfrac{1}{8}\xi_2(\xi_1^2 + \xi_2^2) \frac{\partial}{\partial \xi_2}.
\end{aligned}$$

Note that X_a^2 is S^1 invariant. Pulling $\widetilde{\Omega}_a$ back by the time ε^2 map of the flow of X_a^2 flattens $\widetilde{\Omega}_a$ to fourth order, that is

$$\overline{\Omega}_a = (\exp \varepsilon^2 L_{X_a^2})^* \widetilde{\Omega}_a = \widetilde{\Omega}_a^0 + O(\varepsilon^4) = \omega + O(\varepsilon^4).$$

Since X_a^2 is S^1 invariant, the time ε^2 map of its flow commutes with the S^1 action on $\mathbf{R}^4$. Therefore in the flattened coordinates the new Hamiltonian is S^1 invariant and is given by

$$\begin{aligned}\overline{H}_a &= (\exp \varepsilon^2 L_{X_a^2})^* \widetilde{H}_a = \widetilde{H}_a^2 + \varepsilon^2(\widetilde{H}_a^4 + L_{X_a^2}\widetilde{H}_a^2) + O(\varepsilon^4) \\ &= \overline{H}_a^2 + \varepsilon^2 \overline{H}_a^4 + O(\varepsilon^4),\end{aligned}$$

where

$$\overline{H}_a^2 = \tfrac{1}{2}(\eta_1^2 + \eta_2^2) + \tfrac{1}{8}(a^2 - 4\beta)(\xi_1^2 + \xi_2^2) - \tfrac{1}{2}a(2\gamma - 1)(\xi_1\eta_2 - \xi_2\eta_1)$$

and

$$\begin{aligned}\overline{H}_a^4 &= \tfrac{1}{32}a^2(\xi_1^2 + \xi_2^2)^2 - \tfrac{1}{8}a(\xi_1^2 + \xi_2^2)(\xi_1\eta_2 - \xi_2\eta_1) \\ &\quad + \tfrac{1}{4}(2\gamma - 1)(\xi_1\eta_2 - \xi_2\eta_1)^2 - \tfrac{1}{8}(\xi_1^2 + \xi_2^2)(\eta_1^2 + \eta_2^2).\end{aligned}$$

Flattening $\overline{\Omega}_a$ at order four or higher does not change the terms through second order in $\overline{H}_a$. Since the occurrence of the Hamiltonian Hopf bifurcation depends only on the terms in $\overline{H}_a$ up through second order in $\overline{H}_a$, we will treat $\overline{H}_a$ as a Hamiltonian on $(\mathbf{R}^4, \omega)$.

We now find the normal form of $\overline{H}_a$ to second order using representation theory. The algebra of S^1 invariant polynomials on $(\mathbf{R}^4, \omega)$ is generated by

$$M = \tfrac{1}{2}(\xi_1^2 + \xi_2^2), \quad N = \tfrac{1}{2}(\eta_1^2 + \eta_2^2), \quad T = \xi_1\eta_1 + \xi_2\eta_2, \quad \text{and} \quad S = \xi_1\eta_2 - \xi_2\eta_1.$$

Note that $T^2 + S^2 = 4MN$. Using the Poisson bracket $\{\,,\,\}$ associated to the symplectic form ω, we find that $\{M, N, T, S\}$ form a Lie algebra $\mathscr{L}$ with bracket relations

$$\{T, M\} = -2M, \quad \{T, N\} = 2N, \quad \{M, N\} = T,$$

and

$$\{S, M\} = \{S, N\} = \{S, T\} = 0.$$

In other words, $\mathscr{L}$ is isomorphic to $\mathrm{sl}(2, \mathbf{R}) \times \mathbf{R}$. Let $\mathfrak{Q}$ be the vector space of homogeneous quadratic polynomials on $\mathscr{L}$. On $\mathfrak{Q}$ we have a representation of $\mathrm{sl}(2, \mathbf{R})$ given by

$$\mathrm{ad} : \mathrm{sl}(2, \mathbf{R}) \to \mathrm{gl}(\mathfrak{Q}, \mathbf{R}) : X \to \mathrm{ad}_X .$$

From the theory of representations of $\mathrm{sl}(2, \mathbf{R})$ we know that

$$\mathfrak{Q} = \ker \mathrm{ad}_M \oplus \mathrm{im}\, \mathrm{ad}_N .$$

Explicitly, $\ker \mathrm{ad}_M$ is spanned by $\{M^2, MS, S^2, 4MN - T^2\}$ and $\mathrm{im}\, \mathrm{ad}_N$ is spanned by

$$\{\mathrm{ad}_N M^2, \mathrm{ad}_N^2 M^2, \mathrm{ad}_N^3 M^2, \mathrm{ad}_N^4 M^2, \mathrm{ad}_N(MS), \mathrm{ad}_N^2(MS)\},$$

that is,

$$\{-2MT, 2T^2+4MN, 12NT, 24N^2, 2ST, 2MS\}.$$

Thus

$$6MN = (T^2+2MN)+(4MN-T^2) = -\mathrm{ad}_N(MT)+S^2. \tag{136}$$

Writing $\overline{H}_a$ in terms of $\{M, N, S, T\}$ gives

$$\overline{H}_a^2 = \tfrac{1}{2}N + \tfrac{1}{4}(a^2-4\beta)M - \tfrac{1}{2}a(2\gamma-1)S$$

and

$$\overline{H}_a^4 = \tfrac{1}{8}a^2M^2 - \tfrac{1}{4}MS + \tfrac{1}{4}(2\gamma-1)S^2 - \tfrac{1}{2}MN.$$

To remove the term $-\frac{1}{2}MN$ from $\overline{H}_a$ and thus bring $\overline{H}_a$ into normal form to second order, we apply the change of coordinates $\exp\varepsilon^2\mathrm{ad}_{\frac{1}{6}MT}$ and obtain

$$\begin{aligned} H_a &= (\exp\varepsilon^2\mathrm{ad}_{\frac{1}{6}MT})^*\overline{H}_a \\ &= \overline{H}_a^2 + \varepsilon^2(\overline{H}_a^4 + \mathrm{ad}_{\frac{1}{6}MT}\overline{H}_a^2) + O(\varepsilon^4) \\ &= \overline{H}_a^2 + \varepsilon^2(\overline{H}_a^4 - \tfrac{1}{6}\mathrm{ad}_{\overline{H}_a^2}MT) + O(\varepsilon^4) \\ &= H_a^2 + \varepsilon^2 H_a^4 + O(\varepsilon^4). \end{aligned}$$

Here $H_a^2 = \overline{H}_a^2$ and

$$H_a^4 = \tfrac{1}{24}(a^2+8\beta)M^2 - \tfrac{1}{4}MS + \tfrac{1}{6}(3\gamma-2)S^2.$$

Thus the 4-jet of H_a at the origin is in the proper form for a nonlinear Hamiltonian Hopf bifurcation.

9 Exercises

1. Let (h, j) be a regular value for the energy momentum mapping $\mathcal{EM}$ of the magnetic spherical pendulum. Suppose that

$$\Gamma : \mathbf{R} \to \mathcal{EM}^{-1}(h, j) \subseteq TS^2$$

is an integral curve of the Hamiltonian vector field X_H of the magnetic spherical pendulum. Let $\pi : TS^2 \to S^2$ be the bundle projection.

(a). Show that the curvature of $\gamma = \pi \circ \Gamma$ is a decreasing function of its height on S^2. Deduce that γ has no downward pointing cusps.

(b)*. Does the rotation number $\Theta(h, j)$ of the flow of X_H on $\mathcal{EM}^{-1}(h, j)$ satisfy an estimate of the form $C_1 < \Theta(h, j) < C_2$ where C_i are positive constants which do not depend on h or j?

2. Give an argument (not a calculation) why the coefficient of $(\xi_1^2+\xi_2^2)^2$ in the normal form for the nonlinear Hamiltonian Hopf bifurcation of the Lagrange top does *not* depend on the parameter $\gamma = I_1/I_3$.

3. (a). Show that as the parameter a increases through $a_0 = \sqrt{\frac{4\beta}{\gamma+1}}$ the linearization R_a (121) of the Euler-Poisson equations of the Lagrange top at (e_3, ae_3) has a semisimple eigenvalue 0 of multiplicity 2.

   (b). Let $\mathrm{sp}(4,\mathbf{R})^{S^1}$ be the set of all 4×4 infinitesimally symplectic matrices on $(\mathbf{R}^4, \omega)$ which commute with the S^1 action

$$\varphi : S^1 \times \mathbf{R}^4 \to \mathbf{R}^4 : (s, z) \to \mathrm{diag}(S_s, S_s)z,$$

   where $S_s = \begin{pmatrix} \cos s & -\sin s \\ \sin s & \cos s \end{pmatrix}$. Show that the $\mathrm{sp}(4,\mathbf{R})^{S^1}$ versal normal form of U_a (122) is

$$\mu \to \left(\begin{array}{cc|cc} 0 & 0 & -\alpha(\mu) & 0 \\ 0 & 0 & 0 & -\mu \\ \hline \alpha(\mu) & 0 & 0 & 0 \\ 0 & \mu & 0 & 0 \end{array}\right)$$

   where $\alpha(0) > 0$. Show that the corresponding quadratic Hamiltonian is indefinite when $\mu < 0$ and small and positive definite when $\mu > 0$ and small. This is the linear *sign exchange bifurcation*.

   (c). Find a smooth family of linear symplectic mappings which commute with the S^1 action φ and which bring U_a into versal normal form for all a near a_0.

   (d). Why is there no bifurcation in the topology of the level sets of the energy momentum map of the Lagrange top along the crease when a passes through a_0?

4. Give a geometric explanation how a cusp catastrophe (= the discriminant locus) of the generic cubic polynomial

$$F(u) = au^3 - bu^2 + cu - d^2$$

   where $a > 0$ and becomes a swallowtail surface (= the discriminant locus) of the special quartic

$$G(u) = u^4 - \frac{b}{2a}u^3 + \frac{d}{\sqrt{a}}u + \left(\frac{b}{16a^2} - \frac{c}{4a}\right).$$

   In particular, explain why it is essential that the constant term in F is a square.

5. Lift the S^1 action

$$\begin{gathered} S^1 \times \mathbf{R}^2 \to \mathbf{R}^2 : \\ \big(t, x = (x_1, x_2)\big) \to R_t\begin{pmatrix} x_1 \\ x_2 \end{pmatrix} = \begin{pmatrix} \cos t & -\sin t \\ \sin t & \cos t \end{pmatrix}\begin{pmatrix} x_1 \\ x_2 \end{pmatrix} \end{gathered}$$

   to a Hamiltonian action

$$S^1 \times T^*\mathbf{R}^2 \to T^*\mathbf{R}^2 : \big(t, (x, y)\big) \to (R_t x, R_t y)$$

on $(T^*\mathbf{R}^2, \omega_4 = dx_1 \wedge dy_1 + dx_2 \wedge dy_2)$ with momentum mapping

$$J : T^*\mathbf{R}^2 \to \mathbf{R} : (x, y) \to x_1 y_2 - x_2 y_1.$$

Also consider the symplectic action of $\mathbf{Z}_2$ on $(T^*\mathbf{R}, \omega_2 = dx_1 \wedge dy_1)$ which is generated by $(x_1, y_1) \to (-x_1, -y_1)$. The goal of this exercise is to show that the singular reduced spaces $J^{-1}(0)/S^1$ and $T^*\mathbf{R}/\mathbf{Z}_2$ are isomorphic.

(a). Show that the map

$$\psi : T^*\mathbf{R} \to J^{-1}(0) \subseteq T^*\mathbf{R}^2 : (x_1, y_1) \to (x_1, 0, y_1, 0)$$

induces a homeomorphism between $T^*\mathbf{R}/\mathbf{Z}_2$ and $J^{-1}(0)/S^1$.

(b). Because $\psi^*\omega_4 = \omega_2$, show that

$$\psi^* : C^\infty(T^*\mathbf{R}^2)^{S^1} \to C^\infty(T^*\mathbf{R})^{\mathbf{Z}_2} : f \to f \circ \psi$$

is an injective Poisson map.

(c). Using Schwarz's theorem show that

$$x_1^2, \; y_1^2, \text{ and } x_1 y_1$$

generate $C^\infty(T^*\mathbf{R})^{\mathbf{Z}_2}$. Because

$$\psi^*(x_1^2 + x_2^2) = x_1^2, \; \psi^*(y_1^2 + y_2^2) = y_1^2, \text{ and } \psi^*(x_1 y_1 + x_2 y_2) = x_1 y_1,$$

deduce that ψ^* is surjective and hence is an isomorphism of singular reduced spaces.

6. Consider the Lagrange top after reduction of the left S^1 action, (see section 7.3). In other words we look at the reduced Hamiltonian system (H_a, P_a, ω_a), which is invariant under the Hamiltonian right S^1 action given by (114). The goal of this exercise is to carry out singular reduction of this right S^1 symmetry near the fixed point $p_a = (e_3, ae_3)$.

   Let φ_a be the coordinate change (126), which maps $D_1 \times \mathbf{R}^2 = \{x_1^2 + x_2^2 = 1\} \times \mathbf{R}^2$ into a neighborhood of p_a in P_a. Recall that φ_a intertwines the right S^1 action (114) on P_a with the S^1 action (127) on $\mathbf{R}^4$. In addition, φ_a pulls back the Hamiltonian system (H_a, P_a, ω_a) to the Hamiltonian system $(\widehat{H}_a, \mathbf{R}^4, \widehat{\Omega}_a)$, where $\widehat{H}_a$ is given by (128) and $\widehat{\Omega}_a$ by (129).

   (a). Near p_a show that the pull back by φ_a of the momentum map

$$J_r^a : P_a \subseteq \mathbf{R}^3 \times \mathbf{R}^3 \to \mathbf{R} : (z, w) \to w_3 - a$$

   of the right S^1 action on P_a is the momentum map

$$\widehat{J}_r^a : D_1 \subseteq \mathbf{R}^4 \to \mathbf{R} : (x, y) \to a - (x_1 y_1 + x_2 y_2)\big(1 - (x_1^2 + x_2^2)\big)^{-1/2}$$

   of the S^1 action on $(\mathbf{R}^4, \widehat{\Omega}_a)$.

(b). Let

$$\pi : (\widehat{J}^a)^{-1}(0) \to (\widehat{J}^a)^{-1}(0)/S^1$$

be the reduction map. By exercise 5 the singular reduced space $(\widehat{J}^a)^{-1}(0)/S^1$ is an orbifold near $\pi(0)$, (see appendix B section 4). Consider the symplectic map

$$\begin{aligned} \psi : T^*\mathbf{R} &\to (\widehat{J}^a)^{-1}(0) \subseteq T^*\mathbf{R}^2 : \\ (q,p) &\to \left(q, 0, a(1-(1-q^2)^{1/2})q^{-1}, p\right). \end{aligned}$$

Show that $\pi \circ \psi$ is an orbifold chart for $(\widehat{J}^a)^{-1}(0)/S^1$. In this chart show that the symplectic form on the singular reduced space is

$$\widetilde{\omega} = \psi^*\widehat{\Omega}_a = -(1-q^2)^{-1/2}\, dq \wedge dp.$$

(c). Since $\widehat{H}_a$ is invariant under the S^1 action, there is an induced Hamiltonian $\widetilde{H}_a$ on the singular reduced space $(\widehat{J}^a)^{-1}(0)/S^1$. In the orbifold chart given by π show that

$$\widetilde{H}_a(q,p) = \psi^*\widehat{H}_a(q,p) = \tfrac{1}{2}p^2 + \tfrac{1}{2}a^2\left(1-(1-q^2)^{1/2}\right)^2 q^{-2}.$$

(d). Analyze the behavior of the one parameter family of one degree of freedom Hamiltonian systems $(\widetilde{H}_a, \widetilde{\omega})$ near $(0,0)$.

(e)*. Using the family $(\widetilde{H}_a, \widetilde{\omega})$ can one see that the Lagrange top has monodromy?

Appendix A

Fundamental concepts

In this appendix we describe the basic mathematical structures needed to do Hamiltonian mechanics. We begin with a section on symplectic linear algebra. The motion of a Hamiltonian system takes place on a symplectic manifold, that is, a manifold with a closed nondegenerate 2-form, called a symplectic form. The symplectic form allows one to turn the differential of a function, called a Hamiltonian, into a vector field whose integral curves satisfy Hamilton's equations. An algebraic way of treating Hamiltonian mechanics is via Poisson brackets. When the vector space of smooth functions on a symplectic manifold, which is a Lie algebra under Poisson bracket, is made into an algebra using pointwise multiplication of smooth functions, we obtain a Poisson algebra. The symplectic formulation of mechanics can be recovered from this Poisson algebra.

1 Symplectic linear algebra

In this section we treat the fundamentals of symplectic linear algebra.

Let V be a finite dimensional real vector space. A skew symmetric bilinear form $\sigma : V \times V \rightarrow \mathbf{R}$ is said to be *nondegenerate* if the linear mapping

$$\sigma^{\sharp} : V \rightarrow V^{*} : v \rightarrow \sigma^{\sharp}(v)$$

is bijective. Here $\sigma^{\sharp}(v)$ is the linear map

$$\sigma^{\sharp}(v) : V \rightarrow \mathbf{R} : w \rightarrow \sigma(v, w).$$

A vector space V on which is defined a nondegenerate skew symmetric bilinear form σ is called a *symplectic vector space* and σ is called a *symplectic form* on V.

Example 1. Let W be a finite dimensional real vector space and let $V = W \times W^{*}$. Define a bilinear form σ on V by

$$\sigma\big((w, \alpha), (w', \alpha')\big) = \alpha(w') - \alpha'(w),$$

for every $(w, \alpha), (w', \alpha') \in W \times W^{*}$. (V, σ) is a symplectic vector space. Clearly σ is skew symmetric. We need only show that σ is nondegenerate. Suppose that for some

$(w, \alpha) \in W \times W^*$ we have $0 = \sigma^\sharp\big((w,\alpha)\big)(w',\alpha')$ for every $(w',\alpha') \in W \times W^*$. In particular, for every $w' \in W$ we have

$$0 = \sigma^\sharp\big((w,\alpha)\big)(w',0) = \sigma((w,\alpha),(w',0)) = \alpha(w'),$$

that is, $\alpha = 0$. Similarly, for every $\alpha' \in W^*$ we have $0 = \sigma^\sharp\big((w,\alpha)\big)(0,\alpha') = -\alpha'(w)$, that is, $w = 0$. Hence the linear map $\sigma^\sharp$ is injective. Since $\dim V^* = \dim V$, $\sigma^\sharp$ is surjective and hence bijective. □

Let (V, σ) be a symplectic vector space and let W be a subspace of V. The *symplectic perpendicular* W^σ of W in V is the set of all vectors $v \in V$ such that $\sigma(v, w) = 0$ for every $w \in W$. We assemble below the basic properties of symplectic perpendicularity.

Fact: Let U, W be subspaces of (V, σ).
(a). If $U \subseteq W$, then $W^\sigma \subseteq U^\sigma$.
(b). Let W° be the set of all covectors $\alpha \in V^*$ such that $\alpha|W = 0$.
Then $\sigma^\sharp(W^\sigma) = W^\circ$.
(c). $\dim V = \dim W + \dim W^\sigma$.
(d). $V = W + W^\sigma$.
(e). $W = (W^\sigma)^\sigma$.
(f). $U^\sigma \cap W^\sigma = (U + W)^\sigma$.

(1.1) **Proof**:
(a). Suppose that $v \in W^\sigma$. Then $\sigma(v, w) = 0$ for every $w \in W$ and hence for every $w \in U$, because $U \subseteq W$. Therefore $v \in U^\sigma$.

(b). Suppose that $\alpha \in \sigma^\sharp(W^\sigma)$. Then there is a $v \in W^\sigma$ such that $\alpha = \sigma^\sharp(v)$. Thus for every $w \in W$, we have $\alpha(w) = \sigma^\sharp(v)w = \sigma(v, w) = 0$, that is, $\alpha \in W^\circ$. Consequently, $\sigma^\sharp(W^\sigma) \subseteq W^\circ$. Conversely, if $\alpha \in W^\circ$, then there is a $v \in V$ such that $\alpha = \sigma^\sharp(v)$, because $\sigma^\sharp$ is bijective. Since $\alpha \in W^\circ$, for every $w \in W$ we see that

$$0 = \alpha(w) = \sigma^\sharp(v)w = \sigma(v, w),$$

that is, $v \in W^\sigma$. Hence $\alpha \in \sigma^\sharp(W^\sigma)$. Therefore $\sigma^\sharp(W^\sigma) = W^\circ$.

(c). First we show that

$$\dim V = \dim W^\circ + \dim W.$$

Let $\{w_1, \ldots, w_m\}$ be a basis for W. Extend this to a basis $\{w_1, \ldots, w_m, w_{m+1}, \ldots, w_n\}$ of V. Let $\{w_1^*, \ldots, w_m^*, w_{m+1}^*, \ldots, w_n^*\}$ be the corresponding dual basis of V^*. Let $\alpha \in W^\circ \subseteq V^*$. Then $\alpha = \sum_{j=1}^n \alpha_j w_j^*$. Since $\alpha \in W^\circ$, it follows that

$$0 = \alpha(w_i) = \sum_{j=1}^n \alpha_j w_j^*(w_i) = \alpha_i,$$

for every $1 \le i \le m$. In other words, α lies in the space spanned by the covectors $\{w_i^*\}_{i=m+1}^n$. Since $\{w_i^*\}_{i=m+1}^n$ lie in W° and are independent, they form a basis for W°. Hence $\dim W^\circ = n - m = \dim V - \dim W$. By (b), $\sigma^\sharp(W^\sigma) = W^\circ$. Because $\sigma^\sharp$ is bijective, $\dim W^\sigma = \dim W^\circ$. Hence we obtain (c).

(d). Clearly $W + W^\sigma \subseteq V$. Using (c) it follows that $W + W^\sigma = V$.

(e). Let $w \in W$. For every $v \in W^\sigma$, we have $\sigma(v, w) = 0$. Hence $w \in (W^\sigma)^\sigma$. Consequently, $W \subseteq (W^\sigma)^\sigma$. As $\dim(W^\sigma)^\sigma + \dim W^\sigma = \dim V = \dim W + \dim W^\sigma$, we see that $\dim W = \dim(W^\sigma)^\sigma$ and thus $W = (W^\sigma)^\sigma$.

(f). Since W and U are subspaces of $U + W$, using (a) we obtain $(U + W)^\sigma \subseteq U^\sigma \cap W^\sigma$. From

$$\begin{aligned} \dim V &= \dim U^\sigma + \dim W^\sigma - \dim(U^\sigma \cap W^\sigma) \\ &= \dim V - \dim U + \dim V - \dim W - \dim U^\sigma \cap W^\sigma, \end{aligned}$$

we deduce that $\dim V = \dim U + \dim W + \dim(U^\sigma \cap W^\sigma)$. But

$$\dim V = \dim(U + W) + \dim(U + W)^\sigma \le \dim U + \dim W + \dim(U + W)^\sigma.$$

Therefore $\dim U^\sigma \cap W^\sigma \le \dim(U + W)^\sigma$. Because $(U + W)^\sigma \subseteq U^\sigma \cap W^\sigma$, we conclude that $U^\sigma \cap W^\sigma = (U + W)^\sigma$ □

A subspace W of a symplectic vector space (V, σ) is *isotropic* if and only if $W \subseteq W^\sigma$. In other words, W is isotropic if and only if $\sigma|(W \times W)$ vanishes identically.

Example 2. Let u be a nonzero vector in V and let U be the subspace of V spanned by u. Then U is isotropic, because $\sigma(u, u) = -\sigma(u, u)$ by skew symmetry of σ. Hence $\sigma(u, u) = 0$. □

Let W be an isotropic subspace of the symplectic vector space (V, σ). W is *Lagrangian* if and only if $W = W^\sigma$.

Fact: Every isotropic subspace of a symplectic vector space is contained in a Lagrangian subspace.

(1.2) **Proof**: Let W be an isotropic subspace of (V, σ). If $W = W^\sigma$ we are done. Otherwise, because W is properly contained in W^σ, there is a nonzero vector v in W^σ which is not in W. Let U be the space spanned by v. Then U is isotropic, that is, $U \subseteq U^\sigma$. As $U \subseteq W^\sigma$, we deduce that $U \subseteq U^\sigma \cap W^\sigma$. From ((1.1a)) it follows that $W \subseteq U^\sigma$. Because W is isotropic, $W \subseteq W^\sigma$ and hence $W \subseteq U^\sigma \cap W^\sigma$. Therefore

$$U + W \subseteq U^\sigma \cap W^\sigma = (U + W)^\sigma,$$

that is, $U + W$ is isotropic. In addition, W is properly contained in $U + W$. Thus after a finite number of repetitions of the above argument we have constructed a Lagrangian subspace. □

Example 3. If (V, σ) is the symplectic vector space constructed in example 1, then $W \times \{0\}$ and $\{0\} \times W^*$ are Lagrangian subspaces. □

A *subspace* W of a symplectic vector space (V, σ) is *symplectic* if and only if $\sigma|(W \times W)$ is nondegenerate. Below we collect together some properties of symplectic subspaces.

Fact:

(A). W is symplectic if and only if $W \cap W^{\sigma} = \{0\}$.

(B). If W is a symplectic subspace, then so is W^{σ}.

(C). Every symplectic vector space (V, σ) is the direct sum of σ-perpendicular 2-dimensional symplectic subspaces $(V_i, \sigma|(V_i \times V_i))$, which have a basis $\{e_i, f_i\}$ with respect to which the matrix of $\sigma|(V_i \times V_i)$ is $\begin{pmatrix} 0 & -1 \\ 1 & 0 \end{pmatrix}$.

(D). If Y is a subspace of W which is complementary to $W \cap W^{\sigma}$ in W, then Y is symplectic. In fact, Y is a maximal symplectic subspace of (V, σ) contained in W.

(1.3) **Proof:**

(A). Suppose that $w' \in W \cap W^{\sigma}$. Then $\sigma(w', w) = 0$ for every $w \in W$. In other words, $\sigma^{\sharp}(w') = 0 \in W^*$. As $\sigma|(W \times W)$ is nondegnerate, it follows that $w' = 0$. Hence $W \cap W^{\sigma} = \{0\}$. Conversely, suppose that for some $w' \in W$ we have $\sigma^{\sharp}(w') = 0 \in W^*$. Then for every $w \in W$, we have $0 = \sigma(w', w)$, that is, $w' \in W^{\sigma}$. Hence $w' \in W \cap W^{\sigma} = \{0\}$. Therefore the map $\sigma^{\sharp} : W \rightarrow W^*$ is injective and hence surjective because $\dim W = \dim W^*$. Thus $\sigma|(W \times W)$ is nondegenerate.

(B). Since W is symplectic, $\{0\} = W \cap W^{\sigma} = (W^{\sigma})^{\sigma} \cap W^{\sigma}$, using ((1.1d)). From (A) it follows that W^{σ} is symplectic.

(C). Let v be a nonzero vector in V. Suppose that for every $v' \in V$, $\sigma(v, v') = 0$. Then $v \in V^{\sigma}$. Since σ is nondegenerate, it follows that $V \cap V^{\sigma} = \{0\}$ and hence $v = 0$, which is a contradiction. Thus there is a $v' \in V$ such that $\sigma(v, v') = r \neq 0$. Let V_1 be the space spanned by $\{v, \frac{1}{r}v'\}$. The matrix of $\sigma|(V_1 \times V_1)$ is $\begin{pmatrix} 0 & -1 \\ 1 & 0 \end{pmatrix}$, which is invertible. In other words, $\big(V_1, \sigma|(V_1 \times V_1)\big)$ is a symplectic subspace of (V, σ). Set $W = V_1^{\sigma}$. Repeat the above argument on the symplectic vector space $\big(W, \sigma|(W \times W)\big)$. After a finite number of iterations we find that $V = \sum_{i=1}^{p} \oplus V_i$ where $\big(V_i, \sigma|(V_i \times V_i)\big)$ are two dimensional symplectic subspaces, which are mutually σ-perpendicular and hence the dimension of V is even.

(D). From $(W \cap W^{\sigma})^{\sigma} = W^{\sigma} + W = V$, it follows that for every $w' \in W \cap W^{\sigma}$ and every $v \in V$, $\sigma(w', v) = 0$. Suppose that there is a $y \in Y$ such that for every $y' \in Y$ we have $\sigma(y, y') = 0$. Since $W = W \cap W^{\sigma} \oplus Y$ and $\sigma(w', y'') = 0$ for every $w' \in W \cap W^{\sigma}$ and every $y'' \in Y$, it follows that $0 = \sigma(y, w)$ for every $w \in W$. Because $y \in W$, for every $\widetilde{w} \in W^{\sigma}$ we have $\sigma(y, \widetilde{w}) = 0$. Since $V = W + W^{\sigma}$, we find that $0 = \sigma(y, v)$ for every $v \in V$. But σ is nondegenerate on V. Hence $y = 0$. Consequently, $\sigma|(Y \times Y)$ is nondegenerate. In other words, Y is a symplectic subspace of (V, σ). Suppose that Y' is a symplectic subspace of (V, σ) which is contained in W and properly contains Y. Then $\dim Y' > \dim Y$. Consequently, $Y' \cap W \cap W^{\sigma} \neq \{0\}$. This follows because

$$\dim W = \dim Y' + \dim(W \cap W^{\sigma}) - \dim(Y' \cap W \cap W^{\sigma}) = \dim Y + (\dim W \cap W^{\sigma}),$$

and hence $\dim Y' \cap W \cap W^{\sigma} = \dim Y' - \dim Y > 0$. Let $z \in Y' \cap W \cap W^{\sigma}$. Then $0 = \sigma(z, y')$ for every $y' \in Y'$. Since Y' is symplectic, we deduce that $z = 0$, that is, $Y' \cap W \cap W^{\sigma} = \{0\}$. This is a contradiction. Therefore Y is a maximal symplectic subspace of (V, σ) contained in W. □

The following decomposition, called the *Witt decomposition*, is fundamental in the study of normal forms for momentum mappings in Appendix 2.

Fact: Let (V, σ) be a symplectic vector space and let W be a subspace of V. Then V may be written as $V = X \oplus Y \oplus Z$ where X, Y and Z are σ-perpendicular symplectic subspaces such that

$$W = X \oplus (W \cap W^{\sigma}) \quad \text{and} \quad W^{\sigma} = Y \oplus (W \cap W^{\sigma})$$

and $W \cap W^{\sigma}$ is a Lagrangian subspace of Z.

(1.4) **Proof**: Choose $X \subseteq W$ complementary to $W \cap W^{\sigma}$, and $Y \subseteq W^{\sigma}$ complementary to $W \cap W^{\sigma}$. Then ((1.3D)) shows that both X and Y are symplectic. Let $Z = (X \oplus Y)^{\sigma}$. Then Z is symplectic as well. Hence $V = X \oplus Y \oplus Z$ is a decomposition of V into σ-perpendicular symplectic subspaces. $W \cap W^{\sigma} \subseteq Z$ by construction. Let T be an isotropic subspace of Z with the properties that $T \cap W \cap W^{\sigma} = \{0\}$ and $T \oplus (W \cap W^{\sigma}) = R$ is symplectic. Then $R \oplus X \oplus Y$ is symplectic. To see that $R \oplus X \oplus Y = V$ observe that

$$W = X \oplus (W \cap W^{\sigma}) \subseteq X \oplus R \subseteq X \oplus Y \oplus R$$

and

$$W^{\sigma} = Y \oplus (W \cap W^{\sigma}) \subseteq Y \oplus R \subseteq X \oplus Y \oplus R.$$

Therefore $(R \oplus X \oplus Y)^{\sigma} \subseteq W \cap W^{\sigma} \subseteq R$. As $R \subseteq R \oplus X \oplus Y$, we find that $(R \oplus X \oplus Y)^{\sigma} \subseteq R^{\sigma}$. So $(R \oplus X \oplus Y)^{\sigma} \subseteq R \cap R^{\sigma} = \{0\}$, because R is symplectic. Therefore $V = R \oplus X \oplus Y$. Hence $\dim R = \dim Z$. As $R \subseteq Z$, $R = Z$.

We now show that $W \cap W^{\sigma}$ is a Lagrangian subspace of Z. Since $Z = T \oplus (W \cap W^{\sigma})$, it follows that

$$\dim Z = \dim(W \cap W^{\sigma}) + \dim T \leq \dim(W \cap W^{\sigma}) + \tfrac{1}{2} \dim Z,$$

because T is an isotropic subspace of the symplectic subspace Z. Therefore $\frac{1}{2} \dim Z \leq \dim W \cap W^{\sigma}$. But $W \cap W^{\sigma}$ is also an isotropic subspace of Z. Hence $\dim(W \cap W^{\sigma}) \leq \frac{1}{2} \dim Z$. Thus $\frac{1}{2} \dim Z = \dim(W \cap W^{\sigma})$, that is, $W \cap W^{\sigma}$ is a Lagrangian subspace of Z. □

2 Symplectic manifolds

In this section we define the concept of a symplectic manifold.

A *symplectic manifold* (M, ω) is a pair consisting of a smooth manifold M with a 2-form ω which is

1). closed, that is, $d\omega = 0$;

2). nondegenerate, that is, $\omega(p)$ is an nondegenerate skew symmetric bilinear form on T_pM at each $p \in M$. In other words, $\big(T_pM, \omega(p)\big)$ is a symplectic vector space.

Example 1. Let V be a real vector space and set $M = V \times V^*$. On M define the constant 2-form ω by

$$\omega\big((v,\alpha),(v',\alpha')\big) = \alpha(v') - \alpha'(v),$$

(see example 1 of section 1). Then (M, ω) is a symplectic manifold.

Example 2. Consider the cotangent bundle T^*N of a smooth manifold N. This is the *phase space* of classical mechanics, which is typically the space of all positions and momenta of a physical system. To show that T^*N is a symplectic manifold we first define the *canonical 1-form* θ by

$$\theta_\alpha(X) = \alpha(T\tau X). \tag{1}$$

Here $\tau : T^*N \to N$ is the map which assigns to every covector α in T^*N its point of attachment in N and X is a vector in the tangent space $T_\alpha(T^*N)$ to T^*N at α. The *canonical 2-form* or *symplectic structure* Ω on T^*N is the 2-form $\Omega = -d\theta$. Therefore Ω is closed. We have not yet shown that Ω is nondegenerate. We will return to this point below.

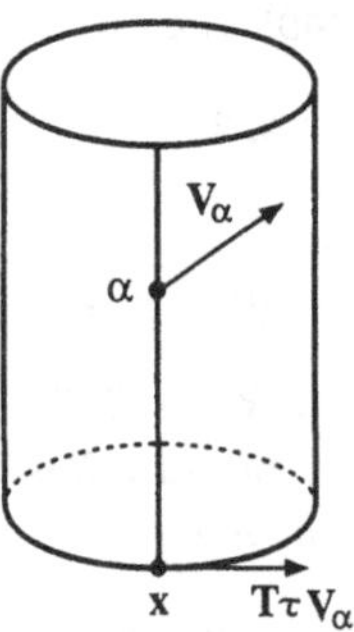

Figure 2.1 The canonical 1-form on T^*N.

Example 2': Suppose that T^*N is the cotangent bundle T^*G of a Lie group G. Let $\mathcal{G}^*$ be the dual to its Lie algebra $\mathcal{G}$. Then $\tau : T^*G \to G : \alpha_g \to g$ is a *trivial* bundle with bundle projection τ. A trivialization of T^*G is given, for example, by left translation

$$\mathcal{L} : G \times \mathcal{G}^* \to T^*G : (g, \alpha) \to (T_gL_{g^{-1}})^t\alpha = \alpha_g, \tag{2}$$

where $L_g : G \to G : h \to gh$ is *left translation* by G. The map $\mathcal{L}$ identifies a covector at a point with a left invariant 1-form, which may be thought of as an element of $\mathcal{G}^*$. Pulling back the canonical 1-form θ on T^*G by the mapping $\mathcal{L}$ gives the 1-form $\vartheta = \mathcal{L}^*\theta$ on $G \times \mathcal{G}^*$. Similarly, pulling back the canonical 2-form Ω on T^*G gives the 2-form $\omega = \mathcal{L}^*\Omega$ on $G \times \mathcal{G}^*$.

We calculate the 1-form ϑ as follows. First we compute the tangent of the mapping $\tau \circ \mathcal{L}$. For $\xi \in \mathcal{G}$ and $\beta \in \mathcal{G}^*$, the mapping

$$\varphi_t^{(\xi,\beta)} : G \times \mathcal{G}^* \to G \times \mathcal{G}^* : (g, \alpha) \to (g \exp t\xi, \alpha + t\beta)$$

is the flow of the vector field $X^{(\xi,\beta)}(g,\alpha) = (T_eL_g\xi, \beta)$ on $G \times \mathcal{G}^*$. Using the left trivialization $\mathcal{L}$, we pull back the action L of left translation to obtain a G-action ℓ on $G \times \mathcal{G}^*$ defined by $\ell_g(h,\alpha) = (L_gh, \alpha)$. Note that $X^{(\xi,\beta)}$ is invariant under the action ℓ because

$$\varphi_t^{(\xi,\beta)}\big(\ell_g(h,\alpha)\big) = \varphi_t^{(\xi,\beta)}(gh,\alpha) = (gh\exp t\xi, \alpha + t\beta) = \ell_g\big(\varphi_t^{(\xi,\beta)}(h,\alpha)\big).$$

Therefore the tangent of $\tau \circ \mathcal{L}$ is

$$T_{(g,\alpha)}(\tau\circ\mathcal{L})(T_eL_g\xi,\beta) = \frac{d}{dt}\Big|_{t=0}\tau\circ\mathcal{L}\big(\varphi_t^{(\xi,\beta)}(g,\alpha)\big) = \frac{d}{dt}\Big|_{t=0} L_g\exp t\xi = T_eL_g\xi.$$

Consequently,

$$\begin{aligned}\vartheta(g,\alpha)(T_eL_g\xi,\beta) &= \theta(\mathcal{L}(g,\alpha))\big(T_{(g,\alpha)}\mathcal{L}(T_eL_g\xi,\beta)\big)\\ &= \alpha_g\big(T_{(g,\alpha)}(\tau\circ\mathcal{L})(T_eL_g\xi,\beta)\big) = \alpha_g(T_eL_g\xi) = \alpha(\xi). \quad (3)\end{aligned}$$

To compute the 2-form ω, we use the formula

$$\begin{aligned}&-d\vartheta(X^{(\xi,\beta)}, X^{(\eta,\gamma)}) =\\ &\quad -X^{(\xi,\beta)} \lrcorner\, d(X^{(\eta,\gamma)}\lrcorner\,\vartheta) + X^{(\eta,\gamma)} \lrcorner\, d(X^{(\xi,\beta)}\lrcorner\,\vartheta) + \big[X^{(\xi,\beta)}, X^{(\eta,\gamma)}\big]\lrcorner\,\vartheta\end{aligned}$$

for the exterior derivative of ϑ. We now calculate each term on the right hand side of the above equation. For the first term, we find that

$$\begin{aligned}\big(X^{(\xi,\beta)}\lrcorner\, d(X^{(\eta,\gamma)}\lrcorner\,\vartheta)\big)(g,\alpha) &= L_{X^{(\xi,\beta)}}(X^{(\eta,\gamma)}\lrcorner\,\vartheta)(g,\alpha)\\ &= \frac{d}{dt}\Big|_{t=0}(\varphi_t^{(\xi,\beta)})^*(X^{(\eta,\gamma)}\lrcorner\,\vartheta)(g,\alpha)\\ &= \frac{d}{dt}\Big|_{t=0}\vartheta(g\exp t\xi, \alpha+t\beta)\big(X^{(\eta,\gamma)}(g\exp t\xi,\alpha+t\beta)\big)\\ &= \frac{d}{dt}\Big|_{t=0}\vartheta(g\exp t\xi,\alpha+t\beta)(T_eL_{g\exp t\xi}\eta,\gamma) = \frac{d}{dt}\Big|_{t=0}(\alpha+t\beta)(\eta) = \beta(\eta).\end{aligned}$$

The second term is calculated similarly. The third term follows once we notice that

$$\begin{aligned}\big[X^{(\xi,\beta)}, X^{(\eta,\gamma)}\big](g,\alpha) &= \frac{d}{dt}\Big|_{t=0}\big(\varphi^{(\eta,\gamma)}_{-\sqrt{t}}\circ\varphi^{(\xi,\beta)}_{-\sqrt{t}}\circ\varphi^{(\eta,\gamma)}_{\sqrt{t}}\circ\varphi^{(\xi,\beta)}_{\sqrt{t}}\big)(g,\alpha)\\ &= \frac{d}{dt}\Big|_{t=0}\big(g\exp\sqrt{t}\xi\exp\sqrt{t}\eta\exp-\sqrt{t}\xi\exp-\sqrt{t}\eta,\alpha\big)\\ &= (T_eL_g[\xi,\eta],0),\end{aligned}$$

as then

$$\vartheta\big(\big[X^{(\xi,\beta)}, X^{(\eta,\gamma)}\big]\big)(g,\alpha) = \vartheta(g,\alpha)(T_eL_g[\xi,\eta],0) = \alpha([\xi,\eta]).$$

Therefore the 2-form ω on $G \times \mathcal{G}^*$ is given by

$$\omega(g,\alpha)\big((T_eL_g\xi,\beta),(T_eL_g\eta,\gamma)\big) = -\beta(\eta)+\gamma(\xi)+\alpha([\xi,\eta]). \tag{4}$$

From (4) it is immediate that ω is a nondegenerate 2-form on $G \times \mathcal{G}^*$. Since $\omega = \mathcal{L}^*\Omega$ and Ω is closed, it follows that ω is closed. Hence ω is symplectic. Because $\mathcal{L}$ is a diffeomorphism, we deduce that Ω is symplectic. □

Now we can show that the canonical 2-form on T^*N is nondegenerate. We may think of the cotangent bundle T^*N as follows. Consider a chart for T^*N to be a chart for the cotangent bundle of the abelian Lie group $\mathbf{R}^n$. In this case the term $\alpha([\xi,\eta])$ in (4) vanishes. Consequently, the canonical 2-form on T^*N is nondegenerate. □

Example 3. Another important example of a symplectic manifold is an orbit $\mathcal{O}_\mu$ of the coadjoint action of a Lie group G on the dual $\mathcal{G}^*$ of its Lie algebra $\mathcal{G}$. In more detail, the *coadjoint action* of G on $\mathcal{G}^*$ is defined by

$$\mathrm{Ad}^* : G \times \mathcal{G}^* \to \mathcal{G}^* : (g,\nu) \to \mathrm{Ad}^t_{g^{-1}}\,\nu \tag{5}$$

where the *adjoint action* Ad of G on its Lie algebra $\mathcal{G}$ is

$$\mathrm{Ad} : G \times \mathcal{G} \to \mathcal{G} : (g,\xi) \to \left.\frac{d}{dt}\right|_{t=0} g\,\exp t\xi\, g^{-1}.$$

▷ The *coadjoint orbit* $\mathcal{O}_\mu$ through $\mu \in \mathcal{G}^*$ is $\{\nu = \mathrm{Ad}^*_g\,\mu \in \mathcal{G}^* \,|\, g \in G\}$. The following calculation shows that the tangent space $T_\nu\mathcal{O}_\mu$ to $\mathcal{O}_\mu$ at ν is $\{\mathrm{ad}^t_\xi\,\nu \,|\, \xi \in \mathcal{G}\}$. It is reasonable that this should be the case, because this equality is obtained by differentiating the defining relation of $\mathcal{O}_\mu$. Here ad_ξ is the tangent to the mapping Ad in the direction ξ, that is,

$$\mathrm{ad}_\xi : \mathcal{G} \to \mathcal{G} : \eta \to \left.\frac{d}{dt}\right|_{t=0} \mathrm{Ad}_{\exp t\xi}\,\eta = [\xi,\eta],$$

where [,] is the Lie bracket on $\mathcal{G}$.

(2.1) **Proof**: For $\xi \in \mathcal{G}$ the curve

$$\gamma^\xi : \mathbf{R} \to \mathcal{O}_\mu : s \to \mathrm{Ad}^t_{\exp -s\xi}\,\nu$$

lies in $\mathcal{O}_\mu$ and passes through ν at $s = 0$. γ^ξ represents the tangent vector $-\mathrm{ad}^t_\xi\nu$ to the coadjoint orbit $\mathcal{O}_\mu$ at the point ν because

$$\left.\frac{d}{ds}\right|_{s=0} \gamma^\xi(s) = \left.\frac{d}{ds}\right|_{s=0} \mathrm{Ad}^t_{\exp -s\xi}\,\nu = -\,\mathrm{ad}^t_\xi\,\nu.$$

Therefore $T_\nu\mathcal{O}_\mu \subseteq \{\mathrm{ad}^t_\xi\nu \,|\, \xi \in \mathcal{G}\}$. To prove the reverse inclusion, for some $\xi \in \mathcal{G}$ consider the curve

$$\Gamma^\xi : \mathbf{R} \to \mathcal{O}_\mu : s \to \exp(s\,\mathrm{ad}^t_\xi)\mu.$$

The image of Γ^ξ is contained in $\mathcal{O}_\mu$ because $\exp(s\,\mathrm{ad}^t_\xi)\mu = \mathrm{Ad}^t_{\exp s\xi}\mu$. Since $\Gamma^\xi(0) = \mu$ and $\frac{d}{ds}\Gamma^\xi(0) = \mathrm{ad}^t_\xi$, we are done. □

For $\xi \in \mathcal{G}$ define the vector field X^{ξ} on $\mathcal{O}_{\mu}$ by

$$X^{\xi}(\nu) = -\,\mathrm{ad}^{t}_{\xi}\,\nu. \tag{6}$$

As is easily checked the flow φ^{ξ}_{s} of X^{ξ} on $\mathcal{O}_{\mu}$ is

$$\varphi^{\xi}_{s}(\nu) = \mathrm{Ad}^{t}_{\exp -s\,\xi}\,\nu. \tag{7}$$

On $\mathcal{O}_{\mu}$ define a 2-form Ω by

$$\Omega(\nu)(X^{\xi}(\nu), X^{\eta}(\nu)) = -\nu([\xi,\eta]). \tag{8}$$

Claim: Ω is a symplectic form on the coadjoint orbit $\mathcal{O}_{\mu}$.

(2.2) **Proof**: First we show that Ω is closed. Recall that the exterior derivative of an n-form Θ is

$$\begin{aligned} d\Theta(X_0, X_1, \ldots, X_n) &= \textstyle\sum_{i=0}^{n}(-1)^{i} d(X_i \lrcorner \Theta)(X_0, \ldots, \widehat{X_i}, \ldots, X_n) \\ &\quad + \textstyle\sum_{0\le i<j\le n}(-1)^{i+j}([X_i, X_j]\lrcorner\Theta)(X_0, \ldots, \widehat{X_i}, \ldots, \widehat{X_j}, \ldots, X_n). \end{aligned} \tag{9}$$

Applying (9) to the 2-form Ω gives

$$\begin{aligned} d\Omega(X^{\xi}, X^{\eta}, X^{\zeta}) &= d(X^{\xi}\lrcorner\Omega)(X^{\eta}, X^{\zeta}) + d(X^{\eta}\lrcorner\Omega)(X^{\zeta}, X^{\xi}) \\ &\quad + d(X^{\zeta}\lrcorner\Omega)(X^{\xi}, X^{\eta}) - \Omega([X^{\xi}, X^{\eta}], X^{\zeta}) - \Omega([X^{\zeta}, X^{\xi}], X^{\eta}) \\ &\quad - \Omega([X^{\eta}, X^{\zeta}], X^{\xi}). \end{aligned} \tag{10}$$

Again using (9), this time to compute the exterior derivative of the 1-form $X^{\xi}\lrcorner\Omega$, we obtain

$$d(X^{\xi}\lrcorner\Omega)(X^{\eta}, X^{\zeta}) = d(X^{\eta}\lrcorner(X^{\xi}\lrcorner\Omega))X^{\zeta} - d(X^{\zeta}\lrcorner(X^{\xi}\lrcorner\Omega))X^{\eta} - [X^{\eta}, X^{\zeta}]\lrcorner(X^{\xi}\lrcorner\Omega).$$

Similarly by cyclically permuting the variables in the above equation gives

$$d(X^{\eta}\lrcorner\Omega)(X^{\zeta}, X^{\xi}) = d(X^{\zeta}\lrcorner(X^{\eta}\lrcorner\Omega))X^{\xi} - d(X^{\xi}\lrcorner(X^{\eta}\lrcorner\Omega))X^{\zeta} - [X^{\zeta}, X^{\xi}]\lrcorner(X^{\eta}\lrcorner\Omega)$$

and

$$d(X^{\zeta}\lrcorner\Omega)(X^{\xi}, X^{\eta}) = d(X^{\xi}\lrcorner(X^{\zeta}\lrcorner\Omega))X^{\eta} - d(X^{\eta}\lrcorner(X^{\zeta}\lrcorner\Omega))X^{\xi} - [X^{\xi}, X^{\eta}]\lrcorner(X^{\zeta}\lrcorner\Omega).$$

Substituting the above three equations into (10) gives

$$\begin{aligned} d\Omega(X^{\xi}, X^{\eta}, X^{\zeta}) &= d(X^{\eta}\lrcorner(X^{\xi}\lrcorner\Omega))X^{\zeta} - d(X^{\zeta}\lrcorner(X^{\xi}\lrcorner\Omega))X^{\eta} \\ &\quad - d(X^{\xi}\lrcorner(X^{\eta}\lrcorner\Omega))X^{\zeta} + d(X^{\zeta}\lrcorner(X^{\eta}\lrcorner\Omega))X^{\xi} \\ &\quad + d(X^{\xi}\lrcorner(X^{\zeta}\lrcorner\Omega))X^{\eta} - d(X^{\eta}\lrcorner(X^{\zeta}\lrcorner\Omega))X^{\xi}. \end{aligned} \tag{11}$$

We calculate the first term in (11) as follows

$$\begin{aligned} d(X^{\eta}\lrcorner X^{\xi}\lrcorner\Omega)(\nu)X^{\zeta}(\nu) &= \big(L_{X^{\zeta}}\Omega(X^{\xi}, X^{\eta})\big)(\nu) \\ &= \frac{d}{ds}\Big|_{s=0}\Omega(\varphi^{\zeta}_{s}(\nu))\big(X^{\xi}(\varphi^{\zeta}_{s}(\nu)), X^{\eta}(\varphi^{\zeta}_{s}(\nu))\big) \\ &= -\frac{d}{ds}\Big|_{s=0}\varphi^{\zeta}_{s}(\nu)([\xi,\eta]), \quad \text{using (8)} \\ &= \mathrm{ad}^{t}_{\zeta}\,\nu([\xi,\eta]), \quad \text{using (7)} \\ &= \nu([\,\zeta, [\xi,\eta]\,]). \end{aligned} \tag{12}$$

Therefore

$$\begin{aligned} d\Omega(X^\xi, X^\eta, X^\zeta) &= \nu([\zeta,[\xi,\eta]]) - \nu([\eta,[\xi,\zeta]]) - \nu([\zeta,[\eta,\xi]]) \\ &\qquad + \nu([\xi,[\eta,\zeta]]) + \nu([\eta,[\zeta,\xi]]) - \nu([\xi,[\zeta,\eta]]) \\ &= 2\nu\big([\zeta,[\xi,\eta]] + [\xi,[\eta,\zeta]] + [\eta,[\zeta,\xi]]\big) \\ &= 0 \end{aligned}$$

by the Jacobi identity. Thus $d\Omega = 0$, which is what we wanted to show.

To see that Ω is nondegenerate, suppose that

$$0 = \Omega(\nu)(X^\xi(\nu), X^\eta(\nu)) \tag{13}$$

for evey $X^\eta(\nu) = -\,\mathrm{ad}^t_\eta\, \nu \in T_\nu \mathcal{O}_\mu$. Then from the definition of Ω it follows that for every $\eta \in \mathcal{G}$, we have

$$0 = \nu([\xi,\eta]) = \nu(\mathrm{ad}_\xi\, \eta) = (\mathrm{ad}^t_\xi\, \nu)(\eta) = -X^\xi(\nu)(\eta),$$

that is, $X^\xi(\nu) = 0$, since $X^\xi(\nu)$ annihilates all of $\mathcal{G}^*$. Consequently Ω is nondegenerate. Note that Ω is invariant under the coadjoint action of G on $\mathcal{O}_\mu$. □

3 Hamilton's equations

In this section we define the concept of Hamiltonian vector field.

On a symplectic manifold (M,ω), the nondegeneracy of the symplectic form ω implies that the mapping $\omega^\sharp(p) : T_pM \to T_p^*M$ defined by

$$\big(\omega^\sharp(p)(v)\big)(w) = \omega(p)(v,w) \quad \text{for } v, w \in T_pM \tag{14}$$

is an isomorphism for every $p \in M$. Denote the inverse of $\omega^\sharp(p)$ by $\omega^\flat(p)$. For each smooth function f on M we may define the *Hamiltonian vector field* X_f of the *Hamiltonian function* f by

$$X_f(p) = \omega^\flat(p)\big(df(p)\big), \tag{15}$$

or equivalently

$$X_f \lrcorner\, \omega = df. \tag{16}$$

Example 1. Consider the *cotangent bundle* T^*N with its canonical symplectic form Ω. Suppose that $f : T^*N \to \mathbf{R}$ is a smooth function. We calculate a local expression for the Hamiltonian vector field X_f as follows. Let U be an open subset of $\mathbf{R}^n$. Then locally T^*N is $T^*U = U \times (\mathbf{R}^n)^*$ with bundle projection $\tau : T^*U \to U : (u,\alpha) \to u$. From the definition of the canonical 2-form Ω it follows that for every $(u,\alpha) \in T^*U$

$$\begin{aligned} \Omega^\sharp(u,\alpha) : T_{(u,\alpha)}(T^*U) = \mathbf{R}^n \times (\mathbf{R}^n)^* \to T^*_{(u,\alpha)}(T^*U) = (\mathbf{R}^n)^* \times \mathbf{R}^n : \\ (v,\beta) \to \big(\Omega^\sharp(u,\alpha)\big)(v,\beta), \end{aligned}$$

where $\Omega^\sharp(u,\alpha)\begin{pmatrix} v \\ \beta \end{pmatrix}$ is the linear function on $T_{(u,\alpha)}(T^*U)$ whose value at (w,γ) is

$$\left(\Omega^\sharp(u,\alpha)\begin{pmatrix} v \\ \beta \end{pmatrix}\right)\begin{pmatrix} w \\ \gamma \end{pmatrix} = -\beta(w) + \gamma(v),$$

that is, $\Omega^\sharp(u,\alpha)\begin{pmatrix} v \\ \beta \end{pmatrix} = (-\beta, v)$. Consequently, the local expression for $\Omega^\flat$ is

$$\Omega^\flat(u,\alpha)(\gamma, w) = \begin{pmatrix} w \\ -\gamma \end{pmatrix}. \tag{17}$$

Locally, the 1-form df is

$$df : T^*U \subseteq \mathbf{R}^n \times (\mathbf{R}^n)^* \to (\mathbf{R}^n)^* \times \mathbf{R}^n : (u,\alpha) \to \big(D_1 f(u,\alpha), D_2 f(u,\alpha)\big).$$

Thus on T^*N the local expression for the Hamiltonian vector field X_f corresponding to f is given by

$$\begin{aligned} &X_f : T^*U \to \mathbf{R}^n \times (\mathbf{R}^n)^* : \\ &\quad (u,\alpha) \to \Omega^\flat(u,\alpha)\big(D_1 f(u,\alpha), D_2 f(u,\alpha)\big) = \big(D_2 f(u,\alpha), -D_1 f(u,\alpha)\big). \end{aligned} \tag{18}$$

An integral curve $\gamma : I \to T^*U$ of X_f satisfies *Hamilton's equations*

$$\frac{d}{dt}\gamma(t) = X_f(\gamma(t)) = \begin{pmatrix} D_2 f(\gamma(t)) \\ -D_1 f(\gamma(t)) \end{pmatrix}. \quad \square \tag{19}$$

Example 2. We derive an expression for the Hamiltonian vector field X_K on TN associated to the Hamiltonian function

$$K : TN \to \mathbf{R} : v \to \tfrac{1}{2} g(x)(v,v),$$

where g is a smooth nondegenerate *metric* on N and $v \in T_x N$. On TN we use the symplectic form $\omega = (g^\sharp)^*\Omega$ where Ω is the canonical 2-form on T^*N. The mapping $g^\sharp(x) : T_x N \to T_x^* N$ is defined by

$$g^\sharp(x)(v) : T_x N \to \mathbf{R} : w \to g(x)(v,w).$$

From the local expression of X_K given in (22) below we see that its integral curves are geodesics for the metric g. Thus X_K is the geodesic vector field associated to the metric g.

We begin by deriving a local expression for the 1-form ϑ on TN defined by

$$\vartheta(v)(w_v) = g(x)(T_v\tau\, w_v, v). \tag{20}$$

Here $v \in T_x N$, $w_v \in T_v N$ and $\tau : TN \to N$ is the bundle projection. A little thought shows that $\vartheta = (g^\sharp)^*\theta$, where θ is the canonical 1-form on T^*N. Let U be an open subset of $\mathbf{R}^n$. Then locally TN is $TU = U \times \mathbf{R}^n$ with bundle projection

$$\tau : TU \to U : (x, v) \to x.$$

Moreover, the tangent of τ is

$$T\tau : T(TU) = (U \times \mathbf{R}^n) \times (\mathbf{R}^n \times \mathbf{R}^n) \to TU = U \times \mathbf{R}^n : \\ \big((x,v),(w,z)\big) \to (x,w).$$

Therefore, ϑ is

$$\vartheta(x,v)(w,z) = g^\sharp(x)(v)w,$$

where $\vartheta(x,v)$ is a linear mapping from $\mathbf{R}^n \times \mathbf{R}^n$ into $\mathbf{R}$. To find the symplectic form ω on TN we note that

$$\omega = (g^\sharp)^*\Omega = -(g^\sharp)^* d\theta = -d(g^\sharp)^*\theta = -d\vartheta.$$

Thus it suffices to compute the exterior derivative of ϑ. This we do as follows. By definition of exterior derivative

$$\begin{aligned}-\big(d\vartheta(x,v)(w_1,z_1)\big)(w_2,z_2) &= -D_1\vartheta(x,v)w_1(w_2,z_2) + D_2\vartheta(x,v)z_1(w_2,z_2) \\ &= -Dg^\sharp(x)w_1(v,w_2) - g^\sharp(x)(z_1)w_2 + Dg^\sharp(x)w_2(v,w_1) + g^\sharp(x)(z_2)w_1.\end{aligned}$$

Hence

$$\begin{aligned}\omega(x,v)\big((w_1,z_1),(w_2,z_2)\big) &= -Dg^\sharp(x)w_1(v,w_2) - g^\sharp(x)(z_1)w_2 \\ &\quad + Dg^\sharp(x)w_2(v,w_1) + g^\sharp(x)(z_2)w_1\end{aligned}$$

is the desired local expression for the symplectic form ω on TN.

Now we compute a local expression for the Hamiltonian vector field X_K. On TU the function K is given by

$$K : U \times \mathbf{R}^n \subseteq \mathbf{R}^n \times \mathbf{R}^n \to \mathbf{R} : (x,v) \to \tfrac{1}{2}g(x)(v,v) = \tfrac{1}{2}g^\sharp(x)(v)v.$$

Thus the 1-form dK is

$$\begin{aligned}dK(x,v) : \mathbf{R}^n \times \mathbf{R}^n \to \mathbf{R} : (w_1,z_1) \to D_1K(x,v)w_1 + D_2K(x,v)z_1 &= \\ = \tfrac{1}{2}Dg^\sharp(x)w_1(v,v) + g^\sharp(x)(v)z_1.&\end{aligned} \tag{21}$$

Let X_K be the vector field on TU given by

$$X_K : U \times \mathbf{R}^n \to \mathbf{R}^n \times \mathbf{R}^n : (x,v) \to \big(X^1(x,v), X^2(x,v)\big).$$

Then for any vector field Y on TU equal to (Y^1, Y^2) we have $\omega(X_K, Y) = dK \cdot Y$, by definition of X_K, that is,

$$\begin{aligned}-Dg^\sharp(x)X^1(v,Y^1) - g^\sharp(x)(X^2)Y^1 + Dg^\sharp(x)Y^1(v,X^1) + g^\sharp(x)(Y^2)X^1 &= \\ = \tfrac{1}{2}Dg^\sharp(x)Y^1(v,v) + g^\sharp(x)(v)Y^2.&\end{aligned}$$

Setting $Y^1 = 0$ in the above equation gives $g^\sharp(x)(Y^2)X^1 = g^\sharp(x)(v)Y^2$. In other words,

$$g(x)(Y^2, X^1) = g(x)(v, Y^2)$$

for every Y^2. Thus by symmetry and nondegeneracy of $g(x)$ we obtain $X^1(x, v) = v$. Setting $Y^2 = 0$ and using $X^1 = v$ gives

$$-Dg^\sharp(x)v(v, Y^1) - g^\sharp(x)(X^2)Y^1 + Dg^\sharp(x)Y^1(v, v) = \tfrac{1}{2}\, Dg^\sharp(x)Y^1(v, v),$$

which rewritten, using the symmetry property

$$Dg^\sharp(x)X(Y, Z) = Dg^\sharp(x)X(Z, Y),$$

becomes

$$\begin{aligned} g(x)(X^2, Y^1) &= \tfrac{1}{2}\big(Dg^\sharp(x)v(Y^1, v) - Dg^\sharp(x)v(v, Y^1) - Dg^\sharp(x)Y^1(v, v)\big) \\ &= -\widetilde{\Gamma}(x, v)Y^1. \end{aligned}$$

Therefore we obtain $X^2(x, v) = -g(x)^\flat\big(\widetilde{\Gamma}(x, v)\big) = -\Gamma(x, v)$. Hence the local expression of the vector field X_K is

$$X_K(x, v) = (v, -\Gamma(x, v)). \tag{22}$$

Since the components of Γ are the usual *Christoffel symbols* of the metric g, the image of the integral curves of the *geodesic vector field* X_K under the bundle projection τ are *geodesics* on N for the metric g. □

Let $\varphi : M \to M$ be a diffeomorphism of the symplectic manifold (M, ω). If φ preserves ω, that is, $\varphi^*\omega = \omega$, then φ is a *symplectic diffeomorphism.*

Claim: The *flow* φ_t^f of a Hamiltonian vector field X_f is a local one parameter group of symplectic diffeomorphism of (M, ω).

(3.1) **Proof**: From the definition of the *flow* of a vector field it follows that $t \to \varphi_t^f$ is a local one parameter group. To show that φ_t^f is symplectic, we calculate

$$\begin{aligned} L_{X_f}\omega &= X_f \lrcorner d\omega + d(X_f \lrcorner \omega) \\ &= d^2 f, \quad \text{using the definition of } X_f \text{ and the fact that } \omega \text{ is closed} \\ &= 0. \end{aligned}$$

Since by definition of *Lie derivative*

$$\frac{d}{dt}(\varphi_t^f)^*\omega = (\varphi_t^f)^*(L_{X_f}\omega) = 0,$$

we obtain $(\varphi_t^f)^*\omega = \omega$. □

A basic property of a symplectic diffeomorphism is that it maps a Hamiltonian vector field into another Hamiltonian vector field. More precisely, for the symplectic diffeomorphism φ of (M, ω) we have

Claim:

$$\varphi^* X_f = X_{\varphi^* f}. \tag{23}$$

(3.2) **Proof**: We calculate

$$X_{\varphi^* f} \lrcorner \omega = d(\varphi^* f) = \varphi^* df = \varphi^*(X_f \lrcorner \omega) = \varphi^* X_f \lrcorner \varphi^* \omega = \varphi^* X_f \lrcorner \omega,$$

since φ is symplectic. Equation (23) follows because ω is nondegenerate. □

Example 3. We derive Hamilton's equations on the cotangent bundle of a Lie group, which has been trivialized using left translation. Let G be a Lie group and (T^*G, Ω) its cotangent bundle with canonical 2-form Ω. Suppose that $\mathcal{H} : T^*G \to \mathbf{R}$ is a smooth function. The Hamiltonian vector field $X_{\mathcal{H}}$ on (T^*G, Ω) corresponding to $\mathcal{H}$ is defined by

$$d\mathcal{H}(\alpha) V_\alpha = \Omega(\alpha)(X_{\mathcal{H}}(\alpha), V_\alpha) \tag{24}$$

for every $V_\alpha \in T_\alpha(T^*G)$. Trivialize T^*G using the mapping $\mathcal{L} : G \times \mathcal{G}^* \to T^*G$ (2) coming from left translation. Pulling the Hamiltonian $\mathcal{H}$ back by $\mathcal{L}$ gives the Hamiltonian

$$H : G \times \mathcal{G}^* \to \mathbf{R} : (g, \alpha) \to \mathcal{H}(\alpha_g).$$

Note that $\mathcal{L}^*\Omega = \omega$ (4). Using ((3.2)) we find that the Hamiltonian vector field X_H on $(G \times \mathcal{G}^*, \omega)$ is given by $X_H = \mathcal{L}^*(X_{\mathcal{H}})$.

We now derive Hamilton's equations, which are satisfied by the integral curves of the vector field X_H on $G \times \mathcal{G}^*$. In this context Hamilton's equations are called the *Euler-Arnol'd equations*. Partially differentiating H in the directions $T_eL_g\eta \in T_gG$ and $\beta \in \mathcal{G}^*$ gives

$$\begin{aligned} dH(g,\alpha)(T_eL_g\eta, \beta) &= D_1H(g,\alpha)T_eL_g\eta + D_2H(g,\alpha)\beta \\ &= D_1H(g,\alpha)T_eL_g\eta + \beta\big(D_2H(g,\alpha)\big), \end{aligned}$$

since $D_2H(g,\alpha) \in \mathcal{G}^{**} = \mathcal{G}$. Because $X_H(g,\alpha) \in T_gG \times \mathcal{G}^*$, we may write

$$X_H(g,\alpha) = \big(T_eL_gX(g,\alpha), \Lambda(g,\alpha)\big). \tag{25}$$

By definition of X_H,

$$dH(g,\alpha)(T_eL_g\eta, \beta) = \omega(g,\alpha)\big((T_eL_gX(g,\alpha), \Lambda(g,\alpha)), (T_eL_g\eta, \beta)\big)$$

for any $\eta \in \mathcal{G}$ and any $\beta \in \mathcal{G}^*$, or using (4)

$$D_1H(g,\alpha)T_eL_g\eta + \beta(D_2H(g,\alpha)) = -\Lambda(g,\alpha)\eta + \beta(X(g,\alpha)) + \alpha([X(g,\alpha), \eta]). \tag{26}$$

Setting $\eta = 0$ in (26) gives

$$\beta(D_2H(g,\alpha)) = \beta(X(g,\alpha))$$

for every $\beta \in \mathcal{G}^*$. Therefore

$$X(g,\alpha) = D_2H(g,\alpha). \tag{27}$$

Setting $\beta = 0$ in (26) gives

$$D_1H(g,\alpha)T_eL_g\eta = -\Lambda(g,\alpha)\eta + (\mathrm{ad}^t_X\,\alpha)\eta,$$

for any $\eta \in \mathcal{G}$. Using (27), we obtain

$$\Lambda(g, \alpha) = -(T_e L_g)^t (D_1 H(g, \alpha)) + \mathrm{ad}^t_{D_2 H(g.\alpha)} \alpha. \tag{28}$$

Substituting (27) and (28) into (25) gives the *Euler-Arnol'd vector field* X_H on $G \times \mathcal{G}^*$. The integral curves of X_H satisfy the Euler-Arnol'd equations

$$\begin{aligned} \dot{g} &= T_e L_g (D_2 H(g, \alpha)) \\ \dot{\alpha} &= -(T_e L_g)^t (D_1 H(g, \alpha)) + \mathrm{ad}^t_{D_2 H(g,\alpha)} \alpha. \end{aligned} \tag{29}$$

If the Hamiltonian $\mathcal{H} : T^*G \to \mathbf{R}$ is left invariant, then so is H. Therefore, $D_1 H = 0$, so the Euler-Arnol'd equations for a left invariant Hamiltonian are

$$\begin{aligned} \dot{g} &= T_e L_g (D_2 H(g, \alpha)) \\ \dot{\alpha} &= \mathrm{ad}^t_{D_2 H(g,\alpha)} \alpha. \qquad \square \end{aligned}$$

4 Poisson algebras and manifolds

In this section we define the notions of a Poisson algebra and a Poisson manifold. This leads to an algebraic formulation of Hamiltonian mechanics.

A *Poisson algebra* $(\mathcal{A}, \{\ ,\ \}, \cdot)$ is a real Lie algebra under the *bracket* $\{\ ,\ \}$ which is also a commutative ring with unit under the multiplication $\cdot$. In addition, for every $f, g, h \in \mathcal{A}$ *Leibniz' rule*

$$\{f, g \cdot h\} = h \cdot \{f, g\} + g \cdot \{f, h\} \tag{30}$$

holds. Leibniz' rule simply states that for every $f \in \mathcal{A}$ the linear mapping

$$\mathrm{ad}_f : \mathcal{A} \to \mathcal{A} : g \to \{f, g\}$$

is a derivation. We call ad_f the *Hamiltonian derivation* associated to f. Thinking of ad_f as a formal vector field, its *formal flow* is given by

$$\varphi^f_t = \exp(t\ \mathrm{ad}_f) = \sum_{n=0}^{\infty} \frac{1}{n!} t^n\, \mathrm{ad}^n_f\,.$$

A map $\varphi : \mathcal{A} \to \mathcal{B}$ is a *homomorphism* of Poisson algebras if it is a homomorphism of the Lie algebra and commutative ring structures. A bijective homomorphism of the Poisson algebra $\mathcal{A}$ is an automorphism of $\mathcal{A}$. It is easy to check that the formal flow φ^f_t is a one parameter group of automorphisms of $\mathcal{A}$, that is,

$$(\varphi^f_t)^*(g \cdot h) = (\varphi^f_t)^* g \cdot (\varphi^f_t)^* h$$

and

$$(\varphi^f_t)^*\{g, h\} = \{(\varphi^f_t)^* g, (\varphi^f_t)^* h\}.$$

An element $f \in \mathcal{A}$ is a *Casimir* if and only if $\{f, g\} = 0$ for every $g \in \mathcal{A}$. Clearly the unit element of $\mathcal{A}$ is a Casimir. If the only Casimir elements are real multiples of the unit element of $\mathcal{A}$, then $\mathcal{A}$ is a *nondegenerate* Poisson algebra.

We now give some examples of Poisson algebras.

Example 1. Let $\mathcal{G}^*$ be the dual of the finite dimensional Lie algebra $\mathcal{G}$ with Lie bracket $[\,,\,]$. For every $f, g \in C^\infty(\mathcal{G}^*)$ and every $\mu \in \mathcal{G}^*$ define a Poisson bracket $\{\,,\,\}_{\mathcal{G}^*}$ by

$$\{f, g\}_{\mathcal{G}^*}(\mu) = \mu([df(\mu), dg(\mu)]). \tag{31}$$

The right hand side of (31) is well defined since the differential $df(\mu)$ of f at μ is a linear form on $\mathcal{G}^*$, which we identify with an element of $\mathcal{G}$. To check that $(C^\infty(\mathcal{G}^*), \{\,,\,\}_{\mathcal{G}^*})$ is a Poisson algebra, where $\cdot$ is the usual product of smooth functions, we first verify Leibniz' rule:

$$\begin{aligned}\{f, g\cdot h\}_{\mathcal{G}^*}(\mu) &= \mu\big([df(\mu), d(g\cdot h)(\mu)]\big) \\ &= \mu\big([df(\mu), g(\mu)dh(\mu) + h(\mu)dg(\mu)]\big) \\ &= \big(g\cdot\{f,h\}_{\mathcal{G}^*} + h\cdot\{f,g\}_{\mathcal{G}^*}\big)(\mu).\end{aligned} \tag{32}$$

We now need only show that $\{\,,\,\}_{\mathcal{G}^*}$ satisfies the Jacobi identity. This is not entirely straightforward. Our argument involves introducing the concept of a structure matrix.

Let $\{x_i\}$, $i = 1, \ldots n$ be a basis for $\mathcal{G}$. Identifying $(\mathcal{G}^*)^*$ with $\mathcal{G}$ we can think of $\{x_i\}$ as coordinates on $\mathcal{G}^*$. Since

$$\mathrm{ad}_f : \mathcal{G}^* \to \mathcal{G}^* : h \to \{f, h\}_{\mathcal{G}^*}$$

is a derivation on $\mathcal{G}^*$, it follows that

$$\mathrm{ad}_f = \sum_{k=1}^{n} f_k\, \partial_k$$

for some $f_k \in C^\infty(\mathcal{G}^*)$. Here $\partial_k = \frac{\partial}{\partial x_k}$. Evaluating the preceding expression for ad_f on x_j and using the fact that $\partial_k\, x_j = \delta^j_k$, we obtain $f_k = \{f, x_k\}_{\mathcal{G}^*}$, that is,

$$\mathrm{ad}_f = \sum_{k=1}^{n} \{f, x_k\}_{\mathcal{G}^*}\, \partial_k. \tag{33}$$

Replacing f in (33) by x_j gives

$$\mathrm{ad}_{x_j} = \sum_{k=1}^{n} \{x_j, x_k\}_{\mathcal{G}^*}\, \partial_k. \tag{34}$$

Hence

$$\begin{aligned}\{f, g\}_{\mathcal{G}^*} &= \mathrm{ad}_f g = \sum_k \{f, x_k\}_{\mathcal{G}^*}\, \partial_k g \\ &= -\sum_k \mathrm{ad}_{x_k} f\, \partial_k g = \sum_{j,k} \{x_j, x_k\}_{\mathcal{G}^*}\, \partial_j f\, \partial_k g.\end{aligned} \tag{35}$$

In other words, $(\{x_j, x_k\}_{\mathcal{G}^*})$ is the *structure matrix* of the Poisson bracket $\{\,,\,\}_{\mathcal{G}^*}$.

Claim: $(C^\infty(\mathcal{G}^*), \{\ ,\ \}_{\mathcal{G}^*})$ is a Lie algebra if and only if the coordinate functions x_i satisfy the *Jacobi identity*, that is, if and only if

$$\{\{x_i, x_j\}_{\mathcal{G}^*}, x_k\}_{\mathcal{G}^*} = \{\{x_i, x_k\}_{\mathcal{G}^*}, x_j\}_{\mathcal{G}^*} + \{x_i, \{x_j, x_k\}_{\mathcal{G}^*}\}_{\mathcal{G}^*} \tag{36}$$

for every $1 \le i, j, k \le n$.

(4.1) **Proof**: Before launching into the proof we first verify the identity

$$\sum_k \{x_k, x_i\}_{\mathcal{G}^*} \{\partial_k f, x_j\}_{\mathcal{G}^*} = \sum_k \{x_k, x_j\}_{\mathcal{G}^*} \{\partial_k f, x_i\}_{\mathcal{G}^*}. \tag{37}$$

We calculate.

$$\begin{aligned}\sum_k \{x_k, x_i\}_{\mathcal{G}^*} \{\partial_k f, x_j\}_{\mathcal{G}^*} &= -\sum_k \{x_k, x_i\}_{\mathcal{G}^*} \operatorname{ad}_{x_j} \partial_k f \\ &= -\sum_{k,\ell} \{x_k, x_i\}_{\mathcal{G}^*} \{x_j, x_\ell\}_{\mathcal{G}^*} \partial_\ell \partial_k f = \sum_k \{x_k, x_j\}_{\mathcal{G}^*} \{\partial_k f, x_i\}_{\mathcal{G}^*}.\end{aligned}$$

The last equality follows by interchanging the order of summation and partial differentiation, and using (34) and the skew symmetry of $\{\ ,\ \}_{\mathcal{G}^*}$.

We now prove the claim. Let $f, g, h \in C^\infty(\mathcal{G}^*)$. Then

$$\begin{aligned}\{\{f, g\}, h\}_{\mathcal{G}^*} &= \{\sum_{j,k} \{x_j, x_k\}_{\mathcal{G}^*} \partial_j f\, \partial_k g, h\}_{\mathcal{G}^*}, \quad \text{using (35)} \\ &= \sum_{j,k,\ell} \big[\{\{x_j, x_k\}_{\mathcal{G}^*}, h\}_{\mathcal{G}^*}\, \partial_j f\, \partial_k g + \{x_j, x_k\}_{\mathcal{G}^*} \{\partial_j f, h\}_{\mathcal{G}^*}\, \partial_k g \\ &\qquad + \{x_j, x_k\}_{\mathcal{G}^*} \{\partial_k g, h\}_{\mathcal{G}^*}\, \partial_j f \big], \quad \text{using Leibniz' rule} \\ &= \sum_{j,k,\ell} \{\{x_j, x_k\}_{\mathcal{G}^*}, x_\ell\}_{\mathcal{G}^*}\, \partial_\ell h\, \partial_j f\, \partial_k g + \overbrace{\sum_{j,k,\ell} \{x_j, x_k\}_{\mathcal{G}^*} \{\partial_j f, x_\ell\}_{\mathcal{G}^*}\, \partial_\ell h\, \partial_k g}^{I} \\ &\qquad + \overbrace{\sum_{j,k,\ell} \{x_j, x_k\}_{\mathcal{G}^*} \{\partial_k g, x_\ell\}_{\mathcal{G}^*}\, \partial_j f\, \partial_\ell h}^{II}, \quad \text{using (33).}\end{aligned}$$

Interchanging h and g in the above formula gives

$$\begin{aligned}\{\{f, g\}, h\}_{\mathcal{G}^*} &= \sum_{j,k,\ell} \{\{x_j, x_k\}_{\mathcal{G}^*}, x_\ell\}_{\mathcal{G}^*}\, \partial_\ell g\, \partial_j f\, \partial_k h \\ &\quad + \overbrace{\sum_{j,k,\ell} \{x_j, x_k\}_{\mathcal{G}^*} \{\partial_j f, x_\ell\}_{\mathcal{G}^*}\, \partial_\ell g\, \partial_k h}^{I'} + \overbrace{\sum_{j,k,\ell} \{x_j, x_k\}_{\mathcal{G}^*} \{\partial_k h, x_\ell\}_{\mathcal{G}^*}\, \partial_j f\, \partial_\ell g}^{III}.\end{aligned}$$

Since $\{f, \{g, h\}\}_{\mathcal{G}^*} = -\{f, \{h, g\}\}_{\mathcal{G}^*} = \{\{h, g\}, f\}_{\mathcal{G}^*}$, interchanging f and h gives

$$\{f, \{g, h\}\}_{\mathcal{G}^*} = \sum_{j,k,\ell} \{\{x_j, x_k\}_{\mathcal{G}^*}, x_\ell\}_{\mathcal{G}^*} \partial_\ell f\, \partial_j h\, \partial_k g$$

$$+ \overbrace{\sum_{j,k,\ell} \{x_j, x_k\}_{\mathcal{G}^*} \{\partial_j h, x_\ell\}_{\mathcal{G}^*} \partial_\ell f\, \partial_k g}^{III'} + \overbrace{\sum_{j,k,\ell} \{x_j, x_k\}_{\mathcal{G}^*} \{\partial_k g, x_\ell\}_{\mathcal{G}^*} \partial_j h\, \partial_\ell f}^{II'} .$$

Using (37) one can show that the terms $I = I'$, $II = II'$, and $III = -III'$. We will prove only the last equality.

$$\begin{aligned} III &= -\sum_{j,\ell}\sum_k \{x_k, x_j\}_{\mathcal{G}^*} \{\partial_k h, x_\ell\}_{\mathcal{G}^*} \partial_j f\, \partial_\ell g \\ &= -\sum_{j,\ell}\sum_k \{x_k, x_\ell\}_{\mathcal{G}^*} \{\partial_k h, x_j\}_{\mathcal{G}^*} \partial_j f\, \partial_\ell g, \quad \text{using (37)} \\ &= -III', \end{aligned}$$

replacing k by j, ℓ by k and j by ℓ. Therefore

$$\{\{f, g\}, h\}_{\mathcal{G}^*} - \{\{f, h\}, g\}_{\mathcal{G}^*} - \{f, \{g, h\}\}_{\mathcal{G}^*} =$$

$$= \sum_{j,k,\ell} \big[\{\{x_j, x_k\}_{\mathcal{G}^*}, x_\ell\}_{\mathcal{G}^*} - \{\{x_j, x_\ell\}_{\mathcal{G}^*}, x_k\}_{\mathcal{G}^*} \{x_j, \{x_\ell, x_k\}_{\mathcal{G}^*}\}_{\mathcal{G}^*}\big] \partial_j f\, \partial_k g\, \partial_\ell h,$$

which proves the claim. □

▷ We now verify that (36) holds.

(4.2) **Proof**: By definition

$$\{x_i, x_j\}_{\mathcal{G}^*}(\mu) = \mu\big([dx_i, dx_j] \big) = \mu([x_i, x_j]), \tag{38}$$

because x_ℓ are linear functions on $\mathcal{G}^*$. Since ($\mathcal{G}$, [,]) is a Lie algebra the Jacobi identity

$$\big[[x_i, x_j], x_k\big] = \big[[x_i, x_k], x_j\big] + \big[x_i, [x_j, x_k]\big]$$

holds. Using (38) it follows that the Jacobi identity

$$\{\{x_j, x_k\}_{\mathcal{G}^*}, x_\ell\}_{\mathcal{G}^*} = \{\{x_j, x_\ell\}_{\mathcal{G}^*}, x_k\}_{\mathcal{G}^*} + \{x_j, \{x_\ell, x_k\}_{\mathcal{G}^*}\}_{\mathcal{G}^*}$$

holds. □

Hence by ((4.1)) $(C^\infty(\mathcal{G}^*), \{\ ,\ \}_{\mathcal{G}^*})$ is a Lie algebra.

▷ When $\mathcal{G}$ is a semisimple Lie algebra, the Poisson algebra $(C^\infty(\mathcal{G}^*), \{\ ,\ \}_{\mathcal{G}^*})$ is *not* nondegenerate.

(4.3) **Proof**: We need only construct a Casimir element of $(C^\infty(\mathcal{G}^*), \{\ ,\ \}_{\mathcal{G}^*}, \cdot)$ to prove the assertion. Recall that $\mathcal{G}$ is *semisimple* if and only if the *Killing form*

$$k(X, Y) = \mathrm{tr}(\mathrm{ad}_X\, \mathrm{ad}_Y) \tag{39}$$

is a nondegenerate symmetric bilinear form on $\mathcal{G}$. The Killing form satisfies

$$k(\mathrm{ad}_Z X, Y) = -k(X, \mathrm{ad}_Z Y) \tag{40}$$

for every $X, Y, Z \in \mathcal{G}$ because

$$\begin{aligned} k(\mathrm{ad}_Z X, Y) &= \mathrm{tr}(\mathrm{ad}_{[Z,X]}\,\mathrm{ad}_Y) = \mathrm{tr}\big(\,\mathrm{ad}_Z\,\mathrm{ad}_X - \mathrm{ad}_X\,\mathrm{ad}_Z)\,\mathrm{ad}_Y\,\big) \\ &= \mathrm{tr}\big(\,\mathrm{ad}_X(\mathrm{ad}_Y\,\mathrm{ad}_Z - \mathrm{ad}_Z\,\mathrm{ad}_Y)\big), \\ &\qquad \text{since } \mathrm{tr}(ABC) = \mathrm{tr}(BCA) \text{ for linear maps } A, B, C. \\ &= \mathrm{tr}(\mathrm{ad}_X\,\mathrm{ad}_{[Y,Z]}) = -k(X, \mathrm{ad}_Z Y). \end{aligned}$$

Using the isomorphism

$$k^\flat = (k^\sharp)^{-1} : \mathcal{G}^* \to \mathcal{G}$$

to identify $\mathcal{G}^*$ with $\mathcal{G}$, the Poisson bracket $\{\,,\,\}_{\mathcal{G}^*}$ (31) on $(C^\infty(\mathcal{G}^*)$ becomes the Poisson bracket $\{\,,\,\}_{\mathcal{G}}$ on $C^\infty(\mathcal{G})$ defined by

$$\{f, g\}_{\mathcal{G}}(X) = k\big(X, \big[k^\flat(df(X)), k^\flat(dg(X))\big]\big).$$

The following calculation shows that the smooth function

$$g : \mathcal{G} \to \mathbf{R} : X \to \tfrac{1}{2} k(X, X)$$

is a Casimir element of $(C^\infty(\mathcal{G}), \{\,,\,\}_{\mathcal{G}}, \cdot)$.

$$\begin{aligned} \{f, g\}_{\mathcal{G}}(X) &= k\big(X, \big[k^\flat(df(X)), X)\big]\big), \quad \text{by definition of } g \text{ and } \{\,,\,\}_{\mathcal{G}} \\ &= -k\big(\,[X, X]\,, k^\flat(df(X))\big), \quad \text{using (40)} \\ &= 0. \qquad \square \end{aligned}$$

Example 2. Let (M, ω) be a smooth symplectic manifold. On the space of smooth functions $C^\infty(M)$ define a *Poisson bracket* $\{\,,\,\}$ by

$$\{f, g\} = \omega(X_f, X_g) \tag{41}$$

for every $f, g \in C^\infty(M)$. Here X_f, X_g are the Hamiltonian vector fields corresponding to the Hamiltonians f, g, respectively. It follows that $\{f, g\} = L_{X_g} f = -L_{X_f} g$. Thus ad_f is the derivation $-L_{X_f}$ which may be identified with the *Hamiltonian vector field* $-X_f$, since M is a smooth manifold.

Claim: $(C^\infty(M), \{\,,\,\}, \cdot)$ is a Poisson algebra.

(4.4) **Proof**: Clearly the bracket $\{\,,\,\}$ is linear in each argument and is skew symmetric. It remains to show that the derivation property and the Jacobi identity hold.

To see that the bracket is a derivation in each slot, note that

$$\{f \cdot g, h\} = L_{X_h}(f \cdot g) = (L_{X_h} f) \cdot g + f \cdot (L_{X_h} g) = \{f, h\} \cdot g + f \cdot \{g, h\}.$$

Thus the derivation property holds because Leibniz' rule holds for the Lie derivative.

To show that the Jacobi identity

$$\{f, \{g, h\}\} = \{\{f, g\}, h\} + \{g, \{f, h\}\} \tag{42}$$

is true we use the fact that the symplectic form ω is closed. Recall (see (9)) that for smooth vector fields X, Y, Z on M

$$\begin{aligned} d\omega(X, Y, Z) = \quad & -d(X \lrcorner \omega)(Y, Z) + d(Y \lrcorner \omega)(Z, X) + d(Z \lrcorner \omega)(X, Y) \\ & -\omega([X, Y], Z) - \omega([Y, Z], X) - \omega([Z, X], Y). \end{aligned} \tag{43}$$

Letting $X = X_f$, $Y = X_g$ and $Z = X_h$, we see that the first three terms in (43) vanish because

$$d(X_f \lrcorner \omega) = d(df) = 0.$$

To deal with the next three terms note that

$$X_{\{f,g\}} = -\left[X_f, X_g\right], \tag{44}$$

because

$$\begin{aligned} -\left[X_f, X_g\right] = L_{X_g} X_f \quad &= \quad \frac{d}{dt}\Big|_{t=0} (\varphi_t^g)^* X_f \ = \ \frac{d}{dt}\Big|_{t=0} X_{(\varphi_t^g)^* f}, \quad \text{by } ((3.2)) \\ &= \quad X_{L_{X_g} f} = X_{\{f,g\}}. \end{aligned}$$

Thus

$$\omega(\left[X_f, X_g\right], X_h) = -\omega(X_{\{f,g\}}, X_h) = -\{\{f, g\}, h\}.$$

So

$$d\omega(X_f, X_g, X_h) = \{\{f, g\}, h\} - \{\{f, h\}, g\} + \{\{g, h\}, f\}. \tag{45}$$

Consequently, the closedness of ω implies that Jacobi's identity holds. □

Corollary: If $\varphi : M \rightarrow M$ is a symplectic diffeomorphism of (M, ω), then

$$\varphi^*\{f, g\} = \{\varphi^* f, \varphi^* g\}$$

for every $f, g \in C^\infty(M)$.

(4.5) **Proof**: For every $m \in M$ we compute

$$\begin{aligned} \{\varphi^* f, \varphi^* g\}(m) \quad &= \quad \omega(m)\big(X_{\varphi^* f}(m), X_{\varphi^* g}(m)\big) \\ &= \quad \omega(m)\big(\varphi^* X_f(m), \varphi^* X_g(m)\big) \quad \text{by } ((3.2)) \\ &= \quad \omega(\varphi^{-1}(\varphi(m)))\big(T\varphi^{-1} X_f(\varphi(m)), T\varphi^{-1} X_g(\varphi(m))\big) \\ &= \quad \big((\varphi^{-1})^* \omega\big)(\varphi(m))\big(X_f(\varphi(m)), X_g(\varphi(m))\big) \ = \ \{f, g\}(\varphi(m)) \\ &= \quad \varphi^*(\{f, g\})(m). \qquad \square \end{aligned}$$

M is a *Poisson manifold* if there is a Poisson bracket $\{\,,\,\}$ on $C^\infty(M)$ such that ▷ $(C^\infty(M), \{\,,\,\}, \cdot)$ is a Poisson algebra. We now show that the Poisson bracket $\{\,,\,\}$ is determined by the brackets of local coordinate functions on M.

(4.6) **Proof**: For a fixed f the map $g \to \{f, g\}$ is a derivation on $C^\infty(M)$, and hence may be represented by a vector field Y_f. Thus in local coordinates we have an expression of the form

$$\{f, g\} = \sum_i Y_f^i \, \partial_i g, \tag{46}$$

where ∂_i is the partial derivative with respect to the ith coordinate function x_i and $Y_f^i = L_{Y_f} x_i$. By the skew symmetry of the Poisson bracket we have a similar expression with f and g interchanged. Therefore we have a local expression

$$\{f, g\} = \sum_{i,j} W_{ij} \, \partial_i f \, \partial_j g. \tag{47}$$

This tells us that the W_{ij} are components of a skew symmetric contravariant tensor W of order 2. Because the Poisson bracket $\{\, ,\, \}$ satisfies the Jacobi identity, the components of W satisfy

$$\sum_\ell \left(\frac{\partial W_{jk}}{\partial x_\ell} W_{i\ell} + \frac{\partial W_{ij}}{\partial x_\ell} W_{k\ell} + \frac{\partial W_{ki}}{\partial x_\ell} W_{j\ell} \right) = 0. \tag{48}$$

If we compute the bracket of our local coordinate functions $(x_1, \ldots, x_n)$ using (47), we find that $\{x_i, x_j\} = W_{ij}$. □

The *rank* of the Poisson structure at a point is the rank of the structure matrix W at the point.

Claim: A Poisson manifold M is symplectic if and only if the matrix of the Poisson structure tensor W is everywhere invertible.

(4.7) **Proof**: Suppose that W is everywhere invertible. Let $X_f = W(\cdot, df)$, and define ω by

$$\omega(X_f, X_g) = W(df, dg) = df \cdot X_g = \{f, g\}. \tag{49}$$

We only need check that ω is closed and nondegenerate. Since at any point $p \in M$

$$\text{span}\left\{ df(p) \middle| f \in C^\infty(M) \right\} = T_p^* M$$

and the matrix (W_{ij}) is invertible, we see that ω is nondegenerate. As a corollary, the vector fields X_f span T_pM as well. They really are Hamiltonian vector fields since (49) implies

$$df = W^{-1}(X_f, \cdot) = X_f \lrcorner \omega.$$

To see that ω is closed, we refer to ((4.3)) where we showed that a symplectic manifold (M, ω) gave a Lie algebra structure to $C^\infty(M)$ such that

$$d\omega(X_f, X_g, X_h) = \{f, \{g, h\}\} + \{\{g, h\}, f\} + \{\{h, f\}, g\}. \tag{50}$$

Because W is a Poisson structure tensor, (48) holds and thus the right hand side of (50) vanishes. Hence $d\omega = 0$.

Now suppose that (M, ω) is a symplectic manifold. Let $\{x_1, \ldots, x_n\}$ be local coordinates on (M, ω). Then the structure matrix W has entries $W_{ij} = \omega(X_{x_i}, X_{x_j})$. For $f, g \in C^\infty(M)$ define

$$\{f, g\} = \sum_{ij} W_{ij}\, \partial_i f\, \partial_i g = W(df, dg).$$

Thus $\{f, g\} = \omega(X_f, X_f)$. Since ω is closed, the Poisson bracket $\{\ ,\ \}$ satisfies the Jacobi identity ((4.3)) and thus (48). Suppose that $W(df, dg) = 0$ for every $g \in C^\infty(M)$. Then $0 = \omega(m)(X_f(m), X_g(m))$ for every $m \in M$. Since

$$\text{span}\left\{ dg(m) \middle| g \in C^\infty(M) \right\} = T^*_m M$$

and $\omega(m)$ is nondegenerate, the vector fields $X_g(m)$ span $T_m M$. From the nondegeneracy of $\omega(m)$ we deduce that $X_f(m) = 0$ for every $m \in M$. Therefore $df = 0$, that is, W is nondegenerate. This implies that the matrix $(W_{ij}(m))$ is invertible for every $m \in M$. □

We now prove a result which gives the local structure of Poisson manifolds

Claim: Suppose that $(M, \{\ ,\ \})$ is a Poisson manifold of dimension n whose structure tensor W in a neighborhood of p has constant rank k. Then k is even, say $k = 2\ell$, and there are local coordinates about p of the form $x = (\xi_1, \ldots, \xi_\ell, \eta_1, \ldots, \eta_\ell, \zeta_{2\ell+1}, \ldots, \zeta_{n-2\ell})$ such that

$$W(x) = \begin{array}{l} \ell\,\{ \\ \ell\,\{ \\ n-2\ell\,\{ \end{array} \left(\begin{array}{cc|c} 0 & I & \\ -I & 0 & \\ \hline & & 0 \end{array} \right).$$

(4.8) **Proof**: Pick a chart so that p is mapped to the origin. If the rank of $W(0)$ is zero, then put $W(x) = 0$ and we are done. If not, then there are smooth functions $\widetilde{\xi}_1$ and η_1 in a neighborhood of 0 such that $\{\widetilde{\xi}_1, \eta_1\}(0) \neq 0$. This implies $L_{X_{\eta_1}} \widetilde{\xi}_1(0) \neq 0$, and in particular, that the vector field $X_{\eta_1}(0) \neq 0$. We may find a system of local coordinates $(\xi_1, \ldots, \xi_n)$ about 0 so that $X_{\eta_1} = \frac{\partial}{\partial \xi_1}$ by rectifying the vector field X_{η_1}. Then in this neighborhood $\{\xi_1, \eta_1\} = 1$. If the rank of W at 0 is 2 we are done. Otherwise, let $N = \{x \,|\, \xi_1 = \eta_1 = 0\}$. In a possibly smaller neighborhood, N is a smooth submanifold containing the origin as $d\xi_1 \wedge d\eta_1 \neq 0$ at 0. Letting $\varphi_t^{\xi_1}$ and $\varphi_s^{\eta_1}$ denote the flows of X_{ξ_1} and X_{η_1}, we see that they commute as $\left[X_{\xi_1}, X_{\eta_1}\right] = -X_{\{\xi_1, \eta_1\}} = 0$. For small enough t and s it follows that

$$\mathcal{U} = \left\{ u \in M \middle| u = \varphi_t^{\xi_1} \circ \varphi_s^{\eta_1}(v),\ v \in N \right\}$$

is an open neighborhood of N about 0. Choosing coordinates $(\overline{x}_1, \ldots, \overline{x}_{n-2})$ on N, we extend them to functions on $\mathcal{U}$ by setting $x_i(u) = \overline{x}_i(v)$. Then $(\xi_1, \eta_1, x_1, \ldots, x_{n-2})$ are coordinates on $\mathcal{U}$. Furthermore, the functions x_i are invariant under the flows $\varphi_t^{\xi_1}$ and $\varphi_s^{\eta_1}$. Therefore $\{\xi_1, x_i\} = 0 = \{\eta_1, x_i\}$. As

$$\frac{\partial}{\partial \xi_1}(\{x_i, x_j\}) = L_{X_{\eta_1}}(\{x_i, x_j\}) = \{L_{X_{\eta_1}} x_i, x_j\} + \{x_i, L_{X_{\eta_1}} x_j\}$$

$$= \{\{x_i, \eta_1\}, x_j\} + \{x_i, \{x_j, \eta_1\}\} = 0,$$

$\{x_i, x_j\}$ does not depend on ξ_1. Similarly, $\{x_i, x_j\}$ does not depend on η_1. Therefore $\{x_i, x_j\}$ depends only on $(x_1, \ldots, x_{n-2})$. To finish the argument we just have to check that we can repeat this argument on N. Identify $C^\infty(N)$ with the smooth functions on $\mathcal{U}$ which are invariant under the flows of X_{ξ_1} and X_{η_1}. For these functions define a Poisson structure $\{\,,\,\}_N$ on N by

$$\{f|N, g|N\}_N = \{f, g\}|N.$$

Thus we may continue the argument inductively on $(C^\infty(N), \{\,,\,\}_N)$. □

When W is invertible, that is, $k = n$, the above result gives a local structure theorem for symplectic manifolds.

An important consequence of ((4.7)) is that in (ξ, η) coordinates about 0 the Hamiltonian derivation $-\operatorname{ad}_f$ for $f \in C^\infty(\mathbf{R}^n)$ has a local flow which satisfies Hamilton's equations

$$\begin{aligned} \dot{\xi}_j &= -\{f, \xi_j\} = \frac{\partial f}{\partial \eta_j} \\ \dot{\eta}_j &= -\{f, \eta_j\} = -\frac{\partial f}{\partial \xi_j} \end{aligned}$$

for $j = 1, \ldots, \ell$. Moreover the flow of $-\operatorname{ad}_f$ is a local one parameter group of symplectic diffeomorphisms of $(\mathbf{R}^n, \omega)$.

Suppose that $(M, \{\,,\,\}_1)$ and $(N, \{\,,\,\}_2)$ are Poisson manifolds. Then $\varphi : M \to N$ is a *Poisson map* if $\varphi^*\{f, h\}_2 = \{\varphi^* f, \varphi^* h\}_1$ for every $f, h \in C^\infty(M)$. In other words, $\varphi^* : C^\infty(N) \to C^\infty(M)$ is a homomorphism of Poisson algebras.

Claim: Let $f \in C^\infty(N)$ and $\varphi : M \to N$ be a Poisson map. If γ is an integral curve of the Hamiltonian vector field $X_{\varphi^* f}$ on M, then $\varphi \circ \gamma$ is an integral curve of X_f.

(4.9) **Proof**: For every $h \in C^\infty(N)$ we have

$$\frac{d}{dt}\varphi^* h(\gamma(t)) = \{\varphi^* h, \varphi^* f\}_1(\gamma(t)), \tag{51}$$

because γ is an integral curve of $X_{\varphi^* f}$. But (51) is equivalent to

$$\begin{aligned} \frac{d}{dt}h(\varphi(\gamma(t))) &= \varphi^*\{h, f\}_2(\gamma(t)), \quad \text{since } \varphi \text{ is a Poisson map} \\ &= \{h, f\}_1(\varphi(\gamma(t))). \end{aligned}$$

Thus $\varphi \circ \gamma$ is an integral curve of X_f. □

Let $(M, \{\,,\,\})$ be a Poisson manifold. A submanifold N of M is a *Poisson submanifold* if we can define a Poisson bracket $\{\,,\,\}_N$ on N by $\{\overline{f}, \overline{g}\}_N = \{f, g\}|N$, where f and g are smooth extensions to M of $\overline{f}, \overline{g} \in C^\infty(N)$. Equivalently, N is a Poisson submanifold of M if the inclusion map $i : N \to M$ is a Poisson map.

An important consequence of the local classification of Poisson manifolds ((4.7)) is the following. Define an equivalence relation $\sim$ on M by declaring $p \sim q$ if q can be reached from p by a piecewise smooth curve each segment of which is an integral curve of a Hamiltonian vector field on M. Then $\sim$ is an equivalence relation and an equivalence class is called a *leaf* of the Poisson structure on M.

Claim: Every leaf on M is a smooth Poisson submanifold of M with a nondegenerate Poisson bracket.

(4.10) **Proof**: Because the rank of a tensor is lower *semicontinuous*, we need only show that the rank of a Poisson structure is invariant under the flow of a Hamiltonian vector field. Then we can use the local classification of Poisson manifolds at each point in the leaf, because the constant rank hypothesis holds on submanifolds. The invariance of the rank follows from the calculation

$$\begin{aligned}\{\{g, f\}, h\} + \{g, \{h, f\}\} &= \{\{g, h\}, f\} = L_{X_f}\{g, h\} = L_{X_f}(W(dg, dh)) \\ &= (L_{X_f} W)(dg, dh) + W(L_{X_f} dg, dh) + W(dg, L_{X_f} dh) \\ &= (L_{X_f} W)(dg, dh) + W(d\{g, f\}, dh) + W(dg, d\{h, f\}) \\ &= (L_{X_f} W)(dg, dh) + \{\{g, f\}, h\} + \{g, \{h, f\}\}.\end{aligned}$$

Since this holds for any f, g, h, it follows that $L_{X_f} W = 0$. Thus the flow of X_f acts by automorphisms of the Poisson structure. Hence the rank of W is invariant under the flow of X_f. □

Example 3. It is often convenient when dealing with constrained systems to be able to work in the ambient space. For example, suppose we wish to examine the motion of a particle *constrained* to the unit sphere. We could work with spherical coordinates (θ, φ) and canonical coordinates $(\theta, \varphi, p_\theta, p_\varphi)$. But suppose that we wanted to work in Cartesian coordinates $(x^1, x^2, x^3, p_1, p_2, p_3)$ with the constraints

$$(x^1)^2 + (x^2)^2 + (x^3)^2 = 1 \qquad \text{and} \qquad x^1 p_1 + x^2 p_2 + x^3 p_3 = 0. \tag{52}$$

What we would like to do is work with the Poisson bracket $\frac{\partial}{\partial \theta} \wedge \frac{\partial}{\partial p_\theta} + \frac{\partial}{\partial \varphi} \wedge \frac{\partial}{\partial p_\varphi}$ in our Cartesian variables. Loosely speaking, we would like to think of the constraints as being given by $r = \text{const.}$ and $p_r = 0$.

Slightly more generally, suppose we have a chart (q, p) on $(T^*\mathbf{R}^3, \omega)$ and a local symplectic submanifold N defined by $q^3 = \text{const.}$ and $p_3 = \text{const.}$ On N we have a Poisson bracket

$$\frac{\partial}{\partial q^1} \wedge \frac{\partial}{\partial p_1} + \frac{\partial}{\partial q^2} \wedge \frac{\partial}{\partial p_2}. \tag{53}$$

Now transform N by the symplectic diffeomorphism

$$\Psi : (T^*\mathbf{R}^3, \omega) \to (T^*\mathbf{R}^3, \omega) : (q, p) \to (Q, P),$$

which is the lift of the diffeomorphism $\psi : \mathbf{R}^3 \to \mathbf{R}^3 : q \to Q$. Then

$$\Psi(q, p) = \left(\psi(q), (T\psi^{-1})^t p\right) = \left((Q_i), (\sum_{k=1}^{3} p_k \frac{\partial q^k}{\partial Q^i})\right),$$

in coordinates. Therefore

$$D\Psi(q,p) = \left(\begin{array}{c|c} \dfrac{\partial Q^k}{\partial q^i} & \displaystyle\sum_k p_k \frac{\partial^2 q^k}{\partial Q^i \partial Q^j} \\ \hline 0 & \dfrac{\partial q^k}{\partial Q^i} \end{array}\right),$$

which takes $(\frac{\partial}{\partial q}, \frac{\partial}{\partial p})$ into $(\frac{\partial}{\partial Q}, \frac{\partial}{\partial P})$. Taking the inverse gives

$$\begin{pmatrix} \frac{\partial}{\partial q} \\ \frac{\partial}{\partial p} \end{pmatrix} = (D\Psi)^{-1} \begin{pmatrix} \frac{\partial}{\partial Q} \\ \frac{\partial}{\partial P} \end{pmatrix} = J^{-1}(D\Psi)^t J \begin{pmatrix} \frac{\partial}{\partial Q} \\ \frac{\partial}{\partial P} \end{pmatrix},$$

since $D\Psi$ is a linear symplectic map with respect to the standard symplectic structure $J = \begin{pmatrix} 0 & -I \\ I & 0 \end{pmatrix}$ on $T^*\mathbf{R}^3$. Therefore

$$\begin{pmatrix} \frac{\partial}{\partial q} \\ \frac{\partial}{\partial p} \end{pmatrix} = \left(\begin{array}{c|c} \dfrac{\partial q^i}{\partial Q^k} & \displaystyle -\sum_{\ell,k} P_\ell \frac{\partial Q^k}{\partial q^\ell} \frac{\partial^2 q^k}{\partial Q^i \partial Q^j} \\ \hline 0 & \dfrac{\partial Q^i}{\partial q^k} \end{array}\right) \begin{pmatrix} \frac{\partial}{\partial Q} \\ \frac{\partial}{\partial P} \end{pmatrix}. \tag{54}$$

Substituting the expression for $\frac{\partial}{\partial q}$ and $\frac{\partial}{\partial p}$ from (54) into the right hand side of (53) gives an expression for the Poisson bracket on the symplectic submanifold $\Psi(N)$, defined by $q^3 = q^3(Q,P)$ and $p_3 = p_3(Q,P)$. In the new variables the functions q^3 and p_3 are Casimirs, that is, functions which Poisson commute with every function. □

We now discuss a generalization of the above example. Let $\{c_i\}_{i=1}^k$ be smooth functions on a symplectic manifold (M,ω) and suppose that $c \in \mathbf{R}^k$ is a regular value of the map

$$\mathscr{C} : M \to \mathbf{R}^k : p \to \big(c_1(p), \ldots, c_k(p)\big).$$

Then $N = \mathscr{C}^{-1}(c)$ is a smooth submanifold of M, called the *constraint manifold* defined by the *constraint functions* $\{c_i\}_{i=1}^k$. If the matrix $\big(\{c_i, c_j\}(p)\big)$ of Poisson brackets of the constraints is invertible for every $p \in N$, then we say that N is a *cosymplectic* submanifold of M.

Claim: If N is a cosymplectic submanifold of a symplectic manifold (M,ω), then $\omega|N$ is a symplectic form on N.

(4.11) **Proof**: For each $p \in N$ let V_p be the space spanned by the vectors $\{X_{c_1}(p), \ldots, X_{c_k}(p)\}$. For every $v \in T_pN$ we have

$$0 = dc_i(p)v = \omega(p)(X_{c_i}(p), v),$$

which implies $X_{c_i}(p) \in (T_pN)^\omega$ and hence $T_pN \subseteq V_p^\omega$. Since the matrix

$$\big(\{c_i, c_j\}(p)\big) = \big(\omega(p)(X_{c_i}(p), X_{c_j}(p))\big)$$

is invertible, V_p is a symplectic subspace of $(T_pM, \omega(p))$ of dimension k, which must be even. But

$$\begin{aligned}\dim T_pN &= \dim T_pM - \dim \mathrm{span}\{dc_1(p), \ldots, dc_k(p)\} \\ &= \dim T_pM - k = \dim T_pM - \dim V_p \\ &= \dim V_p^{\omega}.\end{aligned}$$

Therefore $T_pN = V_p^{\omega}$. Hence T_pN is a symplectic subspace of $(T_pM, \omega(p))$, that is, $\omega|N$ is nondegenerate. Since $d(\omega|N) = (d\omega)|N = 0$, $\omega|N$ is a symplectic form on N. □

Using the symplectic form $\omega|N$ on N, we may define a Poisson bracket $\{\, ,\ \}_N$ on $C^{\infty}(N)$, see (41). We now discuss a way to compute the bracket $\{\, ,\ \}_N$ using the bracket $\{\, ,\ \}$ on M. We make use of *Dirac brackets*, which are defined as follows. Let $C = (C_{ij})$ be the *inverse* of the matrix $(\{c_i, c_j\})$ of Poisson brackets of the constraint functions. Since N is cosymplectic, C is defined in an open neighborhood $\mathcal{U}$ of N. Let $F, G \in C^{\infty}(M)$. For every $u \in \mathcal{U}$ define the Dirac bracket $\{\, ,\ \}^*$ by

$$\{F, G\}^*(u) = \{F, G\}(u) - \sum_{i,j=1}^{k} \{F, c_i\}(u) C_{ij}(u) \{c_j, G\}(u). \tag{55}$$

To keep the notation simple, in what follows we will suppress the variable u. Before we can relate the Dirac bracket to the bracket $\{\, ,\ \}_N$, we make some preliminary observations. First note that $C^{\infty}(N) = C^{\infty}(M)/\mathcal{I}$, where $\mathcal{I}$ is the ideal of smooth functions which vanish on N. In other words, $f \in C^{\infty}(N)$ if there is a function $F \in C^{\infty}(M)$ such that $F|N = f$. Since the ideal $\mathcal{I}$ is generated by the functions $\{c_i\}_{i=1}^{k}$, the function f is represented by

$$F + \sum_{i=1}^{k} \lambda_i c_i = F + \mathcal{I}.$$

Second, observe that the constraint function c_i is a Casimir for the Dirac bracket. This follows because for every $F \in C^{\infty}(M)$

$$\{F, c_k\}^* = \{F, c_k\} - \sum_{ij} \{F, c_i\} C_{ij} \{c_j, c_k\} = \{F, c_k\} - \sum_{i} \{F, c_i\} \delta_{ik} = 0.$$

From the above discussion we see that $\{\, ,\ \}^*|N$ is well defined on $C^{\infty}(N)$ because on N

$$\begin{aligned}\{F + \mathcal{I}, G\}^* &= \{F + \sum_{i=1}^{k} \lambda_i \cdot c_i, G\}^* = \{F, G\}^* + \sum_{i} \{\lambda_i, G\}^* c_i, \\ &\qquad \text{since } \{\, ,\ \}^* \text{ is a derivation and } \{c_i, G\}^* = 0. \\ &= \{F, G\}^* + \mathcal{I}.\end{aligned}$$

▷ The following argument shows that

$$\{F|N, G|N\}_N = \{F, G\}^*|N \tag{56}$$

for $F, G \in C^{\infty}(M)$. This is the basic property of the Dirac bracket.

(4.12) **Proof**: Let

$$F^* = F - \sum_{i,j} \{F, c_i\} C_{ij} c_j.$$

Then

$$\begin{aligned} \{F^*, c_k\} &= \{F, c_k\} - \sum_{i,j} \{F, c_i\} C_{ij} \{c_j, c_k\} + \mathcal{I} \\ &= \{F, c_k\} - \sum_{i} \{F, c_i\} \delta_{ik} + \mathcal{I} = \mathcal{I}, \end{aligned}$$

that is, $\{F^*, c_k\}|N = 0$. Consequently, N is an invariant manifold of X_{F^*}. For every $p \in N$ and every $Y_p \in T_pN$ we have

$$\omega(p)(X_{F^*}(p), Y_p) = dF^*(p)Y_p = dF(p)Y_p = \omega(p)(X_{F^*|N}(p), Y_p).$$

From the nondegeneracy of $\omega|N$ it follows that

$$X_{F^*}|N = X_{F^*|N}. \tag{57}$$

The next calculation shows that $\{F^*, G^*\} = \{F, G\}^* + \mathcal{I}$. Using the definition of F^* and G^* we have

$$\begin{aligned} \{F^*, G^*\} &= \{F - \sum_{i,j} \{F, c_i\} C_{ij} c_j, G - \sum_{k,\ell} \{G, c_k\} C_{k\ell} c_\ell\} \\ &= \{F, G\} - \sum_{k,\ell} \{G, c_k\} C_{k\ell} \{F, c_\ell\} - \sum_{i,j} \{F, c_i\} C_{ij} \{c_j, G\} \\ &\qquad + \sum_{ijk\ell} \{F, c_i\} \{G, c_k\} C_{ij} C_{k\ell} \{c_j, c_\ell\} + \mathcal{I}, \\ &\qquad \text{by bilinearity and Leibniz' rule} \\ &= \{F, G\} - \sum_{i,j} \{F, c_i\} C_{ij} \{c_j, G\} + \mathcal{I}, \\ &\qquad \text{using } \textstyle\sum C_{ij} \{c_j, c_k\} = \delta_{ik}. \\ &= \{F, G\}^* + \mathcal{I}. \end{aligned}$$

After these preliminaries we prove (56) as follows.

$$\begin{aligned} \{F, G\}^*|N &= \{F^*, G^*\}|N \\ &= (\omega|N)(X_{F^*}|N, X_{G^*}|N), \quad \text{by definition of } \{\ ,\ \}. \\ &= (\omega|N)(X_{F^*|N}, X_{G^*|N}), \quad \text{using (57)} \\ &= (\omega|N)(X_{F|N}, X_{G|N}) \\ &= \{F|N, G|N\}_N \quad \text{by definition of } \{\ ,\ \}_N. \qquad \square \end{aligned}$$

Note that in the preceding argument we have *not* used the Jacobi identity for Dirac brackets. Since the Poisson bracket $\{\ ,\ \}_N$ satisfies the Jacobi identity, the Dirac bracket $\{\ ,\ \}^*|N$ also satisfies the Jacobi identity.

The construction of Dirac brackets in ((4.11)) can be modified by letting

$$F^* = F - \sum_{i,j}(\{F, c_i\} + F_i)C_{ij}c_j, \tag{58}$$

where F_i lies in the ideal generated by $\{c_1, \dots, c_k\}$. It is straightforward to check that (57) holds for the modified F^*. This construction we will call *modified* Dirac brackets. In concrete examples it is often necessary to use the freedom in the choice of F_i to simplify F^*.

5 Exercises

1. (Geodesic vector field.) Let g be a Riemannian metric on a smooth manifold M. In local coordinates $x = (x^1, \dots x^n)$ the metric may be written as $g = \sum g_{ij}dx^i \otimes dx^j$. Let $v = (x^1, \dots, x^n, v^1, \dots v^n)$ be natural coordinates on TM. Show that the pullback θ_g by the map $g^\sharp$ to TM of the canonical 1-form θ on T^*M may be written as $\theta_g(v) = \sum g_{ij}v^i dx^j$. Moreover, the pullback Ω_g to TM by $g^\sharp$ of the canonical 2-form Ω on T^*M may be written as

$$\Omega_g(v) = -d\theta_g = -\sum \frac{\partial g_{kj}}{\partial x^i} v^k dx^i \wedge dx^j - \sum g_{ij} dv^i \wedge dx^j.$$

Show that the Hamiltonian vector field Z_g corresponding to the Hamiltonian function $E : TM \to \mathbf{R} : v \to \frac{1}{2} g(v, v)$ is the geodesic vector field.

2. Let $W = \sum_{i,j} W_{ij} \frac{\partial}{\partial x_i} \wedge \frac{\partial}{\partial x_j}$ be the structure tensor for a nondegenerate Poisson structure $\{\,,\,\}$ on $\mathbf{R}^{2n}$ with coordinates $(x_1, \dots, x_{2n})$. In other words $W(dx_i, dx_j) = \{x_i, x_j\}$. Show that $X_{x_j} = \sum_i W_{ij} \frac{\partial}{\partial x_i}$. Define a 2-form by $\Omega = \sum_{k,\ell} W^{k\ell} dx_\ell \wedge dx_k$, where (W^{ij}) is the inverse of the matrix (W_{ij}). Show that $\Omega(X_{x_i}, X_{x_j}) = \{x_i, x_j\}$. Furthermore show that

$$\sum_\ell \frac{\partial W_{jk}}{\partial x_\ell} W_{i\ell} = \sum_\ell \frac{\partial W_{ij}}{\partial x_\ell} W_{\ell k} + \sum_\ell \frac{\partial W_{ik}}{\partial x_\ell} W_{j\ell}. \tag{59}$$

is equivalent to the Jacobi identity for the Poisson structure $\{\,,\,\}$. Let

$$\Gamma_j^{k\ell} = \sum_{mn} \frac{\partial W_{mn}}{\partial x_j} W^{mk} W^{n\ell}.$$

Show that (59) can be written as $\Gamma_j^{k\ell} + \Gamma_k^{\ell j} + \Gamma_\ell^{jk} = 0$. Deduce that Ω is closed.

3. Let G be a Lie group with a nondegenerate metric which is invariant under left and *right* translation. Show that the image of every geodesic on T^*G under the bundle projection is a one parameter subgroup of G.

4. Find the Euler-Arnol'd equations for geodesics of a left invariant metric on the Lie groups $\mathrm{Sl}_2(\mathbf{R})$ and S^3.

Appendix B

Systems with symmetry

In this appendix we discuss Hamiltonian systems with symmetry. By a symmetry of a Hamiltonian system (H, M, ω) we mean an action of a Lie group G on a symplectic manifold (M, ω) which has a momentum mapping $J : M \rightarrow \mathcal{G}^*$ and preserves the Hamiltonian H. We will show that symmetries of a Hamiltonian system give rise to conserved quantities which are constant along integral curves of the Hamiltonian vector field X_H of H. Using the technique of singular reduction we remove a proper G symmetry of (H, M, ω) by constructing a lower dimensional Hamiltonian system $(H_\mu, M_\mu, \{\,,\,\}_\mu)$ for *each* μ in the image of J for which the coadjoint orbit $\mathcal{O}_\mu$ is locally closed.

1 Smooth group actions

In this section we treat some basic properties of smooth group actions on manifolds. We discuss in detail the properties of proper actions.

Let G be a *Lie group*, that is, G is a smooth manifold which is a group such that multiplication is a smooth map. A (left) *action* Φ of G on a smooth manifold M is a smooth mapping

$$\Phi : G \times M \rightarrow M : (g, m) \rightarrow \Phi(g, m) = \Phi_g(m) = \Phi_m(g) \tag{1}$$

such that for every $g, h \in G$ and every $m \in M$ we have $\Phi_{gh}(m) = \Phi_g(\Phi_h(m))$, while for the identity element e in G we have $\Phi_e(m) = m$. A succinct way of expressing these conditions is to say that the mapping

$$G \rightarrow \mathrm{Diff}(M) : g \rightarrow \Phi_g$$

is a homomorphism of G into the group of diffeomorphisms of M.

For $m \in M$ let $\mathcal{O}_m$ be the *orbit* of m under the action of G, that is,

$$\mathcal{O}_m = \left\{ \Phi_g(m) \middle| \; g \in G \right\}.$$

Sometimes, when it is convenient, and not at all confusing, we will write $G \cdot m = \mathcal{O}_m$. Let $\mathcal{G}$ be the Lie algebra of G. Then $T_e\Phi_m(\mathcal{G})$ is the tangent space to the orbit $G \cdot m$ at m.

Denote the *isotropy group* of m by

$$G_m = \left\{ g \in G \middle| \Phi(g, m) = m \right\}.$$

Note that G_m is a Lie group because it is a closed subgroup of G. The Lie algebra of G_m is $\mathcal{G}_m$.

Claim: Let L be a submanifold of G through e such that

$$\mathcal{G} = \mathcal{G}_m \oplus T_e L \tag{2}$$

and let S be a submanifold of M through m such that

$$T_m M = T_m \mathcal{O}_m \oplus T_m S. \tag{3}$$

Then there is an open neighborhood $L_0 \times S_0$ of (e, m) in $L \times S$ such that $\Phi|(L_0 \times S_0)$, the restriction of the action Φ to $L_0 \times S_0$, is a diffeomorphism onto an open neighborhood of m in M.

(1.1) **Proof**: For $(\xi, v) \in T_e L \times T_m S$

$$\begin{aligned} T_{(e,m)}(\Phi|(L \times S))(\xi, v) &= T_e \Phi_m \xi + T_m \Phi_e v, \\ &\qquad \text{by the formula for partial derivatives} \\ &= T_e \Phi_m \xi + v, \quad \text{since } \Phi_e \text{ is the identity map.} \end{aligned}$$

Now $T_{(e,m)}\Phi|(L \times S)$ is injective; for if $0 = T_e \Phi_m \xi + v$, then $T_m \mathcal{O}_m \cap T_m S = \{0\}$ (which follows from (3)) implies that $v = 0$ and thus $T_e \Phi_m \xi = 0$. But by (2) we have $\ker T_e \Phi_m \cap T_e L = \{0\}$. Hence $\xi = 0$. Moreover $T_{(e,m)}\Phi|(L \times S)$ is surjective because

$$T_m M = T_m \mathcal{O}_m + T_m S = T_e \Phi_m \, T_e L + T_m S.$$

Applying the inverse function theorem concludes the argument. □

An immediate consequence of the above result is the following.

Corollary: If $\Phi_\ell(S_0) \cap S_0 \neq \varnothing$ for some $\ell \in L_0$, then $\ell = e$.

(1.2) **Proof**: By hypothesis, there is an $s \in S_0$ such that $\Phi_\ell(s) = s' \in S_0$. But then $\Phi_e(s') = \Phi_\ell(s)$. In other words, $\Phi(\ell, s) = \Phi(e, s')$. Since Φ is a local diffeomorphism on $L_0 \times S_0$, we obtain $\ell = e$. □

In order to insure that orbit spaces, treated in the following section, are reasonably well behaved, some condition must be imposed on the action Φ. A condition which works well is that the action Φ be *proper*, that is, the map

$$G \times M \to M \times M : (g, m) \to (m, g \cdot m)$$

is a *proper map*. As G and M are both manifolds, this means that the inverse image of a compact set under this map is compact. One might think that it would be more natural to require that the map

$$G \times M \to M : (g, m) \to g \cdot m = \Phi(g, m)$$

be proper. But what we need is a condition which insures that nearby points in M may only be mapped into each other by elements of G which are close to the identity. Actually, properness is somewhat stronger than this. For more details see exercise 3.

Claim: If the G action is proper, then the isotropy group G_m is compact.

(1.3) **Proof**: This follows immediately by looking at the inverse image of (m, m). □

A nice feature of a proper action of a Lie group near a fixed point is that it is locally linearizable. In particular, we have

Claim: For any point $m \in M$ there is a G_m-invariant neighborhood U of $m \in M$ and a diffeomorphism $\psi : U \subseteq M \to T_m M$ such that

1. ψ maps the point m to the origin and its tangent at m is the identity.
2. ψ is G_m-equivariant, that is, for every $g \in G_m$ and $p \in U$ we have

$$\psi(\Phi_g(p)) = T_m\Phi_g(\psi(p)).$$

In other words, the diffeomorphism ψ locally identifies the G_m-action on M near m with a linear action of G_m on the vector space $T_m M$.

(1.4) **Proof**: For every chart $(\varphi, \widetilde{U})$ on M with $\varphi(m) = 0$ there is a G_m-invariant open set U containing m with $U \subseteq \widetilde{U}$, (see exercise 2). Working in this chart we may identify the image of the chart map with the tangent space to M at m.

We are reduced to the following situation. There is a compact group G (= isotropy group of 0) acting (not necessarily linearly) on $\mathbf{R}^n$ and the origin is a fixed point. For $p \in \mathbf{R}^n$ define a map $\psi : U \subseteq \mathbf{R}^n \to U' \subseteq \mathbf{R}^n$ by

$$\psi(p) = \int_G D_2\Phi(g^{-1}, 0) \cdot \Phi(g, p)\, dg$$

with dg being Haar measure on G with $\mathrm{vol}(G) = 1$. In other words, dg is a bi-invariant volume form on G. Now for any $h \in G$ and $p \in U$

$$\begin{aligned}
D_2\Phi(h, 0) \cdot \psi(p) &= \int_G D_2\Phi(h, 0) \cdot D_2\Phi(g^{-1}, 0) \cdot \Phi(g, p)\, dg \\
&= \int_G D_2\Phi(hg^{-1}, 0) \cdot \Phi(g, p)\, dg \\
&= \int_G D_2\Phi(k^{-1}, 0) \cdot \Phi(kh, p)\, d(kh), \\
&\qquad \text{changing integration variable by } g = kh \\
&= \int_G D_2\Phi(k^{-1}, 0) \cdot \Phi(kh, p)\, dk \\
&= \int_G D_2\Phi(k^{-1}, 0) \cdot \Phi(k, \Phi(h, p))\, dk \\
&= \psi(\Phi(h, p)).
\end{aligned}$$

Thus we have produced a map ψ which intertwines the action of G with an action by linear maps. Since

$$T_0\psi = \int_G D_2\Phi(g^{-1},0)\, D_2\Phi(g,0)\, dg = id,$$

the map ψ is a local diffeomorphism. □

Corollary 1. Let H be a compact subgroup of G and let

$$M_{[H]} = \left\{ m \in M \middle| H \subseteq G_m \right\}. \tag{4}$$

Then every connected component of $M_{[H]}$ is a smooth submanifold of M. Moreover, the tangent space $T_pM_{[H]}$ at $p \in M_{[H]}$ is the set $(T_pM)^H$ of H-fixed vectors of T_pM, that is,

$$(T_pM)^H = \left\{ v_p \in T_pM \middle| T_p\Phi_h v_p = v_p, \ \text{ for every } h \in H \right\}.$$

(1.5) **Proof**: Let V be an open neighborhood of 0 in T_pM which is mapped diffeomorphically onto the open neighborhood U of p in M by ψ^{-1}. The map ψ^{-1} intertwines the linear H-action

$$\widehat{\Phi} : H \times T_pM \to T_pM : (h, v_p) \to T_p\Phi_h v_p$$

with the H-action $\widetilde{\Phi} = \Phi|(H \times M)$. Since $(T_pM)^H$ is clearly a vector subspace of T_pM, to show that $M_{[H]}$ is a submanifold of M it suffices to verify that

$$\psi^{-1}\big((T_pM)^H \cap V\big) = U \cap M_{[H]}. \tag{5}$$

If $v \in (T_pM)^H \cap V$, then $m = \psi^{-1}(v) \in U$ and $\widehat{\Phi}_h(v) = v$ for every $h \in H$. Since ψ^{-1} intertwines the H-actions $\widehat{\Phi}$ and $\widetilde{\Phi}$, it follows that $\Phi_h(m) = m$ for every $h \in H$, that is, $H \subseteq G_m$. So $m \in M_{[H]}$. Therefore $\psi^{-1}\big((T_pM)^H \cap V\big) \subseteq U \cap M_{[H]}$. Now suppose that $m \in U \cap M_{[H]}$. Then $\psi(m) = v \in V$ and $\Phi_h(m) = m$ for every $h \in H$. Since ψ intertwines the H-actions $\widetilde{\Phi}$ and $\widehat{\Phi}$, we obtain $\widehat{\Phi}_h(v) = v$ for every $h \in H$, that is, $v \in (T_pM)^H$. Therefore $\psi(U \cap M_{[H]}) \subseteq (T_pM)^H \cap V$. This proves (5) and the corollary. □

An important property of proper actions is the existence of a slice. A *slice* at m for the action Φ is a smooth submanifold S of M through m such that

1. S is transverse and complementary to the orbit $\mathcal{O}_m$ of m at the point m, that is,

$$T_mM = T_m\mathcal{O}_m \oplus T_mS.$$

2. For every $p \in S$, S is transverse to $\mathcal{O}_p$, that is,

$$T_pM = T_p\mathcal{O}_p + T_pS.$$

3. S is G_m-invariant.

4. For $p \in S$ and $g \in G$, if $\Phi_g(p) \in S$ then $g \in G_m$.

▷ We now show that slices exist for proper group actions.

(1.6) **Proof**: We begin by constructing a candidate slice S_ε and then show that properties 1) through 4) hold.

For every $k \in G_m$

$$\Phi_k \circ \Phi_g(m) = \Phi_k \circ \Phi_g \circ \Phi_{k^{-1}}(m) = \Phi_{kgk^{-1}}(m).$$

Therefore for every $\xi \in \mathcal{G}$

$$T_m\Phi_k \circ T_e\Phi_m\xi = \left.\frac{d}{dt}\right|_{t=0} \Phi_k\big(\Phi_m(\exp t\xi)\big) = \left.\frac{d}{dt}\right|_{t=0} \Phi_{k(\exp t\xi)k^{-1}}(m).$$

Since the one parameter subgroups $t \to k(\exp t\xi)k^{-1}$ and $t \to \exp t\,\mathrm{Ad}_k\,\xi$ have the same tangent vector $\mathrm{Ad}_k\,\xi$ at $t = 0$, we obtain $T_m\Phi_k \circ T_e\Phi_m\xi = T_e\Phi_m(\mathrm{Ad}_k\,\xi)$. This implies that $T_m\Phi_k$ leaves $T_m\mathcal{O}_m$ invariant. With respect to a G_m-invariant positive definite inner product on T_mM, the orthogonal complement $(T_m\mathcal{O}_m)^\perp$ is a G_m-invariant subspace. Using the local linearizing diffeomorphism ψ of ((1.4)) the submanifold $S_\varepsilon = \psi^{-1}\big((T_m\mathcal{O}_m)^\perp \cap B_\varepsilon\big)$, where B_ε is a ball of radius ε in T_mM, is G_m-invariant. Hence S_ε satisfies property 3). Moreover, $T_mS_\varepsilon = (T_m\mathcal{O}_m)^\perp$. Hence S_ε has property 1). Property 2) is an open condition in S_ε as we see from the following argument. Consider the map

$$\widetilde{\Phi} : G \times S \to M : (g, s) \to \Phi_g(s).$$

Then $T_{(e,s)}\widetilde{\Phi}(\mathcal{G} \times T_sS) = T_s\mathcal{O}_s + T_sS$. By property 1) we see that $T_{(e,m)}\widetilde{\Phi}$ is surjective. This is an open condition on S, that is, for every $s \in S$ near m, the map $T_{(e,s)}\widetilde{\Phi}$ is surjective. Hence $T_sM = T_s\mathcal{O}_s + T_sS$. Thus property 2) holds on S_ε.

Now suppose that property 4) does not hold for any $\varepsilon > 0$. Then there is a sequence $\{m_j\}$ with $m_j \in S_{1/j}$, $m_j \to m$, and a sequence $g_j \notin G_m$ with $g_j \cdot m_j \in S_{1/j}$ (which implies $g_j \cdot m \to m$). By properness, there is a subsequence $g_{j_k} \to g$, which we may assume is just g_j itself. Replacing g_j by $g^{-1}g_j$ we may suppose that $\{g_j\}$ converges to e, but $g_j \notin G_m$. Choosing H to be a submanifold of G which is transverse to G_m, there is an open neighborhood $V \times W$ of (e, e) in $H \times G_m$ such that the multiplication $(h, k) \to hk$ is a diffeomorphism onto an open neighborhood of e in G. Thus we may assume that $g_j = h_jk_j$. As $g_j \notin G_m$ and $k_j \in G_m$, we conclude $h_j \neq e$ for all j. Arguing as above once more with $S_{1/j}$ in place of H, we obtain

$$g_j \cdot m_j = h_j \cdot k_j \cdot m_j \in S_{1/j}.$$

Because $S_{1/j}$ is G_m-invariant, $k_j \cdot m_j \in S_{1/j}$. But ((1.2)) implies that $h_j = e$. This contradiction establishes there is an $\varepsilon > 0$ such that property 4) holds. Hence S_ε is a slice. □

We now discuss some consequences of the existence of a slice for a proper G-action Φ. Let

$$M_H = \Big\{ m \in M \,\Big|\, G_m = H \Big\}. \tag{6}$$

From the definition of $M_{[H]}$ (4) we see that $M_H \subseteq M_{[H]}$.

Claim. Give $M_{[H]}$ the topology induced from that of M. Then M_H is an open subset of $M_{[H]}$. Hence every connected component of M_H is a submanifold of M.

(1.7) **Proof**: Suppose that $p \in M_H$. Let S_p be a slice to the G-action at p. The map

$$\varphi : S_p \times \mathcal{L} \to M : (s, \eta) \to \Phi_{\exp \eta}(s)$$

with $\varphi(p, 0) = p$ and $\mathcal{L}$ a complement to $\mathcal{H}$ in $\mathcal{G}$ is a diffeomorphism of an open neighborhood W of $(p, 0)$ in $S_p \times \mathcal{L}$ onto an open neighborhood U of p in M. Let $q \in M_{[H]} \cap U$. Then there is a $g \in \exp \mathcal{L}$ and an $s \in S_p$ such that $\Phi_g(s) = q$. Therefore

$$g^{-1} G_q \, g = G_s \subseteq H, \tag{7}$$

where the inclusion follows because S_p is a slice. But $q \in M_{[H]}$. Therefore

$$H \subseteq G_q \subseteq g H g^{-1}. \tag{8}$$

Since H and gHg^{-1} have the same Lie algebras, it follows that $H = gHg^{-1}$ if H is connected. Because the map $H \to gHg^{-1}$ is a diffeomorphism, H and gHg^{-1} have the same number of connected components, which is finite since H is compact. From (8) it follows that every connected component of H is a connected component of gHg^{-1}. Therefore H and gHg^{-1} have the same connected components. Hence $H = gHg^{-1}$, which implies that $G_q = H$ using (8). Therefore, $q \in M_H \cap U$, that is, $M_{[H]} \cap U \subseteq M_H \cap U$. But $M_H \subseteq M_{[H]}$. So $M_{[H]} \cap U = M_H \cap U$. □

Let

$$M_{(H)} = G \cdot M_H. \tag{9}$$

Then $M_{(H)}$ is the set of points of M of *orbit type* (H). From the definition of M_H it follows that

$$M_{(H)} = \left\{ m \in M \,\middle|\, \text{there is a } g \in G \text{ such that } gG_m g^{-1} = H \right\}.$$

The following argument shows that $M_{(H)}$ is a submanifold of M.

(1.8) **Proof**: Let $m \in M_{(H)}$. Since $M_{(H)} = G \cdot M_H$, there is a $p \in M_H$ and a $g \in G$ such that $\Phi_g(p) = m$. Because Φ_g is a diffeomorphism, it suffices to show that $M_{(H)}$ is a submanifold through p.

Let $\mathcal{L}$ be a complement to $\mathcal{H}$ in $\mathcal{G}$. Let S_p be a slice to the G-action Φ at p. Then the map

$$\varphi : S_p \times \mathcal{L} \to M : (s, \eta) \to \Phi_{\exp \eta}(s)$$

with $\varphi(0, p) = p$ is a local diffeomorphism at p. Since

$$\vartheta = \psi \times \mathrm{id}_{\mathcal{L}} : T_p S_p \times \mathcal{L} \to S_p \times \mathcal{L}$$

with $\vartheta(0, 0) = (0, p)$ is a local diffeomorphism, where ψ^{-1} is the local diffeomorphism which intertwines the H-action $\widehat{\Phi}$ on $T_p M$ with the H-action $\widetilde{\Phi} = \Phi|(H \times M)$ on M (see ((1.4))), the map

$$\theta : T_p S_p \times \mathcal{L} \to M : (v_p, \eta) \to \Phi_{\exp \eta}(\psi^{-1}(v_p))$$

with $\theta(0,0) = p$ is a local diffeomorphism of an open neighborhood $W \times V \subseteq T_pS_p \times \mathcal{L}$ of $(0,0)$ onto an open neighborhood $U = \theta(W \times V)$ of p in M. Now

$$Y = \psi(M_H \cap S_p \cap U) = \psi(M_{[H]} \cap S_p \cap U)$$

is the intersection of an open neighborhood of 0 in T_pS_p with $T_p(M_{[H]} \cap S_p)$. Since $G \cdot M_H = M_{(H)}$, we obtain

$$\theta(W \times Y) = \Phi_{\exp W}(M_H \cap S_p \cap U) \cap U = M_{(H)} \cap U.$$

Therefore $M_{(H)}$ is a submanifold of M. □

On the set of all closed subgroups of G define a relation $\asymp$ by saying that $K \asymp H$ if and only if K is conjugate to H in G. Clearly $\asymp$ is an equivalence relation on the set of all closed subgroups of G. Thus as H runs over the set of conjugacy classes of closed ▷ subgroups of G, the orbit types $M_{(H)}$ partition M. The *orbit type decomposition* of M is locally finite if the action of G is proper.

(1.9) **Proof**: We prove the local finiteness of the orbit type decomposition by induction on the dimension of the group G and the dimension of the manifold M. If G is a finite group acting on a manifold, the number of conjugacy classes of subgroups is finite. Hence the orbit type decomposition is finite. Suppose that M is a finite number of points, then the orbit type decompostition is finite.

With these special cases out of the way, we can begin the induction argument. Let $G \cdot m$ be the G orbit through m of the proper group action Φ on M. Since Φ is proper there is a G_m-invariant slice S through the point m. Note that the dimension of the isotropy group G_m is at most equal to the dimension of G. Because G_m is compact and leaves m fixed, the G_m-action on S is locally equivalent to the linear isotropy action

$$\widehat{\Phi} : G_m \times T_mS \to T_mS.$$

Suppose that the action $\widehat{\Phi}$ is trivial, that is, $\widehat{\Phi}_h(v_m) = v_m$ for every $h \in G_m$ and every $v_m \in T_mS$. In other words, if $p \in S$ then the isotropy group of a point p on the G-orbit is equal to G_m. Since every G-orbit near $G \cdot m$ intersects S, the isotropy group is conjugate in G to G_m, that is, it lies in the equivalence class of the orbit type decomposition corresponding to (G_m). Hence the orbit type decomposition near $G \cdot m$ is locally finite.

Now suppose that the action $\widehat{\Phi}$ is not trivial. Since G_m is compact, there is a G_m-invariant Euclidean inner product $\langle\, , \,\rangle$ on T_mS. Let $\mathcal{S}_r$ be the sphere of radius r in $(T_mS, \langle\, , \,\rangle)$. Note that the dimension of $\mathcal{S}_r$ is strictly less than the dimension of M. Since the action $\widehat{\Phi}$ is linear, it induces a proper action $\Psi : G_m \times \mathcal{S}_r \to \mathcal{S}_r$.

Repeating the above argument on the G_m-action Ψ either reduces the dimension of the group or the dimension of the manifold on which it acts. This completes the induction step and hence the proof. □

2 Orbit spaces

In this section we describe the space of orbits M/G of a proper group action G on a smooth manifold M. In general, M/G is only a topological space. It need not be a topological manifold. If in addition we assume that the action is free, that is, the isotropy group at each point is the identity element, then the orbit space is a smooth manifold. If we assume that the action is only locally free, that is, the isotropy group at each point is a finite group, then the orbit space is a mildly singular manifold called an orbifold.

2.1 Orbit space of a proper action

In this subsection we define the orbit space of a Lie group acting properly on a smooth manifold and give some of its properties.

Let $\Phi : G \times M \to M$ be an action of a Lie group G on a smooth manifold M. We begin by constructing the orbit space M/G set theoretically. Let $\overline{m} = G \cdot m$ be the orbit of the action Φ through the point $m \in M$. Define the relation $\sim$ on M by saying that $m \sim m'$ if and only if m and m' lie on the same G-orbit of Φ. It is easy to check that $\sim$ is an equivalence relation on M and thus partitions M into G-orbits. The *orbit space* M/G is the set of all G-orbits of Φ on M. Thus the *orbit map*

$$\pi : M \to M/G : m \to \overline{m} \tag{10}$$

is surjective.

We put a topology on M/G by saying that $\mathcal{U} \subseteq M/G$ is *open* if and only if $\pi^{-1}(\mathcal{U})$ is an open subset of M. Let S be a slice at m to the G-action Φ. Then S is transverse to ▷ the G-orbit $G \cdot m$ at m. For a suitable open neighborhood $S_0 \subseteq S$ of m, the set $\pi(S_0)$ is an open subset of M/G.

(2.1) **Proof**: It suffices to show that the union of all G-orbits through S_0, namely, $\pi^{-1}(\pi(S_0)) = G \cdot S_0$, is an open subset of M. From ((1.1)) it follows that there is an open neighborhood G_0 of e in G and a G_m-invariant open neighborhood S_0 in the slice S at m such that $\Phi_{G_0}(S_0)$ is an open neighborhood of m in M. Since G is the union of translates $\{g \cdot G_0 | \, g \in G\}$, the set $G \cdot S_0$ is the union of open sets $\Phi_{g \cdot G_0}(S_0)$ and hence is open. □

▷ Let S be a slice for a proper G-action at the point m in M. Then the orbit spaces $(G \cdot S)/G$ and S/G_m are homeomorphic.

(2.2) **Proof**: Consider the map

$$\varphi : (G \cdot S)/G \to S/G_m : G \cdot s \to (G \cdot s) \cap S.$$

By property 4) of a slice, $(G \cdot s) \cap S = G_m \cdot s$. Hence φ is well defined. Because $G_m \subseteq G$, it follows that the G_m orbit $G_m \cdot s$ is contained in the G-orbit $G \cdot s$. Hence the map

$$\vartheta : S/G_m \to (G \cdot S)/G : G_m \cdot s \to G \cdot s$$

is well defined. It is easy to see that $\vartheta \circ \varphi = id_{(G \cdot S)/G}$ and $\varphi \circ \vartheta = id_{S/G_m}$. Therefore φ is bijective.

We now show that φ is a homeomorphism. Let $\rho : S \to S/G_m$ be the orbit map for the G_m-action on S and let $\pi : (G \cdot S) \to (G \cdot S)/G$ be the orbit map for the G-action on $G \cdot S$. Let $\mathcal{U}$ be an open set in S/G_m containing the orbit $G_m \cdot s$. Then $\rho^{-1}(\mathcal{U}) = U$ is an open set in S containing $G_m \cdot s$. Since $G \cdot U$ is an open set in $G \cdot S$, $\pi(G \cdot U)$ is an open set in $(G \cdot S)/G$ containing $G \cdot s$. Thus φ is continuous. In fact φ is an open map. For suppose that $\mathcal{V}$ is an open set in $(G \cdot S)/G$ containing $G \cdot s$. Then $\pi^{-1}(\mathcal{V}) = V$ is an open set in $G \cdot S$. Hence $U = V \cap S$ is an open set in S containing $G_m \cdot s$. Therefore $\rho(U)$ is an open set in S/G_m. Hence φ is a homeomorphism. □

Claim: If

$$\left\{ (m, \Phi_g(m)) \in M \times M \middle| g \in G \text{ and } m \in M \right\}$$

is closed, then the orbit space M/G is Hausdorff.

(2.3) **Proof**: Suppose that M/G is not Hausdorff, that is, $\overline{m} \neq \overline{m}'$ cannot be separated. Let U_k and V_k be nested neighborhood bases of m and m' respectively. As $\overline{m}$ and $\overline{m}'$ cannot be separated, there is a $p_k \in (G \cdot U_k) \cap (G \cdot V_k)$ for each k. Set

$$p_k = g_k \cdot m_k = h_k \cdot m'_k,$$

where $m_k \in U_k$ and $m'_k \in V_k$. Then $h_k^{-1} g_k m_k \to m'$ and consequently the sequence $\{(m_k, h_k^{-1} g_k m_k)\}$ converges to (m, m'). By hypothesis $m' = \Phi_g(m)$ for some $g \in G$. This implies $\overline{m} = \overline{m}'$, which is a contradiction. □

Corollary: If the G-action Φ is proper, then M/G is Hausdorff.

(2.4) **Proof**: Let $\{(m_k, \Phi_{g_k}(m_k))\}$ be a sequence of points in $M \times M$ which converges to (m, n). By properness of the action there is a subsequence $\{(g_{k_\ell}, m_{k_\ell})\}$ which converges to (g, m) in $G \times M$. By continuity of Φ, $g_{k_\ell} \cdot m_{k_\ell} \to n = g \cdot m$. This implies that $\{(m, g \cdot m)\} \subseteq M \times M$ is a closed set. Hence by the claim, M/G is Hausdorff. □

Example 1. Let H be a closed subgroup of G. Suppose that $(g_k, h_k g_k) \to (g, j)$ in $G \times G$. Then $h_k = (h_k g_k) g_k^{-1} \to j g^{-1}$ since G is a Lie group. Thus the action of the closed subgroup H on G is proper. Consequently, the orbit space G/H is Hausdorff.

Claim: Suppose that G acts properly on M, and that $\{U_\alpha\}$ is a G-invariant open covering of M. Then there is a smooth G-invariant partition of unity subordinate to $\{U_\alpha\}$.

Before giving the proof, it is worthwhile to note two things. First, if we were only concerned with continuous functions, we could argue as follows. As the orbit space M/G is a closed continuous image of a paracompact space, and hence paracompact, we may construct a partition of unity on M/G subordinate to the image of $\{U_\alpha\}$ under the orbit map. Pulling this back to M does the job. Second, we cannot extend this topological result to a smooth one because M/G need *not* be a manifold. With these remarks in mind, we can understand why the proof of the claim proceeds the way it does.

(2.5) **Proof**: As M is locally compact and σ-compact, and M/G is Hausdorff, it follows that M/G is locally compact and paracompact, and hence normal. We can choose a set of points m_k with slices S_k through m_k so that $\{G \cdot S_k\}$ is a locally finite open covering of M subordinate to $\{U_\alpha\}$ and $\pi(S_k)$ is a locally finite open covering of M/G. The local compactness of M/G implies that the slices S_k may be chosen in such a way that there are relatively compact sets $V_k \subseteq S_k$ so that $\{\pi(V_k)\}$ also cover M/G. On each S_k we may construct a smooth function $\widetilde{f_k}$ which is positive on V_k and has compact support in S_k. Define $\overline{f}_k$ to be the average of $\widetilde{f_k}$ over the compact isotropy group G_{m_k}. Let f_k be $f_k(g \cdot m) = \overline{f}_k(m)$ if $m \in S_k$, otherwise $= 0$. Then f_k is a smooth nonnegative function on M. The proof is completed by setting $h_k = f_k / \sum_k f_k$. □

Corollary: Suppose that N is a G-invariant subset of M and f is a smooth function on M such that $f|N$ is constant on G-orbits. Then there is a smooth G-invariant function F on M such that $F|N = f|N$.

(2.6) **Proof**: First we define a function F_k on each $G \cdot S_k$ by the following rule. If $N \cap S_k = \varnothing$, then set $F_k = 0$. Otherwise, set

$$F_k|S_k = \text{average of } f|S_k \text{ over } G_{m_k}.$$

Note that $F_k|(S_k \cap N) = f|(S_k \cap N)$. Extend $F_k|S_k$ to $G \cdot S_k$ by the G-action. Letting $F = \sum_k h_k F_k$ where $\{h_k\}$ is a G-invariant partition of unity subordinate to $\{G \cdot S_k\}$ does the job. □

Corollary: Smooth G-invariant functions on M separate orbits, that is, if m and m' are on distinct G-orbits of M, then there is a smooth G-invariant function F such that $F|\mathcal{O}_m = 1$ and $F|\mathcal{O}_{m'} = 0$.

(2.7) **Proof**: Let S_m be a slice to the G-action at m. Suppose that the G-orbit $G \cdot m'$ does not intersect S_m. Then let $N = G \cdot S_m$ and apply the proof of the above corollary to the function f which is identically 1 on $\mathcal{O}_m = G \cdot m$. This gives a smooth function F on M which is 1 on $\mathcal{O}_m$ and 0 on $\mathcal{O}_{m'}$. Now suppose that $q \in \mathcal{I} = (G \cdot m') \cap S_m$ at q. Then $q \neq m$ since the G-orbits $G \cdot m$ and $G \cdot m'$ are distinct. By property 4) of a slice, $\mathcal{I}$ is just $G_m \cdot q$, which is compact and does not contain m. Hence there is an open set U in S_m which contains m but does not intersect $\mathcal{I}$. Again apply the proof of the above corollary taking $N = G \cdot U$ and letting f be identically equal to 1 on $G \cdot m$. This gives a smooth function F on M which is 1 on $\mathcal{O}_m$ and 0 on $\mathcal{O}_{m'}$. □

2.2 Orbit space of a free action

In this subsection we show that the orbit space of a proper free group action is a smooth manifold. In fact, it is the base of a principal bundle whose total space is the original manifold.

Before proving the result, we pause to consider a germane example: the quotient of a Lie group by a closed subgroup.

Example 1. Let H be a closed subgroup of a Lie group G. Then the orbit space $G/H = \{Hg \mid g \in G\}$ is a smooth manifold.

(2.8) **Proof**: Choose $\mathcal{L} \subseteq \mathcal{G}$ such that $\mathcal{G} = \mathcal{H} \oplus \mathcal{L}$. Then there are open neighborhoods $U^{\mathcal{H}}$ and

$U^{\mathcal{L}}$ of 0 in $\mathcal{H}$ and $\mathcal{L}$ such that

$$\varphi : U^{\mathcal{H}} \times U^{\mathcal{L}} \to G : (h, \ell) \to \exp h \exp \ell$$

with $\varphi(0,0) = e$ is a local diffeomorphism. Since H is a closed subgroup of G, there is an open neighborhood V of e in G such that $V \cap H = \exp U^{\mathcal{H}}$. Let U be an open neighborhood of 0 in $U^{\mathcal{L}}$ with compact closure $\overline{U}$ satisfying $\exp(\overline{U}) \exp(-\overline{U}) \subseteq V$. To show that the H-orbit mapping $\pi : \exp(\overline{U}) \subseteq G \to \pi(\exp(\overline{U})) \subseteq G/H$ is a homeomorphism, it suffices to show that π is one to one, since $\pi(\exp(\overline{U}))$ is compact and Hausdorff. If $\pi(\exp u) = \pi(\exp u')$ for some $u, u' \in \overline{U}$, then there is an $h \in U^{\mathcal{H}}$ such that

$$\exp h \exp u = \exp u' = \exp 0 \exp u'.$$

Because φ is a local diffeomorphism, it follows that $u = u'$ and $h = 0$. Hence $\pi \,|\, \exp(\overline{U})$ is a homeomorphism. From the definition of π it follows that the set $\pi(\exp U^{\mathcal{H}} \exp U) = \pi(\exp U)$ is open and contains $\overline{e} = \pi(e)$ in G/H. Exactly the same argument applies to the map $\varphi_g = R_g \circ \varphi$, where $R_g : G \to G : h \to hg$. This allows us to conclude that every $\overline{g} = \pi(g)$ in G/H has an open neighborhood homeomorphic to an open set in $\mathcal{L}$. Consequently G/H is a topological manifold.

To show that G/H is a smooth manifold, we construct charts as follows. Let $\mathcal{U}_g \subseteq G/H$ be a set such that

$$\mathcal{U}_g = \pi(\exp U\, g) = \overline{e} \exp U\, g.$$

Note that G acts on the *right* on G/H by $(Hg, k) \to Hgk$. Define the mapping ψ_g by

$$\psi_g : \mathcal{U}_g \subseteq G/H \to U \subseteq \mathcal{L} : \overline{e} \exp u\, g \to u.$$

The proof is complete once we have shown that $\{(\mathcal{U}_g, \psi_g) \,|\, g \in G\}$ form an atlas for G/H with smooth transition functions. To do this form the local section maps

$$\sigma_g : \mathcal{U}_g \to G : \overline{e} \exp u\, g \to \exp u\, g.$$

Clearly, $\pi \circ \sigma_g = id_{\mathcal{U}_g}$. So $\sigma_g(\mathcal{U}_g) \subseteq \pi^{-1}(\mathcal{U}_g)$. Furthermore,

$$\sigma_g \circ \psi_g^{-1} : U \to G : u \to \exp u\, g$$

is a diffeomorphism onto its image $\sigma_g(\mathcal{U}_g)$ in G. Now $\psi_g = (\psi_g \circ \sigma_g^{-1}) \circ \sigma_g$. Since $\psi_g \circ \sigma_g^{-1}$ is a diffeomorphism and σ_g is a homeomorphism, we conclude that ψ_g is a homeomorphism.

To prove the smoothness of the transition function

$$\psi_{g'} \circ \psi_g^{-1} : \psi_g(\mathcal{U}_g \cap \mathcal{U}_{g'}) \to \psi_g'(\mathcal{U}_g \cap \mathcal{U}_{g'})$$

we argue as follows. Let $\overline{k} \in \mathcal{U}_g \cap \mathcal{U}_{g'}$. Then there is a unique element $h \in H$ such that $hk = k'$, where $k = \sigma_g(\overline{k})$ and $k' = \sigma_{g'}(\overline{k})$. Since multiplication by h is a smooth map, $h^{-1} \cdot \sigma_{g'}(\mathcal{U}_{g'})$ is a smooth submanifold of G containing k and is transverse to Hk. Using the local diffeomorphism

$$\widetilde{\varphi}_k : \exp U^{\mathcal{H}} \times U \to V : (h, \ell) \to (h \exp \ell)k,$$

we see that the image of the submanifold $h^{-1} \cdot \sigma_{g'}(\mathcal{U}_{g'})$ under $\widetilde{\varphi}_k^{-1}$ is the graph of a smooth mapping $\rho : U \subseteq U^{\mathcal{L}} \to H$. Therefore the transition mapping is

$$\psi_{g'} \circ \psi_g(u) = \pi_2 \circ \widetilde{\varphi}_{k'}^{-1}\big(h\widetilde{\varphi}_k(\rho(u), u)\big),$$

which is smooth. □

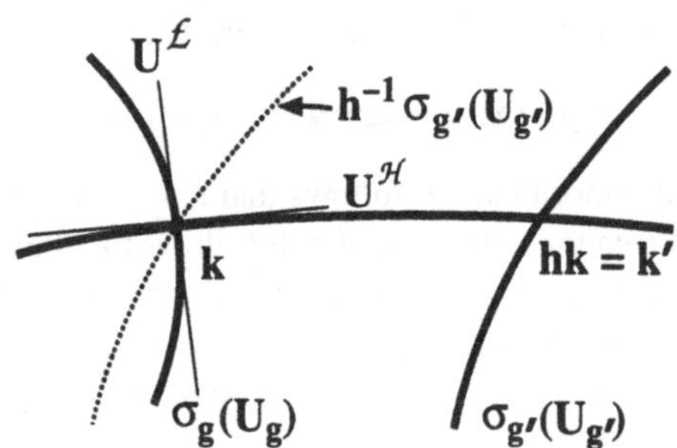

Figure 2.1. The coordinate transition map.

It may help to think of this proof in the following way. In order to show that the transition map $\psi_{g'} \circ \psi_g^{-1}$ is a diffeomorphism we have by design made it a homeomorphism. The multiplication by h is a technical device to get things in a single chart so that we can deduce that the transition map is a local diffeomorphism. A homeomorphism which is everywhere a local diffeomorphism is a diffeomorphism.

In general, to insure that the orbit space M/G of a G-action Φ is a smooth manifold, we need to put very strong conditions on the action. For example, one might hope that if the dimension of each orbit $G \cdot m$ is equal to the dimension of the group G, in particular there are no fixed points, and G acts properly, then M/G is a smooth manifold. Unfortunately this need *not* be true, (see section 4 example 3). Recall that a G-action on M is *free* if and only if for every $m \in M$ the isotropy group G_m is the identity element of G. We now prove

Claim: If the action $\Phi : G \times M \to M$ is free and proper, then the orbit space M/G is a smooth manifold. Moreover, for every $\overline{m} \in M/G$ there is an open neighborhood $\mathcal{U}$ of $\overline{m}$ in M/G and a diffeomorphism

$$\psi_U : U = \pi^{-1}(\mathcal{U}) \subseteq M \to G \times S : m \to \big(\vartheta_U(m), s_U(m)\big) \tag{11}$$

such that for every $g \in G$ and $m \in U$

$$\psi_U(g \cdot m) = \big(g\vartheta_U(m), s_U(m)\big). \tag{12}$$

(2.9) **Proof:**

Step 1

Our first task is construct the diffeomorphisms ψ_U, because these maps will allow us to put a smooth structure on M/G. To do this, let $\widetilde{S}$ be a smooth submanifold through $m \in M$

such that

$$T_m\widetilde{S} \oplus T_e\Phi_m\mathcal{G} = T_m M.$$

As this is an open condition in $\widetilde{S}$, there is an open neighborhood $S \subseteq \widetilde{S}$ about m such that

$$T_s S \oplus T_e\Phi_s\mathcal{G} = T_s M \tag{13}$$

for each $s \in S$.

Claim: For each $s \in S$ and $g \in G$, the action $\Phi|(G \times S)$ has a bijective tangent map at (g, s).

(2.10) **Proof**: To see that $T_{(e,s)}\Phi|(G \times S)$ is injective, suppose that

$$0 = T_{(e,s)}\Phi|(G \times S)(\xi, v) = T_e\Phi_s\xi + v \quad \text{for some } (\xi, v) \in \mathcal{G} \times T_s S.$$

Then $v \in T_e\Phi_s\mathcal{G} \cap T_s S = \{0\}$, so $v = 0$ and $T_e\Phi_s\xi = 0$. Therefore $T_e\Phi_s t\xi = 0$ for all $t \in \mathbf{R}$. This implies that s is a fixed point for Φ restricted to the one parameter subgroup $t \to \exp t\xi$. Because Φ is free, $\exp t\xi = e$ for all $t \in \mathbf{R}$. Therefore $\xi = 0$.

To see that $T_{(e,s)}\Phi|(G \times S)$ is surjective, note that

$$\begin{aligned}
\operatorname{rank} T_{(e,s)}\Phi|(G \times S) &= \dim T_s S + \dim T_e\Phi_s\mathcal{G} - \dim(T_e\Phi_s\mathcal{G} \cap T_s S) \\
&= \dim T_s M, \qquad \text{by (13)}.
\end{aligned}$$

Since $\Phi(g, s) = \Phi_g \circ \Phi(e, s)$, we find that

$$T_{(g,s)}\Phi|(G \times S) = T_s\Phi_g \circ T_{(e,s)}\Phi|(G \times S).$$

However, Φ_g is a diffeomorphism. Thus $T_{(g,s)}\Phi$ is bijective. □

From the inverse function theorem it follows that the mapping $\Phi|(G \times S)$ is a local diffeomorphism of $G \times S$ onto an open subset U of M.

Claim: Shrinking S if necessary, we can arrange that $\Phi|(G \times S)$ is injective and hence is a diffeomorphism.

(2.11) **Proof**: Suppose that for no $S_0 \subseteq S$ the map $\Phi|(G \times S_0)$ is one to one. Let $\{U_k\}$ be a nested neighborhood base of m in S. Since $\Phi|(G \times S_0)$ is not one to one, there is an $s_k \in U_k$ and a $g_k \in G$ bounded away from e such that $g_k \cdot s_k \in U_k$. Therefore $g_k \cdot s_k \to m$. By properness of Φ there is a subsequence $g_{k_\ell} \to g$ and $g \cdot m = m$. This contradicts the freeness of Φ. □

The diffeomorphism

$$\psi_U = (\Phi|(G \times S))^{-1} : U \subseteq M \to G \times S : m \to \big(\vartheta_U(m), s_U(m)\big) \tag{14}$$

intertwines the G-action Φ on U with the G action $\bullet$ on $G \times S$ given by

$$g{\bullet}(h, s) = (gh, s),$$

that is,

$$\psi_U(g \cdot m) = g{\bullet}\big(\vartheta_U(m), s_U(m)\big). \tag{15}$$

Step 2
Now we want to show that M/G is a *topological manifold*. We already know that M/G is paracompact. So it remains to show that each point has a neighborhood homeomorphic to Euclidean space and that the transition maps are continuous. To do this, let

$$\psi_{\mathcal{U}} : \mathcal{U} \subseteq M/G \rightarrow S$$

be the mapping with makes diagram 2.1 commute, that is, $\psi_{\mathcal{U}} \circ \pi = \pi_2 \circ \psi_U$. Here π_2 is projection on the second factor. The map $\psi_{\mathcal{U}}$ is well defined because of the way ψ_U intertwines the G-actions. Now π and π_2 are not only continuous but also open

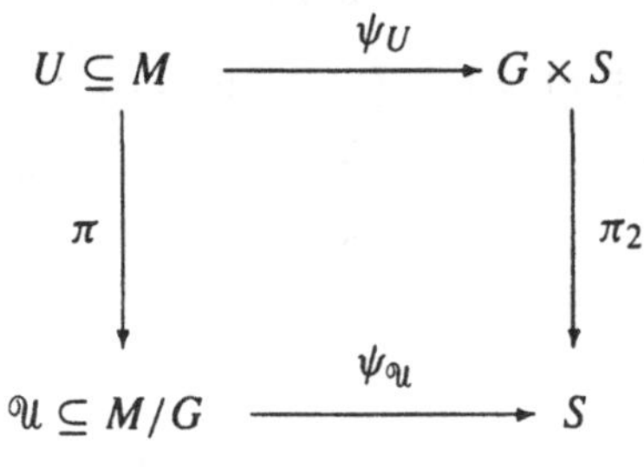

Diagram 2.1

mappings and ψ_U is a diffeomorphism. If V is an open subset of S, then $\psi_{\mathcal{U}}^{-1}(V) = \pi \circ \psi_U^{-1} \circ \pi_2^{-1}(V)$ is open. So $\psi_{\mathcal{U}}$ is continuous. The map $\psi_{\mathcal{U}}$ is one to one because

$$\psi_{\mathcal{U}}^{-1}(\text{pt}) = \pi\big(\psi_U^{-1}(G \times \{\text{pt}\})\big) = \pi(\text{single } G\text{-orbit}) = \text{pt}.$$

Chasing the diagram in the opposite direction, the same reasoning as above shows that $\psi_{\mathcal{U}}^{-1}$ is continuous. Hence $\psi_{\mathcal{U}}$ is a homeomorphism. By our construction of the charts $(\psi_{\mathcal{U}}, \mathcal{U})$, on overlaps the transition map is a homeomorphism.

Step 3
To complete the proof that M/G is a smooth manifold, we only have to check that the transition maps are diffeomorphisms. Let T be a submanifold of M passing through $m' = h \cdot m$ such that the map

$$\psi_V : V \subseteq M \rightarrow G \times T$$

is a diffeomorphism satisfying (14) and $\psi_V^{-1}(e, t) \in T \subseteq M$ for every $t \in T$. Since $U = \Phi(G \times S)$ and $V = \Phi(G \times T)$, there are open subsets $\widehat{S} \subseteq S$, $\widehat{T} \subseteq T$ such that $U \cap V = \Phi(G \times \widehat{S}) = \Phi(G \times \widehat{T})$. We wish to show that the map

$$\theta : \widehat{S} \rightarrow \widehat{T} : s \rightarrow t = \widehat{T} \cap (G \cdot s)$$

is a diffeomorphism. To see this choose s and t so that $s = g \cdot t$. Then $g \cdot \widehat{T}$ intersects $\widehat{S}$ at s and both $\widehat{S}$ and $g \cdot \widehat{T}$ intersect $G \cdot s$ transversely. In a local product chart (U, ψ_U) adapted to (e, s) so that $s \in \widehat{S}$ has coordinates $(e, 0)$, $g \cdot \widehat{T}$ is a smooth manifold which intersects the origin and is transverse to $G \times \{0\}$. This means that $\widehat{S}$ is diffeomorphic by

θ to $g \cdot \widehat{T}$ about s. Hence $\widehat{S}$ is diffeomorphic to $\widehat{T}$. Just like in the example 2, we have a homeomorphism that is everywhere a local diffeomorphism. Hence the mapping θ is a diffeomorphism. This tells us that the transition map between the charts $(\psi_{\mathcal{U}}, \mathcal{U} \cap \mathcal{V})$ and $(\psi_{\mathcal{V}}, \mathcal{U} \cap \mathcal{V})$ of M/G is the smooth map

$$\psi_{\mathcal{V}} \circ \psi_{\mathcal{U}}^{-1} : s \to t = \theta(s)$$

because the charts $(\psi_{\mathcal{U}}, \mathcal{U} \cap \mathcal{V})$ and $(\psi_{\mathcal{V}}, \mathcal{U} \cap \mathcal{V})$ are naturally induced from the charts $(\psi_U, U \cap V)$ and $(\psi_V, U \cap V)$ by restricting the image of ψ_U to $\{e\} \times \widehat{S}$ and the image of ψ_V to $\{e\} \times \widehat{T}$. This concludes the proof that M/G is a smooth manifold. □

Let (P, G, N) be a *left principal G-bundle over N*. This means that

1. G acts smoothly on P and there is a surjective submersion $\pi : P \to N$, called the *bundle projection* such that $\pi(p) = \pi(g \cdot p)$.

2. For each $p \in P$ there is an open neighborhood $\mathcal{U} \subseteq N$ of $\overline{p} = \pi(p)$ and a G-equivariant diffeomorphism of the form

$$\psi_{\mathcal{U}} : \pi^{-1}(\mathcal{U}) \to G \times \mathcal{U} : p \to \left(\vartheta_{\mathcal{U}}(p), \widehat{\psi}_{\mathcal{U}}(\overline{p})\right),$$

   such that $\pi \circ \psi_{\mathcal{U}}^{-1} = id_{\mathcal{U}}$. The group action on $G \times \mathcal{U}$ is multiplication by G on the first factor. The pair $(\psi_{\mathcal{U}}, \pi^{-1}(\mathcal{U}))$ is called a *local trivialization* of the bundle $\pi : P \to N$ at p.

3. For every pair of local trivializations given by $(\psi_{\mathcal{U}}, \pi^{-1}(\mathcal{U}))$ and $(\psi_{\mathcal{V}}, \pi^{-1}(\mathcal{V}))$ the transition mapping $\psi_{\mathcal{V}} \circ \psi_{\mathcal{U}}^{-1} : G \times (\mathcal{U} \cap \mathcal{V}) \to G \times (\mathcal{U} \cap \mathcal{V})$ is a diffeomorphism of the form

$$\psi_{\mathcal{V}} \circ \psi_{\mathcal{U}}^{-1}(g, n) = \left(g \vartheta_{\mathcal{U}\mathcal{V}}(n), n\right) \tag{16}$$

   where $\vartheta_{\mathcal{U}\mathcal{V}} : \mathcal{U} \cap \mathcal{V} \to G$ is smooth.

Note this is a *left* principal G-bundle. The modifications for a *right* principal G-bundle are straightforward.

Example 2. If H is a closed subgroup of the Lie group G and π is the orbit mapping $\pi : G \to G/H : g \to Hg$ then (G, H, π) is a right principal H-bundle. To see this we use the same notation as in example 1. Define the map $\psi_{\mathcal{U}_g}$ by

$$\psi_{\mathcal{U}_g} : \pi^{-1}(\mathcal{U}_g) \to H \times \mathcal{U}_g : k \to \left(k(\sigma_g \circ \pi(k))^{-1}, \pi(k)\right),$$

where $\sigma_g : \mathcal{U}_g \to G : \overline{e} \exp u\, g \to \exp u\, g$ and $u \in U \subseteq \mathcal{L}$. The map $\psi_{\mathcal{U}_g}$ is well defined because Hk intersects $\sigma_g(\mathcal{U}_g)$ exactly once, and consequently is a diffeomorphism. Hence $\pi : G \to G/H$ is a smooth bundle with local trivializations $\{\left(\psi_{\mathcal{U}_g}, \pi^{-1}(\mathcal{U}_g)\right)\}$. From the definition of the local trivialization it follows that the transition maps are

$$\psi_{\mathcal{U}_{g'}} \circ \psi_{\mathcal{U}_g}^{-1} : H \times (\mathcal{U}_g \cap \mathcal{U}_{g'}) \to H \times (\mathcal{U}_g \cap \mathcal{U}_{g'}) :$$
$$(h, u) \to (h', u) = \left(h(\sigma_g(u))(\sigma_{g'}(u))^{-1}, u\right). \quad \square$$

Example 2 should motivate the proof of our next

Claim: Let G be a Lie group which acts freely and properly on a smooth manifold M. Then M is a principal G-bundle over the orbit space M/G with orbit mapping $\pi : M \to M/G$ being the bundle projection.

(2.12) **Proof**: We use the same notation as in ((2.9)). Observe that the map

$$\varphi_U : U = \pi^{-1}(\mathcal{U}) \to G \times M/G : m \to \big(\vartheta_U(m), \psi_{\mathcal{U}}^{-1}(s_U(m))\big)$$

is a diffeomorphism. Here ψ_U is the map defined in step 2 of ((2.11)). So $\varphi_U = (id \times \psi_{\mathcal{U}}^{-1}) \circ \psi_U$. Because it intertwines the G-action, φ_U is a local trivialization. To finish the argument, we need only show that if (φ_U, U) and (φ_V, V) are local trivializations of the bundle $\pi : M \to M/G$, then the transition map $\varphi_V \circ \varphi_U^{-1}$ is of the form (16). The reason for this is that by design there is a canonical diffeomorphism of $S \subseteq M$, with $\mathcal{U} \subseteq M/G$. The computation is

$$\varphi_V \circ \varphi_U^{-1}(u)(g, u) = \varphi_V\big(\Phi_g(\varphi_U^{-1}(e, u))\big) = g \bullet \varphi_V \circ \varphi_U^{-1}(e, s) = \big(g \cdot \vartheta_{UV}(u), u\big).$$

Here the value of the *transition map* $\vartheta_{UV}(u)$ is defined as $\pi_1 \circ \varphi_V \circ \varphi_U^{-1}(e, u)$ where π_1 is the projection onto the first factor of $G \times M/G$. □

Example 3. Let G be a Lie group with a closed subgroup H. Suppose that $\rho : H \to \mathrm{Gl}(V)$ is a homomorphism. Define an action of H on $G \times V$ by

$$\Psi : H \times (G \times V) \to G \times V : (h, (g, v)) \to (gh^{-1}, \rho(h)v).$$

Then Ψ is free and proper. Hence the orbit space $G \times_H V$ is a smooth manifold. The map $\pi_1 : G \times V \to G$ intertwines the H-action Ψ with the action of H on G. Hence it induces a map $\pi : G \times_H V \to G/H$ on the orbit spaces. π is a locally trivial bundle over G/H with fiber V and total space $G \times_H V$. It is an *associated bundle* of the principal bundle $G \to G/H$.

2.3 Orbit space of a locally free action

In this subsection we define the concept of an orbifold and show that the orbit space of a proper locally free action is an orbifold.

Let $\Phi : G \times M \to M$ be an action of a Lie group G on a smooth manifold M. The action Φ is *locally free* if for every $m \in M$ there is an open neighborhood U of e in G such that $U \cap G_m = \{e\}$. We now prove

Fact: The action is locally free if and only if for every $m \in M$ the Lie algebra $\mathcal{G}_m$ of the isotropy group G_m has dimension zero.

(2.13) **Proof**: Suppose that $\dim \mathcal{G}_m \geq 1$ for some $m \in M$. Then there is a nonzero $\xi \in \mathcal{G}_m$. Hence the image of the one parameter subgroup $t \to \exp t\xi$ lies in G_m. Taking a sequence $t_i \to 0$ we see that $\exp t_i \xi \to e$. Thus e is an accumulation point of the one parameter subgroup. This contradicts the assumption that the action Φ is locally free. Therefore $\dim \mathcal{G}_m = 0$.

Conversely, suppose that $\dim \mathcal{G}_m = 0$. Let W be a sufficiently small neighborhood of 0 in $\mathcal{G}$ such that $\exp | W$ is a diffeomorphism. Since $\dim \mathcal{G}_m = 0$, it follows that $W \cap \mathcal{G}_m = \{0\}$. Therefore $(\exp W) \cap G_m = \{e\}$. Consequently, the action Φ is locally free. □

Corollary: Suppose that the G-action Φ is proper. Then Φ is locally free if and only if the isotropy group G_m is finite at every point m in M.

(2.14) **Proof**: Since Φ is a proper action, at each $m \in M$ the isotropy group G_m is compact. Because Φ is locally free, G_m has no accumulation points. Hence G_m is finite. Conversely, if G_m is finite, it has no nontrivial one parameter subgroups. Hence $\dim \mathcal{G}_m = 0$ and thus Φ is locally free. □

Let us look at some examples of orbit spaces of locally free proper actions.

Example 4. Consider the $\mathbf{Z}_2$ action

$$\mathbf{Z}_2 \times \mathbf{R}^2 \to \mathbf{R}^2 : \big(s, (x, y)\big) \to (x, sy),$$

where $s^2 = 1$. Clearly this action is locally free. Its orbit space $\mathbf{R}^2/\mathbf{Z}_2$ is diffeomorphic to the closed upper half plane $\mathbf{R} \times \mathbf{R}_{\geq}$, which is a smooth manifold with boundary.

Example 5. Consider the $\mathbf{Z}_2$ action

$$\mathbf{Z}_2 \times \mathbf{R}^2 \to \mathbf{R}^2 : \big(s, (x, y)\big) \to (sx, sy),$$

where $s^2 = 1$. Clearly this action is locally free. Its orbit space is the closed upper half plane $\{(x, y) \in \mathbf{R}^2 | \, y \geq 0\}$ with the points on its boundary of the form $(x, 0)$ and $(-x, 0)$ identified. Thus the orbit space $\mathbf{R}^2/\mathbf{Z}_2$ is a cone on a circle with vertex at the origin.

Orbit spaces of a proper locally free action form an interesting class of smooth manifolds with mild isolated singularities called orbifolds. We now define precisely what we mean by an *orbifold.* Let $\mathcal{T}$ be a Hausdorff topological space. An *orbifold chart* associated to an open set $\mathcal{U} \in \mathcal{T}$ is a triple (U, G, φ) where

1. U is a connected open subset of $\mathbf{R}^n$.
2. G is a finite group of linear transformations of U onto itself.
3. $\varphi : U \to \mathcal{U}$ is a continuous surjective G-invariant map which induces a homeomorphism $\widetilde{\varphi} : U/G \to \mathcal{U}$.

Let (U, G, φ) and (U', G', φ') be orbifold charts associated to the open sets $\mathcal{U}$ and $\mathcal{U}'$ in $\mathcal{T}$. An *orbifold transition map* $\lambda : (U, G, \varphi) \to (U', G', \varphi')$ is a diffeomorphism $\lambda : U \to \lambda(U) \subseteq U'$ such that

i). $\varphi = \varphi' \circ \lambda$.

ii). For each $g \in G$, there is a $g' \in G'$ such that $\lambda \circ g = g' \circ \lambda$.

The element g' is uniquely determined because $g'|\lambda(U) = \lambda \circ g \circ \big(\lambda^{-1}|\lambda(U)\big)$. Since $\lambda(U)$ is open in U', the map $g'|\lambda(U)$ extends to a unique linear mapping g' of U' into itself. Thus the map $j : G \to G' : g \to g'$ is well defined. In fact, j is an injective homomorphism. It is a homomorphism because

$$j(g\widetilde{g})|\lambda(U) = \lambda \circ g\widetilde{g} \circ \left(\lambda^{-1}|\lambda(U)\right) = \left(j(g)|\lambda(U)\right)\right) \circ \left(j(\widetilde{g})|\lambda(U)\right)$$

and is injective because setting $g' = e$ in *ii*) gives $\lambda \circ g = \lambda$, which implies that $g = e$.

Some basic properties of orbifold transition maps are given in

Claim:

1. If $\lambda, \mu : (U, G, \varphi) \to (U', G', \varphi')$ are orbifold transition maps, then there is a $g' \in G'$ such that $\mu = g' \circ \lambda$.

2. The composition of two orbifold transition maps is an orbifold transition map on a component of its domain with the groups G and G' replaced by subgroups which fix the component.

(2.15) **Proof**: 1) Since $\mu : U \to \mu(U)$ is a diffeomorphism, $\lambda \circ \mu^{-1} : \mu(U) \to \lambda(U)$ is also a diffeomorphism. Furthermore, since $\varphi = \varphi' \circ \lambda = \varphi' \circ \mu$, the point $\lambda \circ \mu^{-1}(w)$ lies on the same G'-orbit as w. Thus there is a $g'_w \in G'$ such that $g'_w(w) = \lambda \circ \mu^{-1}(w)$.

To finish the argument we need to show that we can choose the *same* $g' = g'_w$ for every $w \in \mu(U)$. To verify this first observe that if g'_1, g'_2 are distinct elements of G', then, considered as linear transformations on all of $\mathbf{R}^n \supseteq U$, they agree on at most a hyperplane. Since G' is finite, the set of points $K \subseteq \mu(U)$ on which at least two elements of G' agree is at most a finite union of hyperplanes. Therefore on each path component of $\mu(U) - K$, none of the elements of G' are equal. Moreover, on a fixed path component $\lambda \circ \mu^{-1}$ is a fixed element of G'. To see this for any $g' \in G'$, the set of points in $\mu(U)$ for which $\lambda \circ \mu^{-1} = g'$ is closed by continuity. Thus each path component of $\mu(U) - K$ is partitioned by closed sets (one for every $g' \in G'$ where $\lambda \circ \mu^{-1} = g'$). By connectedness all but one of these closed sets must be empty. Next we show that the element $g' \in G'$ for which $\lambda \circ \mu^{-1} = g'$ does not change from one path component of $\mu(U) - K$ to another. Suppose it did. Let C_1 and C_2 be neighboring components separated by a hyperplane H on which $\lambda \circ \mu^{-1}|C_i = g'_i$ differ. By continuity of g'_i and $\lambda \circ \mu^{-1}$, g'_1 and g'_2 are equal on H. Therefore $g'_1 = g'_2 \circ R$ where R is a reflection in H. Hence $\lambda \circ \mu^{-1}$ cannot be injective near H, which is a contradiction. Therefore $\lambda \circ \mu^{-1} = g'$ for some fixed $g' \in G'$, that is, $\lambda = g' \circ \mu$.

2) The argument is a straightforward consequence of the definition of an orbifold transition map. The details are omitted. □

A collection $\mathcal{A}$ of orbifold charts of the topological space $\mathcal{T}$ is an *orbifold atlas* if the following conditions are satisfied:

1. Let (U, G, φ) and (U', G', φ') be orbifold charts associated to the open sets $\mathcal{U}$ and $\mathcal{U}'$, respectively, of $\mathcal{T}$. If $\mathcal{U} \subseteq \mathcal{U}'$, then there is an orbifold transition map $\lambda : (U, G, \varphi) \to (U', G', \varphi')$.

2. Let (U, G, φ) and (U', G', φ') be orbifold charts associated to the open sets $\mathcal{U}$ and $\mathcal{U}'$, respectively, of $\mathcal{T}$. If $\mathcal{U} \cap \mathcal{U}' \neq \emptyset$, then there is an orbifold chart (U'', G'', φ'') associated to $\mathcal{U}'' \subseteq \mathcal{U} \cap \mathcal{U}'$.

3. To each $(U, G, \varphi) \in \mathcal{A}$, let $\mathcal{U}$ be the associated open set in $\mathcal{T}$. Then $\cup \mathcal{U} = \mathcal{T}$.

Two orbifold atlases $\mathcal{A}$ and $\mathcal{A}'$ are *related* if $\mathcal{A} \cup \mathcal{A}' \subseteq \mathcal{A}''$ for some orbifold atlas $\mathcal{A}''$. Two orbifold atlases $\mathcal{A}$ and $\mathcal{A}'$ are *equivalent* if there are orbifold atlases $\mathcal{A} = \mathcal{A}_1, \mathcal{A}_2, \ldots, \mathcal{A}_m = \mathcal{A}'$ such that $\mathcal{A}_i$ and $\mathcal{A}_{i+1}$ are related for $i = 1, \ldots, m-1$. An equivalence class of orbifold atlases is an *orbifold structure* on $\mathcal{T}$. In other words, $\mathcal{T}$ is an *orbifold* (of dimension n).

The main result of this subsection is the following

Claim: If $\Phi : G \times M \to M$ is a proper locally free action of a Lie group G on a smooth manifold M, then the orbit space M/G is an orbifold.

(2.16) **Proof**: First we construct orbifold charts for M/G. Since the G-action is proper from ((2.4)) it follows that M/G is Hausdorff. Let $m \in M$ and let S be a slice to the G-orbit $G \cdot m$ at m. By ((2.2)) the set of G_m-orbits on S is homeomorphic via the map $\tau : S/G_m \to (G \cdot S)/G$ to the set of G-orbits through S. Because the action Φ is proper, the action of the isotropy group G_m on S is locally the same as the linear action

$$\widehat{\Phi} : G_m \times T_m S \to T_m S : (h, v_m) \to T_m \Phi_h v_m,$$

that is, there is a local diffeomorphism $\psi^{-1} : T_m S \to S$ with $\psi^{-1}(0) = m$ which intertwines the action $\widehat{\Phi}$ with the action Φ. In other words, for every $h \in G_m$ and every $v_m \in T_m S$, we have $\psi^{-1} \circ \widehat{\Phi}_h(v_m) = \Phi_h(\psi^{-1}(v_m))$, (see ((1.4))). To keep the notation simple we will assume that ψ^{-1} is a global diffeomorphism. Associate to the open set $\mathcal{U} = \pi(G \cdot S)$ of the orbit space M/G with orbit map $\pi : M \to M/G$ the triple (U, H, φ) where

1. $U = T_m S$. Actually we should take U to be a ball of radius ε in $T_m S$ for some G_m-invariant metric.

2. $H = G_m$. Since the action Φ is locally free, H is finite ((2.14)). Moreover, H acts linearly on U by $\widehat{\Phi}$.

3. The map $\varphi : U \to \mathcal{U}$ is the composition of the map $\psi^{-1} : T_m S \to S$ followed by the map $\pi|S : S \subseteq M \to \pi(G \cdot S) = \mathcal{U} \subseteq M/G$. Clearly, φ is invariant under the G_m-action $\widehat{\Phi}$. The induced map $\widetilde{\varphi} : U/H \to \mathcal{U}$ is the composition of the map $\widetilde{\psi}^{-1} : (T_m S)/H \to S/H$ followed by the map $\tau : S/H \to \mathcal{U} = (G \cdot S)/G$. Since $\widetilde{\psi}^{-1}$ and τ are homeomorphisms, $\widetilde{\varphi}$ is a homeomorphism.

Thus (U, H, φ) is an orbifold chart associated to the open set $\mathcal{U}$ of M/G.

Let $\mathcal{A}$ be the collection of orbifold charts as constructed above. We now show that $\mathcal{A}$ is an orbifold atlas.

Clearly the open sets in M/G associated to orbifold charts cover M/G. This verifies condition 3 of the definition of orbifold atlas.

Before proceding further with the verification of the definition of an orbifold atlas, consider two open sets $\mathcal{U}$ and $\mathcal{U}'$ in M/G which correspond to the orbifold charts (U, H, φ) and (U', H', φ') in $\mathcal{A}$. We may assume that $\mathcal{U} = \pi(G \cdot S_m)$ and $\mathcal{U}' = \pi(G \cdot S'_{m'})$, where

S_m and $S'_{m'}$ are slices at m and m'. Because the G-action is locally linearizeable, we may identify U with S_m and U' with $S'_{m'}$. Under this identification, $\varphi = \pi | S_m$ and $\varphi' = \pi | S'_{m'}$. To keep the ideas of the argument from being obscured by excessive notation, we will assume that all local diffeomorphisms are diffeomorphisms.

Using these identifications we now verify condition 1. Suppose that $\mathcal{U} \subseteq \mathcal{U}'$. Then we must show that there is an orbifold transition map $\lambda : (U, H, \varphi) \rightarrow (U', H', \varphi')$. We argue as follows. Since $\mathcal{U} \subseteq \mathcal{U}'$, it follows that $G \cdot S_m \subseteq G \cdot S'_{m'}$. Thus the G-orbit through m intersects $S'_{m'}$ say at m''. In particular, there is a $k \in G$ such that $k \cdot m = m''$. Write $\mathcal{G} = \mathcal{H} \oplus \mathcal{L}$ and let $L = \exp \mathcal{L}$. By ((1.1)) the map

$$\theta : L \times S_m \rightarrow U \subseteq M : (\exp \ell, s_m) \rightarrow \exp \ell \cdot s_m$$

is a diffeomorphism. Since $S'_{m'}$ meets G-orbits transversely, $V = (k^{-1} \cdot S'_{m'}) \cap U$ is a smooth section of U, thought of as a trivial bundle over S_m with fiber L. Thus there is a smooth map $\rho : S_m \rightarrow L \subseteq G$ such that $\rho(S_m) \cdot S_m = V$. Hence, the map

$$\lambda : S_m \rightarrow k \cdot V \subseteq S'_{m'} : s_m \rightarrow k\, \rho(s_m) \cdot s_m$$

is a diffeomorphism. By construction of λ it follows that $\varphi' \circ \lambda(s_m) = \varphi(s_m)$ for every $s_m \in S_m$. By construction of the slice S_m, we have $h \cdot S_m = S_m$ for every $h \in H = G_m$. Therefore $\lambda(h \cdot s_m) = k \cdot \rho(h \cdot s_m)\, h \cdot s_m$ for every $s_m \in S_m$. Since

$$\lambda(h \cdot s_m) = k\, \rho(h \cdot s_m)\, h(\rho(s_m))^{-1} k^{-1} \cdot \lambda(s_m),$$

the points $\lambda(s_m)$ and $\lambda(h \cdot s_m)$ lie on the same G-orbit. By property 4) of a slice, $h'(s_m) = k\rho(h \cdot s_m)h(\rho(s_m))^{-1}k^{-1} \in H'$. Hence the map $S_m \rightarrow H' : s_m \rightarrow h'(s_m)$ is continuous. Since S_m is connected and H' is finite, this map is constant, that is, $h'(s_m)$ does not depend on s_m. Consequently, λ is an orbifold transition map.

To verify condition 2, suppose that $\mathcal{U}$ and $\mathcal{U}'$ are open subsets of M/G corresponding to the orbifold charts (U, H, φ) and (U', H', φ') such that $\mathcal{U} \cap \mathcal{U}'$ is a nonempty open subset of M/G. Then $\pi^{-1}(\mathcal{U} \cap \mathcal{U}')$ is the set of points $p \in M$ such that for some $g, \widetilde{g} \in G$, $g \cdot p \in S_m$ and $\widetilde{g} \cdot p \in S'_{m'}$. Let S_p be a slice to the the G-action Φ at $p \in \pi^{-1}(\mathcal{U} \cap \mathcal{U}')$. Since $\pi^{-1}(\mathcal{U} \cap \mathcal{U}')$ is a G-invariant open subset of M containing p and S_p is a slice at p, there is an open subset U'' of S_p containing p such that $G \cdot U'' \subseteq \pi^{-1}(\mathcal{U} \cap \mathcal{U}')$. Let $\mathcal{U}'' = \pi(G \cdot U'')$. Since $G \cdot U''$ is a G-invariant open subset of M and π is an open map, $\mathcal{U}''$ is an open subset of M/G which is contained in $\mathcal{U} \cap \mathcal{U}'$. Identify U'' with S_p and define $\varphi'' = \pi | S_p$. Moreover, let H'' be the isotropy group G_p. By construction (U'', H'', φ'') is an orbifold chart associated to $\mathcal{U}''$.

Hence $\mathcal{A}$ is an orbifold atlas. Thus the orbit space of a proper locally free action is an orbifold. □

3 Momentum mappings

In this section we define the concept of a momentum mapping of a Hamiltonian group action. In addition to proving some general properties, we find a local normal form for an equivariant momentum mapping near a fixed point in its zero level set. The normal form is used to show that every level set of the momentum map is locally arcwise connected. This fact is needed in section 5 when we treat singular reduction.

3.1 General properties

Let Φ be an action of a Lie group G on a smooth manifold M. For every ξ in the Lie algebra $\mathcal{G}$ of G define a vector field X^{ξ} on M by

$$X^{\xi}(m) = \left.\frac{d}{dt}\right|_{t=0} \Phi_m(\exp t\xi) \; = \; T_e\Phi_m\xi, \tag{17}$$

for every $m \in M$. The vector field X^{ξ} is called the *infinitesimal generator* of Φ in the direction ξ.

Claim: The linear map

$$\mathcal{G} \to \mathcal{X}(M) : \xi \to X^{\xi} \tag{18}$$

is an antihomomorphism of the Lie algebra $\mathcal{G}$ to the Lie algebra $\mathcal{X}(M)$ of vector fields on M.

(3.1) **Proof**: Linearity is obvious from (17). To verify that (18) is an antihomomorphism we compute

$$\begin{aligned}
\left[X^{\xi}, X^{\eta}\right](m) &= \left.\frac{d}{dt}\right|_{t=0} \Phi^*_{\exp t\xi} X^{\eta}(m) = \left.\frac{d}{dt}\right|_{t=0} T\Phi_{\exp -t\xi} X^{\eta}(\Phi_{\exp t\xi}(m)) \\
&= \left.\frac{d}{dt}\right|_{t=0} \left.\frac{d}{ds}\right|_{s=0} \Phi_{\exp -t\xi} \circ \Phi_{\exp s\eta} \circ \Phi_{\exp t\xi}(m) \\
&= \left.\frac{d}{dt}\right|_{t=0} \left.\frac{d}{ds}\right|_{s=0} \Phi_m(\exp -t\xi \; \exp s\eta \; \exp t\xi) \\
&= \left.\frac{d}{dt}\right|_{t=0} \left.\frac{d}{ds}\right|_{s=0} \Phi_m\big(\exp s(\mathrm{Ad}_{\exp -t\xi}\eta)\big) \\
&= \left.\frac{d}{dt}\right|_{t=0} X^{\mathrm{Ad}_{\exp -t\xi}\eta}(m) \; = \; X^{-[\xi,\eta]}(m) \; = \; -X^{[\xi,\eta]}(m). \qquad \square
\end{aligned}$$

Let (M, ω) be a symplectic manifold. We say that Φ is a *Hamiltonian G-action* if for every $\xi \in \mathcal{G}$ the infinitesimal generator X^{ξ} is a Hamiltonian vector field on (M, ω). In other words, for every $\xi \in \mathcal{G}$, there is a smooth function $J^{\xi} : M \to \mathbf{R}$ such that

$$X^{\xi} = X_{J^{\xi}}. \tag{19}$$

▷ If G is connected, then Φ_g is a symplectic diffeomorphism for every $g \in G$.

(3.2) **Proof**: Suppose that $\xi \in \mathcal{G}$. Then

$$L_{X^{\xi}}\omega = X^{\xi} \lrcorner\, d\omega + d(X^{\xi} \lrcorner\, \omega) \; = \; X^{\xi} \lrcorner\, d\omega + d(dJ^{\xi}) \; = \; 0.$$

Therefore

$$\frac{d}{dt}\Phi^*_{\exp t\xi}\omega = \Phi^*_{\exp t\xi}\Big(\frac{d}{ds}\Big|_{s=0}\Phi^*_{\exp s\xi}\omega\Big) = \Phi^*_{\exp t\xi}(L_{X^\xi}\omega) = 0,$$

which gives $\Phi^*_{\exp t\xi}\omega = \omega$ for every $\xi \in \mathcal{G}$. Since G is connected, it is generated by an open neighborhood of e. Therefore $\Phi^*_g\omega = \omega$ for every $g \in G$. □

To avoid limiting ourselves to actions of connected Lie groups, we supplement the definition of Hamiltonian G-action Φ by requiring that Φ_g is a symplectic diffeomorphism of (M, ω) for every $g \in G$.

Bringing together the information given in (19) we define a mapping

$$J : M \rightarrow \mathcal{G}^*, \tag{20}$$

by $J(m)\xi = J^\xi(m)$ where $\xi \in \mathcal{G}$ and $m \in M$. This makes sense because $\xi \rightarrow J^\xi(m)$ is a linear form on $\mathcal{G}$ for every fixed $m \in M$. The mapping J is a *momentum map* of the Hamiltonian action Φ provided that

$$J^{[\xi,\eta]} = \{J^\xi, J^\eta\}, \tag{21}$$

for every $\xi, \eta \in \mathcal{G}$. To pass from (19) to (21) requires a mild condition on the G-action that we tacitly assume is satisfied.

Claim: If (M, ω) is connected, then the momentum mapping is determined up to an additive constant $\mu_0 \in \mathcal{G}^*$.

(3.3) **Proof**: Suppose that $\widetilde{J}$ is another momentum map for the Hamiltonian action Φ. Let $\mathcal{J} = J - \widetilde{J}$ and fix $m \in M$. For every $\xi \in \mathcal{G}$ and every $v_m \in T_mM$

$$\begin{aligned}(T_m\mathcal{J})(v_m)\xi &= d\mathcal{J}^\xi(m)v_m = dJ^\xi(m)v_m - d\widetilde{J}^\xi(m)v_m \\ &= \omega(m)\big(X_{J^\xi}(m) - X_{\widetilde{J}^\xi}(m), v_m\big) \\ &= \omega(m)\big(X^\xi(m) - X^\xi(m), v_m\big) = 0.\end{aligned}$$

Hence $T_m\mathcal{J}$ vanishes for every $m \in M$. Since M is connected, it follows that $\mathcal{J} = \mu_0$ for some fixed $\mu_0 \in \mathcal{G}^*$. □

Suppose that the Hamiltonian G-action Φ preserves a Hamiltonian function H on M, that is, for every $g \in G$, we have $\Phi^*_g H = H$. Then Φ is a *symmetry* of the Hamiltonian system (M, ω, H). Symmetries of Hamiltonian systems give rise to conserved quantities.
▷ More precisely, for every $\xi \in \mathcal{G}$ the function J^ξ is an *integral* of X_H, that is,

$$L_{X_H}J^\xi = 0.$$

(3.4) **Proof**: The map Φ preserves the Hamiltonian H, that is, $\Phi^*_{\exp t\xi}H = H$ for every $\xi \in \mathcal{G}$. Therefore

$$\begin{aligned}0 &= \frac{d}{dt}\Big|_{t=0}\Phi^*_{\exp t\xi}H = L_{X^\xi}H, \quad \text{by definition of } X^\xi \\ &= L_{X_{J^\xi}}H, \quad \text{because } \Phi \text{ is a Hamiltonian action} \\ &= \{H, J^\xi\}, \quad \text{by definition of Poisson bracket } \{\,,\,\} \\ &= -\{J^\xi, H\} = -L_{X_H}J^\xi.\end{aligned}$$

□

We now give some examples of momentum mappings.

Example 1. Suppose that ψ is an action of a Lie group G on a configuration space M, which is a smooth manifold. Then ψ *lifts* to an action Ψ of G on the phase space T^*M. Ψ is defined by

$$\Psi_g(\alpha_m) = (T_m\psi_{g^{-1}})^t\alpha_m, \tag{22}$$

for every $\alpha_m \in T^*_mM$. The action Ψ covers the action ψ, that is,

$$\tau(\Psi_g(\alpha_m)) = \psi_g(\tau(\alpha_m)), \tag{23}$$

where $\tau : T^*M \rightarrow M : \alpha_m \rightarrow m$ is the bundle projection. To show that Ψ is a Hamiltonian action we need the following:

Fact: For every $g \in G$, the mapping Ψ_g preserves the canonical 1-form θ on T^*M, (see appendix A section 2).

(3.5) **Proof**: For $v_{\alpha_m} \in T_\alpha(T^*M)$ and $\alpha_m \in T^*_mM$,

$$\begin{aligned}
\big((\Psi_g)^*\theta\big)(\alpha_m)v_{\alpha_m} &= \theta(\Psi_g\alpha_m)T\Psi_g\, v_{\alpha_m} \\
&= (\Psi_g\alpha_m)T\tau\circ T\Psi_g\, v_{\alpha_m}, \quad \text{by definition of } \theta \\
&= (\Psi_g\alpha_m)T\psi_g\circ T\tau\, v_{\alpha_m}, \quad \text{differentiating (23)} \\
&= \alpha_m(T\psi_{g^{-1}}\circ T\psi_g\circ T\tau\, v_{\alpha_m}), \quad \text{by definition of the action } \Psi \\
&= \theta(\alpha_m)v_{\alpha_m}. \qquad \square
\end{aligned}$$

From this follows the *momentum lemma*:

Claim: For every $\xi \in \mathcal{G}$, the infinitesimal generator X^ξ of Ψ in the direction ξ is the Hamiltonian vector field on (T^*M, Ω) corresponding to the Hamiltonian function

$$J^\xi : T^*M \rightarrow \mathbf{R} : \alpha_m \rightarrow \alpha_m(X_\xi(m)). \tag{24}$$

Here Ω is the canonical 2-form on T^*M and $X_\xi(m) = \left.\frac{d}{dt}\right|_{t=0} \psi_{\exp t\xi}(m)$ is the infinitesimal generator of the action ψ on M in the direction ξ.

(3.6) **Proof**: Since the action Ψ preserves the 1-form θ, $L_{X^\xi}\theta = 0$. But

$$L_{X^\xi}\theta = X^\xi \lrcorner d\theta + d(X^\xi \lrcorner \theta).$$

Therefore

$$X^\xi \lrcorner \Omega = -X^\xi \lrcorner d\theta = d(X^\xi \lrcorner \theta).$$

Thus X^ξ is the Hamiltonian vector field on (T^*M, Ω) corresponding to the Hamiltonian function

$$J^\xi : T^*M \rightarrow \mathbf{R} : \alpha_m \rightarrow (X^\xi \lrcorner \theta)(\alpha_m) = \alpha_m(T_{\alpha_m}\tau\, X^\xi(\alpha_m)) = \alpha_m(X_\xi(m)).$$

The last equality above follows by differentiating $\tau(\Psi_{\exp t\xi}(\alpha_m)) = \psi_{\exp t\xi}(\tau(\alpha_m))$ and evaluating the resulting expression at $t = 0$. $\square$

Claim: The Hamiltonian G-action Ψ has a coadjoint equivariant momentum mapping

$$J : T^*M \to \mathcal{G}^* : \alpha_m \to (T\tau \circ T_e\Psi_{\alpha_m})^t\alpha_m. \tag{25}$$

(3.7) **Proof**: First

$$J(\alpha_m)\xi = \alpha_m\big(T_{\alpha_m}\tau\, X^\xi(\alpha_m)\big) = \alpha_m(X_\xi(m)) = J^\xi(\alpha_m). \tag{26}$$

Second, the mapping J intertwines the G-action Ψ on T^*M with the coadjoint action of G on $\mathcal{G}^*$. In other words, J is *coadjoint equivariant*, that is, for every $g \in G$

$$J(\Psi_g(\alpha_m)) = \mathrm{Ad}^t_{g^{-1}}\big(J(\alpha_m)\big). \tag{27}$$

To see this we compute

$$\begin{aligned} J(\Psi_g(\alpha_m))\xi &= \theta(\Psi_g(\alpha_m))X^\xi(\Psi_g(\alpha_m)) = \Psi_g(\alpha_m)\big(T\tau \circ T\Psi_g\, X^{\mathrm{Ad}_{g^{-1}}\xi}(\alpha_m)\big), \\ &\qquad \text{see below} \\ &= \alpha_m\big(T\psi_{g^{-1}} \circ T\psi_g \circ T\tau\, X^{\mathrm{Ad}_{g^{-1}}\xi}(\alpha_m)\big) = J(\alpha_m)\,\mathrm{Ad}_{g^{-1}}\xi \\ &= \mathrm{Ad}^t_{g^{-1}}(J(\alpha_m))\xi. \end{aligned}$$

The second equality above follows from the definition of θ and

$$\begin{aligned} X^\xi(\Psi_g(\alpha_m)) &= \frac{d}{dt}\Big|_{t=0} \Psi_{\exp t\xi} \circ \Psi_g(\alpha_m) \\ &= \frac{d}{dt}\Big|_{t=0} \Psi_g \circ \Psi_{g^{-1}(\exp t\xi)g}(\alpha_m) = T\Psi_g X^{\mathrm{Ad}_{g^{-1}}\xi}(\alpha_m). \end{aligned}$$

To verify $J^{[\xi,\eta]} = \{J^\xi, J^\eta\}$ and thus how that J is a momentum mapping, we calculate

$$\begin{aligned} J^{[\xi,\eta]}(\alpha_m) &= \alpha_m\big(T\tau\, X^{[\xi,\eta]}(\alpha_m)\big) = -\alpha_m\big(T\tau\,[X^\xi, X^\eta](\alpha_m)\big) \\ &= -\alpha_m\big(T\tau\,[X_{J^\xi}, X_{J^\eta}](\alpha_m)\big), \quad \text{using (19)} \\ &= \alpha_m\big(T\tau(L_{X_{J^\eta}} X_{J^\xi})(\alpha_m)\big) = L_{X_{J^\eta}}\big(\alpha_m\big(T\tau\, X_{J^\xi}(\alpha_m)\big)\big) \\ &= \big(L_{X_{J^\eta}} J^\xi\big)(\alpha_m), \quad \text{using (26)} \\ &= \{J^\xi, J^\eta\}(\alpha_m), \quad \text{by definition of Poisson bracket.} \qquad \square \end{aligned}$$

Example 2. Let (V, σ) be a real symplectic vector space. A linear mapping $A : V \to V$ is *symplectic* if it preserves σ, that is, if $A^*\sigma = \sigma$. Note that a linear symplectic map is invertible; for if $Av = 0$, then $0 = \sigma(Av, Aw) = \sigma(v, w)$ for every $w \in V$. Hence $v = 0$, since σ is nondegenerate. The set $\mathrm{Sp}(V, \sigma)$ of all real symplectic linear mappings on (V, σ), is a Lie group because it is a closed subgroup of the group of all invertible linear mappings of V to itself. $\mathrm{Sp}(V, \sigma)$ is called the *real symplectic group*. The Lie algebra $\mathrm{sp}(V, \sigma)$ of $\mathrm{Sp}(V, \sigma)$ is the set of *infinitesimally symplectic* linear mappings $a : V \to V$, that is, $\sigma^\sharp \circ a + a^t \circ \sigma^\sharp = 0$.

Define a linear action Φ of $\mathrm{Sp}(V, \sigma)$ on V by

$$\Phi : \mathrm{Sp}(V, \sigma) \times V \to V : (A, v) \to Av. \tag{28}$$

For $\xi \in \mathrm{sp}(V,\sigma)$ the infinitesimal generator X^{ξ} of Φ in the direction ξ is ξ, because for every $v \in V$

$$X^{\xi}(v) = \left.\frac{d}{dt}\right|_{t=0} \Phi_{\exp t\xi}(v) = \xi(v).$$

Claim: Define the mapping $J : V \rightarrow \mathrm{sp}(V,\sigma)^*$ by

$$J(v)\xi = J^{\xi}(v) = \tfrac{1}{2}\sigma(\xi(v), v), \tag{29}$$

for every $\xi \in \mathrm{sp}(V,\sigma)$. Then J is a coadjoint equivariant momentum mapping for the linear action Φ.

(3.8) **Proof**: By definition J^{ξ} is a homogeneous quadratic function on V. Differentiating (29) gives

$$dJ^{\xi}(v)w = \sigma(\xi(v), w) = \sigma(X^{\xi}(v), w)$$

for every $v, w \in V$. In other words, X^{ξ} is a Hamiltonian vector field on (V,σ) corresponding to the Hamiltonian J^{ξ}. Suppose that $A \in \mathrm{Sp}(V,\sigma)$. Then

$$\begin{aligned} J(Av)\xi &= J^{\xi}(Av) = \tfrac{1}{2}\sigma\big(\xi(Av), Av\big) = \tfrac{1}{2}\sigma\big((A^{-1}\xi A)(v), v\big) \\ &= J^{\mathrm{Ad}_{A^{-1}}\xi}(v) = J(v)\mathrm{Ad}_{A^{-1}}\xi = \big(\mathrm{Ad}^t_{A^{-1}}(J(v))\big)\xi. \end{aligned} \tag{30}$$

Hence J is coadjoint equivariant. From (30) we obtain

$$J^{\mathrm{Ad}_{\exp -t\eta}(\xi)}(v) = J^{\xi}\big((\exp -t\eta)(v)\big),$$

for every $\eta \in \mathrm{sp}(V,\sigma)$ and every $t \in \mathbf{R}$. Taking the derivative with respect to t and evaluating at $t = 0$ gives

$$J^{-[\eta,\xi]}(v) = -dJ^{\xi}(v)\,X^{\eta}(v) = -dJ^{\xi}(v)\,X_{J^{\eta}}(v) = -\{J^{\eta}, J^{\xi}\}(v).$$

Consequently, J is a momentum mapping for the action Φ. □

Example 3. Let G be a Lie group with Lie algebra $\mathcal{G}$. The group G acts naturally on itself by left translation, namely,

$$L : G \times G \rightarrow G : (h, g) \rightarrow L_h g = hg.$$

The action L induces an action Φ on the cotangent bundle T^*G given by

$$\Phi : G \times T^*G \rightarrow T^*G : (h, \alpha_g) \rightarrow \alpha_{hg} = (T_g L_{h^{-1}})^t \alpha_g.$$

Φ preserves the canonical 1-form θ on T^*G and therefore the canonical symplectic ▷ form $\Omega = -d\theta$. For every $\xi \in \mathcal{G}$, the vector field

$$X^{\xi}(\alpha_g) = \left.\frac{d}{dt}\right|_{t=0} \Phi_{\exp t\xi}(\alpha_g)$$

on (T^*G, Ω) is Hamiltonian with Hamiltonian function

$$\mathcal{J}^{\xi} : T^*G \rightarrow \mathbf{R} : \alpha_g \rightarrow (T_e R_g)^t(\alpha_g)\xi.$$

(3.9) **Proof**: To see this, note that for $\xi \in \mathcal{G}$ the infinitesimal generator of the action L in the direction ξ is

$$X_\xi(g) = \frac{d}{dt}\bigg|_{t=0} L_{\exp t\xi} g = \frac{d}{dt}\bigg|_{t=0} R_g \exp t\xi = T_e R_g \xi.$$

Here $R_g : G \to G : h \to hg$ is right translation by g. Therefore, by the momentum lemma ((3.6)),

$$\mathcal{J}^\xi(\alpha_g) = \alpha_g(T_e R_g \xi) = (T_e R_g)^t(\alpha_g)\xi.$$

Thus the function

$$\mathcal{J} : T^*G \to \mathcal{G}^* : \alpha_g \to (T_e R_g)^t(\alpha_g)$$

is the momentum mapping of the action Φ. □

$\mathcal{J}$ intertwines the action Φ on T^*G with the coadjoint action of G on $\mathcal{G}^*$. In other words, $\mathcal{J}(\Phi_h(\alpha_g)) = \mathrm{Ad}^t_{h^{-1}}\, \mathcal{J}(\alpha_g)$.

Using the left trivialization

$$\mathcal{L} : G \times \mathcal{G}^* \to T^*G : (g, \alpha) \to (T_g L_{g^{-1}})^t \alpha = \alpha_g,$$

the G-action Φ on T^*G pulls back to the G-action $\varphi = \mathcal{L}^*\Phi$ on $G \times \mathcal{G}^*$ defined by $\varphi_h(g, \alpha) = (hg, \alpha)$ for $h \in G$. Because φ_h preserves $\vartheta = \mathcal{L}^*\theta$, the vector field

$$X^\xi(g, \alpha) = \frac{d}{dt}\bigg|_{t=0} \varphi_{\exp t\xi}(g, \alpha) = (T_e R_g \xi, 0)$$

is Hamiltonian with corresponding Hamiltonian function

$$\begin{aligned} \vartheta(X^\xi)(g, \alpha) &= \vartheta(g, \alpha) X^\xi(g, \alpha) = \alpha_g(T_e R_g \xi) \\ &= \alpha(T_g L_{g^{-1}} T_e R_g \xi) = \alpha(\mathrm{Ad}_{g^{-1}} \xi) = (\mathrm{Ad}^t_{g^{-1}} \alpha)\xi. \end{aligned}$$

Therefore, the pull back of the momentum mapping $\mathcal{J}$ by left trivialization $\mathcal{L}$ is the momentum mapping

$$J : G \times \mathcal{G}^* \to \mathcal{G}^* : (g, \alpha) \to \mathrm{Ad}^t_{g^{-1}} \alpha$$

of the action φ. Clearly the mapping J intertwines the G-action φ on $G \times \mathcal{G}^*$ with the coadjoint action of G on $\mathcal{G}^*$. □

Example 3′. Let H be a closed subgroup of a Lie group G. Then H is a Lie group. Let H act on G by left translation. The lift of this action to T^*G has a momentum mapping

$$J_H : T^*G \to \mathcal{H}^* : \alpha_g \to (T_e R_g)^t \alpha_g,$$

which is coadjoint equivariant. The lift of left translation by G on itself to T^*G has a coadjoint equivariant momentum mapping

$$J_G : T^*G \to \mathcal{G}^* : \alpha_g \to (T_e R_g)^t \alpha_g.$$

Because H is a Lie subgroup of G, we have an inclusion map $i : \mathcal{H} \to \mathcal{G}$ of Lie algebras. By duality this becomes the projection $\pi = i^t : \mathcal{G}^* \to \mathcal{H}^*$. Clearly, $J_H = \pi \circ J_G$. □

Example 4. Let $\mathcal{O}_\mu$ be the orbit through $\mu \in \mathcal{G}^*$ of the coadjoint action of G on the dual $\mathcal{G}^*$ of its Lie algebra, (see appendix A section 2 example 3). Clearly

$$\Phi : G \times \mathcal{O}_\mu \to \mathcal{O}_\mu : (g, \nu) \to \mathrm{Ad}^t_{g^{-1}}\nu$$

is a G-action. For every $\xi \in \mathcal{G}$, the vector field

$$X^\xi(\nu) = \frac{d}{ds}\Big|_{s=0} \mathrm{Ad}^t_{\exp -s\xi}\nu = -\mathrm{ad}^t_\xi \nu$$

is the infinitesimal generator of the action Φ in the direction ξ. Recall that in appendix A ((2.2)) we have shown that $\mathcal{O}_\mu$ is a symplectic manifold with symplectic form

$$\omega_{\mathcal{O}_\mu}(\nu)\big(X^\xi(\nu), X^\eta(\nu)\big) = -\nu([\xi, \eta]). \tag{31}$$

We now show that Φ is a Hamiltonian action on $(\mathcal{O}_\mu, \omega_{\mathcal{O}_\mu})$. First we verify that Φ_g is a symplectic diffeomorphism. We calculate:

$$\begin{aligned}(\Phi_g^*\omega_{\mathcal{O}_\mu})(\nu)\big(X^\xi(\nu), X^\eta(\nu)\big) &= \omega_{\mathcal{O}_\mu}(\mathrm{Ad}^t_{g^{-1}}\nu)\big(\mathrm{Ad}^t_{g^{-1}}(\mathrm{ad}^t_\xi(\nu)), \mathrm{Ad}^t_{g^{-1}}(\mathrm{ad}^t_\eta(\nu))\big)\\ &= \omega_{\mathcal{O}_\mu}(\mathrm{Ad}^t_{g^{-1}}\nu)\big(\mathrm{ad}^t_{\mathrm{Ad}_g\xi}(\mathrm{Ad}^t_{g^{-1}}\nu), \mathrm{ad}^t_{\mathrm{Ad}_g\eta}(\mathrm{Ad}^t_{g^{-1}}\nu)\big)\\ &= -(\mathrm{Ad}^t_{g^{-1}}\nu)([\mathrm{Ad}_g\xi, \mathrm{Ad}_g\eta]) = -\nu([\xi, \eta])\\ &= \omega_{\mathcal{O}_\mu}(\nu)\big(X^\xi(\nu), X^\eta(\nu)\big).\end{aligned}$$

Next we show that X^ξ is a Hamiltonian vector field on $(\mathcal{O}_\mu, \omega_{\mathcal{O}_\mu})$. Observe that (31) may be rewritten as

$$\omega_{\mathcal{O}_\mu}(\nu)\big(X^\xi(\nu), X^\eta(\nu)\big) = (\mathrm{ad}^t_\eta\nu)\xi = -X^\eta(\nu)(\xi) = -\xi(X^\eta(\nu)), \tag{32}$$

where we have identified $\mathcal{G}^{**}$ with $\mathcal{G}$. Now consider the linear function

$$J^\xi : \mathcal{G}^* \to \mathbf{R} : \alpha \to -\alpha(\xi).$$

Then $dJ^\xi(\alpha)\beta = -\beta(\xi) = -\xi(\beta)$, again identifying $\mathcal{G}^{**}$ with $\mathcal{G}$. Thus (32) may be written as

$$\omega_{\mathcal{O}_\mu}(\nu)\big(X^\xi(\nu), X^\eta(\nu)\big) = dJ^\xi(\nu)X^\eta(\nu). \tag{33}$$

Since $T_\nu\mathcal{O}_\mu$ is spanned by the vectors $X^\eta(\nu)$ as η ranges over $\mathcal{G}$, it follows that X^ξ is the Hamiltonian vector field X_{J^ξ}. To show that Φ has momentum mapping

$$J : \mathcal{O}_\mu \subseteq \mathcal{G}^* \to \mathcal{G}^* : \nu \to -\nu,$$

we must verify that $J^{[\xi,\eta]} = \{J^\xi, J^\eta\}$ for every $\xi, \eta \in \mathcal{G}$. For every $\nu \in \mathcal{O}_\mu$ we have

$$\begin{aligned}J^{[\xi,\eta]}(\nu) &= -\nu([\xi, \eta]), \quad \text{by definition of } J\\ &= \omega_{\mathcal{O}_\mu}(\nu)\big(X^\xi(\nu), X^\eta(\nu)\big), \quad \text{by (31)}\\ &= \omega_{\mathcal{O}_\mu}(\nu)\big(X_{J^\xi}(\nu), X_{J^\eta}(\nu)\big) = \{J^\xi, J^\eta\}. \qquad \square\end{aligned}$$

3.2 Normal form for a momentum mapping

In this subsection we find a local normal form for a coadjoint equivariant momentum mapping of a proper Hamiltonian action near a fixed point in its zero level set. The normal form is used to show that the zero level set of the momentum mapping is locally arcwise connected.

Let $\Phi : G \times M \to M$ be a proper Hamiltonian action of a Lie group G on a symplectic manifold (M, ω) with coadjoint equivariant momentum map $J : M \to \mathcal{G}^*$. For $\mu \in \mathcal{G}^*$ let m be a point in the level set $J^{-1}(\mu)$. Let G_m be the isotropy group of m under the G-action Φ and let G_μ be the isotropy group of μ under the coadjoint action of G on $\mathcal{G}^*$. Note that the linear $\omega(m)$-symplectic action

$$\widehat{\Phi} : G_m \times T_m M \to T_m M : (h, v_m) \to T_m \Phi_h \, v_m \tag{34}$$

has a G_m-coadjoint equivariant momentum mapping $\widehat{J} : T_m M \to \mathcal{G}^*_m$, where

$$\widehat{J}(v_m)\xi = \tfrac{1}{2}\,\omega(m)(X^\xi(v_m), v_m), \tag{35}$$

for $\xi \in \mathcal{G}_m$ and $v_m \in T_m M$, (see example 2). Here the vector field X^ξ is the infinitesimal generator of the G_m-action $\widehat{\Phi}$ in the direction ξ.

Before we can state the local normal form of the momentum mapping J, we need the following decomposition of $T_m M$.

Fact:

$$T_m M = \mathcal{H} \oplus \big(X \oplus T^*_m(G_\mu \cdot m)\big), \tag{36}$$

where X is a complement to $W = (\ker T_m J) \cap (\ker T_m J)^{\omega(m)}$ in $\ker T_m J$ and $\mathcal{H}$ is a subspace of $\mathcal{G}$ which is isomorphic to $\mathcal{G}/\mathcal{G}_m$, that is, $\mathcal{H}$ is isomorphic to $T_m(G \cdot m)$.

(3.10) **Proof**: Using the Witt decomposition ((1.4)) of appendix A starting with the subspace $\ker T_m J$ of the symplectic vector space $(T_m M, \omega(m))$, we obtain the decomposition

$$T_m M = X \oplus Y \oplus Z, \tag{37}$$

where X, Y and Z are the $\omega(m)$-symplectic subspaces of $T_m M$ defined by $\ker T_m J = X \oplus W$, $(\ker T_m J)^{\omega(m)} = Y \oplus W$, and $Z = (X \oplus Y)^{\omega(m)}$. From the Witt decomposition it follows that W is a Lagrangian subspace of Z. Therefore Z is isomorphic to $W \oplus W^*$. Thus we may rewrite (37) as

$$T_m M = (Y \oplus W) \oplus (X \oplus W^*) = (\ker T_m J)^{\omega(m)} \oplus (X \oplus W^*). \tag{38}$$

To finish the proof of (36) it suffices to show

Fact:

$$T_m(G \cdot m) = (\ker T_m J)^{\omega(m)} \tag{39}$$

and

$$T_m(G_\mu \cdot m) = (\ker T_m J) \cap (\ker T_m J)^{\omega(m)} = W. \tag{40}$$

(3.11) **Proof**: We prove (39) as follows. From the definition of the momentum mapping it follows that

$$\omega(m)(X^{\xi}(m), v) = dJ^{\xi}(m)v = (T_mJ(v))\xi, \tag{41}$$

for every $\xi \in \mathcal{G}$. If $v \in \ker T_mJ$, then from (41) we find that $\omega(m)(X^{\xi}(m), v) = 0$ for every $\xi \in \mathcal{G}$. In other words, $v \in (T_m(G \cdot m))^{\omega(m)}$. Conversely, if $v \in (T_m(G \cdot m))^{\omega(m)}$, then $\omega(m)(X^{\xi}(m), v) = 0$ for every $\xi \in \mathcal{G}$. Hence from (41) it follows that $(T_mJ(v))\xi = 0$ for every $\xi \in \mathcal{G}$, that is, $v \in \ker T_mJ$. This establishes (39).

To prove (40) we begin by showing that

$$T_m(G_{\mu} \cdot m) = T_m(G \cdot m) \cap \ker T_mJ. \tag{42}$$

To verify the inculsion $\subseteq$ we argue as follows. Let $v_m \in T_m(G_{\mu} \cdot m)$. Then for some ξ in the Lie algebra $\mathcal{G}_{\mu}$ of G_{μ}, $v_m = X^{\xi}(m)$. Differentiating the relation $J(\Phi_{\exp s\,\xi}(m)) = \mathrm{Ad}^t_{\exp -s\,\xi}(J(m))$ with respect to s and then setting $s = 0$ gives

$$T_mJ\,X^{\xi}(m) = X^{\xi}_{\mathcal{G}^*}(J(m)) = X^{\xi}_{\mathcal{G}^*}(\mu) = 0, \tag{43}$$

where $X^{\xi}_{\mathcal{G}^*}(\mu) = \left.\frac{d}{ds}\right|_{s=0} \mathrm{Ad}^t_{\exp -s\xi}\,\mu$ and the last equality follows because $\xi \in \mathcal{G}_{\mu}$. Therefore $X^{\xi}(m) \in \ker T_mJ$, which proves $T_m(G_{\mu} \cdot m) \subseteq \ker T_mJ$. Since $G_{\mu} \subseteq G$, $T_m(G_{\mu} \cdot m) \subseteq T_m(G \cdot m)$. Therefore the inclusion $\subseteq$ in (42) holds. To prove $\supseteq$ suppose that $v_m \in T_m(G \cdot m) \cap \ker T_mJ$. Then there is a ξ in the Lie algebra of $\mathcal{G}$ such that $v_m = X^{\xi}(m) = T_e\Phi_m\xi$. Because $v_m \in \ker T_mJ$, it follows from $T_mJ\,X^{\xi}(m) = X^{\xi}_{\mathcal{G}^*}(\mu)$ that $X^{\xi}_{\mathcal{G}^*}(\mu) = 0$, that is, ξ is an element of the Lie algebra $\mathcal{G}_{\mu}$ of G_{μ}. Consequently, $v_m \in T_m(G_{\mu} \cdot m)$. This proves (42). Substituting (39) into (42) gives (40). □

This completes the proof of the decomposition (36). □

We now prove

Corollary 1: When $\mu = 0$ the decomposition (36) reads

$$T_mM = \mathcal{H} \oplus X \oplus W^*, \tag{44}$$

where $\mathcal{H} = W = T_m(G \cdot m)$.

(3.12) **Proof**: When $\mu = 0$ the isotropy group G_{μ} equals G. Therefore from (39) and (40) we obtain

$$T_m(G \cdot m) = (\ker T_mJ)^{\omega(m)} = (\ker T_mJ) \cap (\ker T_mJ)^{\omega(m)} = W. \tag{45}$$

Hence by definition $Y = \{0\}$ in (38). □

Corollary 2: The spaces X, Y, W and W^* in the Witt decomposition

$$T_mM = X \oplus Y \oplus (W \oplus W^*) \tag{46}$$

can be choosen to be G_m-invariant.

(3.13) **Proof**: Since $J^{-1}(\mu)$ is invariant under the action $\Phi|(G_m \times M)$, $\ker T_m J = T_m J^{-1}(\mu)$ is invariant under the linear $\omega(m)$-symplectic action $\widehat{\Phi}$ (34). Consequently, $(\ker T_m J)^{\omega(m)}$ and $W = \ker T_m J \cap (\ker T_m J)^{\omega(m)}$ are also invariant under $\widehat{\Phi}$.

Let γ be an inner product on $T_m M$. Then averaging over G_m, which is compact because the G-action is proper, we may assume that γ is $\widehat{\Phi}$-invariant. Let X, Y be the orthogonal complement of W in $\ker T_m J$ and $(\ker T_m J)^{\omega(m)}$, respectively. Then X and Y are $\widehat{\Phi}$-invariant $\omega(m)$-symplectic subspaces of $(T_m M, \omega(m))$. Therefore Z, which is the orthogonal complement of $X \oplus Y$ in $T_m M$, is a $\widehat{\Phi}$-invariant $\omega(m)$-symplectic subspace. Hence we have obtained the $\widehat{\Phi}$-invariant Witt decomposition $T_m M = X \oplus Y \oplus Z$. Since W is a $\widehat{\Phi}$-invariant Lagrangian subspace of Z, its orthogonal complement $W^{\perp}$ in Z is $\widehat{\Phi}$-invariant. Since $W^{\perp}$ is isotropic and hence Lagrangian, it is isomorphic to W^*. This proves (46). □

We now are in a position to state the local normal form theorem.

Claim: Let $m \in J^{-1}(0)$. Using the decomposition (44) we write the tangent space $T_m M$ to M at m as the sum $\mathcal{H} \oplus X \oplus W^*$. Let (η, x, α) be coordinates on $T_m M$ with respect to this decomposition. Then there is a local diffeomorphism $\vartheta : T_m M \to M$ with $\vartheta(0) = m$ and $T_0\vartheta = \mathrm{id}_{T_m M}$ such that for every $v_m = (\eta, x, \alpha)$ sufficiently close to 0,

$$\vartheta^* J(\eta, x, \alpha) = \mathrm{Ad}^t_{\exp -\eta}(\widehat{J}(x) + \alpha). \tag{47}$$

Outline of the proof: We prove the claim in four steps. In step 1 we symplectically identify a neighborhood of 0 in $(T_m M, \omega(m))$ with a neighborhood of m in (M, ω) in such a way that the G_m-action $\Phi|(G_m \times M)$ becomes the linear $\omega(m)$-symplectic G_m-action $\widehat{\Phi}$. In step 2, after identifying the momentum mapping J with a locally defined G_m-coadjoint equivariant mapping $\mathcal{J} : T_m M \to \mathcal{G}^*$, we split $\mathcal{J}$ into a sum of two locally defined G_m-coadjoint equivariant maps $\mathcal{J}' : T_m M \to \mathcal{G}^*_m$ and $\mathcal{J}'' : T_m M \to \mathcal{H}^*$ with $\mathcal{J}'(0) = 0$, $\mathcal{J}''(0) = 0$. In step 3 we analyze the mapping $\mathcal{J}'$. Because $\mathcal{J}'(0) = 0$, $\mathcal{J}' = \widehat{J}$ (35). We then use the decomposition

$$T_m M = X \oplus \left(W \oplus W^*\right), \tag{48}$$

where the summands are G_m-invariant and symplectically perpendicular symplectic subspaces of $(T_m M, \omega(m))$, to show that $W^* \subseteq \ker \widehat{J}$. Since $\mathcal{H} = W$, it follows that $\mathcal{H}^* = W^*$. In step 4 we show that the map $\mathcal{J}''$ is a local submersion. Changing coordinates on $T_m M$ by a local diffeomorphism θ such that $\theta^*\mathcal{J}''$ is locally a projection onto $\mathcal{H}^*$ and $\theta|(X \oplus W)$ is the identity map, we see that for $(x, 0, \alpha) \in T_m M = X \oplus W \oplus W^*$ near $(0, 0, 0)$ we have

$$\theta^*\mathcal{J}(x, 0, \alpha) = \widehat{J}(x) + \alpha. \tag{49}$$

Using the exponential mapping from $\mathcal{H}$ to G, the G-action Φ, and the local diffeomorphism of step 1, we bring the momentum mapping J into the local normal form (47).

(3.14) **Proof**: We now prove the claim.

Step 1.
Since the G-action Φ is proper, the isotropy group G_m is compact. Let γ be a G_m-invariant Riemannian metric on M. Then the *exponential map* $\mathrm{Exp} : T_m M \to M$ associated to

the metric γ is a local diffeomorphism with $\mathrm{Exp}(0) = m$ such that $T_0 \,\mathrm{Exp} = \mathrm{id}_{T_m M}$. Because γ is G_m-invariant, the map Exp intertwines the linear G_m-action $\widehat{\Phi}$ (34) on $T_m M$ with the G_m-action $\Phi|(G_m \times M)$ on M. Pulling back the symplectic form ω on M by the exponential map gives a symplectic form $\widehat{\sigma}$ on a neighborhood of 0 in $T_m M$ with $\widehat{\sigma}(0) = \omega(m)$. Since the map $\Phi_h : M \to M$ is a symplectic diffeomorphism for every $h \in G_m$, it follows that $\widehat{\Phi}_h : T_m M \to T_m M$ is a linear symplectic mapping of $(T_m M, \widehat{\sigma})$. The equivariant Darboux theorem (see ((3.15))) shows that there is a local diffeomorphism $\psi : T_m M \to T_m M$ with $\psi(0) = 0$, which pulls back the symplectic form $\widehat{\sigma}$ to the *constant* symplectic form $\omega(m)$, commutes with the linear G_m action $\widehat{\Phi}$, and has $T_0 \psi = \mathrm{id}_{T_m M}$. Thus the local diffeomorphism $\varphi = \psi \circ \mathrm{Exp} : T_m M \to M$ symplectically identifies a neighborhood of 0 in $T_m M$ with a neighborhood of m in M so that the ω-symplectic G_m-action $\Phi|(G_m \times M)$ becomes the linear $\omega(m)$-symplectic G_m-action $\widehat{\Phi}$ on $T_m M$.

To finish the first step we prove the *equivariant Darboux theorem.*

Fact: Let $G \times V \to V$ be a linear action of a compact Lie group G on a vector space V. If this action preserves the symplectic forms ω_0 and ω_1 on V, where $\omega_0(0) = \omega_1(0)$, then there are G-invariant neighborhoods $\mathcal{U}_0$ and $\mathcal{U}_1$ of 0 and a diffeomorphism $\varphi : \mathcal{U}_0 \to \mathcal{U}_1$ which maps 0 to 0 and commutes with the G-action such that $\varphi^* \omega_1 = \omega_0$ on $\mathcal{U}_0$.

(3.15) **Proof**: Suppose that on a G-invariant neighborhood $\mathcal{U}_0$ of 0 we can find a G-invariant 1-form ζ for which $\zeta(0) = 0$ and $d\zeta = \omega_0 - \omega_1$. Then on a possibly smaller G-invariant neighborhood of 0 define a G-invariant time dependent vector field ξ where ξ_t is the vector field $\big(t\omega_1 + (1-t)\omega_0\big)^{\flat} \zeta$ with t held fixed. Let ϑ be the flow of ξ. By shrinking the G-invariant neighborhood of 0, we can arrange that $\mathcal{U}_0 \times [0,1]$ is in the domain of ϑ. The next calculation shows that for every $t \in [0,1]$ the curve $t \to \vartheta_t^*\big(t\omega_1 + (1-t)\omega_0\big)$ is constant. It suffices to show that its derivative is zero. We compute:

$$
\begin{aligned}
\frac{d}{dt}\vartheta_t^*\big(t\omega_1 + (1-t)\omega_0\big) &= \vartheta_t^*\big(L_\xi(t\omega_1 + (1-t)\omega_0)\big) \\
&= \vartheta_t^*\big(L_{\xi_t + \frac{\partial}{\partial t}}(t\omega_1 + (1-t)\omega_0)\big), \quad \text{since } \xi = \xi_t + \tfrac{\partial}{\partial t} \\
&= \vartheta_t^*\big(L_{\xi_t}(t\omega_1 + (1-t)\omega_0) + \omega_1 - \omega_0\big) \\
&= \vartheta_t^*\big(d(\xi_t \,\lrcorner\, (t\omega_1 + (1-t)\omega_0)) + \xi_t \,\lrcorner\, d(t\omega_1 + (1-t)\omega_0) + \omega_1 - \omega_0\big) \\
&= \vartheta_t^*(d\zeta + \omega_1 - \omega_0), \quad \text{since } d\omega_0 = d\omega_1 = 0 \\
&= 0, \quad \text{by construction of } \zeta.
\end{aligned}
$$

Therefore on $\mathcal{U}_0$ we have $\omega_0 = \vartheta_1^*(\omega_1)$. Since ξ_t is G-invariant, its flow ϑ_t commutes with the G-action on V. Setting $\varphi = \vartheta_1$, we see that φ, which maps $\mathcal{U}_0$ onto $\mathcal{U}_1 = \varphi(\mathcal{U}_0)$, is the desired diffeomorphism.

To finish the proof it remains to construct the 1-form ζ with the desired properties. For every $v \in V$ let $\psi_t(v) = (1-t)v$ be a one parameter group of radial contractions on V. Then $\psi_0 = \mathrm{id}_V$, $\psi_1 = 0$, $\psi_t(0) = 0$, and ψ_t commutes with the linear action of G on V. We calculate

$$
-(\omega_1 - \omega_0) = \psi_1^*(\omega_1 - \omega_0) - \psi_0^*(\omega_1 - \omega_0)
$$

$$= \int_0^1 \frac{d}{dt}\big(\psi_t^*(\omega_1 - \omega_0\big)\, dt$$

$$= \int_0^1 \psi_t^*\big(L_\eta(\omega_1 - \omega_0)\big)\, dt, \quad \text{where } \eta \text{ is the vector field infinitesimally generated by } \psi_t.$$

$$= \int_0^1 \psi_t^*\big(d(\eta \lrcorner (\omega_1 - \omega_0))\big)\, dt, \quad \text{since } d(\omega_1 - \omega_0) = 0$$

$$= d\Big(\int_0^1 \psi_t^*(\eta \lrcorner (\omega_1 - \omega_0))\, dt\Big).$$

Set $\zeta = \int_0^1 \psi_t^*\big(\eta \lrcorner (\omega_1 - \omega_0)\big)\, dt$. Since $(\omega_1 - \omega_0)(0) = 0$, it follows that $\zeta(0) = 0$. Because the integrand is G-invariant, it follows that ζ is also. Thus the 1-form ζ has the right properties. □

Step 2.
Using the local diffeomorphism $\varphi : T_m M \to M$ constructed in step 1, pull back the G-coadjoint equivariant momentum map $J : M \to \mathcal{G}^*$ with $J(m) = 0$ to obtain a locally defined G_m-equivariant map $\mathcal{J} = \varphi^*(J) : T_m M \to \mathcal{G}^*$ with $\mathcal{J}(0) = 0$. Split $\mathcal{J}$ into the sum of two maps as follows. Since the isotropy group G_m is closed, it is a Lie subgroup of G. Thus the Lie algebra $\mathcal{G}_m$ of G_m is a subalgebra of $\mathcal{G}$. Hence we have an inclusion map $i' : \mathcal{G}_m \to \mathcal{G}$. By duality we obtain a projection map $\pi' = (i')^t : \mathcal{G}^* \to \mathcal{G}_m^*$. Since G_m is compact, there is a G_m-coadjoint invariant inner product on $\mathcal{G}^*$. Let $\mathcal{H}^* = (\mathcal{G}_m^*)^\perp$. Then $\mathcal{G}^* = \mathcal{G}_m^* \oplus \mathcal{H}^*$ and there is a projection $\pi'' : \mathcal{G}^* \to \mathcal{H}^*$ such that $\pi' + \pi'' = \mathrm{id}_{\mathcal{G}^*}$. Write $\mathcal{J} = \pi' \circ \mathcal{J} + \pi'' \circ \mathcal{J} = \mathcal{J}' + \mathcal{J}''$. Then $\mathcal{J}' : T_m M \to \mathcal{G}_m^*$ and $\mathcal{J}'' : T_m M \to \mathcal{H}^*$ are locally defined G_m-coadjoint equivariant maps with $\mathcal{J}'(0) = \pi'(0) = 0$ and $\mathcal{J}''(0) = \pi''(0) = 0$.

Step 3.
We now analyze the mapping $\mathcal{J}'$. Since $\mathcal{J}'$ is G_m-coadjoint equivariant and $\mathcal{J}'(0) = 0$, it follows that $\mathcal{J}'$ is the canonical momentum map $\widehat{J}$ (35) of the linear symplectic action $\widehat{\Phi}$ on $(T_m M, \omega(m))$.

Note that the mapping $\rho : \mathcal{G}_m \to \mathrm{sp}(T_m M, \omega(m)) : \xi \to X^\xi$ is an antihomomorphism of Lie algebras. We now study ρ more closely. Because the Witt decomposition $T_m M = X \oplus (W \oplus W^*)$ ((3.12)) has summands which are invariant under the linear G_m-action $\widehat{\Phi}$ and because the linear Hamiltonian vector field X^ξ is the infinitesimalization of this action in the direction ξ, it follows that each summand of the Witt decomposition is *invariant* under X^ξ for every $\xi \in \mathcal{G}_m$. Since W and W^* are $\widehat{\Phi}$-invariant Lagrangian subspaces of Z, W and W^* are invariant Lagrangian subspaces of $X^\xi | Z$ for every $\xi \in \mathcal{G}_m$.

We are now in position to prove

Claim: For every $\alpha \in W^*$, $\widehat{J}(\alpha) = 0$.

(3.16) **Proof**: Since W^* is an X^ξ-invariant subspace for every $\xi \in \mathcal{G}_m$, it follows that $X^\xi(\alpha) \in W^*$ for every $\xi \in \mathcal{G}_m$. Because W^* is Lagrangian $\omega(m)(X^\xi(\alpha), \alpha) = 0$. Hence $\widehat{J}(\alpha)(\xi) = 0$ for every $\xi \in \mathcal{G}_m$, which proves the claim. □

Corollary: Let $\alpha \in W^*$ and $x \in X$, then

$$\widehat{J}(\alpha + x) = \widehat{J}(x). \tag{50}$$

(3.17) **Proof**: Let $\xi \in \mathcal{G}_m$. We compute:

$$\begin{aligned} 2\,\widehat{J}(\alpha + x)\xi &= \omega(m)(X^\xi(\alpha), \alpha) + \omega(m)(X^\xi(\alpha), x) \\ &\qquad + \omega(m)(X^\xi(x), \alpha) + \omega(m)(X^\xi(x), x) \\ &= \omega(m)(X^\xi(x), x), \\ &\qquad \text{since } X^\xi(\alpha) \in W^*,\ X^\xi(x) \in X, \text{ and } W^* \subseteq X^{\omega(m)} \\ &= 2\,\widehat{J}(x). \quad \square \end{aligned}$$

Step 4.
Next we analyze the mapping $\mathcal{J}'' : T_m M \to \mathcal{H}^*$.

Claim: The map $\mathcal{J}''$ with $\mathcal{J}''(0) = 0$ is a local submersion at 0.

(3.18) **Proof**: It suffices to show that $T_0\mathcal{J}''|W^*$ is injective, because $\mathcal{H}^* = W^*$. Suppose that there is an $\alpha \in W^*$ such that

$$0 = T_0\mathcal{J}''(\alpha) = \pi''(T_0\mathcal{J}(\alpha)) = \pi''(T_m J(\alpha)).$$

In other words, $T_m J(\alpha) \in \mathcal{G}_m^*$. Suppose that $T_m J(\alpha)$ is nonzero, then there is a $\zeta \in \mathcal{G}_m$ such that

$$1 = T_m J(\alpha)\zeta = \omega(m)(X^\zeta(\alpha), \alpha).$$

But $X^\zeta(\alpha) = 0$, since $\zeta \in \mathcal{G}_m$. This is a contradiction. Therefore $T_m J(\alpha) = 0$, that is, $\alpha \in \ker\, T_m J = X \oplus W$. From the decomposition (44) it follows that

$$(X \oplus W) \cap W^* = \{0\}.$$

Therefore $\alpha = 0$. Hence $T_0\mathcal{J}''|W^*$ is injective. $\square$

Let (x, w, α) be coordinates on $T_m M = X \oplus W \oplus W^*$. From the above claim and the implicit function theorem it follows that there is a local diffeomorphism $\theta : T_m M \to T_m M$ with $\theta(0) = 0$ such that $\mathcal{J}'' \circ \theta$ is the projection $T_m M \to \mathcal{H}^* = W^* : (x, w, \alpha) \to \alpha$ and $\theta|(X \oplus W) = id_{X \oplus W}$. In other words

$$\mathcal{J}''\big(\theta(x, w, \alpha)\big) = \alpha \tag{51}$$

and

$$\theta(x, w, \alpha) = \big(x, w, \theta_3(x, w, \alpha)\big) \tag{52}$$

for every $(x, w, \alpha) \in X \oplus W \oplus W^*$. Therefore

$$\begin{aligned} \theta^*\mathcal{J}(x, 0, \alpha) &= \mathcal{J}'(\theta(x, 0, \alpha)) + \mathcal{J}''(\theta(x, 0, \alpha)) \\ &= \mathcal{J}'(x, 0, \theta_3(x, 0, \alpha)) + \alpha, \quad \text{using (51) and (52)} \\ &= \widehat{J}(x) + \alpha, \quad \text{using the corollary of ((3.16)).} \end{aligned}$$

Since $X \oplus W^*$ is transverse to the G-orbit $G \cdot m$ at m and $\mathcal{H}$ is its tangent space $T_m(G \cdot m)$, using the map φ constructed in step 1 define

$$\vartheta : (X \oplus W^*) \times \mathcal{H} \to M : \big((x, \alpha), \eta\big) \to \Phi_{\exp \eta}\big(\varphi(\theta(x, 0, \alpha))\big).$$

Note that $\vartheta(0,0,0) = m$. Moreover, ϑ is a local diffeomorphism because $T_{(0,0,0)}\vartheta = \mathrm{id}_{T_m M}$. Therefore, for every u in M near m there is a unique point (x, α, η) which lies in $(X \oplus W) \times \mathcal{H}$ such that $u = \vartheta(x, \alpha, \eta)$. Hence

$$J(u) = J\big(\exp \eta \cdot \varphi(\theta(x, \alpha))\big) = \mathrm{Ad}^t_{\exp -\eta}\big(\theta^* \mathcal{J}(x, 0, \alpha)\big) = \mathrm{Ad}^t_{\exp -\eta}(\widehat{J}(x) + \alpha),$$

which is the desired normal form. □

Corollary: For every $\mu \in \mathcal{G}^*$ the level set $J^{-1}(\mu)$ is locally arcwise connected.

Using the following device, called the *shifting trick*, we reduce the proof of the corollary to the case of the 0-level of a G coadjoint equivariant momentum mapping $\mathcal{K}$ of a proper G-action. Consider the symplectic manifold $M \times \mathcal{O}_\mu$ with symplectic form $\Omega = \pi_1^*\omega + \pi_2^*\omega_{\mathcal{O}_\mu}$, where ω is a symplectic form on M and $\omega_{\mathcal{O}_\mu}$ is the symplectic form on the G coadjoint orbit $\mathcal{O}_\mu$ given in example 3 of section 2 of appendix A. Here π_i is the projection on the ith factor of $M \times \mathcal{O}_\mu$. Define a G-action on $M \times \mathcal{O}_\mu$ by

$$\big((g, (m, \nu))\big) \to (\Phi_g(m), \mathrm{Ad}^t_{g^{-1}}\nu).$$

This action is proper and has coadjoint equivariant momentum mapping $\mathcal{K} : M \times \mathcal{O}_\mu \to \mathcal{G}^*$ given by $\mathcal{K}(m, \nu)\xi = J^\xi(m) - \nu(\xi)$ for every $\xi \in \mathcal{G}$.

Fact. The level sets $J^{-1}(\mu) \times \mathcal{O}_\mu$ and $\mathcal{K}^{-1}(0)$ are locally diffeomorphic.

(3.19) **Proof**: Let $\mathcal{U}$ be a neighborhood of $\mu \in \mathcal{O}_\mu$. Suppose that $\sigma : \mathcal{U} \to G$ is a local section of the bundle $G \to G/G_\mu$ such that $\sigma(\mu) = e$ and $\mathrm{Ad}^t_{\sigma(\nu)^{-1}}\mu = \nu$, (see section 2 example 1). Observe that the the map

$$\alpha : M \times \mathcal{U} \to M \times \mathcal{U} : (m, \nu) \to \big(\Phi_{\sigma(\nu)}(m), \nu\big) = \big(\sigma(\nu) \cdot m, \nu\big)$$

is a local diffeomorphism. For every $\xi \in \mathcal{G}^*$ we have

$$\begin{aligned} \mathcal{K}(\alpha(m, \nu))\xi &= \mathcal{K}(\sigma(\nu) \cdot m, \nu)\xi \\ &= (J(\sigma(\nu) \cdot m) - \nu)(\xi) = \big(\mathrm{Ad}^t_{\sigma(\nu)^{-1}} J(m) - \nu\big)\xi. \end{aligned}$$

Thus $\mathcal{K}(\alpha(m, \nu)) = 0$ if and only if $J(m) = \mathrm{Ad}^t_{\sigma(\nu)}\nu = \mu$ and $\nu \in \mathcal{U}$. □

(3.20) **Proof**: We now prove the corollary. Since $\mathcal{K}^{-1}(0)$ and $J^{-1}(\mu) \times \mathcal{O}_\mu$ are locally diffeomorphic and $\mathcal{O}_\mu$ is locally arcwise connected, it follows that $\mathcal{K}^{-1}(0)$ is locally arcwise connected if and only if $J^{-1}(\mu)$ is. Thus it suffices to prove the corollary when $\mu = 0$. Applying the normal form ((3.14)) to the value 0 of the coadjoint equivariant momentum mapping J, we see that

$$\begin{aligned} u &= J(x, \alpha, \eta) \in J^{-1}(0) \Leftrightarrow 0 = \mathrm{Ad}^t_{\exp -\eta}\big(\widehat{J}|X)(x) + \alpha\big) \\ &\Leftrightarrow 0 = (\widehat{J}|X)(x) \text{ and } \alpha = 0. \end{aligned}$$

But $\widehat{J}$ is the canonical quadratic momentum map of a linear symplectic G_m-action. Hence $(\widehat{J}|X)^{-1}(0)$ is a cone in X, which is locally arcwise connected. Therefore $J^{-1}(0)$ is locally arcwise connected. □

4 Reduction: the regular case

Reduction is the basic technique in symplectic geometry for removing symmetry from a Hamiltonian system.

In the regular case the *reduction theorem* reads

Theorem: Let Φ be a free proper action of the Lie group G on the symplectic manifold (M, ω), which has a coadjoint equivariant momentum mapping $J : M \to \mathcal{G}^*$. Suppose that $\mu \in \mathcal{G}^*$ is a *regular value* of J. Then the *reduced space* $M_\mu = J^{-1}(\mu)/G_\mu$ is a smooth symplectic manifold with symplectic form ω_μ defined by $\pi_\mu^* \omega_\mu = i^*\omega$. Here $\pi_\mu : J^{-1}(\mu) \to M_\mu$ is the orbit map (= *reduction map*) of the G_μ-action $\Phi|(G_\mu \times J^{-1}(\mu))$ and $i : J^{-1}(\mu) \to M$ is the inclusion.

(4.1) **Proof**: The reduced space M_μ is a smooth manifold, because G_μ acts freely and properly on the smooth manifold $J^{-1}(\mu)$, (see ((2.9))). Thus we need only construct the symplectic form ω_μ on M_μ. Let $m \in J^{-1}(\mu)$. Since μ is a regular value of J, we have $T_m J^{-1}(\mu) = \ker T_m J$. Therefore

$$W_m = T_m J^{-1}(\mu) \cap (T_m J^{-1}(\mu))^{\omega(m)} = \ker \big(\omega(m)|T_m J^{-1}(\mu)\big)$$

is an $\omega(m)$-isotropic subspace of the symplectic vector space $(T_m M, \omega(m))$. Let X_m be a complement to W_m in $T_m J^{-1}(\mu)$. From ((1.3D)) of appendix A, it follows that X_m is an $\omega(m)$-symplectic subspace of $(T_m M, \omega(m))$. Using ((3.11)) we see that $W_m = T_m(G_\mu \cdot m)$. Thus X_m is isomorphic to $T_m J^{-1}(\mu)/W_m = T_{\pi_\mu(m)} M_\mu$. Since $\ker T_m \pi_\mu = W_m$, the symplectic form $\omega(m)|X_m$ pushes down under $T_m \pi_\mu$ to a symplectic form $\omega_\mu(\pi_\mu(m))$ on $T_{\pi_\mu(m)} M_\mu$. In other words, for every $m \in J^{-1}(\mu)$ we have $(\pi_\mu^* \omega_\mu)(m) = i^*\omega(m)$. Now

$$\pi_\mu^*(d\omega_\mu) = d(\pi_\mu^* \omega_\mu) = d(i^*\omega) = i^*(d\omega) = 0,$$

because ω is closed. Hence $d\omega_\mu = 0$, since π_μ is a submersion. Therefore, ω_μ is a symplectic form on M_μ. □

Example 1. Consider the Hamiltonian G action

$$\Phi : G \times T^*G \to T^*G : (h, \alpha_g) \to \alpha_{hg} = (T_e L_{h^{-1}})^t \alpha_g$$

on the cotangent bundle T^*G of G with its canonical symplectic form Ω. The action Φ is the lift to T^*G of the action

$$L : G \times G \to G : (h, g) \to L_h g = hg$$

of left multiplication on G. Using example 3 of section 3 we see that Φ has a coadjoint equivariant momentum mapping

$$\mathcal{J} : T^*G \to \mathcal{G}^* : \alpha_g \to (T_e R_g)^t(\alpha_g).$$

Pulling back by the left trivialization

$$\mathcal{L} : G \times \mathcal{G}^* \to T^*G : (g, \alpha) \to (T_e L_{g^{-1}})^t \alpha = \alpha_g, \tag{53}$$

which is a symplectic diffeomorphism of $(G \times \mathcal{G}^*, \omega = \mathcal{L}^*\Omega)$ with (T^*G, Ω), the action Φ becomes the G-action

$$\varphi : G \times (G \times \mathcal{G}^*) \to G \times \mathcal{G}^* : \big(h, (g, \alpha)\big) \to (hg, \alpha),$$

on $G \times \mathcal{G}^*$. The action φ is Hamiltonian with momentum mapping

$$J = \mathcal{L}^* \mathcal{J} : G \times \mathcal{G}^* \to \mathcal{G}^* : (g, \alpha) \to \mathrm{Ad}^t_{g^{-1}} \alpha.$$

We now apply the regular reduction theorem to φ. To begin with, observe that for every $\mu \in \mathcal{G}^*$ the μ-level set of the momentum mapping J is the smooth manifold

$$\Big\{(g, \mathrm{Ad}^t_g \mu) \in G \times \mathcal{G}^* \big| g \in G\Big\}.$$

Indeed, $J^{-1}(\mu)$ is diffeomorphic to G, because it is the graph of the smooth mapping $g \to \mathrm{Ad}^t_g \mu$. Let

$$G_\mu = \Big\{h \in G \big| \mathrm{Ad}^t_{h^{-1}} \mu = \mu\Big\}$$

be the isotropy group of the coadjoint action of G on $\mathcal{G}^*$ at μ. Note that $J^{-1}(\mu)$ is G_μ-invariant by the equivariance of J. Since G_μ is a closed subgroup of G, it is a smooth submanifold of G. Moreover, the orbit space G/G_μ is a smooth manifold, (see section 2 example 1). Because the diffeomorphism

$$\chi : G \to J^{-1}(\mu) : g \to (g, \mathrm{Ad}^t_g \mu)$$

intertwines the G_μ-action on G with the G_μ-action on $J^{-1}(\mu)$, the orbit spaces $J^{-1}(\mu)/G_\mu$ and G/G_μ are diffeomorphic. In particular the orbit space $J^{-1}(\mu)/G_\mu$ is a smooth manifold. We now show that $J^{-1}(\mu)/G_\mu$ is *symplectically* diffeomorphic to the coadjoint orbit $\mathcal{O}_\mu$.

(4.2) **Proof**: Consider the diagram

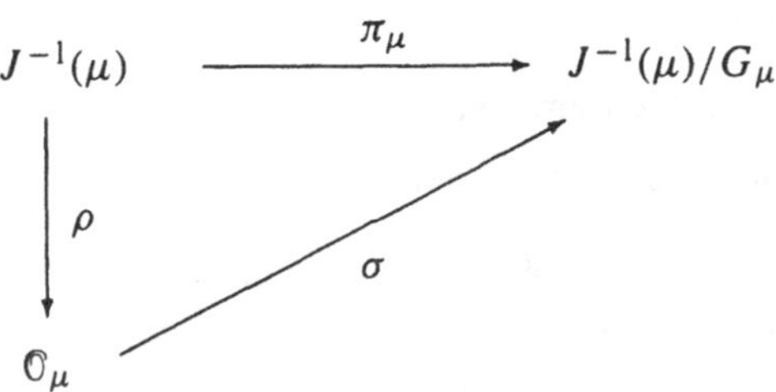

Diagram 4.1

Here σ is the map induced by π_μ and ρ is the restriction of the projection $(g, \alpha) \to \alpha$ to $J^{-1}(\mu)$. More explicitly, $\rho(g, \mathrm{Ad}^t_g \mu) = \mathrm{Ad}^t_g \mu$. The image of $J^{-1}(\mu)$ under ρ is the coadjoint orbit $\mathcal{O}_\mu$ through μ. The map σ is injective because every fiber of the mapping ρ is a single G_μ-orbit. To see this, suppose that $\rho(g, \mathrm{Ad}^t_g \mu) = \rho(h, \mathrm{Ad}^t_h \mu)$. Then $\mathrm{Ad}^t_g \mu = \mathrm{Ad}^t_h \mu$ or $\mu = \mathrm{Ad}^t_{g^{-1}} \mathrm{Ad}^t_h \mu = \mathrm{Ad}^t_{hg^{-1}} \mu$. Whereupon $hg^{-1} \in G_\mu$. Consequently, $\varphi_{hg^{-1}}(g, \mathrm{Ad}^t_g \mu) = (h, \mathrm{Ad}^t_g \mu)$, by definition of φ action. Thus $(g, \mathrm{Ad}^t_g \mu)$ and $(h, \mathrm{Ad}^t_g \mu)$ lie on the same G_μ-orbit. Since π_μ is surjective, σ is bijective. Because π_μ and ρ are smooth and because the bundle ρ has a smooth local cross section, it follows that σ is smooth. Since $J^{-1}(\mu)$ is a principal G_μ-bundle ((2.12)), it has a smooth local cross section. Therefore σ^{-1} is smooth. Hence σ is a diffeomorphism.

Next we show that the symplectic form $\sigma^*\omega_\mu$ on the reduced space $\mathcal{O}_\mu$ is $\omega_{\mathcal{O}_\mu}$. First we calculate the tangent of the mapping ρ. Differentiating the curve

$$s \to \rho(g \exp s\xi, \mathrm{Ad}^t_{g \exp s\xi} \mu) = \mathrm{Ad}^t_{g \exp s\xi} \mu$$

and evaluating the derivative at $s = 0$ gives

$$T_{(g,\nu)}\rho(T_eL_g\xi, \mathrm{ad}^t_\xi \nu) = \mathrm{ad}^t_\xi \nu$$

with $\nu = \mathrm{Ad}^t_g\mu$, since $\mathrm{Ad}^t_{hg} = \mathrm{Ad}^t_g \mathrm{Ad}^t_h$ and $\left.\frac{d}{dt}\right|_{t=0} \mathrm{Ad}^t_{\exp t\xi}\, \eta = \mathrm{ad}^t_\xi\, \eta$. We now compute $\sigma^*\omega_\mu$:

$$\begin{aligned}
(\sigma^*\omega_\mu)(\nu)\big(\mathrm{ad}^t_\xi \nu, \mathrm{ad}^t_\eta \nu\big) &= \\
&= (\sigma^*\omega_\mu)(\rho(g,\nu))\big(T_{(g,\nu)}\rho(T_eL_g\xi, \mathrm{ad}^t_\xi \nu), T_{(g,\nu)}\rho(T_eL_g\eta, \mathrm{ad}^t_\eta \nu)\big) \\
&= \big((\sigma\circ\rho)^*\omega_\mu\big)(g,\nu)\big((T_eL_g\xi, \mathrm{ad}^t_\xi \nu), (T_eL_g\eta, \mathrm{ad}^t_\eta \nu)\big) \\
&= \omega(g,\nu)\big((T_eL_g\xi, \mathrm{ad}^t_\xi \nu), (T_eL_g\eta, \mathrm{ad}^t_\eta \nu)\big), \\
&\qquad \text{because } \sigma\circ\rho = \pi_\mu \text{ and } \pi^*_\mu\omega_\mu = \omega|J^{-1}(\mu) \\
&= -(\mathrm{ad}^t_\xi \nu)\eta + (\mathrm{ad}^t_\eta \nu)\xi + \nu([\xi,\eta]), \\
&\qquad \text{by construction of } \omega, \text{ (see appendix A section 1 equation (4))} \\
&= -\nu([\xi,\eta]) + \nu([\eta,\xi]) + \nu([\xi,\eta]) \\
&= -\nu([\xi,\eta]).
\end{aligned}$$

Hence the reduced phase space is symplectically diffeomorphic to $(\mathcal{O}_\mu, \omega_{\mathcal{O}_\mu})$. □

One of the main motivations for the reduction theorem is to remove symmetries from a Hamiltonian system (H, M, ω). We say that G is a symmetry of (H, M, ω) if H is invariant under a proper Hamiltonian action of G on the symplectic manifold (M, ω) with coadjoint equivariant momentum map $J : M \to \mathcal{G}^*$. When μ is a regular value of J, $H|J^{-1}(\mu)$ induces a smooth function H_μ on the reduced space M_μ, called the *reduced Hamiltonian*, which satisfies

$$\pi^*_\mu H_\mu = i^* H. \tag{54}$$

Thus we have constructed a *reduced Hamiltonian system* $(H_\mu, M_\mu, \omega_\mu)$. Its importance lies in the fact that the vector field X_H on $J^{-1}(\mu)$ is π_μ-related to the *reduced vector field* X_{H_μ}, that is,

$$T\pi_\mu \circ X_H = X_{H_\mu} \circ \pi_\mu. \tag{55}$$

This is a geometric way of saying what it means to use the symmetry to reduce the number of variables in a problem.

(4.3) **Proof**: To see that (55) holds and that the reduced equations are Hamiltonian first look at the given equation of motion on $J^{-1}(\mu)$, namely,

$$X_H \rfloor \omega = dH. \tag{56}$$

As everything is G_μ-invariant, we can push X_H and H down to M_μ. To push ω down to M_μ, we need to know that ω is a pull back. Hence ω must vanish on vectors tangent to G_μ-orbits. From ((3.11)) we know that $\ker \omega | J^{-1}(\mu) = T_m(G_\mu \cdot m)$. Thus we can push ω down to M_μ. Consequently, the equations of motion (56) pass to the quotient, because both sides of (56) vanish on vectors tangent to G_μ-orbits. □

Example 2: An interesting special case of the regular reduction theorem occurs when the Hamiltonian $\mathcal{H}$ on T^*G is a quadratic function associated to a left invariant metric ρ on G. In this case the image of the integral curves of the Hamiltonian system under the bundle projection are geodesics on G, (see appendix A section 3 example 2).

More explicitly, if ρ^* is the metric dual to ρ, it is left invariant. The Hamiltonian

$$\mathcal{H} : T^*G \to \mathbf{R} : \alpha_g = T_e L^t_{g^{-1}} \alpha \to \tfrac{1}{2}\rho^*(g)(\alpha_g, \alpha_g) = \tfrac{1}{2}\rho^*(e)(\alpha, \alpha)$$

is left invariant. Pulling $\mathcal{H}$ back by the left trivialization $\mathcal{L}$ (53) gives the Hamiltonian

$$H : G \times \mathcal{G}^* \to \mathbf{R} : (g, \alpha) \to \tfrac{1}{2} r^*(e)(\alpha, \alpha),$$

where $r^*(g) : \mathcal{G}^* \times \mathcal{G}^* \to \mathbf{R}$ is the dual metric on $\mathcal{G}^*$ defined by

$$r^*(g)(\alpha, \beta) = \rho^*(g)(\alpha_g, \beta_g).$$

Since the dual metric ρ^* is left invariant, so is the dual metric r^*, that is, $r^*(g)(\alpha, \beta) = r^*(e)(\alpha, \beta)$. Therefore the Hamiltonian H is left invariant as well. The integral curves of X_H are the pull back of integral curves of $X_{\mathcal{H}}$ by the left trivialization $\mathcal{L}$. On $G \times \mathcal{G}^*$ the integral curves of X_H satisfy the Euler-Arnol'd equations

$$\begin{aligned} \dot{g} &= T_e L_g (r^*(e))^\sharp(\alpha) \\ \dot{\alpha} &= \mathrm{ad}^t_{(r^*(e))^\sharp(\alpha)} \alpha, \end{aligned} \tag{57}$$

since

$$D_2 H(g, \alpha)\beta = r^*(e)(\alpha, \beta) = (r^*(e))^\sharp(\alpha)\beta,$$

(see appendix A section 3 example 3).

Because H is left invariant, we may remove the symmetry given by the left action of G on T^*G using reduction, (see example 1). We obtain a reduced phase space $(\mathcal{O}_\mu, \omega_{\mathcal{O}_\mu})$ with a reduced Hamiltonian

$$H_\mu : \mathcal{O}_\mu \subseteq \mathcal{G}^* \to \mathbf{R} : \nu = \mathrm{Ad}^t_g \mu \to H_\mu(\nu)$$

where $H_{\mu}(\nu) = H(e, \nu) = \frac{1}{2} r^{*}(e)(\nu, \nu)$. Since

$$dH_{\mu}(\nu)\mathrm{ad}^{t}_{\xi}\nu = (r^{*}(e))^{\sharp}(\nu)\mathrm{ad}^{t}_{\xi}\nu,$$

the integral curves of the reduced Hamiltonian vector field $X_{H_{\mu}}$ on the reduced space $(\mathcal{O}_{\mu}, \omega_{\mathcal{O}_{\mu}})$ satisfy *Euler's equations*

$$\dot{\nu} = \mathrm{ad}^{t}_{(r^{*}(e))^{\sharp}(\nu)}\nu. \quad \square \tag{58}$$

We now reexamine the hypothesis in the regular reduction theorem ((4.1)) that μ is a regular value of the momentum mapping J.

Claim: If $\mu \in \mathcal{G}^{*}$ is a regular value of the coadjoint equivariant momentum mapping $J : M \to \mathcal{G}^{*}$, then G_{μ} acts locally freely on $J^{-1}(\mu)$.

(4.4) **Proof**: The following reasoning shows that $\dim \mathcal{G}_{\mu} = 0$ and hence G_{μ} acts locally freely on $J^{-1}(\mu)$ by ((2.13)). Since μ is a regular value of J, for every $m \in J^{-1}(\mu)$

$$\dim \ker T_{m}J = \dim T_{m}M - \dim \mathcal{G}^{*} = \dim T_{m}M - \dim \mathcal{G}. \tag{59}$$

Because $T_{m}(G \cdot m) = (\ker T_{m}J)^{\omega(m)}$ (39), it follows that

$$\dim(\ker T_{m}J)^{\omega(m)} = \dim \mathcal{G} - \dim \mathcal{G}_{m}. \tag{60}$$

In addition, since $T_{m}(G_{\mu} \cdot m) = \ker T_{m}J \cap (\ker T_{m}J)^{\omega(m)}$ (40), we obtain

$$\dim\left(\ker T_{m}J \cap (\ker T_{m}J)^{\omega(m)}\right) = \dim \mathcal{G}_{\mu} - \dim \mathcal{G}_{m}. \tag{61}$$

Therefore

$$\begin{aligned}\dim T_{m}M &= \dim \ker T_{m}J + \dim(\ker T_{m}J)^{\omega(m)} \\ &\qquad - \dim\left(\ker T_{m}J \cap (\ker T_{m}J)^{\omega(m)}\right) \\ &= \dim T_{m}M - \dim \mathcal{G} + \dim \mathcal{G} - \dim \mathcal{G}_{m} + \dim \mathcal{G}_{m} - \dim \mathcal{G}_{\mu},\end{aligned}$$

using (59), (60) and (61). Hence $\dim \mathcal{G}_{\mu} = 0$. $\square$

We now prove an orbifold version of the regular reduction theorem.

Theorem 2: Suppose that $J : M \to \mathcal{G}^{*}$ is a coadjoint equivariant momentum mapping of a proper Hamiltonian action of a Lie group G on a symplectic manifold (M, ω). If $\mu \in \mathcal{G}^{*}$ is a regular value of J, then the reduced space $M_{\mu} = J^{-1}(\mu)/G_{\mu}$ is a symplectic orbifold.

Because M_{μ} is not necessarily a smooth manifold, we need to define what we mean by a symplectic form on M_{μ}. We say that ω_{μ} is a symplectic form on M_{μ} if it is the push forward under the G_{μ}-orbit map $\pi_{\mu} : J^{-1}(\mu) \to M_{\mu}$ of a symplectic form Ω_{μ} on M restricted to $J^{-1}(\mu)$, In other words, if $\pi^{*}_{\mu}\omega_{\mu} = \Omega_{\mu}|J^{-1}(\mu)$.

(4.5) **Proof**: Since μ is a regular value of J, using ((4.4)) we see that G_{μ} acts locally freely on $J^{-1}(\mu)$. Hence M_{μ} is an orbifold by ((2.16)).

To construct a symplectic form on M_μ we first construct a special orbifold atlas for M_μ. Let $m \in J^{-1}(\mu)$. Then there is a local diffeomorphism $\varphi_m : T_m M \to M$ with $\varphi_m(0) = m$ which pulls back the symplectic form ω on M to the constant symplectic form $\sigma_m = \omega(m)$ on $T_m M$ and intertwines the linear σ_m-symplectic G_m-action

$$\widehat{\Phi} : G_m \times T_m M \to T_m M : (h, v_m) \to T_m \Phi_h v_m$$

with the G_m-action $\Phi|(G_m \times M)$ on (M, ω), (see step 1 of ((3.14))). Under the mapping φ_m the momentum mapping $J : M \to \mathcal{G}^*$ pulls back to a locally defined G_m-equivariant map $\mathcal{J} : T_m M \to \mathcal{G}^*$. Clearly, $\varphi_m|\mathcal{J}^{-1}(\mu)$ is a local diffeomorphism from a neighborhood of 0 in $\mathcal{J}^{-1}(\mu)$ to a neighborhood of m in $J^{-1}(\mu)$. Since μ is a regular value of J, from ((4.4)) it follows that $\dim \mathcal{G}_\mu = 0$. Therefore $T_m(G_\mu \cdot m) = \{0\}$. Hence $(T_m M, \sigma_m)$ is transverse to $G_\mu \cdot m$ at m and is G_m-invariant. Consequently there is a G_m-invariant open neighborhood U_m of 0 in $T_m M$ such that $\varphi_m(U_m)$ is a slice to the G_μ-action $\Phi|(G_\mu \times J^{-1}(\mu))$ at m. Since $\pi_\mu \circ \varphi_m$ is G_m-invariant and $\big(G_\mu \cdot \varphi_m(U_m)\big)/G_\mu$ is homeomorphic to U_m/G_m by ((2.2)), the diffeomorphism φ_m induces a homeomorphism from U_m/G_m to $\pi_\mu(\varphi_m(U_m))$. Therefore $(U_m, G_m, \pi_\mu \circ \varphi_m)$ is an orbifold chart for M_μ at $\pi_\mu(m)$.

On the G_μ-invariant set $\mathcal{U}_m = G_\mu \cdot \varphi_m(U_m)$ define a G_μ-invariant symplectic form Ω_m by

$$\Omega_m\big(\Phi_h(\varphi_m(u_m))\big) = (T_{\varphi_m(u_m)}\Phi_{h^{-1}})^t \sigma_m$$

for every $h \in G_\mu$ and every $u_m \in U_m$. Since $\{(U_m, G_m, \pi_\mu \circ \varphi_m)\}_{m \in J^{-1}(\mu)}$ is an orbifold atlas for M_μ, $\{\mathcal{U}_m\}_{m \in J^{-1}(\mu)}$ is a G_μ-invariant open covering of $J^{-1}(\mu)$. By ((2.5)) there is a smooth G_μ-invariant partition of unity $\{\psi_m\}$ subordinate to the covering $\{\mathcal{U}_m\}$. Therefore $\Omega_\mu = \sum_m \psi_m \cdot \Omega_m$ is the desired smooth G_μ-invariant symplectic form on $J^{-1}(\mu)$. By construction Ω_μ pushes down to a symplectic form ω_μ on M_μ. □

Example 3 (The 2:1 resonance). On $(\mathbf{R}^4, \omega = \sum_{i=1}^2 dx_i \wedge dy_i)$ consider the Hamiltonian $S^1 = \mathbf{R}/2\pi\mathbf{Z}$-action

$$\Phi : S^1 \times \mathbf{R}^4 \to \mathbf{R}^4 : \big(t, (x, y)\big) \to \left(\begin{array}{cc|cc} \cos t & 0 & \sin t & 0 \\ 0 & \cos 2t & 0 & \sin 2t \\ \hline -\sin t & 0 & \cos t & 0 \\ 0 & -\sin 2t & 0 & \cos 2t \end{array}\right) \begin{pmatrix} x \\ y \end{pmatrix}, \tag{62}$$

which has a momentum mapping

$$H : \mathbf{R}^4 \to \mathbf{R} : (x, y) \to \tfrac{1}{2}\left[(y_1^2 + x_1^2) + 2\,(y_2^2 + x_2^2)\right]. \tag{63}$$

Here we have identified the dual of the Lie algebra of S^1 with $\mathbf{R}$. The action Φ restricted to $S^1 \times H^{-1}(h)$ is proper since S^1 is compact. However, it is not free, because for $h > 0$ the isotropy group

$$\begin{aligned} S^1_{(0,\sqrt{h},0,0)} &= \left\{t \in \mathbf{R}/2\pi\mathbf{Z} \,\middle|\, (0, \sqrt{h}, 0, 0) = (0, \sqrt{h}\cos 2t, 0, \sqrt{h}\sin 2t)\right\} \\ &= \left\{0 \bmod 2\pi,\ \pi \bmod 2\pi\right\} \end{aligned}$$

is isomorphic to $\mathbf{Z}_2$. Since $h > 0$ is a regular value of H, the S^1-action $\Phi|(S^1 \times H^{-1}(h))$ is locally free. Hence the reduced space $M_h = H^{-1}(h)/S^1$ is a symplectic orbifold.

To exhibit M_h as a semialgebraic variety we use invariant theory. Introducing complex conjugate coordinates $z_j = x_j + i\,y_j$, $\overline{z}_j = x_j - i\,y_j$ for $j = 1, 2$, the action Φ becomes the S^1-action

$$\widetilde{\Phi} : S^1 \times \mathbf{C}^4 \to \mathbf{C}^4 : \big(t, (z, \overline{z})\big) \to \big(e^{it} z_1, e^{2it} z_2, e^{-it} \overline{z}_1, e^{-2it} \overline{z}_2\big).$$

The Hermitian monomial $z^j \overline{z}^k = z_1^{j_1} z_2^{j_2} \overline{z}_1^{k_1} \overline{z}_2^{k_2}$ is invariant under $\widetilde{\Phi}$ if and only if

$$j_1 + 2j_2 - k_1 - 2k_2 = 0. \tag{64}$$

Claim: The algebra of Hermitian polynomials $\sum c_{jk} z^j \overline{z}^k$ (where $c_{jk} = \overline{c_{kj}} \in \mathbf{C}$) which are invariant under $\widetilde{\Phi}$ is generated by the Hermitian monomials

$$z_1\overline{z}_1, \quad z_2\overline{z}_2, \quad z_1^2\overline{z}_2, \quad z_2\overline{z}_1^2. \tag{65}$$

These monomials satisfy the relation

$$(z_1^2\overline{z}_2)(z_2\overline{z}_1^2) = (z_1\overline{z}_1)^2(z_2\overline{z}_2) \tag{66}$$

and the inequalities

$$z_1\overline{z}_1 \ge 0 \qquad z_2\overline{z}_2 \ge 0. \tag{67}$$

(4.6) **Proof**: To show that the monomials in (65) generate the algebra of invariant Hermitian polynomials, we need to determine all $j, k \in \mathbf{Z}^2_{\ge}$ such that (64) holds. Let $n_1 = j_1 - k_1$ and $n_2 = j_2 - k_2$. Then (64) becomes

$$n_1 + 2n_2 = 0. \tag{68}$$

Suppose that $n_2 \ge 0$. Then $n_2 = j_2 - k_2$ and $2n_2 = -n_1 = k_1 - j_1$. Hence the monomial

$$z_1^{j_1} z_2^{k_2+n_2} \overline{z}_1^{j_1+2n_2} \overline{z}_2^{k_2} = (z_1\overline{z}_1)^{j_1} (z_2\overline{z}_2)^{k_2} (z_2\overline{z}_1^2)^{n_2}$$

is invariant. Thus $z_1\overline{z}_1$, $z_2\overline{z}_2$ and $z_2\overline{z}_1^2$ are generators. Now suppose that $n_2 \le 0$. Then $n = -n_2 \ge 0$, $n = -n_2 = k_2 - j_2$ and $2n = -2n_2 = n_1 = j_1 - k_1$. Hence the monomial

$$z_1^{2n+k_1} z_2^{j_2} \overline{z}_1^{k_1} \overline{z}_2^{n+j_2} = (z_1\overline{z}_1)^{k_1} (z_2\overline{z}_2)^{j_2} (z_1^2\overline{z}_2)^n$$

is invariant. Thus $z_1\overline{z}_1$, $z_2\overline{z}_2$ and $z_1^2\overline{z}_2$ are generators. Consequently, the algebra of $\widetilde{\Phi}$-invariant Hermitian polynomials is generated by (65). □

Translating the above result into real coordinates, we have shown that the algebra of real polynomials in x, y which are invariant under the S^1-action Φ (62) is generated by the monomials

$$\begin{aligned} \sigma_1 &= \operatorname{Re} z_1^2\overline{z}_2 = x_2(x_1^2 - y_1^2) + y_2(2x_1y_1) \\ \sigma_2 &= \operatorname{Im} z_1^2\overline{z}_2 = x_2(2x_1y_1) - y_2(x_1^2 - y_1^2) \\ \sigma_3 &= y_2^2 + x_2^2 \\ \sigma_4 &= y_1^2 + x_1^2, \end{aligned} \tag{69}$$

which satisfy the relation

$$\sigma_1^2 + \sigma_2^2 = \sigma_3 \sigma_4^2 \qquad \sigma_3 \geq 0,\ \sigma_4 \geq 0. \tag{70}$$

Claim: The orbit space $\mathbf{R}^4/S^1$ of the S^1-action Φ is the semialgebraic variety $\mathcal{V}$ defined by (70).

(4.7) **Proof**: Consider the Hilbert map

$$\Psi : \mathbf{R}^4 \to \mathbf{R}^4 : (x, y) \to \big(\sigma_1(x, y), \sigma_2(x, y), \sigma_3(x, y), \sigma_4(x, y)\big).$$

It suffices to show that for every $\xi = (\xi_1, \xi_2, \xi_3, \xi_4) \in \mathcal{V}$ the fiber $\Psi^{-1}(\xi)$ is a single S^1 orbit of Φ. Suppose that $\xi_4 > 0$. Then solving the linear equations

$$\begin{pmatrix} \xi_1 \\ \xi_2 \end{pmatrix} = \begin{pmatrix} x_1^2 - y_1^2 & 2x_1y_1 \\ 2x_1y_1 & -(x_1^2 - y_1^2) \end{pmatrix} \begin{pmatrix} x_2 \\ y_2 \end{pmatrix}$$

gives

$$\begin{aligned} x_2(x_1, y_1) &= \frac{1}{\xi_4^2}\left[(x_1^2 - y_1^2)\xi_1 + (2x_1y_1)\xi_2\right] \\ y_2(x_1, y_1) &= \frac{1}{\xi_4^2}\left[(2x_1y_1)\xi_1 - (x_1^2 - y_1^2)\xi_2\right]. \end{aligned} \tag{71}$$

Thus

$$\Psi^{-1}(\xi) = \left\{ \big(x_1, y_1, x_2(x_1, y_1), y_2(x_1, y_1)\big) \middle| x_1^2 + y_1^2 = \xi_4 \right\},$$

since

$$(x_2(x_1, y_1))^2 + (y_2(x_1, y_1))^2 = \frac{1}{\xi_4^2}(\xi_1^2 + \xi_2^2) = \xi_3,$$

by (70). Suppose that $\xi_4 = 0$, then $x_1 = y_1 = 0$. Hence $\xi_1 = \xi_2 = 0$ and

$$\Psi^{-1}(0, 0, \xi_3, 0) = \left\{ (0, 0, x_2, y_2) \middle| x_2^2 + y_2^2 = \xi_3 \right\}.$$

Therefore for every $\xi \in \mathcal{V}$, the fiber $\Psi^{-1}(\xi)$ is a single S^1 orbit of Φ. □

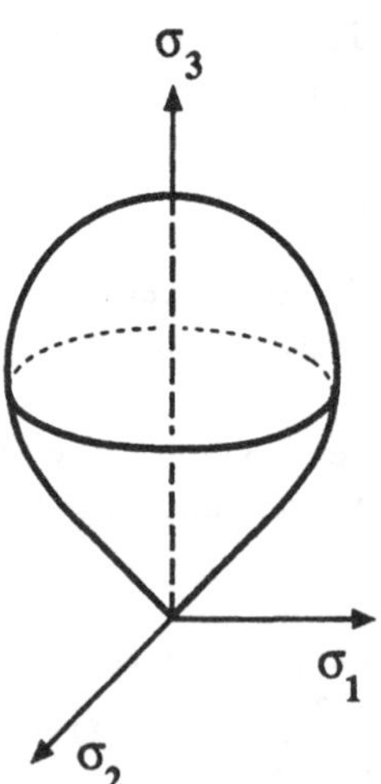

Figure 4.1. The reduced space M_h.

Since the momentum mapping H (63) is S^1-invariant, the reduced space $M_h = H^{-1}(h)/S^1$ is defined by (70) and

$$\sigma_4 + 2\sigma_3 = 2h. \tag{72}$$

After eliminating σ_4 from (70) and (72), we find that M_h is the semialgebraic variety given by

$$\sigma_1^2 + \sigma_2^2 = 4\sigma_3(h-\sigma_3)^2 \qquad 0 \le \sigma_3 \le h, \tag{73}$$

M_h is a compact surface, which is homeomorphic to S^2, with a conical singularity at $(0, 0, h)$, see figure 4.1. □

After this digression we return to discussing regular reduction. Let Φ be a free proper G-action on M. As a warm up to our treatment of singular reduction in section 5, we construct another model for the reduced space M_μ given by regular reduction. Let $\mathcal{O}_\mu$ be the G-coadjoint orbit through $\mu \in \mathcal{G}^*$. The following discussion shows that the G-orbit space $J^{-1}(\mathcal{O}_\mu)/G$ is diffeomorphic to the regular reduced space M_μ.

We begin by observing that the coadjoint equivariance of the momentum mapping J implies

$$J^{-1}(\mathcal{O}_\mu) = G \cdot J^{-1}(\mu). \tag{74}$$

▷ If $\mu \in \mathcal{G}^*$ is a regular value of J, then $J^{-1}(\mathcal{O}_\mu)$ is a smooth submanifold of M.

(4.8) **Proof**: Since μ is a regular value of J, the level set $J^{-1}(\mu)$ is a submanifold of M. Because J is coadjoint equivariant, the induced action $\Phi|(G_\mu \times J^{-1}(\mu))$ is defined and is proper. Let $m \in J^{-1}(\mu)$. Then there is a slice S_m to $\Phi|(G_\mu \times J^{-1}(\mu))$ at m. Choose $\mathcal{L}$ so that $\mathcal{L} \oplus \mathcal{G}_m = \mathcal{G}_\mu$ and let $L = \exp \mathcal{L}$. By definition of a slice, the map $\Psi : L \times S_m \to J^{-1}(\mu)$ with $\Psi(e, m) = m$ is a local diffeomorphism. Because $J^{-1}(\mathcal{O}_\mu) = G \cdot J^{-1}(\mu)$, the map

$$\Theta : G \times (L \times S_m) \to J^{-1}(\mathcal{O}_\mu) : \big(g, (\ell, s_m)\big) \to \Phi_g\big(\Psi(\ell, s_m)\big)$$

with $\Theta\big(e, (e, m)\big) = m$ is a local diffeomorphism. Hence Θ is a local parametrization of $J^{-1}(\mathcal{O}_\mu)$. □

The restricted action $\Phi|(G \times J^{-1}(\mathcal{O}_\mu))$ is free and proper because the action Φ is free and proper. Therefore by ((2.9)) the orbit space $\mathcal{M}_\mu = J^{-1}(\mathcal{O}_\mu)/G$ is a smooth manifold. **Claim**: $\mathcal{M}_\mu = J^{-1}(\mathcal{O}_\mu)/G$ is diffeomorphic to the reduced space $M_\mu = J^{-1}(\mu)/G_\mu$.

(4.9) **Proof**: Consider diagram 4.2 where i is the inclusion and π, π_μ are orbit maps. Let $g \in G_\mu$. For $q \in J^{-1}(\mu)$,

$$\pi\big(i(\Phi_g(q))\big) = \pi\big(\Phi_g(i(q))\big) = \pi(i(q)).$$

Hence the mapping $\pi \circ i$ is G_μ-invariant and thus it induces a smooth mapping

$$\sigma : J^{-1}(\mu)/G_\mu \to J^{-1}(\mathcal{O}_\mu)/G.$$

The claim is established if we show that σ is bijective and that $T\sigma$ is bijective. The map

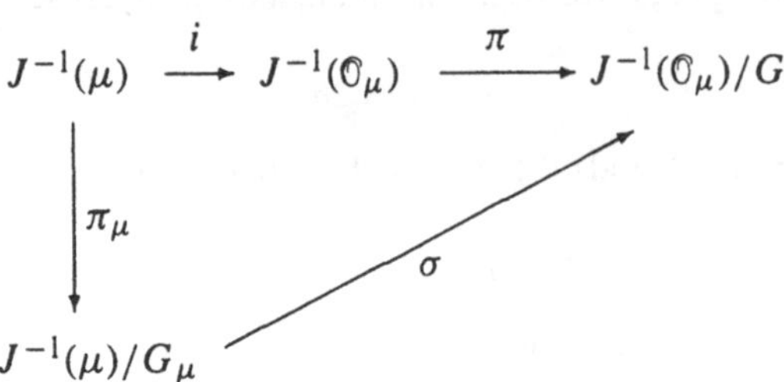

Diagram 4.2

σ is one to one, for if $q, q' \in J^{-1}(\mu)$ with $\pi(i(q)) = \pi(i(q'))$, then there is a $g \in G$ such that $\Phi_g(i(q)) = i(q')$. Since

$$\mu = J(i(q')) = \mathrm{Ad}^t_{g^{-1}} J(i(q)) = \mathrm{Ad}^t_{g^{-1}} \mu,$$

it follows that $g \in G_\mu$. Thus every fiber of the mapping $\pi \circ i$ is a single G_μ-orbit, which shows that σ is injective. The following argument shows that $\pi \circ i$ is surjective. For $n \in J^{-1}(\mathcal{O}_\mu)$ consider the orbit $G \cdot n$ in $J^{-1}(\mathcal{O}_\mu)/G$. Since $J^{-1}(\mathcal{O}_\mu) = G \cdot J^{-1}(\mu)$, there is an $m \in J^{-1}(\mu)$ and a $g \in G$ such that $n = g \cdot m$. Therefore $\pi \circ i(m) = G \cdot m = G \cdot n$. Thus σ is surjective and hence is bijective. We know that the tangent map of σ

$$T_{\pi_\mu(q)}\sigma : T_{\pi_\mu(q)}\big(J^{-1}(\mu)/G_\mu\big) \to \big(T_{\pi \circ i(q)} J^{-1}(\mathcal{O}_\mu)/G\big)$$

is bijective once we show that it is injective and

$$\dim T_{\pi_\mu(q)}\big(J^{-1}(\mu)/G_\mu\big) = \dim T_{\pi \circ i(q)}\big(J^{-1}(\mathcal{O}_\mu)/G\big).$$

To see that $T_{\pi_\mu(q)}\sigma$ is injective suppose that $\overline{v} \in T_{\pi_\mu(q)}\big(J^{-1}(\mu)/G_\mu\big)$ lies in $\ker T_{\pi_\mu(q)}\sigma$. Because π_μ is a submersion, there is a $v \in J^{-1}(\mu)$ such that $T_q \pi_\mu v = \overline{v}$. Hence $T_q(\pi_\mu \circ \sigma)v = 0$. But $\pi_\mu \circ \sigma = \pi \circ i$, so $v \in \ker T_q(\pi \circ i)$. Because the fibers of $\pi \circ i$ are G_μ-orbits in $J^{-1}(\mu)$, we obtain $\ker T_q(\pi \circ i) = T_q(G_\mu \cdot q) = \ker T_q \pi_\mu$. Hence $\overline{v} = 0$. In addition,

$$\begin{aligned}
\dim T_{\pi_\mu(q)}\big(J^{-1}(\mu)/G_\mu\big) &= \dim T_q J^{-1}(\mu) - \dim \mathcal{G}_\mu \\
&= \dim T_q J^{-1}(\mu) + \dim T_\mu \mathcal{O}_\mu - \dim T_\mu \mathcal{O}_\mu - \dim \mathcal{G}_\mu \\
&= \dim T_q J^{-1}(\mathcal{O}_\mu) - \dim \mathcal{G} \\
&= \dim T_{\pi \circ i(q)}\big(J^{-1}(\mathcal{O}_\mu)/G\big).
\end{aligned}$$

Thus $\mathcal{M}_\mu$ is another model for the reduced space M_μ. □

We can define a symplectic structure on $\mathcal{M}_\mu$ by noting that the form $\omega|(J^{-1}(\mu)) + J^*\omega_{\mathcal{O}_\mu}$ passes to the quotient $J^{-1}(\mathcal{O}_\mu)/G$. Here $\omega_{\mathcal{O}_\mu}$ is the form on $\mathcal{O}_\mu$ given in example 2 of section 2 of appendix A. However, with an eye towards singular reduction (see section 5), we will obtain this form by constructing a Poisson bracket on $\mathcal{M}_\mu$. We begin by constructing a Poisson bracket on M_μ. Because (M_μ, ω_μ) is a smooth symplectic manifold, the Poisson bracket $\{\, ,\, \}$ on $C^\infty(M_\mu)$ is defined in the standard way, namely: for $f, h \in C^\infty(M_\mu)$

$$\{f, h\}(\overline{q}) = \omega_\mu(\overline{q})(X_f(\overline{q}), X_h(\overline{q}))$$

for every $\overline{q} \in M_\mu$. For $\mathcal{M}_\mu$ we use a different approach. For technical reasons from now on we shall assume that $J^{-1}(\mathcal{O}_\mu)$ is a *closed* subset of M. On $C^\infty(\mathcal{M}_\mu)$, we define a Poisson bracket $\{\, ,\, \}_\mu$ as follows. Let $f, h \in C^\infty(\mathcal{M}_\mu)$. Then $\pi^* f, \pi^* h$ are smooth G-invariant functions on $J^{-1}(\mathcal{O}_\mu)$. Since $J^{-1}(\mathcal{O}_\mu)$ is a closed G-invariant subset of M, there are smooth G-invariant functions F and H on M such that $F|J^{-1}(\mathcal{O}_\mu) = \pi^* f$ and $H|J^{-1}(\mathcal{O}_\mu) = \pi^* h$, see ((2.6)). For every $q \in J^{-1}(\mathcal{O}_\mu) \subseteq M$ let $\overline{q} = \pi(q) \in \mathcal{M}_\mu$. Then define $\{\, ,\, \}_\mu$ by

$$\{f, h\}_\mu(\overline{q}) = \omega(q)(X_F(q), X_H(q)) = \big(\{F, H\}_M | J^{-1}(\mathcal{O}_\mu)\big)(q),$$

▷ where $\{\, ,\, \}_M$ is the standard Poisson bracket on $C^\infty(M)$. Before we can show that the bracket $\{\, ,\, \}_\mu$ is well defined, we need to know that $J^{-1}(\mathcal{O}_\mu)$ is an invariant manifold of X_F.

(4.10) **Proof**: First we show that $J^{-1}(\mu)$ is invariant under the flow of X_F. For $\xi \in \mathcal{G}$ and $q \in J^{-1}(\mu)$ we have

$$\{J^\xi, F\}_M(q) = -L_{X_{J^\xi}} F(q) = -L_{X^\xi} F(q) = 0,$$

since F is G-invariant. Therefore J^ξ is constant along the integral curves of X_F. Hence $J^{-1}(\mu)$ is an invariant manifold of X_F. Because $J^{-1}(\mathcal{O}_\mu) = G \cdot J^{-1}(\mu)$ and the flow of X_F commutes with the action of G, it follows that $J^{-1}(\mathcal{O}_\mu)$ is an invariant manifold of X_F. □

Claim: The Poisson bracket $\{\, ,\, \}_\mu$ on $C^\infty(\mathcal{M}_\mu)$ is well defined.

(4.11) **Proof**: Suppose that $\widetilde{F}$ is a smooth G-invariant function on M such that $\widetilde{F}|J^{-1}(\mathcal{O}_\mu) = \pi^* f$. Then $(F - \widetilde{F})|J^{-1}(\mathcal{O}_\mu) = \pi^* f - \pi^* f = 0$. Therefore

$$\{H, F - \widetilde{F}\}|J^{-1}(\mathcal{O}_\mu) = -L_{X_H}\big((F - \widetilde{F})|J^{-1}(\mathcal{O}_\mu)\big) = -L_{X_H} 0 = 0,$$

since $X_H(q) \in T_q J^{-1}(\mathcal{O}_\mu)$. (*Warning*: $F|J^{-1}(\mathcal{O}_\mu) = 0$ does *not* imply that $X_F|J^{-1}(\mathcal{O}_\mu) = 0$.) Therefore $\{f, h\}_\mu$ does not depend on the choice of G-invariant extension of $\pi^* f$. Since the bracket $\{\, ,\, \}_\mu$ is skew symmetric, it does not depend on the choice of G-invariant extension of $\pi^* h$ either. Thus the Poisson bracket $\{\, ,\, \}_\mu$ is well defined. □

We now want to show that the Poisson bracket $\{\, ,\, \}_\mu$ is nondegenerate. First we prove

Claim: For every $q \in J^{-1}(\mu)$, the tangent space $T_q J^{-1}(\mu)$ is spanned by Hamiltonian vector fields of smooth G-invariant functions.

(4.12) **Proof**: Suppose that F is a smooth G-invariant function on M. Then for every $\xi \in \mathcal{G}$, $F(\Phi_{\exp t\xi}(q)) = F(q)$, which differentiated with respect to t and then evaluated at $t = 0$ gives $X^\xi \lrcorner\, dF = 0$. Since $T_q(G \cdot q)$ is spanned by $\{X^\xi(q) \mid \xi \in \mathcal{G}\}$, it follows that $dF(q)$

lies in the annihilator $(T_q(G \cdot q))^\circ$ of $T_q(G \cdot q)$. However by ((1.1b)) of appendix A we have

$$\omega(q)^\flat (T_q(G \cdot q))^\circ = (T_q(G \cdot q))^{\omega(q)}. \tag{75}$$

Since $T_q(G \cdot q)^{\omega(q)} = T_q J^{-1}(\mu)$, we obtain

$$X_F(q) = \omega^\flat(q) dF(q) \in T_q J^{-1}(\mu)$$

by (75). Conversely, given a vector $v \in T_q J^{-1}(\mu)$, there is a smooth G-invariant function F on M such that $X_F(q) = v$. To see this observe that

$$\text{span} \left\{ df(q) \middle| f \in C^\infty(M)^G \right\} = \left(T_q(G \cdot q)\right)^\circ,$$

where $C^\infty(M)^G = \{f \in C^\infty(M) | \ f \text{ is } G\text{-invariant}\}$. Therefore

$$\text{span} \left\{ X_f(q) \middle| \text{for every } f \in C^\infty(M)^G \right\} = (T_q(G \cdot q))^{\omega(q)}.$$

But $T_q(G \cdot q)^{\omega(q)} = T_q J^{-1}(\mu)$ by (39), so we are done. □

Now we are in a position to show that

Claim: The Poisson bracket $\{\,,\,\}_\mu$ on $C^\infty(\mathcal{M}_\mu)$ is nondegenerate.

(4.13) **Proof**: Let F be a smooth G-invariant function on M. Suppose that $0 = \{F, H\}$ on $J^{-1}(\mathcal{O}_\mu)$ for every $H \in C^\infty(M)^G$. Then for every $q \in J^{-1}(\mu)$

$$0 = L_{X_H} F(q) = dF(q) X_H(q) = dF(q) v,$$

for every $v \in T_q J^{-1}(\mu)$ by ((4.11)). Therefore $dF = 0$ on $J^{-1}(\mu)$. Since $J^{-1}(\mu)$ is a smooth manifold, it is locally connected. Hence F is locally constant on $J^{-1}(\mu)$ and by G-invariance, locally constant on $J^{-1}(\mathcal{O}_\mu)$. Therefore, if $0 = \{f, h\}_\mu$ for every $h \in C^\infty(\mathcal{M}_\mu)$, the function f is locally constant. Thus the Poisson bracket $\{\,,\,\}_\mu$ is nondegenerate. □

The next argument proves

Claim: The Poisson algebras $(C^\infty(\mathcal{M}_\mu), \{\,,\,\}_\mu, \cdot)$ and $(C^\infty(M_\mu), \{\,,\,\}, \cdot)$ are isomorphic.

(4.14) **Proof**: Consider the map $\sigma : M_\mu \to \mathcal{M}_\mu$ which makes diagram 4.3 commutative. Here

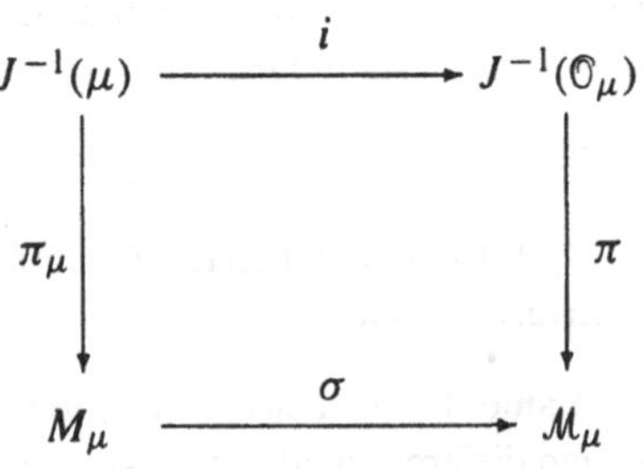

Diagram 4.3

the map i is the inclusion and π_μ and π are the G_μ and G orbit maps, respectively. From diagram 4.3 we obtain the commutative diagram 4.4. We assert that σ^* is the desired isomorphism of Poisson algebras. For $f, h \in C^\infty(\mathcal{M}_\mu)$ it is straightforward to check that

$$\sigma^* f \cdot \sigma^* h = \sigma^*(f \cdot h).$$

Therefore we need only show that σ^* is a Poisson map, that is,

$$\sigma^*(\{f, h\}_\mu) = \{\sigma^* f, \sigma^* h\}. \tag{76}$$

To prove (76) we make some preliminary observations. Since $\sigma^* f \in C^\infty(M_\mu)$, the Hamiltonian vector field $X_{\sigma^* f}$ on (M_μ, ω_μ) makes sense. However, the Hamiltonian vector field $X_{\pi^* f}$ does not make sense until we have extended $\pi^* f \in C^\infty(J^{-1}(\mathcal{O}_\mu))$ to a smooth G-invariant function F on *all* of M. This we can do using a G-invariant partition of unity ((2.5)), because $J^{-1}(\mathcal{O}_\mu)$ is *closed*. The vector field $X_{\pi^* f}$ does not depend on the choice of extension. To check this observe that $C^\infty(J^{-1}(\mathcal{O}_\mu)) = C^\infty(M)^G/\mathcal{J}$, where $\mathcal{J}$ is the ideal of smooth G-invariant functions which vanish on $J^{-1}(\mathcal{O}_\mu)$. $\mathcal{J}$ is a closed

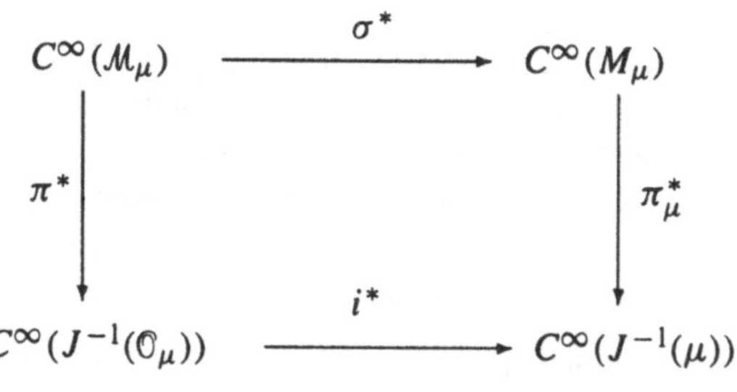

Diagram 4.4

subspace of $C^\infty(M)^G$ since $J^{-1}(\mathcal{O}_\mu)$ is closed. Because $J^{-1}(\mathcal{O}_\mu)$ is invariant under the flow of every Hamiltonian vector field corresponding to a smooth G-invariant function on M, $\mathcal{J}$ is also invariant. Hence $X_{\pi^* f}$ thought of as a derivation on $C^\infty(J^{-1}(\mathcal{O}_\mu))$ is independent of the extension. We now show that for every $q \in J^{-1}(\mu)$

$$T\pi_\mu \circ X_F(q) = X_{\sigma^* f}(\pi_\mu(q)). \tag{77}$$

Since $F \in C^\infty(M)^G$, the vector field X_F is tangent to $J^{-1}(\mu)$. Hence for every $Y(q) \in T_q J^{-1}(\mu)$ we have

$$\begin{aligned}
\omega_\mu(\pi_\mu(q))\big(T\pi_\mu X_F(q), T\pi_\mu Y(q)\big) &= \\
&= \omega(q)(X_F(q), Y(q)), \quad \text{by ((4.1))} \\
&= dF(q)Y(q) \\
&= d\big(\pi_\mu^* \sigma^*(f)\big)(q)Y(q), \\
&\qquad\qquad \text{since } F|J^{-1}(\mu) = i^*\pi^*(f) = \pi_\mu^*\sigma^*(f). \\
&= \omega_\mu(\pi_\mu(q))\big(X_{\sigma^* f}(\pi_\mu(q), T\pi_\mu Y(q)\big).
\end{aligned}$$

From this (77) follows, because ω_μ is nondegenerate and π_μ is a submersion.

For $h \in C^\infty(\mathcal{M}_\mu)$ let $H \in C^\infty(M)^G$ such that $H|J^{-1}(\mu) = \pi^* h$. Let $q \in J^{-1}(\mu)$. The following calculation proves (76).

$$
\begin{aligned}
\{\sigma^* f, \sigma^* h\}(\pi_\mu(q)) &= \\
&= \omega_\mu(\pi_\mu(q))\big(X_{\sigma^* f}(\pi_\mu(q)), X_{\sigma^* h}(\pi_\mu(q))\big), \text{ by definition of } \{\,,\,\} \\
&= \omega_\mu(\pi_\mu(q))(T\pi_\mu X_F(q), T\pi_\mu X_H(q)), \quad \text{using (77)} \\
&= \omega(q)(X_F(q), X_H(q)), \quad \text{because } \pi_\mu^* \omega_\mu = \omega|J^{-1}(\mu) \\
&= \{F, H\}(q) = \{f, h\}_\mu(\pi(q)), \quad \text{by definition of } \{\,,\,\}_\mu \\
&= \{f, h\}_\mu\big(\sigma(\pi_\mu(q))\big), \quad \text{since diagram 4.3 commutes} \\
&= \sigma^*\{f, h\}_\mu(\pi(q)). \quad \square
\end{aligned}
$$

This shows that $\mathcal{M}_\mu$ is symplectically diffeomorphic to M_μ, provided that $J^{-1}(\mathcal{O}_\mu)$ is a closed subset of M.

5 Reduction: the singular case

In this section we examine the reduction of symmetry of a Hamiltonian system when we relax the hypotheses of the regular reduction theorem of the previous section. In particular, we only assume that the Hamiltonian action of the symmetry group is proper and that the coadjoint orbit through the value of the momentum mapping is locally closed. We make *no other* hypotheses on the values of the momentum map. Singular reduction gives rise to reduced spaces which are not smooth manifolds.

The main goal of this section is to prove the *singular reduction theorem.*

Theorem. Let $\Phi : G \times M \to M$ be an action of a Lie group G on a symplectic manifold (M, ω). Suppose that this action is proper, Hamiltonian, and has a coadjoint equivariant momentum mapping $J : M \to \mathcal{G}^*$. Furthermore, suppose that the coadjoint orbit $\mathcal{O}_\mu$ through $\mu \in \mathcal{G}^*$ is locally closed. Then on the singular reduced space $\mathcal{M}_\mu = J^{-1}(\mathcal{O}_\mu)/G$ there is a nondegenerate Poisson algebra $(C^\infty(\mathcal{M}_\mu), \{\,,\,\}_\mu, \cdot)$. In addition, $\mathcal{M}_\mu$ is a locally finite union of symplectic manifolds called symplectic pieces. The flow of a Hamiltonian derivation corresponding to a smooth function on $\mathcal{M}_\mu$ preserves the decomposition of $\mathcal{M}_\mu$ into symplectic pieces and the inclusion map of a symplectic piece into $\mathcal{M}_\mu$ is a Poisson map.

First we construct the singular reduced space. Let $\pi : M \to M/G$ be the orbit mapping of the G-action Φ. The topological space

$$\mathcal{M}_\mu = J^{-1}(\mathcal{O}_\mu)/G = \pi(J^{-1}(\mathcal{O}_\mu))$$

is the *singular reduced space* corresponding to μ. Because the action of G is proper, $\mathcal{M}_\mu$ is Hausdorff ((2.3)).

Next we define $C^\infty(\mathcal{M}_\mu)$, the space of smooth functions on $\mathcal{M}_\mu$. A continuous function $f_\mu : \mathcal{M}_\mu \to \mathbf{R}$ is said to be *smooth* if there is a smooth G-invariant function $f : M \to \mathbf{R}$ such that

$$\pi^* f_\mu = f|J^{-1}(\mathcal{O}_\mu).$$

Let $C^\infty(M)^G$ be the space of smooth G-invariant functions on M. From the definition of smooth function it follows that $C^\infty(\mathcal{M}_\mu)$ is $C^\infty(M)^G|J^{-1}(\mathcal{O}_\mu)$, the space of smooth G-invariant functions on M restricted to $J^{-1}(\mathcal{O}_\mu)$. In other words,

$$C^\infty(\mathcal{M}_\mu) = C^\infty(M)^G/\mathcal{J},$$

where $\mathcal{J} = \mathcal{J}(J^{-1}(\mathcal{O}_\mu))$ is the ideal of smooth G-invariant functions which vanish identically on $J^{-1}(\mathcal{O}_\mu)$. Because $\mathcal{O}_\mu$ is locally closed, it is a submanifold of $\mathcal{G}^*$. Hence $J^{-1}(\mathcal{O}_\mu)$ is closed. Thus $\mathcal{J}$ is a *closed* subspace of $C^\infty(M)^G$ and we may give $C^\infty(\mathcal{M}_\mu)$ the quotient topology. In addition, because $J^{-1}(\mathcal{O}_\mu)$ is closed and G-invariant, we can use a G-invariant partition of unity ((2.5)) to extend every smooth G-invariant function on $J^{-1}(\mathcal{O}_\mu)$ to a smooth G-invariant function on M.

Before we can define a Poisson bracket on $C^\infty(\mathcal{M}_\mu)$, we need some preparation. Recall that ▷ $(C^\infty(M), \{\, ,\, \}, \cdot)$ is a Poisson algebra, (see appendix A section 4). The following argument shows that $(C^\infty(M)^G, \{\, ,\, \}, \cdot)$ is a Poisson subalgebra of $(C^\infty(M), \{\, ,\, \}, \cdot)$.

(5.1) **Proof**: For every $g \in G$ the mapping $\Phi_g : M \to M : (g, m) \to g \cdot m$ is a symplectic diffeomorphism of (M, ω). For every $f, h \in C^\infty(M)^G$

$$\begin{aligned} \Phi_g^*\{f, h\} &= \{\Phi_g^* f, \Phi_g^* h\} \quad \text{by ((4.4)) of appendix A} \\ &= \{f, h\}, \text{ since } f, h \text{ are } G\text{-invariant.} \end{aligned}$$

Therefore $\{f, h\} \in C^\infty(M)^G$. Clearly

$$\Phi_g^*(f \cdot h) = \Phi_g^* f \cdot \Phi_g^* h = f \cdot h.$$

So $f \cdot h \in C^\infty(M)^G$. □

To construct a Poisson bracket on $C^\infty(\mathcal{M}_\mu)$ we verify that the ideal $\mathcal{J}$ is a *Poisson ideal* of $(C^\infty(M)^G, \{\, ,\, \}, \cdot)$. In other words, $\mathcal{J}$ is an ideal of the commutative ring $(C^\infty(M)^G, \cdot)$ *and* also an ideal of the Lie algebra $(C^\infty(M)^G, \{\, ,\, \})$. This follows from

Claim: Let $\mathcal{A}$ be a Poisson subalgebra of $(C^\infty(M), \{\, ,\, \}, \cdot)$. Suppose that for every $f \in \mathcal{A}$ the flow of the Hamiltonian vector field X_f preserves a subset N of M. Then the ideal $\mathcal{J}(N) = \{f \in \mathcal{A} | \, f|N = 0\}$ is a Poisson ideal of $\mathcal{A}$.

(5.2) **Proof**: First note that it is easy to see that $\mathcal{J}$ is an ideal in $(C^\infty(M), \cdot)$. To show that $\mathcal{J}$ is an ideal in $(C^\infty(M), \{\, ,\, \})$, let $f \in \mathcal{A}$, $n \in N$ and $h \in \mathcal{J}(N)$. Suppose that $t \to \gamma(t)$ is an integral curve of X_f with $\gamma(0) = n$. Then $\gamma(t) \in N$ for all t in some open interval I_n containing 0. Consequently, $h(\gamma(t)) = 0$ for all $t \in I_n$. Differentiating with respect to t and evaluating at $t = 0$ gives

$$0 = \left.\frac{d}{dt}\right|_{t=0} h(\gamma(t)) = \{f, h\}(n).$$

Thus for every $n \in N$ we have $\{f, h\}(n) = 0$, that is, $\{f, h\} \in \mathcal{J}(N)$. □

By ((4.10)) the flow of a Hamiltonian vector field corresponding to G-invariant Hamiltonian function on M preserves $J^{-1}(\mathcal{O}_\mu)$. Applying ((5.2)) with $\mathcal{A} = (C^\infty(M)^G, \{\, ,\, \}, \cdot)$

and $N = J^{-1}(\mathcal{O}_\mu)$ shows that $\mathcal{J} = \mathcal{J}(J^{-1}(\mathcal{O}_\mu))$ is a Poisson ideal of $\mathcal{A}$. On $C^\infty(\mathcal{M}_\mu)$ define a *Poisson bracket* $\{\, ,\,\}_\mu$ by

$$\pi^*(\{f_\mu, h_\mu\}_\mu) = \{f, h\}|J^{-1}(\mathcal{O}_\mu),$$

where $\pi^* f_\mu = f|J^{-1}(\mathcal{O}_\mu)$ and $\pi^* h_\mu = h|J^{-1}(\mathcal{O}_\mu)$.

Claim: The Poisson bracket $\{\, ,\,\}_\mu$ on $C^\infty(\mathcal{M}_\mu)$ is well defined.

(5.3) **Proof**: Let $h' \in C^\infty(M)^G$ such that $\pi^* h_\mu = h'|J^{-1}(\mathcal{O}_\mu)$. Then $h - h' \in \mathcal{J}$. Since $\mathcal{J}$ is a Poisson ideal, $\{f, h - h'\} \in \mathcal{J}$, that is , $\{f, h - h'\}|J^{-1}(\mathcal{O}_\mu) = 0$. Thus by bilinearity, $\{\, ,\,\}_\mu$ is well defined. □

Because a Poisson algebra modulo a Poisson ideal is a Poisson algebra, we find that

$$(C^\infty(\mathcal{M}_\mu), \{\, ,\,\}_\mu, \cdot) = (C^\infty(M)^G, \{\, ,\,\}, \cdot)/\mathcal{J}$$

is a Poisson algebra.

The Poisson bracket $\{\, ,\,\}_\mu$ on $C^\infty(\mathcal{M}_\mu)$ is quite useful because its only Casimir functions are locally constant. In other words,

Claim: The Poisson bracket $\{\, ,\,\}_\mu$ on $C^\infty(\mathcal{M}_\mu)$ is nondegenerate.

(5.4) **Proof**: It suffices to show that if $h \in C^\infty(M)^G$ such that

$$\{f, h\}(p) = 0 \tag{78}$$

for every $f \in C^\infty(M)^G$ and every $p \in J^{-1}(\mathcal{O}_\mu)$, then $h|J^{-1}(\mathcal{O}_\mu)$ is locally constant. Note that by G-invariance of $\{f, h\}$ and the fact that $J^{-1}(\mathcal{O}_\mu) = G \cdot J^{-1}(\mu)$, we need only show that if

$$\{f, h\}(q) = 0 \tag{79}$$

for every $f \in C^\infty(M)^G$ and every $q \in J^{-1}(\mu)$, then $h|J^{-1}(\mu)$ is locally constant. Recall that $J^{-1}(\mu)$ is locally arcwise connected, (see the corollary to ((3.20))). Thus to verify that $h|J^{-1}(\mu)$ is locally constant it suffices to show that

$$\frac{dh}{dt}(\gamma(t)) = 0, \tag{80}$$

for every locally defined C^1 curve $t \to \gamma(t)$ with $\gamma(0) = q$ and $\gamma(t) \in J^{-1}(\mu)$. Since $\{f, h\}(q) = df(q)X_h(q)$ and

$$\text{span}\,\{df(q) \mid f \in C^\infty(M)^G\} = T_q(G \cdot q)^\circ,$$

from (79) it follows that

$$X_h(q) \in (T_q(G \cdot q)^\circ)^\circ = T_q(G \cdot q) \tag{81}$$

for every $q \in J^{-1}(\mu)$. By definition of the momentum map J, we have

$$\omega(\gamma(t))\Big(\frac{d\gamma}{dt}, X^\xi(\gamma(t))\Big) = dJ^\xi(\gamma(t))\frac{d\gamma}{dt} = \frac{dJ^\xi}{dt}(\gamma(t)) = 0,$$

for every $\xi \in \mathcal{G}$. The last equality in the above equation follows because $\gamma(t) \in J^{-1}(\mu)$. Therefore

$$\frac{d\gamma}{dt} \in T_{\gamma(t)}\big(G \cdot \gamma(t)\big)^{\omega(\gamma(t))}. \tag{82}$$

Using (81) and (82) we have

$$0 = \omega(\gamma(t))\big(X_h(\gamma(t)), \frac{d\gamma}{dt}\big) = dh(\gamma(t))\frac{d\gamma}{dt} = \frac{dh}{dt}(\gamma(t)),$$

which proves (80). Thus the Poisson algebra $(C^\infty(\mathcal{M}_\mu), \{\,,\,\}_\mu, \cdot)$ is nondegenerate. □

We now turn to constructing the symplectic pieces of the singular reduced space $\mathcal{M}_\mu$. Because we work locally, we can use the shifting trick ((3.19)) to reduce to the case where $\mu = 0$. First we partition $J^{-1}(0)$ by orbit types. More formally, let H be a compact subgroup of G and let $M_{(H)}$ be the corresponding orbit type, (see (9)). Define $J_{(H)} = J^{-1}(0) \cap M_{(H)}$. The collection of orbit types $M_{(H)}$ partitions M. Therefore, $J_{(H)}$ partitions $J^{-1}(0)$. Because of ((1.9)) the partition of M into orbit types is locally finite. Hence the partition of $J^{-1}(0)$ by $J_{(H)}$ is locally finite.

The following proposition is the key result in constructing the symplectic pieces of $\mathcal{M}_0$.

Proposition. For every compact subgroup H of G, the set $J_{(H)}$ is a submanifold of (M, ω). Moreover, for every $q \in J_{(H)}$ we have $T_q J_{(H)} \cap (T_q J_{(H)})^{\omega(q)} = T_q \mathcal{O}_q$.

We prove this proposition in a series of lemmas and claims. From now on we assume that $p \in M$, $H = G_p$ and $\mathcal{H}$ is the Lie algebra of H.

Lemma 1. For any $p \in M$, $\operatorname{im} T_p J = \mathcal{H}^\circ = \{\alpha \in \mathcal{G}^* \,|\, \alpha|\mathcal{H} = 0\}$.

(5.5) **Proof**: For every $\xi \in \mathcal{G}$ we have

$$dJ^\xi(p) = X^\xi(p) \,\lrcorner\, \omega(p). \tag{83}$$

Now

$$\begin{aligned} \xi \in \mathcal{H} \quad &\Leftrightarrow \quad X^\xi(p) = 0 \Leftrightarrow (T_pJ\, v_p)\xi = 0 \quad \forall\, v_p \in T_pM, \quad \text{using (83) and the nondegeneracy of } \omega \\ &\Leftrightarrow \quad \xi \in (\operatorname{im} T_pJ)^\circ. \end{aligned}$$

Therefore

$$\mathcal{H} = (\operatorname{im} T_pJ)^\circ. \tag{84}$$

Taking the annihilator of both sides of (84) gives (83). □

Lemma 2. Let $p \in M_H \cap J^{-1}(0)$ and let M_H° be the connected component of M_H containing p. Then $J(M_H^\circ) \subseteq \mathcal{H}^\circ$.

(5.6) **Proof**: Since M_H° is an open connected subset of the submanifold M_H (see ((1.7))), it is arcwise connected. Thus for every $q \in M_H^\circ$ there is a piecewise smooth curve $\gamma : [0,1] \to M_H^\circ$ with $\gamma(0) = p$ and $\gamma(1) = q$. For every $\widetilde{q} \in M_H^\circ$ and for every $\xi \in \mathcal{H}$, we have $X^\xi(\widetilde{q}) = 0$. Therefore

$$dJ^\xi(\widetilde{q}) = X^\xi(\widetilde{q}) \,\lrcorner\, \omega(\widetilde{q}) = 0,$$

which implies that $\frac{dJ^{\xi}}{dt}(\gamma(t)) = 0$. Hence

$$J^{\xi}(q) - J^{\xi}(p) = \int_0^1 \frac{dJ^{\xi}}{dt}(\gamma(t))\,dt = 0$$

for every $q \in M_H^{\circ}$ and every $\xi \in \mathcal{H}$. But $p \in J^{-1}(0)$. Therefore $J^{\xi}(q) = 0$. In other words, $J(M_H^{\circ}) \subseteq \mathcal{H}^{\circ}$. □

Lemma 3. Let $q \in J^{-1}(0)$. Then $T_q \mathcal{O}_q \subseteq \ker T_q J$.

(5.7) **Proof**: By hypothesis

$$0 = J^{[\xi,\eta]}(q) = \{J^{\xi}, J^{\eta}\}(q) = dJ^{\eta}(q) X^{\xi}(q),$$

for every $\xi, \eta \in \mathcal{G}$. Therefore $X^{\xi}(q) \in \ker T_q J$ for every $\xi \in \mathcal{G}$. But $T_q \mathcal{O}_q = \text{span}\{X^{\xi}(q) \in T_q M | \xi \in \mathcal{G}\}$. Hence $T_q \mathcal{O}_q \subseteq \ker T_q J$. □

Let

$$\mathcal{H}_H^{\circ} = \left\{ \alpha \in \mathcal{H}^{\circ} \middle| \text{Ad}^t_{h^{-1}} \alpha = \alpha \text{ for every } h \in H \right\}$$

and write $\mathcal{G}^* = \mathcal{H}_H^{\circ} \oplus (\mathcal{H}_H^{\circ})^{\perp}$ for some inner product on $\mathcal{G}^*$ which is invariant under the coadjoint action of H. Let $\pi_H : \mathcal{G}^* \to \mathcal{H}_H^{\circ}$ be the projection along $(\mathcal{H}_H^{\circ})^{\perp}$ onto $\mathcal{H}_H^{\circ}$. Define the H-coadjoint equivariant map $J_H : M \to \mathcal{H}_H^{\circ}$ by $J_H = \pi_H \circ J$.

Claim. Let $p \in J^{-1}(0) \cap M_H$. Then there is an open neighborhood U of p in M such that $J^{-1}(0) \cap U \cap M_H = J_H^{-1}(0) \cap U \cap M_H$.

(5.8) **Proof**: Since $J^{-1}(0) \subseteq J_H^{-1}(0)$, we obtain $J^{-1}(0) \cap V \cap M_H \subseteq J_H^{-1}(0) \cap V \cap M_H$ for every open neighborhood V of p in M. Because $p \in J^{-1}(0) \cap M_H$, from ((5.6)) we find that there is an open neighborhood U of p in M such that $J(q) \in \mathcal{H}^{\circ}$ for every $q \in J_H^{-1}(0) \cap U \cap M_H$. Since $q \in M_H$ and J is coadjoint equivariant, $J(q) \in \mathcal{H}_H^{\circ}$. But $0 = J_H(q) = \pi_H(J(q))$, that is, $J(q) \in \ker \pi_H = (\mathcal{H}_H^{\circ})^{\perp}$. Hence $J(q) = 0$, that is, $q \in J^{-1}(0) \cap U \cap M_H$. □

Claim. Let $p \in J^{-1}(0) \cap M_H$ and let S_p be a slice to the G-action Φ at p. Then $J_H^{-1}(0) \cap M_H \cap S_p$ is a submanifold of M through p.

(5.9) **Proof**: Let γ be an H-invariant Riemannian metric on M. From ((5.5)) we have $\text{im}\, T_p J = \mathcal{H}^{\circ}$. Since $\mathcal{H}_H^{\circ} \subseteq \mathcal{H}^{\circ}$, it follows that for every $\mu \in \mathcal{H}_H^{\circ}$ there is a unique $v_{\mu} \in (\ker T_p J)^{\perp}$ such that

$$T_p J\, v_{\mu} = \mu, \tag{85}$$

which is equivalent to

$$T_p J_H\, v_{\mu} = \mu, \tag{86}$$

because $\mu \in \mathcal{H}_H^{\circ}$.

The following argument shows that v_{μ} is left fixed by the linear action $T_m \Phi$ of H on $T_p M$. Differentiating the relation $J(\Phi_h(p)) = \text{Ad}^t_{h^{-1}}(J(p))$ for every $h \in H$, which expresses the fact that J is H-coadjoint equivariant, we obtain

$$T_p J \circ T_p \Phi_h v_{\mu} = T_{\Phi_h(p)} J \circ T_p \Phi_h v_{\mu} = \text{Ad}^t_{h^{-1}}(T_p J\, v_{\mu}) = T_p J\, v_{\mu} = \mu. \tag{87}$$

The first equality in (87) follows because $h \in G_p = H$, while the third equality holds because $T_pJ\,v_\mu \in \mathcal{H}_H^\circ$. Since $v_\mu \in (\ker T_pJ)^\perp$, we have

$$0 = \gamma(p)(v_\mu, w) = \gamma(\Phi_h(p))(T_p\Phi_h v_\mu, T_p\Phi_h w),$$

for every $w \in \ker T_pJ$. Because J is coadjoint equivariant, $T_p\Phi_h(\ker T_pJ) = \ker T_pJ$. Therefore $T_p\Phi_h v_\mu \in (\ker T_pJ)^\perp$. From the uniqueness of solutions of (85), we obtain $T_p\Phi_h v_\mu = v_\mu$ for every $h \in H$, that is, $v_\mu \in T_pM_{[H]}$, (see ((1.5))).

From ((5.7)) and the construction of the slice S_p ((1.6)) it follows that $v_\mu \in (\ker T_pJ)^\perp \subseteq T_p\mathcal{O}_p{}^\perp = T_pS_p$. Therefore $v_\mu \in T_pM_{[H]} \cap T_pS_p = T_p(M_{[H]} \cap S_p)$.

Thus we have shown that T_pJ_H is a surjective linear map from $T_p(M_{[H]} \cap S_p)$ onto $\mathcal{H}_H^\circ$. In other words, p is a regular point of $J_H|(M_{[H]} \cap S_p)$. Consequently,

$$J_H^{-1}(0) \cap M_{[H]} \cap S_p = (J_H|(M_{[H]} \cap S_p))^{-1}(0)$$

is a submanifold of $M_{[H]} \cap S_p$ through p. Because M_H is open in $M_{[H]}$, we deduce that $J_H^{-1}(0) \cap M_H \cap S_p$ is a submanifold of M through p. □

Corollary.

$$T_p\big(J_H^{-1}(0) \cap M_{[H]} \cap S_p\big) = (\ker T_pJ) \cap T_p(M_{[H]} \cap S_p),$$

for $p \in M_H \cap J^{-1}(0)$.

(5.10) **Proof**: Since $M_H \cap J^{-1}(0) \subseteq M_{[H]} \cap J_H^{-1}(0)$, it follows that $p \in J_H^{-1}(0) \cap M_{[H]} \cap S_p$. From the proof of the claim, p is a regular point of $J_H|(M_{[H]} \cap S_p)$. Therefore

$$T_p\big(J_H^{-1}(0) \cap M_H \cap S_p\big) = T_p\big(J_H^{-1}(0) \cap M_{[H]} \cap S_p\big) = (\ker T_pJ_H) \cap T_p(M_{[H]} \cap S_p). \quad (88)$$

Again, from the proof of the claim, we have $T_pJ\,v \in \mathcal{H}_H^\circ$, for every $v \in T_p(M_{[H]} \cap S_p)$. Consequently,

$$T_pJ_H\,v = \pi_H(T_pJ\,v) = T_pJ\,v$$

for every $v \in T_p(M_{[H]} \cap S_p)$. Using (88) we obtain the corollary. □

We now are in a position to prove the first assertion of the proposition.

Claim. The set $J_{(H)} = J^{-1}(0) \cap M_{(H)}$ is a submanifold of M.

(5.11) **Proof**: Since $J_{(H)} = G \cdot \big(J^{-1}(0) \cap M_H\big)$, it suffices to show that there is a neighborhood U of $p \in J^{-1}(0) \cap M_H$ such that $U \cap J^{-1}(0) \cap M_H$ is a submanifold of M through p. First we show that $J_{(H)} \cap S_p \subseteq M_H$. Let $q \in J_{(H)} \cap S_p$. Then for some $g \in G$

$$gHg^{-1} = G_q \subseteq H, \quad (89)$$

where the equality follows because $q \in M_{(H)}$ and the inclusion because q lies in the slice S_p. If H is connected, then $gHg^{-1} = H$ because both H and gHg^{-1} have the same Lie algebra. If H is not connected, then H and gHg^{-1} have the same finite number of connected components. From (89) it follows that every component of gHg^{-1} is contained in a component of H. Therefore $H = gHg^{-1}$. In other words, $q \in M_H$. Hence $J_{(H)} \cap S_p \subseteq M_H$.

Next there is an open neighborhood U of p in M such that

$$\begin{aligned} J_{(H)} \cap S_p \cap U &= J^{-1}(0) \cap M_{(H)} \cap S_p \cap U, \quad \text{by definition of } J_{(H)} \\ &= J^{-1}(0) \cap M_H \cap S_p \cap U, \quad \text{since } J_{(H)} \cap S_p \subseteq M_H \\ &= J_H^{-1}(0) \cap M_H \cap S_p \cap U, \quad \text{by ((5.8)).} \end{aligned} \tag{90}$$

From ((5.9)) we know that $J_H^{-1}(0) \cap M_H \cap S_p \cap U$ is a submanifold of M. From (90) we deduce that $J_{(H)} \cap S_p \cap U$ is a submanifold of M and hence is a submanifold of S_p. Let $\mathcal{L}$ be a complement to $\mathcal{H}$ in $\mathcal{G}$. From the properties of a slice it follows that the map

$$\varphi : S_p \times \mathcal{L} \to M : (s, \eta) \to \Phi_{\exp \eta}(s)$$

with $\varphi(p, 0) = p$ is a local diffeomorphism. Because $J_{(H)} \cap S_p \cap U$ is a submanifold of S_p, there is a local diffeomorphism

$$\theta : (J_{(H)} \cap S_p) \times Y \to S_p$$

such that $\theta | \big((J_{(H)} \cap S_p) \times \{0\} \big) = \mathrm{id}_{J_{(H)} \cap S_p}$. Here Y is a space which is complementary to $T_p(J_{(H)} \cap S_p)$ in $T_p S_p$. Therefore

$$\Psi : (J_{(H)} \cap S_p) \times Y \times \mathcal{L} \to M : (q, y, \eta) \to \Phi_{\exp \eta}\big(\theta(q, y)\big)$$

with $\Psi(q, 0, 0) = q$ is a local diffeomorphism at $(p, 0, 0)$. Because S_p is a slice and $J_{(H)}$ is G-invariant, it follows that Ψ maps an open subset of $(J_{(H)} \cap S_p) \times \{0\} \times \mathcal{L}$ containing $(p, 0, 0)$ diffeomorphically onto an open subset of $J_{(H)}$ containing p. Therefore $J_{(H)}$ is a submanifold of M. □

In a neighborhood of p in (M, ω) we have a $G_p \,(= H)$-invariant Riemannian metric γ and a G-invariant symplectic form ω. Because H is compact, we can use the equivariant Darboux theorem to obtain an H-invariant constant symplectic form $\widetilde{\omega} = \omega(p)$ and an H-invariant inner product Γ on T_pM. Let $A = \Gamma^{\flat} \circ \widetilde{\omega}^{\sharp}$. Then $j = A(-A^2)^{-1/2}$ is an H-invariant almost complex structure on T_pM such that $j^2 = -1$ and $\Omega(v, w) = \Gamma(j(v), w)$ for every $v, w \in T_pM$ is an H-invariant symplectic form. An important consequence of the construction of j is

Lemma 4. For every subspace V of T_pM, $(V^{\perp})^{\Omega} = j(V)$.

(5.12) **Proof**: The lemma follows from

$$\begin{aligned} (V^{\perp})^{\Omega} &= \left\{ u \in T_pM \,\middle|\, \Omega(u, v) = 0, \ \forall\, v \in V^{\perp} \right\} \\ &= \left\{ u \in T_pM \,\middle|\, j(u) \in (V^{\perp})^{\perp} \right\}, \quad \text{since } \Gamma(j(u), v) = \Omega(u, v). \\ &= j(V), \quad \text{since } (V^{\perp})^{\perp} = V \text{ and } j^2 = -1. \quad \square \end{aligned}$$

Lemma 5. The symplectic form $\omega(p)$ on T_pM restricted to $T_p(J_{(H)} \cap S_p)$ is nondegenerate.

(5.13) **Proof**: First we show that $T_p(J_{(H)} \cap S_p)$ is invariant under the almost complex structure j. From the construction of the slice S_p, $T_pS_p = (T_p\mathcal{O}_p)^\perp$. Using ((5.7)) we obtain

$$j(\ker T_pJ) = ((\ker T_pJ)^\perp)^\Omega = ((\ker T_pJ)^\Omega)^\perp = (T_p\mathcal{O}_p)^\perp = T_pS_p. \tag{91}$$

Since $j^2 = -1$, it follows that $j(T_pS_p) = \ker T_pJ$. Therefore $\ker T_pJ \cap T_pS_p$ is j-invariant. Because j commutes with the H-action, $j(T_pM_{[H]}) = T_pM_{[H]}$. Using (90) and ((5.10)), we find that

$$T_p(J_{(H)} \cap S_p) = (\ker T_pJ) \cap T_p(M_{[H]} \cap S_p) = (\ker T_pJ) \cap T_pS_p) \cap T_pM_{[H]}. \tag{92}$$

Therefore $T_p(J_{(H)} \cap S_p)$ is j-invariant, because it is the intersection of two j-invariant spaces.

Suppose that for some $v \in T_p(J_{(H)} \cap S_p)$ we have $\omega(p)(v,u) = 0$ for every $u \in T_p(J_{(H)} \cap S_p)$. Then writing $u = j(u')$, we obtain $0 = \Gamma(p)(v,u')$ for every $u' \in T_p(J_{(H)} \cap S_p)$. But $\Gamma(p)$ is a positive definite inner product on T_pM. Therefore $v = 0$, that is, $\omega(p)$ is nondegenerate on $T_p(J_{(H)} \cap S_p)$. □

We now prove the second assertion in the proposition.

Claim: $T_qJ_{(H)} \cap (T_qJ_{(H)})^{\omega(q)} = T_q\mathcal{O}_q$, for every $q \in J_{(H)}$.

(5.14) **Proof**: From the proof of ((5.11)) and the definition of the slice it follows that

$$T_{g\cdot p}\Phi_{g^{-1}}\big(T_{g\cdot p}J_{(H)}\big) = T_p(J_{(H)} \cap S_p) \oplus T_p\mathcal{O}_p \tag{93}$$

for every $p \in J_{(H)} \cap S_p$ and every $g \in G$. Note that $g \cdot p \in J_{(H)}$ because $J_{(H)} \cap S_p \subseteq M_H \cap J^{-1}(0) \cap S_p$ by (90) and $J_{(H)} = G \cdot (M_H \cap J^{-1}(0))$. Therefore

$$T_p(J_{(H)} \cap S_p) \subseteq T_p(M_H \cap J^{-1}(0) \cap S_p) = (\ker T_pJ) \cap T_p(M_{[H]} \cap S_p) \subseteq \ker T_pJ, \tag{94}$$

where the equality in (94) follows from the corollary of ((5.9)). Consequently

$$T_p\mathcal{O}_p \subseteq (\ker T_pJ)^\omega \subseteq T_p(J_{(H)} \cap S_p)^\omega,$$

where the first inclusion above follows from (39) and the second by taking the symplectic perpendicular of (94). Because $T_p\mathcal{O}_p \subseteq \ker T_pJ = T_p\mathcal{O}_p{}^\omega$ (using ((5.7)) and (39)), $T_p\mathcal{O}_p$ is isotropic. Since $T_p(J_{(H)} \cap S_p)$ is symplectic ((5.13)), the subspace $T_p(J_{(H)} \cap S_p)^\omega$ is symplectic. Thus there is an isotropic subspace W of $T_p(J_{(H)} \cap S_p)^\omega$ such that $T_p\mathcal{O}_p \oplus W$ is a symplectic subspace of $T_p(J_{(H)} \cap S_p)^\omega$. Hence we obtain the decompositions

$$T_p(J_{(H)} \cap S_p)^\omega = (T_p\mathcal{O}_p \oplus W) \oplus (T_p\mathcal{O}_p \oplus W)^\omega \tag{95}$$

and

$$\begin{aligned} T_pM &= T_p(J_{(H)} \cap S_p) \oplus T_p(J_{(H)} \cap S_p)^\omega \\ &= T_p(J_{(H)} \cap S_p) \oplus (T_p\mathcal{O}_p \oplus W) \oplus (T_p\mathcal{O}_p \oplus W)^\omega \end{aligned} \tag{96}$$

into symplectically perpendicular subspaces. Again because $T_p\mathcal{O}_p$ is isotropic, from (96) we deduce

$$T_p\mathcal{O}_p{}^\omega = T_p\mathcal{O}_p \oplus T_p(J_{(H)} \cap S_p) \oplus (T_p\mathcal{O}_p \oplus W)^\omega. \tag{97}$$

Therefore

$$\begin{aligned} T_{g\cdot p}\Phi_{g^{-1}}\big((T_{g\cdot p}J_{(H)})^{\omega}\big) &= \big(T_{g\cdot p}\Phi_{g^{-1}}T_{g\cdot p}J_{(H)}\big)^{\omega} = T_p(J_{(H)}\cap S_p)^{\omega}\cap T_p\mathcal{O}_p{}^{\omega} \\ &= T_p\mathcal{O}_p \oplus (T_p\mathcal{O}_p\oplus W)^{\omega}. \end{aligned}$$

The first equality above holds because $\Phi_{g^{-1}}$ is symplectic, the second by (93) and the third is a consequence of (95) and (97). Consequently,

$$T_{g\cdot p}\Phi_{g^{-1}}(T_{g\cdot p}J_{(H)}\cap (T_{g\cdot p}J_{(H)})^{\omega}) = T_p\mathcal{O}_p.$$

In other words,

$$T_{g\cdot p}J_{(H)}\cap (T_{g\cdot p}J_{(H)})^{\omega} = T_{g\cdot p}\mathcal{O}_{g\cdot p},$$

which proves the claim. □

Having verified the claims ((5.11)) and ((5.14)), we have proved the proposition. □

We now use this proposition to decompose the singular reduced space $\mathcal{M}_0 = J^{-1}(0)/G$ into symplectic pieces. For every compact subgroup H of G let

$$\mathcal{M}_{(H)} = \pi\big(M_{(H)}\cap J^{-1}(0),\big).$$

where $\pi : M \to M/G$ is the G-orbit map. By the proposition and the proof of the regular reduction theorem ((4.1)), $\mathcal{M}_{(H)}$ is a smooth manifold equipped with a symplectic form $\omega_{(H)}$ such that $\pi^*\omega_{(H)} = \omega|M_{(H)}$. Clearly, $\mathcal{M}_0 = \bigcup_{(H)}\mathcal{M}_{(H)}$. Thus we have obtained a partition of the singular reduced space $\mathcal{M}_0$ into symplectic manifolds. Using the local diffeomorphism in the shifting trick ((3.19)) we can transport this partition of $\mathcal{M}_0$ to a partition of the singular reduced space $\mathcal{M}_\mu$ for $\mu \neq 0$ into smooth symplectic manifolds. This latter partition is entirely determined by the Poisson structure $\{\, ,\, \}_\mu$ on $C^\infty(\mathcal{M}_\mu)$.

To verify the last claim of the singular reduction theorem, we need a notion of a Hamiltonian vector field on the singular reduced space $\mathcal{M}_\mu$. Because $\mathcal{M}_\mu$ is in general singular, it does *not* make sense to talk about a Hamiltonian vector field on $\mathcal{M}_\mu$. But recall: if M is a smooth manifold, then every derivation $\mathcal{X}$ of the algebra $(C^\infty(M), \cdot)$ of smooth functions on M is the Lie derivative of a unique smooth vector field X on M. Thus we may identify a derivation on $C^\infty(M)$ with a vector field on M. Hence on the algebra $(C^\infty(\mathcal{M}_\mu), \cdot)$ of smooth functions of the singular reduced space $\mathcal{M}_\mu$ it makes sense to talk about a *Hamiltonian derivation*

$$-\mathrm{ad}_{f_\mu} : C^\infty(\mathcal{M}_\mu) \to C^\infty(\mathcal{M}_\mu) : h_\mu \to -\mathrm{ad}_{f_\mu}h_\mu = \{h_\mu, f_\mu\}_\mu$$

corresponding to the smooth Hamiltonian function $f_\mu \in C^\infty(\mathcal{M}_\mu)$. When $\mathcal{M}_\mu$ is nonsingular, then $-\mathrm{ad}_{f_\mu}$ can be identified with the Hamiltonian vector field X_{f_μ}, (see appendix A section 4). Before we can talk about a flow for a Hamiltonian derivation we need to define some terms. A continuous function $\gamma : I \to \mathcal{M}_\mu$, defined on an open interval I, is said to be *smooth* if for every $h_\mu \in C^\infty(\mathcal{M}_\mu)$ the function $h_\mu\circ\gamma$ is smooth. A flow φ_t is said to be *smooth* if the following conditions hold:

1. $\varphi_t : \mathcal{M}_\mu \to \mathcal{M}_\mu$ is a one parameter group of homeomorphisms such that for every $h_\mu \in C^\infty(\mathcal{M}_\mu)$ and every $t \in \mathbf{R}$, the function $h_\mu\circ\varphi_t \in C^\infty(\mathcal{M}_\mu)$.

2. for every $m_\mu \in \mathcal{M}_\mu$ the curve $t \to \varphi_t(m_\mu)$ is smooth.

A smooth flow φ_t is a *flow of a Hamiltonian derivation* $-\mathrm{ad}_{f_\mu}$ if for every $m_\mu \in \mathcal{M}_\mu$ and every $h_\mu \in C^\infty(\mathcal{M}_\mu)$ we have

$$\frac{d}{dt} h_\mu(\varphi_t(m_\mu)) = \{f_\mu, h_\mu\}_\mu(\varphi_t(m_\mu)). \tag{98}$$

In general (98) is *not* a system of differential equations written in a coordinate free way, ▷ since the reduced space $\mathcal{M}_\mu$ need not be locally Euclidean. Therefore we have to prove the existence of solutions of (98).

(5.15) **Proof**: Let $f \in C^\infty(M)^G$. Then the flow φ_t^f of X_f commutes with the G-action Φ on M, that is, $\Phi_g \circ \varphi_t^f = \varphi_t^f \circ \Phi_g$ for every $g \in G$. Since φ_t^f maps $J^{-1}(\mathcal{O}_\mu)$ diffeomorphically onto itself, φ_t^f descends to a flow ψ_t on $\mathcal{M}_\mu$, that is, $\pi \circ \varphi_t^f = \psi_t \circ \pi$. To show that ψ_t is smooth, note first that ψ_t is a one parameter group of homeomorphisms. To show that $h_\mu \circ \psi_t \in C^\infty(\mathcal{M}_\mu)$ for every $h_\mu \in C^\infty(\mathcal{M}_\mu)$ we observe that by definition of smooth function on $\mathcal{M}_\mu$ there is an $h \in C^\infty(M)^G$ such that $\pi^* h_\mu = h|J^{-1}(\mathcal{O}_\mu)$. Therefore on $J^{-1}(\mathcal{O}_\mu)$

$$\begin{aligned} \pi^*(h_\mu \circ \psi_t) &= h_\mu \circ \psi_t \circ \pi = h_\mu \circ \pi \circ \varphi_t^f \\ &= (h|J^{-1}(\mathcal{O}_\mu)) \circ \varphi_t^f = (h \circ \varphi_t^f)|J^{-1}(\mathcal{O}_\mu) \end{aligned} \tag{99}$$

and $h_\mu \circ \psi_t$ is G-invariant, since

$$h \circ \varphi_t^f \circ \Phi_g = h \circ \Phi_g \circ \varphi_t^f = h \circ \varphi_t^f.$$

Therefore $h_\mu \circ \psi_t \in C^\infty(\mathcal{M}_\mu)$. For fixed $m_\mu \in \mathcal{M}_\mu$ the smoothness of the curve

$$t \to h_\mu \circ \psi_t(m_\mu)$$

follows from (99). Finally we have to show that (98) holds for ψ_t, that is, ψ_t is Hamiltonian flow for $-\mathrm{ad}_{f_\mu}$. For $m \in J^{-1}(\mathcal{O}_\mu)$ and $h \in C^\infty(M)^G$ we have

$$\begin{aligned} \frac{d}{dt} h(\varphi_t^f(m)) &= \{f, h\}(\varphi_t^f(m)) = \pi^*\{f_\mu, h_\mu\}_\mu(\varphi_t^f(m)) \\ &= \{f_\mu, h_\mu\}_\mu(\pi(\varphi_t^f(m))) = \{f_\mu, h_\mu\}_\mu\big(\psi_t(\pi(m))\big). \end{aligned}$$

But $\pi^*(h_\mu \circ \psi_t)(m) = h(\varphi_t^f(m))$. Therefore

$$\frac{d}{dt}(h_\mu \circ \psi_t)(m_\mu) = \{f_\mu, h_\mu\}_\mu(\psi_t(m_\mu)),$$

for every $m_\mu = \pi(m) \in \mathcal{M}_\mu$. □

▷ The uniqueness of the flow of a Hamiltonian derivation on $\mathcal{M}_\mu$ follows because the smooth invariant functions of a proper action separate G-orbits.

(5.16) **Proof**: In more detail, suppose that φ_t and ψ_t are smooth Hamiltonian flows on $\mathcal{M}_\mu$ of $-\mathrm{ad}_{f_\mu}$. Then φ_{-t} is a smooth flow of ad_{f_μ}. Since the action of G on M is proper, smooth G-invariant functions separate G-orbits ((2.7)). Therefore smooth functions on $\mathcal{M}_\mu$ separate points. Thus it is enough to show that for every $m_\mu \in \mathcal{M}_\mu$ and every $h_\mu \in C^\infty(\mathcal{M}_\mu)$ we have $h_\mu(\psi_t \circ \varphi_{-t}(m_\mu)) = h_\mu(m_\mu)$. But

$$\frac{d}{dt}(\psi_t \circ \varphi_{-t}(m_\mu)) = \{h, f\}(\psi_t(\varphi_{-t}(m_\mu)) + \{h, -f\}(\psi_t(\varphi_{-t}(m_\mu)) = 0. \quad \square$$

▷ We now show that the flow of a Hamiltonian derivation corresponding to a smooth function on the singular reduced space $\mathcal{M}_0$ preserves the decomposition of $\mathcal{M}_0$ into symplectic pieces.

(5.17) **Proof**: Let f_0 be a smooth function on the symplectic piece $\mathcal{M}_{(H)}$ of $\mathcal{M}_0$. Then $\pi^* f_0 = f|J_{(H)}$ for some smooth G-invariant function f on M. Let φ_t^f be the flow of the vector field X_f. First we verify that $\varphi_t^f(J_{(H)}) \subseteq J_{(H)}$. Since $J_{(H)} = G \cdot (J^{-1}(0) \cap M_H)$, it suffices to show that $\varphi_t^f(J^{-1}(0)) \subseteq J^{-1}(0)$ and $\varphi_t^f(M_H) \subseteq M_H$. The proof of the former follows because $f \in C^\infty(M)^G$, (see ((4.10))). To prove the latter, suppose that $p \in M_H$. Because $f \in C^\infty(M)^G$, φ_t^f and Φ_g commute for every $g \in G$. Thus for every $h \in H$

$$\Phi_h(\varphi_t^f(p)) = \varphi_t^f(\Phi_h(p)) = \varphi_t^f(p). \tag{100}$$

In other words, $\varphi_t^f(p) \in M_{[H]}$. Because M_H is relatively open in $M_{[H]}$ ((1.7)), there is an open neighborhood U of p in M such that $U \cap M_H = U \cap M_{[H]}$. Let $q \in U \cap M_H$. Then repeating the argument in (100) shows that $\varphi_t^f(q) \in U \cap M_{[H]}$, that is, $\varphi_t^f(U \cap M_H) \subseteq U \cap M_{[H]}$. Because φ_t^f is a homeomorphism of M, it induces a homeomorphism of M_H in the relative topology. Therefore $\varphi_t^f(U \cap M_H)$ is a relative open subset of $M_{[H]}$ containing $\varphi_t^f(p)$. Consequently, $\varphi_t^f(p) \in M_H$. Thus we obtain $\varphi_t^f(J_{(H)}) \subseteq J_{(H)}$. Hence the flow of the Hamiltonian derivation $-\mathrm{ad}_{f_0}$ maps the symplectic piece $\mathcal{M}_{(H)}$ of $\mathcal{M}_0$ into itself. □

▷ We now complete the proof of the singular reduction theorem by showing that the inclusion map $\tilde{\imath} : \mathcal{M}_{(H)} \to \mathcal{M}_0$ is a Poisson map.

(5.18) **Proof**: Since $J_{(H)}$ is a G-invariant symplectic submanifold of (M, ω), it follows that the inclusion map $i : J_{(H)} \to M$ induces a homomorphism of Poisson algebras

$$\imath^* : (C^\infty(M)^G, \{\, ,\, \}|J_{(H)}, \cdot) \to (C^\infty(M)^G|J_{(H)}, \{\, ,\, \}, \cdot). \tag{101}$$

Suppose that $f_0, g_0 \in C^\infty(\mathcal{M}_0)$, then there are functions $f, g \in C^\infty(M)^G$ such that $\pi^*(f_0|\mathcal{M}_{(H)}) = f|J_{(H)}$ and $\pi^*(g_0|\mathcal{M}_{(H)}) = g|J_{(H)}$. Therefore

$$\begin{aligned}
\{f_0|\mathcal{M}_{(H)}, g_0|\mathcal{M}_{(H)}\}_0(\pi(q)) &= \{f|J_{(H)}, g|J_{(H)}\}(q), \quad \text{by definition of } \{\, ,\, \}_0. \\
&= \{f, g\}(q), \quad \text{using (101)} \\
&= \{f_0, g_0\}_0(\pi(q)), \quad \text{by definition of } \{\, ,\, \}_0,
\end{aligned}$$

for every $q \in J_{(H)}$. Thus the map $\tilde{\imath}^*$ is a Poisson map. □

▷ An important consequence of the singular reduction theorem is that the decomposition of the singular reduced space $\mathcal{M}_\mu$ into symplectic pieces is encoded in the Poisson algebra $(C^\infty(\mathcal{M}_\mu), \{\ ,\ \}_\mu, \cdot)$.

(5.19) **Proof**: To see this recall that on a connected symplectic manifold the group generated by the time one maps of the flows of a Hamiltonian vector fields corresponding to smooth Hamiltonian functions acts transitively. From this and the fact that flows of Hamiltonian derivations corresponding to smooth Hamiltonian functions on $\mathcal{M}_\mu$ preserve the decomposition of $\mathcal{M}_\mu$ into symplectic pieces, it follows that the connected components of symplectic pieces of $\mathcal{M}_\mu$ are equivalence classes of the relation: m_μ is equivalent to m'_μ if and only if there are smooth functions $f^1_\mu, \ldots, f^n_\mu$ on $\mathcal{M}_\mu$ such that the composition of the time one maps of the flows of $-\mathrm{ad}_{f^i_\mu}$ maps m_μ to m'_μ. □

More precisely we have proved:

Corollary: Suppose that $\mathcal{M}_\mu$ and $\mathcal{M}'_\mu$ are singular reduced spaces and that $\varphi : \mathcal{M}_\mu \to \mathcal{M}'_\mu$ is a homeomorphism. If $\varphi^* : C^\infty(\mathcal{M}'_\mu) \to C^\infty(\mathcal{M}_\mu)$ is an isomorphism of the Poisson algebra $(C^\infty(\mathcal{M}'_\mu), \{\ ,\ \}_\mu, \cdot)$ onto the Poisson algebra $(C^\infty(\mathcal{M}_\mu), \{\ ,\ \}_\mu, \cdot)$, then φ maps a symplectic piece of $\mathcal{M}_\mu$ to a symplectic piece of $\mathcal{M}'_\mu$.

6 Exercises

1. Look up the definition of a slice in Palais' paper [168] and show that it is equivalent to ours.

2. Let G be a compact group acting on $\mathbf{R}^n$ with the origin 0 fixed by the subgroup G_0. Show that for each open set $\widetilde{U}$ containing 0 there is an open set U containing 0 such that $U \subseteq \widetilde{U}$ and U is G_0-invariant.

3. (a). Show that the adjoint action of $\mathrm{Sl}(2, \mathbf{R})$ on $\mathrm{sl}(2, \mathbf{R})$ is not proper.

   (b). Give an example of a Lie group G acting freely on a manifold M with the following two properties: 1) for every $m \in M$ the orbit $G \cdot m$ is a closed subset of M; 2) the orbit space M/G is nowhere locally Hausdorff.

   (c). Show that if a Lie group G acts isometrically on a Riemannian manifold M and each orbit $G \cdot m$, $m \in M$, is closed, then the quotient space M/G is a metric space in natural way.

   *(d)**. If a Lie group G acts effectively and isometrically on Riemannian manifold M and each orbit $G \cdot m$, $m \in M$, is a closed submanfold, is the action necessarily proper?

4. Let G be a compact Lie group which acts linearly on $\mathbf{R}^n$. For $y, z \in \mathbf{R}^n$ suppose that the G-orbit through y is disjoint from the G-orbit through z. Show that there is a G-invariant polynomial P on $\mathbf{R}^n$ such that $P(y) \neq P(z)$.

5. Let G be a compact Lie group acting linearly on $\mathbf{R}^n$. Suppose that $\sigma_1, \ldots, \sigma_r$ generate the algebra of G-invariant polynomials on $\mathbf{R}^n$. Then the fiber of the Hilbert map $\Phi : \mathbf{R}^n \to \mathbf{R}^r : x \to \big(\sigma_1(x), \ldots, \sigma_r(x)\big)$ is a single G-orbit.

6. Give an example of a proper action of a Lie group on a smooth manifold whose fixed point set is a union of submanifolds of *different* dimension.

7. If G is a compact Lie group acting by linear symplectic mappings on the symplectic vector space (V, ω), show that the set of vectors V^G fixed by *every* element of G is a symplectic vector space.

8. Let $\Phi : G \times M \to M$ be a proper action of a Lie group G on a symplectic manifold (M, ω). Suppose that the action Φ is symplectic, that is, Φ_g is a symplectic diffeomorphism for every $g \in G$.

   (a). Show that every connected component of $M_{[H]}$ is a symplectic submanifold of M.

   (b). If Φ is locally free, deduce that $M_{[H]}$ has codimension at least two and hence symplectic orbifolds have no topological boundary.

9. Give a counterexample to the statement: For every symplectic form Ω on $T^*\mathbf{R}^n$, which is invariant under an action of a compact Lie group G, there is a local diffeomorphism φ about 0, which commutes with the G-action, such that $\varphi^*\Omega$ is the *standard* symplectic form on $T^*\mathbf{R}^n$ near 0.

10. (Linear Hamiltonian vector fields.) (a). Let (V, ω) be a real symplectic vector space. Let $\mathrm{Sp}(\omega, \mathbf{R})$ be the group of all linear symplectic isomorphisms of (V, ω). In other words, $P \in \mathrm{Sp}(\omega, \mathbf{R})$ if and only if $P \in \mathrm{Gl}(V, \mathbf{R})$ and $P^*\omega = \omega$. Show that $\mathrm{Sp}(\omega, \mathbf{R})$ is a closed subgroup of $\mathrm{Gl}(V, \mathbf{R})$ and hence is a Lie group. Let $\mathrm{sp}(\omega, \mathbf{R})$ be the set of all $p \in \mathrm{gl}(V, \mathbf{R})$ such that $\omega(pv, w) + \omega(v, pw) = 0$ for every $v, w \in V$. Show that $\mathrm{sp}(\omega, \mathbf{R})$ is the Lie algebra of $\mathrm{Sp}(\omega, \mathbf{R})$ with Lie bracket $[\xi, \eta] = \xi\eta - \eta\xi$. Let $\{e_1, \ldots, e_n, f_1, \ldots, f_n\}$ be a basis of V such that the matrix of ω is $J_{2n} = \begin{pmatrix} 0 & -I_n \\ I_n & 0 \end{pmatrix}$. Show that $R = \begin{pmatrix} a & b \\ c & d \end{pmatrix} \in \mathrm{Sp}(\omega, \mathbf{R}) = \mathrm{Sp}(2n, \mathbf{R})$ if and only if $R^t J_{2n} R = J_{2n}$ if and only if $a^t c = c^t a$, $b^t d = d^t b$ and $a^t d - c^t b = 1$. Similarly show that $r = \begin{pmatrix} a & b \\ c & d \end{pmatrix} \in \mathrm{sp}(\omega, \mathbf{R}) = \mathrm{sp}(2n, \mathbf{R})$ if and only if $r^t J_{2n} + J_{2n} r = 0$ if and only if $d = -a^t$, $b = b^t$ and $c = c^t$.

    (b). Let $H : V \to \mathbf{R}$ be a homogeneous quadratic function on V, that is, there is a symmetric bilinear form $\widetilde{H}$ on V such that $H(v) = \frac{1}{2}\widetilde{H}(v, v)$. Let X_H be the *linear Hamiltonian vector field* associated to the Hamiltonian H on (V, ω). Show that $X_H \in \mathrm{sp}(\omega, \mathbf{R})$. Conversely, show that $\xi \in \mathrm{sp}(\omega, \mathbf{R})$ is the linear Hamiltonian vector field associated to the homogeneous quadratic function $v \to \frac{1}{2}\omega(\xi v, v)$. Let $\mathcal{Q}$ be the vector space of homogeneous quadratic functions on V. For $F, G \in \mathcal{Q}$ define their *Poisson bracket* by $\{F, G\} = \omega(X_F, X_G)$. Show that $(\mathcal{Q}, \{\,,\,\})$ is a Lie algebra which is isomorphic to $(\mathrm{sp}(\omega, \mathbf{R}), [\,,\,])$.

    (c). The *linear symmetry group* G of the Hamiltonian system (H, V, ω) where $H \in \mathcal{Q}$ is the set of all $Q \in \mathrm{Sp}(\omega, \mathbf{R})$ such that $Q^*H = H$. Show that G is a closed subgroup of $\mathrm{Sp}(\omega, \mathbf{R})$ and hence is a Lie group. Show that the Lie algebra $\mathcal{G}$ of G is the set of all $\xi \in \mathrm{sp}(\omega, \mathbf{R})$ such that $[\xi, X_H] = 0$. In other words, $\mathcal{G}$ is the Lie subalgebra of all $F \in \mathcal{Q}$ such that $\{F, H\} = 0$.

(d). Show that the linear action $\Phi : G \times V \rightarrow V$: $(Q, v) \rightarrow Qv$ is Hamiltonian with *momentum mapping* $J : V \rightarrow \mathcal{G}^*$ where $J(v)\xi = \frac{1}{2}\,\omega(\xi v, v)$ for every $\xi \in \mathcal{G}$. Show that J is coadjoint equivariant. Verify that for every $\xi \in \mathcal{G}$ the function $J^{\xi} : V \rightarrow \mathbf{R}$: $v \rightarrow J(v)\xi$ is an integral of X_H.

11. (The $1^p : -1^q$ semisimple resonance.) Consider the quadratic Hamiltonian

$$H : \mathbf{R}^{2n} \rightarrow \mathbf{R} : (x, y) \rightarrow \frac{1}{2}\left\langle \begin{pmatrix} \mathcal{I} & 0 \\ 0 & \mathcal{I} \end{pmatrix} \begin{pmatrix} x \\ y \end{pmatrix}, \begin{pmatrix} x \\ y \end{pmatrix} \right\rangle,$$

where $\mathcal{I} = \mathrm{diag}(I_p, -I_q)$.

(a). Show that the linear symmetry group G of the Hamiltonian system $(H, \mathbf{R}^{2n}, \omega)$ is the set of all $R \in \mathrm{Sp}(2n, \mathbf{R})$ such that $R = \begin{pmatrix} a & b \\ -\mathcal{I}b\mathcal{I} & \mathcal{I}a\mathcal{I} \end{pmatrix}$, where $a, b \in \mathrm{gl}(n, \mathbf{R})$ satisfy

$$\begin{aligned} (a\mathcal{I})^t(\mathcal{I}a) + (\mathcal{I}b)^t(b\mathcal{I}) &= I_n \\ (b\mathcal{I})^t(\mathcal{I}a) &= (\mathcal{I}a)^t(b\mathcal{I}). \end{aligned}$$

(b). Let $\mathcal{I}$ be the matrix of a Hermitian inner product on $\mathbf{C}^n$ with respect to the standard basis. Let $\mathrm{U}(p, q)$ be the set of all $S \in \mathrm{gl}(n, \mathbf{C})$ such that $\overline{S}^t\mathcal{I}S = \mathcal{I}$. Show that the matrix $S = \begin{pmatrix} a & b \\ c & d \end{pmatrix}$, where $a \in \mathrm{gl}(p, \mathbf{C})$, $d \in \mathrm{gl}(q, \mathbf{C})$, b is a $q \times p$ and c is a $p \times q$ complex matrix, lies in $\mathrm{U}(p, q)$ if and only if

$$\begin{aligned} \overline{a}^t a - \overline{c}^t c &= I_p \\ \overline{d}^t d - \overline{d}^t d &= I_q \\ \overline{a}^t b &= \overline{c}^t d. \end{aligned}$$

Verify that the map

$$\Theta : G \rightarrow \mathrm{U}(p, q) : R = \begin{pmatrix} a & b \\ -\mathcal{I}b\mathcal{I} & \mathcal{I}a\mathcal{I} \end{pmatrix} \rightarrow S = a + i\,b\mathcal{I}$$

is an isomorphism of Lie groups.

(c). Let $\mathrm{u}(p, q)$ be the set of all $s \in \mathrm{gl}(n, \mathbf{C})$ such that $\overline{s}^t\mathcal{I} + \mathcal{I}s = 0$. Show that $\mathrm{u}(p, q)$ is the Lie algebra of $\mathrm{U}(p, q)$. Show that the matrix $s = \begin{pmatrix} a & b \\ c & d \end{pmatrix}$, where $a \in \mathrm{gl}(p, \mathbf{C})$, $d \in \mathrm{gl}(q, \mathbf{C})$, b is a $q \times p$ and c is a $p \times q$ complex matrix, lies in $\mathrm{u}(p, q)$ if and only if $a = -\overline{a}^t$, $c = -\overline{b}^t$, and $d = -\overline{d}^t$. Show that the Lie algebra $\mathcal{G}$ of the symmetry group G is the set of all $r \in \mathrm{sp}(2n, \mathbf{R})$ such that $r = \begin{pmatrix} \mathcal{I}a & b \\ -b & \mathcal{I}a \end{pmatrix}$, where $a, b \in \mathrm{gl}(n, \mathbf{R})$ satisfy $(\mathcal{I}a)^t = -\mathcal{I}a$ and $b = b^t$. Show that the map

$$\theta : \mathcal{G} \rightarrow \mathrm{u}(p, q) : r = \begin{pmatrix} \mathcal{I}a & b \\ -b & \mathcal{I}a \end{pmatrix} \rightarrow a + i\,b\mathcal{I}$$

is an isomorphism of Lie algebras.

(d). Define a Hermitian inner product $\langle\, ,\, \rangle$ on $\mathrm{u}(p,q)$ by $\langle u, w\rangle = -\frac{1}{2}\operatorname{tr}(\mathcal{I}\,\overline{w}^t\,\mathcal{I}u)$. Show that $\langle\, ,\, \rangle$ is invariant under Ad_U for every $U \in \mathrm{U}(p,q)$ and that it is nondegenerate. Show that the map

$$J : \mathbf{C}^n = \mathbf{C}^p \times \mathbf{C}^q \to \mathrm{u}(p,q) : (z, w) \to \begin{pmatrix} i\,(z \otimes \overline{z}^t) & z \otimes \overline{w}^t \\ -w \otimes \overline{z}^t & i\,(w \otimes \overline{w}^t) \end{pmatrix}$$

is a Hermitian momentum mapping for the linear $\mathrm{U}(p,q)$ action

$$\Phi : \mathrm{U}(p,q) \times \mathbf{C}^n \to \mathbf{C}^n : (U, \zeta) \to U\zeta.$$

Here we have identified $\overline{\mathrm{u}(p,q)}^*$ with $\mathrm{u}(p,q)$ using $\langle\, ,\, \rangle$. Writing

$$J(z, w) = i\,(z, iw) \otimes \overline{(z, iw)}^t$$

deduce that the rank of $J(z,w)$ is at most 1. Hence

$$(z_j\overline{z}_k)(w_\ell\overline{w}_m) = (z_j\overline{w}_m)(\overline{z}_k w_\ell)$$

are the only relations among the quadratic Hermitian integrals

$$\begin{cases} z_j\overline{z}_k, & 1 \le j,k \le p; \\ z_j\overline{w}_\ell,\ \overline{z}_j w_\ell, & 1 \le j \le p \ \&\ 1 \le \ell \le q; \\ w_\ell\overline{w}_m, & 1 \le \ell, m \le q \end{cases}$$

of the Hermitian vector field X_H. Show that the $\mathrm{U}(p,q)$-adjoint orbit through $J(z,0)$, $z \ne 0$, $J(0,w)$, $w \ne 0$ and $J(z,w)$, $z \ne 0 \,\&\, w \ne 0$ is diffeomorphic to $\mathbf{CP}^{p-1}$, $\mathbf{CP}^{q-1}$, $\mathbf{CP}^{n-1}$, respectively. Find an expression for the symplectic form on each of these adjoint orbits.

12. Let $S^2_\ell = \{x \in \mathbf{R}^3 \,|\, \langle x, x\rangle = \ell^2,\ \ell > 0\}$ be the 2-sphere of radius ℓ in $\mathbf{R}^3$ with Euclidean inner product $\langle\, ,\, \rangle$. Consider the action

$$\Phi : S^1 \times (S^2_\ell \times S^2_\ell) \to S^2_\ell \times S^2_\ell : \big(t, (x, y)\big) \to (R_t x, R_t y),$$

where $R_t = \begin{pmatrix} \cos t & -\sin t & 0 \\ \sin t & \cos t & 0 \\ 0 & 0 & 1 \end{pmatrix}$. Let $\Omega = \pi_1^*\sigma_2 + \pi_2^*\sigma_2$, where π_i for $i = 1, 2$ is the projection onto the ith factor and σ_2 is the standard volume form on S^2_ℓ.

(a). Show that Ω is a symplectic form on $S^2_\ell \times S^2_\ell$ and the S^1-action Φ is Hamiltonian with momentum map

$$J : S^2_\ell \times S^2_\ell \to \mathbf{R} : (x, y) \to x_3 + y_3.$$

Show that ℓ is a regular value of J. Using Morse theory show that $J^{-1}(\ell)$ is homeomorphic to a 3-sphere.

(b). Show that $\pm\ell(e_3, e_3)$ are fixed points of the action Φ restricted to $J^{-1}(\ell)$. Thus the reduced space $\mathcal{M}_0 = J^{-1}(\mu)/S^1$ has singularities. Use invariant theory to construct a concrete model of $\mathcal{M}_\ell$ as a semialgebraic variety in $\mathbf{R}^3$. (Hint: first find the generators of the algebra of polynomials which are invariant under the action Φ. Next show that they satisfy one relation and two inequalities. Using the fact that Φ is an action on $S^2_\ell \times S^2_\ell$ obtain additional relations.)

(c). Find the structure matrix of the Poisson bracket on $\mathcal{M}_\ell$ and show that it is induced from a Poisson bracket on $\mathbf{R}^3$. Write out the Hamiltonian derivation of a smooth function on $\mathcal{M}_\ell$.

13. (The $p:q$ resonance). Let $p, q \in \mathbf{Z}_\geq$ with $\gcd(p,q) = 1$. Consider the resonant harmonic oscillator on $(T\mathbf{R}^2, \omega)$ given by the Hamiltonian

$$H_2 : T\mathbf{R}^2 \to \mathbf{R} : (x, y) \to \tfrac{1}{2}\left[p(y_2^2 + x_2^2) + q(y_1^2 + x_1^2)\right]. \tag{102}$$

(a). Show that all orbits of X_{H_2} are periodic of period $2\pi pq$.

(b). Show that the projection of every orbit of energy $1/2$ on configuration space $\mathbf{R}^2$ is a closed curve, contained in the rectangle $\mathfrak{R} : |x_1| \leq 1/q$ & $|x_2| \leq 1/p$, and is tangent to the horizontal sides of $\mathfrak{R}$ q-times and the vertical sides of $\mathfrak{R}$ p times.

(c). Show that the algebra of polynomials of $T\mathbf{R}^2$ which are invariant under the flow φ_t^H of X_{H_2} is generated by the polynomials

$$\begin{cases} \sigma_1 = y_1^2 + x_1^2 & \sigma_3 = \mathrm{Re}\,[(x_1 + i\,y_1)^p (x_2 - i\,y_2)^q] \\ \sigma_2 = y_2^2 + x_2^2 & \sigma_4 = \mathrm{Im}\,[(x_1 + i\,y_1)^p (x_2 - i\,y_2)^q], \end{cases}$$

subject to the relation

$$\sigma_3^2 + \sigma_4^2 = \sigma_1^p \sigma_2^q \qquad \sigma_1 \geq 0, \;\&\; \sigma_2 \geq 0. \tag{103}$$

(d). Show that the S^1 action generated by X_{H_2} is locally free on $H^{-1}(0)$ $h > 0$ and that the orbit space $P_h^{p,q} = H^{-1}(h)/S^1$ is an orbifold defined by (103) and $2h = p\sigma_2 + q\sigma_1$. In other words, $P_h^{p,q}$ is defined by

$$\sigma_3^2 + \sigma_4^2 = \left(\frac{2h - \sigma_1}{p}\right)^q \sigma_1^p \qquad 0 \leq \sigma_1 \leq 2h.$$

Draw a picture of $P_h^{p,q}$ for $(p,q) = (1,3), (2,3)$ and $(1,4)$. Find the structure matrix for the Poisson algebra of smooth functions on $P_h^{p,q}$.

(e). Suppose that H is a smooth Hamiltonian function on $(T^*\mathbf{R}^2, \omega)$ whose Taylor series at the origin starts with the quadratic terms H_2 (102). Suppose that H commutes with the S^1 action generated by the flow of X_{H_2}. Using the orbifold reduction theorem appendix 2 ((4.3)), write out the Hamiltonian derivation on the singular reduced space $P_h^{p,q}$ corresponding to the Hamiltonian form by the reducing H.

14. (a). Let $\Phi : G \times V \to V$ be a linear Hamiltonian action of a compact Lie group on a symplectic vector space (V, ω), (see exercise 10). Let $J : V \to \mathcal{G}^*$ be the quadratic momentum mapping of Φ. Choose a set of generators $\{\sigma_1, \ldots, \sigma_k\}$ of the

algebra of invariant polynomials. Embed the reduced space $\mathcal{M}_0 = J^{-1}(0)/G$ into $\mathbf{R}^k$ using the Hilbert map $\pi : V \to \mathbf{R}^k : x \to \big(\sigma_1(x), \ldots, \sigma_k(x)\big)$. Then there exists a Poisson bracket on $\mathbf{R}^k$ that restricts to the Poisson structure on $\mathcal{M}_0$.

(b). Show that the conclusion of (a) holds for the linear Hamiltonian action on $(\mathbf{R}^4, \omega)$ given by the $p : q$-resonance, (see exercise 13).

(c)*. Is this also true for the $p_1 : p_2 : \ldots : p_n$-resonance?

(d)*. Consider the linear Hamiltonian action on $(T^*(\mathbf{R}^3)^n, \omega)$ given by lifting the diagonal action of SO(3) on n-copies of $\mathbf{R}^3$. Let $J : T^*(\mathbf{R}^3)^n \to \mathrm{so}(3)^*$ be the momentum mapping. Does the conclusion of (a) hold for the Poisson structure on the reduced space $J^{-1}(0)/\,\mathrm{SO}(3)$?

15. (Reduction in stages). (a). Define the notion of a momentum mapping of a smooth Poisson action of a Lie group on a singular reduced space.

    (b). Let G_1 and G_2 be compact Lie groups acting in a Hamiltonian fashion on the smooth symplectic manifold (M, ω) with corresponding momentum mappings $J_i : M \to \mathcal{G}^*$ for $i = 1, 2$. Suppose that the actions commute, that is, the group $G_1 \times G_2$ acts in a Hamiltonian way on (M, ω) with momentum mapping

$$J : M \to \mathcal{G}_1^* \times \mathcal{G}_2^* : m \to (J_1(m), J_2(m)).$$

    Show that we may assume that J is $G_1 \times G_2$ coadjoint equivariant. Show that $J^{-1}(0)/(G_1 \times G_2) = \widetilde{J}_2^{-1}(0)/G_2$ where $\widetilde{J}_2$ is the restriction of the function J_2 to $J_1^{-1}(0)/G_1$. First try proving the special case when all the actions are free and proper and all the reduced spaces are smooth.

16. Show that an integral for a Hamiltonian vector field on $(T^*\mathbf{R}^n, \omega)$ comes from an action on the configuration space $\mathbf{R}^n$ lifted to an action on the cotangent bundle $T^*\mathbf{R}^n$ if and only if it is linear in the momenta (affine if the symplectic form has magnetic terms).

Appendix C

Ehresmann connections

1 Basic Properties

In this section we define the notion of an Ehresmann connection for a surjective submersion. A connection permits a curve in N to be *locally lifted* to a horizontal curve in M. An Ehresmann connection is good if every smooth curve in N has a global horizontal lift. For good connections we define the notions of parallel translation and monodromy.

Let $\pi : M \rightarrow N$ be a submersion. Consider two smooth distributions on M called vertical and horizontal. The *vertical* distribution is defined as $\mathrm{Vert} : m \rightarrow \mathrm{Vert}_m = \ker T_m\pi \subseteq T_mM$, while the *horizontal* distribution $\mathrm{Horz} : m \rightarrow \mathrm{Horz}_m \subseteq T_mM$, is defined by a subspace Horz_m of T_mM which is complementary to Vert_m. These distributions give a smooth splitting

$$m \rightarrow \mathrm{Horz}_m \oplus \mathrm{Vert}_m = T_mM.$$

of TM, which is called an *Ehresmann connection* $\mathcal{H}$ associated to the submersion π provided that for every $m \in M$,

$$T_m\pi(\mathrm{Horz}_m) = T_{\pi(m)}N.$$

To see how an Ehresmann connection allows smooth curves in N to be locally lifted to smooth curves in M, we define the notion of a local horizontal lift of a curve. Let $J = [0, \varepsilon)$, $\varepsilon > 0$ and let $\gamma : J \rightarrow N$ be a smooth curve in N. Suppose that $m_0 \in \pi^{-1}(\gamma(0))$. A smooth curve $\widetilde{\gamma} : J \rightarrow M$ is called the *local horizontal lift* of γ starting at m_0 if and only if $\widetilde{\gamma}(0) = m_0$, $(\pi \circ \widetilde{\gamma})(t) = \gamma(t)$, and $\frac{d}{dt}\widetilde{\gamma}(t) \in \mathrm{Horz}_{\widetilde{\gamma}(t)}$ for every $t \in J$. From these conditions it follows that the local horizontal lift $\widetilde{\gamma}$ satisfies a system of ordinary differential equations with a given initial condition and hence is unique. ▷

(1.1) **Proof**: Let $(p, v_p) \in TM$. Choose local coordinates $(x_1, \ldots, x_m, \frac{\partial}{\partial x_1}, \ldots, \frac{\partial}{\partial x_m})$ on TM near (p, v_p) such that $\{\frac{\partial}{\partial x_{n+1}}, \ldots, \frac{\partial}{\partial x_m}\}$ span $\ker T_x\pi$. Let $(y_1, \ldots, y_n, \frac{\partial}{\partial y_1}, \ldots, \frac{\partial}{\partial y_n})$ be local coordinates on TN near $\big(\pi(p), T_p\pi\, v_p\big)$. In these coordinates

$$\gamma : J \rightarrow N : t \rightarrow \big(\gamma_1(t), \ldots, \gamma_n(t)\big) \quad \text{and} \quad \widetilde{\gamma} : J \rightarrow M : t \rightarrow \big(\widetilde{\gamma}_1(t), \ldots, \widetilde{\gamma}_m(t)\big).$$

There is an $m \times n$ matrix $(a_{ij}(y))$ such that for every $1 \le i \le m$ we have

$$T\pi\left(\frac{\partial}{\partial x_i}\right) = \sum_{j=1}^{n} a_{ij}(y)\,\frac{\partial}{\partial y_j}.$$

Because π is a submersion, the matrix $(a_{ij}(y))$ has rank n. In order that $\pi(\widetilde{\gamma}(t)) = \gamma(t)$ for $t \in J$, we need $T\pi\left(\frac{d\widetilde{\gamma}}{dt}(t)\right) = \frac{d\gamma}{dt}(t)$. Equivalently,

$$\sum_{i=1}^{m}\sum_{j=1}^{n} \widetilde{\gamma}_i{}'(t)\, a_{ij}(\gamma(t))\,\frac{\partial}{\partial y_j} = \sum_{j=1}^{n} \gamma_j{}'(t)\,\frac{\partial}{\partial y_j},$$

that is,

$$\sum_{i=1}^{n} \widetilde{\gamma}_i{}'(t)\, a_{ij}(\gamma(t)) = \gamma_j{}'(t) \tag{1}$$

for every $1 \le j \le n$. Because the projection

$$\rho : TM \to \ker T\pi : \left(\frac{\partial}{\partial x_1}, \ldots, \frac{\partial}{\partial x_m}\right) \to \left(\frac{\partial}{\partial x_{n+1}}, \ldots, \frac{\partial}{\partial x_m}\right)$$

is surjective, there is an $m \times (m-n)$ matrix $(b_{ij}(y))$ of rank $m-n$ such that

$$\rho\left(\frac{\partial}{\partial x_i}\right) = \sum_{k=1}^{m-n} b_{ik}(y)\,\frac{\partial}{\partial x_k}.$$

If $\widetilde{\gamma}$ is locally horizontal, then $\rho\left(\frac{d\widetilde{\gamma}}{dt}(t)\right) = 0$ for every $t \in J$, that is,

$$\sum_{i=1}^{m} \widetilde{\gamma}_i{}'(t)\, b_{ik}(\gamma(t)) = 0 \tag{2}$$

for every $1 \le k \le m-n$. Hence (1) and (2) may be written as

$$\sum_{i=1}^{m} \widetilde{\gamma}_i{}'(t)\, c_{ij}(\gamma(t)) = \begin{cases} \gamma_j{}'(t), & 1 \le j \le n \\ 0, & n+1 \le j \le m \end{cases} \tag{3}$$

where $c_{ij}(y) = \begin{cases} a_{ij}(y), & 1 \le j \le n \\ b_{ij-n}(y), & n+1 \le j \le m \end{cases}$. Since the $m \times m$ matrix $(c_{ij}(y))$ has rank m, we may solve (3) and obtain

$$\widetilde{\gamma}_i{}'(t) = \sum_{j=1}^{n} c^{ij}(\gamma(t))\,\gamma_j{}'(t) \tag{4}$$

for $1 \le i \le m$ and $t \in J$. Here $(c^{ij}(y))$ is the inverse of $(c_{ij}(y))$. Equation (4) is the desired differential equation for the local horizontal lift. Given the initial condition $\widetilde{\gamma}(0)$, equation (4) has a unique solution. □

A smooth curve $\widetilde{\gamma} : [0,1] \to M$ is a *global horizontal lift* of a curve $\gamma : [0,1] \to N$ if it is a local horizontal lift for every $t \in [0,1]$.

Example 1: Consider the Ehresmann connection associated to the surjective submersion $\pi : \mathbf{R}^2 \to \mathbf{R} : (x, y) \to x$ defined by the vertical distribution

$$\mathrm{Vert} : \mathbf{R}^2 \to T_{(x,y)}\mathbf{R}^2 : (x, y) \to \mathrm{span}\,\{(0,1)\}$$

and the horizontal distribution

$$\mathrm{Horz} : \mathbf{R}^2 \to T_{(x,y)}\mathbf{R}^2 : (x, y) \to \mathrm{span}\,\{(1, y^2)\}.$$

Suppose that $\gamma : [0,1] \to \mathbf{R} : t \to t$. Then the local horizontal lift

$$\widetilde{\gamma} : J \to \mathbf{R}^2 : t \to \big(\widetilde{\gamma}_1(t), \widetilde{\gamma}_2(t)\big)$$

of γ starting at $(2,0)$ satisfies

$$\widetilde{\gamma}_1(t) = \pi(\widetilde{\gamma}(t)) = \gamma(t) = t, \tag{5}$$

$\widetilde{\gamma}_1(0) = 0$, $\widetilde{\gamma}_2(0) = 2$, and

$$\frac{d\widetilde{\gamma}(t)}{dt} \in \mathrm{span}\,\Big\{(1, (\widetilde{\gamma}_2(t))^2\Big\}.$$

This last condition is equivalent to

$$\frac{d\widetilde{\gamma}_1}{dt} = \lambda(t) \quad \text{and} \quad \frac{d\widetilde{\gamma}_2}{dt} = \lambda(t)\,\big(\widetilde{\gamma}_2(t)\big)^2 \tag{6}$$

for some smooth function $\lambda : J \to \mathbf{R}$. Differentiating (5) gives $1 = \dfrac{d\widetilde{\gamma}_1}{dt} = \lambda(t)$. Hence the second equation in (6) is

$$\frac{d\widetilde{\gamma}_2}{dt}(t) = \big(\widetilde{\gamma}_2(t)\big)^2 \tag{7}$$

with initial condition $\widetilde{\gamma}_2(0) = 2$. Integrating (7) gives $\widetilde{\gamma}_2(t) = \frac{1}{\frac{1}{2} - t}$. Thus the local horizontal lift $\widetilde{\gamma}$ of γ exists only on $[0, \frac{1}{2})$. Hence γ does *not* have a global horizontal lift, since $\widetilde{\gamma}$ is not defined on all of $[0,1]$. □.

We say that an Ehresmann connection $\mathcal{H}$ associated to the surjective submersion $\pi : M \to N$ is *good* if every smooth curve $\gamma : [0,1] \to N$ has a horizontal lift $\widetilde{\gamma} : [0,1] \to M$ starting at any $m_0 \in \pi^{-1}(\gamma(0))$. The following claim gives a criterion when there is a good Ehresmann connection.

Claim: Let M be a connected smooth manifold. If $\pi : M \to N$ is a proper surjective submersion, then there is a good Ehresmann connection $\mathcal{H}$ associated to π.

(1.2) **Proof**: Since M has a Riemannian metric we may define the horizontal distribution at $m \in M$ as the orthogonal complement of the vertical distribution in T_mM. In other words, $m \to \mathrm{Horz}_m = (\ker T_m\pi)^{\perp}$, where $m \to \mathrm{Vert}_m = \ker T_m\pi$. Clearly this defines an Ehresmann connection associated to π.

To show that this connection is good, let $\gamma : [0,1] \rightarrow N$ be a smooth curve. Consider the set

$$\mathcal{I} = \left\{ t \in [0,1] \,\middle|\, \begin{array}{l} \gamma \,|\, [0,t] \text{ has a horizontal lift } \tilde{\gamma} \text{ defined on } [0,t] \text{ for every} \\ \text{starting point } m \in \pi^{-1}(\gamma(0)). \end{array} \right\}$$

First we show that $\mathcal{I}$ is nonempty. Let $m_0 \in \pi^{-1}(\gamma(0))$. Because the local horizontal lift $\tilde{\gamma}$ of γ starting at m_0 satisfies a smooth set of differential equations, there is an open neighborhood $U_{m_0} \subseteq \pi^{-1}(\gamma(0))$ of m_0 whose closure is compact and a positive time T_{m_0} such that for every $m \in U_{m_0}$ the local horizontal lift of γ starting at m is defined on $\left[0, T_{m_0}\right]$. Since π is a proper map, the fiber $\pi^{-1}(\gamma(0))$ is compact. Therefore the open covering $\{U_m \,|\, m \in \pi^{-1}(\gamma(0))\}$ has a finite subcovering $\{U_{m_i}\}_{i=1}^{r}$. Let $T = \min_i \{T_{m_i}\} > 0$. Then the local horizontal lift of γ starting at any point of the fiber $\pi^{-1}(\gamma(0))$ is defined on $[0, T]$. In other words, $0 \in \mathcal{I}$.

Let τ be the least upper bound for $\mathcal{I}$. Because $\mathcal{I}$ is closed, τ lies in $\mathcal{I}$. Suppose that $\tau < 1$. Then repeating the argument of the above paragraph with the fiber $\pi^{-1}(\gamma(0))$ replaced by the fiber $\pi^{-1}(\gamma(\tau))$, we find a $T' > 0$ such that every local horizontal lift of γ starting at any $m_\tau \in \pi^{-1}(\gamma(\tau))$ is defined on $\left[0, T'\right]$. For every $m_0 \in \pi^{-1}(\gamma(0))$ the horizontal lift of γ starting at m_0 has left end point at some $m_\tau \in \pi^{-1}(\gamma(\tau))$. This horizontal lift joins smoothly to the local horizontal lift of γ at m_τ, because the domains of the lifts overlap and local horizontal lifts are unique. The new horizontal lift of γ formed from this joining process starts at m_0 and has left end point in $\pi^{-1}(\gamma(T' + \tau))$. But this contradicts the definition of τ. Hence $\mathcal{I} = [0, 1]$. □

In what follows we will assume that $\mathcal{H}$ is a good Ehresmann connection associated to the submersion π. Let $\gamma : [0,1] \rightarrow N$ be a smooth path and let $\tilde{\gamma}_m$ be a horizontal lift of γ starting at $m \in \pi^{-1}(\gamma(0))$ with respect to the connection $\mathcal{H}$. The mapping

$$\mathcal{P}_\gamma : \pi^{-1}(\gamma(0)) \rightarrow \pi^{-1}(\gamma(1)) : m \rightarrow \tilde{\gamma}_m(1)$$

is called *parallel translation* along γ with respect to $\mathcal{H}$. Note that for the Ehresmann connection defined in example 1 no fiber of π can be even locally parallel transported along any curve. From smooth dependence of solutions of differential equations on initial conditions, we have

Claim: If $\pi : M \rightarrow N$ is a surjective submersion with a good Ehresmann connection $\mathcal{H}$, then along γ the parallel translation map $\mathcal{P}_\gamma : \pi^{-1}(\gamma(0)) \rightarrow \pi^{-1}(\gamma(1))$ is a diffeomorphism.

If γ is a closed path, that is, $\gamma(0) = \gamma(1)$, then the map $\mathcal{P}_\gamma$ is the *monodromy map* of γ. Fix a point $n \in N$. Then the set of diffeomorphisms

$$G_n = \left\{ \mathcal{P}_\gamma \,\middle|\, \gamma \text{ a smooth closed curve in } N \text{ with } \gamma(0) = n \right\}$$

is called the *holonomy group* of the Ehresmann connection $\mathcal{H}$ at n.

Example 2. Consider the mapping

$$\pi : \mathbf{R}^4 \to \mathbf{R}^3 :$$
$$(x_1, x_2, x_3, x_4) \to (w_1, w_2, w_3) = \big(x_1^2 + x_2^2 - x_3^2 - x_4^2, 2(x_1x_4 + x_2x_3), 2(x_2x_4 - x_1x_3)\big).$$

Then the image of the unit 3-sphere S^3 under π is the unit 2-sphere S^2. Moreover, the mapping $\mathcal{F} = \pi|S^3 : S^3 \to S^2$ is the Hopf fibration, (see chapter 1 section 4). Define an Ehresmann connection $\mathcal{H}$ associated to $\mathcal{F}$ as follows. The vertical distribution of $\mathcal{H}$ at $x \in S^3$ is

$$\begin{aligned}
\mathrm{Vert}_x &= (\ker T_x\pi) \cap T_xS^3 = \ker \begin{pmatrix} x_1 & x_2 & -x_3 & -x_4 \\ x_4 & x_3 & x_2 & x_1 \\ -x_3 & x_4 & -x_1 & x_2 \\ x_1 & x_2 & x_3 & x_4 \end{pmatrix} \\
&= \mathrm{span}\Big\{(-x_2, x_1, x_4, -x_3)\Big\}.
\end{aligned}$$

The horizontal distribution of $\mathcal{H}$ at $x \in S^3$ is

$$\mathrm{Horz}_x = \Big(\mathrm{span}\Big\{(-x_2, x_1, x_4, -x_3)\Big\}\Big)^{\perp} \cap T_xS^3,$$

where $X^{\perp}$ is the orthogonal complement of the subspace X of $\mathbf{R}^4$ with respect to the Euclidean inner product on $\mathbf{R}^4$. Therefore

$$\mathrm{Horz}_x = \mathrm{span}\Big\{(x_3, x_4, -x_1, -x_2), (-x_4, x_3, -x_2, x_1)\Big\}.$$

It follows from the definition of Vert_x and Horz_x that

$$\mathrm{Vert}_x \oplus \mathrm{Horz}_x = T_xS^3$$

for every $x \in S^3$. Because the rank of $D\pi(x)$ is 2 at every $x \in \mathbf{R}^4 - \{0\}$, we have $T_x\pi(\mathrm{Horz}_x) = T_{\mathcal{F}(x)}S^2$ for every $x \in S^3$. Therefore $\mathcal{H}$ is an Ehresmann connection associated to $\mathcal{F}$. Since $\mathcal{F}$ is a proper submersion, the connection is good. Let

$$\gamma : [0, T] \to S^2 : t \to w(t)$$

be a smooth curve with $\gamma(0) = w$. Then γ lifts to a horizontal curve

$$\widetilde{\gamma} : [0, T] \to S^3 : t \to x(t)$$

with $\widetilde{\gamma}(0) = x$. In other words,

$$\mathcal{F}(x(t)) = \mathcal{F}(\widetilde{\gamma}(t)) = \gamma(t) = w(t) \quad \text{and} \quad \frac{d\widetilde{\gamma}(t)}{dt} \in \mathrm{Horz}_{\widetilde{\gamma}(t)}$$

for every $t \in [0, T]$. The second condition for a horizontal lift may be rewritten as

$$\begin{cases} \dot{x}_1 = \lambda_1 x_3 - \lambda_2 x_4 \\ \dot{x}_2 = \lambda_1 x_4 + \lambda_2 x_3 \\ \dot{x}_3 = -\lambda_1 x_1 - \lambda_2 x_2 \\ \dot{x}_4 = \lambda_2 x_1 - \lambda_1 x_2, \end{cases} \tag{8}$$

for some smooth functions $\lambda_1 = \lambda_1(t)$ and $\lambda_2 = \lambda_2(t)$.

Assume that x lies in the open set

$$U_1 = \left\{ x \in S^3 \middle| x_3^2 + x_4^2 > 0 \right\} = S^3 - \mathcal{F}^{-1}(1, 0, 0).$$

Using $x_3^2 + x_4^2 = \frac{1}{2}(1 - w_1)$ and the definition of $\mathcal{F}$, we eliminate λ_1 and λ_2 from (8), and obtain

$$\begin{cases} \dot{x}_3 = \dfrac{1}{(1 - w_1)} (w_3 \dot{x}_1 - w_2 \dot{x}_2) \\ \dot{x}_4 = -\dfrac{1}{(1 - w_1)} (w_2 \dot{x}_1 + w_3 \dot{x}_2). \end{cases} \tag{9}$$

From the definition of $\mathcal{F}$, we have $\begin{pmatrix} x_4 & x_3 \\ -x_3 & x_4 \end{pmatrix} \begin{pmatrix} x_1 \\ x_2 \end{pmatrix} = \frac{1}{2} \begin{pmatrix} w_2 \\ w_3 \end{pmatrix}$. Because $x \in U_1$, these equations may be solved to give

$$\begin{cases} x_1 = x_1(x_3, x_4) = \dfrac{1}{(1 - w_1)} (-w_3 x_3 + w_2 x_4) \\ x_2 = x_2(x_3, x_4) = \dfrac{1}{(1 - w_1)} (w_2 x_3 + w_3 x_4). \end{cases} \tag{10}$$

Note that

$$\tau_1 : \mathcal{F}(U_1) \times S^1 = (S^2 - \{w_1 = 1\}) \times \{x_3^2 + x_4^2 = \tfrac{1}{2}(1 - w_1)\} \to U_1 :$$
$$\big((w_1, w_2, w_3), (x_3, x_4)\big) \to \big(x_1(x_3, x_4), x_2(x_3, x_4), x_3, x_4\big)$$

is a local trivialization of the Hopf fibration. Differentiating (10) and substituting the result into (9) gives

$$\begin{aligned} 2(1 - w_1)\dot{x}_3 &= -\big[(w_2 \dot{w}_2 + w_3 \dot{w}_3) x_3 - (w_3 \dot{w}_2 - w_2 \dot{w}_3) x_4 + (1 + w_1)\dot{w}_1 x_3\big] \\ 2(1 - w_1)\dot{x}_4 &= \big[(w_2 \dot{w}_3 - w_3 \dot{w}_2) x_3 + (w_2 \dot{w}_2 + w_3 \dot{w}_3) x_4 + (1 + w_1)\dot{w}_1 x_4\big]. \end{aligned} \tag{11}$$

These are the equations satisfied by the horizontal lift of γ with respect to the connection $\mathcal{H}$ in the local trivialization τ_1.

Using (11) we compute the monodromy map of $\mathcal{F}^{-1}(0, 0, 1)$ along the equator

$$\gamma : [0, 2\pi] \to S^2 - \{w_1 = 1\} : t \to w(t) = (0, \cos t, \sin t),$$

traversed in a *counterclockwise* direction. Since $\dot{w}_1 = 0$, $w_2 \dot{w}_2 + w_3 \dot{w}_3 = 0$, and $w_2 \dot{w}_3 - w_3 \dot{w}_2 = 1$, equation (11) becomes

$$\begin{cases} \dot{x}_3 = \frac{1}{2} x_4 \\ \dot{x}_4 = -\frac{1}{2} x_3. \end{cases} \tag{12}$$

Using the fact that $x_3^2 + x_4^2 = \frac{1}{2}$ on γ, equation (12) integrates to

$$\begin{pmatrix} x_3(t) \\ x_4(t) \end{pmatrix} = \frac{1}{\sqrt{2}} \begin{pmatrix} \cos t/2 & \sin t/2 \\ -\sin t/2 & \cos t/2 \end{pmatrix} \begin{pmatrix} x_3(0) \\ x_4(0) \end{pmatrix}.$$

Therefore parallel transport of $\mathcal{F}^{-1}(0,0,1)$ along the closed curve γ is

$$\mathcal{P}_t^1 : \mathcal{F}^{-1}(0,0,1) \to \mathcal{F}^{-1}(\gamma(t)) : \tau_1\Big(0,0,1,x_3(0),x_4(0)\Big) = \begin{pmatrix} x_4(0) \\ x_3(0) \\ x_3(0) \\ x_4(0) \end{pmatrix} \to$$

$$\tau_1\Big(0,\cos t,\sin t,x_3(t),x_4(t)\Big) = \frac{1}{\sqrt{2}} \begin{pmatrix} \cos 3t/2 & -\sin 3t/2 & 0 & 0 \\ \sin 3t/2 & \cos 3t/2 & 0 & 0 \\ 0 & 0 & \cos t/2 & \sin t/2 \\ 0 & 0 & -\sin t/2 & \cos t/2 \end{pmatrix} \begin{pmatrix} x_4(0) \\ x_3(0) \\ x_3(0) \\ x_4(0) \end{pmatrix}.$$

In other words, $\mathcal{P}_t^1$ is a clockwise rotation of the circle $\mathcal{F}^{-1}(0,0,1)$ through an angle $t/2$. Consequently, the monodromy map $\mathcal{P}_{2\pi}^1 = -id$ is a clockwise rotation through an angle π.

Let

$$U_2 = \Big\{x \in S^3 \big| x_1^2 + x_2^2 > 0\Big\} = S^3 - \mathcal{F}^{-1}(-1,0,0).$$

Then $\mathcal{F}(U_2) = S^2 - \{w_1 = -1\}$. The map

$$\tau_2 : \mathcal{F}(U_2) \times S^1 = (S^3 - \{w_1 = -1\}) \times \{x_1^2 + x_2^2 = \tfrac{1}{2}(1+w_1)\} \to U_2 :$$
$$\Big((w_1,w_2,w_3),(x_1,x_2)\Big) \to \Big(x_1,x_2,x_3(x_1,x_2),x_4(x_1,x_2)\Big),$$

where

$$\begin{cases} x_3(x_1,x_2) = \dfrac{1}{1+w_1}(-w_3x_1 + w_2x_2) \\[2ex] x_4(x_1,x_2) = \dfrac{1}{1+w_1}(w_2x_1 + w_3x_2), \end{cases}$$

is a local trivialization of the Hopf fibration. The horizontal lift with respect to $\mathcal{H}$ of the equator $\{w_1 = 0\} \cap S^2$ traversed in the *clockwise* direction, that is, the horizontal lift of the curve $-\gamma : [0,2\pi] \to S^2 : t \to \gamma(2\pi - t) = (0,\cos t,-\sin t)$, satisfies

$$\begin{cases} \dot{x}_1 = -\tfrac{1}{2}x_2 \\ \dot{x}_2 = \tfrac{1}{2}x_1. \end{cases}$$

Thus the parallel translation of $\mathcal{F}^{-1}(0,0,1)$ along the curve $-\gamma$ is the map

$$\mathcal{P}_t^2 : \mathcal{F}^{-1}(0,0,1) \to \mathcal{F}^{-1}\big(-\gamma(t)\big) = \mathcal{F}^{-1}\big(\gamma(2\pi - t)\big)$$

which is a clockwise rotation through an angle $2\pi - t/2$. Thus the monodromy map $\mathcal{P}_{2\pi}^2 = -id$ on $\mathcal{F}^{-1}(0,0,1)$ is a counterclockwise rotation through an angle $-\pi$.

We note that the classifying map of the Hopf fibration is

$$\chi : S^1 = \{w_1 = 0\} \cap S^2 \to S^1 = \mathcal{F}^{-1}(0,0,1) : t \to (\mathcal{P}^2_{-2\pi+t})^{-1} \circ \mathcal{P}^1_{-t},$$

which is a counterclockwise rotation of $\mathcal{F}^{-1}(0,0,1)$ through an angle t. Hence $\deg\chi = 1$, (see also chapter 1 section 4). □

2 The Ehresmann theorems

In this section we prove the Ehresmann fibration and trivialization theorems.

We begin with the *fibration theorem.*

Claim: Let $\pi : M \rightarrow N$ be a proper surjective submersion. Then π is a *locally trivial* fibration.

(2.1) **Proof**: Because π is a proper surjective submersion, there is a good Ehresmann connection $\mathcal{H}$ associated to π. Give N a Riemannian metric. For each $n \in N$, let B_n be an open ball in T_nN about 0 where the exponential map $\exp_n : B_n \subseteq T_nN \rightarrow N$ is a diffeomorphism. Let $U_n = \exp_n B_n$. Then U_n is a open neighborhood of n in N. For $n' \in U_n$ let

$$\gamma_{n,n'} : [0,1] \rightarrow U_n \subseteq N : t \rightarrow \exp_n t v_{n'},$$

be the geodesic joining n to n'. Because $\exp_n$ is a diffeomorphism, the vector $v_{n'} \in B_n \subseteq T_nN$ is uniquely determined. Let

$$\mathcal{P}_{\gamma_{n,n'}} : \pi^{-1}(n) \rightarrow \pi^{-1}(n')$$

be parallel translation along $\gamma_{n,n'}$ using the Ehresmann connection $\mathcal{H}$.

Consider the mapping

$$\tau : U_n \times \pi^{-1}(n) \rightarrow \pi^{-1}(U_n) : (n', m) \rightarrow \mathcal{P}_{\gamma_{n,n'}}(m).$$

and the projection mapping

$$\pi_1 = \pi|\pi^{-1}(U_n) : \pi^{-1}(U_n) \rightarrow U_n : m \rightarrow \pi(m).$$

We want to show that τ is a local trivialization of the fibration π_1. Towards this goal define the smooth mappings

$$\rho : U_n \times \pi^{-1}(U_n) \rightarrow \pi^{-1}(U_n)$$

where for $m \in \pi^{-1}(n')$ we have $\rho(m) = \mathcal{P}_{-\gamma_{n,n'}}(m)$ and

$$\sigma : \pi^{-1}(U_n) \rightarrow U_n \times \pi^{-1}(n) : m \rightarrow \big(\pi_1(m), \rho(m)\big).$$

Then $\sigma \circ \tau = id_{U_n \times \pi^{-1}(n)}$. Since

$$\tau\big(\pi_1(m), \rho(m)\big) = \mathcal{P}_{\gamma_{n,\pi_1(m)}} \circ \mathcal{P}_{-\gamma_{n,\pi_1(m)}}(m) = m,$$

for every $m \in \pi^{-1}(U_n)$, we have $\tau \circ \sigma = id_{\pi^{-1}(U_n)}$. Thus τ is a local trivialization of the surjective submersion π_1. □

We now prove the *trivialization theorem.*

Claim: Suppose that $\pi : M \rightarrow N$ is a proper surjective submersion and that N is smoothly contractible, that is, there is a one parameter family of mappings $F : [0,1] \times N \rightarrow N$ such that

1. F_t is a diffeomorphism for every $t \in [0, 1]$.
2. $F_0 = id_N$.
3. For every $n \in N$, $F_1(n) = n_0$ for some fixed n_0 in N.

Then the fibration π is trivial, that is, M is diffeomorphic to $N \times \pi^{-1}(n_0)$ and this diffeomorphism maps fibers of π onto fibers of

$$\pi_1 : N \times \pi^{-1}(n_0) \to N : (n, m) \to n$$

(2.2) **Proof**: For $n \in N$ consider the curve $\gamma_n : [0, 1] \to N : t \to F_t(n)$, which smoothly joins the point n to n_0. Now define the map

$$\varphi : N \times \pi^{-1}(n_0) \to M : (n, m) \to \mathcal{P}_{\gamma_n}(m).$$

Following the same argument used in the proof of the Ehresmann fibration theorem, we see that φ is a diffeomorphism which is fiber preserving. □

3 Exercises

1. (Levi-Civita connection.) Let g be a Riemannian metric on M. Let $\tau_M : TM \to M$ be the tangent bundle of M with natural coordinates (x^i, v^i). On the second tangent bundle $\tau_{TM} : T(TM) \to TM$ with natural coordinates (x^i, v^i, X^i, V^i) define two distributions

$$\mathrm{Horz}^g : TM \to T(TM) : (x^i, v^i) \to \mathrm{span}\left\{ (x^i, v^i, X^i, -\sum_{j,k} \Gamma^i_{jk} v^j X^k) \right\}$$

and

$$\mathrm{Vert}^g : TM \to T(TM) : (x^i, v^i) \to \mathrm{span}\left\{ (x^i, v^i, 0, V^i + \sum_{j,k} \Gamma^i_{jk} v^j X^k) \right\}.$$

Show that Horz^g and Vert^g define an Ehresmann connection $\mathcal{H}^g$ on the bundle τ_{TM}. $\mathcal{H}^g$ is called *Levi-Civita connection.* Prove the following properties of $\mathcal{H}^g$.

(a). The geodesic vector field Z_g (see appendix A exercise 1) lies in Horz^g, that is, for every $(x, v) \in TM$ we have $Z_g(x, v) \in \mathrm{Horz}^g_{(x,v)}$.

(b). $\mathcal{H}^g$ is symmetric, that is, $j \circ \mathrm{Horz}^g = \mathrm{Horz}^g$, where

$$j(x, v, X, V) = (x, X, v, V)$$

is the canonical involution on $T(TM)$.

(c). Let

$$K_g : T(TM) \to TM : (x^i, v^i, X^i, V^i) \to (x^i, V^i + \sum_{j,k} \Gamma^i_{jk} v^j X^k).$$

For every $X_1, X_2 \in T_{(x,v)}(TM)$

$$\Omega_g(x,v)(X_1,X_2) = g(x)\big(K_g(X_2), T\tau_M(X_1)\big) - g(x)\big(K_g(X_1), T\tau_M(X_2)\big).$$

$\mathrm{Horz}^g_{(x,v)}$ and $\mathrm{Vert}^g_{(x,v)}$ are *Lagrangian* subspaces of $T_{(x,v)}(TM)$, that is, they are $\Omega_g(x,v)$-isotropic and have half the dimension of $T_{(x,v)}(TM)$.

2. (Connections on a principal bundle.) (a). Let

$$\Phi : G \times P \to P : (g,m) \to \Phi_g(m) = g \cdot m$$

be a free and proper action of a Lie group G on a manifold P. Then P is a (left) *principal G-bundle* over the G-orbit space M with bundle projection $\pi : P \to M$ given by the orbit map. The bundle π is locally trivial, that is, for every $m \in M$ there is an open neighborhood U and a diffeomorphism

$$\tau : \pi^{-1}(U) \to G \times U : p \to \big(\varphi(p), \pi(p)\big)$$

such that $\varphi(g \cdot p) = g\varphi(p)$. Show that

$$\pi^{-1}(\pi(p)) = \big\{ g \cdot p \in P \big| g \in G \big\}.$$

Let $\mathcal{G}$ be the Lie algebra of G. For a fixed $p \in P$, let

$$\sigma_p : \mathcal{G} \to T_pP : \xi \to X_\xi(p),$$

where $X_\xi(p) = \frac{d}{dt}\Big|_{t=0} \exp t\xi \cdot p$. Show that $\mathrm{im}\,\sigma_p = T_p\big(\pi^{-1}(p)\big)$.

(b). A smooth $\mathcal{G}$-valued 1-form ϑ on P is a smooth section of the bundle $L(TP, \mathcal{G}) = \coprod_{p\in P} L(T_pP, \mathcal{G})$ with bundle projection $\rho(L(T_pP, \mathcal{G})) = p$. A *connection* on a (left) principal G-bundle $\pi : P \to M$ is a $\mathcal{G}$-valued 1-form ϑ such that

1). $\vartheta(p)(\sigma_p(\xi)) = \xi$ for every $\xi \in \mathcal{G}$ and every $p \in P$.

2). For every $g \in G$ and every $v_p \in T_pP$,

$$\vartheta\big(\Phi_g(p)\big)(T_e\Phi_g v_p) = \mathrm{Ad}_g\big(\vartheta(p)(v_p)\big).$$

Consider the distributions

$$\mathrm{Vert}^\vartheta : P \to TP : p \to \mathrm{span}\big\{\sigma_p(\xi) \in T_pP \big| \xi \in \mathcal{G}\big\}$$

and

$$\mathrm{Horz}^\vartheta : P \to TP : p \to \ker \vartheta(p).$$

Show that Vert^ϑ and Horz^ϑ define an Ehresmann connection on the bundle π. Show that Horz^ϑ is G-invariant, that is, for every $p \in P$ and every $g \in G$,

$$T_e\Phi_g(\mathrm{Horz}^\vartheta_p) = \mathrm{Horz}^\vartheta_{\Phi_g(p)}.$$

Conversely, show that an Ehresmann connection on a principal G-bundle π whose horizontal distribution is G-invariant determines a $\mathcal{G}$-valued 1-form ϑ which satisfies 1) and 2) above.

(c)*. Show that every connection 1-form on a principal bundle defines an Ehresmann connection where every curve has a horizontal lift.

3. (Connections on S^1 principal bundles.) Let $\psi : S^1 \times P \to P$ be a proper free action on a smooth manifold P and let X be the infinitesimal generator of ψ. The orbit space M of this action is a smooth manifold. The orbit map $\pi : P \to M$ gives P the structure of an S^1 principal bundle. Let θ be a 1-form on P which is ψ-invariant and for which $X \lrcorner \theta = 1$. Then θ is a *connection 1-form* on P.

(a). Show that $X \lrcorner d\theta = 0$. Since $\ker T\pi = \mathrm{span}\{X\} \subseteq \ker d\theta$ and $d\theta$ is ψ-invariant, deduce that there is a 2-form Ω on M such that $\pi^*\Omega = d\theta$. Ω is called the *curvature form* of θ.

(b). (Infinitesimal holonomy formula.) Let $\varphi : V \subseteq M \to U \subseteq \mathbf{R}^n$ be a chart for M such that $\tau : \pi^{-1}(V) \subseteq P \to U \times S^1$ is a trivialization of the principal bundle π. Let

$$\gamma : I \times [0,1] \to U : (s,t) \to \gamma(s,t) = \gamma_s(t)$$

be a family of smooth curves with $\gamma_s(0) = u$ for every $s \in I = (-\varepsilon, \varepsilon)$. For every $s \in I$ let $\Gamma_s : [0,1] \to \pi^{-1}(U)$ be the horizontal lift of γ_0 with respect to the connection 1-form θ with $\Gamma_s(0) = p \in \pi^{-1}(u)$ fixed. In other words, $\pi(\Gamma_s(t)) = \gamma_s(t)$ and $\frac{\partial \Gamma_s}{\partial t}(t) \lrcorner \theta(\Gamma_s(t)) = 0$ for every $(s,t) \in I \times [0,1]$. Suppose that $\Gamma_s(1) \in \pi^{-1}(u)$ for every $s \in I$. Then the *holonomy of θ along γ_s as measured from the point p* is the element $\mathrm{hol}_p^{\gamma_s}$ of $S^1 = \mathbf{R}/2\pi\mathbf{Z}$ such that $\psi_{\mathrm{hol}_p^{\gamma_s}}(p) = \Gamma_s(1)$. Let $\frac{\partial}{\partial s}\mathrm{hol}_p^{\gamma_s}$ be the *infinitesimal holonomy vector field* on the fiber $\pi^{-1}(u)$. Then

$$\frac{\partial}{\partial s}\mathrm{hol}_p^{\gamma_s}(s) = \int_0^1 \Omega(\gamma_s(t)) \left(\frac{\partial \gamma_s}{\partial t}(t), \frac{\partial \gamma_s}{\partial s}(t) \right) dt. \tag{13}$$

Each step of the following calculation, which proves (13), takes place in the image of the trivialization τ. Supply a justification.

$$\begin{aligned}
0 &= \frac{d}{ds}\big(\theta(\Gamma_s(t))\frac{\partial \Gamma_s}{\partial t}\big) = \big(D\theta(\Gamma_s(t))\frac{\partial \Gamma_s}{\partial s}\big)\frac{\partial \Gamma_s}{\partial t} + \theta(\Gamma_s(t))\frac{\partial^2 \Gamma_s}{\partial s\, \partial t} \\
&= \big(D\theta(\Gamma_s(t))\frac{\partial \Gamma_s}{\partial s}\big)\frac{\partial \Gamma_s}{\partial t} - \big(D\theta(\Gamma_s(t))\frac{\partial \Gamma_s}{\partial t}\big)\frac{\partial \Gamma_s}{\partial s} + \frac{d}{dt}\big(\theta(\Gamma_s(t))\frac{\partial \Gamma_s}{\partial s}\big) \\
&= -d\theta(\Gamma_s(t))\big(\frac{\partial \Gamma_s}{\partial t}, \frac{\partial \Gamma_s}{\partial s}\big) + \frac{d}{dt}\big(\theta(\Gamma_s(t))\frac{\partial \Gamma_s}{\partial s}\big) \\
&= -\Omega(\gamma_s(t))\big(\frac{\partial \gamma_s}{\partial t}, \frac{\partial \gamma_s}{\partial s}\big) + \frac{d}{dt}\big(\theta(\Gamma_s(t))\frac{\partial \Gamma_s}{\partial s}\big).
\end{aligned}$$

Integrating over $t \in [0,1]$ gives

$$\theta(\Gamma_s(1))\frac{\partial \Gamma_s}{\partial s}(1) = \int_0^1 \Omega(\gamma_s(t))\big(\frac{\partial \gamma_s}{\partial t}, \frac{\partial \gamma_s}{\partial s}\big)\, dt.$$

But $\frac{\partial \Gamma_s}{\partial s}(1) = \frac{\partial}{\partial s}\mathrm{hol}_p^{\gamma_s} X(\Gamma_s(1))$ and $X(\Gamma_s(1)) \lrcorner \theta(\Gamma_s(1)) = 1$. This proves (13).

(c). (Hopf bundle.) Let

$$S^3 = \Big\{ (z_1, z_2) = (x_1 + iy_1, x_2 + iy_2) \in \mathbf{C}^2 \big| |z_1|^2 + |z_2|^2 = 1 \Big\}.$$

Consider the proper free $S^1 = \mathbf{R}/2\pi\mathbf{Z}$ action

$$\psi : S^1 \times S^3 \to S^3 : \big(t, (z_1, z_2)\big) \to (\mathrm{e}^{-it} z_1, \mathrm{e}^{-it} z_2).$$

Its orbit space is

$$S^2 = \Big\{ (w_1, w_2, w_3) \in \mathbf{R}^3 \big| w_1^2 + w_2^2 + w_3^2 = 1 \Big\}$$

and its orbit map

$$\begin{aligned} h : S^3 \to S^2 : (z_1, z_2) \to \big(2\,\mathrm{Im}\,\overline{z}_1 z_2, 2\,\mathrm{Re}\,\overline{z}_1 z_2, |z_1|^2 - |z_2|^2\big) = \\ = \big(2(x_1y_2 - x_2y_1), 2(x_1x_2 + y_1y_2), x_1^2 + y_1^2 - x_2^2 - y_2^2\big) \end{aligned}$$

is the Hopf fibration. On $\mathbf{C}^2 = \mathbf{R}^4$ consider the 1-form

$$\begin{aligned} \theta &= -\frac{1}{2i}(\overline{z}_1 dz_1 + \overline{z}_1 dz_2 - z_1 d\overline{z}_1 - z_2 d\overline{z}_2) \\ &= y_1 dx_1 - x_1 dy_1 + y_2 dx_2 - x_2 dy_2. \end{aligned}$$

Show that $\vartheta = \theta | S^3$ is a connection 1-form on the principal S^1 bundle h. Show that the distributions

$$\mathrm{Horz} : S^3 \to TS^3 : z \to \ker \vartheta(z) \quad \text{and} \quad \mathrm{Vert} : S^3 \to TS^3 : z \to \mathrm{span}\{X(z)\},$$

where $\ker \vartheta(z)$ is spanned by

$$\begin{cases} X_1(z) = -y_2 \frac{\partial}{\partial x_1} - x_2 \frac{\partial}{\partial y_1} + y_1 \frac{\partial}{\partial x_2} + x_1 \frac{\partial}{\partial y_2} \\ X_2(z) = x_2 \frac{\partial}{\partial x_1} - y_2 \frac{\partial}{\partial y_1} - x_1 \frac{\partial}{\partial x_2} + y_1 \frac{\partial}{\partial y_2} \end{cases}$$

and

$$X(z) = y_1 \frac{\partial}{\partial x_1} - x_1 \frac{\partial}{\partial y_1} + y_2 \frac{\partial}{\partial x_2} - x_2 \frac{\partial}{\partial y_2},$$

define the Ehresmann connection associated to ϑ.

The following argument shows that the curvature Ω of the connection 1-form ϑ is $\frac{1}{2}\,\mathrm{vol}_{S^2}$. Here vol_{S^2} is the standard volume form on S^2 defined by $\mathrm{vol}_{S^2}(w)(u, v) = (w, u \times v)$, where $w \in S^2$ and $u, v \in T_w S^2$. Note that $\int_{S^2} \mathrm{vol}_{S^2} = 4\pi$. Since $H^3(S^2) = 0$, Ω is a closed 2-form. Because $\dim H^2(S^2) = 1$, there is a $\lambda \in \mathbf{R}$ such that $\Omega = \lambda\,\mathrm{vol}_{S^2}$. To determine λ we must orient S^3. Let

$$n(z) = x_1 \frac{\partial}{\partial x_1} + y_1 \frac{\partial}{\partial y_1} + x_2 \frac{\partial}{\partial x_2} + y_2 \frac{\partial}{\partial y_2}$$

be the outward normal to S^3 in $\mathbf{R}^4$ at $z = (x_1, y_1, x_2, y_2)$. Show that $n \lrcorner \mathrm{vol}_{\mathbf{R}^4} = -\frac{1}{2}\theta \wedge d\theta$. From the fact that

$$-\tfrac{1}{2}(\theta \wedge d\theta)(z)\big(X(z), X_1(z), X_2(z)\big) = 1,$$

it follows that $\mathrm{vol}_{S^3} = -\frac{1}{2}\vartheta \wedge d\vartheta$. Since

$$\det\big(n(z), X(z), X_1(z), X_2(z)\big) = 1,$$

the basis $\{X(z), X_1(z), X_2(z)\}$ of T_zS^3 is positively oriented with respect to vol_{S^3}. Show that T_zh maps $\{X(z), X_1(z), X_2(z)\}$ onto the frame $\{2v_1, 2v_2\}$ on $T_{h(z)}S^2$ where

$$v_1 = (x_1^2 + x_2^2 - y_1^2 - y_2^2)\frac{\partial}{\partial w_1} + 2(x_1y_1 - x_2y_2)\frac{\partial}{\partial w_2} - 2(x_1y_2 + x_2y_1)\frac{\partial}{\partial w_3}$$

and

$$v_2 = 2(x_1y_1 + x_2y_2)\frac{\partial}{\partial w_1} + 2(x_1x_2 + y_1y_2)\frac{\partial}{\partial w_2} + (x_1^2 + y_1^2 - x_2^2 - y_2^2)\frac{\partial}{\partial w_3}.$$

Since $\det(h(z), v_1, v_2) = -1$, the frame $\{2v_1, 2v_2\}$ is *negatively* oriented with respect to vol_{S^2}. Hence $(h^*\mathrm{vol}_{S^2})(z)(X_1(z), X_2(z)) = -4$. From the fact that $d\vartheta(z)(X_1(z), X_2(z)) = -2$, it follows that $\lambda = \frac{1}{2}$. Therefore $\Omega = \frac{1}{2}\,\mathrm{vol}_{S^2}$.

The following argument computes the holonomy of the connection 1-form ϑ along the equator $\gamma_{\pi/2}$ of S^2. Let D be the closed upper hemisphere, which is positively oriented with respect to vol_{S^2} and is bounded by the positively oriented equator $\gamma_{\pi/2}$. For each $s \in [0, \pi/2]$ let Π_s be the plane in $\mathbf{R}^3$ passing through $u = (1, 0, 0) \in S^2$ with normal vector $(\cos s, 0, \sin s)$. For $s \in (0, \pi/2]$ let $\gamma_s : [0, 1] \to D$ be a parametrization of the ellipse $\Pi_s \cap D$. Note that $\gamma_0(t) = u$ for every $t \in [0, 1]$. Thus we have a smooth family of curves $\gamma : [0, \pi/2] \times [0, 1] \to D : (s, t) \to \gamma_s(t)$. Since the horizontal lift Γ_s of γ_s with respect to the connection ϑ starts at $p \in h^{-1}(u)$, the holonomy $\mathrm{hol}_p^{\gamma_0}$ along γ_0 is 0. Justify each step of the following calculation.

$$\begin{aligned}\mathrm{hol}_p^{\gamma_{\pi/2}} &= \int_0^{\pi/2} \frac{d}{ds}\mathrm{hol}_p^{\gamma_s}\, ds = \int_0^{\pi/2}\int_0^1 \Omega(\gamma_s(t))\Big(\frac{\partial\gamma_s(t)}{\partial t}, \frac{\partial\gamma_s(t)}{\partial s}\Big)\, dt\, ds \\ &= \int_{[0,\pi/2]\times[0,1]} \gamma^*\Omega = \int_D \Omega = \tfrac{1}{2}\,\mathrm{vol}_{S^2}(D) = \pi.\end{aligned}$$

Generalize this argument to show that the holonomy along any closed positively oriented curve γ on S^2 is equal to one half the area on S^2 of the 2-disc bounded by γ.

(d). Consider the S^1 principal bundle

$$\pi : T_1S^2 \subseteq T\mathbf{R}^3 \to S^2 \subseteq \mathbf{R}^3 : (x, y) \to x$$

with S^1 action

$$\psi : S^1 \times T_1S^2 \to T_1S^2 : \big(t, (x, y)\big) \to \big(x, y\cos t + (x \times y)\sin t\big).$$

Let $\theta(x, y) = (x \times y, \frac{\partial}{\partial y})$ be a 1-form on $T\mathbf{R}^3$ where $(\ ,\ )$ is the Euclidean inner product on $\mathbf{R}^3$. Show that $\vartheta = \theta | T_1S^2$ is a connection 1-form for the principal bundle π. The vector fields

$$X_1(x, y) = (x, \frac{\partial}{\partial y}) - (y, \frac{\partial}{\partial x}) \quad \text{and} \quad X_2(x, y) = (x \times y, \frac{\partial}{\partial x}),$$

when restricted to T_1S^2 span $\ker \vartheta(x, y)$. Let ω be the symplectic form on TS^2 given by restricting the standard symplectic form on $T\mathbf{R}^3$. Orient TS^2 using the volume form $\mathrm{vol}_{TS^2} = \omega \wedge \omega$. With respect to the outward normal $n(x, y) = (y, \frac{\partial}{\partial y})$ at $(x, y) \in T_1S^2$ in TS^2, we have $n \lrcorner \mathrm{vol}_{TS^2} = \mathrm{vol}_{T_1S^2}$. Show that $T_{(x,y)}\pi$ maps the positively oriented frame $\{(x \times y, \frac{\partial}{\partial y}), X_1(x, y), X_2(x, y)\}$ of $T_{(x,y)}(T_1S^2)$ onto the positively oriented frame $\{(y, \frac{\partial}{\partial x}), (x \times y, \frac{\partial}{\partial x})\}$ of T_xS^2. Here we are using the standard volume form vol_{S^2} to orient S^2. Show that the curvature 2-form Ω of ϑ is equal to vol_{S^2}. Hence the holonomy around the positively oriented equator, as measured from a point in the fiber over a fixed point on the equator, is 2π. Generalize this argument to show that the holonomy along any closed positively oriented curve on S^2 is equal to the area on S^2 of the 2-disc bounded by γ.

Appendix D

Action Angle Coordinates

In this appendix we prove the existence of local action angle coordinates for a Liouville integrable Hamiltonian system in a neighborhood of a compact connected fiber of its integral mapping. Under the additional assumption that all the smooth fibers of the integral map are compact and connected, we discuss the monodromy obstruction to the existence of global action angle coordinates.

1 Local action angle coordinates

Here we give a proof of the construction of action angle variables for certain Liouville integrable Hamiltonian systems.

A Hamiltonian system (f_1, M, ω) where $\dim M = 2n$ is a *Liouville integrable* system with n-degrees of freedom if there are *Poisson commuting* functions $(f_1, \ldots, f_n)$, that is, $\{f_i, f_j\} = 0$, whose differentials are linearly independent on an open dense subset W of M and whose Hamiltonian vector fields X_{f_i} are complete.

An important result for Liouville integrable systems is the following.

Claim: If (f_1, M, ω) is an n-degree of freedom Liouville integrable Hamiltonian system with integrals $(f_1, \ldots, f_n)$, then every connected component of a regular fiber of the integral map

$$f : M \to \mathbf{R}^n : p \to \big(f_1(p), \ldots, f_n(p)\big)$$

is diffeomorphic to a product of a k-torus $\mathbf{T}^k$ and $\mathbf{R}^{n-k}$ for some $0 \le k \le n$.

We postpone the proof of this claim until after we proved the existence of local action angle coordinates.

Claim: Let (M, ω) be a symplectic manifold of dimension $2n$. Suppose that $(f_1, \ldots, f_n)$ are Poisson commuting functions. Consider the integral map

$$f : M \to \mathbf{R}^n : p \to \big(f_1(p), \ldots, f_n(p)\big).$$

Suppose that q is a regular value in the image of f and that $f^{-1}(q)$ has a compact connected component F_q. Then there is a neighborhood V of F_q, an open set U of $\mathbf{R}^n$ and a diffeomorphism

$$V \to U \times \mathbf{T}^n : p \to (j_1, \dots j_n, \phi_1, \dots \phi_n),$$

where $\mathbf{T}^n$ is the n-torus $\mathbf{R}^n/\mathbf{Z}^n$. The coordinates j_i, called the *actions*, are smooth functions of the integrals f_i. The coordinates ϕ_i are called the *angles*. In *action angle coordinates* the symplectic form ω may be written as

$$\omega = \sum_{i=1}^{n} d\phi_i \wedge dj_i.$$

Note that for fixed $j = (j_1, \dots, j_n)$ the symplectic form ω vanishes identically on the n-torus $\{j\} \times \mathbf{T}^n$. Because the dimension of this *torus* is half the dimension of M it is a *Lagrangian* manifold.

In other words, V has the structure of a symplectic principal bundle with structure group $\mathbf{T}^n$, Lagrangian fibers, and a Hamiltonian action of the structure group with momentum map the projection map of the bundle. Furthermore, a section may be chosen to be Lagrangian so that the action angle variables form a symplectic chart.

The proof of the claim takes several steps.

1. Show that F_q has a neighborhood V diffeomorphic to $U \times F_q$.
2. Show that F_q is diffeomorphic to $\mathbf{T}^n$.
3. Define the group action.
4. Define the action variables and show that the action variables are a momentum map for the group action. This shows that the group action is Hamiltonian.
5. Construct a Lagrangian section of V.

(1.1) **Proof**: The steps of the proof are numbered as in the outline above.

Step 1.
Since q is a regular value of f and F_q is a compact component of $f^{-1}(q)$, there is an open ball $U \subseteq \mathbf{R}^n$ containing q which is contained in the set of regular values of f. Because F_q is a smooth compact submanifold of M, there is an open neighborhood V of F_q whose closure is compact such that $f(\overline{V}) \subseteq U$. Therefore

$$\tilde{f} = f|V : V \to U$$

is a proper submersion. Applying the Ehresmann fibration and trivialization theorems (see appendix C ((2.1)) and ((2.2))) and shrinking U and V if necessary, it follows that $\tilde{f}$ is a trivial smooth fibration, that is,

$$\tau : V \to U \times \tilde{f}^{-1}(u_0),$$

where $F_q = \widetilde{f}^{-1}(u_0)$ with $u_0 = q \in U$ such that

$$\tau^{-1}\big(\{u\} \times \widetilde{f}^{-1}(u_0)\big) = \widetilde{f}^{-1}(u).$$

Step 2.
Let ψ_t^i be the flow of the Hamiltonian vector field X_{f_i}. This flow leaves the fibers of f and hence those of $\widetilde{f}$ invariant. Because $\widetilde{f}^{-1}(u)$ is compact, the flow ψ_t^i on $\widetilde{f}^{-1}(u)$ is complete. Since $\{f_i, f_j\} = 0$, the flows ψ_t^i and ψ_t^j commute. Thus we have an $\mathbf{R}^n$ action

$$\Psi : \mathbf{R}^n \times \widetilde{f}^{-1}(u) \to \widetilde{f}^{-1}(u) : (t, p) = \big((t_1, \ldots, t_n), p\big) \to \psi_{t_1}^1 \circ \cdots \circ \psi_{t_n}^n(p).$$

Since the vector fields $X_{f_i} | \widetilde{f}^{-1}(u)$ are linearly independent, $D_1\Psi$ has rank n. Hence the orbits of Ψ are open. Since $\widetilde{f}^{-1}(u)$ is foliated by the orbits and is connected, there is only a single orbit, that is, Ψ is *transitive*. Let $\widetilde{p} \in \widetilde{f}^{-1}(u)$ and let

$$I_{\widetilde{p}} = \Big\{ t \in \mathbf{R}^n \Big| \Psi(t, \widetilde{p}) = \widetilde{p} \Big\},$$

be the isotropy group of the action Ψ at $\widetilde{p}$. Because $\widetilde{f}^{-1}(u)$ is compact $I_{\widetilde{p}} \neq \{0\}$. By transitivity, the isotropy groups at any two points in $\widetilde{f}^{-1}(u)$ are conjugate and hence are equal since Ψ is an abelian action. Denote this isotropy group by $\mathcal{P}_f(u)$ and call it the *period lattice* of Ψ at u. Because $D_1\Psi$ has rank n, the period lattice $\mathcal{P}_f(u)$ is a zero dimensional Lie subgroup of $\mathbf{R}^n$ and hence is *discrete*. We claim that $\mathcal{P}_f(u)$ is a *lattice*, that is, it is a free $\mathbf{Z}$-module with n generators. To see this we prove

Lemma: Let Λ be a nonzero discrete subgroup of $\mathbf{R}^n$. Then for some $1 \le k \le n$ there are k linearly independent vectors $\lambda_1, \ldots, \lambda_k$, called *generators*, such that

$$\Lambda = \mathbf{Z}\lambda_1 + \cdots + \mathbf{Z}\lambda_k. \tag{1}$$

(1.2) **Proof**: Since $\Lambda \neq 0$, there is a nonzero vector $\overline{\lambda}_1 \in \Lambda$. Let span $\overline{\lambda}_1$ be the subspace of $\mathbf{R}^n$ spanned by $\overline{\lambda}_1$. Since $\Lambda \cap \overline{\lambda}_1$ is a discrete group, there is a $\widehat{\lambda}_1 \in \Lambda \cap \operatorname{span} \overline{\lambda}_1$ such that the open line segment $\{t\widehat{\lambda}_1 \,|\, t \in (0, 1)\}$ does *not* intersect $\Lambda \cap \overline{\lambda}_1$. Set $\lambda_1 = \widehat{\lambda}_1$. Then $\mathbf{Z}\lambda_1 = \Lambda \cap \operatorname{span}\{\lambda_1\}$. If $\Lambda = \mathbf{Z}\lambda_1$, then $k = 1$ and we are done.

Otherwise, let $\overline{\lambda}_2 \in \Lambda - \mathbf{Z}\lambda_1$. As before let $\widehat{\lambda}_2$ be a generator of $\Lambda \cap \operatorname{span} \overline{\lambda}_2$ and let

$$P_{\lambda_1\widehat{\lambda}_2} = \Big\{ s_1\lambda_1 + s_2\widehat{\lambda}_2 \Big| s_1, s_2 \in [0, 1] \Big\}$$

be the *parallelogram* spanned by λ_1 and $\widehat{\lambda}_2$. Suppose that $P_{\lambda_1\widehat{\lambda}_2} \cap \Lambda$ contains an infinite number of points $\widetilde{\lambda}_i$. Select one, say $\widetilde{\lambda}_{i_0}$. A finite number of $\mathbf{Z}^2$-translates of $P_{\lambda_1\widetilde{\lambda}_{i_0}}$ cover the parallelogram $P_{\lambda_1\widehat{\lambda}_2}$. Hence one of these translates contains an infinite number of points of Λ. Since Λ is closed under $\mathbf{Z}^2$-translations, we conclude that $P_{\lambda_1\widetilde{\lambda}_{i_0}}$ contains an infinite number of points of $P_{\lambda_1\widehat{\lambda}_2} \cap \Lambda$. Repeat the above argument on the parallelogram $P_{\lambda_1\widetilde{\lambda}_{i_0}}$. We obtain a sequence of points $\widehat{\lambda}_i \in P_{\lambda_1\widehat{\lambda}_2} \cap \Lambda$ which converges to zero. This contradicts the discreteness of Λ. Hence the parallelogram $P_{\lambda_1\widehat{\lambda}_2}$ contains only *finitely* many points of Λ. Therefore it contains a parallelogram $P_{\lambda_1\lambda_2}$ such that $P_{\lambda_1\lambda_2} \cap \Lambda$ contains only the vertices $\{0, \lambda_1, \lambda_2, \lambda_1 + \lambda_2\}$. Suppose that $\mathbf{Z}\lambda_1 + \mathbf{Z}\lambda_2$ is not equal to $\mathrm{span}\{\lambda_1, \lambda_2\} \cap \Lambda$. Then there is a $\lambda = r_1\lambda_1 + r_2\lambda_2 \in \Lambda$ with $r_i \in \mathbf{R} - \mathbf{Z}$. Hence for some $n_i \in \mathbf{R} - \mathbf{Z}$ with $r_i - n_i \in (0, 1)$, the vector $\mu = \lambda - n_1\lambda_1 - n_2\lambda_2$ lies in $P_{\lambda_1\lambda_2}$ and is not a vertex. This contradicts the definition of $P_{\lambda_1\lambda_2}$. Therefore $\mathbf{Z}\lambda_1 + \mathbf{Z}\lambda_2 = \mathrm{span}\{\lambda_1, \lambda_2\} \cap \Lambda$. If $\Lambda = \mathbf{Z}\lambda_1 + \mathbf{Z}\lambda_2$, then $k = 2$ and we are done.

Otherwise repeat a similar argument as above with parallelograms replaced by parallelopipeds. Eventually we obtain a linearly independent set of vectors $\{\lambda_1, \ldots, \lambda_k\}$ such that (1) holds. □

Corollary: If $\mathbf{R}^n/\Lambda$ is compact, then $k = n$.

(1.3) **Proof**: Suppose that $1 \le k < n$. Then there is a nonzero vector $\lambda \in \mathbf{R}^n$ such that $\lambda \notin \mathrm{span}\{\lambda_1, \ldots, \lambda_k\}$. Thus for every $t \in \mathbf{R} - \{0\}$

$$\mu_t = t\lambda + \mathbf{Z}\lambda_1 + \cdots + \mathbf{Z}\lambda_k$$

is a nonzero element of $\mathbf{R}^n/\Lambda$. Since $\mathbf{R}^n/\Lambda$ is compact, there is a subsequence $\{t_i\}$ in $[1, 2]$ such that $\{\mu_{t_i}\}$ converges to $\mu \in \mathbf{R}^n/\Lambda$. Taking another subsequence we may assume that t_i converges to $t' \in [1, 2]$ and that μ_{t_i} converges to

$$\mu = \sum_{i=1}^{k} r_i\lambda_i + \mathbf{Z}\lambda_1 + \cdots + \mathbf{Z}\lambda_k$$

for some $r_i \in [0, 1)$. Therefore

$$\lambda = \frac{1}{t'}\big(\mu + \mathbf{Z}\lambda_1 + \cdots + \mathbf{Z}\lambda_k\big) \in \mathrm{span}\{\lambda_1, \ldots, \lambda_k\},$$

which contradicts the fact that $\lambda \notin \Lambda$. Conseqently $k = n$. □

From the corollary it follows that the period lattice $\mathcal{P}_f(u)$ is a lattice. Therefore $\widetilde{f}^{-1}(u)$ is diffeomorphic to an n-torus $\mathbf{R}^n/\mathcal{P}_f(u)$.

Step 3.
Because the fibration $\widetilde{f}$ is trivial, for every $p_0 \in \widetilde{f}^{-1}(u_0)$ there is a smooth section

$$\sigma : U \to \widetilde{f}^{-1}(U) \subseteq V$$

with $\sigma(u_0) = p_0$, which is a diffeomorphism onto its image. We now show that the isotropy group varies smoothly with u. Consider the function

$$\Theta : \mathcal{W} \times U \to \mathcal{U} \subseteq \mathbf{R}^n : (t, u) \to \sigma^{-1} \circ \Psi(t, \sigma(u)) - u,$$

where $\mathcal{W}$ is an open subset of $\mathbf{R}^n$ about $t_0 \in I_{p_0} - \{0\}$ and $\mathcal{U}$ is an open subset of $\mathbf{R}^n$. Since $\Psi(t_0, p_0) = p_0$, it follows from the continuity of σ and Ψ that both $\mathcal{W}$ and $\mathcal{U}$ can be chosen so that Θ is well defined. Since σ is a diffeomorphism onto its image and $D_1\Psi$ is surjective, $D_1\Theta$ is surjective. Thus 0 is a regular value of Θ and $(t_0, u_0) \in \Theta^{-1}(0)$. Therefore, near (t_0, u_0) the set $\Theta^{-1}(0)$ is the graph of a smooth function $T_0 : U \to \mathbf{R}^n$.

Let t_0 run through a basis $\{t_0^i\}$ of the period lattice $\mathcal{P}_f(u_0)$. Using the preceeding argument for $i = 1, \ldots, n$, we obtain smooth functions

$$T^i : U \to \mathbf{R}^n : u \to (T_1^i(u), \ldots, T_n^i(u))$$

such that $\{T^1(u), \ldots, T^n(u)\}$ form a basis of $\mathcal{P}_f(u)$ which depends smoothly on $u \in U$ and $\{T^i(u_0)\} = \{t_0^i\}$.

Let Y_i be a smooth vector field on V whose flow is

$$\Phi^i : \mathbf{R} \times V \to V : (t, p) \to \Psi\big((tT_1^i(u), \ldots, tT_n^i(u)), p\big).$$

Then

$$Y_i(p) = \sum_{j=1}^{n} T_j^i(f(p)) X_{f_i}(p).$$

The flow of Y_i is periodic with period one. The vector fields $\{Y_i\}$ are linearly independent and $[Y_i, Y_j] = 0$ since $\{f_i, f_j\} = 0$. Thus the $\{\Phi^i\}$ are n commuting flows and define an action of the n-dimensional torus group $\mathbf{T}^n = \mathbf{R}^n/\mathbf{Z}^n$ on $\widetilde{f}^{-1}(U)$.

By restricting U if necessary, we can arrange that

$$0 = [\omega] \in H^2(\widetilde{f}^{-1}(U), \mathbf{R}).$$

To see this we argue as follows. In a neighborhood V of $\widetilde{f}^{-1}(u_0)$ which retracts to $\widetilde{f}^{-1}(u_0)$, (this exists by virtue of the previous statements about the local fibration) any two-cycle Σ is homotopic to a cycle σ in the fiber $\widetilde{f}^{-1}(u_0)$. We have

$$\int_\Sigma \omega = \int_\sigma \omega = 0$$

the first equality holds because ω is closed and the second because $\widetilde{f}^{-1}(u_0)$ is isotropic. This shows that the cohomology class of ω vanishes, so by the *de Rham theorem* there exists a smooth 1-form ζ on V such that $\omega = -d\zeta$.

Step 4.
Define

$$j_a(p) = \int_0^1 (\zeta(Y_a)) \circ \Phi_a^t(p)\, dt. \tag{2}$$

We assert that the torus action $\Phi = (\Phi_1, \ldots, \Phi_n)$ is symplectic and has a momentum map $j = (j_1, \ldots, j_n)$.

To establish the assertion, it is enough to demonstrate this for each Φ_a individually because the torus group is abelian. This is established if we can show

$$dj_a(p)\, Z(p) = \omega(p)(Y_a(p), Z(p)) \tag{3}$$

for each $p \in V$ and each $Z(p) \in T_pV$. In other words, the vector field Y_a is the Hamiltonian vector field X_{j_a} corresponding to the action j_a.

Extend Z to a vector field, also denoted by Z. We can arrange things so that $\Phi_* Z = Z$. In particular, $[Y_a, Z] = 0$. Differentiating (2) and remembering that Z is Φ-invariant gives

$$dj_a(Z) = \int_0^1 \big(d(\zeta(Y_a))Z\big) \circ \Phi_a^t \, dt.$$

Here the reference to p has been dropped. However,

$$d(\zeta(Y_a))Z = L_Z(\zeta(Y_a)) = (L_Z\zeta)Y_a + \zeta(L_Z Y_a). \tag{4}$$

The second term in (4) vanishes since $[Z, Y_a] = 0$. Now

$$(L_Z\zeta)(Y_a) = \big(d(Z \lrcorner \zeta)\big)Y_a + (Z \lrcorner d\zeta)Y_a = L_{Y_a}(\zeta(Z)) + \omega(Y_a, Z).$$

Since

$$\int_0^1 L_{Y_a}(\zeta(Z)) \circ \Phi_a^t \, dt = \int_0^1 \frac{d}{dt}(\zeta(Z)) \circ \Phi_a^t \, dt = 0,$$

we have

$$dj_a(Z) = \int_0^1 \omega(Y_a, Z) \circ \Phi_a^t \, dt. \tag{5}$$

Now we show that the integrand in (5) is invariant under the flow Φ_a. We calculate

$$L_{Y_a}\big(\omega(Y_a, Z)\big) = L_{Y_a}\big((Y_a \lrcorner \omega)(Z)\big) = \big(Y_a \lrcorner L_{Y_a}\omega\big)(Z)$$

since $[Y_a, Z] = 0$. Furthermore

$$\begin{aligned} Y_a \lrcorner L_{Y_a}\omega &= Y_a \lrcorner \big(Y_a \lrcorner d\omega + d(Y_a \lrcorner \omega)\big) = Y_a \lrcorner d(Y_a \lrcorner \omega) \\ &= Y_a \lrcorner d\Big(\sum_{c=1}^n T_a^c X_{f_c} \lrcorner \omega\Big) = Y_a \lrcorner \Big(\sum_{c=1}^n dT_a^c \wedge df_c\Big) \\ &= 0, \end{aligned}$$

because both the T_a^c and f_c are constant on the fibers to which Y_a is tangent. Since the integrand is invariant, $dj_a(Z) = \omega(Y_a, Z)$, which establishes (3) and is precisely what is required to show that we have a momentum map.

Note that this shows that the momentum map has rank n because

$$(dj_1 \wedge \ldots \wedge dj_n)(Z_1, \ldots, Z_n) = (\omega \wedge \ldots \wedge \omega)(Y_1, \ldots, Y_n, Z_1, \ldots, Z_n),$$

and we can certainly choose vector fields $Z_1, \ldots, Z_n$ so that the vectors $Y_a(p)$ and $Z_a(p)$ are a basis for $T_p V$.

Step 5.
In order to choose angle variables correctly, it is useful to review first the situation so far. By restricting U if necessary, the results of the previous sections show that V has the structure of a symplectic principal bundle. The fibers of the bundle are Lagrangian, and the action of the structure group is Hamiltonian, with momentum map locally identified with the projection map of the bundle.

About a point $j(p)$ in the base, we choose a section s of the bundle, and compute the symplectic form in these coordinates. This yields

$$\omega = \begin{pmatrix} F & I \\ -I & 0 \end{pmatrix}.$$

The off diagonal blocks are the identity because j is the momentum map for the torus group, and the group action expressed in angle coordinates (call them θ^a) is just addition. Since ω is closed, so is

$$F = \tfrac{1}{2} \sum_{a,b} F^{ab} dj_a \wedge dj_b.$$

This implies that the functions F^{ab} depend only on the actions and not on the angles. By restricting the domain of the section s if necessary, we may arrange that $F = -dA$ for some 1-from A. We move the section s by defining new angle coordinates ϕ^a using the map

$$\phi_a = \theta^a - A^a.$$

This moves s to a Lagrangian section, and we conclude that in the coordinates (j_a, ϕ^b) the symplectic form ω becomes

$$\sum_{a=1}^{n} d\phi^a \wedge dj_a.$$

This concludes the proof of local action angle coordinates. □

(1.4) **Proof**: We now turn to proving the claim about Liouville integrable systems. The proof follows along the lines of step 2 of the proof of the existence of local action angle coordinates. The $\mathbf{R}^n$ action Ψ on F_q exists because the flows of X_{f_i} on F_q are complete. Because this action is transitive, the isotropy groups are all the same group and are discrete. Using the lemma we see that F_q is diffeomorphic to $\mathbf{T}^{n-k} \times \mathbf{R}^k$ for some $1 \le k \le n$. □

2 Monodromy

Note that the construction of action angle variables is in general only a *local* one, and it is of some interest to see whether it can be extended to a global one. We do not describe all the obstructions to doing this. In this section we will discuss the crudest obstruction to the existence of global action angle coordinates called *monodromy*.

Suppose that we are in the following situation. Let M be a symplectic manifold foliated by connected Lagrangian n-tori which are the fibers of a smooth map

$$F : M \to B.$$

Then F is a fiber bundle over B with total space M and fiber an n-torus. We wish to determine if F is a product bundle. This we do by constructing a *bundle of period lattices* associated to the bundle F. The total space of the associated bundle is $\mathcal{P}$, which is the disjoint union $\coprod_{b \in B} \mathcal{P}_F(b)$ of the period lattice $\mathcal{P}_F(b)$ at b of the bundle F. The base of the bundle is B and the projection mapping is

$$\mathcal{F} : \mathcal{P} \to B : \mathcal{P}_b = \mathcal{P}_F(b) \to b.$$

▷ We construct local trivializations for $\mathcal{F}$ as follows.

(2.1) **Proof**: Let $\{U_\alpha\}$ be an open covering of B such that on $F^{-1}(U_\alpha)$ local action angle coordinates

$$(j_\alpha^1, \ldots j_\alpha^n, \phi_\alpha^1, \ldots \phi_\alpha^n)$$

exist. In addition, assume that $\{U_\alpha\}$ is a good covering of B in the sense of Leray, that is,

1). U_α and $U_\alpha \cap U_\beta$ are connected and contractible.

2). $U_\alpha \cap U_\beta \cap U_\gamma = \varnothing$.

The flows $\psi_t^{j_\alpha^i}$ of the Hamiltonian vector fields $X_{j_\alpha^i}$ on $F^{-1}(U_\alpha)$ generate a $\mathbf{T}^n = \mathbf{R}^n/\mathbf{Z}^n$ action

$$\Psi^\alpha : \mathbf{T}^n \times F^{-1}(U_\alpha) \to F^{-1}(U_\alpha) : \big((t_1, \ldots, t_n), p\big) \to \psi_{t_1}^{j_\alpha^1} \circ \cdots \circ \psi_{t_n}^{j_\alpha^n}(p),$$

which is free and proper. Therefore, $F^{-1}(U_\alpha)$ is a principal $\mathbf{T}^n$ bundle, (see appendix B section 2). Moreover, $F^{-1}(U_\alpha)$ is trivial because U_α is contractible, (see appendix C section 2). The trivialization

$$\tau_\alpha : F^{-1}(U_\alpha) \to U_\alpha \times \mathbf{T}^n$$

intertwines the $\mathbf{T}^n$ action Ψ^α with the affine action of $\mathbf{T}^n$ on itself given by

$$\mathbf{T}^n \times \mathbf{T}^n \to \mathbf{T}^n : (\bar{s}, \bar{t}) \to \overline{s+t}.$$

Here $\bar{t} = t \bmod 1$. From this intertwining property it follows that for fixed $u_\alpha \in U_\alpha$ the tangent of

$$\tau_{\alpha,u_\alpha}^{-1} : \mathbf{T}^n \to F^{-1}(u_\alpha) : t \to \tau_\alpha^{-1}(u_\alpha, t) = p$$

maps the lattice $\mathbf{Z}^n \subseteq T_t\mathbf{T}^n$ into the lattice in $T_pF^{-1}(u_\alpha)$ spanned by the vectors

$$\Big\{ X_{j_\alpha^1}(p), \ldots, X_{j_\alpha^n}(p) \Big\}.$$

This latter lattice is the period lattice $\mathcal{P}_F(u_\alpha)$ of F at $F(p) = u_\alpha$. Since $\mathcal{P}_F(u_\alpha)$ does *not* depend on the choice of $p \in F^{-1}(u_\alpha)$, the tangent of the map $\tau_{\alpha.u_\alpha}^{-1}$ does *not* depend on the point in $\mathbf{T}^n$. Thus a local trivialization of $\mathcal{F}|\mathcal{F}^{-1}(U_\alpha)$ is given by

$$\rho_\alpha : \bigcup_{p \in F^{-1}(U_\alpha)} \mathcal{P}_F(F(p)) \to U_\alpha \times \mathbf{Z}^n : p \to \big(F(p), T_p\tau_{\alpha.F(p)}\mathcal{P}_F(F(p))\big).$$

To complete the construction of the bundle $\mathcal{F}$, we find its transition mappings. Towards this end let

$$\tau_\beta \circ \tau_\alpha^{-1} : (U_\alpha \cap U_\beta) \times \mathbf{T}^n \to (U_\alpha \cap U_\beta) \times \mathbf{T}^n : (u, t) \to \big(\varphi_{\alpha\beta}(u), \tau^{\alpha\beta}(u, t)\big)$$

be the transition map for the bundle $F|F^{-1}(U_\alpha \cap U_\beta)$. For $u \in U_\alpha \cap U_\beta$, the partial transition mapping

$$\tau_u^{\alpha\beta} : \mathbf{T}^n \to \mathbf{T}^n : t \to \tau^{\alpha\beta}(u, t)$$

has a tangent map $T_t\tau_u^{\alpha\beta}$ which does *not* depend on the point $t \in \mathbf{T}^n$. Moreover, $T\tau_u^{\alpha\beta}$ is a linear isomorphism of $\mathbf{R}^n$, which preserves the lattice $\mathbf{Z}^n$. Since $U_\alpha \cap U_\beta$ is connected, we may assume that this isomorphism is orientation preserving. The collection of all such isomorphisms forms the group $\mathrm{Sl}(n, \mathbf{Z})$, which is a *discrete* subgroup of $\mathrm{Sl}(n, \mathbf{R})$. Because $U_\alpha \cap U_\beta$ is connected, it follows from discreteness and continuity that the map

$$U_\alpha \cap U_\beta \to \mathrm{Sl}(n, \mathbf{Z}) : u \to T\tau_u^{\alpha\beta}$$

is constant, say $g^{\alpha\beta}$. Consequently, the transition map for the period lattice bundle $\mathcal{F}|\mathcal{F}^{-1}(U_\alpha \cap U_\beta)$ is

$$\rho_\beta \circ \rho_\alpha^{-1} : (U_\alpha \cap U_\beta) \times \mathbf{Z}^n \to (U_\alpha \cap U_\beta) \times \mathbf{Z}^n : (u, z) \to \big(\varphi_{\alpha\beta}(u), g^{\alpha\beta}(z)\big).$$

This completes the construction of the bundle $\mathcal{P}$ of period lattices associated to F. □

Next we describe the monodromy representation associated to the bundle of period lattices. Let $\gamma : [0, 1] \to B$ be a smooth closed curve in B with starting point b. Restricting the bundle $\mathcal{F}$ to γ

$$\mathcal{F}|\gamma : \mathcal{P}|\gamma = \bigcup_{u \in \gamma[0,1]} \mathcal{P}_F(u) \to \gamma : \mathcal{P}_F(u) \to u$$

which is a smooth bundle of period lattices over γ. Since γ is diffeomorphic to a circle, the isomorphism class of the bundle $\mathcal{F}|\gamma$ is determined by an automorphism $g(b)$ of the fiber $\mathcal{P}_b$ over b. This isomorphism class depends only on the homotopy class of γ in B. Therefore we have the map

$$\pi : \pi_1(B) \to \mathrm{Sl}(n, \mathbf{Z}) : b \to g(b),$$

which is the *monodromy representation* of the fundamental group of B associated to the bundle $\mathcal{F}$. If π is *not* the identity, that is, if for some smooth loop γ in B the bundle $\mathcal{F}|\gamma$ is *not* trivial, then the bundle of period lattice $\mathcal{F}$ is said to have *monodromy.*

The bundle of period lattices is isomorphic to the bundle $H_1(F^{-1}(B), \mathbf{Z})$ over B whose fiber at $b \in B$ is the first homology group of the fiber $F^{-1}(b)$ of the n-torus bundle F. The proof of this assertion is left as an exercise, (see chapter 4 section 5 for a special case).

We now prove

Claim: The n-torus bundle $F : M \to B$ is not trivial if the associated period lattice bundle $\mathcal{F} : \mathcal{P} \to B$ has monodromy.

(2.2) **Proof**: We prove the contrapositive. Suppose that the bundle F is trivial. Then over every smooth closed loop γ in B, the bundle $F|\gamma$ is trivial. Hence the classifying map

$$\chi : F^{-1}(\gamma(0)) \to F^{-1}(\gamma(0))$$

of $F|\gamma$ is homotopic to the identity map. Hence χ induces the map

$$\chi_* : H_1(F^{-1}(\gamma(0)), \mathbf{Z}) \to H_1(F^{-1}(\gamma(0)), \mathbf{Z}),$$

which is the identity. Therefore the monodromy representation of the period lattice bundle $\mathcal{F} : \mathcal{P} \to B$ associated to the n-torus bundle $F : M \to B$ is the identity. In other words, the bundle $\mathcal{F}$ has no monodromy. □

3 Exercises

1. (Particle in a champagne bottle potential.) On $(T\mathbf{R}^2, \omega)$ consider the Hamiltonian
$$H : T\mathbf{R}^2 \to \mathbf{R} : (x, y) \to \tfrac{1}{2}(y_1^2 + y_2^2) - \tfrac{1}{2}(x_1^2 + x_2^2) + \tfrac{1}{2}(x_1^2 + x_2^2)^2.$$
The Hamiltonian vector field X_H describes the motion of a particle in a circularly symmetric double well potential. The Hamiltonian H is invariant under the lift of this S^1 action to $T\mathbf{R}^2$. Show that $(0, 0)$ is an isolated critical value of the energy momentum mapping $\mathcal{EM}$. Let Γ be a circle in the range of $\mathcal{EM}$ which bounds a 2-disc containing $(0, 0)$ in its interior. Suppose that Γ lies in the set of regular values of $\mathcal{EM}$. Show that the 2-torus bundle $\mathcal{EM}^{-1}(\Gamma)$ has monodromy $\begin{pmatrix} 1 & 0 \\ 1 & 1 \end{pmatrix}$.

2. (Quadratic spherical pendulum.) On $TS^2 \subseteq T\mathbf{R}^3$ with symplectic form $\omega|TS^2$ consider the Hamiltonian
$$H : TS^2 \subseteq T\mathbf{R}^3 \to \mathbf{R} : (x, y) \to \tfrac{1}{2}(y_1^2 + y_2^2 + y_3^2) + x_3^2.$$
The Hamiltonian vector field X_H describes the motion of a particle on a 2-sphere S^2 under a force coming from a potential which is the square of the height. Show that $(1, 0)$ is an isolated critical value of the energy momentum mapping $\mathcal{EM}$ and show that the 2-torus bundle $\mathcal{EM}^{-1}(\Gamma)$ over a circle Γ about $(1, 0)$ which lies in the set of regular values of $\mathcal{EM}$ has monodromy $\begin{pmatrix} 1 & 0 \\ 2 & 1 \end{pmatrix}$.

3. (Monodromy and Seifert manifolds.) Consider the circle S defined by $j^2 + h^2 = 1/4$ in the image of the energy momentum map $\mathcal{EM}$ of the particle in a champagne bottle potential, (see exercise 1). Show that S lies in the set of regular values of $\mathcal{EM}$.

   (a). Show that $\mathcal{EM}^{-1}(S)$ is a smooth compact connected orientable 3-manifold M^3.

   (b)*. Let $\pi : V^2 \to M^3 \to S^1$ be a bundle with fiber V^2 a compact orientable surface of genus g, total space M^3 a compact orientable 3-manifold and base space a circle S^1. Let A be the $2g \times 2g$ matrix induced on $H_1(M^3, \mathbf{Z})$ by the monodromy map of the bundle. Show that $H_1(V^2, \mathbf{Z})$ is isomorphic to $\mathbf{Z} \oplus \operatorname{coker}(A - I)$. When π is the bundle $V^2 \to M^3 \overset{\mathcal{EM}}{\to} S$ with $\mathcal{EM}$ given in exercise 1, show that V^2 is diffeomorphic to a 2-torus.

   (c)*. Show that the flow of the angular momentum vector field in exercise 1 allows one to realize $\mathcal{EM}^{-1}(S)$ as the total space M^3 of a principal circle bundle
$$\rho : S^1 \to M^3 \to W^2$$
   with base space a smooth compact connected 2-manifold W^2. Show that $H_1(W^2, \mathbf{Z})$ is isomorphic to $\mathbf{Z}/e\mathbf{Z} \oplus \mathbf{Z}$. Compute the Euler class e of the bundle ρ. Show that W^2 is diffeomorphic to a 2-torus. Obtain this result using invariant theory.

   (d). Identify M^3 as a Siefert manifold.

4. Give an example of a T^3-bundle π over S^2 which is not trivial. Is there an integrable Hamiltonian system on $(T\mathbf{R}^3, \omega)$, whose momentum mapping has $0 \in \mathbf{R}^3$ as an isolated critical value, which gives rise to the bundle π?

5. (a). (Burke's egg.) This exercise presents an attempt to find the simplest example of a mechanical system with symmetry which has monodromy. Consider a particle of mass m moving on the plane under the influence of a potential $-r^2 = -(x_1^2 + x_2^2)$. Prove the following assertions.

   1. Burke's egg is Liouville integrable.

   2. The origin is an isolated critical value of the energy momentum map of Burke's egg.

   3. The fiber of a regular value of the energy momentum map is topologically $S^1 \times \mathbf{R}$. From this deduce that the monodromy is *trivial.*

   (b). (Burke's egg (poached).) The problem is the same as in part (a) except that we add a magnetic term to the symplectic structure. More explicitly, the phase space of the problem is $T^*\mathbf{R}^2$ with coordinates (x_1, x_2, y_1, y_2) and symplectic form

$$\Omega = (x_1^2 + x_2^2)\, dx_1 \wedge dx_2 + dx_1 \wedge dy_1 + dx_2 \wedge dy_2.$$

   The Hamiltonian is $H(x, y) = \frac{1}{2}\left(y_1^2 + y_2^2\right) - (x_1^2 + x_2^2)$. Prove the following assertions.

   1. The problem is Liouville integrable.

   2. The origin is an isolated critical value of the energy momentum mapping.

   3. The topology of the fiber of a regular value of the energy momentum mapping is a 2-torus T^2. Show that every fiber of the energy momentum mapping is *compact.* Deduce that the projection of every orbit of X_H onto configuration space is *bounded.*

   4. Show that the monodromy is *nontrivial* even though the topology of the enegry surfaces does *not* change when the value of the energy passes through 0.

6. (Kerr-Newman black hole). Consider the motion of a particle about a Kerr-Newman black hole. It is known that this problem is Liouville integrable because there is a "hidden "integral of motion.

   (a)*. Describe the image of the energy momentum mapping.

   (b)*. Describe the topology of the fibers of the energy momentum mapping.

   (c)*. Do global action angle coordinates exist for this problem?

7. Let P^{2n} be the phase space of a Liouville integrable system with energy momentum map $\mathscr{EM} : P \to \mathbf{R}^n$.

   (a). Let $f : U \to \mathbf{R}^n$ be a local set of action variables on the open set $U \subseteq \mathbf{R}^n$. Pull back the standard flat connection

$$\nabla_X Y = DY \cdot X$$

   on $\mathbf{R}^n$ by the mapping f. Show that the pulled back connection ∇_X^* in U is

$$\nabla_X^* Y = DY \cdot X + Df^{-1} \cdot D^2 f(X, Y).$$

   Deduce that $\nabla_X^* Y$ defines a covariant derivative in U.

(b). Let $g : V \to \mathbf{R}^n$ be another set of local action variables. If we define ∇ in V as in part (a), show that the two covariant derivatives agree on $U \cap V$. (Hint: how are the two sets of action variables related?)

(c). Show that the construction in (a) and (b) defines a flat connection on the set $\mathcal{R}$ of regular values of $\mathcal{EM}$ (where one can define local action variables).

(d). Show that the holonomy of the connection defined in part (c) has the same matrix representation (in a suitable basis) as the monodromy of the bundle

$$\mathcal{EM}|\mathcal{EM}^{-1}(\mathcal{R}) : \mathcal{EM}^{-1}(\mathcal{R}) \to \mathcal{R}.$$

Appendix E

Basic Morse Theory

In this appendix we give an introduction to basic Morse theory. We define the notion of the Hessian of a smooth function at a critical point on a smooth manifold and show that if it is nondegenerate then there are local coordinates in which the function is equal to its second derivative. We also prove the Morse isotopy lemma which gives a criterion when two suitable level sets of a smooth function are diffeomorphic. We conclude the appendix by extending the notion of nondegenerate critical point to nondegenerate critical submanifold.

1 Preliminaries

In this section we review the concepts of a smooth function and a smooth manifold (in $\mathbf{R}^n$). We show that if 0 is a regular value of a smooth function then its zero level set is a smooth manifold. We prove the Lagrange multiplier criterion for finding the critical points of a real valued smooth function restricted to a manifold which is a level set of a smooth function.

Let $U \subseteq \mathbf{R}^m$ be an open set. The function $f : U \subseteq \mathbf{R}^m \rightarrow \mathbf{R}^k$ is *smooth* if for $x \in U$ and every $n \in \mathbf{Z}_{\geq}$, the nth derivative $D^n f(x)$ of f at x exists. If $V \subseteq \mathbf{R}^m$ is a closed set, then $f : V \subseteq \mathbf{R}^m \rightarrow \mathbf{R}^k$ is *smooth* if there is an open set U containing V and a smooth function $F : U \subseteq \mathbf{R}^m \rightarrow \mathbf{R}^k$ such that $F|V = f$. A smooth function $f : V \subseteq \mathbf{R}^m \rightarrow \mathbf{R}^k$ is a *diffeomorphism* if it is one to one and its inverse is smooth on $f(V)$.

A set $M \subseteq \mathbf{R}^n$ is a k-dimensional *smooth manifold* if for every $m \in M$ there is an open set $U \subseteq \mathbf{R}^n$ containing m and a diffeomorphism $\varphi : U \cap M \rightarrow V \subseteq \mathbf{R}^k$. The pair (U, φ) is called a *coordinate chart at* m and the pair (V, φ^{-1}) is called a *local parametrization of* M *at* m.

Claim: Suppose that $F : \mathbf{R}^m \times \mathbf{R}^k \rightarrow \mathbf{R}^k$ is a smooth mapping with $F(0,0) = 0$ and $D_2F(0,0)$ is an isomorphism. Then there is an open set U in $\mathbf{R}^m$ containing 0, an open set V in $\mathbf{R}^k$ containing 0, and a smooth function $g : U \subseteq \mathbf{R}^m \rightarrow V \subseteq \mathbf{R}^k$ such that

$$F^{-1}(0) \cap (U \times V) = (\operatorname{gr} g)(U) = \left\{ (u, g(u)) \in \mathbf{R}^m \times \mathbf{R}^k \middle| u \in U \right\}.$$

Here gr g is the graph of g. Restated, the conclusion says that $F(u, g(u)) = 0$ for every $u \in U$, that is, locally $F^{-1}(0)$ is the graph of a smooth function.

(1.1) **Proof**: Define the mapping

$$H : \mathbf{R}^m \times \mathbf{R}^k \to \mathbf{R}^m \times \mathbf{R}^k : (z, w) \to \big(z, F(z, w)\big) = (x, y).$$

Then $H(0, 0) = (0, 0)$ and

$$DH(x, y) = \begin{pmatrix} I & 0 \\ D_1F(x, y) & D_2F(x, y) \end{pmatrix}.$$

Since $D_2F(0, 0)$ is a linear isomorphism, $DH(0, 0)$ is a linear isomorphism of $\mathbf{R}^m \times \mathbf{R}^k$. By the inverse function theorem, H is a local diffeomorphism of $\mathbf{R}^m \times \mathbf{R}^k$, that is, there is an open set U' in $\mathbf{R}^m$ containing 0 and an open set V' in $\mathbf{R}^k$ containing 0 such that the open set $U' \times V'$ is mapped onto $U \times V$ by H and on $U \times V$ the function H has a smooth inverse. Let

$$\pi_2 : \mathbf{R}^m \times \mathbf{R}^k \to \mathbf{R}^k : (x, y) \to y$$

be projection on the second factor. By definition of H we have $\pi_2 \circ H(z, w) = F(z, w)$. For $(x, y) \in U \times V$ the mapping H^{-1} is defined. Therefore

$$y = \pi_2(x, y) = F \circ H^{-1}(x, y), \tag{1}$$

where $H^{-1}(x, y) = (z, w)$. In other words, after changing coordinates by H^{-1}, the function F is locally the projection π_2. Write $H^{-1}(x, y) = \big(h_1(x, y), h_2(x, y)\big)$. Using the definition of H it follows that

$$(x, y) = H \circ H^{-1}(x, y) = \big(h_1(x, y), F(h_1(x, y), h_2(x, y))\big). \tag{2}$$

Hence $h_1(x, y) = x \in U$. Setting $y = 0$ in (2) we obtain $0 = F\big(x, h_2(x, 0)\big)$ for every $x \in U$. Let

$$g : U \subseteq \mathbf{R}^m \to V \subseteq \mathbf{R}^k : u \to h_2(u, 0).$$

Then for every $u \in U$ we have $F(u, g(u)) = 0$, that is, $F^{-1}(0) \cap (U \times V) = \text{gr}\, g$. Since H^{-1} is differentiable, g is also. □

Corollary 1: If $F : \mathbf{R}^{m+k} \to \mathbf{R}^k$ is smooth with $F(0) = 0$ and $DF(0) : \mathbf{R}^{m+k} \to \mathbf{R}^k$ is surjective, then near 0 the set $F^{-1}(0)$ is the graph of a smooth function.

Another way to state the conclusion of the above corollary is that the zero level set of F near 0 is an m-dimensional manifold, if $DF(0)$ is surjective.

Let $F : \mathbf{R}^{m+k} \to \mathbf{R}^k$ be a smooth mapping. If $DF(x)$ is not surjective, then x is a *critical point* of F. If for every $x \in F^{-1}(z)$, the derivative $DF(x)$ is surjective, then z is a *regular value* of F. Note that if z is not in the image of F, then z is a regular value. If z is *not* a regular value, then z is a *critical value* of F.

Corollary 2: If z is a regular value of the smooth mapping $F : \mathbf{R}^{m+k} \to \mathbf{R}^k$, then $F^{-1}(z)$ is an m-dimensional manifold.

Example 1: Let $f : \mathbf{R}^3 \rightarrow \mathbf{R} : (x, y, z) \rightarrow x^2 + y^2 + z^2 - 1$. Then

$$Df : \mathbf{R}^3 \rightarrow (\mathbf{R}^3)^* : (x, y, z) \rightarrow 2(x, y, z)$$

is not surjective if and only if $(x, y, z) = (0, 0, 0)$. But $(0, 0, 0) \notin f^{-1}(0)$. Thus 0 is a regular value of f. Hence $f^{-1}(0)$ is a 2-dimensional manifold called the unit *2-sphere*. □

Example 2: Let $O(n)$ be the set of all $n \times n$ real matrices A such that $AA^t = I$. Then $O(n)$ is the set of all orthogonal matrices. We now show that $O(n)$ is a smooth manifold. Let

$$F : \mathrm{gl}(n, \mathbf{R}) \rightarrow \mathrm{Sym}_n(\mathbf{R}) : A \rightarrow AA^t - I,$$

where $\mathrm{Sym}_n(\mathbf{R})$ is the set of all $n \times n$ real *symmetric* matrices. Differentiating F gives

$$DF : \mathrm{gl}(n, \mathbf{R}) \rightarrow L\big(\mathrm{gl}(n, \mathbf{R}), \mathrm{Sym}_n(\mathbf{R})\big)$$

with $DF(A)B = AB^t + BA^t$, where $B \in \mathrm{gl}(n, \mathbf{R})$. Suppose that $F(A) = 0$ and let $C \in \mathrm{Sym}_n(\mathbf{R})$. Then $B = \frac{1}{2}CA$ solves the equation $DF(A)B = C$. Therefore, for every $A \in F^{-1}(0)$ the derivative of F at A is surjective, that is, 0 is a regular value of the mapping F. Thus $O(n)$ is a manifold of dimension $\frac{1}{2}n(n-1)$. □

Claim: Let $g : \mathbf{R}^m \times \mathbf{R}^k \rightarrow \mathbf{R}$ be a smooth function and let 0 be a regular value of the smooth function $f : \mathbf{R}^m \times \mathbf{R}^k \rightarrow \mathbf{R}^k$. If $x \in f^{-1}(0)$ is a critical point of $g|f^{-1}(0)$, then for some linear function $\lambda : \mathbf{R}^k \rightarrow \mathbf{R}$,

$$Dg(x) + \lambda Df(x) = 0. \tag{3}$$

(1.2) **Proof**: We assume that $D_2f(x)$ is an isomorphism. Let $\varphi : \mathbf{R}^m \rightarrow \mathbf{R}^k$ be a smooth function such that

$$\mathrm{gr}\,\varphi : U \subseteq \mathbf{R}^m \rightarrow f^{-1}(0) \cap (U \times \varphi(U)) \subseteq \mathbf{R}^m \times \mathbf{R}^k$$

with $\mathrm{gr}\,\varphi(0) = (0, \varphi(0)) = x$ is a local parametrization of $f^{-1}(0)$ at x. Look at the function

$$g \circ \mathrm{gr}\,\varphi : U \subseteq \mathbf{R}^m \rightarrow \mathbf{R} : y \rightarrow g(y, \varphi(y)).$$

Because $\mathrm{gr}\,\varphi$ is a local diffeomorphism and $Dg(x)$ is not surjective, the function $g \circ \mathrm{gr}\,\varphi$ has a critical point at 0, that is,

$$0 = D(g \circ \mathrm{gr}\,\varphi)(0) = D_1g(0, \varphi(0)) + D_2g(0, \varphi(0))D\varphi(0) = D_1g(x) + D_2g(x)\,D\varphi(0). \tag{4}$$

For every $y \in U$, we have $f(y, \varphi(y)) = 0$, which differentiated and evaluated at 0 gives

$$D_1f(x) + D_2f(x)D\varphi(0) = 0.$$

But $D_2f(x)$ is an isomorphism. Hence

$$D\varphi(0) = -\big(D_2f(x)\big)^{-1} \circ D_1f(x). \tag{5}$$

Substituting (5) into (4) gives

$$0 = D_1 g(x) - \big(D_2 g(x) \circ (D_2 f(x))^{-1}\big) D_1 f(x). \tag{6}$$

Define a linear map $\lambda : \mathbf{R}^k \to \mathbf{R}$, by $\lambda = -D_2 g(x) \circ \big(D_2 f(x)\big)^{-1}$ called the *Lagrange multiplier*. Then (6) may be written as $0 = Dg(x) + \lambda Df(x)$. □

Example 3. Let $(\, ,\,)$ be the Euclidean inner product on $\mathbf{R}^{n+1}$. and let

$$S : \mathbf{R}^{n+1} \to \mathbf{R} : x \to (x, x) - 1.$$

Because 0 is a regular value of S, the unit n-sphere S^n is the manifold $S^{-1}(0)$. Consider the function

$$Q : \mathbf{R}^{n+1} \to \mathbf{R} : x \to (x, Ax),$$

where $A : \mathbf{R}^{n+1} \to \mathbf{R}^{n+1}$ is linear and symmetric, that is, for every $y, z \in \mathbf{R}^{n+1}$, $(y, Az) = (Ay, z)$. Because $S^{-1}(0)$ is compact, $Q|S^n$ has a critical point, say x. Using Lagrange multipliers ((1.2)), we find that x satisfies

$$\begin{cases} 0 = DQ(x) - \lambda DS(x) \\ 0 = S(x). \end{cases}$$

Since $DQ(x)h = 2(x, Ah)$ and $DS(x)h = 2(x, h)$, we see that for every $h \in \mathbf{R}^{n+1}$

$$\begin{cases} 0 = (x, Ah) - \lambda(x, h) \;=\; ((A - \lambda)x, h) \\ 0 = (x, x) - 1, \end{cases}$$

which implies

$$\begin{cases} (A - \lambda)x &=& 0 \\ (x, x) &=& 1. \end{cases}$$

Thus x is an eigenvector of A of unit length. If x is a unit length eigenvector of A corresponding to a simple real eigenvalue λ, $\pm x$ is a critical point of $Q|S^n$. More generally, if λ is an eigenvalue of A of geometric multiplicity $m + 1 \le n + 1$, that is, the dimension of the space of eigenvectors V_λ corresponding to the eigenvalue λ is $m + 1$, then $Q|S^n$ has a critical set which is a unit m-sphere $S^m \subseteq V_\lambda$.

2 The Morse Lemma

In this section, we prove the Morse lemma. We also define the Hessian of a function on a manifold at a critical point and give a formula for computing it using Lagrange multipliers.

Suppose that x is a critical point of a smooth function $f : \mathbf{R}^m \to \mathbf{R}$. Then the *Hessian* of f at x is the symmetric bilinear mapping $D^2 f(x)$ where

$$D^2 f : \mathbf{R}^m \to \mathrm{Sym}_m(\mathbf{R}) : x \to \Big(\frac{\partial^2 f}{\partial x_i \partial x_j}\Big).$$

If the $m \times m$ matrix $D^2 f(x)$ is invertible, then x is a *nondegenerate* critical point of f. If all the critical points of f are nondegenerate, then f is a *Morse function*. We now prove the *Morse lemma*.

Claim: Suppose that 0 is a nondegenerate critical point of a smooth function $f : \mathbf{R}^m \rightarrow \mathbf{R}$ with $f(0) = 0$. Then there is an open set $U \subseteq \mathbf{R}^m$ with $0 \in U$ and a diffeomorphism $\varphi : U \subseteq \mathbf{R}^m \rightarrow \mathbf{R}^m$ with $\varphi(0) = 0$ such that $f \circ \varphi^{-1}(y) = \frac{1}{2} D^2 f(0)(y, y)$ for every $y \in \varphi(U)$.

(2.2) **Proof**: We start by computing the Taylor series of f at 0 to second order with integral remainder:

$$\begin{aligned} f(x) &= \int_0^1 \frac{d}{dt} f(tx)\, dt, \quad \text{by the fundamental theorem of calculus and the fact that } f(0) = 0 \\ &= -\frac{d}{dt} f(tx)(1-t)\Big|_0^1 + \int_0^1 (1-t) \frac{d^2}{dt^2} f(tx)\, dt, \\ &\qquad \text{by partial integration} \\ &= Df(0)x + \Big(\int_0^1 (1-t) D^2 f(tx)\, dt \Big)(x, x) \\ &= B(x)(x, x), \end{aligned}$$

where $B(x)$ is the quadratic function $\int_0^1 (1-t) D^2 f(tx)\, dt$. The last equality above follows because $Df(0) = 0$. Note that $B(0) = \frac{1}{2} D^2 f(0)$ is invertible because 0 is a nondegenerate critical point of the function f.

Let $\varphi(x) = R(x)x$, where $R(x)$ is an $m \times m$ matrix depending smoothly on x such that $R(0) = I$. We must determine $R(x)$ so that

$$B(x)(x, x) = f(x) = B(0)\big(R(x)x, R(x)x \big) \tag{7}$$

for all x in some open neighborhood U of 0. Towards this goal let

$$F : \mathbf{R}^m \times \mathrm{gl}(m, \mathbf{R}) \rightarrow \mathrm{Sym}_m(\mathbf{R}) : (x, R) \rightarrow R^t B(0) R - B(x). \tag{8}$$

Then $F(0, I) = 0$. Differentiating F with respect to R and evaluating at $(0, I)$ gives

$$D_2 F(0, I)S = S^t B(0) + B(0)S. \tag{9}$$

$D_2 F(0, I)$ is surjective because for any given $C \in \mathrm{Sym}_m(\mathbf{R})$, choose $S = \frac{1}{2} B(0)^{-1} C$. From (9) we see that $D_2 F(0, I)S = C$. Therefore, by ((1.1)), there is an open neighborhood U of 0 in $\mathbf{R}^m$ and a differentiable function

$$R : U \subseteq \mathbf{R}^m \rightarrow \mathrm{gl}(m, \mathbf{R}) : x \rightarrow R(x)$$

with $R(0) = I$ such that $F(x, R(x)) = 0$ for every $x \in U$. Shrinking U if necessary, we may assume that $R(x)$ is invertible for every $x \in U$. Thus the map φ is a diffeomorphism of U. □

Let M be a smooth manifold and $F : M \to \mathbf{R}$ a smooth function. Suppose that $p \in M$ is a critical point of F. Define the Hessian of F at p as follows. Let $\varphi : U \subseteq \mathbf{R}^{n-k} \to M$ with $\varphi(0) = p$ be a local parametrization of M. Then the *Hessian* of F at p is the symmetric bilinear form on the *tangent space* T_pM given by

$$(\mathrm{Hess}_p F)(v_p, w_p) = D^2(F \circ \varphi)(0)(v, w) \tag{10}$$

Here the curves $t \to \varphi(tv)$ and $t \to \varphi(tw)$ represent the *tangent vectors* v_p and w_p to M at p.

We now look at what happens to (10) when we change local parametrizations. Suppose that $\widetilde{\varphi} : U \subseteq \mathbf{R}^m \to M$ with $\widetilde{\varphi}(0) = p$ is another local parametrization for M. Let $\psi = \varphi^{-1} \circ \widetilde{\varphi}$. Then

$$\begin{aligned} D^2(F \circ \widetilde{\varphi})(0) &= D^2\big((F \circ \varphi) \circ \psi\big)(0) = D\big(D(F \circ \varphi)(\psi(0)) \cdot D\psi(0)\big) \\ &= D^2(F \circ \varphi)(\psi(0))\big(D\psi(0), D\psi(0)\big) + D(F \circ \varphi)(\psi(0)) \cdot D^2\psi(0) \quad (11) \\ &= D^2(F \circ \varphi)(0)\big(D\psi(0), D\psi(0)\big), \end{aligned}$$

where the second equality above follows by the chain rule, the third equality because $\cdot$ is bilinear and the last equality because $\psi(0) = 0$ and $D(F \circ \varphi)(\psi(0)) = 0$, since p is a critical point of F. Thus

$$D^2(F \circ \widetilde{\varphi})(0)(v, w) = D^2(F \circ \varphi)(0)\big(D\psi(0)v, D\psi(0)w\big). \tag{12}$$

Here the curves $t \to \varphi(t\, D\psi(0)v)$ and $t \to \varphi(t\, D\psi(0)w)$ represent the tangent vectors v_p and w_p in the local parametrization $\widetilde{\varphi}$. From (12) we see that the invariants of the Hessian such as its nondegeneracy (= its invertibility), its *Morse index* (= its number of negative eigenvalues) and *signature* (= the difference of its number of positive and negative eigenvalues), do not depend on the choice of parametrization.

Claim: Let 0 be a regular value of the smooth function $f : \mathbf{R}^n \to \mathbf{R}^k$ and suppose that $g : \mathbf{R}^n \to \mathbf{R}$ is a smooth function. If $x \in f^{-1}(0)$ is a critical point of

$$g|f^{-1}(0) : f^{-1}(0) \subseteq \mathbf{R}^n \to \mathbf{R}$$

with Lagrange multiplier $\lambda : \mathbf{R}^k \to \mathbf{R}$, that is,

$$Dg(x) + \lambda \circ Df(x) = 0, \tag{13}$$

then the Hessian of $g|f^{-1}(0)$ at x is

$$\big(D^2g(x) + \lambda \circ D^2f(x)\big)\Big|T_x f^{-1}(0), \tag{14}$$

where $T_x\big(f^{-1}(0)\big) = \ker Df(x)$.

(2.3) **Proof**: Let $\varphi : U \subseteq \mathbf{R}^{n-k} \to f^{-1}(0)$ with $\varphi(0) = x$ be a local parametrization of $f^{-1}(0)$ at x. Then

$$D^2(g \circ \varphi)(0) = D^2g(x)\big(D\varphi(0), D\varphi(0)\big) + Dg(x)D^2\varphi(0), \tag{15}$$

using (11). Since φ is a local parametrization of $f^{-1}(0)$, for every $y \in U$ we have

$$\lambda \circ f(\varphi(y)) = 0. \tag{16}$$

Differentiating (16) twice and evaluating at 0 gives

$$\begin{aligned} 0 &= D^2((\lambda \circ f) \circ \varphi)(0) \\ &= D^2(\lambda \circ f)(x)\big(D\varphi(0), D\varphi(0)\big) + D(\lambda \circ f)(x)\, D^2\varphi(0), \\ &= D^2(\lambda \circ f)(x)\big(D\varphi(0), D\varphi(0)\big) - Dg(x)\, D^2\varphi(0). \end{aligned} \tag{17}$$

The second equality above follows from (11). Equation (17) follows because the map λ is linear, which implies that $D(\lambda \circ f)(x) = \lambda \circ Df(x)$, and because of (13). Adding (15) and (17) gives

$$D^2(g \circ \varphi)(0)(v, w) = D^2(g + \lambda \circ f)(x)(D\varphi(0)v, D\varphi(0)w).$$

Since the linear map $D\varphi(0) : T_0\mathbf{R}^{n-k} \to T_x\big(f^{-1}(0)\big)$ is an isomorphism, we obtain

$$\mathrm{Hess}_x\, g | f^{-1}(0) = D^2(g + \lambda \circ f)(x)\Big| T_x\big(f^{-1}(0)\big). \qquad \square$$

Example 3 (continued): From (14) it follows that the Hessian of $Q|S^n$ at the critical point x is

$$\big(D^2Q(x) + \lambda D^2S(x)\big)\Big| \ker DS(x) = 2(A - \lambda)|\,\langle x\rangle^{\perp},$$

where $\langle x\rangle^{\perp} = \{y \in \mathbf{R}^{n+1} | \,(y, x) = 0\}$. Since $\ker(A - \lambda)|\,\langle x\rangle^{\perp}$ has dimension $\dim V_{\lambda} - 1$, x is a nondegenerate critical point of $Q|S^n$ if and only if λ is a simple real eigenvalue of A. Suppose that all the eigenvalues λ_i for $i = 1, \ldots, n+1$ of A are real and simple and that $\lambda_{n+1} > \lambda_n > \cdots > \lambda_1$. Let x_i be unit eigenvectors corresponding to the eigenvalues λ_i. The Hessian of $Q|S^n$ at $\pm x_i$ with respect to the basis $\{e_1, e_2, \ldots \widehat{e_i}, \ldots, e_{n+1}\}$ is the $n \times n$ matrix $\mathrm{diag}\,(\lambda_{n+1} - \lambda_i, \ldots, \lambda_{i+1} - \lambda_i,\ \lambda_{i-1} - \lambda_i,\ \ldots, \lambda_1 - \lambda_i)$, which is invertible and has $i - 1$ negative eigenvalues. In other words, $\pm x_i$ is a nondegenerate critical point of $Q|S^n$ of Morse index $i - 1$. Hence $Q|S^n$ is a Morse function with $2(n+1)$ critical points. $\square$

Suppose that $f : M \subseteq \mathbf{R}^m \to \mathbf{R}$ is a smooth function with a critical set P which is a manifold. We say that P is a *nondegenerate critical manifold* of f if

1. $T_pP \subseteq \ker D^2f(p)$ for every $p \in P$.

2. $D^2f(p)|\, N_pP$ is nondegenerate for every $p \in P$, where N_pP is any subspace of T_pM such that $T_pM = T_pP \oplus N_pP$.

In other words, the tangent space to P at the point p is contained in the kernel of the Hessian of f at p for every $p \in P$ and the Hessian of f restricted to a normal space at $p \in P$ in M is nondegenerate. A smooth function whose critical set is made up of nondegenerate critical submanifolds is called a *Bott-Morse function*.

Example 3 (continued). Suppose that λ_i is an eigenvalue of A of multiplicity $m+1 < n+1$. List the eigenvalues of A in a nonincreasing order with a mulitiple eigenvalue being listed as many times as its multiplicity. Then the m-sphere

$$\left\{x \in \mathbf{R}^{n+1} \middle| (A - \lambda_i)x = 0 \quad \text{and} \quad (x, x) = 1\right\}$$

is a critical manifold of $Q|S^n$. For $x \in S^m$ let $N_x S^m = V_{\lambda_i}{}^{\perp}$. Then the Hessian of $Q|N_x S^m$ is the $(n-m) \times (n-m)$ matrix diag $(\lambda_{n+1} - \lambda_i, \ldots, \lambda_{i+1} - \lambda_i, \lambda_{i-1} - \lambda_i, \ldots \lambda_1 - \lambda_i)$, which is invertible and has Morse index $n - m - j$ where j is the first occurrence of the eigenvalue λ_i. Therefore S^m is a nondegenerate critical submanifold of Q. Hence Q is a Bott-Morse function on S^n.

3 The Morse isotopy lemma

In this section we prove the Morse isotopy lemma, which gives a useful criterion when two regular level sets of a smooth function are diffeomorphic.

Claim: Let $f : M \rightarrow \mathbf{R}$ be a proper smooth function. If $[a, b]$ is a nonempty closed interval contained in the set of regular values in the image of f, then $f^{-1}(a)$ is smoothly *isotopic* to $f^{-1}(b)$, that is, there is a smooth one parameter family of diffeomorphisms $\psi : [0, 1] \times M \rightarrow M$ such that for every $t \in [0, 1]$, $\psi_t : f^{-1}(a) \rightarrow f^{-1}(a + t(b-a))$.

(3.1) **Proof**: The idea of the proof is to push the level set $f^{-1}(a)$ to the level set $f^{-1}(b)$ along the integral curves of the gradient vector field associated to f.

First we define the gradient vector field associated to f. Using a partition of unity, we can define a smooth Riemannian metric $\langle\, , \,\rangle$ on the manifold M. Since the metric is nondegenerate, the mapping $\langle\, , \,\rangle^{\sharp} : T_m M \rightarrow T_m^* M : v_m \rightarrow \langle v_m, \cdot \rangle$ is an isomorphism for every $m \in M$. Making use $\langle\, , \,\rangle^{\sharp}$ we convert the 1-form $df : M \rightarrow T^*M$ to a vector field X^f on M, called the *gradient vector field* associated to f. More explicitly, X^f is defined by

$$df(m)v_m = \langle X^f(m), v_m \rangle$$

for every $m \in M$ and $v_m \in T_m M$. Suppose that a is a regular value of f. Then at every $m \in f^{-1}(a)$, we have $X^f(m) \in T_m\big(f^{-1}(a)\big)^{\perp}$. This follows because a is a regular value of f implies that $T_m f^{-1}(a) = \ker df(m)$. Hence

$$0 = df(m)v_m = \langle X^f(m), v_m \rangle$$

for every $v_m \in T_m f^{-1}(a)$.
Because f is a proper mapping, $f^{-1}([a, b]) = K$ is a compact subset of M. Since $[a, b]$ contains no critical values of f, the gradient vector field X^f does not vanish on K. Let

$$\widetilde{X}(m) = \frac{1}{\langle X^f(m), X^f(m) \rangle} X^f(m).$$

Let U be an open set containing K whose closure is compact. Using a partition of unity, there is a smooth function $\rho : M \rightarrow \mathbf{R}$, which is 1 on K and 0 outside U. Consider the vector field $X = \rho \widetilde{X}$. Since X is a smooth vector field on M which vanishes outside a compact subset, it is *complete*, that is, its flow $\varphi : \mathbf{R} \times M \to M$ is a one parameter group of diffeomorphisms of M. Now

$$\begin{aligned} \frac{d}{dt} f(\varphi_t(m)) &= df(\varphi_t(m))X(\varphi_t(m)) = df(\varphi_t(m))\rho(\varphi_t(m))\widetilde{X}(\varphi_t(m)) \\ &= \rho(\varphi_t(m)). \end{aligned} \tag{18}$$

Suppose that $\varphi_t(m) \in K$, that is, $f(\varphi_t(m)) \in [a, b]$. From the definition of ρ it follows that

$$\frac{d}{dt} f(\varphi_t(m)) = 1. \tag{19}$$

Integrating (19) gives $t \to f(\varphi_t(m)) = t + f(m)$. Therefore $\varphi_{b-a} : M \to M$ is a diffeomorphism which maps $f^{-1}(a)$ onto $f^{-1}(b)$. Moreover

$$\psi : [0, 1] \times M \to M : (t, m) \to \varphi_{a+t(b-a)}(m)$$

is a smooth isotopy of $f^{-1}(a)$ onto $f^{-1}(b)$. □

We now apply the Morse lemma and the Morse isotopy lemma to prove a characterization of the sphere.

Claim: Let M be a compact smooth n-dimensional manifold without a boundary. Suppose that f is a Morse function on M with two nondegenerate critical points. Then M is homeomorphic to the n-sphere S^n.

(3.2) **Proof**: Let $p_\pm$ be the critical points of f. We may assume that f has a maximum value c_+ at the point p_+ and a minimum value c_- at the point p_-. By the Morse lemma, there are coordinates $(x_1, \ldots, x_n)$ in an open neighborhood U_+ of p_+ with 0 corresponding to p_+ such that

$$f|U_+(x_1, \ldots, x_n) = c_+ - x_1^2 - \cdots - x_n^2.$$

Thus there is a $b < c_+$ such that $f^{-1}([b, c_+]) = \overline{D}_+$ is a neighborhood of p_+ which is diffeomorphic to a closed n-disc. Similarly, there is an $a > c_-$ such that $f^{-1}([c_-, a]) = \overline{D_-}$ is diffeomorphic to a closed n-disc containing p_-. We may assume that $a < b$. Note that $\partial \overline{D}_+$ and $\partial \overline{D}_-$ are diffeomorphic to an $(n-1)$-sphere S^{n-1}. Because the function f restricted to $f^{-1}([a, b])$ is proper and has no critical values, it follows from the Morse isotopy lemma that $f^{-1}([a, b])$ is diffeomorphic to $S^{n-1} \times [a, b]$.

Let $S^n \subseteq \mathbf{R}^{n+1}$ be the standard unit n-sphere in Euclidean $n+1$-space with $q_\pm$ its north and south poles respectively. Let $V_\pm$ be polar cap open neighborhoods of $q_\pm$ which are diffeomorphic to open n-discs $B_\pm$, respectively. Set $C = S^n - (B_+ \cup B_-)$. Then C is diffeomorphic to $S^{n-1} \times [0, 1]$. Note that $\partial C = \partial \overline{B}_+ \cup \partial \overline{B}_-$. Let $h_0 : \overline{D}_+ \to \overline{B}_+$ be a diffeomorphism of the closed n-disc. Extend the diffeomorphism $h_0|\partial \overline{D}_+ : \partial \overline{D}_+ \to \partial \overline{B}_+$ of $(n-1)$-spheres to a diffeomorphism $\widetilde{h}_0 : \partial \overline{D}_+ \times [a, b] \to \partial \overline{B}_+ \times [0, 1]$ of cylinders on S^{n-1}. This gives rise to the diffeomorphism $\widehat{h}_0 : f^{-1}([a, b]) \to C$, which extends to a diffeomorphism

$$h_1 : D_+ \cup f^{-1}([a, b]) \to B_+ \cup C.$$

Identify $f^{-1}(a)$ with S^{n-1}. Hence we are left with extending the diffeomorphism

$$\widetilde{h}_2 = h_1 | f^{-1}(a) : f^{-1}(a) \rightarrow \partial \overline{B}_-$$

of S^{n-1} into itself to a homeomorphism h_2 of the closed n-disc $\overline{D}^n$ to $\overline{B}_-$ where $\partial \overline{D}^n = S^{n-1}$. This is done by defining

$$h_2(x) = \begin{cases} |x| \, \widetilde{h}_2(x/|x|), \text{ if } x \neq 0 \\ 0, \text{ if } x = 0. \end{cases}$$

The function h_2 maps the radial segment $[0, y]$ where $y \in S^{n-1}$ linearly onto the line segment $[0, \widetilde{h}_2(y)]$ in $\overline{D}^n$. The map h_2 extends the diffeomorphism h_1 to a *homeomorphism* between M and S^{n-1}. □

The example of an exotic 7-sphere shows that it is *not* always possible to extend a diffeomorphism of a 7-sphere to a *diffeomorphism* of the 8-disc.

4 Exercise

1. (Equivariant Morse lemma). Let G be a compact connected Lie group which acts linearly on $\mathbf{R}^n$. Suppose that $f : \mathbf{R}^n \rightarrow \mathbf{R}$ is a G-invariant smooth function with $f(0) = 0$ and $Df(0) = 0$. If $D^2 f(0)$ is nondegenerate, show that there is a local diffeomorphism φ such that $\varphi(0) = 0$; φ commutes with the action of G, and $\varphi^* f(x) = \frac{1}{2} D^2 f(0)(x, x)$.

2. (a). Find a Morse function f on real projective 2-space $\mathbf{RP}^2$ with three critical points, one of index 0, 1 and 2.

   (b). How many orbits of the gradient vector field of f join critical points of index 0 or 2 with critical points of index 1? Take a sphere in the local unstable manifold of the critical point of index 1. Use the flow of the gradient vector field of f to transport this sphere until it intersects a sphere in the local stable manifold of the critical point of index 2. What is its intersection number? Answer a similar question for a sphere in the local stable manifold of the critical point of index 1.

   (c)*. Using ideas from b) find a decomposition of $\mathbf{RP}^2$ into cells together with the degrees of their attaching maps. From this compute the cohomology ring of $\mathbf{RP}^2$ with integer coefficients.

Notes

Foreword

p.v par.1 *Three body problem.* See for instance Xia [224] or [225].
p.xi par.4 *three critical points.* For a construction of a function on a two dimensional torus with three critical points see Banchoff and Takens [18].

Harmonic oscillator

p.4 par.0 *energy level set.* This program for analyzing the energy momentum mapping was started by Smale [189] who suggested that studying the regular values was enough. The importance and physical significance of the critical values was missed. Indeed by studying popular texts such as Goldstein [78] one realizes that many physically interesting motions occur precisely at the critical values.
p.9 par.4 *quadratic integrals.* Our treatment of the U(2) momentum mapping follows Cushman [45] and Cushman and Rod [53]. In [110] Jauch and Hill observed that in quantum mechanics the degeneracy of the spectrum of the harmonic oscillator Hamiltonian is due to the U(2) symmetry.
p.14 par.3 *Hopf mapping.* Dulock and McIntosh [67] observed that there was a mapping from integral curves of the harmonic oscillator of a fixed energy onto a 2-sphere, each point of which was on a unique such integral curve. They also noted that this orbit map was the Hopf fibration.
p.15 par.1 *Hopf fibration.* The first paper on the Hopf fibration is that of Hopf [101].
p.17 par.1 *linking number.* There are many equivalent definitions of linking number. For one, which is different from the one used in the text, see Milnor[153] and exercises 3, 5 and 6.
p.21 par.2 *classifying map.* For the definition of the classifying map of a bundle with base space a sphere, see Steenrod [201].
p.24 par.2 *invariant theory.* Arnol'd [11] seems to be the first one to have seen that a Hamiltonian invariant under the flow of the harmonic oscillator induces dynamics on the reduced space which are given by a generalized Euler equation. See Cushman and Rod [53] for a detailed treatment of reduction for such Hamiltonian functions. This result is basic to studying *perturbations* of the harmonic oscillator. Some papers on the use of normal form and reduction to study the Hénon-Heiles Hamiltonian are Hénon-Heiles [97], Gustavson [91], Moser [158], Kummer [120], Cushman [46], Churchill, Kummer and Rod [42], and Cotter [44].
p.25 par.3 *claim.* The proof that there is only one relation follows that of Billera, Cushman and Sanders [31].
p.28 par.3 *prime ideal.* For definitions of the terms from commutative algebra used in this and the next paragraph see Hungerford [104].
p.28 par.3 *Hilbert Nullstellensatz.* For a proof of the Nullstellensatz see Mumford [161].
p.28 par.3 *Schwarz.* For a proof of Schwarz' theorem on smooth invariants see Schwarz [186] or Mather [149]. In Bierstone [30] one can find a transparent proof.
p.32 par.3 *linking number.* This exercise is taken from Bott and Tu [36, p. 230] .
p.33 par.0 *Hopf invariant.* This exercise is taken from Bott and Tu [36, p. 228] .

p.34 par.1 *Hopf invariant of Hopf map.* This exercise is a modified version of one in Bott and Tu [36, p. 238] .
p.36 par.1 *Fubini-Study.* See Arnol'd [13, p. 343] for a derivation of the Fubini-Study Hermitian metric on $\mathbf{CP}^{n-1}$.

Geodesics on S^3

p.38 par.1 *constrained.* Our discussion of constrained motion follows Moser [160] and Dirac [63].
p.47 par.2 *dual pairs.* The term dual pairs was introduced by Howe [103]. See Cushman and Sanders [55] for an example of the use of dual pairs in studying perturbations of the geodesic flow on the 3-sphere. The subject of *perturbations* of the geodesic flow on the 3-sphere has a vast literature being essentially the main topic of *celestial mechanics* (see section 4 of the text for an explanation of the relation between the Kepler problem and the geodesic flow). Some classic references are Delaunay [62], Hill [99], Poincaré [173], Levi-Civita [133] and Birkhoff [32]. For a survey of normalization techniques in the study of perturbations of the geodesic flow see Cushman [49].
p.51 par.1 *smooth function.* This is the smooth analogue of the Hilbert Nullstellensatz of algebraic geometry.
p.51 par.1 *partition of unity.* See Sternberg [203] pp.55–57, for the construction of a partition of unity.
p.53 par.3 *Kepler's problem.* For Kepler's treatment see [114]. A good historical study of Kepler's astronomical work may be found in Stevenson [204]. For Newton's treatment of the Kepler problem see his *Principia* [163] Book 1, paragraph 3, propositions XI, XII and XIII.
p.55 par.2 *eccentricity vector.* The term eccentricity vector is due to Duistermaat. It is also called the Laplace-Runge-Lenz vector [127], but actually was discovered by J. Hermann [98] and J. Bernoulli [29]. See Goldstein [79] for a historical discussion. The eccentricity vector was first written down as a vector using quaternions by Hamilton [96] in 1845.
p.62 par.2 *where on the orbit.* Our discussion of where the particle is at a given time on its elliptical orbit follows Pollard [175]. For another interesting treatment see Souriau [199] who uses an integrated version of the equation defining the eccentric anomaly. This reduces the integration of the Kepler problem to a four dimensional system of linear differential equations, see exercise 8.
p.65 par.1 *regularization.* The process of removing the incompleteness of the flow of the Kepler vector field has a long history. It begins with Levi-Civita [134] in the context of removing binary collisions in the restricted three body problem and is followed by the *spinor* regularization of Kustaanheimo and Stiefel [123]. More recently Moser [159] used the tangent of the stereographic projection map to regularize the bounded Keplerian motion energy level by energy level. This works because of a theorem of Hamilton [96], see also Milnor [154]. Moser's method is essentially the same as the one used in quantum mechanics by Fock [75], Bargmann [20] and Pauli [171], which explained the SO(4) symmetry in the spectrum of the *hydrogen atom.* The regularization of the Keplerian motion for all negative energies at once was first done by Souriau [198] and independently by Ligon and Schaaf [139]. The text follows Cushman and Duistermaat [50]. Other references relating the Kepler problem to the geodesic flow are Belbruno [28] and Osipov [167].

p.65 par.1 *Ligon–Schaaf map.* The Ligon–Schaaf map is due to Ligon and Schaaf [139], who provided no motivation for how they found it. The fact that their map has a simple characterization is due to Cushman and Duistermaat [50].
p.74 par.2 *geometrically.* An early treatment of the topology of the Kelper problem may be found in Kaplan [113]. Important references for the global geometry of the Kepler problem are Souriau [199], [198], and [197] and Ligon and Schaaf [139].
p.74 par.2 *symmetry.* For a discussion of the SO(4, 1) and SO(4, 2) symmetries in the Kepler problem see Guillemin and Sternberg [86], [88], [89], Bacry [17], Györgyi [92], Iosifescu and Scutaru [107], Prince and Eliezer [177] and a review article by Bander and Itzykson [19] .
p.75 par.4 *third law.* A comprehensive discussion of Kepler's third law may be found in Stephenson [202].
p.76 par.2 *Bessel.* Our treatment of Kepler's equation is taken from Colwell [43] who has references to the historical literature.
p.78 par.1 *map to.* We note that analogous formulæ (up to $\pm$ signs) hold for the positive energy case as well as the case of a repulsive potential. When you get stuck some good references to consult are Gyorgyi [93] and Souriau [199].
p.80 par.1 *Kustaanheimo-Stiefel.* Our approach to Kustaanheimo-Stiefel regularization of the Kepler problem follows the treatment of Kummer [122]. See also Kustaanheimo and Stiefel [123] and Baumgarte [27].

Euler top

p.83 par.1 *Euler top.* Euler [74] was the first to treat the motion of a force free rigid body with one point fixed. We call this rigid body problem the Euler top. More modern treatments can be found in Whittaker [218, p.144–155] and Pars [170, p.216–224].
p.83 par.1 *geodesics.* The model of the Euler top as a geodesic flow of a left invariant metric on the rotation group is due to Arnol'd [11]. This chapter can be considered as an attempt to make his ideas explicit and provide more detail than Iacob [106]. For generalizations of the Euler top to Lie groups other than the rotation group see Mischenko and Fomenko [155].
p.87 par.3 *derivative of the exponential function.* The proof of the formula for the derivative of the exponential map is taken from Freudenthal and de Vries [76]. See also Tuynman [208].
p.89 par.3 *solid ball.* We know of no good reference for the solid ball model of the rotation group, even though every topologist knows it.
p.91 par.3 *sphere bundle model.* The structure of the sphere bundle model of the rotation group as a Lie group is not the standard one of a closed subgroup of Gl(3).
p.95 par.2 *double covered.* The fact that the bundle projection from the unit tangent sphere bundle of the 2-sphere to the 2-sphere is double covered by the Hopf fibration is vital for understanding the qualitative properties of the energy momentum mapping of the Euler top.
p.102 par.2 *principal axes.* The Lie theoretic meaning of the principal axis transformation seems to be new.
p.103 par.2 *sphere bundle.* The derivation of the Euler-Arnol'd equations in the sphere bundle model seems to be new.

p.121 par.1 *level sets of H*. The picture in figure 5.2 of the foliation of a level set of angular momentum by energy level sets is new.
p.123 par.2 *integrate.* The integration of the Euler-Arnol'd equations, which describe the motion of the Euler top in space, is usually done using Euler angles (see Whittaker [218] or Pars [170]. See also exercise 4. A discussion of the various conventions used for matrices, angular velocities, and Euler angles can be found in Synge [205]). When properly restricted, Euler angles form a chart for the rotation group. At least three such charts are needed to form an atlas for the rotation group. The trouble with the usual approach toward integrating the Euler-Arnol'd equations is that it is not obvious which motions of the top are left out when one uses only one chart. This difficulty is avoided when one uses the sphere bundle model.
p.133 par.2 *second.* This geometric interpretation of the rotation number is basically that given by Arnol'd [11] and Iacob [106]. They both overlooked the fact that two isotropy orbits project onto a single boundary component of the annulus. Thus the rotation number is the angle between every *second* point of contact of the projected integral curve with a fixed boundary component of the annulus and not every point of contact. The projected integral curve is called the herpolhode, see exercise 7.
p.134 par.2 *half twist.* This twisting phenomenon is well known to physicists, although its first published explanation seems to be Ashbaugh, Chicone, and Cushman [16]. Our exposition follows theirs quite closely, except we do not use Euler angles.
p.137 par.2 *Jacobi elliptic functions.* This exercise is taken from Tricomi [207].
p.139 par.1 *Euler's equations.* This exercise is taken from Lawden [128].
p.143 par.4 *herpolhode.* See Whittaker [218, p.154–55].
p.144 par.3 *phenomenon.* This exercise is taken from Ashbaugh, Chicone, and Cushman [16]. For some nice pictures see Cushman and Hoveijn [51].

Spherical pendulum

p.147 par.2 *spherical pendulum.* According to Whittaker [218, p.104], the spherical pendulum was first treated by Lagrange [124] and was first integrated using elliptic functions by Jacobi [109] in 1839. This seems to overlook the fact that Huygens [105] understood the relative equilibria (see the discussion in Yoder [226]) many years before 1673 when he published his results about them. A modern analytic treatment of the spherical pendulum can be found in Whittaker [218, p.104–105]. Nice pictures of the orbits may be found in Webster.
p.150 par.2 *singular reduction.* Arms, Cushman and Gotay [7] were the first to apply the technique of singular reduction to the spherical pendulum.
p.151 par.1 *Claim.* This claim is a special case of the general fact [172, p.20–21.] that the orbit space of a linear action of a compact Lie group on some Euclidean space is homeomorphic to the image of its Hilbert map. See exercises 4 and 5.
p.154 par.2 *orbit space.* The model of the zero level set of angular momentum as an orbit space of a two element group on TS^1 may be found in Lerman, Montgomery, and Sjamaar [132].
p.156 par.2 *energy momentum mapping.* The geometry of the energy momentum map follows the treatment in Duistermaat [66] and Cushman [47]. The *relative equilibria* were found by Huygens [105] and a description of the closed orbits on regular tori by Emch

[72].
p.171 par.2 *Hartog's theorem.* For a proof of Hartog's theorem see Gunning [90, p.15].
p.172 par.3 *Cauchy's theorem.* For a proof of Cauchy's theorem see Ahlfors [3].
p.172 par.3 *residue theorem.* For a proof of the residue theorem see Ahlfors [3].
p.175 par.0 *monodromy map.* We give two proofs for the existence of monodromy for the spherical pendulum: an analytic one due to Duistermaat [66] and a geometric one due to Cushman [47]. For another one using *Picard Lefschetz* techniques see Zou [227]. Monodromy is a topological phenomenon which was classically unknown. Presumably this is a consequence of the linguistic curiosity that notions of global geometry require the language of modern geometry. Monodromy can *not* be observed by examining a single trajectory since it is a property of the way in which trajectories fit together. It can be observed in a series of experiments where the angular momentum and energy vary along a loop in the energy angular momentum plane which encloses the isolated critical value $(h, j) = (1, 0)$. One sees the monodromy as the increase in the angle $\vartheta(h, j)$ by 2π.
p.180 par.3 *rotation number.* The inequality $\frac{1}{2} < \theta(h, \ell)/2\pi < 1$ for the rotation number in the spherical pendulum was first proved by Puiseux in [178]. Our proof of $\theta(h, \ell) < 2\pi$ follows that of M.-A. de Saint-Germain [60], which in turn is based on an idea of Hadamard [94]. Our proof of $\theta(h, \ell) > \pi$ follows that of Alex Weinstein [217].
p.182 par.1 *coordinates.* This *theorem* is due to Horozov [102] and was unknown classically. Our proof follows his with some modifications.
p.184 par.2 *properties.* For an idea how to prove i) see Cushman and Sanders [54].

Lagrange top

p.187 par.2 *Lagrange top.* The Lagrange top was studied by Lagrange [124, p.261]. Since then studies of the top have been a legion. Some standard works are Golubev [80], Leimanis [129], Routh [183], and Klein and Sommerfeld [116]. Whittaker [218, p.155–63] gives a very analytic treatment. Goldstein [78] has a nice (but incomplete) qualitative treatment. Pars [170, p.113–117 and 152–57] has a thorough discussion which is a mixture of analysis and qualitative argument.
p.197 par.2 *magnetic spherical pendulum.* The fact that reducing the right S^1 action for the Lagrange top leads to the magnetic spherical pendulum was first shown by Novikov [165, p.30–31]. The identification of the symplectic form of the reduced Lagrange top as the standard symplectic form on TS^2 plus a magnetic term is a special case of the cotangent reduction theorem of Kummer [121]. This theorem does *not* provide an explicit expression for the equivalence mapping.
p.201 par.1 *Hamiltonian system.* Our treatment of the reduction of the Lagrange top to a system with one degree of freedom follows the treatment of reduction of the S^1 symmetry of the magnetic spherical pendulum given by Bates and Cushman [23].
p.205 par.1 *tangent cone* For the definition of the tangent cone to an algebraic variety see Brieskorn and Knörrer [38].
p.209 par.3 *smooth function.* Again we have used the theorem of Schwarz [186].
p.223 par.3 *motion of the axis.* The proof that the classification of the possible motions of the axis is complete does not seem to be available in the classical literature on tops and gyroscopes. Many authors such as Goldstein [78] draw pictures sans supporting arguments. Goldstein's pictures do not give all the possibilities as Pars [170, p.152–57]

shows in figures 9b)-f) on page 157. See also Webster [210]. Deimel [61, p. 84] cites Hadamard [94] for the proof of the nonexistence of a downward loop. Deimel's physical argument for the nonexistence of downward cusps is too soft for us.
p.226 par.1 *blow up map.* For more details on the geometry of the blow up map see Brieskorn and Knörrer [38, p. 486–89].
p.231 par.1 *rotation number.* Our proof that the rotation number is positive follows that of Hadamard [94].
p.236 par.2 *reconstructing the topology.* The reconstruction of the topology of the singular energy momentum level sets is new. It completes the analysis of Iacob [106] and Tatarinov [206].
p.245 par.2 *discriminant of cubic and quartic.* For the proofs of these facts about discriminants see Brieskorn and Knörrer [38, p.186–190].
p.246 par.2 *swallowtail surface.* For more information about the swallowtail surface see Poston and Stewart [176]. Our discriminant locus consists of the real points of the complex discriminant and *not* the locus of multiple real roots of the real quartic. This means that our swallowtail surface has a whisker of multiple purely imaginary roots whereas the locus of multiple real roots does not.
p.253 par.2 *monodromy.* The fact that the energy momentum map of the Lagrange top has monodromy was first proved in Cushman and Knörrer [52].
p.255 par.3 *gyroscopically stabilized.* See Klein and Sommerfeld [116] for a discussion of gyroscopic stabilization.
p.256 par.2 *Linearizing.* A complete linear analysis of gyroscopic stability can not be found in such classical sources as Routh [183] or Webster [210], because the necessary symplectic linear algebra was not known until 1937, see Williamson [219] .
p.258 par.1 *normal form.* For more details on normal forms for real infinitesimally symplectic linear maps see Burgoyne and Cushman [39] and [40].
p.258 par.1 *smooth normal form.* Arnol'd in [12] introduced the concept of smooth (versal) normal form for a smooth family of matrices.
p.261 par.2 *Hamiltonian Hopf bifurcation.* Guillemin [85] observed that there was a Hamiltonian Hopf bifurcation in the Lagrange top. His unpublished notes inspired the treatment in Cushman and van der Meer [57]. For a full treatment of the Hamiltonian Hopf bifurcation see van der Meer [209].
p.263 par.1 *Poincaré lemma.* Our proof of the Poincaré lemma follows Duistermaat [64].
p.268 par.2 *Lift.* This exercise is due to E. Lerman [130].
p.269 par.1 *goal.* This exercise is due to E. Lerman [131].

Fundamental concepts

p.275 par.3 *Witt.* In [221] Witt proved the Witt decomposition for symmetric bilinear forms. Artin [15, p. 120-1] noted that Witt's proof also works for symplectic forms.
p.275 par.3 *symplectic manifold.* It is a vexing historical question trying to answer who was the first to recognize explicitly that symplectic geometry was the basic underlying mathematical structure in Hamiltonian mechanics. For linear Hamiltonian systems the credit seems due to Wintner [220]. For nonlinear Hamiltonian systems the question is not so easily answered. In [41] Cartan gives a modern treatment of Hamilton's equations from the variational point of view. Even though he used exterior differential calculus,

there is no mention of the symplectic structure of phase space. Later symplectic structures were studied by Ehresmann and his school: Liebermann [70], Gallisot [77], Lichnerowicz [136], Souriau [193] and Reeb [180], [181], [182]. The term "symplectic geometry" seems to have been coined by Souriau [194] . All this activity did not bring symplectic geometry and its relation to mechanics to the attention of the working mathematician. Mackey [143] makes it clear that the symplectic formulation of mechanics was "understood" by 1963. This formulation was used extensively by Sternberg in his 1964 book [203]. In a widely circulated but unpublished letter in 1965 [169], Palais explained the symplectic formulation of Hamiltonian mechanics. Shortly after this the classic books by Abraham and Marsden [1], Arnol'd [13] and Souriau [197] appeared popularizing the symplectic geometric treatment of mechanics.

p.278 par.1 *symplectic form.* The symplectic form on the left trivialization of the cotangent bundle of a Lie group is perhaps the basic formula of Hamiltonian mechanics.

p.279 par.1 *symplectic form.* Lagrange [124] was the first to use the symplectic form in mechanics. In his case, the form was pulled back to the tangent bundle. He also proved its invariance under the flow of the Hamiltonian vector field. This was pointed out to us by J. Duistermaat. The symplectic form on a coadjoint orbit is often called the Kostant-Kirillov-Souriau form after Kostant [119], Kirillov [115] and Souriau [197]. However, it was discovered by Lie [138].

p.280 par.3 *Hamiltonian vector field.* Lagrange [125] was the first one to write down Hamilton's equations for Hamiltonians of the form kinetic plus potential. He also wrote the equations of motion for perturbations of the Kepler problem in Hamiltonian form. However, he did not write down Hamilton's equations for a general Hamiltonian. For an interesting discussion of this see Weinstein [213]. Poisson [174, p.343] wrote down the action corresponding to a Lagrangian and then obtained one half of Hamilton's equations. Hamilton [95] wrote the equations of motion in Hamiltonian form.

p.281 par.1 *geodesic vector field.* Our derivation of the equations satisfied by the integral curves of the geodesic vector field follows Lang [126].

p.283 par.2 *Christoffel symbol* For more details on the Christoffel symbols see Spivak [200].

p.285 par.3 *Poisson manifold.* The term "Poisson manifold" is due to Lichnerowicz [137]. Weinstein proved most of their basic properties in [214] and [216]. The concept of a Poisson manifold is due to Lie [138, section 62 volume 2]. For a discussion of Lie's contributions to symplectic geometry see Weinstein [215] and Bloch and Ratiu [33].

p.289 par.2 *Poisson bracket.* Poisson brackets were first written down by Poisson in [174, p.266]. Again it was Lie [138] who discovered that they gave a Lie algebra structure on the space of smooth functions.

p.290 par.1 *because.* Our proof follows that of Duistermaat [64].

p.290 par.1 *Jacobi identity.* The fact that the closedness of the 2-form ω is equivalent to the Jacobi identity was observed by Jost [111].

p.293 par.5 *continue the argument.* The purely local character of this result was established by Gromov [84].

p.293 par.5 *Darboux's theorem* This result is called Darboux's theorem after Darboux [58], even though it was known to Liouville. For a discussion of this see Lützen [142]. Our proof follows that of Arnol'd [13] and Weinstein [214].

p.297 par.3 *identity.* Actually the requirement that c is a regular value of the constraint mapping $\mathscr{C}$ is not needed to show that the Dirac bracket satisfies the Jacobi identity. It is only necessary that the matrix C is invertible for every $n \in \mathscr{C}^{-1}(0)$. For details we refer to Dirac [63].

Systems with symmetry

p.300 par.2 *closed subgroup.* For a proof of the fact that a closed subgroup of a Lie group is itself a Lie group, see Adams [2].
p.301 par.2 *linearizability.* The local linearizablity of a proper action is called Bochner's lemma, see Bochner [34].
p.302 par.1 *slice.* In [168] Palais proved the slice theorem. Our definition of a slice is slightly different from Palais', see exercise 1. Our proof follows that of Duistermaat and Kolk [65].
p.305 par.6 *local finiteness.* Our proof of the local finiteness of the orbit type decomposition follows Palais [168].
p.308 par.2 *partition of unity.* The proof of the existence of a G-invariant partition of unity subordinate to a G-invariant open covering follows that given in Palais [168].
p.310 par.2 *free and proper.* The proof that the orbit space of a proper free action follows along the lines of Duistermaat and Kolk [65].
p.313 par.3 *principal G-bundle.* For another definition see Steenrod [201].
p.315 par.4 *orbifold.* Our definition of orbifold follows that given by Satake [184] except that we do not assume that the codimension of the fixed point set of the action of G is greater than or equal to two. This means that our orbifolds can have a topological boundary.
p.317 par.5 *locally free.* There seems to be no proof in the literature that the orbit space of a locally free action is an orbifold, although this result was certainly known to Satake [184].
p.320 par.3 *momentum map.* The notion of a momentum mapping goes back at least to Lie [138]. In volume 2 he shows that the momentum map is canonical (page 300), equivariant with respect to some linear action (page 329), and that its image is invariant under the coadjoint action, provided that the rank of the momentum mapping is constant (page 338). The momentum map was formally defined by Souriau in [196] who studied its equivariance properties in [195]. Souriau discussed the momentum mapping at length in [197] as did Kostant in [119, p.187]. For a historical discussion see Bloch and Ratiu [33]. The French term of Souriau for momentum map is "moment". Hence the alternative English term "moment map". The term "momentum mapping" seems to have been first used by Cushman [45].
p.320 par.3 *mild condition.* See Woodhouse [223], Abraham and Marsden [1], and Souriau [197] for a more exhaustive discussion of conditions which insure that a momentum mapping is coadjoint equivariant.
p.320 par.3 *integral.* The fact that the G-momentum mapping of a G-invariant Hamiltonian is invariant under the flow of the Hamiltonian vector field is often called Noether's theorem because of [164].
p.322 par.1 *coadjoint equivariance.* The coadjoint equivariance of the momentum mapping was known to Lie [138, volume 2, chapter 20] but its importance for understanding the topological structure of integral manifolds of mechanical systems was emphasized by Smale [189].

p.326 par.1 *normal form.* Our proof of the normal form of a momentum mapping follows that given by Arms, Gotay and Jennings [8], which in turn is based on that of Arms, Marsden, and Moncrief [9]. For other proofs see Marle [146] and Guillemin and Sternberg [87].
p.332 par.1 *shifting trick.* This standard construction seems to have first appeared in print in Guillemin and Sternberg [88]. However, it was certainly "known "to experts before that. See the discussion at the end of Arms [5].
p.333 par.2 *regular reduction.* The regular reduction theorem has a long history going back to at least Jacobi [108] when he eliminated the node in the three body problem. The modern global formulation is due to Meyer [150] and Marsden and Weinstein [148]. For a sampling of the many applications of regular reduction see Marsden and Ratiu [147].
p.337 par.3 *regular value.* For another proof that the isotropy group at μ acts locally freely on $J^{-1}(\mu)$ if μ is a regular value see Guillemin and Sternberg [88].
p.337 par.3 *orbifold reduction.* Certainly the orbifold reduction theorem was known to Weinstein [213]. However, we could not find a detailed proof of it in the literature. See also Baily and Emmerich and Römer [73].
p.346 par.5 *singular reduction.* The first paper on singular reduction was Arms, Marsden and Moncrief [9] who considered the singular reduced space as a *set* and studied its decomposition into symplectic pieces. Our proof of the decomposition into symplectic pieces follows theirs. Our argument contains simplifications and clarifications due to Śniatycki, Schwarz and Bates [191] and uses only the fact that the group action is proper and Hamiltonian. For another proof see Bates and Lerman [24]. Another treatment for a Hamiltonian action of a *compact* group can be found in Sjamaar and Lerman [188]. The construction of a *smoothness structure* for the singular reduced space and a Poisson bracket on its smooth functions was first given by Arms, Cushman, and Gotay [7]. Examples using the technique of singular reduction (but with no proofs) had been given *before* the appearance of [7] by Cushman [48] and van der Meer [209]. Recent articles with additional examples are Cushman [49], Cushman and Sjamaar [56] and Lerman, Montgomery, and Sjamaar [132]. A rather complete survey and comparison of early theories of singular reduction may be found in Arms, Gotay, and Jennings [8]. Of these early papers we note the following: Śniatycki [190], Śniatycki and Weinstein [192], Gotay [81], Gotay and Bos [83] Arms [6], Gotay [82], and Arms and Wilbour [10].
p.348 par.2 *nondegenerate.* Our proof of the nondegeneracy of the Poisson bracket follows Arms, Cushman and Gotay [7] and uses clarifications suggested by Lerman.
p.355 par.0 *flow.* The existence and uniqueness of the flow of a Hamiltonian derivation is taken from Sjamaar and Lerman [188].
p.356 par.3 *encoded.* The result that the stratification of the singular reduced space is determined by its Poisson structure is due to Sjamaar and Lerman [188].
p.361 par.2 *true.* See Egilsson [68] for a counterexample.

Ehresmann connections

p.363 par.3 *Ehresmann connection.* The basic notion of Ehresmann connection was defined by Ehresmann in [69]. We follow the exposition of Zou [227]. It is amazing that no detailed arguments establishing the basic properties of an Ehresmann connection are available in texts on differential geometry.

p.363 par.3 *local horizontal lift.* Our proof of the existence of a local horizontal lift follows that indicated by Wolf in [222].
p.366 par.3 *parallel translation.* Local parallel translation need not exist for every Ehresmann connection. Some authors finesse this by making it part of the definition.
p.371 par.4 *claim* The Ehresmann trivialization theorem is a special case of the basic property of bundles called covering homotopy theorem, which is due to Steenrod [201].
p.373 par.3 *holonomy.* This formula is due to Schlesinger [185] although it is often credited to Ambrose and Singer [4].

Action angle coordinates

p.377 par.4 *Liouville integrable.* The earliest version of this theorem appears in Jacobi [109, p. 252] in the sense that n independent commuting integrals provide a complete solution to the Hamilton-Jacobi equation and thus would reduce the integration of Hamilton's equations to quadratures. This is a purely local procedure. Early authors did not discuss tori, although ones sees quasiperiodic motions in the special cases they treated. It seems that the conclusion that in every integrable system where the integral map has compact fibers all near by solutions are quasiperiodic is modern and is due to Arnol'd and Avez [14]. Jacobi's result was rediscovered by Liouville [140]. For a more detailed discussion of this see Lützen [142]. The first modern formulation of action angle coordinates also appears to be due to Arnol'd and Avez [14]. Some hypotheses in the theorem, equivalent to the openness of the set of compact components, the functional independence of the actions and the exactness of the symplectic form were later removed by Jost [112] as well as Markus and Meyer [144]. The proof of Markus and Meyer was streamlined by Duistermaat [66] whose proof in turn was shortened by Bates and Śniatycki [25]. Other proofs of varying degrees of completeness appear in Abraham and Marsden [1], Guillemin and Sternberg [88] and Liebermann and Marle [135]. Generalizations to singular integrable systems may be found in Eliasson [71] and Knörrer [117] and [118]. Generalizations to partially integrable systems in Nehorošev [162].
p.378 par.2 *Proof.* Our proof of the existence of local action angle coordinates follows Bates and Śniatycki [25].
p.381 par.4 *de Rham's theorem.* For a proof of de Rham's theorem see de Rham [59].
p.383 par.2 *obstructions.* All the obstructions for finding global action angle coordinates are described by Duistermaat in [66]. An example of the *Chern class obstruction* is given in Bates [21].
p.383 par.2 *monodromy.* Our treatment of the *monodromy obstruction* follows the period lattice construction of Duistermaat [66]. A Čech approach has been given by Bates [22]. There are many interesting Liouville integrable systems which have been proved to have monodromy, namely, the spherical pendulum by Duistermaat [66] and Cushman [47]; the Lagrange top by Cushman and Knörrer [52]; the Hamiltonian Hopf bifurcation by van der Meer [209]; the champagne bottle potential by Bates [22]; the Kirchhoff top by Bates and Zou [26], see also Zou [227] and the quadratic spherical pendulum, see exercise 2.
p.386 par.4 *Monodromy.* See Bates [22] for more details about the motion in the *champagne bottle potential.*
p.386 par.4 *Monodromy.* For a detailed treatment of the *quadratic spherical pendulum* see Bates and Zou [26]. .

p.386 par.4 *isomorphic.* See Shastri and Zvengrowski [187, proposition 3.6, p.474] for a proof of this fact.
p.386 par.4 *isomorphic.* See proposition 3.5 of Shastri and Zvengrowski [187, p.474].
p.386 par.4 *Euler class.* For more background on the *Euler class* see Montesinos[157].
p.386 par.4 *Siefert manifold.* For more information on *Seifert manifolds* see [166].
p.387 par.2 *Burke's egg.* This example is due to W. Burke who discovered it in the spring of 1985.
p.387 par.2 *Kerr-Newman black hole.* For a discussion of the Kerr-Newman black hole see Misner, Thorne, and Wheeler [156, p.899–901].

Morse theory

p.390 par.4 *inverse function.* For a proof of the inverse function theorem see Loomis and Sternberg [141].
p.391 par.1 *Claim.* This claim is usually referred to as the *Lagrange multiplier theorem.*
p.392 par.2 *Morse lemma.* The proof we give here is attributed to Hörmander. For other proofs see Milnor [152] and Weinstein [211]
p.395 par.1 *nondegenerate.* The definition of a nondegenerate critical manifold was given by Bott [35].
p.396 par.3 *Morse isotopy lemma.* The proof of the Morse isotopy lemma follows that given by Milnor in [152, p.12].
p.397 par.2 *one parameter group.* For a proof of the existence of a flow for a vector field see Loomis and Sternberg [141].
p.397 par.2 *two nondegenerate critical points.* This result is attributed by Milnor [151] to Reeb [179]. Our proof follows Hirsch [100, p. 154–55].
p.398 par.4 *exotic 7-sphere.* J. Milnor [151] gave an example of a 7-sphere which was not diffeomorphic to the standard 7-sphere. Brieskorn [37] gives explicit polynomials which give rise to every exotic 7-sphere.

Bibliography

[1] R. Abraham and J. Marsden. *Foundations of Mechanics, 2nd edition*. Addison-Wesley, Reading, Mass., 1978.

[2] J. Adams. *Lectures on Lie groups*. Benjamin, New York, 1969.

[3] L.V. Ahlfors. *Complex Analysis*. 2ed McGraw-Hill Book Company, New York, 1966.

[4] W. Ambrose and I. Singer. A theorem on holonomy. *Transactions of the American Mathematical Society*, 75:428–443, 1953.

[5] J. Arms. Symmetry and solution set singularities in Hamiltonian field theories. *Acta Physica Polonica*, B17:499–523, 1986.

[6] J. Arms. Reduction of Hamiltonian systems for singular values of momentum. *Contemporary Mathematics*, 81:99–110, 1988.

[7] J. Arms, R. Cushman, and M. Gotay. A universal reduction procedure for Hamiltonian group actions. In T. Ratiu, editor, *The Geometry of Hamiltonian Systems*, pages 33–51. Springer Verlag, New York, 1991.

[8] J. Arms, M. Gotay, and G. Jennings. Geometric and algebraic reduction for singular momentum mappings. *Advances in Mathematics*, 74:43–103, 1990.

[9] J. Arms, J. Marsden, and V. Moncrief. Symmetry and bifurcation of momentum mappings. *Communications in Mathematical Physics*, 78:455–478, 1981.

[10] J. Arms and D. Wilbour. Reduction procedures for Poisson manifolds. In P. Donato et al., editors, *Symplectic Geometry and Mathematical Pysics*, pages 462–475. Birkhäuser, Boston, 1991.

[11] V. Arnol'd. Sur le géométrie différentielle des groupes de Lie de dimension infinie et ses applications a l'hydrodynamique des fluides parfaits. *Annales de l'Insitut Fourier, Grenoble*, 6:319–361, 1966.

[12] V. Arnol'd. On matrices depending on parameters. *Russian Mathematics Surveys*, 26:29–41, 1971.

[13] V. Arnol'd. *Mathematical Methods of Classical Mechanics*. Springer Verlag, New York, 1978.

[14] V. Arnol'd and A. Avez. *Ergodic Problems of Classical Mechanics*. Benjamin, New York, 1967.

[15] E. Artin. *Geometric Algebra*. Interscience, New York, 1963.

[16] M. Ashbaugh, C. Chicone, and R. Cushman. The twisting tennis racket. *Journal of Dynamics and Differential Equations*, 3:67–86, 1991.

[17] H. Bacry. The de Sitter group $l_{4,1}$ and bound states of the hydrogen atom. *Nuovo Cimento*, 41 A:222–34, 1966.

[18] T. Banchoff and F. Takens. Height functions on surfaces with three critical points. *Illinois Journal of Mathematics*, 19:325–335, 1972.

[19] M. Bander and C. Itzykson. Group theory and the hydrogen atom (I). *Reviews of Modern Physics*, 38:330–345, 1966.

[20] V. Bargmann. Zur Theorie des Wasserstoffatoms. Bemerkungen zur gleichnamigen Arbeit von V. Fock. *Zeitschrift für Physik*, 99:576–582, 1936.

[21] L. Bates. Examples for obstructions to action angle coordinates. *Proceedings of the Royal Society at Edinburgh*, 110A:27–30, 1989.

[22] L. Bates. Monodromy in the champagne bottle. *Journal of Applied Mathematics and Physics (ZAMP)*, 42:837–847, 1991.

[23] L. Bates and R. Cushman. The magnetic spherical pendulum. *Meccanica*, 30:271–289, 1995.

[24] L. Bates and E. Lerman. Proper actions and symplectic stratified spaces. *(to appear in Pacific Journal of Mathematics)*, 1997.

[25] L. Bates and J. Śniatycki. On action angle variables. *Archive for Rational Mechanics and Analysis*, 120:337–343, 1992.

[26] L. Bates and M. Zou. Degeneration of Hamiltonian monodromy cycles. *Nonlinearity*, 6:313–335, 1993.

[27] J. Baumgarte. Das Oszillator-Kepler Problem und die Lie Algebra. *Journal für die reine und angewandte Mathematik*, 301:50–76, 1978.

[28] E. Belbruno. Two body motion under the inverse square central force and equivalent geodesic flows. *Celestial Mechanics*, 15:467–476, 1977.

[29] J. Bernoulli. *Histoires de L'Académie Royale des Sciences avec Mémoires de Mathématique et Physique*, pages 523–544. Paris, 1712. reprint: Amsterdam, 1713, pages 635–703 = Opera Omnia, volume 1, pages 470–480.

[30] E. Bierstone. *The Structure of orbit spaces and the singularities of equivariant mappings*. monograph 35, Instituto de Matemática Pura e Aplicada, Rio de Janeiro, Brazil, 1980.

[31] L. Billera, R. Cushman, and J. Sanders. The Stanley decomposition of the harmonic oscillator. *Proceedings of the Koninklijke Nederlandse Akademie van Wetenschappen*, series A, 91(4):375–393, 1988.

[32] G.D. Birkhoff. Sur le probléme restreint des trois corps I. *Annali della Scuola Normale Superiora de Pisa*, 4:267–306, 1935. II, Annali della Scuola Normale Superiora de Pisa, volume 5, 1936, 1–72.

[33] A. Bloch and T. Ratiu. Convexity and integrability. In P. Donato et al., editors, *Symplectic Geometry and Mathematical Physics*, pages 48–79. Birkhäuser, Boston, 1991.

[34] S. Bochner. Compact groups of differentiable transformations. *Annals of Mathematics*, 46:372–381, 1945.

[35] R. Bott. Nondegenerate critical manifolds. *Annals of Mathematics*, 60:249–261, 1954.

[36] R. Bott and R. Tu. *Differential forms in Algebraic Topology*. Graduate Texts in Mathematics, vol 82, Springer Verlag, New York, 1982.

[37] E. Brieskorn. Beispiele zur Differentialtopologie von Singularitäten. *Inventiones Mathematicae*, 2:1–14, 1966.

[38] E. Brieskorn and H. Knörrer. *Plane Algebraic Curves*. Birkhäuser, Boston, 1986.

[39] N. Burgoyne and R. Cushman. Normal forms for real linear Hamiltonian systems with purely imaginary eigenvalues. *Celestial Mechanics*, 8:435–443, 1974.

[40] N. Burgoyne and R. Cushman. Normal forms for linear Hamiltonian systems. In C. Martin and R. Hermann, editors, *The 1976 Ames Research Center (NASA) conference on geometric control theory*, pages 483–529. Math Sci Press, Brookline, Mass., 1977.

[41] E. Cartan. *Leçons sur les invariants intégraux*. Hermann, Paris, 1922. 3rd printing, 1971.

[42] R. Churchill, M. Kummer, and D. Rod. On averaging, reduction and symmetry in Hamiltonian systems. *Journal of Differential Equations*, 49:359–414, 1983.

[43] P. Colwell. Bessel functions and Kepler's equation. *American Mathematical Monthly*, 99:44–48, 1992.

[44] C. Cotter. *The 1:1 semisimple resonance*. PhD thesis, University of California at Santa Cruz, 1986.

[45] R. Cushman. The momentum mapping of the harmonic oscillator. *Symposia Mathematica*, 14:323–342, 1974.

[46] R. Cushman. Geometry of the bifurcations of the Henon-Heiles family. *Proceedings of the Royal Society, London, series A*, 382:361–371, 1982.

[47] R. Cushman. Geometry of the energy momentum mapping of the spherical pendulum. *Centrum voor Wiskunde en Informatica Newsletter*, 1:4–18, 1983.

[48] R. Cushman. Normal form for Hamiltonian vector fields with periodic flow. In S. Sternberg, editor, *Differential geometric methods in mathematical physics*, pages 125–144. Reidel, Dordrecht, 1984.

[49] R. Cushman. A survey of normalization techniques applied to perturbed Keplerian systems. In K. Jones et al., editors, *Dynamics Reported, volume 1, new series*, pages 54–112. Springer Verlag, New York, 1991.

[50] R. Cushman and J.J. Duistermaat. A characterization of the Ligon-Schaaf regularization map. preprint, University of Utrecht, 1996.

[51] R. Cushman and I. Hoveijn. Visualizing special motions of the Euler top. In R.P. Agarwal, editor, *Dynamical Systems and Applications*, pages 153–167. World Scientific, 1995.

[52] R. Cushman and H. Knörrer. The momentum mapping of the Lagrange top. *Lecture Notes in Mathematics*, 1139:12–24, 1985.

[53] R. Cushman and D. Rod. Reduction of the semisimple 1:1 resonance. *Physica D*, 6:105–112, 1982.

[54] R. Cushman and J. Sanders. Abelian integrals and global Hopf bifurcations. *Lecture Notes in Mathematics*, 1125:87–98, 1985.

[55] R. Cushman and J. Sanders. Constrained normal form. *Celestial Mechanics*, 45:181–187, 1989.

[56] R. Cushman and R. Sjamaar. On singular reduction of Hamiltonian spaces. In P. Donato et al., editors, *Symplectic Geometry and Mathematical Physics*, pages 114–128. Birkhäuser, Boston, 1991.

[57] R. Cushman and J.-C. van der Meer. The Hamiltonian Hopf bifurcation in the Lagrange top. *Lecture Notes in Mathematics*, 1416:26–38, 1991.

[58] G. Darboux. Sur le problème de Pfaff. *Bulletin des Sciences Mathématique*, 6:14–36 and 48–68, 1882.

[59] G. de Rham. *Variétés Différentiables*. Hermann, Paris, 1955. English translation Springer Verlag, New York, 1984.

[60] M. A. de Saint-Germain. Note sur le pendule sphérique. *Bulletin des Sciences Mathématique*, 20:114–16, 1896. Bulletin des Sciences Mathématique, 22, 1898, pages 95–96.

[61] R. Deimel. *Mechanics of the gyroscope*. Dover, New York, 1950.

[62] C. Delaunay. Théorie du mouvement de la lune. Memoire 28, Académie des Sciences de France, Paris, 1860. Memoire 29, 1867.

[63] P. Dirac. Generalized Hamiltonian dynamics. *Canadian Journal of Mathematics*, 2:129–148, 1950.

[64] J. Duistermaat. *Fourier Integral Operators*. Courant Institute of Mathematical Sciences, New York, 1973.

[65] J. Duistermaat and J. Kolk. Chapter IV of an unpublished manuscript on Lie groups, University of Utrecht. 1990.

[66] J.J. Duistermaat. On global action angle coordinates. *Communications on Pure and Applied Mathematics*, 33:687–706, 1980.

[67] V.A. Dulock and H.V. McIntosh. On the degeneracy of the two dimensional harmonic oscillator. *American Journal of Physics*, 33:109–118, 1965.

[68] S. Egilsson. On embedding the 1 : 1 : 2 resonance space in a Poisson manifold. *preprint, Department of Mathematics, University of California at Berkeley*, 1995.

[69] C. Ehresmann. Les connections infinitésimales dans un espace fibré differentiable. *Colloque de Topologie, CBRM*, pages 29–35, 1950.

[70] C. Ehresmann and P. Liebermann. Sur les formes differentielles exterieurs de degré 2. *Comptes Rendus Académie des Sciences*, 227:420–21, 1948.

[71] H. Eliasson. Normal form for Hamiltonian systems with Poisson commuting integrals – elliptic case. *Commentari Mathematica Helvitica*, 65:4–35, 1990.

[72] A. Emch. On closed curves described by a spherical pendulum. *American Mathematical Monthly*, pages 218–221, 1918.

[73] C. Emmerich and Römer. Orbifolds as configuration spaces of systems with gauge symmetries. *Communications in Mathematical Physics*, 129:69–94, 1990.

[74] L. Euler. Du mouvement de rotation des corps solides autour d'un axe variable. *Mémories de l'académie des sciences de Berlin*, 14:154–193, 1758.

[75] V. Fock. Zur Theorie des Wasserstoffatoms. *Zeitschrift für Physik*, 98:145–154, 1935.

[76] H. Freudenthal and H. de Vries. *Linear Lie Groups*. Academic Press, New York, 1969.

[77] F. Gallisot. Formes extérieures en méchanique. *Annales de l'Institut Fourier, Grenoble*, 4:145–297, 1952.

[78] H. Goldstein. *Classical Mechanics*. Addison-Wesley, Reading, 1950.

[79] H. Goldstein. More on the prehistory of the Laplace or Runge-Lenz vector. *American Journal of Physics*, 44:1123–24, 1976.

[80] V. V. Golubev. *Lectures on integration of the equations of motion of a rigid body about a fixed point*. Israel program for scientific translations, Jerusalem, 1960.

[81] M. Gotay. Poisson reduction and quantization for the $(n + 1)$-photon. *Journal of Mathematical Physics*, 25:2154–2159, 1984.

[82] M. Gotay. Reduction of homogeneous Yang-Mills fields. *Journal of Geometry and Physics*, 6:349–365, 1989.

[83] M. Gotay and L. Bos. Singular angular momentum mappings. *Journal of Differential Geometry*, 24:181–203, 1986.

[84] M. Gromov. Pseudo holomorphic curves in symplectic manifolds. *Inventiones Mathematicae*, 82:207–347, 1985.

[85] V. Guillemin. Unpublished manuscript. MIT, 1987.

[86] V. Guillemin and S. Sternberg. *Geometric Asymptotics*. American Mathematical Society, Providence, R.I., 1977.

[87] V. Guillemin and S. Sternberg. A normal form for the moment map. In S. Sternberg, editor, *Differential Geometric Methods in Physics*, pages 161–175. D. Reidel, Dordrecht, 1984.

[88] V. Guillemin and S. Sternberg. *Symplectic Techniques in Physics*. Cambridge University Press, Cambridge, U.K., 1984.

[89] V. Guillemin and S. Sternberg. *Variations on a Theme by Kepler, Colloquium Publications, 42*. American Mathematical Society, Providence, R.I., 1991.

[90] R.C. Gunning. *Introduction to holomorphic functions of several variables, volume 1: Function Theory*. Wadsworth and Brooks/Cole, Belmont, California, 1990.

[91] F. Gustavson. On constructing formal integrals of a Hamiltonian system near an equilibrium point. *The Astrophysical Journal*, 71:670–686, 1966.

[92] G. Györgyi. Kepler's equation, Fock variables, Bacry's generators and Dirac brackets. *Nuovo Cimento*, 53 A:717–735, 1968. II, Nouvo Cimento, 62 A, 1969, pages 343–360.

[93] G. Györgyi. Csoportelméleti dinamika 1. *Magyar Fizikai Folyórat*, 17:39–69, 1969.

[94] M. Hadamard. Sur la précession dans le mouvement d'un corps pesant de révolution fixé par un point de son axe. *Bulletin des Sciences Mathématique*, 19:228–230, 1895.

[95] W. Hamilton. On a general method in dynamics by which the study of the motions of all free systems of attracting or repelling points is reduced to the search and differentiation of one central characteristic function I. *Philosophical Transactions*, pages 247–308, 1834. = Mathematical Papers, volume I, pages 103–161; II Philosophical Transactions, 1835, 95–114 = Mathematical Papers, volume II, pages 162-211.

[96] W. Hamilton. Application of quaternions to some dynamical questions. *Proceedings of the Royal Irish Academy,*, 3:appendix, xxxvi–1, 1847. = Collected Mathematical Papers, vol. III, Algebra, no. XLI, p. 441, Cambridge University Press, Cambridge, UK., 1967.

[97] M. Hénon and C. Heiles. The applicability of the third integral of motion. *The Astronomical Journal*, 69:73–79, 1964.

[98] J. Hermann. *Histoires de L'Academie Royale des Sciences avec Mémoires de Mathématique et Physique*, pages 102–103 and 519–523. Paris, 1712. reprint: Amsterdam, 1713, pages 135–36 and 686–682 = Johannis Bernoulli, Opera Omnia, volume 1, pages 469–470.

[99] G. Hill. Researches in lunar theory. *American Journal of Mathematics*, 1:5–26, 129–147, 245–260, 1883.

[100] M.W. Hirsch. *Differential Topology*. Springer Verlag, New York, 1976.

[101] H. Hopf. Über die Abbildungen der dreidimensionalen Sphären auf der Kugelfläche. *Mathematische Annalen*, 104:637–665, 1931.

[102] E. Horozov. Perturbations of the spherical pendulum and abelian integrals. *Journal für die reine und angewandte Mathematik*, 408:114–135, 1990.

[103] R. Howe. Remarks on classical invariant theory. *Transactions of the American Mathematical Society*, 313:539–570, 1989.

[104] T.W. Hungerford. *Algebra*. Holt, Rinehart, and Winston, New York, 1974.

[105] C. Huygens. *l'Horloge à Pendule*. Paris, 1673. Oeuvres Complétes, volume 18, 69–368. (See section Sur la force centrifugal) = English translation by R.J. Blackwell, Christiaan Huygens' the pendulum clock, Iowa State University Press, Ames, Iowa, 1985.

[106] A. Iacob. Invariant manifolds in the motion of a rigid body about a fixed point. *Revue roumaine de mathématiques pures et appliquées*, 16:1497–1521, 1971.

[107] I. Iosifescu and H. Scutaru. On six dimensional canonical realization of the $so(4, 2)$ algebra. *Journal of Mathematical Physics*, 21:2033 –2045, 1980.

[108] C.G.J. Jacobi. Sur l'elimination des noeuds dans le problème des trois corps. *Journal für die reine und angewandte Mathematik*, 26:115–131, 1843.

[109] C.G.J. Jacobi. *Vorlesungen über Dynamik*. G. Reimer, Berlin 1891. reprint: Chelsea, New York, 1967.

[110] J.M. Jauch and E.I. Hill. On the problem of degeneracy in quantum mechanics. *Physical Review*, 57:641–645, 1940.

[111] R. Jost. Poisson brackets (an unpedogogical lecture). *Reviews of Modern Physics*, 36:572–579, 1964.

[112] R. Jost. Winkel und Wirkungsvariable für allgemeine mechanische Systeme. *Helvetica Physica Acta*, 41:965–968, 1968.

[113] W. Kaplan. Topology of the 2-body problem. *American Mathematical Monthly*, 49:316–323, 1942.

[114] J. Kepler. Astronomia nova. Prague, 1609. also "Neue Astronomie ", Berlin 1929 = Gesammelte Werke, volume 3, Munich, 1937.

[115] A. Kirillov. *Methods of Representation Theory*. Springer Verlag, New York, 1976.

[116] F. Klein and A. Sommerfeld. *Über die Theorie des Kreisels*. Teubner, Berlin, 1897. reprint: Johnson Reprint Company, New York, 1965.

[117] H. Knörrer. Singular fibers of the momentum mapping for integrable Hamiltonian systems. *Journal für die reine und angewandte Mathematik*, 355:67–107, 1985.

[118] H. Knörrer. Unpublished lecture notes on mechanics. ETH, Zürich, 1987.

[119] B. Kostant. Quantization and unitary representations. *Lecture Notes in Mathematics*, 170:87–208, 1970.

[120] M. Kummer. On resonant nonlinearly coupled oscillators with two equal frequencies. *Communications in Mathematical Physics*, 48:53–79, 1976.

[121] M. Kummer. On the construction of the reduced phase space of a Hamiltonian system with symmetry. *Indiana University Mathematics Journal*, 30:281–291, 1981.

[122] M. Kummer. On the 3-dimensional lunar problem and other perturbations of the Kepler problem. *Journal of Mathematical Analysis and Applications*, 101:142–194, 1983.

[123] P. Kustaanheimo and E. Stiefel. Perturbation theory of Kepler motion based on spinor regularization. *Journal für die reine und angewandte Mathematik*, 219:204–219, 1965.

[124] J. Lagrange. *Mécanique Analytique*. Paris, 1788. reprint. 4th edition Librairie scientifique et technique, ed. A. Blanchard, Paris, 1965 = Oeuvres, volume 12, Gauthier-Villars, Paris, 1867–92.

[125] J. Lagrange. Mémoire sur la théorie des variations des éléments des planétes et en particulier des variations des grands axes de leurs orbites. *Memoire de la premiere Classe de l'Institut de France*, 1808. = Oeuvres, volume 6, pages 711–816.

[126] S. Lang. *Differentiable Manifolds, 2nd edition*. Springer Verlag, New York, 1985.

[127] P. Laplace. *Traité de Mecanique Céleste*. Paris, 1798. reprint: Celestial Mechanics, Chelsea, New York, 1966.

[128] D.F. Lawden. *Elliptic Functions and Applications*. Applied Mathematical Sciences, vol. 80, Springer Verlag, New York, 1989.

[129] E. Leimanis. *The general problem of the motion of coupled rigid bodies about a fixed point*. Springer Verlag, New York, 1965.

[130] E. Lerman. Hamiltonian Hopf bifurcation in symmetric tops. preprint, MIT, 1994.

[131] E. Lerman. Private communication. July, 1996.

[132] R. Lerman, E. Montgomery and R. Sjamaar. Examples of singular reduction. In D. Salamon, editor, *Symplectic Geometry*, pages 127–155. Cambridge University Press, Cambridge, 1993.

[133] T. Levi-Civita. Traiettorie singolari ed urti net problema ristretto dei tre corpi. *Annali di Matemàtica, serie III*, 9:1, 1903.

[134] T. Levi-Civita. Sur la régularisation du problème des trois corps. *Acta Mathematica*, 42:99–144, 1920.

[135] P. Libermann and C. Marle. *Symplectic Geometry and Analytical Mechanics*. D. Reidel, Dordrecht, 1987.

[136] A. Lichnerowicz. Sur les varietés symplectique. *Comptes Rendus Academie des Sciences*, 233:723–26, 1951.

[137] A. Lichnerowicz. Les varietés de Poisson et leur algébres de Lie associés. *Journal of Differential Geometry*, 12:253–257, 1977.

[138] S. Lie. *Theorie der Transformationsgruppen,* Unter Mitwerkung von F. Engel, Band 1,2,3. B.G. Teubner, Leipzig, 1888,1890,1893. reprint: 2d ed., Chelsea, New York, 1970.

[139] T. Ligon and M. Schaaf. On the global symmetry of the classical Kepler problem. *Reports on Mathematical Physics*, 9:281–300, 1976.

[140] J. Liouville. Note sur l'intégration des équations differentielles de la dynamique, présentée au bureau des longitudes le 29 juin 1853. *Journal de Mathématiques pures et appliquées*, 20:137–138, 1855.

[141] L. Loomis and S. Sternberg. *Advanced Calculus*. Addison-Wesley, Reading, Mass., 1968.

[142] J. Lützen. *Joseph Liouville*. Springer Verlag, New York, 1990.

[143] G. Mackey. *Mathematical Foundations of Quantum Mechanics*. Benjamin Cummings, Reading, Mass., 1963.

[144] L. Markus and K. Meyer. *Generic Hamiltonian dynamical systems are neither integrable nor ergodic, Memoir # 144*. American Mathematical Society, Providence, R.I., 1974.

[145] A.I. Markushevich. *Theory of functions of a complex variable*. Prentice-Hall, Englewood Cliffs, N.J., 1965.

[146] C.M. Marle. Modèle d'action hamiltonienne d'un groupe de Lie sur une variété symplectique. *Rendiconte del Seminario Mathematico Universita e Politechnico Torino*, 43:227–251, 1985.

[147] J. Marsden and T. Ratiu. *Introduction to Mechanics and Symmetry*. Springer, Verlag, New York, 1994.

[148] J. Marsden and A. Weinstein. Reduction of symplectic manifolds with symmetry. *Reports on Mathematical Physics*, 5:121–130, 1974.

[149] J. Mather. Differentiable invariants. *Topology*, 16:145–156, 1977.

[150] K. Meyer. Symmetries and integrals in mechanics. In M. Piexoto, editor, *Dynamical Systems*, pages 259–272. Academic Press, New York, 1973.

[151] J. Milnor. On manifolds homeomorphic to the 7-sphere. *Annals of Mathematics*, 64:399–405, 1956.

[152] J. Milnor. *Morse Theory*. Princeton University Press, Princeton, N.J., 1963.

[153] J. Milnor. *Topology from the Differentiable Point of View*. University Press of Virginia, Charlottesville, 1965.

[154] J. Milnor. On the geometry of the Kepler problem. *American Mathematical Monthly*, 90:353–365, 1983.

[155] A. Mischenko and A. Fomenko. Euler's equations on finite dimensional Lie groups. *Mathematika Izvestia*, 12:371–389, 1978.

[156] K.S. Misner, C.W. Thorne and J.A. Wheeler. *Gravitation*. W.H. Freeman and Company, San Francisco, Calf., 1973.

[157] J. Montesinos. *Classical Tesselations and Three Manifolds*. Springer Verlag, New York, 1987.

[158] J. Moser. *Lectures on Hamiltonian systems, Memoir # 81*. American Mathematical Society, Providence, R.I., 1968.

[159] J. Moser. Regularization of Kepler's problem and the averaging method on a manifold. *Communications on Pure and Applied Mathematics*, 23:609–636, 1970.

[160] J. Moser. Various aspects of integrable Hamiltonian systems. In J. Guckenheimer et al., editors, *Dynamical Systems, Progress in Mathematics, volume 8*, pages 197–220. Birkhäuser, Boston, 1975.

[161] D. Mumford. *Algebraic Geometry I: Complex Projective Varieties*. Springer Verlag, New York, 1979.

[162] L. Nehorošev. Action angle coordinates and their generalizations. *Transactions of the Moscow Mathematical Society*, 26:180–198, 1972.

[163] I. Newton. *Philosophiae Naturalis Principia Mathematica*. London, 1687. English edition, F. Cajori, University of California Press, Berkeley, 1946.

[164] E. Noether. *Invariante Variationsprobleme*. Nachrichten von der Gesellschaft der Wissenschaften zu Göttingen Mathematisch-physikalische Klasse, pages 235–258, 1918.

[165] S. Novikov. The Hamiltonian formalism and a many-valued analog of Morse theory. *Russian Mathematics Surveys*, 37:1–56, 1982.

[166] P. Orlik. *Seifert Manifolds.* Lecture Notes in Mathematics, vol. 281, Springer Verlag, New York, 1972.

[167] Yu. S. Osipov. The Kepler problem and geodesic flows in spaces of constant curvature. *Celestial Mechanics*, 16:191–208, 1977.

[168] R. Palais. On the existence of slices for actions of noncompact Lie groups. *Annals of Mathematics*, 73:295–323, 1961.

[169] R. Palais. unpublished letter. 1965.

[170] L.A. Pars. *A Treatise on Analytical Dynamics*. Heinemann, London, 1965. reprint: Oxbow Press, Woodbridge, Conn., 1979.

[171] W. Pauli. Über das Wasserstoffspektrum vom Standpunkt der neuen Quantenmechanik. *Zeitschrift für Physik*, 36:336–363, 1926.

[172] V. Poènaru. Singularities C^∞ en présence de symétrie. *Lecture Notes in Mathematics*, 1278, 1987.

[173] H. Poincaré. *Les Méthodes Nouvelles de la Mécanique Céleste*, volume 1–3. Gauthier-Villars, Paris, 1892/1893/1898. reprint: Dover, New York, 1957.

[174] L. Poisson. Mémoire sur la variation des constantes arbitraires dans les questions de la mécanique. *Journal de l'Ecole Polytechnique*, 8 (15 cahier), 1809.

[175] H. Pollard. *Introduction to Celestial Mechanics*. Mathematical Association of America, Buffalo, New York, 1976.

[176] T. Poston and I. Stewart. *Taylor Expansions and Catastrophes*. Pitman, Publishing Company, London, 1976.

[177] G. Prince and C. Eliezer. On the Lie symmetries of the classical Kepler problem. *Journal of Physics A*, 14:587–596, 1981.

[178] M. Puiseux. Sur le mouvement d'un point matériel pesant sur un sphère. *Journal de mathématique pures et appliquées*, 7:517–520, 1842.

[179] G. Reeb. *Sur certain propriétés feuilletées*. Actualités Scientifiques et Industrielles, # 1183, Hermann et Cie, Paris, 1952.

[180] G. Reeb. Variétés symplectiques, variétés presque-complex et systémes dynamique. *Comptes Rendus Academie des Sciences*, 235:776–778, 1952.

[181] G. Reeb. Sur les espaces de Finsler et les espaces de Cartan. In *Colloque Géométrie differentielle, Strasbourg*, pages 35–40. Centre Nationale de Recherche de Science, Paris, 1953.

[182] G. Reeb. Sur certaines problémes relatif aux variétés presque-symplectiques et systémes dynamique. In *Convegno Internationale di Geometria Differenziale*, pages 104–106. Cromomese, Rome, 1954.

[183] E. Routh. *Advanced Rigid Body Dynamics*. Macmillian and Co., London, 1884. reprint: "A treatise on the dynamics of a system of rigid bodies", parts I and II, Dover, New York, 1960.

[184] I. Satake. On a generalization of the notion of a manifold. *Proceedings of the National Academy of Science USA*, 42:359–363, 1956.

[185] L. Schlesinger. Parallelverschiebung und Krümmungstensor. *Mathematische Annalen*, 99:413–434, 1928.

[186] G. Schwarz. Smooth functions invariant under the action of a compact Lie group. *Topology*, 14:63–68, 1975.

[187] A.R. Shastri and P. Zvengrowski. Type of 3-manifolds and addition of relativistic kinks. *Reviews in Mathematical Physics*, 3:467–478, 1991.

[188] R. Sjamaar and E. Lerman. Stratified symplectic spaces and reduction. *Annals of Mathematics*, 134:375–422, 1991.

[189] S. Smale. Topology and mechanics I. *Inventiones Mathematicae*, 10:305–331, 1970. II, Inventiones Mathematicae, 11, 45–64, 1970.

[190] J. Śniatycki. Constraints and reduction. *Lecture Notes in Mathematics*, 1037:301–344, 1983.

[191] J. Śniatycki, G. Schwarz, and L. Bates. Yang-Mills and Dirac fields in a bag, constraints and reduction. *Communications in Mathematical Physics*, 176:95–115, 1995.

[192] J. Śniatycki and A. Weinstein. Reduction and quantization for singular momentum mappings. *Letters in Mathematical Physics*, 7:155–161, 1983.

[193] J.M. Souriau. Géométrie symplectique différentielle-applications. In *Colloque Géometrie Différentielle, Strasbourg*, pages 53–59. Centre Nationale de Recherche de Science, Paris, 1953.

[194] J.M. Souriau. Equations canoniques et géometrie symplectique. *Alger Mathématiques, Series A*, 1, no. 2:239–265, 1954.

[195] J.M. Souriau. Modèles classiques quantifiables pour les particules élémentaires II. *Comptes Rendus Académie des Sciences*, 265:165–167, 1967.

[196] J.M. Souriau. Quantification géométrique. Applications. *Annales de l'Institut Henri Poincaré*, 6:311–341, 1967.

[197] J.M. Souriau. *Structure des Systèmes Dynamiques*. Dunod, Paris, 1970.

[198] J.M. Souriau. Sur la varieté de Kepler. *Symposia Mathematica*, 14:343–360, 1974.

[199] J.M. Souriau. Géométrie globale du problème à deux corps. In S. Benenti et al., editors, *Proceedings IUTAM-ISIMM: Modern Developments in Analytical Mechanics, supplement to volume 117*, pages 609–636, 1983.

[200] M. Spivak. *A Comprehensive Introduction to Differential Geometry*. Publish or Perish, Berkeley, Calif., 1970.

[201] N. Steenrod. *The Topology of Fibre Bundles*. Princeton University Press, Princeton, N.J., 1951.

[202] B. Stephenson. *Kepler's physical astronomy, Studies in the history of mathematics and physical sciences, volume #13*. Springer Verlag, New York, 1987.

[203] S. Sternberg. *Lectures on Differential Geometry*. Prentice Hall, Englewood, Cliffs, N.J., 1964.

[204] B. Stevenson. *Kepler's Physical Astronomy*. Springer Verlag, New York, 1987.

[205] J. Synge. Classical Dynamics. In S. Flügge, editor, *Encyclopedia of Physics, volume III/1*, pages 1–223. Springer Verlag, Berlin, 1960.

[206] Ia. V. Tatarinov. Portraits of the classical integrals of the problem of rotation of a rigid body about a fixed point. *Vestnik Moscow University*, 6:99–105, 1974. (Russian).

[207] F. G. Tricomi. *Differential Equations*. Blackie and Son, London, 1953.

[208] G.M. Tuynman. The derivative of the exponential map of matrices. *American Mathematical Monthly*, 102:818–820, 1995.

[209] J-C. van der Meer. The Hamiltonian Hopf bifurcation. *Lecture Notes in Mathematics*, 1160, 1985.

[210] A. G. Webster. *Dynamics of particles and rigid elastic and fluid bodies, 2nd edition*. reprint. Dover, New York, 1959.

[211] A. Weinstein. The spherical pendulum and complex integration. *American Mathematical Monthly*, 49:521–523, 1942.

[212] A. Weinstein. Symplectic manifolds and their Lagrangian submanifolds. *Advances in Mathematics*, 6:329–346, 1971.

[213] A. Weinstein. Symplectic V-manifolds, periodic orbits of Hamiltonian systems, and the volume of certain Riemannian manifolds. *Communications on Pure and Applied Mathematics*, 30:265–271, 1977.

[214] A. Weinstein. Symplectic geometry. *Bulletin of the American Mathematical Society*, 5:1–13, 1981.

[215] A. Weinstein. Local structure of Poisson manifolds. *Journal of Differential Geometry*, 18:523–557, 1983.

[216] A. Weinstein. Sophus Lie and symplectic geometry. *Expositiones Mathematicae*, 1:95–96, 1983.

[217] A. Weinstein. Poisson structures and Lie algebras. *The Mathematical Heritage of Élie Cartan, Asterisque, Numéro hors series*, pages 421–434, 1985.

[218] E. Whittaker. *A treatise on the Analytical Dynamics of Particles and Rigid Bodies, 4th edition*. Cambridge University Press, Cambridge, U.K., 1937.

[219] J. Williamson. On the normal forms of linear canonical transformations in dynamics. *American Journal of Mathematics*, 59:599–617, 1937.

[220] A. Wintner. On linear conservative dynamical systems. *Annali di Mathematica*, 13:105–112, 1934.

[221] E. Witt. Theorie der quadratischen Formen in beliebigen Körpern. *Journal für die reine und angewandte Mathematik*, 176:31–44, 1937.

[222] J. Wolf. Differentiable fibre spaces and mappings compatible with Riemannian metrics. *Michigan Mathematics Journal*, 11:65–70, 1964.

[223] N. Woodhouse. *Geometric Quantization, 2nd edition*. Oxford University Press, Oxford, 1991.

[224] Z. Xia. Melnikov method and transversal homoclinic orbits in the restricted three body problem. *Journal of Differential Equations*, 96:170–184, 1992.

[225] Z. Xia. Arnold diffusion in the elliptic restricted three body problem. *Journal of Dynamics and Differential Equations*, 5:219–240, 1993.

[226] J. Yoder. *Unrolling time*. Cambridge University Press, New York,, 1988.

[227] M. Zou. *Geometry of two degree of freedom integrable Hamiltonian systems*. PhD thesis, University of Arizona, Tucson, Arizona, 1992.

Acknowledgements

This book has been a long time in the making. Because I have been writing it for years, I owe many people and institutions a thank you. First, I thank my coauthor for challenging me in the summer of 1987 to get busy and start writing. His financial support through his NSERC grant has been essential for seeing this project through. Now that it is done, I promise, not monopolize the computer anymore. Second, I would like to thank my colleagues J. Śniatycki, and D. Rod in Calgary and H. Duistermaat in Utrecht for their support (both moral and financial) in this project. Especially I would like to thank J. Arms in Seattle and E. Lerman in Champaign-Urbana for their critique of an early version of appendix B. I hope I have corrected all the mistakes they found. Also I would like to thank S. Maltby and K. Foltinek for their labors in finding and removing errors in various preliminary versions of this book. Their efforts are visible throughout the text. Certainly, I would like to thank C. Saull and M. Lamoureux in Calgary and I. Hoveijn in Groningen for their help in drawing the figures. Finally I would like to thank the Mathematics Departments of the University of Utrecht, the University of California at Santa Cruz, the University of Calgary and the Fields Institute for their financial support during the course of writing this book.

I would appreciate hearing from readers about errors or obscurities in the text. In spite of my best effort I know that they exist but I do not know their location. I can be reached via email at cushman@math.ruu.nl.

It goes without saying that my wife Corrie has made this all possible. Perhaps I will be a different person when I do not have to spend most of my waking hours working on this book.

Index

SPRINGER NATURE

GPSR Compliance

The European Union's (EU) General Product Safety Regulation (GPSR) is a set of rules that requires consumer products to be safe and our obligations to ensure this.

If you have any concerns about our products, you can contact us on ProductSafety@springernature.com

In case Publisher is established outside the EU, the EU authorized representative is:

Springer Nature Customer Service Center GmbH
Europaplatz 3
69115 Heidelberg, Germany

Zeitfracht Medien GmbH
Ferdinand-Jühlke-Straße 7
99095 Erfurt, Deutschland
produktsicherheit@kolibri360.de